Navigation
Land, Sea, Air, & Space

Edited and with comprehensive introductions by
Myron Kayton
President
Kayton Engineering Company

IEEE PRESS

A volume in the IEEE PRESS Selected Reprint Series, prepared under the sponsorship of the IEEE Aerospace and Electronic Systems Society

The Institute of Electrical and Electronics Engineers. Inc., New York

PRINTED IN THE UNITED STATES OF AMERICA

IEEE Order Number: PC0246-9

Library of Congress Cataloging-in-Publication Data

Navigation : land, sea, air, and space / edited by Myron Kayton.
p. cm.—(IEEE Press selected reprint series)
"Prepared under the sponsorship of the IEEE Aerospace and Electronic Systems Society."
ISBN 0-87942-257-2
1. Navigation. 2. Aids to Navigation. 3. Navigation (Aeronautics). 4. Navigation (Astronautics). I. Kayton, Myron. II. IEEE Aerospace and Electronic Systems Society.
VK555.N345 1989
623.89—dc20 89-39737
CIP

Contents

Preface

THIS is a new kind of book, a comprehensive tutorial on navigation for all vehicles. It consists of 8 Introductions that I wrote specifically for this book and 39 reprints that I selected to illustrate modern navigation technology. The Introductions are easily identified by a black tab at the edge of each page. The book is intended for experienced engineers who want to step away from the details of their job and learn about adjacent navigation technologies, and for young engineers who want to know what they are getting into. I wish I had had this book when I entered the navigation and avionics field 30 years ago. I wish I had had it again when I expanded into new vehicle-electronic technologies several times in my career. This is the first book I know of that covers the navigation of vehicles on land, at sea, in the air, and in space.

My goals have been:

—to give a complete picture of vehicle navigation as it is practiced today;

—to describe past technologies and show how they led to the most recent integrated systems, satellite navaids, and expert systems; and

—to combine on-board navigation and externally derived navigation in one book, perhaps for the first time in regard to the navigation of spacecraft.

This personal view of navigation narrates how today's technology evolved, describes the present state of the art, and predicts the future. Though it is inevitably colored by my own experience, I have tried to balance the various technologies and types of vehicle.

In Part 1, I describe the attributes of navigation systems: cost, accuracy, autonomy, time-delay, global coverage, and human interfaces. Part 1 distinguishes among surveying, position–location, navigation, and guidance. I confine "navigation" to the measurement of a vehicle's position or velocity, whether on-board or by an external station (the latter sometimes called "position location"), whether in real-time or calculated several days later. I do not address the "conning" of ships, the piloting of aircraft, or the attitude-control of spacecraft; they are technologies in their own right.

Parts 2 through 7 are organized in the same way. Each part has:

—an introduction to past and present technology, including the present state of the art. It includes ample citations of reprints and references.

—reprints of important articles on key aspects of each technology. I selected them from publications of the Institute of Electrical and Electronics Engineers (IEEE), the American Astronautical Society (AAS), the Institute of Navigation (ION), the American Institute of Aeronautics and Astronautics (AIAA), and other sources of information about modern navigation.

In the Introduction to Part 2, I describe the intertwined history of surveying, astronomy, time measurement, and map-making. Navigation can be no more accurate than the underlying geodesy, whose data are derived from astronomic, surveying, and gravimetric information, obtained in the early days on daring expeditions without maps.

In the Introduction to Part 3, I describe the navigation of land vehicles, whether camel caravans, railroads, tanks, or automobiles. I include descriptions of PLRS, JTIDS, GPS, and the Sarsat rescue system, though some of them can also be used on ships, aircraft, and spacecraft.

In the Introduction to Part 4, I describe the navigation of ships, starting from its primitive origins during the settling of the Earth's islands, and continuing to its modern electronic forms. I discuss the applications of astronomy to navigation after AD. 1500, the start of the great age of European exploration. Part 4 includes descriptions of Loran, Transit, and the magnetic compass, though these instruments can also be used on land vehicles and aircraft.

In the Introductions to Parts 5 and 6, I describe the navigation of aircraft and spacecraft, both of which evolved in the electronic era. The Introduction to Part 5 includes a description of Omega, which is used by aircraft and ships. The Introduction to Part 6 is the first presentation I know of that addresses both the ground-tracked and on-board navigation of spacecraft. Part 6 describes the navigation of Earth-orbiting spacecraft and planetary explorers.

The Introduction to Part 7 is my assessment of animal navigation and my speculations on its application by man. Animals are the ultimate neural-net navigators; they modify their data base and algorithms as they gain experience about the world. The study of animal navigation forces us to examine our definitions of "artificial" intelligence. Part 7 is an expansion of an article I wrote for the *IEEE Aerospace and Electronic Systems Magazine* in March 1989.

Part 8 describes my projections of all forms of navigation into the next century. Section 8.1 contains my predictions for navigation and closely related communications that are common to all vehicles. Section 8.2 predicts trends in digital cartography. Sections 8.3–8.6 contain predictions for each type of vehicle. Predictions are feasible, 30 years ahead, because the large investment in ground-based and space-based navigation systems precludes rapid change. Radio systems tend to remain operational for fifty years, though the vehicle-borne receivers are constantly improved by using the latest computers, analog chips, and digital microcircuits to reduce cost, add features, and increase reliability.

Within each Introduction, sections are differentiated with a period (e.g., N.N.), figures with hyphens (e.g., N-N), and tables with colons (e.g., N:N), for the reader's convenience. I use *nm* for nautical miles, *mi* for statute miles, and *m* for meters, to avoid three-character units, though some purists may object. Within the reprints, I left the acronyms, units, and notation as the authors wrote them.

Seven of the 8 Introductions began as a tutorial article I wrote for the *IEEE Transactions on Aerospace and Electronic Systems* (September 1988). I have added many historical anecdotes, more detail about current navigation aids, and more references. I chose the reprints to illustrate important aspects of modern navigation-system engineering. They illustrate performance, error modeling, interfaces, networking, software, and multi-sensor processing. The internal details of the navaids are outlined in the Introductions and are discussed in detail in the references and in a few reprints. I regret that I had to omit many worthy books and papers from the reprints and Bibliography to save space.

I hope nontechnical readers will also use this book. It will be of interest to any reader who has wondered how we find our way on this planet and in the solar system. Students of the history of science will see how a real technology grew from its roots in prehistoric boat travel and ancient astronomy to modern electronic navigation aids. I explain navigation technology as a continuing process, each stage of which is built on the one before. The nontechnical reader will be able to follow the historical explanations and most of the modern navigation material, but he or she may miss some of the subtleties without some knowledge of mechanics, electromagnetics, and multisensor filters.

I have assembled a comprehensive Bibliography and Subject Index. The Bibliographies at the end of each Introduction are annotated so the reader can decide whether to look up the works. When describing modern navigation, I often cite references to guide the reader. However, when describing historical navigation, I do not interrupt the narrative with citations; the annotated Bibliographies show where my data (other than personal experience) came from. Besides the Bibliographies in the Introductions, each reprint has its own bibliography.

The thorough Subject Index covers the Introductions and all reprints. It includes technical terms, people, places, and companies. The authors of reprints are listed separately on page 441. Authors of bibliography entries are not indexed. Companies are indexed under their usual name, not necessarily that of the parent. Some no longer exist. I did not index ''United States'' and a few other common headings. I have few entries under ''computer'' because that topic pervades the entire book. The Subject Index refers the reader to the page number of the Introductions (in italics) and to the first page number of the reprints.

The Glossary on page xi lists most acronyms and abbreviations used in the Introductions and reprints, for the reader's convenience. The Glossary contains the section numbers of the Introductions (in italics) and the first page number of the reprints in which the acronym appears. Acronyms are not repeated in the Subject Index. The following acknowledgments list colleagues who gave me information for this book or for its predecessor article in the *IEEE Transactions*. I thank them for the time they spent educating me.

I would be happy to receive comments from readers, in care of the IEEE PRESS, about suggested additions and corrections that I might use in future editions, with appropriate credit.

Acknowledgments

The author wishes to acknowledge helpful comments by the following colleagues, some of whom provided photographs:

Prof. Thomas Adams of Brown University for his lecture on Page and Mount and his discussion afterward.

Mr. Peter Arnold of Sperry Marine Systems for photos of World War I gyrocompasses.

Mr. James R. Barnett of the FAA Maintenance Division for information about Vortac servicing intervals.

Mr. Dan A. Bathker of NASA's Jet Propulsion Laboratory for photographs and information about the upgrade to the Deep Space network in 1988.

Mr. Harry Black of NASA Johnson Space Center for information about the Shuttle's uplink.

Chief Petty Officer Neil Bryson of the U.S. Coast Guard for photographs of Omega and Loran transmitter antennas and for current manuals.

Prof. C. R. Carter of McMasters University for information about Sarsat-Cospas.

Prof. Robert Chilton of Texas A&M (formerly with NASA's Johnson Space Center) for his comments on Mercury navigation.

Dr. Calvin Culver of Micrologic for his comments on Loran and Omega.

Mr. David Dobson, Administrative Editor of *IEEE Transactions on Aerospace and Electronic Systems*, for suggestions on format and publication of the Sept. 1988 Transactions article.

Mr. Sven Dodington of ITT (retired) for comments on early drafts of the Introductions to Parts 1, 5, and 8 and for part of Table 1:1.

Mr. Don Edward of ITT for information about Mercury tracking.

Mr. Walter R. Fried of Hughes Aircraft Company for his comments on JTIDS and PLRS.

Mr. Tom Gaudion of Rosemount Inc. for photographs of the air-data sensors on the B-767.

Mr. David Gilbert of NASA's Johnson Space Center (retired) for information about Shuttle flight controls.

Prof. Patricia Greenfield of UCLA for comments on the Introduction to Part 7.

Mr. John Griffin of NASA Johnson Space Center for information about the Shuttle's K-band radar and tracking.

Mr. Huey Hernandez of NASA Johnson Space Center for information about the Shuttle's telemetry format.

Mr. Mark Hess of NASA's Office of Space Station for a photograph of Soyuz.

Mr. Lester Isaacs of McDonnell-Douglas for aircraft antenna and cockpit photographs.

Mr. Frank C. Jarvis of Raytheon's Submarine Signal Division for information about Doppler sonar logs.

Dr. Peter D. Joseph of TRW for his review of early drafts of the Introductions to Parts 5 and 6.

Mr. Jerry Klayman of TRW for information about the GEODSS tracking system.

Prof. Richard McFarland of Ohio University for information about ILS antennas.

Mr. Larry McWhorter of NASA's Johnson Space Center for information about Shuttle entry guidance.

Prof. Josh T. Nessmith of the Georgia Institute of Technology for information about space tracking radars.

Mr. Eric Palmer of McDonnell-Douglas for information about landing aids for MD aircraft.

Mr. Don Piehler of TRW for comments about the Combined Space Operations Center.

Mr. Joseph Portney of Litton's Guidance and Control Division for historical information.

Mr. Ian Pryke of the European Space Agency for information about their tracking networks.

Dr. William Redisch of NASA Goddard Space Flight Center for information about NASCOM expansion and the origins of Sarsat.

Dr. Macgregor Reid of NASA's Jet Propulsion Laboratory, for information about early space tracking networks.

Dr. Jesse C. Ryles of the Air Force Avionics Laboratory for information about navigation systems in development.

Mr. Emil Schiesser of NASA's Johnson Space Center for information about Shuttle's on-board gravity model.

Mr. David Scull of the FAA for the Federal Radionavigation Plan.

Mr. Paul Shack for information about the Shuttle's S-band communications.

Mr. Robert Sharples of TRW for allowing me to photograph his Micronesian stick map.

Mr. Merrill Skolnik for his comments on World War II radar.

Mr. Vince Tampio of LTV-Sierra for information about the APN-169 stationkeeping equipment.

Mr. Robert C. Thomson of the International Omega Association for data on Omega availability.

Ms. Lisa Vasquez-Morrison of Media Services Corporation for photographs of the Shuttle's head-up display.

Mr. Al Vilpors of NAVAIR for information about carrier landing systems.

Mr. Mike Wade of Rockwell for photographs of the Navstar satellite.

Mr. Bill Watson of NASA Goddard for STDN and TDRS users' guides.

Mr. Larry Whitcomb of Magnavox Nav-Com Inc. for information about the GPS satellite and his comments on the Introduction to Part 3.

Mr. Bob Wigginton of the Naval Electronic Systems Center for information about SPN-46.

Dr. I. Joe Williams of TRW for his comments on the Introduction to Part 6.

Mr. Jerry Wittles of Aerospace Corporation for information about radio guidance and the Satellite Control Network, including a photograph of an RTS.

Prof. Walter Wrigley of MIT (deceased) for his comments on the Introductions to Parts 4 and 5.

Ms. Marilyn Zotos of the C. S. Draper Laboratory for photos of the Apollo IMU and computer and of the SPIRE gimbal set.

The following people submitted corrections to the *IEEE Transactions* article, September 1988, on which the Introductions to each Part are based:

Mr. Gil Lieberman of Israel Military Industries corrected an error in my description of GPS.

Prof. Dale Pollack of Tuskegee University discovered an error in Part 6.

Prof. Walter Wrigley of MIT (deceased) corrected a historical error in Part 2.

Myron Kayton
Kayton Engineering Company
Santa Monica, California

Glossary

Italicized numbers refer to the page of appearance in the Introductions to each Part. Other numbers refer to the page of the reprint in which the abbreviation first appears. Other terms are indexed on page 443.

A/D	Analog-to-digital, 194
ADI	Attitude-director indicator, *240*, *330*, 374
ADTA	Air data transducer assembly, 396
AFC	Automatic frequency control, 110
AFCS	Automatic flight control system, 296
AGC	Automatic gain control, 359
A-N Range	U.S. aircraft guidance system, *229*
ARCS	Automatic route control system, 101
ARIA	Apollo Range Instrumentation Aircraft, *319*
ARINC	Aeronautical Radio, Inc., 108, *158*, *234*, 285, *437*
ARTCC	Air route traffic control center, *231*, 303
AS	Anti-spoofing, 108, 354
ASCII	American Standard Code for Information Interchange, 108
ASDR	Airport surface detection radar, *238*
ASIC	Application-specific integrated circuit, 357
ATCRBS	Air Traffic Control Radar Beacon System, 285
ATR	Air transport rack, 108
ATRAN	Automatic terrain recognition and navigation, *236*
AVLS	Automatic vehicle location system, 90
AWACS	Airborne Warning and Control System, command center in a Boeing 707, *68*
AZ	Azimuth angle, 313, 390
BAZ	Back-azimuth angle, 314
BCPS	Battery charger power supply, 303
BIH	Bureau International de l'Heure, 53, 58
BIT	Built-in test, 360
BITE	Built-in test equipment, 192, *244*
BLEU	Blind landing experimental unit, *238*
BPF	Band-pass filter, 109
C/A	Coarse-acquisition code, 61, *68*, 136
CAD	Computer-aided design, 91
CAINS	Carrier aircraft INS, 278
CAS	Collision-Avoidance System, *236*
C-band	4–7 GHz
CCD	Charge-coupled device, 406
CCITT	International Consultive Committee for Telephone and Telegraph, 86
CD	Compact disk, 103, *436*
CEP	Circle of equal probability, also circular error probable, *156*, 195, 288
CHAIKA	Soviet navigation system, like U.S. LORAN-C, *156*
CICADA	Soviet satellite navigation system, 156, 322
CID	Continuously integrated Doppler, 44
CLASS	Chrysler Laser Atlas and Satellite System, 101
CMOS	Complementary metal-oxide on silicon, 357
CNES	Centre National des Etudes Spatiales, 353
C/NV	Critical/nonvolatile memory, 121
COAS	Crewman optical alignment sight, 389
COSPAS-SARSAT	Combined Soviet COSPAS and western SARSAT system, *70*, *436*
CRT	Cathode-ray tube, *68*, 104, *240*, 316, *330*
CSOC	Consolidated Space Operations Center, *317*
CSTC	Consolidated Satellite Test Center, *317*
CW	Continuous wave
C-W	Clohessy-Wiltshire, 390
DABS	Discrete Address Beacon System (Mode-S), *243*
DAIS	Digital avionics integration system, *240*
DAP	Digital autopilot, 377
DDA	Digital differential analyzer, 198
DEC	Digital Equipment Corporation, 90
DECCA	British hyperbolic navigation system, *156*
DF	Direction finder, *229*
DH	Decision height, *238*, 296
DIME	Dual independent map encoding, 102
DMA	U.S. defense mapping agency, 33, *436*
DME	Distance-measuring equipment, *1*, *12*, *231*, 250, 304, 313, *330*
DOD	U.S. Department of Defense
DR	Dead-reckoning, 73
DSIF	Deep-space instrumentation facility, *318*
DSVL	Doppler sonar velocity log, 189
ECEF	Earth-centered, Earth-fixed coordinates, 127, 142, 348
ECU	Sonar electronics control unit, 189
EFIS	Electronic flight instrumentation system, 316
EL	Elevation angle, 318, 390
EM	Electro-magnetic, 185, 188
EMA	Electro-magnetic accelerometer, 204
EMI	Electromagnetic interference, 204
EMS	Emergency medical services, 90
EO	Electro-optical, 408
EPLRS	Enhanced PLRS, *68*, 82, *437*
EPUU	Enhanced PLRS user unit, 82
ERGS	Electronic Route Guidance System, 104
EROS	Eliminate range zero system, *236*
ESGM	Electrostatically supported gyro monitor, 196
ET	Shuttle external tank, 374
FA	Final approach, 314
FAA	U.S. Federal Aviation Administration
FBM	Fleet ballistic missile, 194
FCC	Federal Communications Commission (U.S.), *234*
FCPU	Facility central processing unit, 303
FDI	Failure detection and isolation, 267
FEBE	Early inertial navigator, *234*
FILS	Ferranti inertial land surveyor, 35
FO-FO-FS	Fail-operational, fail-operational, fail-safe, 267
FSK	Frequency shift key, 99
FSS	Flight service station, 303
GBF	Geographic base file, 105
GEODSS	Ground-based electro-optical deep-space surveillance system, *321*
GDOP	Geometric dilution of precision, 126, 351
GFY	Government fiscal year, 208
GHz	Gigahertz (10^9 Hz)
GLONASS	Soviet Global Navigation Satellite System, *69*, *322*, 410
GN&C	Guidance, navigation, and control, 374
GPC	Shuttle's general-purpose computer, 374
GPS	U.S. Global Positioning System
GPSDR	GPS demonstration receiver, 353
GPSPAC	GPS package, 347, 410
GSC	Geodetic Survey of Canada, 44
GSFC	NASA Goddard Space Center, 352
GST	Greenwich sidereal time, 22
GSTDN	Ground STDN, 350
GT	Ground transmitter
h	Hour
HDOP	Horizontal dilution of position, 126
HDUE	High-dynamic user equipment, 133
HEAO	High-energy astronomy observatory, 408
HF	High frequency (3–30 MHz), 67, 211, *229*, *318*

HSI	Horizontal Situation Indicator, *240*, *330*
HUD	Head-up display, *238*, *330*
ICAO	International Civil Aviation Organization
IF	Intermediate frequency (in demodulation), 108
IFF	Identification friend or foe, *231*
IID	Intermittently integrated Doppler, 50
ILS	Instrument landing system, *3*, 232, 297, 313, *437*
IMU	Inertial measurement unit
INMARSAT	International Maritime Satellite Organization, *234*, *438*
in	Inch
INS	Inertial navigation system, 250
IOPCP	GPS input–output computer program, 116
IOT	Input-output terminal, 303
IPA	Intermediate power amplifier, 303
IPS	Inertial positioning system, 44
IRS	Inertial reference system, 258
IRU	Inertial reference unit = IMU, 258, 367
ISO	International Standards Organization, 82
Isp	Specific impulse, 374
IUS	Inertial upper stage, *331*
JPL	NASA Jet Propulsion Laboratory, 353
J/S	Jamming-to-signal ratio, 108
JTIDS	Joint tactical information distribution system
K-band	16–28 GHz
Ka-band	26–40 GHz
Ku-band	12–18 GHz
km	Kilometer
KSC	NASA Kennedy Space Center, 375
kt	Knot = nm/h
LASS	Litton automatic surveyor system, 33
Lb	pound
L-band	1–3 GHz
LCD	Liquid crystal display, 104
L/D	Lift-to-drag ratio, 371
LDEF	Long-Duration Exposure Facility, 388
LED	Light-emitting diode, 109
LF	Low frequency (30–300 kHz), 218
LHA	Local hour angle, 15
LM	Lunar module, *325*
LO	Local oscillator, 355
LOP	Line of position, 167, 248
LORAN	Long range navigation
LSO	Landing signal officer, *239*
LST	Local sidereal time, 22
L1, L2	GPS carrier frequencies, *68*, 347, 354
m	Meter
MACE	MGM-13B ground-launched cruise missile, *236*
MC	Midcourse correction, 390
MDA	Minimum descent altitude, 297
MDT	Mobile data terminal, 90
MECO	Main-engine cut-off, 376
MET	Mission elapsed time, 390
MHz	Megahertz (10^6 Hz)
MINS	Marine integrated navigation system, 161
MIRV	Multiple independent re-entry vehicles, 194
MLS	Microwave landing system, *232*, 313, *330*, *438*
MOPS	Minimum operating performance standards, 285, 316
MPS	Shuttle main propulsion system, 374
MPS	Maintenance processor subsystem, 303
MSBLS	Microwave scanning-beam landing system for Shuttle, *240*, *330*, 395
MSFN	Manned space flight network, 318
MTBF	Mean time between failures, 110
MTBO	Mean time between outages, 303
NASA	U.S. National Aeronautics and Space Administration
NASCOM	NASA communication network, *319*
NATO	North Atlantic Treaty Organization, *156*, 192, *231*
NAVCP	GPS navigation computer program, 115
NAVSEA	U.S. Navy's Sea Systems Command, 188
NAVSTAR-GPS	U.S. Global Positioning System
NCS	Network control station, 82
NIST	U.S. National Institute of Standards and Technology, 250
nm	Nautical mile = 1852 m
NNSS	U.S. Navy navigation satellite system, 49
NRZ	Non-return to zero modulation, 87
NVM	Non-volatile memory, 260
OMEGA	Worldwide radio navigation system
OMS	Shuttle orbital maneuvering system, *319*, 383
OPNET	Transit ground operations network, 48
PA	Power amplifier, 303
PADS	Position and azimuth determination system, 33
PAR	Precision approach radar, *237*
PAU	GPS preamplifier unit, 109
PCD	Preferred compass direction, 428
PDOP	Position dilution of precision, 356
PJHI	PLRS–JTIDS hybrid interface, 86
PLANS	Primary land arctic navigation system, 73
PLANS	IEEE Position Location and Navigation Symposium
PLRS	Position location reporting system
P-N	Pseudonoise, *320*, 347
PPC	Omega phase propagation correction, 162
PRN	Pseudo-random noise = P–N, 142, *320*
PSF	Pounds per square foot, 374
PT	PLRS terminal, 82
PTT	Government-owned post, telephone, telegraph agency, 70
PTT	Push to talk, 303
PTTI	Precise time and time interval, 58
RA	TCAS resolution advisory, 287
RAM	Random access memory, *327*, 355
RCA	Radio Corporation of America, now General Electric, *318*
RCO	Remote communication outlet receiver, 303
RCP	GPS receiver computer program, 114
RCS	Reaction–control system, 394
RF	Radio frequency
RFP	Request for proposal, 188
RGSS	Rapid geodetic-surveyor system, 33
RHC	Rotation hand controller, 391
RM	GPS receiver manager, 120
RM	Redundancy management, 123, 396
RMC	Remote monitor and control unit, 303
RMM	Remote maintenance and monitoring, 303
RMS	Root mean square, *157*, 348
RNG	Range, 390
ROM	Read-only memory, 355
R/PA	GPS receiver/processor assembly, 348
RR	Rendezvous radar, 389
RTCA	Radio Technical Commission for Aeronautics, 313
RTS	Remote tracking station, *317*
RVR	Runway visual range, *238*
RXP	Receiver-processor, 108
s	Second
SA	GPS selective availability, 109, 354
SARPs	FAA standards and recommended practices, 314
SARSAT	Search and rescue satellite-aided tracking, *70*, *436*
S-band	2–4 GHz
SCF	Satellite control facility, *317*
SCN	Satellite control network, *317*
SDBM	GPS satellite data-base manager, 120
SDI	Strategic defense initiative, 403
SEM	Standard electronic module, 192, 206
SEP	Separation, 367
SEP	Spherical error probable, 331
SGLS	Space ground link system, *317*
SINS	Ship's inertial navigation system, *157*, 194, *437*

SKE	Stationkeeping equipment, *236*
SMM	Solar maximum mission, 388
S/N	Signal-to-noise ratio, 221
SONAR	Sound navigation and ranging, *153*, 171, 188
SOOS	Stacked Oscars on Scout, 208
SPARS	Space precision attitude reference system, 403
SPIRE	Space inertial reference equipment, *234*
SRB	Solid-rocket booster, 375
SSGS	Standardized space guidance system, 403
SSR	Secondary surveillance radar, *231*
ST	Star tracker, 389
STADAN	Space Tracking and Data Network
STANAG	Standard NATO agreement, 192
STC	Space test center
STDN	Spaceflight Tracking and Data Network, *318*, *438*
STOL	Short takeoff and landing, 315
STS	Space transportation system, *329*, 374
SV	Space vehicle, 347
t	Ton
TA	TCAS traffic alert, 287
TACAN	Tactical air navigation
TCAS	Traffic-alert collision-avoidance system, *231*, 285
TDRS	Tracking and data relay satellite
Techeval	U.S. Navy technical evaluation, 192
TERCOM	Terrain contour mapping, *236*
TFTA	Terrain following/terrain avoidance, *237*
THC	Translation hand controller, 391
TIA	Type inspection authorization, 301
TIG	Time of ignition, 391
TLS	Tracking-loop software, 353, 358
TOC	Times of coincidence, 60, 218
TRANSIT	A U.S. satellite navigation system
TTD	Total timing delay, 60
TVC	Thrust vector control, 375
TXCO	Temperature-compensated crystal oscillator, 307
UHF	Ultrahigh frequency, 300–900 MHz, *66*, *237*, *317*
UK	United Kingdom (England), 221
USAF	U.S. Air Force
USB	Unified S-band, *319*
USCG	U.S. Coast Guard, 221
USCGS	U.S. Coast and Geodetic Survey, 34
USGS	U.S. Geological Survey, 34
USNO	U.S. Naval Observatory, 58, 143, 211
UTC	Universal Time Coordinated
UTM	Universal transverse mercator, *11*, *436*
μs	Microsecond = 10^{-6} s
VAX	Minicomputer made by DEC, 90
VERLORT	Very long range tracking radar, *318*
VHF	Very high frequency, 30–300 MHz, *10*, *67*, *317*
VHSIC	Very-high-speed integrated circuits, 88
VL	Viking lander, 366
VLF	Very-low frequency, 3–30 kHz, 58, 217, *232*, 245
VLF-OMEGA	Very low frequency OMEGA navigation system, *438*
VO	Viking orbiter, 371
VOR	Very high frequency omnirange
VORTAC	VOR–TACAN combined station, *230*, 303, *437*
VSWR	Voltage standing wave ratio, 308
VTS	Vehicle tracking system, 94
W	Watt
WGS	World geodetic system, *11*, *158*, 210
WWV	Radio station of NIST, *10*, *152*, *233*, *319*, 248
X-band	6–12 GHz
YPG	U.S. Army's Yuma proving ground, 133
ZUPT	Zero-velocity update, *2*, *10*, 33, *66*

Part 1
Introduction to Navigation

SINCE primitive man first wandered away from his hut, people have asked "Where am I?" (navigation) and "Which way leads to my destination?" (guidance). Early man had no concept of absolute position on an Earth, of whose size and shape he could not conceive.

In modern times, man has ventured forth on ships, land vehicles, aircraft, and spacecraft. Each vehicle has evolved a distinctive complement of navigation devices. Parts 3–6 of this book describe the navigation devices that have been used on various vehicles throughout history and how the modern complement of equipment evolved. Navigation devices are chosen for the following reasons:

Cost. Cost of the on-board equipment is the most important attribute of most navigation systems. Figure 1-1 shows the cost versus accuracy of several popular navigation receivers. VOR (Part 5.2, pg. 229) sets are produced in large volume using custom chips and are the least costly of the usual aircraft navigators. Transit (Part 4.10) and Loran (Part 4.8) sets are the least costly, to the user, of all radio navigation sets. The cost of Navstar-GPS (Part 3.6) sets is still decreasing as the system approaches its operational date.

The total cost of a navigation system to society includes the construction and maintenance of transmitters, the preparation and dissemination of charts and correction tables, and the construction and maintenance of monitor/control stations. Because most transmitters are government owned, expenses other than the cost of receivers are paid from general tax revenues. Hence, users are not motivated to choose systems with the lowest social cost.

Accuracy. Accuracy is the next most important attribute of most systems. It has steadily improved from hundreds of miles in the 18th century to fractions of a meter with navigation satellites today. The most accurate and the least expensive (for the user) global systems are satellite based (Fig. 1-1). Several radio systems use "differential" techniques in which a stationary surveyed station broadcasts its measured offsets that are applied by nearby receivers to correct their own measurements. This technique often improves accuracy by a factor of 5.

Autonomy. This book subdivides autonomy into the following five classes:

1. *passive self-contained systems* whose users neither receive nor transmit electromagnetic signals. They comprise dead-reckoning systems such as inertial navigators. They emit no radiation that would betray their presence and require no external stations. Failures are detected and corrected on-board.

2. *active self-contained systems* whose users radiate but do not receive externally generated signals. Examples are radar and sonar. They do not depend on the existence of navigation stations.

3. *receivers of natural radiation* whose users measure naturally emitted electromagnetic radiation. Examples are magnetic compasses, optical observation of stars, and optical observation of landmarks. They do not announce their presence by emitting nor do they need external navigation stations.

4. *receivers of man-made radiation* whose users measure man-made electromagnetic radiation from navaids. Users receive radiation emitted by Earth-based or space-based transmitters but they do not themselves transmit. Examples are radio aids such as Loran, Omega (Part 5.3), VOR, Navstar-GPS, and Transit. They are not usually considered autonomous because they require external transmitting stations.

5. *active radio navaids* whose users exchange signals with navigation stations. These include DME (Part 5.2), JTIDS, PLRS (Part 3.5), and collision-avoidance systems. The vehicle betrays its presence by emitting.

Time delay. Some navigation aids give continuous indications of position and velocity; others give intermittent indications. Time delay ("latency") can be due to computer delays, scanning of a radar beam, use of a single radio set to receive signals from several satellites consecutively, or gaps in satellite coverage. Only 20 years ago, the computation of a marine fix took 30 min. Today, navigation calculations are completed in fractions of a second on all vehicles, in an electronic computer. Sub-meter, survey-quality accuracy can be achieved with GPS and Transit after several hours of data reduction.

Global coverage. Radio aids are classified according to their geographic extent of coverage. Line-of-sight radio aids (those that operate at frequencies above 100 MHz) do not propagate beyond the horizon, whereas low-frequency systems (those that operate below 100 kHz) have coverages of several thousand miles. With satellite-based systems, millions of square miles of Earth are covered by each of several satellites.

Human interface. Navigation has become automated as the on-board computer selects modes, stations, and waypoints, and calculates position without human intervention. Hence, instead of requiring a skilled navigator on each vehicle, a smaller number of even-more-skilled engineers is required to design the automated systems. In some systems, humans select the navigation stations; in others, the selection is automatic. In

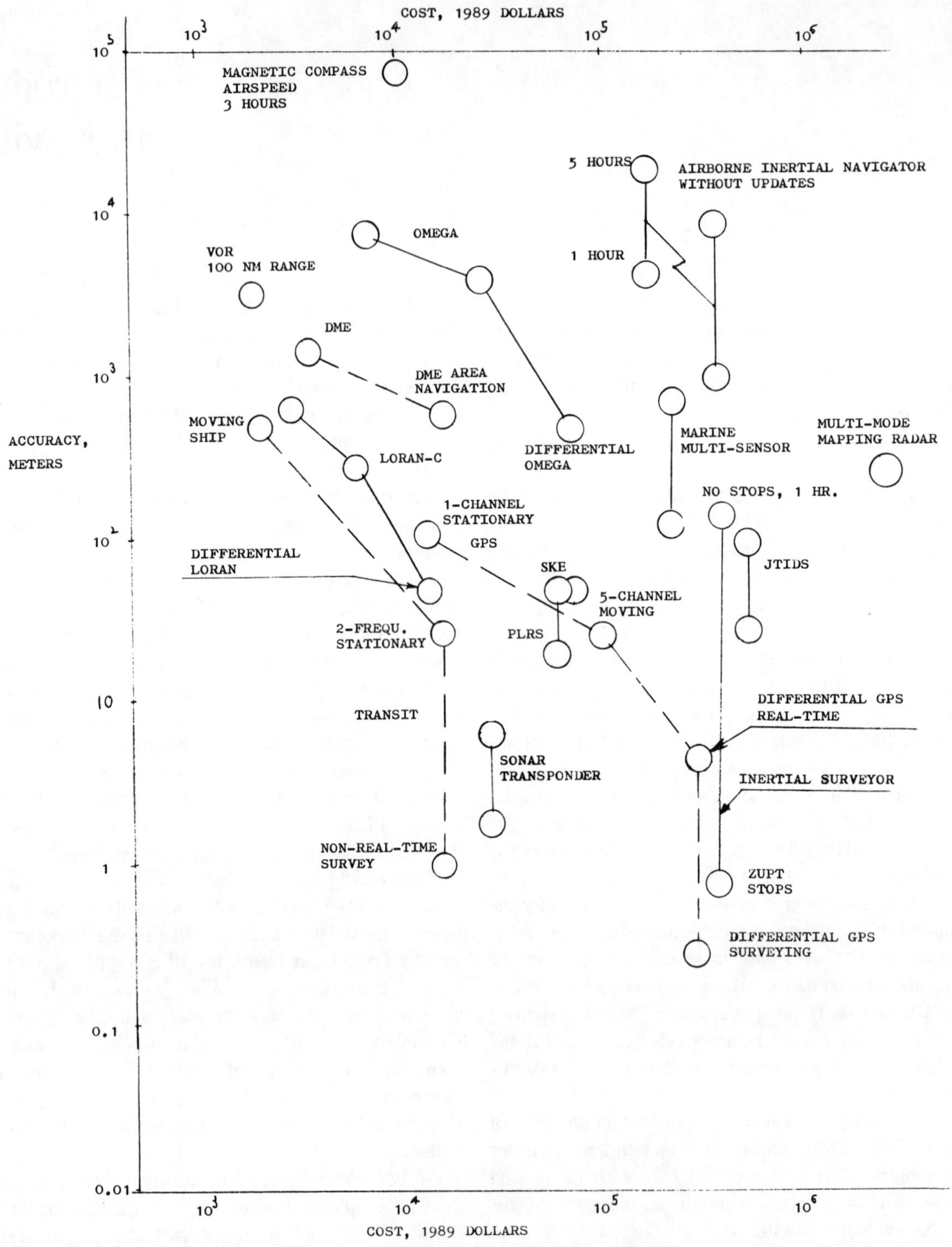

Fig. 1-1. Cost versus accuracy of user equipment.

most manned vehicles, humans select the waypoints that define a route and humans monitor equipment failures.

Navigation sensors are portable and the calculations are done in real time, the time delay being one of the contributors to system error (e.g., a half-second unpredictable delay at a vehicle speed of 1000 ft/s gives a 500-ft position error). When the position of a fixed point is measured, the process is called ''surveying.'' Surveying differs from navigation because:

—surveying sensors may be erected on the site,
—several days may be allowed to calculate the position,
—much greater accuracy is usually demanded.

Some Army survey systems (e.g., for cannons) require such rapid results that the distinction between navigation and surveying is blurred.

Table 1:1 lists the Western radio navigation aids operating in 1989: each is described in the Introductions to Parts 3 through 6. The table shows the number of transmitting stations

TABLE 1:1

1989 RADIO NAVIGATION AIDS, SHOWING THE APPROXIMATE NUMBER OF TRANSMITTING STATIONS AND USERS*

System	Described in Part	Frequency		Number of Stations	Number of Users		
		MHz	Band		Air	Marine	Space
OMEGA	5	0.010–0.013	VLF	8	12,000	8,000	—
LORAN-C	4	0.100	LF	50	60,000	400,000	—
DECCA	4	0.070–0.130	LF	150	2,000	40,000	—
Beacons†	5	0.200–1.6	MF	3000	200,000	500,000	—
ILS†	5	108–112, 329–335	VHF, UHF	2000	200,000	—	—
VOR†	5	108–118	VHF	2000	250,000	—	—
SARSAT/COSPAS	3	121.5 243, 406	VHF UHF	5 satellites	100,000	20,000	—
TRANSIT, CICADA	4	150, 400	VHF, UHF	6 satellites	—	100,000	—
PLRS	3	420–450	UHF			In development	—
JTIDS	3	962–1213	L			In development	—
DME†	5	962–1213	L	1000	100,000	—	4
TACAN	5	962–1213	L	1500	20,000	—	4
SSR/IFF/TCAS†	5	1030, 1090	L	800	250,000	—	4
GPS	3	1227, 1575	L	24 satellites, 1992		Being deployed	6
GLONASS	3	1240–1260 1597–1617	L	24 satellites, 1992		Being deployed	
SCN	6	1760–1850 2200–2300	S	7		—	200
STDN	6	2025–2150 2200–2300	S	3 satellites 10 ground		—	50
SKE	5	3370–3510	S	None	1,000	0	0
Radar Altimeter	5	4200	C	None	10,000	—	—
MLS†	5	5031–5091	C	5		Being deployed	—
FPQ-6/FPS-16	6	5.4–5.9	C	10	—	—	
Weather/Map Radar	5	~10	X	None	3,000	—	3
Shuttle Rendezvous Radar	6	13.9	Ku	None	—	—	4
Airborne Doppler Radar	5	13–16	Ku	None	5,000	—	—
SPN-41	5	15.6	Ku	50	500	—	4
SPN-42/46 radar	5	33	Ka	50	500	—	—

* By Sven Dodington and Myron Kayton.

† ICAO standard format.

and the number of marine, air, and space users of each system. It also shows the letter designations of the navigation frequency bands that correspond to regulatory assignments (see IEEE Standard 521-1984). Regrettably, another series of letters, in alphabetical order, has been arbitrarily assigned to radio frequencies by the military electronic-warfare community. Those bands are concerned with larger issues than navigation and regulatory assignments (jammers do not respect legal niceties). The navigation community hopes that the electronic-warfare community will change its rationally inspired designations, perhaps to numbered bands, to avoid confusion with the older, lettered bands.

Guidance and navigation are closely related. A guidance system determines the direction to steer without necessarily calculating absolute position. The simplest guidance system points the vehicle toward the destination at all times. If the vehicle moves faster than the destination (which may itself be moving), the vehicle will eventually get there, perhaps by a convoluted route. More complex guidance systems "lead" the destination by pointing ahead of it. Typical guidance sensors are radars, infrared telescopes, and radio-homing receivers. One example of a guidance system is the Instrument Landing System (ILS, Part 5.4), which is installed at every major airport in the western world. During the landing approach, the aircraft's receiver measures up–down and left–right deviations from a radio beam that emanates from antennas near the runway. The pilot or autopilot flies the aircraft so that the deviation is zero, thereby flying a straight path almost to touchdown. The ILS is called a "guidance system" because there is no explicit calculation of the airplane's position or velocity relative to the airport. Guidance systems are usually much less expensive than navigation systems.

During the last two millennia, whole navigation technologies were created and then made obsolete (e.g., sandglasses, log-trig computations, the calculation of longitude by observing magnetic variation, and celestial fixes at sea). Other technologies were perfected, ceased to advance, then revived under the impetus of microcomputers (e.g., the compensation of the magnetic compass). Others, still, are completely new (e.g., satellites). I try to explain what happened and why, in the context of new technology and in the changing priorities of each historical period.

I have proposed "Kayton's Law of Constant Travel Time," which says man will make casual, frequent trips that take 1 day (perhaps 10 h). Until the domestication of the horse, this gave man a normal range of 10 to 20 mi; after 2000 B.C., it meant 50 mi on horseback or wagon. Man's range of travel did not change for 4000 years until the invention of the railroad, when casual trips of 200 mi became possible. The invention of the automobile in 1900 did not increase his radius but again put the vehicle under the traveler's control (as was the horse). The development of 75 mi/h steam trains during the 1930s doubled man's range, as did fast automobiles and paved highways. The introduction of pressurized, propeller-driven commercial aircraft after World War II raised man's radius to 2000 mi and the long-range, wide-body jet raised it to 5000 mi, in 1975.

References and Bibliography

Institute of Electrical and Electronics Engineers, "Definition of navigation aid terms," IEEE Standard 172-1983, 30 pp. Abstruse definitions of more than 600 terms.

Institute of Electrical and Electronics Engineers, "Standard letter designations for radar frequency bands," IEEE Standard 521-1984. Nominal range of each band and specific frequencies assigned by the World Administrative Radio Conference. To be updated in 1990.

Philipot, T., *An Historical Discourse of the First Inventions of Navigation.* London: W. Godid, 1661. Desert compass on camelback, 1500. Speculations on making a true-north-pointing needle.

Sherrett, A. (Ed.), *Cambridge Encyclopedia of Archaeology.* Cambridge: Cambridge University Press, 1980. Boats, populations, agriculture, horse.

Skolnik, M. I. (Ed.) *Radar Applications.* New York, NY: IEEE Press, 1988. Extraordinary book of well-chosen reprints on radar system design with perceptive introductions by the editor, who is a true master. Emphasizes U.S. and U.K. technology. History, air-traffic control, frequency selection, military air defense, early warning, airborne radar including moving-target detection, terrain following radar, and space-based radar. Reprints several papers cited in the bibliography of this book. Little on naval radar.

Swords, S. S., *Technical History of the Beginnings of Radar.* London: P. Peregrinus Ltd., 325 pgs., 1986. Pre-World War II VHF radar at 100–500 MHz. Details of World War II Chain Home bistatic radar.

Wrigley, W., "History of inertial navigation," *J. Inst. Navigation*, Spring 1977, pp. 1–6. Excellent treatment, emphasizing ballistic missiles and SINS. Good bibliography.

Part 2
Introduction to Cartography

NAVIGATION can be no better than the underlying knowledge of the shape of the Earth and of the position of landmarks and radio transmitters. The demands of navigators have often impelled advances in map-making and geodesy. Conversely, map improvements have sometimes motivated the invention of new navigation techniques. The history of maps and geodesy itself is tied to the collection of data by astronomers, surveyors, and explorers.

2.1 Ancient Astronomy

It is often said that shepherds and farmers everywhere (who daily observed the Sun, Moon, and stars) learned astronomy. True, they discovered the direction of north and learned the azimuths of the rising and setting of heavenly bodies at various seasons. However, the discovery of long-term astronomic phenomena required the collection of data for decades, even centuries. In preliterate societies, a year's worth of data was a lot to remember since information on planet motions and star constellations was passed from generation to generation in the form of poems, some of which survive from Greece.

In 8000 B.C., long before the introduction of agriculture, a few Europeans may have known the length of the year, as shown by a 3.5-year record of incised marks on an animal's shoulder blade. Among the Earth's many preliterate societies (preliterate in the sense of not writing narratives), very few organized the collection of data. In fact, the value of astronomical records may have shown them the value of literacy. Thus some societies (Mid-East, Egypt, Central America, and China) saved inscribed bones and sticks, thereby organizing the collection of written data, probably for religious/astrological reasons. I suspect that they discovered the Moon's 29-day cycle first, since it was easiest to count. By keeping long-term records, the primitive people living in the north (where the annual path of the Sun changes markedly) must have learned the movement of the Sun in elevation. They would therefore have discovered the approximate length of the year and correlated it with the seasons. Their brethren nearer the equator (like the Egyptians) are known to have correlated the rising of stars with the length of the year and with the seasons.

The Babylonians (1800 B.C.) and Egyptians (1500 B.C.) invented much of our basic astronomy, recording the position of the Sun in the ecliptic and dividing the horizon circle into 360° (they invented myths about why the year was 5 or 6 days longer than the ideal 360 days). They could probably predict some eclipses. They did not use their knowledge to define position on the Earth; that is, they did not create the concepts of latitude and longitude.

Both civilizations knew how to measure azimuth relative to astronomic observations. Some Egyptian temples of 1000 B.C. are apparently aligned to true north, to the solar solstices, and to the heliacal rising of Sirius.

By 2000 B.C., they had learned to add occasional days to account for the length of the year (actually 365.2422 days). They measured the Moon's synodic period (actually 29.531 days between full moons). In 380 B.C., the Greeks correctly identified 235 lunar months with 19 solar years. Only a few civilizations discovered that the Moon did not return to the same stars after 19 years, due to precession of its orbit (precisely locating the Moon required written observations based on a systematic way of naming background constellations and stars). The 12 Islamic lunar months drift rapidly relative to the solar calendar but the Koran explicitly forbids adding an intercalary month from time to time. To synchronize the lunar months with the solar year, 7 months must be added every 19 solar years, as in the Hebrew calendar. The biblical-age Hebrews (12th to 6th century B.C.) added months *ad hoc* by observation of the Moon; they could not predict when a new month would begin nor when an extra month was needed to retain synchronism with the solar calendar. The Moon has rarely been used for navigation anywhere, even in the 20th century when its movement can be predicted for decades ahead.

Digressing from navigation for a moment, the "week" has no astronomical significance. Many primitive people (such as the early Romans and the Polynesians) named each day of the lunar month (e.g., Day-1, Day-2, the Ides of March) so that weeks did not exist. The ancient Greeks had a 10-day week, the Republican Romans had an 8-day week, and the Mayans had a 13-day week. Other civilizations used weeks of 5 to 10 days, often tied to the schedules of itinerant markets. The French revolutionary government tried, unsuccessfully, to institute a 10-day week as an anticlerical, higher productivity measure. I think the Russian revolutionaries tried the same thing. Our 7-day week is a chance legacy of the Semitic people of western Asia that is at least as old as the ancient Hebrews.

Most primitive people did not collect written records and therefore did not learn quantitative astronomy. The Pacific Islanders, though they lived close to the sea, are a good example. A few early civilizations did collect astronomic data and analyzed them numerically for repetitions of orbits, eclipses, and planetary conjunctions. Mesopotamia and China are good examples. I suspect that to force the collection of data over many centuries, astronomic measurements were made into a religious ritual. The many societies that did not institutionalize the collection of data could not learn astronomy. Even of the data-collecting civilizations, only ancient Greece and some of its intellectual successors (e.g., western Europe) created geometric models to explain the numbers. Everywhere, the consumers of astronomical data were priests and astrologers.

Northwestern Europe became civilized much later than did the Mid-East in that farming arrived about 4000 B.C., the start of their Neolithic Age (versus 8000 B.C. in the Mid-East). Many Neolithic burial mounds, called "barrows" in England, are oriented to true north. Some stone circles (called "henges") and monumental stones (called "menhirs" in French) of 2000 B.C. are clearly oriented to a north–south axis. These aligned monuments are found throughout the British Isles, in western France, and in Spain. Late Neolithic henges may contain more complex sightlines based on moonrise and moonset. Some scholars have suggested that henges are analog computers for calculating eclipses.

Predicting eclipses without a knowledge of orbit mechanics requires centuries of patient observation and written records. The first civilization known to collect such data was Babylon, approximately 2000 B.C. The Chinese recorded eclipses, novas, and comets for astrological purposes from approximately 1300 B.C. but could not predict eclipses within a few hours by A.D. 1200. They formed no geometric picture of the Earth's orbit; they merely analyzed past records for periodicities.

I doubt that people just transitioning from hunter–gatherers to villagers would devote enough time to collecting astronomical data. I suggested, 15 years ago, that western European henges may have been aids to determine the proper planting time in a cold climate where the growing season is short and where overcast skies often obliterate the Sun. Henges appear to be outgrowths of early Neolithic "causewayed camps" built as palisaded circles, perhaps as places to expose corpses. I suspect that sightlines to sunrises and sunsets were observed and recorded with sticks in clear weather, then marked by stones. I suspect that the sightlines on favorable agricultural dates were identified after many generations and that the whole structure was used as a giant calendar during extended periods of sustained cloud-cover.

The Neolithic Age began in the Americas about 2500 B.C. when farm villages were created in Central America and on the west coast of South America. By A.D. 500, buildings had been constructed at some Mayan ceremonial centers in which the priests kept track of the 365-day solar year and the 260-day religious year (20 "weeks" of 13 days, each day numbered and each "week" given a sacred name). Figure 2-1 shows the author standing atop a building, usually identified as an observatory, in the now-abandoned Maya city of Chichén Itzá. We call it an observatory because it is a conical building. If the Indians used it as an observatory, they probably observed stars in the horizontal viewing slits, a few of which still survive. The large size of the slits (2 ft square) limited the accuracy of measurement; if the slits were used for astronomic observations, reed or wooden markers were probably placed in them. They did not, of course, have gimballed instruments, being Stone-Age people. Even in tropical Central America, the prediction of planting dates, tied to the onset of the rainy season, may have been the motivation for collecting astronomical data.

Most North American Indians were wanderers, a millenium behind their Central American cousins. Nevertheless, ceremonial centers existed. Burial mounds are found on a north–south alignment at Cahokia, Illinois (A.D. 1000). "Medicine circles" are found in the vicinity of Wyoming and Alberta, dating from A.D. 1500, that may have north-aligned radials and other radials oriented to the solstices. There is no evidence of celestial position determination as an outgrowth of North American, Central American, or South American astronomy. In this respect, their astronomy did not advance beyond that of the Babylonians and Egyptians.

Farmers and herdsmen did not travel more than 30 mi from their homes in their entire lifetimes and would therefore not have known that the elevations of the stars change with latitude. The concept of latitude probably arose among people who traveled on long north–south rivers. This could have occurred in North America, European Russia, Siberia, or Egypt; in fact, it apparently occurred only in Egypt (I suspect because ceremonial centers existed to collect the data). In 1000 B.C., no bright star was within 7° of the North Pole. Hence, people observed the star that just grazed the horizon (e.g., did not set). Travelers (principally sailors) undoubtedly discovered that as they moved north, fewer stars set. Mediterranean travelers were at a disadvantage since their trips were primarily east–west. Still, sharp-eyed sailors would have seen the difference between the stars that set on the north shore versus those that set on the south shore, 9° of latitude apart.

The concept of latitude was known in the eastern Mediterranean by 200 B.C., where Greek-speaking Alexandrians measured absolute latitude from the polestar and measured the latitude difference between places by observing the Sun's peak elevation (at local noon) on the same day in each place. Their scholars knew that the Earth was spherical and estimated its circumference as 17,000 to 25,000 mi (Columbus believed the smaller circumference). They understood the concept of longitude but could not measure the longitude difference between two points since they would have had to make simultaneous observations of the Sun at both locations and there was no way for two observers to synchronize their measurements within, for example, the minute of time needed for a 15-mi error in longitude. Hence, east–west distances were measured in travel time (walking and sailing) and were often in error by 50% on long distances. Hipparchus (150 B.C.) understood the spherical Earth so well that he proposed using lunar eclipses as a means of measuring longitude differences. Columbus did just that on his second voyage to the Americas (Part 4.4, pg. 152). Measurements of latitude after the time of Columbus are discussed in Part 4.3 and measurements of time are discussed in section 2.6 below and in Part 4.4.

2.2 Ancient Maps

The Phoenician sailors (1500–200 B.C.) must have had shoreline maps of the entire Mediterranean, built up over 1300 years of trading, but none apparently survived. The well-organized Romans must have had maps but, though tantalizing literary references exist, none have survived. The first known

Fig. 2-1. Chichén Itzá Observatory, Mexico, 1966. Photo by Paula Kayton.

world maps were published in A.D. 150 by Ptolemy in Alexandria, who began the convention that north is "up" (a convention that did not stick until 1500 years later). As expected, his maps had excellent north–south detail in the eastern Mediterranean (because latitude could be measured), poorer east–west detail there, and poor detail in all directions far from Egypt.

In 12th-century Europe, the *mappa mundi* was fashionable as a royal decoration. They were all stylistically similar; Jerusalem was at the center and east was on top. The eastern half of the map was Asia, the northwest was Europe, and the southwest was Africa. The *mappa mundi* was not drawn to scale. It showed mythical kingdoms and sketches of fanciful monsters and imagined inhabitants.

The first known modern map appeared in the 13th century (Fig. 2-2). It was called a "Portolan" by the Iberians who produced it by hand on sheepskin parchment. It showed the seacoasts and rivers of the Mediterranean in great detail, to scale, but with little detail inland. Figure 2-2 shows the Mediterranean with north on top and Jerusalem at the right. Shorelines are identifiable by the closely spaced names of ports, written perpendicular to the shore. A Portolan was constructed with two large circles, on the circumference of each of which were 8 or 16 equally spaced rose-centers. Part of one circle with 9 visible centers (out of 16) appears in Fig. 2-2. From each rose-center, 16 or 32 (in the figure) radials emerged, which were used for measuring steering directions. Most Portolans had no meridian lines but on some, meridians

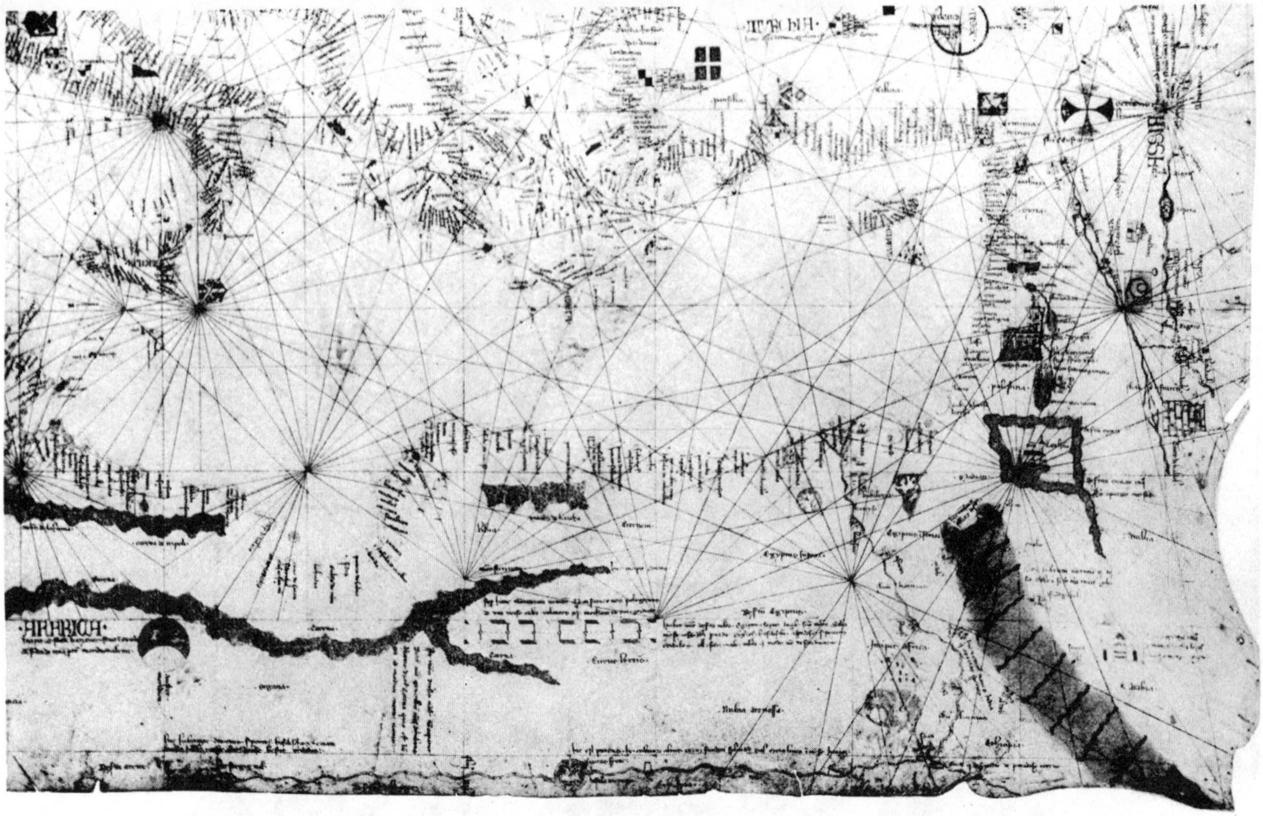

'ortolan Chart of Angellino de Dalorto in the Collection of Prince Corsini at Florence
Sheet 4. R. Geog. Soc. London, 1929

Fig. 2-2. Portolan chart of the eastern Mediterranean [from (The Portolan Chart of Angellino De Dalorto, 1325)].

were drawn parallel and north-pointing. Mercator's innovation, of distorting the vertical scale to match the secant-of-latitude distortion of the horizontal scale, is not present in Ptolemaic, Portolan, or Chinese maps. There is some speculation that the Portolans trace their history to presently lost Phoenician sailing charts. The earliest-known Portolans were hand-drawn in Catalonia, Mallorca, Genoa, and Venice by families of cartographers. Portolans changed little during their 300-year period of use because little new information was being discovered about the world. They disappeared in the 16th century as printed Mercator maps came into use and as new data were collected by explorers.

The invention of the magnetic compass in the Middle Ages (Part 4.2) reinforced the convention of placing north at the top of the map. North-oriented maps, however, were not universally adopted until the 17th century. For example, a large collection of maps of Italy, dated in the 1580s and prominently displayed in the Vatican Museum's Gallery of Maps, are mostly oriented with bodies of water on top. Hence, on the east coast of Italy, northeast is on top, whereas on the west coast, southwest is on top. A survey map of the Rhone River delta, in the Camargue Museum, has southwest on top for no apparent reason.

Chinese maps are known from 200 B.C. They were local survey maps kept by the civil bureaucracy for tax-collecting purposes. The oldest surviving countrywide map of China is dated A.D. 1137, with east up. Though the Chinese were using a north-east grid by that time, they never learned to distort the north–south scale, as did Mercator. Many Chinese maps were drawn with south, an auspicious direction (barbarians invaded from the north), at the top. By 1550, some Chinese maps, drawn by visiting European missionaries, had west on top. Arab "wheel-maps" exist from A.D. 1000. They resemble the European *mappa mundi;* they do not represent real terrain and were probably made for religious purposes. A Turkish map dated 1513 survives, attributed to Admiral Piri Re'is. It contains details of the Americas and Antarctica at a time when no data were available. The map is unexplainable and is probably a forgery.

European maritime maps were protected as secret information until printing technology made dissemination inevitable. Maritime maps from the 13th to 16th century were often weighted, to be tossed overboard rather than allowed to fall into the hands of competing traders. Even in the 20th century, the Soviet Union considers its own maps secret. Two German generals, writing about their World War II attack on the Soviet Union, said, "We were not prepared for what we found because our maps in no way corresponded to reality . . . . Supposed main roads, and there seemed to be many, often proved to be merely sandy tracks. Railways that we were supposed to use simply didn't exist.'' Even in the 1980s, the Soviet Union still deliberately printed erroneous maps on which a few cities and rivers are said to be displaced as much as 200 mi. For a further discussion of maps used by navigators, please see section 2.5 below and Parts 4.1–4.4 starting on pg. 147.

2.3 Surveying

The ancient Egyptians and Babylonians surveyed their land to measure boundaries, to establish irrigation-canal routes, and to construct monumental buildings. The Egyptians had the problem of retrieving land boundaries after Nile floods. This was probably done relative to landmarks or stone boundary markers, though the accuracy of their surveying is unknown. In both civilizations, land was apparently owned by the state and assigned to farmers in return for taxes. Hence, there was a motivation to erect boundary posts and retrieve their location when destroyed. The first surveying record, I believe, was a standard cubit found in the Snefru pyramid, 2500 B.C. Later Egyptian paintings show surveying sticks and squares. Chinese drawings of A.D. 200 show surveying ropes, plumb lines, and leveling rods. Prof. Thom, in 1967, proposed that the megalith builders of England (2000 B.C.) had a ''Neolithic yard'' of 2.72 ft, which was used at many construction sites. This would imply that there was a class of itinerant construction supervisors but I doubt that the rate of construction would have supported such specialists so early in man's history nor would rival clans have employed them.

In feudal days, each local lord, whether in Europe, the Mid-East, China, or Japan, claimed to own all the land he could control by force. Some surveying skill was required to place markers at the boundaries of land assigned to serfs but the serfs worked the land at the lord's pleasure so permanent boundaries may have been unimportant. Surveying the borders of the lord's land was unnecessary since these were regions of permanent war with neighboring lords and the least favored serfs probably lived there.

The existence of European surveying instruments was recorded in medieval times, when the peasants had been freed and small private tracts came to be prized. The first book on surveying appeared anonymously in 1523. Length standards were entirely local, as they had been earlier. During the 300-year period of construction of Gothic cathedrals in northwestern Europe, itinerant construction crews may have carried their own standards and construction drawings from site to site. Ample evidence shows that markets and churches standardized weights and measures locally. Itinerant merchants calibrated their utensils relative to the standard bowls and lengths in each town. In England, nationwide standardization was attempted in Norman times (A.D. 1300) but was resisted by local barons. Length units were standardized throughout England in the 1550s and the metric system was invented and implemented in France in 1790 as an aftermath of their revolution. It is said that 112 different length units were used in the German principality of Baden by 1800. Six different length units, all called ''ch'ih,'' were used in Beijing in 1939 and two others in Shanghai. Their length varied from 12.3 to 14.1 in. In all countries, it was an industrial revolution that enforced nationwide standardization.

The Chinese, Mayans, Babylonians, and Europeans collected vast amounts of naked-eye data for court astrologers and priests. For example, the Nuremberg observatory in Germany, founded in 1472, used calibrated rings to measure azimuth and elevation. Lines of sight were measured along sticks and tubes, without telescopes. For example, Tycho Brahe's observatory in Denmark was nontelescopic. A nontelescopic Chinese observatory of A.D. 1276, equipped with naked-eye measuring rings, was derived from Persian predecessors.

Magnifying glasses, made of naturally occurring rock crystal or glass (a man-made precious material until about A.D. 1100) existed in antiquity. The first recorded mounting of magnifying lenses in eyeglass frames occurred in Florence about A.D. 1300. The first telescopes were apparently used for magic shows in late–16th-century Europe, based on the new technology of two coaxial lenses. In 1609, Galileo trained his telescope on the Moon and Jupiter. Within 20 years, telescopes had been adopted by European army officers and by astronomers. A few local monarchs squandered their taxpayers' money by subsidizing telescopic observatories. By 1670, astronomical refractor telescopes had reached a length of 17 ft. In Europe, naked-eye observatories were retrofitted with telescopes while the Chinese continued to use naked-eye data. We must again note that before the age of navigation, astronomy was funded entirely by astrologers, to predict the future, and by churches to fix the exact dates of festivals.

In Portugal, the kings were merchants (dealing in gold and slaves) and did not disdain trade, as did most other kings. Hence in 1418, Prince Henry the Navigator founded the world's first known navigation ''academy'' at Sagres, Portugal, to collect navigation information and create secret navigational charts for his ship captains who were exploring Africa, and, later, the Americas. In our terms, his academy was a state-owned central data bank. Returning crews were debriefed by a *junta dos matematicos*. The academy did not include an astronomical observatory but some historians credit it with inventing the cross-staff (see Part 4.3). When international shipping became an important industry after 1700, the ruler of each maritime country founded an observatory that served as the zero meridian for his nation's ships (see Part 4.4) and as the origin of its national survey.

In 1635, the circumference of the Earth was measured, in England, by dividing the distance traveled along a land meridian by the latitude difference (obtained from star elevations). A similar measurement was made in France in 1669. In the very late 1600s, the newly created French national observatory (still located just south of the Jardin du Luxembourg in Paris) began to collect latitudes and longitudes of cities around the world. Longitude was measured by observing star transits, time being measured with a pendulum clock whose period and epoch were calibrated at each site by the eclipses of Jupiter's moons. That method of measuring longitude had been suggested by Galileo in 1636 when he realized that Jupiter's great distance from Earth would cause the observations of Jovian satellites to be the same everywhere on Earth. Galileo's method was impractical at sea, as explained in Part 4.4, though many astronomers of the time kept trying. Thanks to the land-based data, the map of the world, as known to western Europeans, began to assume a realistic appearance. The best map, Cassini's 1693 French "planisphere," showed California as an imprecisely located island, because Cabrillo, Drake, and Vizcaino explored the coast before the precise measurement of longitude was possible.

In 1736 the Frenchman, Maupertuis, led expeditions to Peru and Lapland that measured the radius of curvature along the meridian in tropic and arctic regions (by measuring distance and latitude difference), thus establishing the ellipticity of the Earth in the meridian plane for the first time in history. The joint pressure of surveying and navigation accelerated the creation of national observatories all over western Europe and in the United States to collect data and establish "reference ellipsoids" that best matched their local territory. The Clarke Ellipsoid of 1866 is still used in North American surveys though data throughout the Western world are being converted to the 1984 World Geodetic System (see section 2.4).

The measurement of longitude differences was easy after the American radio station, WWV, started broadcasting time in 1923. The radio signal allowed celestial measurements to be made at a known time and, hence, the precise measurement of longitude (see Part 4.4).

Prior to 1960, precise land survey consisted of measuring the angles in a network of triangles. The length of occasional sides was measured with hand-held tapes. Heights were measured between vertices with telescopic "levels" and rods. In the late 1950s, optical and microwave ranging devices began to augment the hand-held tapes and more reliance was placed on the length measurements. Astrogeodetic and gravimetric measurements were always made at occasional vertices.

The reprint by Ewing and Mitchell, pp. 15–32 (from their textbook on geodesy) describes the traditional techniques of surveying with tape and theodolite, and the methods of establishing the astrogeodetic stations. These eminent geodesists show how to calculate a celestial fix, including corrections for refraction and deflection of the vertical.

In 1989 the United States was crossed by a few high-precision traverses (lengths measured to one part in 10^6). A denser net of first-order monuments exists. The lengths of the legs in the net are accurate to one part in 10^5, based on astronomical and satellite measurements. A still-denser net of second-order survey monuments and markers exists whose lengths are accurate to one part in 20,000 to 50,000. Builders and land-developers use third-order surveys (accurate to one part in 5,000 to 10,000), propagated from the second-order markers. Aerial surveys are widely used to make third-order maps, especially in difficult terrain. Hiran and Shiran are sometimes used to calibrate the aircraft's location relative to known ground stations, where the transponders are placed. Most navigators are content with third-order map accuracy but want it over long distances on a world-wide basis.

GPS satellite receivers (Part 3.5) have already demonstrated surveying accuracy to one part in 10^8 after a few hours of observation at each end of an arc [see (Allison *et al.,* 1988), (Bertiger *et al.,* 1988), and (Remondi, 1988); and the reprint by Carson, *et al.* in Part 6]. Hlasny and Schleppe (1988) have described the daily routine of a GPS survey in rugged terrain. The reprint by Kouba, pp. 44–57, shows how Transit satellite fixes are incorporated into the geodetic net. He uses differential Transit (Part 4.10) measurements to a claimed accuracy of half a meter.

In the 1980s, inertial navigators were used as surveying aids where rapid setup was required, as in erecting field artillery. Inertial surveying devices have achieved accuracies of 10–20 cm (5 to 20 parts per million) and measurements of the gravity deflection of the vertical accurate to 0.3 arcsec, after lengthy data reduction. In order to reduce the divergence of velocity errors, the vehicle must stop periodically so velocity can be zeroed. These "zero velocity updates" (ZUPTs) typically occur for 1 to 5 min every 5 to 20 min. Huddle's reprint, pp. 33–43, describes the design of an inertial surveyor and the test results. Wong and Schwarz (1988) have also described an inertial surveyor and Caspary and Konig (1986) have described test results on three inertial surveyors showing sub-meter accuracy. In the next generation, survey accuracy and speed will be improved by combining GPS and inertial data.

2.4 Global Geodesy

Long-distance travelers want to navigate on a worldwide ellipsoid, not on a succession of national grids. A navigator wants an ellipsoid that is symmetrical about the polar axis and that reasonably approximates the geoid. On the other hand, a geophysicist wants to know the exact shape of the geoid, by fitting a triaxial ellipsoid, whose equatorial radii differ by approximately 150 m.

Before the electronic era, the only way to tie national grids together was to observe lunar occultations simultaneously (by mechanical clock) from each grid. That method also determined the longitude of islands. It suffered from clock error, parallax, and uncertainty in the exact time of occultation because each observer saw the occultation behind a different portion of the rough terrain of the Moon. In 1940, the unmeasurable offsets among major grids were estimated to be less than 2000 ft, though some islands were later found to be a mile in error. In the 1950s, the ballistic missile program in the United States introduced a new era in geodesy by demanding more accurate relative positions and orientations of the national ellipsoids. The first solution was to conduct a traverse between national grids using airborne VHF ranging equipment and ground transponders (called Shoran, Hiran, and Shiran). One traverse, in 1953–56, connected the American and European grids from Newfoundland to Scotland via Greenland, Iceland, and the Faroes Island with an error of one part in 2×10^5. Another Hiran traverse circumnavigated the Caribbean Sea with a 25-m closure. A third traverse connected South Africa to the western European grid at Crete. Ewing and Mitchell (1970) and Laurilla (1976) have described these surveys and the electronic equipment that was used. In a Shiran survey, an aircraft interrogates two ground stations simultaneously, in each of two as-yet-unconnected nets, 10 times per second.

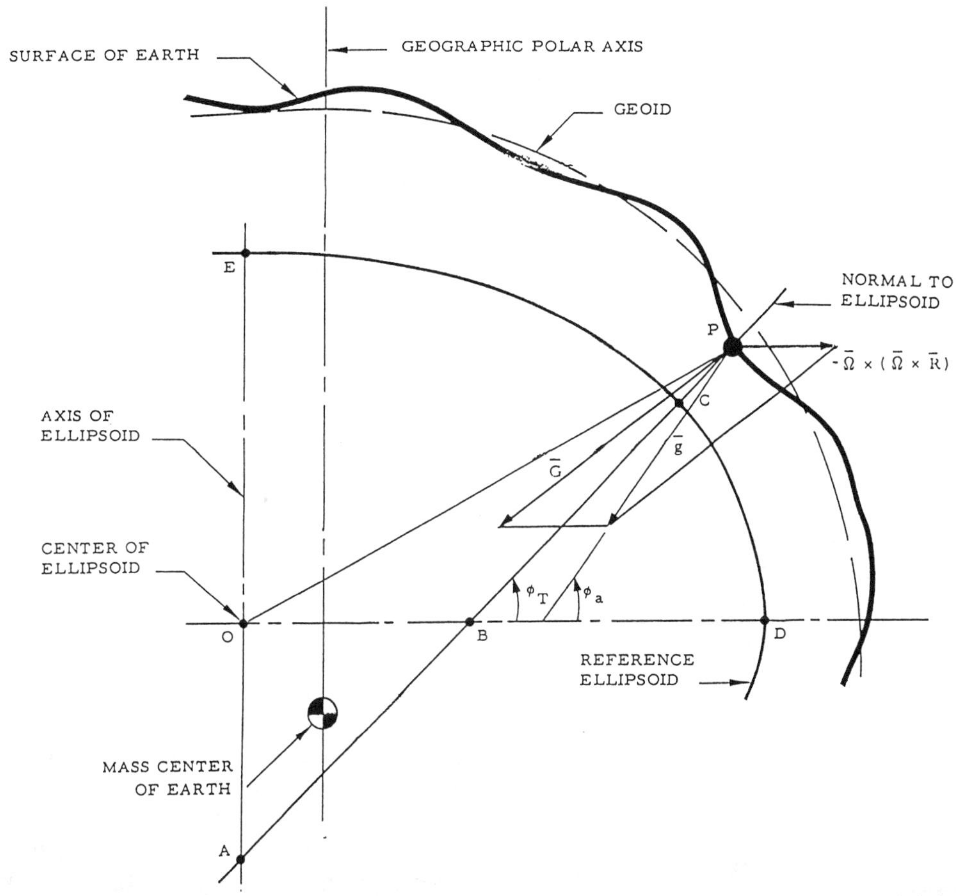

Fig. 2-3. Shape of the Earth [from (Kayton, 1960)]. ϕ_T = geodetic latitude; ϕ_a = observed astronomic latitude.

As a result, the first worldwide ellipsoid, the 1956 Hough, was calculated based on arcs longer than 100°. Figure 2-3 shows a "best-fitting" ellipsoid. The two-dimensional angle between the normal to the ellipsoid at any point and the direction of gravity at the same point is called the "deflection of the vertical." A best-fitting ellipsoid minimizes the mean-square deflection of the vertical which is the same as minimizing the mean-square geoidal height above the reference ellipsoid.

A parallel effort created the 1957 Columbus ellipsoid, based on precise gravity measurements (in 5° × 5° squares, mostly in the Northern Hemisphere). In 1957, Fischer combined the long-arcs and gravity measurements to get a worldwide best fit with an ellipsoid of rotation whose inverse flattening was preselected as 298.3. NASA used that ellipsoid to survey its space tracking stations.

After 1960, a major effort was undertaken to use Earth satellites to calculate the fine structure of the shape of the Earth. The earliest satellite geodetic measurements were made by photographing sunlit satellites against the background stars. Later, flashing lights on satellites were photographed by ground-based, gimballed "Baker–Nunn" cameras. Page 298 of the American Geophysical Union reference (1977) shows sample photographs from which geodetic measurements were made. Later in the 1960s, small satellites were launched with an electronic beacon or transponder at VHF, which permitted accurate ranging. Among those geodetic satellites were Anna, Beacon Explorer, and Geos. In the 1970s, Transit satellite information was used for surveying. Satellite geodesy, including the use of GPS, is discussed further in Part 6.3, pg. 317. The present world standard is the U.S. World Geodetic System (WGS-84) ellipsoid, which is symmetric around the polar axis. Its semi-major axis is 6,378,137 m, its semi-minor axis is 21.385 km shorter. WGS-84 is a modification of the U.S. Navy's Transit satellite Doppler Reference Frame. Gravity harmonics are derived from it to 180th order, corresponding to 1° × 1° squares on the surface of the Earth. The Defense Mapping Agency reference (1987) describes it, including information about transforming older datums to WGS-84. Parker and Bartholomew (1988) have discussed the worldwide ellipsoids that are stored in the standard military GPS navigation receiver.

2.5 Modern Maps

Hard-copy maps are projections of portions of the reference ellipsoid onto flat surfaces. Dozens of such projections have been invented. The Mercator projection, used by mariners for 400 years, is described in Part 4.2. The Lambert Conformal Conic projection is often used for aeronautical charts, while the Transverse Mercator projection is the standard for Army maps. Universal Transverse Mercator (UTM) maps are divided into "grid zones" and are numbered in meters of "eastings" and "northings." Figure 2-4 shows the Lambert Conformal Conic and Transverse Mercator projections picto-

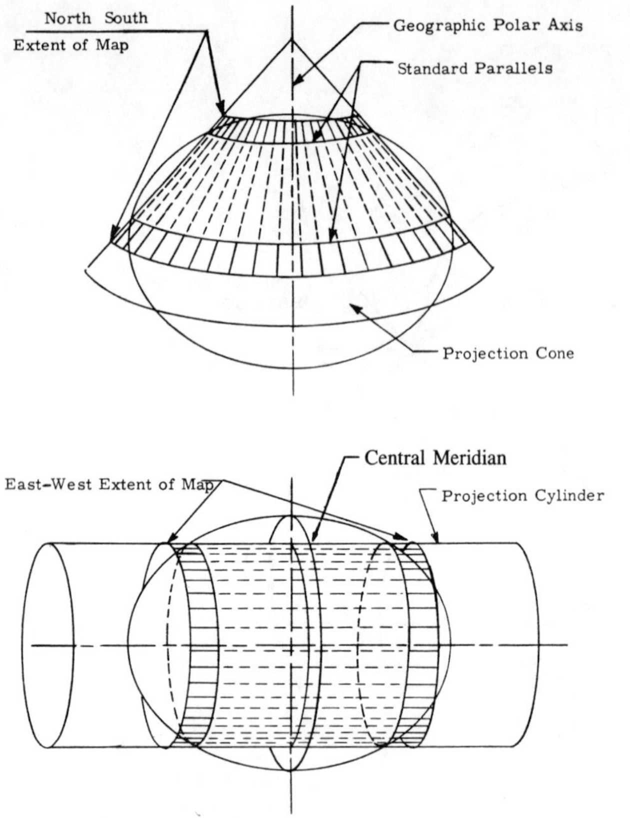

Fig. 2-4. Common map projections [from (Kayton, 1960)].
Top: Lambert Conformal Conical Projection
Bottom: Transverse Mercator Projector.

rially. The precise transformation from an ellipsoidal Earth to a flat *x–y* map grid is a mathematical power series in latitude and longitude.

Digital maps store the exact position and height of points on Earth relative to a reference ellipsoid. The data are usually stored on magnetic media but are now being delivered to users on compact disk (see also Parts 3.3 and 8.2).

In the 1980s, maps were available in traditional paper copy and in digital form from government organizations such as:

—the U.S. Geological Survey (topographic, natural resources),

—the Census Bureau (streets, incomes),

—the Department of Agriculture (crops, yields, forest cover, soils),

—the Environmental Protection Agency (pollution)

—the Defense Mapping Agency (airports, targets, navigation aids),

—county offices (real-estate plats, assessments)

and many other sources in all major nations. The digital data include latitude, longitude, altitude, magnetic variation, gravity deflection of the vertical, and identification of radio navigation stations, for example. The U.S. Navy has extensive digital maps of the ocean floor (useful for position fixing) and digital records of ocean temperature and pressure versus depth (useful for calculating sonar propagation speed and attenuation). Most of the data are freely available. The dissemination of digital data is slow (perhaps a week to order data that are a month old), but I foresee a new industry of commercial digital-map "wholesalers" who are permitted real-time read-only access to the government data bases and who retail the data in specialized formats on tape or optical disk to end-users, such as private map makers, or by radio to ships at sea. Commercial users of digital maps include real-estate developers, mass-marketers, delivery services, and an ever-expanding community that is finding innovative uses of maps, well beyond navigation. The reference by Rogoff (1988) is a plea to standardize the format of digital maps, worldwide. Crews would no longer board ships and aircraft with heavy briefcases full of charts.

2.6 Time

The legal definition of *time intervals* is based on a narrow spectral line of the cesium atom, measured by averaging the frequency of a cluster of (typically 20 to 25) cesium clocks. A single cesium clock holds frequency within one part in 10^{12} to 10^{13} relative to the average for several days, or one part in 10^{10} to 10^{11} for a few minutes. *Absolute time* is the interval that has elapsed since a past event. Astronomical absolute time, "ephemeris time," is usually measured from noon Greenwich time on January 1, 1900, the start of the first Julian day of this century. NASA measures orbits from the crossing of the vernal equinox in 1950. Celestial navigators measure Universal Time, the time between consecutive transits of any star. Since the Earth's rotation is noticeably irregular (milliseconds/day) relative to accumulated atomic-clock counts, Universal Time Coordinated (UTC) is defined. UTC is Universal Time corrected for polar migration and adjusted to an integral second twice per year. Hence, UTC is the accumulated atomic-clock counts since the last leap second was added. To avoid the discontinuity of a leap second, GPS satellites maintain continuous UTC (called UTCG).

In order to measure position within 0.25 nm, a navigator must know time:

—within a second absolute, to calculate position from star measurements;

—within 0.1 s absolute, to calculate position from low-altitude satellites; or

—within a microsecond to calculate position from electronic ranging devices. If two-way ranging signals are exchanged (as in TDRS or DME), the microsecond accuracy must be maintained during the round-trip travel time of the signal; if a one-way signal is measured (as in GPS or one-way Loran or Omega), the accuracy must be 1 μs absolute.

The oldest time-measuring instruments are the sundial and the water clock. Water clocks were found in Egypt (1500 B.C. at Karnak), ancient Greece, and China. They measured the volume of water drained from an orifice. The more advanced clocks kept a constant head above the orifice so the rate of drainage was always the same.

The earliest subdivisions of a day were "hours" that were equal fractions of the sunlit day, hence of variable length throughout the year, especially in northern countries. Before artificial light, astrologers and monks were among the few people who subdivided the night. In Christian and Moslem countries, the night was divided into equal "hours" for the

Fig. 2-5. Yemenite astrolabe in brass, A.D. 1296. Diameter = 6.2 in. Courtesy of the Metropolitan Museum of Art.

purpose of determining the proper times for prayer. The intervals were measured by burning standard candles. Chinese astrologers and navigators measured time intervals by burning joss sticks. In the 12th or 13th century, Europeans invented processes to make inexpensive glass, whereupon sandglasses replaced candles at sea (Parts 4.4 and 4.5).

The astrolabe (Fig. 2-5) was in use before A.D. 600 but the oldest surviving instrument is Arabic, from A.D. 930 (in the Kuwait National Museum). The astrolabe was an analog computer that calculated local solar time from a measurement of peak Sun elevation ("noon") or peak star elevation ("meridian transit"). The measurement was plotted on scales of constant elevation (called "almucantars") on a rotatable plate. In daytime, the plate was rotated to match the measured elevation with the known position of the Sun on an ecliptic scale. At night, the plate was rotated to match the measured elevation with star marks (of which 19 are visible in Fig. 2-5 as points within the upper circle). Day or night, the local time was then read from a scale on the rim. All astrolabes were constructed by projecting stars onto the plane of the instrument in a stereographic projection described by Ptolemy but known at least since 150 B.C. By A.D. 1600, the most complex astrolabes allowed for the eccentricity of the Sun's orbit so mean local solar time could be measured. Astrolabe technology was highly developed in Moslem countries as the preferred means to determine the five required times of prayer each day. Al-Aqsa Mosque in Jerusalem, for example, displays several sacred astrolabes. The modern counterpart of these sacred astrolabes is a hand-held electronic calculator that displays the name and time of day of each prayer, once the user's latitude and longitude are entered on the keyboard.

Mechanical clocks were invented in Europe in the late 1200s and installed in public buildings, usually in churches that were rich enough to pay for the clock (which was built on site) and to hire a human clock "governor," probably a clever blacksmith. These clocks announced the hours by a bell. An hour hand was added in the mid-1300s to display time visually. Clock accuracy did not improve enough to add a minute hand until just before 1700, when the pendulum and anchor escapement were invented. Those clocks were not portable; travelers carried a sundial or astrolabe until costly pocket watches slowly made their appearance after 1600. The accuracy of measuring time steadily improved, by star transits and better mechanical clocks, until the invention of atomic clocks in the 1950s.

Today, portable clocks can be reset within 30 μs of UTC by Transit satellite (Part 4.10) and within 2 to 10 μs of UTC by Loran-C, soon to be improved to 0.1 μs. Loran synchronization is discussed in Part 4.8. Optimistic metrologists are predicting 0.01 μs accuracy with differential GPS (Part 3.6).

A vehicle that knows absolute time within 1 μs can measure one-way range to Loran and Omega stations, within 300-m error, by measuring the arrival time of a radio wave emitted at a known time. The propagation velocity is compensated for ionospheric height (allowing for solar flares) and terrain. Clocks on board ship are discussed in Part 4.4. To synchronize an orbiting clock (e.g., on GPS) relative to an Earth-bound clock better than one part in 10^{10}, a relativistic correction must be included since time appears to run slower in orbit.

The method of disseminating time has progressed from watching the village clock (1300), to carrying a pocket watch (1600), to using the telegraph (1850), to the radio broadcasting of time (1923), to telemetered time (1950), and to using radio navaids (1980).

The reprint by Klepczynski, pp. 58–63, describes the time services of the U.S. Naval Observatory and the effect of timing errors on navigation accuracy. Fruehauf (1982) has described the many quartz and atomic clocks that can be purchased as time references for ground stations and vehicles. Timing errors are also discussed in conjunction with each radio navigation aid in Parts 3–6.

References and Bibliography

Allison, T., *et al.*, "C/A code dual-frequency surveying receiver," in *Rec. IEEE Position Location and Navigation Symp.*, 1988, pp. 434–441.Detailed description of 2-frequency, 10-channel Trimble receiver for surveying. Surveying results on 6- and 269-km baselines.

American Geophysical Union, "National Geodetic Satellite Program, Parts 1, and 2," NASA SP-365, 1977, Thorough description of Baker–Nunn photograpy and electronic ranging to GEOS, ANNA, PEOLE, Beacon-Explorer, etc.

Anon., *Explanatory Supplement to the Astronomical Ephemeris.* London: Her Majesty's Stationery Office, published annually. Excellent observational astronomy.

Bertiger, W. I., *et al.*, "Demonstration of sub-meter GPS orbit determination and high-precision user positioning,'' in *Rec. IEEE Position Location and Navigation Symp.,* 1988, pp. 185–194. Measured survey arcs for 8-h observation periods. Accurate to 1–4 parts in 10^8; several centimeters.

Brown, L. A., ''The longitude,'' in *The World of Mathematics,* J. R. Newman, Ed. New York, NY: Simon & Schuster, 1956, pp. 780–819. 1636 Galileo proposal to use four Jovian satellites for longitude. Academie Royale des Sciences chartered in 1666 to make maps. French use of Jovian eclipses to measure longitude in Europe and in their colonies.

Caspary, W. F. and R. Konig, ''Comparative study of Ferranti, Honeywell, and Litton inertial survey systems,'' *Bulletin Géodésique,* vol. 60, no. 2, pp. 143–153, 1986. Tests of three systems on a Munich range showing 1-cm accuracy.

Defense Mapping Agency, ''Department of Defense World Geodetic System 1984,'' DMA Tech. Rep. 8350.2, Sept. 1987. Dimensions, 364 constants of 18th-order Earth's gravity model, contour maps of geoid height, transformation equations, and parameters to other coordinate frames. Major axis is from 1980 Geodetic Reference System; center is from U.S. Navy Transit Doppler Reference frame.

Eddy, J. A., ''Medicine wheels and plains Indian astronomy,'' *Technol. Rev.,* Dec. 1977, pp. 18–31. North American alignments.

Ewing, C. E. and M. M. Mitchell, *Introduction to Geodesy.* New York, NY: Elsevier, 1970. Excellent treatment of classical geodetic surveying and mapping. Describes photography of satellites against stars and use of Secor ranging from satellites. Shows reduction of lines of sight to ellipsoid surface. Their Chapter 5 is reprinted in Part 2, pp. 15–32.

Falkenberg, W., *et al.,* ''GPS differential positioning technologies for hydrographic surveying,'' in *Rec. IEEE Position Location and Navigation Symp.,* 1988, pp. 310–317. Developed a method to combine pseudorange with carrier phase for 1- to 5-m error. Much of the paper is on multipath compensation for heave. Built oscillating mast and ran experiments.

Fell, P. J., ''TRANSIT and GPS—A report on geodetic positioning,'' *Bulletin Géodésique,* vol. 60, no. 3, pp. 181–192, 1986. Detailed report on geodetic use of Transit in the United States, China, and Japan.

Fruehauf, H., ''Precision oscillators and their role and performance in navigation systems,'' in *Rec. IEEE Position Location and Navigation Symp.,* 1982, pp. 118–125. Excellent descriptions of quartz and atomic clock devices with sizes, weights, costs, and accuracy.

Gingerich, ''Islamic astronomy,'' *Sci. Amer.,* Apr. 1986, vol. 258, pp. 74–83. Excellent, includes discussion of astrolabes.

Guye, S. and H. Michel, *Time and Space: Measuring Instruments from the 15th to 19th Century.* New York, NY: Praeger Publishers, 1972. 289 pp. (Published in French as *Mesures du temps et de l'espace,* 1970.) The best history of European clocks. Some material on ancient weights and measures.

Harley, J. B. and D. Woodward, (Eds.), *The History of Cartography,* vol. 1: *Ancient World and Medieval Europe.* Chicago. IL: University of Chicago Press, 1987. First of a projected six-volume work.

Hlasny, B. and J. B. Schleppe, ''Control surveys for British Columbia,'' in *Rec. IEEE Position Location and Navigation Symp.,* 1988, pp. 201–206. Use of five-channel Trimble GPS receiver, helicopter-delivered to remote sites. Details of survey operations and results.

Huddle, J. R., ''Historical perspective on estimation techniques for position and gravity survey with inertial systems,'' *AIAA J. Guidance,* vol. 9, no. 3, pp. 257–267, May–June 1986. Reprinted in Part 2, pp. 33–43.

Kayton, M., ''Coordinate frames in inertial navigation,'' *Progress in Astronautics and Rocketry: Vol. 8, Guidance and Control.* New York: Academic, 1962, pp. 367–394. Limitations on inertial navigation. Global geodesy; effect of deflection of the vertical.

Klepczynski, W. J., ''Modern navigation systems and their relation to timekeeping,'' *Proc. IEEE,* vol. 71, no. 10, pp. 1193–1198, Oct. 1983. Reprinted in Part 2, pp. 58–63.

Kouba, J., ''A review of geodetic and geodynamic satellite Doppler positioning,'' *Rev. Geophys. Space Phys.,* vol. 21, no. 1, pp. 27–40, Feb. 1983. Reprinted in Part 2, pp. 44–57.

Krupp, E. C., *in Search of Ancient Astronomies.* New York, NY: Doubleday, 1977. 300 pp. Review of Thom speculations, Stonehenge, North American Indian circles, Mesoamerican astronomy, and Egyptian alignments.

Laurilla, S. H., *Electronic Surveying and Navigation.* New York, NY: Wiley, 1976. Strong treatment of ranging devices, including data reduction. Includes Stansell's chapter on Transit (see Part 4).

Logan, K. P. ''Improvements in advanced electronic navigation systems,'' *IEEE AES Mag.,* pp. 20–22, Sept. 1987. Private shore station would access DMS map data base and transmit revisions to subscribers.

Luse, J. D. and R. Malla, ''Geodesy from Astrolabe to GPS,'' *Navigation,* pp. 101–113, Summer 1985. Short review from Eratosthenes to bubble sextant.

Melbourne, W. G., *et al.,* ''GPS-based geodesy in California, Mexico, and the Caribbean,'' in *Rec. IEEE Position Location and Navigation Symp.,* 1988, pp. 219–229. Centimeter-accuracy station costs $350,000; data in a few hours, by differential GPS.

Messmore, J. A., ''Tactical terrain data—Spatial data for the 1990s,'' in *Rec. IEEE Position Location and Navigation Symp.,* 1988, pp. 104–109. Describes a tri-service project to create tactical terrain data.

Needham, J., *Science and Civilization in China,* vol. 3: *Mathematics and the Sciences of the Heavens and the Earth.* Cambridge: Cambridge University Press, 1959. The definitive work on astronomy and cartography in China.

North, J. D., ''The astrolabe,'' *Sci. Amer.,* Jan. 1974, vol. 246, pp. 96–106. Excellent description of construction of the instrument.

Parker, W. S. and R. G. Bartholomew, ''Mapping, charting and geodesy in the DoD standard GPS receivers,'' in *Rec. IEEE Position Location and Navigation Symp.,* 1988, pp. 145–152. Stores 11 ellipsoids (including WGS-84) and 47 datums with translation vectors from each to the worldwide WGS-84 datum.

Phillips, J. O., *Specifications to Support Classification, Standards of Accuracy, and General Specifications of Geodetic Control Surveys.* Washington, D.C., U.S. Dept. of Commerce, 1978. Horizontal and vertical control errors allowed for each class of survey. Techniques.

The Portolan Chart of Angellino De Dalorto, 1325. Reproduced in 1929 by the Royal Geographic Society from the maps of Prince Corsini, Florence.

Remondi, B. W., ''High Precision Survey Results using the Ashtech XII GPS Receiver,'' in *Rec. IEEE Position Location and Navigation Symp.,* 1988, pp. 195–200. Market-oriented test results on short baselines comparing two receivers in differential GPS mode.

Renfrew, C., ''The social archaeology of megalithic monuments,'' *Sci. Amer.,* Nov 1983, vol. 255, pp. 152–163. Evolution of henges from earlier megaliths.

Rogoff, M., ''Electronic charts—At the heart of 21st century navigation,'' in *Rec. IEEE Position Location and Navigation Symp.,* 1988, pp. 88–94. A plea for the creation of an international ''bureau'' to buy government map data and publish them in a standard format, chiefly for marine use.

Wong, R. V. C. and K-P Schwarz, ''Development and field testing a RLG strapdown inertial survey system,'' in *Rec. IEEE Position Location and Navigation Symp.,* 1988, pp. 480–487. Litton ring-laser aircraft inertial navigator modified for added I/O and with personal computer to collect data. Sub-meter accuracy versus GPS-surveyed truth.

Wynter, W. and A. Turner, *Scientific Instruments.* New York, NY: Scribner's, 1975, 239 pp. Good description of old instruments.

Yunck, T. P., *et al.,* ''Role of GPS in precise Earth observations,'' in *Rec. IEEE Position Location and Navigation Symp.,* 1988, pp. 251–258. Describes orbit prediction and ground-station survey to centimeter accuracy using ground-based and orbiting GPS receivers.

Geodetic Astronomy

C. E. EWING AND M. M. MITCHELL

Geodetic astronomy is the application of astronomy to the precise determination of latitude, longitude, and azimuth on the earth.

A distinction must be made between the two terms "geodetic positions and directions" and "astronomical positions and directions." All the discussion about latitude, longitude, and azimuth encountered so far in this book* has been geodetic in nature because the three quantities have been referred to the ellipsoid, which is the mathematical figure of the earth. Astronomical positions and directions are determined with reference to the geoid. The difference between the two is a function of the deflection of the vertical which has been described very briefly earlier and is discussed in detail in Chapter 8. The manner in which the geodetic and the astronomical quantities are related will be shown later in this chapter. In order to avoid confusion, the geodetic latitude, longitude, and azimuth are shown as ϕ, λ, and α, respectively, which has been the practice heretofore in this book. The astronomical latitude, longitude, and azimuth bear the corresponding capital Greek letters Φ, Λ, and A.

The determination of astronomical positions and directions is essentially a study in applied spherical trigonometry. The basic equations of spherical trigonometry are shown in Appendix A. In contrast with geodesy, which has complicated spheroidal triangles, astronomy deals with true spherical triangles. The celestial sphere is considered to be a vast sphere of infinite dimensions with the observer at the center.

Before detailed derivations for latitude, longitude, and azimuth are attempted, it is necessary to have an understanding of some of the fundamental properties associated with geodetic astronomy. The definitions which follow will have more meaning as they are related to appropriate figures and derivations.

Of the four astronomical systems of coordinates in common use, two are important to the geodesist. These two are the horizon system and the equator system. They differ in the position of the fundamental circle, an arbitrary great circle of the celestial sphere.

The horizon system is shown in Fig. 5–1. The fundamental circle is the horizon. Small circles parallel to the horizon are called almucantars. The zenith, designated Z, is the point at which a plumbline produced upward from the observer's position O intersects the celestial sphere. The diametrically opposite point on the celestial sphere is the nadir. The zenith and the nadir are the poles in the horizon system. Great circles which pass through the zenith and the nadir are called vertical circles; they are perpendicular to the horizon. The vertical circle through the star R is ZRC. The astronomical azimuth A of the star is the angular distance measured clockwise on the horizon from the north point to point C which is the foot of the vertical circle through R. The altitude h of the star is the angular distance measured upward from the horizon on the vertical circle through the body. It is the arc CR. The zenith distance z of the star R is the arc ZR; it is the complement of the altitude, or $z = 90° - h$. The two quantities azimuth and

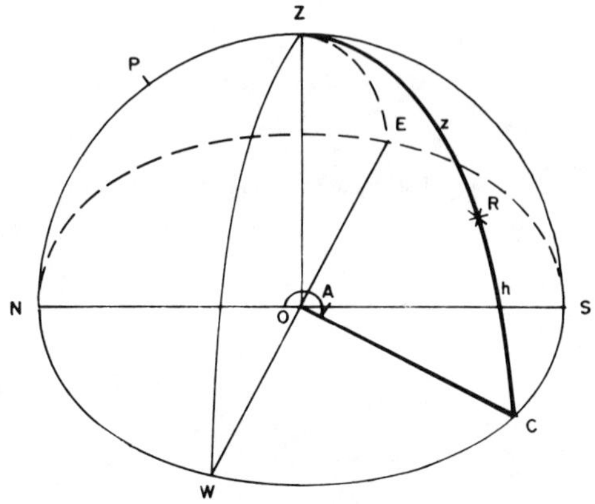

Fig. 5-1. The horizon system of coordinates.

* In this paper, all references to "this book" or specific sections pertain to *Introduction to Geodesy*.

Reprinted with permission from C. E. Ewing and M. M. Mitchell, *Introduction to Geodesy,* New York: Elsevier Science Publishing Company, Inc., 1970, ch. 5, pp. 55–90.

altitude (or zenith distance) determine the position of a heavenly body in the horizon system at any given instant.

The horizon system is used for some geodetic purposes; however, it has its weak points. The coordinates of a heavenly body are not constant. The altitude and azimuth of a star continuously change because of diurnal motion. Furthermore, the system is local. The altitude and azimuth of a star are not the same for two observers at different locations.

The second astronomical coordinate system of geodetic significance is the equator system. It is shown in Fig. 5–2. The celestial poles, hereafter referred to simply as the poles, are the two points where the axis of rotation of the earth, if extended, would intersect the celestial sphere. The celestial equator is the intersection of the celestial sphere and the plane of the earth's equator. It is the fundamental circle of the equator system. Small circles parallel to the celestial equator are known as parallels of declination and are equivalent to parallels of latitude on the earth.

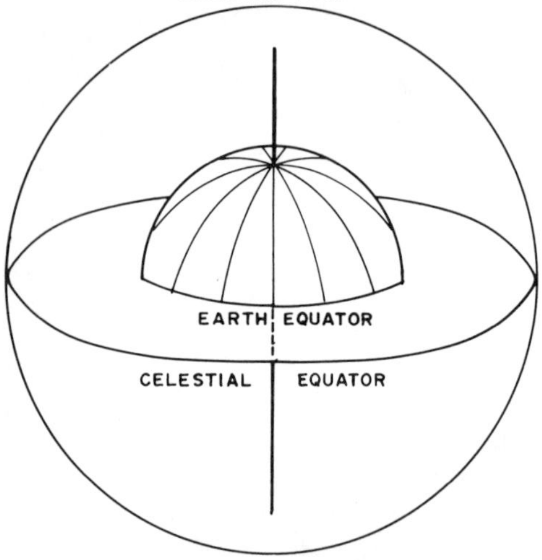

Fig. 5-2. Fundamental properties of the equator system of coordinates.

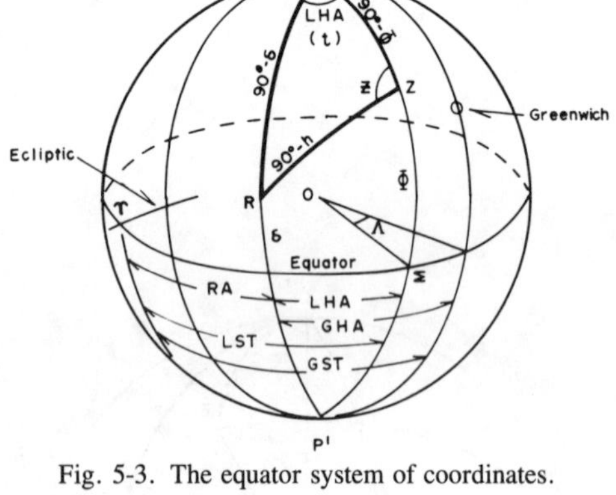

Fig. 5-3. The equator system of coordinates.

The vertical circle which passes through the celestial poles is known as the celestial meridian or, more simply, as the meridian. The prime vertical is the vertical circle at right angles to the meridian. The great circles which are perpendicular to the equator are known as hour circles. The hour circle through the zenith is the meridian of the observer. It is shown by PZP' in Fig. 5–3. Thus the meridian of the observer, which is a vertical circle in the horizon system, is also an hour circle in the equator system. The meridian zenith distance, z_m, of a heavenly body is the angular distance from the zenith to the body, measured along the hour circle through the body.

Figure 5–3 also shows additional properties of the equator system. The local hour angle (LHA) of a heavenly body R is the angular distance measured westward on the equator from the point of intersection of the equator and the meridian of the observer (the Σ point), to the foot of the hour circle through the body. What is exactly the same thing, it is the angle at the pole from the observer's meridian westward to the hour circle through the body. The meridian angle t of a heavenly body is the angular distance measured either eastward or westward, whichever is the shorter distance, on the equator from the Σ point to the foot of the hour circle through the body. LHA and t are similar. The former is measured only in the westward direction; the latter is measured either east or west from 0° to 180° and carries the suffix E or W. In the case of star R in Fig. 5–3, LHA and t have the same value. If R had been to the east of the observer's meridian, t would have been measured in that direction, whereas the LHA would have been measured westward from the meridian of the observer all the way around the equator to the foot of the hour circle through R.

The Greenwich hour angle (GHA) of a heavenly body is the angular distance measured westward on the equator from the meridian of Greenwich to the foot of the hour circle through the body.

A point of great importance in geodetic astronomy is the vernal equinox, designated ♈ (the Ram's head), and often called the first point of Aries. It is that intersection of the ecliptic with the equator at which the sun crosses the equator from south to north. This occurs only once each year, on or near 21 March, hence the name vernal.

The right ascension (RA) of a body is the angular distance measured eastward on the equator from the vernal equinox to the foot of the hour circle through the body. Right ascension is usually expressed in units of time: hours, minutes, and seconds.

Local sidereal time is the local hour angle of the vernal equinox. By reference to Fig. 5–3, it can be seen that LST = LHA + RA. Greenwich sidereal time is the Greenwich hour angle of the vernal equinox, or GST = GHA + RA.

The declineation δ is the angular distance from the equator to a heavenly body measured on the hour circle through the body. It carries a positive sign when the body is north of the equator, and a negative sign when it is south of the equator.

The azimuth angle of a heavenly body is the angular distance measured at the observer's position eastward or westward on the horizon to the foot of the vertical circle through the body. It is designated by the symbol $\mathcal{Z}$ and should not be confused with Z, which is the zenith, or with z, which is zenith distance.

In the equator system, the coordinates of a heavenly body are expressed either by hour angle and declination, or by right ascension and declination. Because right ascension and declination change very slowly through the years. they are especially useful for identifying the position of a body on the celestial sphere.

Knowledge of the fundamental properties of astronomy, as defined above and as illustrated by Figs. 5–1, 5–2, and 5–3, is absolutely essential to a proper understanding of the means by which latitude, longitude, and azimuth are determined. The end products of geodetic astronomy are latitude, longitude, and azimuth. Some instruments are peculiarly adapted to the determination of one quantity, such as the transit instrument for longitude and the zenith telescope for latitude; some are adapted to the simultaneous determination of both latitude and longitude, such as the zenith camera and the astrolabe; others, particularly the theodolite, can be used for the determination of all three quantities.

THE DETERMINATION OF LATITUDE

The astronomical latitude Φ of an observer is the angle between the direction of a plumbline at that point and the plane of the earth's equator. It may be either north or south depending on the position of the observer with respect to the equator. Another definition is that latitude is the declination of the zenith.

Figure 5–4 is taken in the plane of the meridian. The observer is at point O. The north and south points on the horizon are N and S, respectively. P, the celestial pole, is at right angles to the equator, designated E. The zenith of the observer is Z. By definition, the latitude Φ is the declination of the zenith, or the angular distance from the equator to the zenith measured along the hour circle. Since ZO is perpendicular to the horizon NS, and OP is perpendicular to the equator OE, it follows that $\angle NOP$ is equal to Φ. From this basic geometry comes the fundamental fact that the altitude of the pole is equal to the latitude of the observer. This fact is often used in simple determinations of latitude from sightings on Polaris followed by a correction for the local hour angle of observation, which is found in the American Ephemeris and Nautical Almanac [4], hereafter referred to simply as the Almanac. A correction for atmospheric refraction is necessary also. The result is not of geodetic quality.

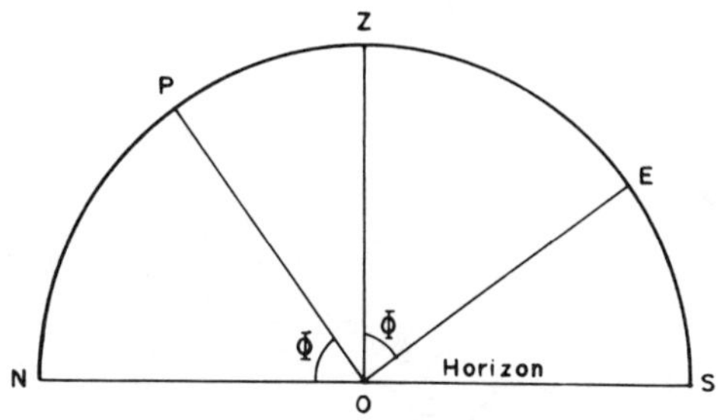

Fig. 5-4. Astronomical latitude.

Another method for determining latitude is to measure the meridian altitude or its complement, the meridian zenith distance, of a star. Figure 5–5 is drawn in the plane of the meridian of the observer. R is a star at the instant of crossing the meridian. The Σ point is the intersection of the observer's meridian and the equator. From this sketch, we see that $\Sigma Z = \Phi$, the astronomical latitude of the observer; $RZ = z_m$, the meridian zenith distance of the star; and $\Sigma R = \delta$, the declination of the star.

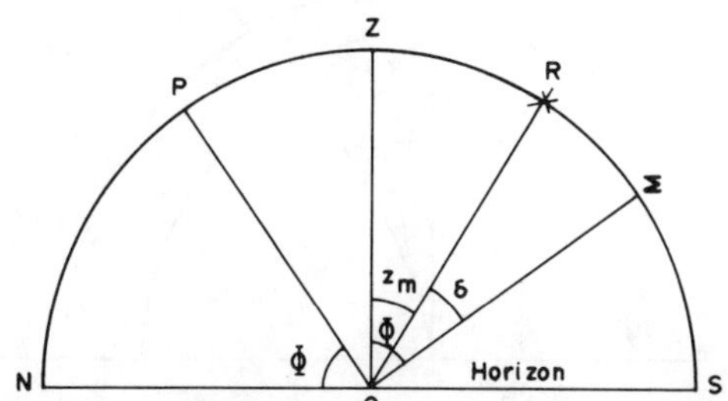

Fig. 5-5. Relationships among latitude, meridian zenith distance, and declination for a body south of the zenith.

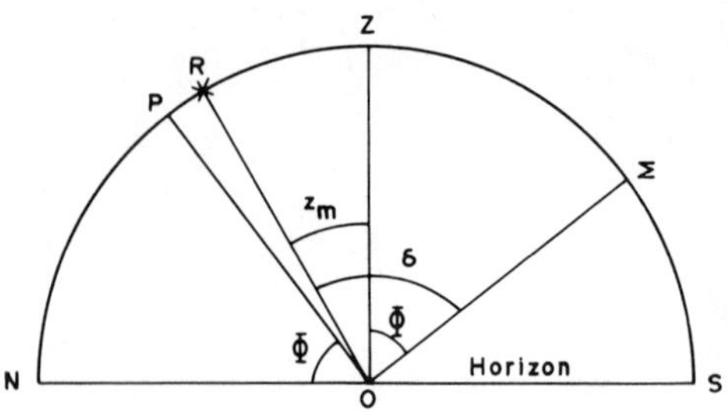

Fig. 5-6. Relationships among latitude, meridian zenith distance, and declination for a body north of the zenith and above the pole.

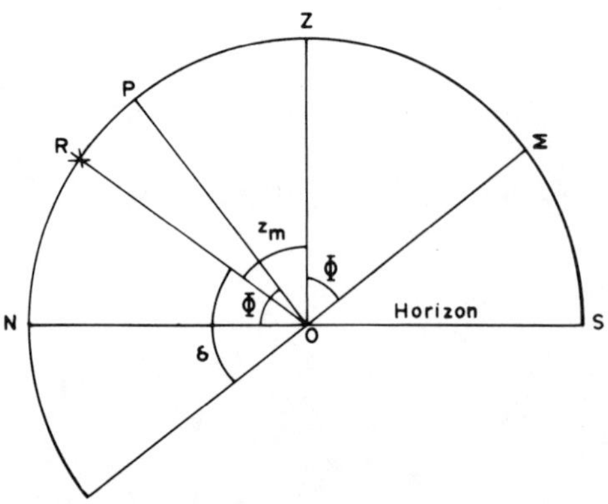

Fig. 5-7. Relationships among latitude, meridian zenith distance, and declination for a body north of the zenith and below the pole.

From these relationships, we can write

$$\Phi = z_m + \delta \quad \text{(for stars south of the zenith).} \tag{5–1}$$

When a star is chosen which has a declination greater than the observer's latitude, it will cross the meridian north of the zenith. Figure 5–6 shows the conditions when the star is above the pole and north of the zenith. From the figure, we may write

$$\Phi = \delta - z_m \quad \text{(for upper transits north of the zenith).} \tag{5–2}$$

When the star crosses the meridian of the observer below the pole, as in the case of circumpolar stars at lower transit, Fig. 5–7 shows the geometry. In this case, the latitude is given by

$$\Phi = 180° - z_m - \delta \quad \text{(for stars at lower transit).} \tag{5–3}$$

These three equations (5–1), (5–2), and (5–3) also hold true when the observer is in the southern hemisphere except that in Eq. (5–1) the word south should be replaced by north, and in Eq. (5–2) the word north should be repaced by south. Equation (5–3) holds true as written. These relationships may be proved by drawing simple sketches of the southern hemisphere.

Care must be used in choosing the sign of δ when these equations are used. When the declination is in the latitude of the observer, its sign is positive; when the declination is in the latitude opposite the observer, its sign is negative.

In applying this method, the selected star is tracked for several minutes before transit, which is defined as its passage across the meridian of the observer. The tangent screw of the vertical circle is used to keep the horizontal wire on the star. When the star appears to move along the horizontal wire with no apparent vertical motion, its meridian zenith distance is recorded. If the meridian altitude is read instead of the meridian zenith distance, it is subtracted from 90° to obtain the meridian zenith distance z_m. The declination of the star being observed is found in the Almanac. Finally, one of Eqs. (5–1), (5–2), or (5–3) is used, depending on the conditions of observation. Of course a correction for atmospheric refraction must be applied.

Several other methods are available for determining latitude, such as observing the altitude of a star not on the meridian when the time is known, and using circummeridian altitudes. From any of these methods it is possible to determine the precise latitude of a place provided we know the proper atmospheric refraction correction to apply. Herein lies the weakeness of these methods discussed previously. Because of atmospheric refraction, a ray of light from a celestial body undergoes a change in direction as it penetrates the atmosphere. When the ray of light from the heavenly body R in Fig. 5–8 arrives at the observer's position, it will appear to come from R', which is the final direction of the light as it enters the eye of the observer. Therefore, any object will appear at a higher altitude than it actually is. The amount of this angular displacement in altitude is the refraction correction r which must be subtracted from the observed altitude, or added to the observed zenith distance, in this manner:

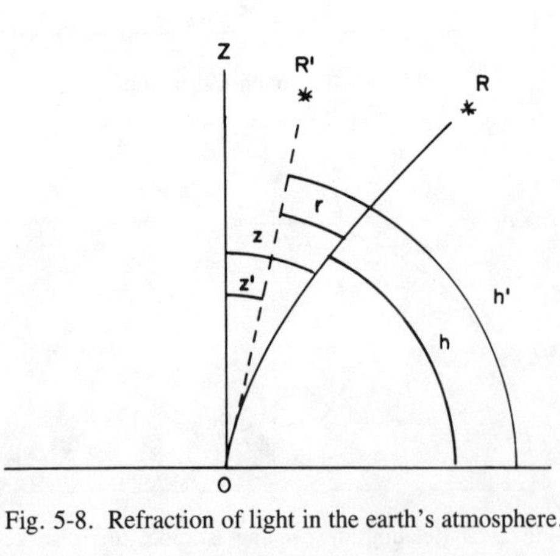

Fig. 5-8. Refraction of light in the earth's atmosphere.

$$h = h' - r \tag{5–4a}$$

or

$$z = z' + r, \tag{5–4b}$$

where h' is the observed altitude and z' is the observed zenith distance.

The magnitude of r may be obtained from tables and formulas. One such correction is

$$r = 60.6 \cot h', \tag{5-5}$$

where r is the refraction correction in seconds of arc.

The error in this correction increases as the altitude decreases and may reach 10 seconds of arc at low altitudes. For this reason, a more accurate correction is needed. Because the index of refraction varies with the temperature and pressure of the atmosphere, it is necessary to use an equation written in terms of these quantities. One equation, known as Comstock's empirical formula, is

$$r = \frac{983b}{460 + t} \cot h' \tag{5-6}$$

where r is the refraction correction in seconds of arc; b is the barometric pressure in inches of mercury; and t is the temperature in degrees Fahrenheit.

For altitudes greater than 15° above the horizon, the error produced by this equation will seldom exceed 1 second of arc. For some types of surveying, this error is not unreasonable. In geodetic surveying, however, it is intolerable. One second of arc in latitude corresponds to slightly more than 100 feet on the earth's surface. Therefore, for geodetic purposes, we must seek a better method of determining latitude than those previously discussed.

This better method is the Horrebow-Talcott method. In this procedure, we observe a series of stars to the north and to the south of the zenith. Each pair of stars consists of a north star and a south star of known declination, at approximately equal zenith distances and transiting the meridian at about the same time. Since Eqs. (5–1) and (5–2) were written for stars to the south of the zenith and stars to the north of the zenith, respectively, their combination is used in the Horrebow-Talcott method.

Figure 5–9 shows the plane of the observer's meridian. R_s is the star transiting the meridian to the south of the zenith. Its declination is δ' and its meridian zenith distance z'_m. R_n is the star which transits the meridian to the north of the zenith. Its declination is δ'' and its meridian zenith distance z''_m. The superscripts are prime and double-prime and not minutes and seconds of arc. Equations (5–1) and (5–2) may be rewritten for this special case as follows:

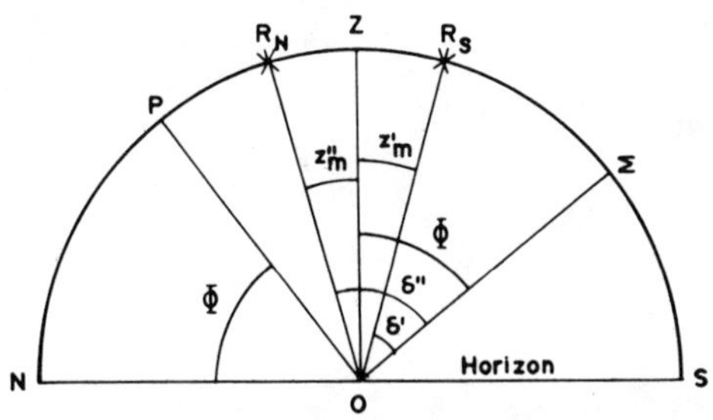

Fig. 5-9. Determination of latitude by the Horrebow-Talcott method.

$$\Phi = \delta' + z'_m \qquad \text{and} \qquad \Phi = \delta'' - z''_m.$$

Adding these two equations gives

$$2\,\Phi = \delta' + \delta'' + z'_m - z''_m \tag{5-7}$$

or, expressed differently,

$$\Phi = \tfrac{1}{2}(\delta' + \delta'') + \tfrac{1}{2}(z'_m - z''_m). \tag{5-8}$$

This equation indicates that the only observed quantity needed is the small difference in meridian zenith distances between the pair of stars and not the absolute meridian zenith distances of the two stars. The declinations of the two stars, needed to complete the information for Eq. (5–8), are found in the Almanac.

This method may be used with several different instruments, among which are the observatory-mounted zenith telescope, the Bamberg broken-telescope transit, shown in Fig. 5–10, and the Wild T–4 Universal Instrument, which is also a broken-telescope instrument, shown in Fig. 5–11. The term "broken telescope" is used to describe a telescope having a prism in the optical axis which directs the rays of

Fig. 5-10. Bamberg broken-telescope transit (U.S. Coast and Geodetic Survey).

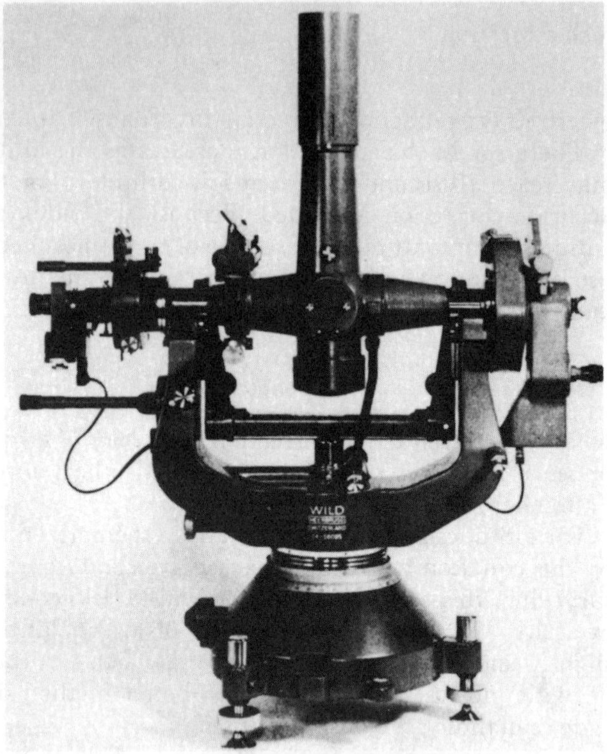

Fig. 5-11. Wild T–4 Universal Instrument (Wild Heerbrugg Instruments, Inc.).

light at right angles through the hollow axis of the instrument to the eyepiece. With this arrangement, the observer remains in the same position no matter what the zenith distance of the observed body may be.

These instruments are equipped with an eyepiece micrometer having a movable reticle. This mechanism is designed to measure to a fraction of a second of arc the difference in meridian zenith distances of a pair of stars. The field of view depends on the instrument used; in the case of the Wild T–4, it is 30 minutes of arc. Because only the difference of two meridian zenith distances is read instead of the absolute quantities, this small field of view is adequate.

The horizontal axis of the telescope is oriented in a true east-west direction so that the line of sight coincides with the meridian when the telescope is rotated about that axis. Reversing the telescope 180° about the vertical axis again orients the instrument so that rotation about the horizontal axis describes the meridian on the other side of the zenith. Without resetting the telescope, and merely by reversing the instrument about its vertical axis, one may observe another star of approximately the same meridian zenith distance on the opposite side of the zenith.

Prior to the field observations, an observing list must be compiled. The limited number of stars shown in the Almanac is inadequate. A general star catalogue is needed. One catalogue is the *Apparent Places of Fundamental Stars* [3], which contains the positions of 1535 stars, with 26 of these being north circumpolar stars and 26 south circumpolar stars. Another listing is the *Boss Star Catalogue* [1], which provides the mean positions of 33,342 stars for a given date. The mean position of a star for the given date must be reduced to its apparent place at any given instant by use of equations provided in the Almanac. It is important to remember that in geodetic work the apparent place of a star is always used.

It would, of course, be a long and tedious process to search through a star catalogue to find six to eight pairs of stars which are constrained to meeting certain definite conditions. The difference in time between the transit of each star must be at least 4 minutes to allow for reading the first star, reversing the instrument, and reading the second star. The time should not exceed 20 minutes because the constants of the instrument might change if the observing period is unduly long. Stars having meridian zenith distances greater than 30° should not be used. The extent of the field of the micrometer is generally about 30 minutes of arc and so the difference in the meridian zenith distance of a pair cannot exceed that amount. Stars fainter than eighth magnitude should be disregarded.

With these constraints, the selection of an observing list must be made methodically. Since we know the time period of our observation, we must choose stars having a right ascension corresponding to this time. The declination must, as stated above, be chosen so that the meridian zenith distance does not exceed 30°. Since we have some prior knowledge of the latitude of the observer, we can use this best estimate and rearrange Eq. (5–7) into the form

$$(\delta' + \delta'') - 2\Phi = z_m'' - z_m'. \tag{5-9}$$

This equation says that the sum of the declinations of the two stars minus twice the latitude is equal to the difference in the meridian zenith distances. This difference is limited by the field of view of the telescope and the extent of the micrometer, which in the case of the Wild T–4 is 30 minutes of arc.

Entering the catalogue with the selected period of right ascension, we select one star brighter than eighth magnitude which has a declination so that the meridian zenith distance does not exceed 30°. Then searching in the catalogue not less than 4 minutes nor more than 20 minutes of right ascension later, we choose a second star of sufficient brightness which fulfills the conditions of Eq. (5–9). A little practice will enable the observer to compile a list of six or eight pairs of stars without difficulty.

Prior to observation of the first star in a pair, the telescope is set to the average meridian zenith distance of the two stars. A few minutes before the time of transit, the first star will come into the field of view of the telescope and appear to travel parallel to the movable wire in the micrometer. The micrometer head must then be turned until the movable wire bisects the star as it transits the meridian. The micrometer reading is recorded; also, the time of transit is recorded merely as a check to insure that the proper star was observed. Without disturbing the setting of the telescope, merely reverse it about the vertical axis until it hits the 180° stop. Proceed with the second star in an identical manner. Knowing the value of one turn of the screw for the specific instrument enables the difference in micrometer readings between the first and the second observations to be translated directly into seconds of arc. This information completes the needs for solution of the astronomical latitude by use of Eq. (5–8).

Refraction, the nemesis of all astronomical observations, is eliminated effectively by this method. For one thing, stars having meridian zenith distances greater than 30° are not used. More importantly, only the small difference in meridian zenith distances of a pair of stars at nearly equal distances north and south of the zenith enters into the computations. Therefore, the refraction effect is very nearly canceled. Accuracies of 0.1 second of arc can be obtained.

Of the several methods of latitude determination described above, only the Horrebow-Talcott method employing the zenith telescope or similar instrument furnishes the precision needed in geodetic work. The other methods have been included to provide a background for the reader and to suggest procedures which may be used when a less precise preliminary determination is needed.

THE DETERMINATION OF LONGITUDE

Geodetic longitude is the angular distance measured on the terrestrial equator from the intersection of a fixed meridian and the equator to the foot of the meridian through the observer's position. The meridian through Greenwich, England, is taken as the fixed meridian. Longitude is reckoned positive westward and negative eastward from the meridian of Greenwich. The standard designation of geodetic longitude is λ.

Astronomical longitude differs from geodetic longitude by the longitudinal component of the deflection of the vertical. The two are related in a manner which will be explained at the end of this chapter. Astronomical longitude is designated Λ.

Longitude is correlated directly with time. The difference in longitude between any two places is equal to the difference in their local times. This local time may be sidereal, apparent, or civil; in geodetic observations, sidereal time is normally used.

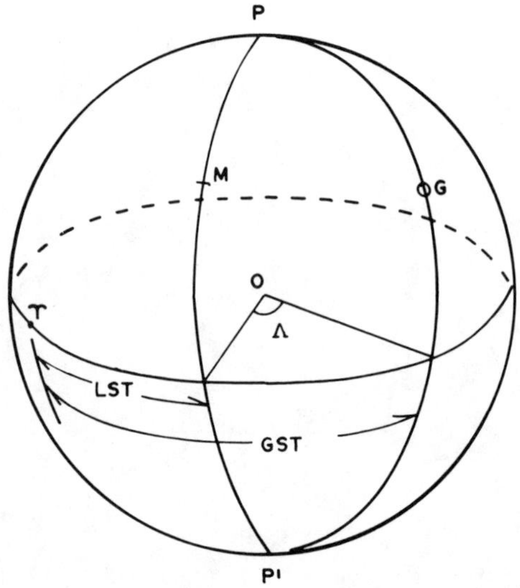

Fig. 5-12. The relationship between time and longitude.

In Fig. 5–12, G designates Greenwich. The point M, the observer's position, is west of Greenwich and is the point for which the longitude is desired. The angle Λ is the difference in astronomical longitude between the Greenwich meridian and the meridian through M. With the vernal equinox at ♈, the sidereal clock at Greenwich reads Greenwich sidereal time (GST); the sidereal clock at M reads local sidereal time (LST). The angle Λ is the difference between the two, or

$$\Lambda = \text{GST} - \text{LST}. \tag{5–10}$$

This equation expresses Λ in terms of time. The following relationships are used to convert units of time to units of arc:

$$\begin{aligned} 1 \text{ hour of time} &= 15^\circ \text{ of arc,} \\ 1 \text{ minute of time} &= 15' \text{ of arc,} \\ 1 \text{ second of time} &= 15'' \text{ of arc.} \end{aligned}$$

In theory, the determination of longitude is quite simple. In practice, it is a difficult and painstaking task, particularly in the precise measurement of time. The determination of the longitude of a place is a three-step solution:

1. *Determine Greenwich sidereal time at the instant of transit of a star.* This is done by use of a sidereal chronometer which is set to Greenwich sidereal time. At the instant of transit, a chronographic record is made.

2. *Determine the local sidereal time at the point M where the longitude is desired.* This is done by obtaining the right ascension for the star which was observed in transit in step 1. Since local sidereal time is defined as the right ascension of the observer's meridian, it follows that, when the star is on the meridian, the local sidereal time is equal to the right ascension of the star. This can be seen from the basic equation of sidereal time,

$$\text{LST} = \text{RA} + \text{LHA}, \tag{5–11}$$

where LST = local sidereal time, RA = right ascension, and LHA = local hour angle. In the case of a star on the meridian, the local hour angle is zero. Therefore,

$$\text{LST} = \text{RA}. \tag{5–12}$$

The RA of a star, and hence its LST, is taken from the Almanac or a star catalogue.

3. *Compare the Greenwich sidereal time as determined by step* 1 *with the local sidereal time as explained in step* 2. The difference is the longitude of the place, as shown by a return to Eq. (5–10).

The instrument most generally used in the precise determination of longitude is the transit instrument mounted in an observatory or on a concrete pedestal. This precise instrument should not be confused with the ordinary surveyor's transit used in plane surveying. It consists basically of a telescope, a horizontal axis perpendicular to the telescope, and a reticle in the focal plane. The horizontal axis is the axis of rotation and is mounted in an east-west direction which causes the telescope to move in the plane of the meridian. With the instrument in perfect adjustment, the mean wire of the reticle traces the meridian when the telescope is rotated about its horizontal axis. A graduated circle is attached so that the proper meridian zenith distance may be set to assist in locating the stars in an observing list.

Other instruments used extensively in the determination of longitude are the Bamberg broken-telescope transit shown in Fig. 5–10 and the Wild T–4 Universal Instrument shown in Fig. 5–11. These instruments are similar to the transit instrument but are more convenient to use. The horizontal axis is hollow. A 90° prism in the optical axis of the telescope directs the rays through the hollow axis to the eye. All observations may be made from one position.

A time signal from the U.S. Bureau of Standards through station WWV is recorded before the observations, again after 6 stars have been observed, and a third time after the operation is completed. From these time signals, the rate of the chronometer is determined. The chronometer drives a cylinder so that the beat of the clock is transmitted by an electric circuit to an electromagnet which releases a pen and marks the seconds on the sheet of graph paper which is wrapped around the cylinder. Other chronographs use a reel and paper tape in lieu of the cylinder and graph paper. As a star moves across the field of the telescope, it is continuously bisected with a movable wire by turning the handwheels of an impersonal micrometer. As the wheels are turned, electrical contacts are broken at regular intervals. These contacts are recorded on the chronograph. The readings of the chronograph, which can be estimated to 0.01 seconds of time, are averaged. The time thus obtained is the instant of transit of the star over the fictitious mean wire. Figure 5–13 shows a combination time-signal receiver, chronograph, and chronometer.

The observed time must be corrected for diurnal aberration. Since the velocity of light is finite compared with the rotational velocity of the observer on the earth's surface, a star is apparently displaced. This displacement, known as diurnal aberration, is designated K. Its magnitude is

$$K = 0.021 \sec \delta \cos \phi, \qquad (5\text{–}13)$$

where K is the diurnal aberration in seconds of time; δ is the declination of the star observed; and ϕ is the latitude of the observer. This correction must be subtracted from the observed time of the meridian transit of the star.

A single determination of longitude consists of the transit observa-

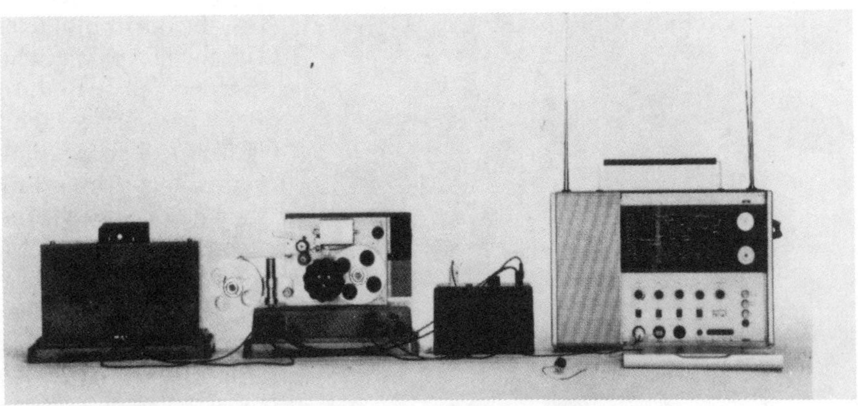

Fig. 5-13. From left to right: Chronometer, chronograph, and time-signal receiver (Wild Heerbrugg Instruments, Inc.).

tions of 12 stars of which 6 are south and 6 north of the zenith. One in each set of 6 should be within 8° of the zenith. The stars should be chosen and arranged in an observing list so there are at least several minutes of time between successive stars. As the star appears in the field, it is bisected by the vertical wire of the impersonal micrometer as described previously. This procedure is continued until the entire list of 6 stars is finished. The telescope is then reversed 180° about its vertical axis and the second set of 6 stars is observed.

As in an ordinary surveyor's transit, it is necessary to keep the instrument in as nearly perfect adjustment as possible. The observing list and the observing procedures are designed to minimize the errors, such as selecting the same number of stars north and south of the zenith and reversing the instrument in order to eliminate the azimuth error. Since these errors in the transit instrument can be minimized, the main element contributing to error in longitude determination is that of time. The velocity of rotation of a point on the equator is approximately 1500 feet per second. At latitude 45°, it is over 1000 feet per second. When time is interpolated to 0.01 second from the chronograph, this means a possible error of about 10 to 15 feet in longitude. The experience of the authors is that an overall accuracy of 50 to 100 feet in longitude is more realistic when the transit instrument is used with the procedures which have been outlined.

THE SIMULTANEOUS DETERMINATION OF LATITUDE AND LONGITUDE

So far in this chapter we have considered the independent determination of latitude and longitude by use of instruments designed for those specific purposes. Several instruments are available for the simultaneous determination of latitude and longitude. One of these is the astrolabe; another is the zenith camera. The astrolabe will be considered in this chapter.

For years it has been common practice at sea and in the air for navigators to use a sextant or an octant in determining the latitude and longitude of a place by observing the altitudes and corresponding times of two or more stars. Accuracies in the order of 2 minutes of arc are usually achieved. Another method, similar to that used by navigators, is often used in geodetic surveys. It is based on determining the time when a star attains a fixed altitude. Accuracies of 1 to 2 seconds of arc are obtainable. The astrolabe is the instrument normally used. It is a rather simple device for observation on the almucantar of 30° zenith distance. There are two types of astrolabes: the prismatic astrolabe and the pendulum astrolabe.

The prismatic astrolabe may be an independent unit specifically designed as such, or it may be an attachment for use on a theodolite. An attachment for use on the Wild T–3 Precision Theodolite is shown in Fig. 5–14. The prismatic astrolabe has three principal parts:

(1) the horizontal telescope, at the focus of which is a reticle;

(2) the 60° prism, which is located in front of the objective with one face perpendicular to the optical axis of the telescope;

(3) a mercury trough located just below the lower face of the prism.

Figure 5–15 shows the arrangement of the components of a typical astrolabe. Two parallel rays S and S' from a star of zenith distance 30° are shown. The ray S' enters the prism at right angles to the surface AB, is reflected at surface AC, and emerges perpendicular to surface BC. The ray S is reflected by the mercury surface, enters the prism at right angles to surface AC, and then emerges perpendicular to surface BC. Therefore, the two rays emerging from surface BC are parallel and produce a single image of the star at F *when the zenith distance of the star is* 30°. At zenith distances other than 30°, the observer will see two images.

The eyepiece *H* is used for finding the star; it has a field of 1°.5 and a magnification of 30 diameters. The eyepiece *G* is used for normal observations and has a field of 0°.5 and a magnification of 80 diameters.

The view through the eyepiece *G* is shown in sequence in Fig. 5–16. First is the view before the star attains 30° zenith distance; second, the instant the star is at 30° zenith distance; and third, after the star passes 30° zenith distance. It is better practice to set the astrolabe so that the stars do not come to coincidence but appear to pass each other a small distance apart. This can be done by rotating the telescope about its optical axis.

The pendulum astrolabe does not employ the double-image principle used in the prismatic astrolabe. With the pendulum astrolabe, or "bent telescope," only one image of a star is observed as it crosses a fixed wire of the reticle. This fixed wire represents a constant zenith distance of 30°. The nonpendulus portion is brought to within 1 minute of arc of being perfectly level. Then four thumbscrews are loosened, allowing a metallic mirror, which is attached to a pendulum, to be suspended by sensitive springs. The pendulum comes to rest within ± 0″.1 from the direction of local gravity. The mirror is then perpendicular to the pendulum and therefore rests in the local horizontal plane. Rays of light enter the objective, are reflected from the mirror, and then observed through the eyepiece.

In theory, determinations of the position of an observer are identical regardless of the type of astrolabe. The theory is quite simple. From any point on the surface of the earth, the zenith distance of a star expresses the angular distance on the surface of the earth from the substellar point of the star to the observer. If the zenith distance of a star is determined, the observer is on the circumference of a circle with its center at the substellar point of the star and angular radius equal to the determined zenith distance. Observations of two stars will yield two circles which intersect. One point of intersection is the location of the observer. An observation of a third star will resolve the ambiguity.

In practice, this method of drawing circles would not be accurate enough for geodetic purposes. Therefore, either a mathematical solution or a graphical solution employing lines of position is used. In the

Fig. 5-14. Prismatic astrolabe attachment for use on the Wild T-3 Precision Theodolite (Wild Heerbrugg Instruments, Inc.).

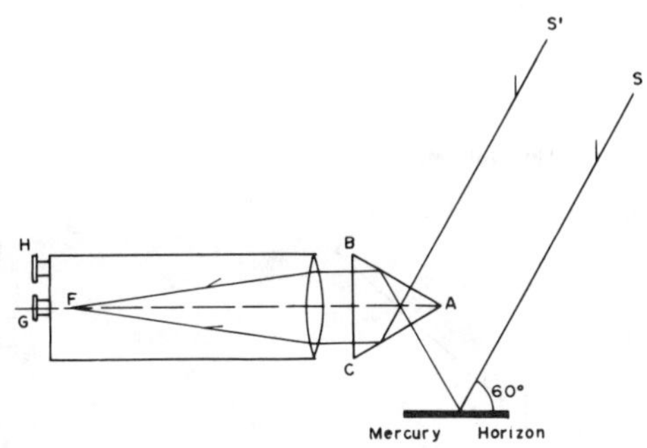

Fig. 5-15. Diagram showing the optical system of the prismatic astrolabe.

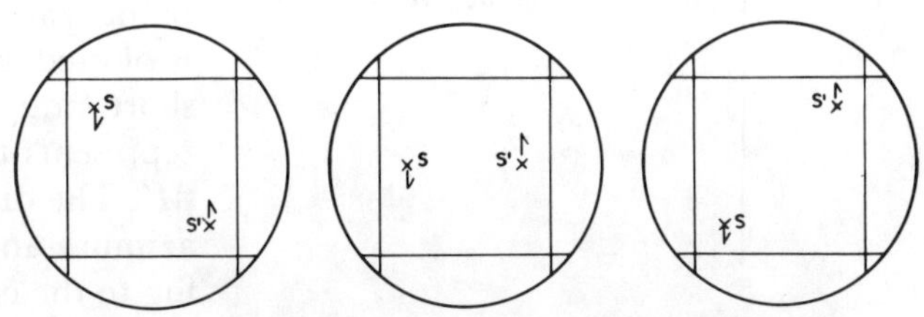

Fig. 5-16. View through the eyepiece of an astrolabe showing three positions of the double image of a star.

authors' experience the graphical solution is less tedious and gives comparable results to that obtained mathematically. Furthermore, the graphical solution enables the reader to follow better the successive steps in the theory. For these reasons, the graphical solution will be explained.

Figure 5–17 shows the geometric conditions on the surface of the earth. Point M is the position of the observer. His position is approximately known and these assumed coordinates are known as Φ and Λ. The right ascension and declination of the star are taken from the Almanac or star catalogue for the instant of observation. The substellar point is designated R. In the spherical triangle PMR, we know the following:

$$PM = 90° - \Phi,$$
$$PR = 90° - \delta,$$
$$t = \text{the meridian angle of the star.}$$

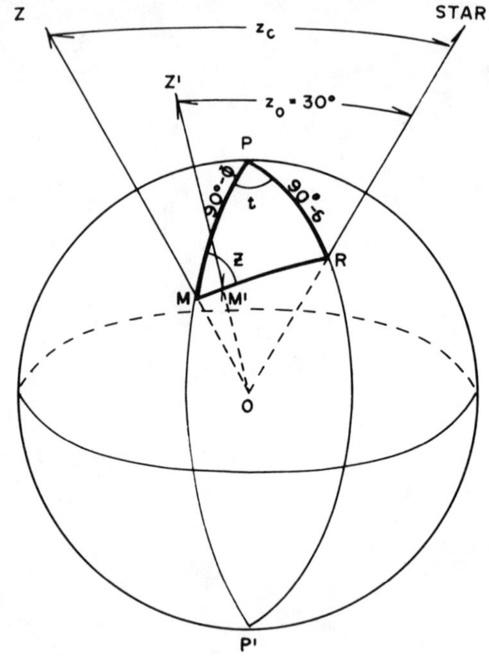

Fig. 5-17. Surface of the earth showing the substellar point of the star at R and the assumed position of the observer at M.

The local chronograph is set to LST by subtracting the assumed longitude Λ from GST. Then LHA = LST − RA, from which t is found.

In the same spherical triangle PMR, we wish to find

$$MR = z_c,$$

which is the computed zenith distance, and

$$\angle PMR = Z$$

the azimuth angle, which is reckoned either east or west of true north as the star is east or west of the meridian.

From the law of cosines of a spherical triangle, we may write

$$\cos z_c = \cos(90° - \Phi)\cos(90° - \delta) + \sin(90° - \Phi)\sin(90° - \delta)\cos t$$

or, simplified,

$$\cos z_c = \sin\Phi\sin\delta + \cos\Phi\cos\delta\cos t. \qquad (5\text{–}14)$$

From the law of sines of a spherical triangle, we obtain

$$\frac{\sin Z}{\sin(90° - \delta)} = \frac{\sin t}{\sin z_c},$$

or

$$\sin Z = \frac{\sin t \cos\delta}{\sin z_c}. \qquad (5\text{–}15)$$

The computation of Eqs. (5–14) and (5–15) furnishes the computed zenith distance z_c and its corresponding azimuth angle Z. The zenith distance as actually observed is called z_o and is designed to be 30° in the case of astrolabes. The observer is situated on the circumference of a circle with its center at the substellar point R and radius z_o. A small arc of this circle is shown intersecting MR at M'. This arc would have intersected M if the assumed position had been the actual position of the observer. This distance MM' is equal to z_c minus z_o and is known as the intercept.

Next, this intercept is plotted on a graph to a convenient scale. The y axis of the graph represents the meridian through point M. Since the intercept is defined as the computed zenith distance z_c minus the observed zenith distance z_o, the intercept is plotted toward the substellar point when the intercept is positive; when the intercept is negative, it is plotted away from the substellar point. At the end of the intercept, a short straight line is drawn perpendicular to the intercept. This line represents the line of position; it is a short segment of the arc through M'. The direction to the substellar point is provided by the computed azimuth angle Z. Figure 5–18 indicates the line of position corresponding to the conditions of Fig. 5–17. This illustrates the procedure for one star.

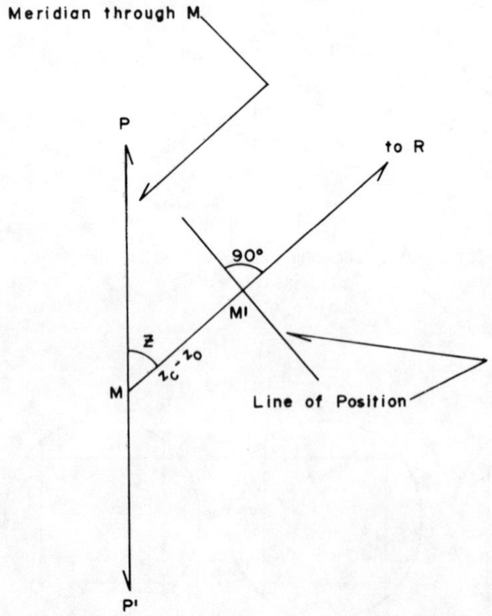

Fig. 5-18. A line of position for a single star.

An observing list consists of 20 stars, 5 in each quadrant. The stars must be selected so they will intersect the circle of constant 30° zenith distance within the period of time selected for the observations. Sometimes the same star may be used twice, once when it intersects the circle of constant zenith distance east of the meridian and again at the intersection west of the meridian. For any latitude Φ of the observer, the stars which will attain a zenith distance of 30° must have a declination within the limits of Φ plus 30° and Φ minus 30°. The selection of 20 stars which fulfill these conditions and the computation of the azimuth angle and local sidereal time is a laborious process when done without some short cut. This assistance has been provided in the form of tables [2] prepared for each 1° of latitude from 60°S to 60°N. These tables give the LST and azimuth at which stars attain the 30° zenith distance. An interpolation for the assumed latitude of the observer is sufficiently accurate to provide the approximate information for identifying each selected star during field observations. By use of these tables, an observing list can be compiled and the stars arranged in order of increasing LST.

Once the observing list has been compiled, the astrolabe is set so there is a clear field of view for 30° around the zenith. The azimuth circle of the instrument is oriented to true north by means of a predetermined azimuth mark, by use of the magnetic needle, or by pointing the telescope tube at one of the readily identifiable stars in the observing list whose approximate azimuth has been determined. When each star in the observing list intersects the circle of constant zenith distance, the observer presses a key which records the instant of time on the chronograph.

By use of Eqs. (5–14) and (5–15), the intercept distance and azimuth angle are computed for each star observed. The same plotting procedure which was explained for Fig. 5–18 is used for the entire list. Figure 5–19 illustrates this procedure. The origin of the axes is the assumed position M of the observer. A convenient scale for the graph is determined to accommodate all the intercepts. Once all the lines of position have been plotted, a circle is drawn which is most nearly tangent to all the lines of position. This is a matter of common sense and judgment. The center of this best-fitting circle is O and represents the true position of the observer. The values of Δx and Δy in seconds of arc are then scaled from the graph. The corrections in latitude and longitude then become

$$\Delta\Phi = \Delta y \tag{5–16}$$

and

$$\Delta\Lambda = \Delta x \sec \Phi. \tag{5–17}$$

These corrections are then applied properly to the assumed latitude and longitude. In the case of Fig. 5–19, the correct position O is north of the assumed position and the correction must be added. The correction in longitude must be subtracted because the actual position O is east of the assumed position. This, of course, is the case when the observer is west of Greenwich and north of the equator. For other conditions, the corrections must be applied properly.

As a matter of convenience, the value of z_0 may be assumed as any convenient value and not held rigidly to 30°. This avoids the use of negative intercepts. The physical location of each line of position is changed, but the center O remains the same. The astrolabe is a relatively simple and reliable instrument. It is light in weight and requires few adjustments. Refraction corrections need not be applied: if the observing list is selected with an equal number of stars in each quadrant, the effect of refraction is canceled. Accuracies of a few hundred feet in position may be attained. The principal weakness, in the authors'

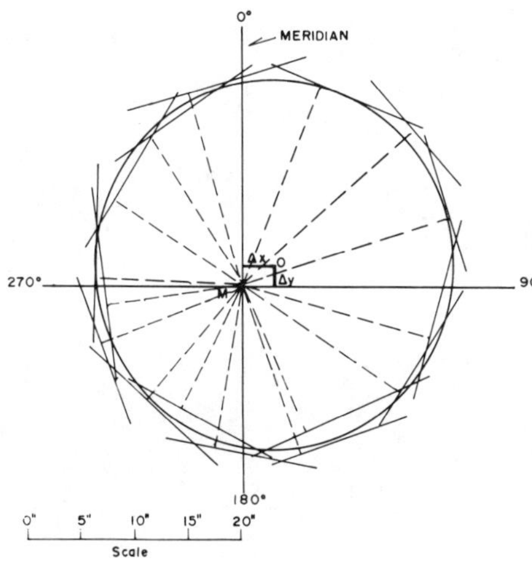

Fig. 5-19. Lines of position for an entire observation list.

experience, is the sensitivity of the prismatic astrolabe to vibrations. When it is used on top of buildings in which air conditioners, motors, or elevators are operating, or near rail or road traffic, the mercury bath tends to tremble and the image of the reflected star is often fuzzy and may disappear entirely.

THE DETERMINATION OF AZIMUTH

The azimuth of a line is the horizontal angle which the line makes with the meridian, measured clockwise from true north. The problem of finding the azimuth of a line consists of two parts:

(1) determining the azimuth angle of the heavenly body R, and

(2) measuring the horizontal angle between the azimuth line and the heavenly body.

The problem is illustrated by Fig. 5–20. The azimuth angle $\mathcal{Z}$ of the star is measured either eastward or westward from true north. The angle K is the angle measured clockwise from the star to the line OT whose azimuth is desired. For stars east of the meridian, the astronomical azimuth A of the line OT is

$$A = \mathcal{Z} + K; \tag{5–18}$$

for stars west of the meridian, the astronomical azimuth is

$$A = 360° - \mathcal{Z} + K. \tag{5–19}$$

The first part of the problem consists of the solution of the spherical triangle formed on the celestial sphere by the pole, the zenith, and the star. The second part of the problem involves simply turning the angle between the azimuth line and the heavenly body.

At this point, a review of spherical trigonometry in Appendix A is suggested, particularly the law of sines, the law of cosines, and the relation between two angles and three sides. The solution of the azimuth problem makes use of these spherical trigonometric relationships. Figure 5–3 shows the spherical triangle $P\mathcal{Z}R$ which must be solved. Figure 5–21 shows the same triangle as viewed from the zenith. The azimuth angle $\mathcal{Z}$ is the quantity we desire. It should not be confused with the zenith, which is designated $\mathcal{Z}$.

With reference to both Figs. 5–3 and 5–21, the notation is as follows: P is the celestial pole; $\mathcal{Z}$ is the zenith of the observer; R is the heavenly body observed; $P\mathcal{Z} = 90° - \Phi$; $PR = 90° - \delta$; $R\mathcal{Z} = 90° - h$; t is the meridian angle of R; and $\mathcal{Z}$ is the azimuth angle of R, the quantity to be computed.

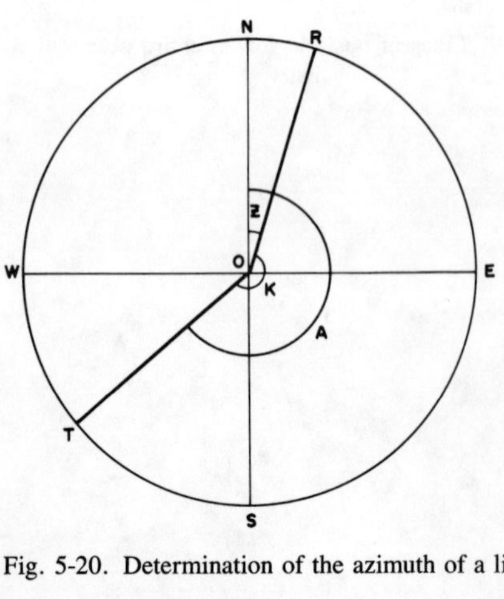

Fig. 5-20. Determination of the azimuth of a line.

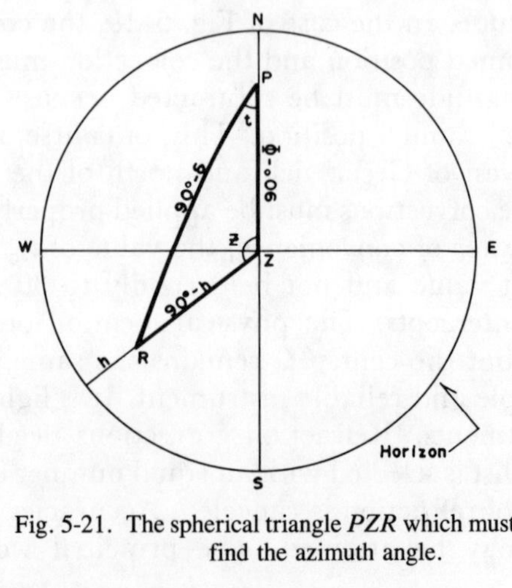

Fig. 5-21. The spherical triangle PZR which must be solved to find the azimuth angle.

There are numerous methods for azimuth determination making use of the sun, moon, or stars. Three of the most common methods will be presented.

Case I. A star is observed at any hour angle, the time and altitude being measured during the observation. The following elements of the celestial triangle are known after the observation:

$$PZ = 90° - \Phi,$$
$$PR = 90° - \delta,$$
$$RZ = 90° - h,$$

and

$$t = \text{RA} - \text{LST} \quad \text{(for stars east of the meridian)},$$

or

$$t = \text{LST} - \text{RA} \quad \text{(for stars west of the meridian)}.$$

It is assumed that the local clock is set to local sidereal time, which is obtained from LST $=$ GST $- \Lambda$.

By the use of the spherical law of sines, we have

$$\frac{\sin t}{\sin(90° - h)} = \frac{\sin Z}{\sin(90° - \delta)},$$

or

$$\cos h \sin Z = \sin t \cos \delta, \tag{5–20}$$

from which

$$\sin Z = \frac{\sin t \cos \delta}{\cos h}. \tag{5–21}$$

Case II. A star is observed at any hour angle, the time (but not the altitude) being recorded during the observation. The elements of the celestial triangle which are known after the observation are PZ, PR, and t.

By use of the relationship between two angles and three sides of a spherical triangle, we may write

$$\sin(90° - h) \cos Z = \sin(90° - \Phi) \cos(90° - \delta) - \cos(90° - \Phi) \sin(90° - \delta) \cos t.$$

This equation may be simplified to

$$\cos h \cos Z = \cos \Phi \sin \delta - \sin \Phi \cos \delta \cos t. \tag{5–22}$$

Dividing Eq. (5–20) by Eq. (5–22) produces

$$\frac{\cos h \sin Z}{\cos h \cos Z} = \frac{\sin t \cos \delta}{\cos \Phi \sin \delta - \sin \Phi \cos \delta \cos t},$$

which finally becomes

$$\tan Z = \frac{\sin t}{\cos \Phi \tan \delta - \sin \Phi \cos t}. \tag{5–23}$$

Case III. A circumpolar star is observed at elongation. Figure 5–22, drawn in the horizon system, shows a circumpolar star at two positions R and R' at the points of eastern and western elongation about the pole P. These are points of tangency with two vertical circles ZL and ZL'; therefore, the spherical triangles PZR and PZR' have right angles at R and R', respectively. The azimuth angle is the same size whether at eastern or western elongation and is designated Z_e. It can be obtained from the law of sines,

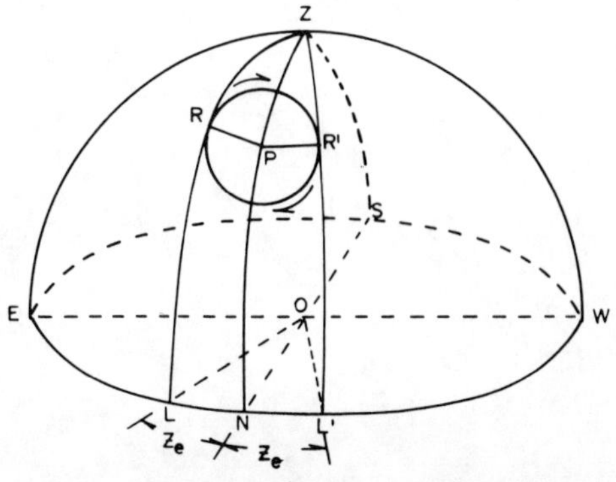

Fig. 5-22. The determination of azimuth by observing a circumpolar star at elongation.

$$\frac{\sin Z_e}{\sin(90° - \delta)} = \frac{\sin 90°}{\sin(90° - \Phi)},$$

or

$$\sin Z_e = \frac{\cos \delta}{\cos \Phi} = \cos \delta \sec \Phi. \tag{5–24}$$

Near elongation, the azimuth of a star changes slowly. For example, the azimuth of Polaris changes less than 1 second of arc from about 4 minutes before elongation to 4 minutes after elongation. In this time period, two sets of observations can be made between the star and the azimuth mark, once with the telescope direct and once with the telescope reversed. If Polaris is used as the circumpolar star, tables in the Almanac may be used for obtaining the azimuth at elongation. Many tedious computations can be avoided by the proper use of tables.

If the theodolite, such as the one shown in Fig. 5–23, is in proper adjustment—particularly, the horizontal axis must be truly horizontal—and good observing techniques are used, the only correction that need be made to the azimuth angle is the aberration correction. Because of the motion of the observer in the eastward direction, there is an effect of the diurnal aberration on the azimuth angle. The amount of this correction in seconds of arc is

$$\Delta Z = +0.319 \frac{\cos \Phi \cos Z}{\cos h}. \tag{5–25}$$

When the star is east of the meridian, the correction is plus; when the star is west of the meridian, the correction is minus. Since the altitude of the pole is equal to the latitude of the observer, Eq. (5–25) may be simplified in the case of a star near the pole. In seconds of arc, the correction becomes

$$\Delta Z = \pm 0.32 \cos Z. \tag{5–26}$$

Because the azimuth angle Z is reckoned either eastward or westward from the north point, whereas azimuth is measured only clockwise from north, the azimuth angle Z may be converted to astronomical azimuth A measured from north by these rules:

$$A = Z \quad \text{(for a star east of the meridian)},$$

and

$$A = 360° - Z \quad \text{(for a star west of the meridian)}.$$

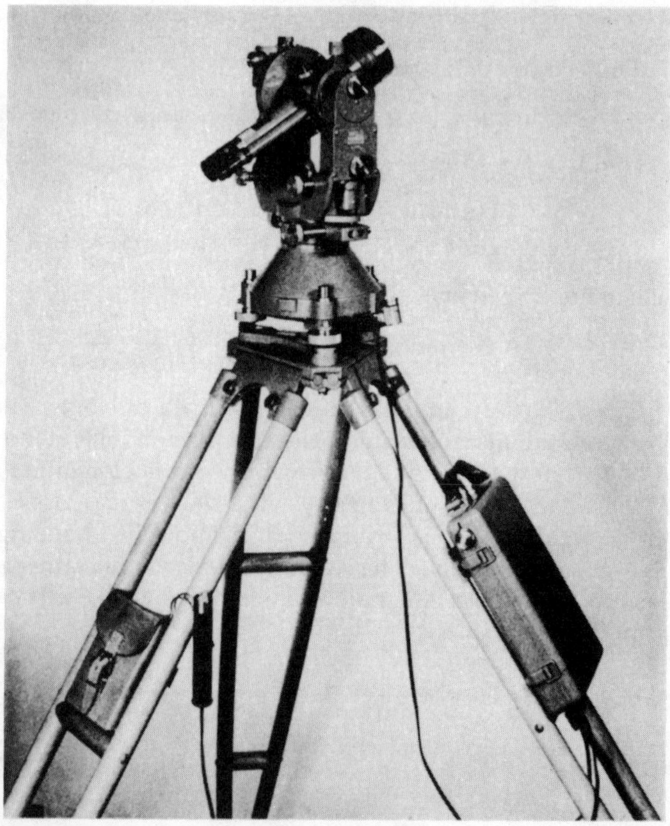

Fig. 5-23. Wild T-3 Precision Theodolite (Wild Heerbrugg Instruments, Inc.).

If an angle from the heavenly body to an azimuth mark has been measured in the procedure, its azimuth is determined by a return to Eqs. (5–18) and (5–19).

The methods outlined in Case II and Case III are those most often used in geodetic surveys. Case II should yield an accuracy of 0.″5 of arc; Case III, 0.″1 of arc.

RELATIONSHIP BETWEEN GEODETIC AND ASTRONOMICAL VALUES

Throughout this chapter, the authors have been very careful to point out that astronomical latitude, longitude, and azimuth are not the same as geodetic latitude, longitude, and azimuth. The capital Greek letters Φ, Λ, and A have consistently been used to distinguish the astronomical quantities from the geodetic quantities, which are designated by the corresponding small Greek letters ϕ, λ, and α.

The fact that the two are not the same is difficult for the beginning geodesist to grasp. It comes as a distinct shock to seasoned mariners who are proud of their abilities with the sextant or some other astronomical instrument. If we accept their claims of ship-positioning accuracy, or even go a little farther and grant them an error-free astronomical determination of their position, then the fact remains that the ship might be as far as $1\frac{1}{2}$ nautical miles from its true geodetic position. This fact has been alluded to several times in this book and is treated more thoroughly in Chapter 8. The mass anomalies within the earth produce corresponding gravity anomalies which deflect the plumbline away from the true vertical. This effect is known as the deflection of the vertical and has a maximum value of about 85 seconds of arc. It is called variously "deviation from the vertical", "deflection of the plumbline," and similar expressions, but the meaning is the same. Because the level bubbles on astronomical instruments are perpendicular to the plumbline, it follows that any astronomical instrument which uses level bubbles for orientation is affected by the deflection of the vertical.

The deflection of the vertical has an effect on latitude, longitude, and azimuth. The famous deflection-of-the-vertical equations relate the geodetic and the astronomical values. These equations are

$$\xi = \Phi - \phi, \tag{5–27}$$

$$\eta = (\Lambda - \lambda) \cos \phi, \tag{5–28}$$

and

$$\eta = (A - \alpha) \cot \phi, \tag{5–29}$$

where ξ is the deflection of the vertical in the plane of the meridian; Φ is the astronomical latitude; ϕ is the geodetic latitude; η is the deflection of the vertical in the plane of the prime vertical; Λ is the astronomical longitude; λ is the geodetic longitude; A is the astronomical azimuth; and α is the geodetic azimuth.

The reader should be aware of a pitfall in consulting other reference books: The symbols ξ and η are sometimes reversed. At least one textbook on geodesy reverses the above definitions of ξ and η.

The three equations (5–27), (5–28), and (5–29) are generally used in a slightly different form because the geodetic position and azimuth are normally required after the astronomical position and azimuth have been determined. Equation (5–27) can be rewritten as

$$\phi = \Phi - \xi. \tag{5–30}$$

If Eq. (5–28) is multiplied by sec ϕ, the result is

$$\eta \sec \phi = \Lambda - \lambda$$

or

$$\lambda = \Lambda - \eta \sec \phi. \tag{5–31}$$

If Eq. (5–29) is multiplied by tan ϕ, the result is

$$\eta \tan \phi = A - \alpha$$

or

$$\alpha = A - \eta \tan \phi. \tag{5–32}$$

From Eqs. (5–30), (5–31), and (5–32), it can be seen that any astronomical value can be converted to its corresponding geodetic value if we know the components of the deflection of the vertical. This is one goal of the worldwide gravity program.

Furthermore, Eqs. (5–29) and (5–28) may be equated:

$$(A - \alpha) \cot \phi = (\Lambda - \lambda) \cos \phi.$$

Multiplying by tan ϕ produces

$$A - \alpha = (\Lambda - \lambda) \sin \phi,$$

and finally

$$\alpha = A - (\Lambda - \lambda) \sin \phi. \tag{5–33}$$

Equation (5–33) is the famous Laplace equation which links differences of astronomical and geodetic longitudes with differences ot astronomical and geodetic azimuths. A station where simultaneous determinations of longitude and azimuth are accomplished is known as a Laplace station.

REFERENCES

1. B. Boss, and collaborators, *General Catalogue*, Vols. I–V. Carnegie Institution of Washington, Washington, D.C.
2. Institute of Geographical Explorations, *Complete 60° Star Lists for the Position Finding by Equal Altitude Method*, Climax Stationery Co., New York.
3. International Astronomical Union, *Apparent Places of Fundamental Stars*. York House, Kingsway, London.
4. Nautical Almanac Office, U.S. Naval Observatory, *The American Ephemeris and Nautical Almanac*. Government Printing Office, Washington, D.C.

Historical Perspective on Estimation Techniques for Position and Gravity Survey with Inertial Systems

James R. Huddle
Litton Guidance and Control Systems Division, Woodland Hills, California

Introduction

INVESTIGATIONS into the application of inertial technology to geodetic surveys have covered a span of at least 25 years, involving a number of private companies and government and university laboratories. The principal agency sponsoring this work in the United States has been the U.S. Army Engineer Topographic Laboratories (USAETL), whose original efforts began in the early 1960s.[1] In 1965 the Guidance and Control Systems Division of Litton Industries began its initial tradeoff studies on the Position and Azimuth Determining System (PADS) under USAETL sponsorship.[2-4]

The original estimation problem posed for the PADS was to design software to process discrete measurements from an odometer, or a laser-velocimeter, of the error in system-computed velocity observable at vehicle stops (subsequently called Zero-Velocity Updates or ZUPTS), so that accuracies of 20 m in level position, 10 m in elevation, and 0.3 mrad in azimuth were preserved over a 6-h open-traverse mission. The solution obtained was that of two independent Kalman filters, one for control of the errors associated with the local-level navigation axes and the other for control of errors associated with the vertical axis. Periodic vehicle stops were required, depending on the mode of system operation, to make system corrections with the Kalman filter(s) of the observed velocity error. The design was verified for all three modes of operation via simulation with extensive system error models and successfully field-tested in 1972. The first paper available to the general public on the basic principles for control of system error in the design of the PADS, along with simulation and test results was published in the same year. As in practice the stopping of the vehicle every 10 min to make ZUPT corrections is not excessively inconvenient, the principal operational mode for the PADS has become the unaided, free-inertial mode during vehicle travel. Also contributing to this result was the fact that the laser-velocimeter was an expensive sensor which gave the vehicle a distinctive signature, although it allowed the travel period to be extended to 1 h. Utilization of the odometer has also diminished as it extended the travel period to only 20 min.

The success of the PADS prototype hardware-testing program for the free-inertial mode in 1972[6-8] stimulated interest by a number of groups on the utilization of this inertial equipment for a higher-accuracy position, elevation, and gravity-vector survey capability in helicopters as well as land vehicles. The higher accuracy was easily obtained by relaxing the constraints imposed by the real-time artillery survey mission. The techniques employed were longer pre-mission alignment and calibration time, shorter travel periods between vehicle stops, longer stop periods during which system corrections of the observed error in system-computed velocity were made, and simple adjustment of the real-time position and elevation survey values, using the errors in these quantities available (only) at traverse closure.[33,35,62]

Early Work in Inertial Position Surveying

This latter interest resulted in the delivery beginning in 1974 of a number of systems based on the prototype PADS hardware, but with minor software and hardware modifications. Users included the U.S. Defense Mapping Agency (DMA), Canadian Department of Energy Mines and Resources (CEMR), U.S. Bureau of Land Management (BLM), and Span, Inc. These users began to report the rather accurate test results they were obtaining in 1975.[11-14] The specific inertial surveyors within this collection of systems have been given various names depending on the user, such as the Inertial Positioning System (IPS), Inertial Survey System (ISS), Spanmark,® and Litton Auto Surveyor System® (LASS I).

These user groups were well versed in the use of adjustment techniques for conventional surveys, and shortly papers began to appear on error models and procedures which could be used to adjust the real-time inertial survey position measurements.[32,36,37] Since the Kalman filter design employed for the original PADS provided real-time position measurements of adequate accuracy for these initial applications, little motivation existed for design change when used with the procedures adopted for enhanced accuracy. For cost and schedule-risk reasons, the real-time implementation of higher dimension, more complex error-coupling mechanizations of the Kalman filter did not find financial support.

Early Work in Inertial Gravity Surveying

Also at the time of the 1972 prototype PADS tests, USAETL expressed interest in the development of a more complete geodetic survey system capable of recovering all the measurable geodetic parameters of level position, elevation, intensity of gravity disturbance, and deflection of the vertical.

The design was constrained to be a reasonably low cost modification of the prototype PADS mechanization. A study performed in 1973[9] established a rather simple modification to the operational level axes real-time Kalman filter to achieve this purpose, along with estimates of the potential accuracy

Received Nov. 25, 1985.

Reprinted with permission from *AIAA J. Guidance,* vol. 9, no. 3, pp. 257–267, May–June 1986.

that the system could obtain for single traverse surveys with post-mission smoothing. This design change deleted the gyro-drift rate states from the Kalman filter so that change in the direction of the gravity vector relative to the mathematical figure of the Earth observed at vehicle stops was contaminated primarily by integrated gyro drift and (necessarily) the accelerometer noise. Correction for these errors in the real-time deflection-change estimates is then made, using the difference between reference astronomic measurements at the initial and closure points of the traverse.[34,38,63] No modification was necessary for the vertical axis real-time Kalman filter to measure the intensity of the gravity disturbance. Reports on the initial testing of the Rapid Geodetic Survey System (RGSS) began in early 1975[10] and improvements in accuracy obtained with various hardware and real-time software modifications have continued to be discussed.[15,16,22,26]

Recent Hardware Developments

During the period that initial deliveries of LASS I equipment were being made and investigations proceeded on the potential for gravity vector survey, the requirements of the U.S. Army dictated that certain changes be made to the prototype hardware to obtain the production model for the PADS. These changes were made, and the resulting systems successfully completed testing in 1979. The Litton Auto Surveyor System II (LASS II) based on the PADS production hardware was created. Photographs depicting this equipment are shown in Figs. 1 and 2. Table 1 presents an overview of the development of survey systems at Litton where Litton Auto Surveyor System I (LASS I) refers to the set of systems based on the original PADS prototype hardware. Users of the newer LASS II equipment include CEMR, Canadian Forces (CF), USAETL/DMA, and ITECH. Some minor variations of this system exist, depending on the interest of the user. For highest accuracy, all three channels (as opposed only to the vertical channel) use the low-noise A1000 accelerometer. The present version of the RGSS uses this equipment.

At the Third International Symposium on Inertial Technology for Surveying and Geodesy at Banff, Canada, in 1985, high-accuracy positioning-test results were reported by different user groups.[28-31] In particular, Rueger[29] has obtained accuracies of a few centimeters in a few kilometers or roughly one part in 100,000 of the relative position change. These accuracies approach theoretical accuracies of one to two parts per million predicted a few years ago in Huddle[38] as obtained from a study of limitations accuracy due to system error instabilities and the gravity disturbance.

Estimation Work Related to the Principal Inertial Surveyors

The First International Symposium on Inertial Technology for Surveying and Geodesy at Ottawa, Canada, in 1977 provided the first major public forum for presentation of various techniques for error estimation in intertial systems for survey. Table 2 provides a summary of estimation techniques associated with the four specific systems which were discussed then and continue to be reported upon.

The first entry in the table is the Aerial Profiling of Terrain System (APTS) being developed by Draper Labs for the U.S. Geological Survey (USGS). This system, described conceptually at the 1977 symposium,[21] discussed a 20-state, real-time Kalman filter to process the difference between laser-tracker range and angular measurements to presurveyed retroreflectors and the inertial system estimates of these variables to control-system error propagation. A later paper[25] on system architecture proposed only simple real-time resets to the inertial system position and velocity so that the navigation error would be bounded in real time, allowing the retroreflectors to be acquired in a timely manner. It had been decided to move the substantial estimation problem to a post-flight processor. In this scheme, a significant amount of real-time recorded

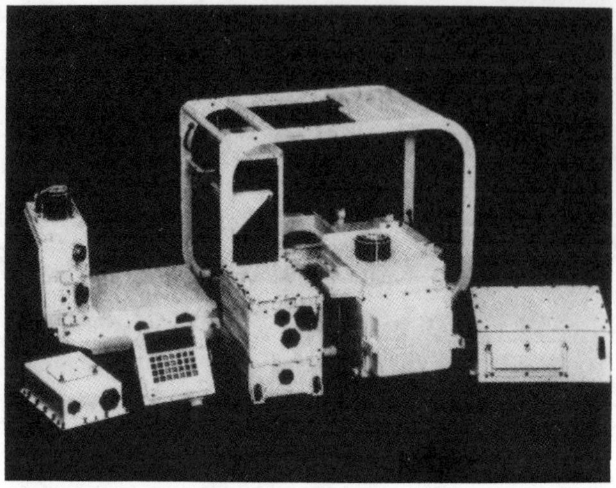

Fig. 1 LASS II hardware elements.

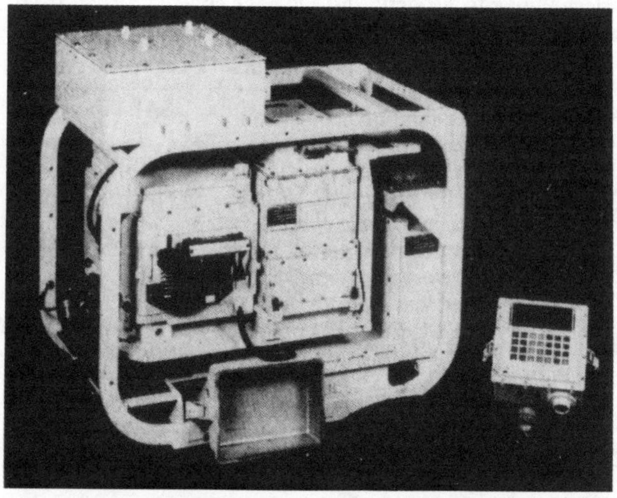

Fig. 2 LASS II hardware assembled in installation fixture.

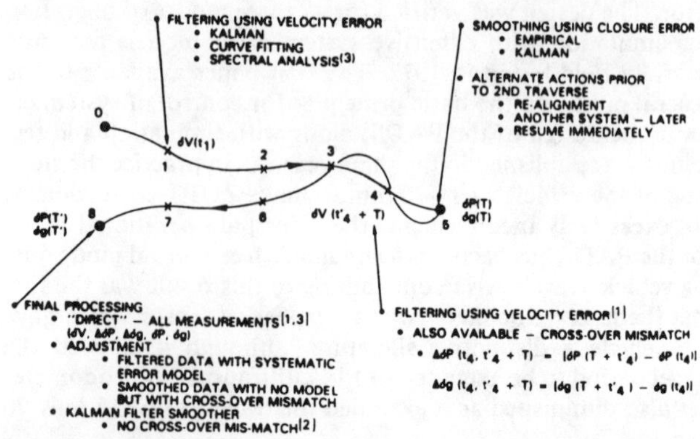

Fig. 3 Summary of various techniques for processing inertial survey measurements.*

---

*1) Kalman filter presumes $\delta g(P) \rightarrow \delta g(t)$; may not be a problem in the filtering step, as the function of the filter is to provide correction gains to bring the inertial system to an equilibrium state at each stopping point; an unknown in the final adjustment is the effect of correlation of error between spatially near traverses induced by the spatially correlated gravity disturbance.

2) Mismatch of position and gravity estimates at crossover points currently processed by empirical smoothers or multitraverse adjustment programs; a possible technique is Kalman filter with "suspended" states.

3) Discussed in Schwarz[54] and Schwarz et al.[59]

Table 1 Summary of Litton inertial survey system hardware

Designation	User	Units delivered (Sept. 1985)
PADS	U.S. Army	207 + 35 spares (402 to be delivered at 10/month)
IPS-1	DMA[a]	
ISS	CEMR[a]	LASS I: 14 (deliveries since 1974)
	CF[a]	
Litton	BLM	LASS II: 13 (deliveries since 1982)
Auto Surveyor®	ITECH[a]	
Spanmark®	SPAN	
RGSS (3 A-1000s)	USAETL/DMA[a]	

[a]Obtained delivery of LASS II equipment.

Table 2 Evolution of estimation approaches in four inertial survey systems

System	1977 papers	1981 papers	Recent
APTS (Draper)	20 state real-time Kalman filter[21]	Simple, real-time position and velocity resets[25] 29 + 11 per retro. states post-flight processor 10-15 megabytes storage	Post flight processor with 90 or more states[27]
GEO-SPIN (Honeywell)	27 state post-survey batch maximum-likelihood for ZUPTS and position differences at repeat points on travcrsc[39]	21 state real-time Kalman with smoothing at each control point[24]	
FILS (Ferranti)	Quadratic-polynominal fits to velocity error at stops to remove accumulated position error during travel in real-time[19]	10 state Kalman post-survey filter-smoother[45] Two-stage post-survey program[46] Functional fits to velocity error to correct position, attitude Repeat points on traverse yield unstable parameters Least-square adjustment Experiments with a six-parameter adjustment model[55]	
LASS (Litton)	14-state Kalman real-time level axes error control[17,38] 4 state for vertical axis Individual single-traverse (only) smoothers Position Elevation Gravity disturbance Deflection	Regional adjustment program with collocation for adjustment of filtered data for many traverses[47,52]	Simulation results with RAP improvements[66]

data (10 to 15 megabytes for a 3-h profile) was to be processed by a 29-state and higher dimensional Kalman filter-smoother, depending upon how many retroreflectors (up to 11 states per retro) were to be considered. Recently, this post-flight processor, which is capable of obtaining estimates of 80 or more error states, has been described in more detail.[27]

The last three entries in the table are inertial survey systems that operate in the free-inertial mode between stopping points at which the error in the system-computed velocity vector is observed. The velocity error includes "acceleration error" implicitly, in that a sequence of velocity measurements at a stopping point is usually obtained. Additional measurements typically available to these systems are the reference position and gravity vector at the beginning and end points of surveys traverses. For survey procedures which result in repetition of points (e.g., direct and reverse) or crossing of the traverses, the difference of the computed position and gravity vector obtained at the common encounter points provides further information for a correction process.

The Honeywell Geo-Spin® discussed at the 1977 symposium,[18] proposed the use of a post-mission, batch maximum-likelihood[39] technique to process the velocity errors at vehicle stops, the reference position at traverse end points, and the difference in computed positions at common points of a direct and reverse traverse. The 27 states in the system-error model employed included the customary nine navigation error states plus nine states for both the gyro and accelerometers (three each). Subsequently,[23,24] a 21-state, real-time Kalman filter was proposed for processing velocity errors at stops, with a Kalman smoother to be employed at the closure of each traverse.

The Ferranti FILS[20] discussed at the symposium[19] incorporated quadratic-polynomial fits used to estimate the velocity error between vehicle stops to then estimate the accrued position error in real time. Subsequently, three papers on post-mission processing of real-time survey measurements were presented. One of these papers described a post-mission 10 state (9 navigation error states plus azimuth drift rate) Kalman filter-smoother.[45] The design was tested for three different cases using forward and reverse traverse data for an L-shaped test course near Calgary. The other paper[46] described a post-processing software package that is comprised of two stages. The first stage obtains functional fits of the three-axis velocity error observed at stops and employs this data to obtain corrections to accrued position, platform tilt, and azimuth misalignment. If a forward-and-reverse traverse has been made, differences in position at each common point of the traverse are then employed to obtain estimates of the time variation in azimuth error and the level accelerometer scale factor errors. The second stage is a least-squares adjustment for horizontal position using reference position data to solve for the two fixed-level accelerometer scale factor errors and azimuth misalignment.

An alternative adjustment model for the FILS horizontal position after velocity correction has been investigated.[55] The parameters in this model include two level accelerometer scale factor errors, two azimuth misalignments, and two azimuth drift rates for each level accelerometer.

The Litton LASS I[17,38] has been described briefly above. This system employs two real-time Kalman filters of 14 states and four states for horizontal and vertical axis-error control, with the indicated simple modification for gravity vector survey. Four smoothing programs for correction of error in real-time survey measurements were available at that time which employed only the errors observed at a single traverse closure point. These included:

1) Horizontal position and azimuth misalignment correction, using closure values in horizontal position and azimuth.[62] The adjustment model included the survey point level position error and azimuth misalignment, level accelerometer scale factor errors, initial azimuth misalignment and azimuth drift rate, nonorthogonality of the level accelerometers, and integrated white noise.

2) Elevation correction using the initial and traverse closure value of elevation with the adjustment model including, in addition to the survey point elevation error, the vertical accelerometer misalignments and integrated white noise.[62] A modified version of this smoother, allowing time variation in the vertical accelerometer misalignments, existed for the RGSS mechanization.[65]

3) Deflection of the vertical correction using the reference deflection component change over the traverse.[63] The adjustment model included the survey point deflection errors and the platform gyro-drift rates.

4) Gravity disturbance correction using corrected elevation from the elevation smoother and the magnitude of the gravity disturbance values at the initial and traverse closure points.[64] The adjustment model included trend in the vertical accelerometer bias.

Table 3 Summary of reported work on systematic error models for level position survey

Authors		Accel. scale factor	Initial azimuth	Initial tilts	Time (t) × distance	$t^{1/2}$	t	$t^{2/3}$	t^2	PE,N
Early models										
Gregerson and Carriere[32]		2, ②	②		2, ②			2	②	
Ball and Voorhees[37]		2	2	2						
Kouba[36]		3	1			2				
Huddle/Litton[33-35,62-65]		2	2			2				2
Later models—via least-squares fit										
Schwarz and Gonthier[49]										
To eliminate systematic error	Forward	2	2							
	Reverse	2	2				2	2		
Hannah[52]		2	2						2	

Table 4 Summary of examinations of systematic error modes using test data

Authors	Remarks	Test Data
Milbert[53]	t^2 coefficients not needed	
(For 0 Gregerson model of Table 3)	Systematic error remains in longitude	SPAN SW Arizona Helicopter 66 and 80 km
Schwarz[41]	A-priori variances for Litton smoother can be selected to obtain near-optimal performance for a specific system.	GSC Ottawa, 100 km L-shaped test course
Schwarz and Gonthier[49]	Examined 6 error models with 4 to 8 parameters by least-squares (L.S.) fit Don't perform position update at forward traverse closure before reverse as it upsets error pattern	GSC Ottawa
Hannah and Pavlis[42]	Examined early error models (G,B,K,L) for 2 control points and concluded performance was similar	USAETL New Mexico, Wash., DC GSC Ottawa
Hannah[52]	Examined 7 error models of 2 to 4 parameters for position, elevation and gravity disturbance and 3 error models of 1 to 3 parameters for deflection by L.S. fit. 15-parameter model for all 6 geodetic variables determined as best	USAETL RGSS New Mexico helicopter and ground vehicle data (Subset of Todd[22] plus more helicopter data)

In 1981 the concept for a more sophisticated post-mission error-correction software, called the Regional Adjustment Program (RAP), was presented with experiments on the application of collocation for improved gravity vector estimation.[47] This program,[57] which incorporates the individual models discussed above into a single program, has been implemented and optimized.[66] In concept, the RAP is capable of processing survey position, elevation, and gravity-vector measurements from a number of traverses throughout a region that have been collected for any number of system realignments and with different inertial survey systems. Further, the differences in the position, elevation, and gravity-vector measurements at all traverse crossover points are processed to obtain excellent self-calibration of the inertial survey systems, as discussed further below.

Later Work in Estimation for Inertial Survey

Beyond the work cited above, there has been a great increase in contributions to the estimation area, beginning with the Second International Symposium on Inertial Technology for Surveying and Geodesy at Banff, Canada, in 1981. A significant amount of work has been carried out by three groups, which include USAETL/DMA, the University of Calgary (UC), and the Ohio State University (OSU). Tables 3-6 identify such work and summarize their contribution in one or more of the following three categories:

1) Appropriateness of parametric models for single-traverse (forward or forward-and-reverse) smoothing or the adjustment of numerous intersecting traverses throughout an area.

2) Performance and properties of various techniques for processing, in addition to the observations for the simple single-traverse category discussed below, the differences of measured positions, elevations, gravity disturbances, and deflections of the vertical at the same physical point, as determined by perhaps different inertial systems after possible new alignments on traverses originating from perhaps different control points. This question therefore deals with techniques for dealing with networks of inertial traverses.

3) Appropriateness of the techniques for processing a subset of the available observations. The types of observations available include the error in system-computed velocity at the vehicle stopping points and the error in system measured position, elevation, gravity disturbance, and deflection of the vertical (as available) at a traverse termination control point. This question, therefore, deals with the simple single-traverse case. This approach is somewhat academic, as in practice the traverse is usually repeated in the reverse direction to ensure reliability of the forward-traverse measurements and to enhance accuracy. Processing of measurement mismatches at the repeated points is sometimes done, obtaining a simplified case of network adjustment as discussed above.

A number of published papers have dealt with more than one of these issues, resulting in relevance to more than one category. This occurred particularly for papers on modeling and network adjustment, as the equality constraint on geodetic parameters at common network points permits higher-dimension systematic-error models.

The first category deals with the initial models being used in the early work of Gregerson and Carriere,[32] Ball,[37] Kouba,[36] Elms and Huddle,[33] Huddle and Lentz,[34] and Huddle[35,62-65] for post-survey smoothing of measurements from a single traverse or from a forward and reverse traverse, as shown in Table 3. Most of the modeling work has been restricted to sets of traverses obtained for a particular test area with a specific system. Conclusions of these various experiments vary and become more divergent as the time of the survey increases. Also, the different test areas and systems used to provide the data may contribute to the diversity of opinion.

Work that has examined these early models and extensions of them is summarized in Table 4. Milbert[53] examined a 12-parameter model ascribed to Gregerson for position smoothing (two level accelerometer scale factor errors and two azimuth misalignments and drift rates, and two coefficients of t^2) and elevation smoothing (two vertical accelerometer misalignments and drifts) and concluded that terms proportional to time-squared (t^2) were not needed and that systematic errors remained in longitude, after smoothing for a forward-and-reverse traverse. The data employed was provided by Span, Inc., for 65- and 80-km test line in southwest Arizona and consisted of 18 forward-and-reverse helicopter traverses.

Schwarz[41] examined the Litton smoothing and found that by changing the a priori variances for the systematic errors, near-optimal performance in position was obtainable. The data employed was provided by the Geodetic Survey of Canada (GSC) for the 100-km L-shaped Ottawa test net and consisted of nine forward and reverse L-shaped traverses and 10 straight traverses. This correponds to an optimal "tuning" of the smoothing software for a specific system. Further, for the same data, Schwarz and Gonthier[49] examined six parametric models by a least-squares fit. They concluded that a simple four-parameter model fit the position error data very well for the forward traverse, but seven or eight parameters were required for the reverse traverse, which was performed after a position update that disrupted the linearity of the error pattern, indicating clearly the presence of initial azimuth misalignment. Empirical smoothing performed with the models indicated the four-parameter model was as good as the others. The weights employed in this model corresponded to level-axes accelerometer scale factor errors of .0027% and .0058%, nonorthogonality of the accelerometer axes of 15 arc-s and initial azimuth misalignments of 38 arc-s. Subsequent empirical analysis of a large amount of data reported recently in Schwarz et al.[59] for the LASS I, yields 1 σ values appropriate for an adjustment model of .0022% and .0010% for the level-accelerometer scale factor errors, 14 arc-s for azimuth misalignment, 5.9 arc-s for the level accelerometer nonorthogonality, and .0007 deg/h for azimuth drift rate.

Hannah and Pavlis[42] examined the Gregerson, Ball, Kouba, and Litton [G,B,K,L] models cited above for position and elevation smoothing for two points on a traverse and concluded that all models performed about the same, but that Kouba's had a slight edge for position and Litton's for elevation. The data employed here was varied, being provided by USAETL from tests in New Mexico and Washington, DC, and from the GSC for Ottawa. For traverses with three control points, a five-parameter (three accelerometer scale factor errors, a common scale factor error for the level accelerometers, and azimuth misalignment) model of Kouba was recommended.

Alternate techniques for processing measurements obtained in a network of intersecting inertial traverses are summarized in Table 5. For a set of RGSS data taken by USAETL in New Mexico, comprising of 11 forward and 11 reverse traverses with alignments at each end and taken with both a ground ($\sim$2.18-h traverses) vehicle and a ($\sim$.77-h) helicopter (22 total traverses per vehicle), Todd[22] examined the performance for Litton's empirically smoothed position, elevation, gravity disturbance, and deflection of the vertical for different survey procedures. These procedures included single pass over a survey point, averaging double passes using the reverse traverse, averaging two single-crossing passes from different traverses, and averaging two double-pass crossings (from two forward-and-reverse surveys crossing the point). Performance improved approximately by the square-root of the number of passes. The best results were obtained in the helicopter mode, due to the shorter time for error accumulation. The same empirically smoothed data was processed by Zimmerman and Harris[48] by least-squares for the entire network of points, minimizing the differences between the forward-and-reverse traverses while simultaneously minimizing the differences at all traverse crossing points. Also in Todd[22] using this least-squares technique, the RMS errors for the helicopter mode were 10 cm and 21 cm for position, 19 cm for elevation, 0.4 and 0.5 arc-s for the deflection components, and 1.1 mgal for the gravity disturbance. This same multiple-traverse position data, prior to empirical smoothing (Kalman filtered only), was then examined by Todd and Tindall[44] in a least-squares adjustment with 48- and 27-parameter models (four plus and minus accelerometer scale factor errors, 22 initial azimuth misalignments [ground vehicle model only], and 22 or 1 azimuth drift rates). The 48-parameter model performed slightly better than the 27-parameter model, yielding results similar to that for the least-squares adjustment performed on the smoothed data.

Subsequently, Hannah[52] performed an exhaustive analysis of a subset of 12 forward traverses from the USAETL RGSS Kalman filtered ground vehicle data, supplemented with Kalman filtered data from a new set of helicopter traverses comprised of 16 (~.5-h) forward traverses, 8 (~1-h) L-shaped traverses, and 6 (~2-h) L-shaped forward-and-reverse traverses (where the reverse traverse was made immediately without position update). He examined seven different parametric models of two to four parameters per axis for position, elevation, and gravity disturbance and three different parameteric models of one to three parameters for each deflection component. A 15-independent-parameter model in total was determined to be best and was then employed in a least-squares network adjustment for the helicopter traverses. This model employed three independent parameters for each position component (accelerometer scale factor error, azimuth misalignment, and a t^2 coefficient), four parameters for elevation (two vertical accelerometer misalignments and a scale factor error, and a t coefficient), four parameters for the gravity disturbance (three in common with the elevation model but also an independent vertical accelerometer bias trend), four parameters for each deflection component (two in common with the corresponding position error, a level axis drift rate, and an independent t^2 coefficient). Previous work by Hannah[51] with the varied IPS mechanization data (a different Kalman filter mechanization than that for the RGSS data above) of Hannah and Pavlis[42] examined a similarly extensive set of candidate parametric models and arrived at a similar model, but without the vertical-accelerometer scale-factor error (14 parameters) and a parameter proportional to the

Table 5 Summary of reported work on processing of networks of inertial traverses

Paper	Procedure/Remarks	Data
Todd[22]	Compared empirically smoothed: Single pass and average 2 single passes Average forward and reverse traverses Average 2 different forward and reverse traverses Error reduces by $\sqrt{N}$ Also least squares (L.S.) (Zimmerman & Harris[48]) Best Performance for L.S. for helicopter runs Position 10 and 21 cm, RMS Elevation 19 cm Deflections 0.4 and 0.5 s Disturbance 1.1 mgal	11 forward and 11 reverse traverses with both helicopter and ground vehicle as obtained by USAETL in New Mexico
Todd & Tindall[44]	Least-squares adjustment of filtered data 27-parameter model with 22 azimuth errors 4 + and-accelerometer scale factors 1 azimuth drift rate 48-parameter model Above 22 azimuth drift rates Performance of 48 slightly better than 27-parameter model but similar to above least squares. On empirically smoothed data empirically smoothed data	Above USAETL ground vehicle test data
Hannah & Mueller[43]	Least-squares adjustment with 5-parameter model Discovered cross-over points of traverse, obtained excellent calibration of inertial system revealing error in underlying control	GSC southern Manitoba 24 intersecting forward and reverse traverses
Hannah[52]	Least-squares adjustment with 15-parameter model Performance —Position 10 cm —Disturbance 0.9 mgal —Deflection 0.4 arc-s Again discovered potential error in control-elevation	USAETL RGSS 24 roughly straight interlocking helicopter traverses
Arden and Schwarz[56]	Examined procedures for efficient network surveys Concluded strong relative accuracy of interlocking traverses can uncover error in peripheral control Further, peripheral control should not be held fixed—assign proper covariances	None
Schwarz, Arden and English[59,60]	Tested four procedures for network surveys Adjustment of Kalman-filtered measurements obtained superior performance	GSC—expanded Manitoba LASS I Data Set

longitudinal $(\Delta\lambda)^2$ difference-squared (accounting for the long-term effects of azimuth drift) in the latitude model instead of a parameter for t^2.

Additionally, in Hannah and Mueller,[43] position (with a simple five-parameter model), and in Hannah,[51] position and elevation [with the nine-parameter model of Ref. 42 but without the $(\Delta\lambda)^2$ parameter], test results for a network of intersecting inertial traverses were reported. The data was provided by the GSC and comprised of 24 forward-and-reverse approximately 100-km (1.5-h) production traverses in southern Manitoba. In this case, systematic error trends of the internal points of the mesh after adjustment were related to potential systematic error trends in the underlying control. The premise was that the crossover points of the traverses were in themselves sufficient to obtain excellent calibration of the inertial data, thereby revealing deficiencies in the underlying control.

Subsequently, with 24 interlocking helicopter traverses (12 roughly straight, forward, and reverse traverses) from the USAETL RGSS data set, Hannah[52] performed another network adjustment experiment with the 15-parameter model. A potential systematic error was discovered in elevation, making control again suspect. Otherwise, performance of the network adjustment was excellent, approaching the residual error level of the least-squares procedure for model identification, indicating the effectiveness of the network approach. The RMS error performance approached that of the control, being less than 10 cm for position, .9 mgal for gravity disturbance, and 0.4 s for the deflection of the vertical.

Arden and Schwarz[56] have examined alternative network field procedures relative to the standard used by the GSC with the LASS. This procedure includes double forward-and-reverse traverses between four corner control stations of an 80×100 km grid. Northerly directed forward-and-reverse traverses were repeated every 10 to 20 km in the east direction, and there was a single easterly-directed forward-and-reverse traverse at the mid-point of the grid. Consequently, the peripheral and east-directed mid-grid lines of points are encountered four times, while all other points are encountered twice. They found that no advantage was obtained by either executing a criss-cross network of single traverses (encountering all points interior to the periphery twice) or adding a control station at the center of the grid. They did find that the double forward-and-reverse traverses on the periphery could be reduced to single forward and reverse traverses without significant performance loss, hence reducing costs. Finally, they concluded, similarly to Hannah above, that a network of interlocking inertial traverses is so strong in relative accuracy that it will uncover relative error in the peripheral control. Hence, control should not be held fixed in the adjustment but should be assigned proper a priori variances to allow the entire network to move in the adjustment.

Reflecting upon the above body of modeling work and some of that to be discussed below under empirical smoothing, the differences obtained by various contributors seem to lie in the method of dealing with azimuth drift rate for traverses which extend over significant time periods (>1.5 h). The premise in the original smoothing work[62] was that survey position errors $(\delta D_{E,N})$ could be approximately related to a constant azimuth drift rate by the following equations:

$$\delta D(t)_E = \int_o^t \mathrm{d}\mu \int_o^\mu a_N(\nu)\nu b_z \mathrm{d}\nu = b_z P_E(t)$$

$$\delta D(t)_N = -\int_o^t \mathrm{d}\mu \int_o^\mu a_E(\nu)\nu b_z \mathrm{d}\nu = -b_z P_N(t)$$

where $a_{E,N}(\nu)$ are the east and north components of vehicle acceleration; $P_{E,N}(t)$ are the dynamic coefficients relating azimuth drift rate to the accelerations that can be generated in real-time in the digital computer and stored for the post-survey adjustment; and b_z is the assumed constant azimuth drift rate over the period of interest. Clearly, the P coefficients can be represented approximately in different ways.

Table 6 summarizes investigations of the appropriateness of methods of processing the available observations for a single traverse. Kouba[36] originally investigated the problem of least-squares adjustment of Kalman-filtered inertial position measurements, using the four-parameter systematic error model cited above. The integrated white-noise model[35] for the accumulation of residual position error induced by (only) system noise and stochastic gravity disturbances after zero-velocity updates, was employed to construct the weight covariance matrix for these errors. Kouba indicated variations required for the multiple-traverse case. Schwarz[40] extended this analysis, deriving the error covariance matrices between different survey points on the traverse for filtered survey measurements and discussed the filter-smoother which can be optimal for the case that position-dependent error sources (e.g., gravity disturbances) can be represented properly as

Table 6 Summary of investigations into various techniques for adustment and smoothing

Authors	Subject
Kouba[36]	Derived a "total" covariance matrix between errors at different survey points, using systematic error plus integrated white noise as the model for filtered survey measurements to be used in a least-squares adjustment: $$\delta P(t) = \Phi(t)X + \int_0^t \xi(\mu)\mathrm{d}\mu$$
Schwarz[41]	Derived the total error covariance for (Kalman) filtered survey measurements Presumes gravity disturbance representable as time-dependent process
Schwarz and Gonthier[50]	Discusses the direct (batch) and two-step methods of measurement processing (velocity error, cross-over mismatch and closure errors) Notes lack of formulas for total covariance for optimal smoothed traverses
Gonthier[61]	Showed from simulation analysis that a suboptimal Kalman filter in conjunction with an empirical smoother performs as well as an optimal Kalman filter-smoother for straight, L-shaped, and Z-shaped forward traverses without the added storage and computation costs
Vassiliou[58]	Examined spectral decomposition approach for velocity error processing (suggested by Schwarz[54] in conjunction with empirical smoothing with good performance)

time-dependent functions. This can be achieved by avoiding commonality or proximity of points along the traverses. Schwarz and Gonthier[50] subsequently pointed out a difficulty in obtaining the total error covariance matrix between all the survey points on a traverse that have been obtained by optimal filtering and smoothing. This error covariance matrix is required to integrate properly the survey results with measurements obtained by other survey methods.

Gonthier[61] examined by simulation the interesting question of performance degradation in position error for:

1) A survey system mechanized with a "simplified" 12-state Kalman filter to control the effects of stochastic noise sources with ZUPTS, in combination with a six-parameter empirical smoother (two level-axis accelerometer scale factor errors and two misalignments, and an azimuth misalignment and drift rate) to eliminate these systematic errors, which are not effectively observed by the Kalman filter with the zero-velocity measurements. This is similar in concept to the LASS mechanization.

2) An "optimal" 16-state Kalman filter-smoother (the seven navigation errors, three [variable] gyro drifts two [variable] acceleration errors and the four systematic errors of accelerometer scale factor error and mechanical misalignment, as cited above).

Simulations were performed for straight, L-shaped, and Z-shaped traverses. The conclusion was that the two mechanizations produced almost identical performance over the traverses. The simplified filter and empirical smoother is attractive from a practical point of view, as it obtains this performance without the extensive storage of data and computation required to implement the optimal smoother.

Vassiliou[58] has examined off-line for three forward-and-reverse traverses (~2 h) of FILS data for the 45-km, L-shaped test course near Calgary, the position and elevation performance of the method of spectral decomposition suggested by Schwarz[54] for the processing of the zero-velocity measurements in combination with a 10-parameter model (the three accelerometer scale factor errors, the six possible misalignments of their sensing axes, and a common drift parameter for all three axes). Spectral decomposition first obtains a least-squares fit of the error in system-computed velocity to a set of theoretically derived basis functions involving sinusoids of the Schuler frequency and Earth rate. Residual systematic error in the velocity samples are then represented as a summation of low-frequency harmonic sinusoids, as determined from a discrete Fourier transform applied to the residual data. The unfiltered position data is then corrected with the integrals of the obtained velocity error functions. In these tests, position error (prior to smoothing) ranged from 5 to 25 m. Error after correction based on the mismatch of position measurements between the forward-and-reverse traverse at intermediate points, and after empirical smoothing based on the position misclosures at the end of the forward-and-reverse traverses, were 15, 30, and 10 cm, respectively, for latitude, longitude, and height. These errors were less than the errors obtained with the suboptimal 10-state Kalman filter-smoother[45] which processed (necessarily) a subset of the same data.

Clearly from the above discussion, a proliferation of techniques has resulted for processing the measurements available from an inertial surveyor. For the observed error in system computed velocity at vehicle stops, the techniques include 1) Kalman filtering, 2) curve filtering, and 3) spectral decomposition. For the observed single traverse mis-closure in the composition and gravity vector at control points, the techniques include 1) Kalman smoothing, 2) empirical smoothing, or 3) no smoothing if further processing of multiple traverses is to be performed. Further, for the observed differences in the position and gravity vector at common points of multiple traverses, the techniques include 1) empirical smoothing or again 2) no smoothing. Finally, for multiple traverses these pre-processed data are subject to subsequent adjustment with a least-squares procedure which either incorporates a systematic error model if only filtered data are employed, or else a least-squares procedure which incorproates no systematic error model if the data have been pre-smoothed. In either of these latter cases, the differences of the measurements at common traverse points and the mis-closures at control points are minimized.

Schwarz[54] has provided a framework from which to consider this diverse combination of methods relative to nearly-rigorous reference method of simulatneous processing of all the available velocity error, traverse cross-over point mismatches, and closure errors. This reference method is referred to as the "direct" approach and rests on the assumption that the gravity-disturbance vector, which is a position-dependent function, can be transformed to a time-dependent function with appropriate stochastic description. This becomes difficult for forward-and-reverse traverses and networks of intersecting traverses. The practical significance of the correlation between survey errors due to the gravity disturbance vector on "spatially near" traverse points not reflected through such a time function representation remains to be ascertained. This could be significant, as the gravity disturbance vector is a major source of survey measurement error, especially for high-accuracy survey systems. In Schwarz et al.,[59] further detail on this framework has been developed along with the results of experiments performed for a LASS I horizontal position and elevation data set provided by the GSC for survey traverses in southern Manitoba. The study reports on the relative merits of:

1) Adjustment of filtered measurements.

2) Free-adjustment of filtered measurements, with subsequent transformation to control.

3) Adjustment of empirically smoothed measurements.

4) Averaging of empirically smoothed measurements.

They tentatively concluded that the first and third techniques are superior and similar in performance, but that final resolution had to await improvements in the available control. Subsequently,[60] better control-point values to a relative accuracy of 5 cm (1 σ) were obtained via the use of the interferometric GPS technique. With the availability of this information, the first technique of simultaneous adjustment of all filtered measurements was clearly superior for latitude and longitude. The poorer performance of the second method remains problematic and will be investigated further by those authors.

Illustration of Estimation Methods

To illustrate the various estimation techniques discussed above for processing inertial survey measurements, Fig. 3 has been constructed. Beginning at the left side of the figure, the inertial system is aligned and initialized with reference data. The surveyor is then moved along the forward traverse, stopping periodically to observe the error in system-computed velocity, which can be used by various filtering techniques to reduce error propagation. At the closure of the forward traverse at the right side of the figure, additional reference data becomes available. The error in the surveyor estimated position, elevation, and gravity-disturbance vector observed at closure can be used by an empirical or Kalman smoother to correct the survey measurements made at intermediate points along the forward traverse. Azimuth data may also be available but is seldom used in operational surveys today.

At the traverse closure point, the inertial system may be realigned before proceeding with the further collection of survey measurements, or proceed immediately with the survey. An alternate possibility is that another surveyor may be used to collect measurements from the traverse termination point at some different time.

On the second traverse, velocity error is observed at vehicle stopping points and used to control error as in the forward traverse. Further, however, as common points are encountered on the reverse portion of the traverse, mismatches between the position, elevation, and gravity-disturbance measurements obtained on the forward traverse provide additional measurements for error control. These measurements are used by some empirical smoothers to solve for error instabilities when the same inertial system is being used, and in all adjustment methods as the adjustment model dictates.

Note that the processing of these differences is not accommodated by a standard Kalman filter-smoother, as covariance information between errors at different points in time is not available. This could be overcome by adding error states to the filter for the position, elevation, and gravity-disturbance measurements at the survey points on the forward traverse,[9] referred to as "suspended" states below, and then processing the mismatches on the reverse portion of the traverse. This presumes that the same system is used on the reverse portion of the traverse and that the spatially correlated gravity-disturbance vector can be represented sufficiently well for our purposes as a time-correlated stochastic process.

At the lower left side of the figure, a final traverse closure point is encountered which completes a very simple network of inertial traverses in our illustration. Here the same reference data are potentially available as at the closure point of the forward traverse. The various procedures noted above can now be employed to obtain final values for the positions and gravity disturbances at the vehicle stopping points.

Conclusions from Literature Review

In reviewing the contributions cited as references for this paper, the following general conclusions are drawn:

1) Suboptimal filtering of various types, whose objective is to control the time-variant sources of system noise using the error in system-computed velocity observed at vehicle stops, can achieve essentially optimal performance.

2) Empirical smoothing, whose objective is to control the effect of systematic error in the survey measurements which are only weakly observable by the velocity error filtering above, can be based upon fairly simple parametric models and obtain essentially optimal performance. Such a model includes accelerometer scale factor errors and misalignments and the instabilitites in these misalignments. Some variation in opinion appears as to how to represent the instability in the accelerometer alignment, which may be aggravated by unknown effects of the gravity disturbance.

3) A network of inertial traverses throughout a region with numerous crossover points obtains high relative accuracy in an adjustment with simple parametric models for systematic error.

4) Control should be introduced into network adjustments with proper covariances, due to the high relative accuracy of the inertial measurements.

5) The relative accuracies obtainable in nominal network surveys with inertial systems currently appear to be from 10 to 30 cm in level position and elevation and from .3 to 4 arc-s for the deflection of the vertical. In fact, accuracies relative to existing control at the upper end of this range have tended to cast doubts on the accuracy of the control.

Future Investigations

Given the above conclusions, it appears that investigations into the following issues would be useful to obtain performance enhancement for the user of inertial survey equipment. For the user whose work does not allow the use of a network approach but collects measurements with forward and reverse traverses:

1) Determine any performance enhancement obtainable by the use of "suspended" states for errors in the estimates of position and the gravity disturbance at survey points on the forward traverse which can then be processed with the comparison measurements at the same points on the reverse traverse. As a Kalman filter-smoother problem, such a study would provide a near-optimal reference for deterministic empirical smoothers, which today are processing these measurements to reduce the effects of error instabilities due to various sources. Recall, however, that the effect of the spatially correlated gravity disturbance is not truly represented as time-correlated stochastic process in a simple manner here.

2) Determine the performance enhancement of an ad hoc technique in which the estimates of the gravity disturbance vector obtained on the forward traverse, and enhanced by traverse closure measurements, can be used to reduce the error effect of the gravity disturbance vector on the reverse traverse.

For the user whose work does allow the use of network techniques, answers to the above questions would also be useful in that they could improve the quality of the data to be used in the network adjustment. Further, however, the error effect of the underlying regionally correlated gravity-disturbance vector has not yet been clearly quantified. Specifically, the effect on the final adjustment results of the correlation of error between survey measurements obtained on spatially near traverses, due to the gravity disturbance, needs to be addressed.

Conclusion

It is clear from a review of the literature on the field of inertial survey that this is an enterprise of significant commercial and military importance where optimal estimation techniques have made a major contribution to success. The experimental and theoretical work performed over the last five years has indicated that some of the original estimation approaches are near optimal in performance, while remaining attractive in their implementation cost. For a single traverse, theoretical relative position accuracies of a few parts per million are being approached by some users. For regional surveys comprised of a number of intersecting traverses, however, some theoretical issues remain as indicated above, which when resolved should lead to further improvements in performance. Further, as we look to the future it is clear that more accurate inertial equipment will be available for survey use. This equipment, in conjunction with developments in other areas of gravity gradiometry and satellite range and doppler measurement, will in some as-yet-to-be-determined combination lead to substantially improved accuracy, while retaining the high production rates available today with inertial survey systems.

References

Investigations and tests of position, elevation, and azimuth determining inertial survey systems for artillery applications.

[1]"Final Technical Report on Inertial Survey Equipment (GEISHA)," USAETL, Fort Belvoir, VA, 1963.

[2]Clark, W.J., "Final Technical Report on the Application of Inertial Technique to Surveying," USAETL AQ22107E66, Litton, Woodland Hills, CA, Nov. 1966.

[3]Maughmer, R.W. and Yamamoto, J.R., "Advanced Study of Position and Azimuth Determing Systems," USAETL Final Rept., AQ51073J68, Litton, Dec. 1968.

[4]Maughmer, R. W. and Byrd, D. M., "Supplement to Final Report, Advanced Study of a Position and Azimuth Determining System," USAETL AQ51073J685, Litton, July 1969.

[5]Huddle, J. R. and Maughmer, R. W., "The Application of Error Control Techniques in the Design of an Advanced Augmented Inertial Surveying System," *28th Annual Meeting of the Institute of Navigation,* West Point, N.Y., Litton, Woodland Hills, CA, June 1972, pp. 1-44.

[6]Ellms, S. R., "The Design, Fabrication, and Test of a Position and Azimuth Determining System (PADS) Final Report," USAETL-CR-73-6, Litton, June 1973.

[7]Kitchens, C. W. Sr., "The Position and Azimuth Determining System," *American Congress on Surveying and Mapping,* Falls Church, VA, Oct. 1973.

[8]Perrin, J. L., "The Prototype Jeep-Mounted Position and Azimuth Determining System (PADS)," *Proceedings of the 8th Data Exchange Conference for Inertial Systems,* Massachusetts Institute of Technology, Cambridge, MA, Aug. 1974, USAETL-CR-73-12, Fort Belvoir, VA, pp. 214-236.

Investigations and tests of high-accuracy position, elevation, and gravity vector survey systems.

[9]Huddle, J. R., "A Study of the Position and Azimuth Determining System for Mapping, Charting and Geodesy Applications," USAETL-CR-73-12, Fort Belvoir, VA, Aug. 1973.

[10]Huddle, J. R. and Mancini, A., "Gravimetric and Position Determinations Using Land-Based Inertial Systems," *Proceedings of the 35th Annual Meeting of the American Congress on Surveying and Mapping*, Washington, DC, March 1975, pp. 93-106.

[11]Ball, W. E., "Testing an Airborne Inertial Survey System for Bureau of Land Management Cadastral Survey Applications in Alaska," *Proceedings of the 35th Annual Meeting of the American Congress on Surveying and Mapping*, Washington, DC, March 1975, pp. 107-137.

[12]Gregerson, L. F., "Inertial Instrumentation at the Geodetic Survey of Canada," *Commonwealth Survey Officers Conference*, Cambridge, England, Aug. 1975.

[13]Barr, J. R., "Inertial Surveys in Private Practice," *Alaska Surveying and Mapping Association Convention*, Anchorage, AK, Jan. 1976.

[14]Harris, H. C., "IPS Field Test Results," (DMA) *Proceedings of the 36th Annual Meeting of the American Conference on Surveying and Mapping*, Washington, DC, Feb. 1976.

[15]Fishel, N. and Roof, E., "Results of Tests Using an Inertial Rapid Geodetic Survey System (RGSS)," *Proceedings of the 37th Annual Meeting of the American Congress on Surveying and Mapping*, Washington, DC, March 1977, pp. 100-153.

[16]Roof, E. F. and Todd, M.S., "The Development of the Inertial Rapid Geodetic Survey System at USAETL," *Proceedings of the 1st International Symposium on Inertial Technology for Surveying and Geodesy*, Ottawa, Canada, Oct. 1977, pp. 113-120.

[17]Ellms, S. R., "A Description of the Auto-Surveyor System and Its Sub-System Accuracies," *Proceedings of the 1st International Symposium on Inertial Technology for Surveying and Geodesy*, Ottawa, Canada, Oct. 1977, p. 78-89.

[18]Hadfield, M. J., "Critical Inertial System Characteristics for Land Surveying," *Proceedings of the 1st International Symposium on Inertial Technology for Surveying and Geodesy*, Ottawa, Canada, Oct. 1977, pp. 90-101.

[19]Brown, K. R., Crawford, C. I., and Rhind, W. G., "Versatility of the Inertial System for Survey Applications," *Proceedings of the 1st International Symposium on Inertial Technology for Surveying and Geodesy*, Ottawa, Canada, Oct. 1977, pp. 102-112.

[20]Hittel, A. and Hagglund, J.E., "Evolution of the Ferranti Inertial Survey System," *Proceedings of the 1st Symposium on Inertial Technology for Surveying and geodesy*, Ottawa, Canada, Oct. 1977, p. 257-277.

[21]Hursh, J. N., Mamon, G., and Soltz, J. A. "Aerial Profiling of Terrain," *Proceedings of the 1st Symposium on Inertial Technology for Surveying and Geodesy*, Ottawa, Canada, Oct. 1977, pp. 121-130.

[22]Todd, M.S. "Rapid Geodetic Survey System (RGSS) White Sands Tests for Position, Height, and the Anomalous Gravity Vector Components" *Proceedings of the 41st Annual Meeting, American Congress on Surveying and Mapping*, Washington, DC, Feb. 1981.

[23]Harris, H. C., "Status of DMA Testing of the Honeywell GEO-SPIN System," *Proceedings of the 2nd International Symposium on Inertial Technology for Surveying and Geodesy*, Banff, Canada, June 1981, p. 359-372.

[24]Adams, G. W. and Hadfield, M. J. "GEO-SPINS/IPS-2 Improvements for Precision Gravity Measurement," *Proceedings of the 2nd International Symposium on Inertial Technology for Surveying and Geodesy*, Banff, Canada, June 1981, p. 235-252.

[25]Soltz, J. A., Mamon, G., and Chapman, W. "Aerial Profiling of Terrain System Implementation," *Proceedings of the 2nd International Symposium on Inertial Technology for Surveying and Geodesy*, Banff, Canada, June 1981, pp. 269-270.

[26]Roof, E. F., "Inertial Survey Systems," *Journal of Surveying Engineering*, Vol. 109, Aug. 1983, pp. 116-135.

[27]Donna, J., "APT System Post-Processor," *Proceedings of the 3rd International Symposium on Inertial Technology for Surveying and Geodesy*, Banff, Canada, Sept. 1985.

[28]Rueger, J. M., "Evaluation of an Inertial Surveying System," *The Australian Surveyor*, Vol. 32, June 1984, pp. 78-98.

[29]Rueger, J. M., "High Accuracy in Short ISS Missions," *Proceedings of the 3rd International Symposium on Inertial Technology for Surveying and Geodesy*, Banff, Canada, Sept. 1985.

[30]Hein, G. W., Eissfeller, B., and Landau, H., "Comparative Study of Ferranti, Honeywell and Litton Inertial Surveying Systems: Test Net Werdenfelser Land," *Proceedings of the 3rd International Symposium on Inertial Technology for Surveying and Geodesy*, Banff, Canada, Sept. 1985.

[31]Hadfield, M. J., "High Precision Position and Gravity Measurement with GEO-SPIN [TM]," *Proceedings of the 3rd International Symposium on Inertial Technology for Surveying and Geodesy*, Banff, Canada, Sept. 1985.

[32]Gregerson, L. F. and Carriere, R. J., "Inertial Surveying System Experiments in Canada," *International Association of Geodesy XVI General Assembly of the IUGG*, Grenoble, France, Aug. 1975, available as Publ. 13414, Litton, Woodland Hills, CA..

[33]Ellms, S. R. and Huddle, J. R., "The Application of Inertial Navigation Systems to Precision Land Survey," *Journal of the Institute of Navigation*, Vol. 23, Summer 1976, pp. 93-105.

[34]Huddle, J. R. and Lentz, R. H., "Post Mission Smoothing and Analysis of the Measurements of the Change in the Deflection of the Vertical Obtained by the Rapid Geodetic Survey System (RGSS)," USAETL-0065, Fort Belvoir, VA, Sept. 1976.

[35]Huddle, J. R., "Navigation to Surveying Accuracy with an Inertial System," *Proceedings of the Bicentennial National Aerospace Symposium*, Warminster, PA, April 1976, pp. 25-32.

[36]Kouba, J., "Geodetic Adjustment of Inertial Surveys," *1977 Annual Meeting of the Canadian Geophysical Union*, Vancouver, Canada, April 1977.

[37]Ball, W. E. Jr. and Voorhees, G. D., "Adjustment of Inertial Survey System Errors," *Proceedings of the 1st International Symposium on Inertial Technology for Surveying and Geodesy*, Ottawa, Canada, Oct. 1977, p. 370-379.

[38]Huddle, J. R., "Theoretical Principles for Design of the Inertial Surveyor for Position and Gravity Determinations," *Journal of Guidance and Control*, Vol. 1, May-June 1978, p. 183-188.

[39]Adams, G. W., "Intertial Survey Data Reduction Using Maximum Likelihood Estimation," *Proceedings of the 1st International Symposium on Inertial Technology for Surveying and Geodesy*, Ottawa, Canada, Oct. 1977, p. 380-387.

[40]Schwarz, K. P. "Error Propagation in Inertial Positioning," *Canadian Surveyor*, Vol. 34, Sept. 1980, pp. 265-276.

[41]Schwarz, K. P. "Inertial Surveying Systems—Experience and Prognosis," *The Canadian Surveyor*, Vol. 34, March 1980, pp. 41-53.

[42]Hannah, J. and Pavlis, D. E., "Post-Mission Adjustment Techniques for Inertial Surveys," Ohio State Geodetic Science Rept. 305, Columbus, OH, Oct. 1980.

[43]Hannah, J. and Mueller, I. I., "Improvement of Inertial Surveys Through Post-Mission Network Adjustment and Self-Calibration," *Proceedings of the 2nd International Symposium on Inertial Technology for Surveying and Geodesy*, Banff, Canada, June 1981, p. 177-190.

[44]Todd, M. S. and Tindall, T. O., "Inertial Survey Adjustment," *Proceedings of the 2nd International Symposium on Inertial Technology for Surveying and Geodesy*, Banff, Canada, June 1981, p. 191-198.

[45]Wong, R.V.C., "A Kalman Filter/Smoother for the Ferranti Inertial Land Surveyor System," *Proceedings of the 2nd International Symposium on Inertial Technology for Surveying and Geodesy*, Banff, Canada, June 1981, p. 133.-146.

[46]Cross, P. A. and Webb, J. P., "Post-Processing Experiments with FILS2 Test Data," *Proceedings of the 2nd International Symposium on Inertial Technology for Surveying and Geodesy*, Banff, Canada, June 1981, p. 199-214.

[47]Bose, S. C. and Huddle, J. R., "Regional Adjustment of Inertial Gravity Disturbance Vector Measurements by Optimal Two-Dimensional Smoothing," *Proceedings of the 2nd International Symposium on Inertial Technology for Surveying and Geodesy*, Banff, Canada, June 1981, p. 147-160.

[48]Zimmerman, D. S. and Harris, H. C., "General Purpose Least-Squares Adjustment of IPS Data," *Proceedings of the 2nd International Symposium on Inertial Technology for Surveying and Geodesy*, Banff, Canada, June 1981, p. 215-227.

[49]Schwarz, K. P. and Gonthier, M., "Analysis of the Heading Sensitivity of the Litton Inertial Survey System," *Proceedings of the 2nd International Symposium on Inertial Technology for Surveying and Geodesy*, Banff, Canada, June 1981, p. 399-414.

[50]Schwarz, K. P. and Gonthier, M., "Adjustment Problems in Inertial Positioning," Symposium on Geodetic Networks and Computation, Munich, Germany, Aug. 1981.

[51]Hannah, J., "The Development of Comprehesive Error Models and Network Adjustment Techniques for Inertial Surveys," Rept. 330, Department of Geodetic Science and Surveying, Ohio State University, Columbus, OH, March 1982.

[52]Hannah, J., "Inertial Rapid Geodetic Survey System (RGSS) Error Models and Network Adjustment," Rept. 332, Department of

Geodetic Science and Surveying, Ohio State University, Columbus, OH, Sept. 1982.

[53]Milbert, D. G., "An Inertial Survey Adjustment Program: Implementation and Validation," National Oceanic and Atmospheric Administration Technical Rept. 97NGS26, Rockville, MD, Nov. 1982.

[54]Schwarz, K.P., "A Unified Approach to Post-Mission Processing of Inertial Data," presented at XVIII General Assembly of International Union of Geodesy and Geophysics, Hamburg, Germany, Aug. 1983.

[55]Van den Herrewegen, M., "Post-Mission Adjustment in Inertial Surveying," presented at the XVIII General Assembly of International Union of Geodesy and Geophysics, Hamburg, Germany, Aug. 1983.

[56]Arden, D. A. G. and Schwarz K. P. "Optimizing Field Procedures for Inertial Networks," presented at the XVIII General Assembly of International Union of Geodesy and Geophysics, Hamburg, Germany, Aug. 1983.

[57]Wei, S. Y. and Brockstein, A. J., "Offline Optimal Adjustment of Inertial System Survey Data," *Proceedings of the Fall Convention, American Congress on Surveying and Mapping-American Society of Photogrammetry,* Salt Lake City, UT, Sept. 1983, pp. 236-246.

[58]Vassiliou, A. A., "Processing of Unfiltered Inertial Data," UCSE Rept. 20006, University of Calgary, Surveying Engineering, Canada, Feb. 1984.

[59]Schwarz, K. P., Arden, D. A. G., and English, J. J. H., "Comparison of Adjustment and Smoothing Methods for Inertial Networks," UCSE Rept. 30006, University of Calgary, Canada, June 1984.

[60]Schwarz, K. P. and Arden, D. B., "A Comparison of Adjustment and Smoothing Methods for Inertial Networks," *Proceedings of the 3rd International Symposium on Inertial Technology for Surveying and Geodesy,* Banff, Canada, Sept. 1985.

[61]Gonthier, M., "Smoothing Procedures for Inertial Survey Systems of Local Level Type," University Calgary Surveying Engineering Rept. 20008, University of Calgary, Canada, June 1984.

[62]Huddle, J. R., "IPS Smoothing Program Mechanization Equations," Litton: JRH-7405-028, Woodland Hills, CA, May 1974.

[63]Huddle, J. R., "Definition of Post-Mission Smoothing Program for Platform Drift Rate Estimation and Deflection Change Estimate Improvement," Litton: JRH-7509-070, Woodland, Hills, CA, Sept. 1975.

[64]Huddle, J. R., "Post-Mission Smoothing Program for Recovery of the Magnitude of the Gravity Distrubance Change with the RGSS," Litton: JRH-7509-072, Woodland Hills, CA, Sept. 1975.

[65]Huddle, J. R., "Modifications to the IPS Elevation Smoother to Obtain the RGSS Configuration," JRH-7509-073, Woodland Hills, CA, Sept. 1975.

[66]Wei, S. Y. and Brockstein, A. J., "Optimization of the Regional Adjustment Program and Its Field Application," *Proceedings of the 3rd International Symposium on Inertial Technology for Surveying and Geodesy,* Banff, Canada, Sept. 1985.

A REVIEW OF GEODETIC AND GEODYNAMIC SATELLITE DOPPLER POSITIONING

J. Kouba

Gravity and Geodynamics Division, Earth Physics Branch, Department of Energy, Mines and Resources
Ottawa, Ontario K1A 0Y3

Abstract. Since 1967, when the U.S. Navy Navigation System was released to the general public, the accuracy of geodetic and geodynamic satellite Doppler positioning using satellite Doppler has improved steadily. This improvement is due mainly to refinements in hardware, refraction, and orbital modeling. Point positioning, using a minimum of 30 passes and the U.S. Defense Mapping Agency (DMA) precise ephemeris, has reached submeter accuracy. Relative positioning among several simultaneously observing stations separated by distances up to 250 km and using a minimum of 30 passes has a typical accuracy below 0.5 m (1σ). Current hardware design and errors in modeling of delays in the wet portion of the troposphere limit the achievable accuracy in relative positioning at the 0.1- to 0.5-m (1σ) level, unless errors in receiver phase and tropospheric refraction delays are continuously measured (calibrated). Also, a crystal oscillator of lower short-term stability ($\geq 5 \times 10^{-12}$/100 s) increases the Doppler data noise and causes a positive correlation persisting over several minutes. Geodetic and geodynamic Doppler data reductions usually involve meticulous modeling for instrumental, atmospheric, and orbital errors as well as careful Doppler data editing. A short-arc orbit computation employed in such satellite Doppler reductions requires a gravitational potential expanded to at least degree and order 10 if the orbit shape accuracy is to be comparable to DMA precise ephemeris. Satellite Doppler positioning has had a significant impact on geodetic datum definition and the establishment of national and continental networks of unprecedented accuracy. Satellite Doppler methods, apart from providing one of the most accurate continuous methods of polar motion monitoring, have also been found useful in other geodynamical and ice motion studies. This review concentrates mainly on system performance, models, and computational methods.

Contents

Paper number 2R1773.

1. Introduction

Measurements of the Doppler shift on radio signals received from the first artificial earth satellites proved useful for determination of the satellite orbital parameters. It was soon realized that Doppler shift measurements could be used to determine positions if the satellite orbits were known [McClure, 1958]. This led to the development of the U.S. Navy Navigational Satellite System (NNSS) for all-weather worldwide marine navigation with the primary objective to support Polaris submarine operations. The system, also known as TRANSIT, was introduced for military applications in 1964 and released for nonmilitary use in 1967. The navigational capabilities of the system have been described and analyzed extensively as it has evolved [Guier and Weiffenbach, 1958, 1960; Kershner and Newton, 1962; Newton, 1967; Piscane et al., 1973; Applied Physics Laboratory Staff, 1976; Black et al., 1976; Stansell, 1978; Black, 1980; Piscane, 1981; Seeber and Egge, 1981]. For users on land the TRANSIT system has become a standard for global surveying. Methods have been developed to facilitate high-precision positioning in a three dimensional terrestrial reference frame which can satisfy global as well as regional geodetic and geodynamic requirements.

The treatment of satellite Doppler measurements and the computational approach needed to obtain useful geodetic and geophysical data differ in a number of ways from those used for navigation. Geodetic solutions require careful modeling of instrumental and environmental effects on Doppler measurements and involve multipass, often multistation, data sets assembled during postprocessing predominantly for earth-fixed stations. This meticulous approach has resulted in significant improvements of our knowledge of the size and shape of the earth and its gravitational field and has led to the establishment of a truly global, nearly geocentric, reference system.

This review will concentrate on models and computational methods employed for precise positioning by means of Doppler measurements on the TRANSIT system satellite signals.

2. Satellite Doppler Measurements

Geodetic Doppler instrumentation typically integrates cycles of a beat frequency ($f_o - f_r$) which results from mixing the local

Reprinted with permission from *Rev. Geophys. Space Phys.*, vol. 21, no. 1, pp. 27–40, Feb. 1983.

reference frequency f_o and the received satellite signal f_r. The basic measurements are integrated counts D between reception times τ_1 and τ_2:

$$D = \int_{\tau_1}^{\tau_2} (f_o - f_r) d\tau \qquad (1)$$

The integration interval $\tau_2 - \tau_1$ depends on instrumentation and usually is between 4.6 and 120 s. Often it is chosen between 20 and 30 s, which is optimal with respect to geometrical strength and data-handling economy.

When f_o is considered constant in the integration of (1), then

$$D = f_o(\tau_2 - \tau_1) - \int_{\tau_1}^{\tau_2} f_r d\tau$$

The remaining integral can be evaluated by arguing that the number of cycles of f_r received between τ_1 and τ_2 has to be equal to the number of cycles of the satellite frequency f_s emitted between the corresponding satellite time marks $t_1 = \tau_1 - S_1/c$ and $t_2 = \tau_2 - S_2/c$, where S_1 and S_2 are satellite-to-receiver distances and c is the vacuum velocity of light:

$$\int_{\tau_1}^{\tau_2} f_r d\tau = \int_{t_1}^{t_2} f_s dt$$

Here we are neglecting propagation, relativistic, and equipment effects, which will be addressed later.

When f_s is also considered constant, the expression for the integrated Doppler count then becomes

$$D = (f_o - f_s)(\tau_2 - \tau_1) + (S_2 - S_1) f_s/c \qquad (2a)$$

An equivalent expression but in terms of ground frequency cycles f_o/c and emission times is obtained from (2a) when substituting $\tau_i = t_i + S_i/c$:

$$D = (f_o - f_s)(t_2 - t_1) + (S_2 - S_1) f_o/c \qquad (2b)$$

The distances

$$S_i = |\underline{r}_s(t_i) - \underline{r}_g(\tau_i)|$$

are distances between a ground station with the radius vector $\underline{r}_g$ at τ_i and a satellite position $\underline{r}_s$ at t_i.

Instrumental Effects

The Doppler ground instrumentation typically consists of an omnidirectional antenna, dual-channel phase-locked receiver, data recorder, and precision clock/oscillator assembly. Each equipment segment contributes to Doppler measurement errors. The most significant instrumental errors are: (1) - antenna phase errors, (2) - receiver phase errors, (3) - receiver time delay variations, and (4) - clock and oscillator errors.

Antenna phase errors are mainly caused by ground reflections (multipath) and can cause significant systematic errors mapping mostly into height solutions. Such height errors can reach several decimeters even for modern Doppler equipment [Kouba, 1976]. The antenna errors can be eliminated either by using a directional antenna or by a proper design of an omni-directional antenna with zero sensitivity below the horizon.

The receiver component, apart from white phase noise, introduces a phase error (also called Doppler delay) which can change during a pass. Such slow phase variations due to changes in signal strength and temperature can reach up to 180° of the Doppler signal (beat frequency) even for modern geodetic receivers and can significantly affect the solution for position parameters. Ideally, this error source should be eliminated by improved receiver design and/or real-time receiver calibrations.

The knowledge of total receiver time delay (in addition to the Doppler phase error) is required for synchronization of the local and satellite clocks. This delay is the time required for time mark propagation and detection within the receiver/decoder assembly. It can vary from receiver to receiver and is usually between 500 and 1200 μs. Ideally, it should be constant or measured continuously. However, variations in the receiver delay occur both during a pass and from one pass to another. Even if the variations during a pass are eliminated, as is the case with some new hardware [Icenbice and Loiler, 1979], the more significant variations between passes still remain (for example, as a function of ambient temperature, signal strength, etc.). These variations can cause systematic synchronization and position biases.

If we more precisely define t_i and τ_i as the epochs at which a particular wave front has zero phase at the satellite and at the detection circuitry of the receiver, respectively, then, taking into account the above instrument sources, they can be related as follows:

$$\tau_i = t_i + S_i/c + d_i + P_i + R_i + \Delta t_a + \delta_i \qquad (3)$$

where d_i denotes instantaneous receiver time delay, P_i represents the combination of receiver and antenna phase error, R_i denotes a small relativistic correction (see section 4), Δt_a is the atmospheric (tropospheric and ionospheric) delay (see next subsection), and δ_i represents local clock errors. Δt_a, d_i, and P_i will be similar for the same receiver types simultaneously observing points with small separations (<100 km); consequently, their effect on relative positioning will be reduced.

In (3) the clock errors δ_i are small and mostly due to clock resolution because other systematic clock errors (excluding clock drift) are negligible for a typical pass of less than 1000 s. On the other hand, instability of the reference frequency standard can contribute significantly to the integrated Doppler counts. The oscillator error contribution is directly related to the two sample Allan variance $\sigma_y^2(\Delta\tau)$ [Allan et al., 1974] conventionally used to characterize frequency stability in time domain, that is,

$$\sigma_D^2(\Delta\tau)[1 - \rho(\Delta\tau)] = \sigma_y^2(\Delta\tau) c^2 \Delta\tau^2 \qquad (4)$$

where $\sigma_D^2(\Delta\tau)$ and $\rho(\Delta\tau)$ are the Doppler count variance and the correlation coefficient between adjacent Doppler counts, respectively, $\Delta\tau$ is the integration interval, and c is the speed of light. The Allan variance for up to 1000 s typically has the following time dependence:

$$\sigma_y^2(\Delta\tau) = \sigma_0^2 \Delta\tau^{-q}$$

where $\sigma_0 = \sigma_y$ (1 s) and the exponent q depends on the frequency standard type. Crystal oscillators (drift removed) typically show no time dependence for $\Delta\tau$ up to 1000 s (i.e., q = 0), cesium and rubidium ($\Delta\tau$ < 100 s) standards exhibit white frequency noise (i.e., q = 1) and hydrogen maser standards contribute only white phase noise (q = 2) for $\Delta\tau$ < 100 s. Cesium and rubidium error contributions (q = 1) are statistically independent from one Doppler count to another; consequently, the correlation $\rho(\Delta\tau)$ between adjacent Doppler counts is zero and according to (4) the Doppler variance is then

$$\sigma_D^2(\Delta\tau) = \sigma_0^2 c^2 \Delta\tau \qquad (5)$$

The evaluation of variances and correlation of crystal oscillators requires error modeling [e.g., Fell, 1980]. Such modeling reveals that crystal oscillator error contributions to Doppler counting have strong positive correlation persisting over several minutes [Kouba and Orosz, 1982].

Atmospheric Delays

Both ionosphere (the ionized portion of the atmosphere extending above 50 km) and troposphere (the neutral atmosphere below 50 km) introduce significant delays in radio wave propagations.

Ionospheric delays, the larger of the two, strongly depend on solar activity, having daily variations (minimum at night and maximum at midafternoon) and annual variations (minimum in winter and maximum in summer). The ionospheric refractive index is a function of the frequency f and can be approximated by a series expansion [Clynch and Altenburg, 1979]

$$n = 1 - A_1 N_e/f^2 - A_2 N_e^2/f^4 + \ldots \qquad (6)$$

where A_1, A_2 are constants and N_e is the electron density. The ionospheric refraction causes the measured ranges S^o to be shorter than the geometrical ranges S, that is,

$$S^o = S - B_1/f^2 - B_2/f^4 + \ldots \qquad (7)$$

where the coefficient $B_q = A_q \int N_e^q dS$, q = 1, 2, ... By measuring ranges S_H^o and S_L^o (or integrated Doppler measurements) on two coherent frequencies f_H and f_L one can eliminate from (7) the most significant term with B_1:

$$S = (f_H^2 S_H^o - f_L^2 S_L^o)/(f_H^2 - f_L^2)$$

It can be easily verified that the above equation is also valid when Doppler counts D_H^o and D_L^o ((2a) or (2b)) integrated on f_H and f_L are substituted for S_H^o and S_L^o.

In the case of NNSS Doppler observations, f_H = 400 MHz, f_L = 150 MHz, and the above equation becomes

$$D_{400} = (64\, D_{400}^o - 24\, D_{150}^o)/55 \qquad (8)$$

where D_{400}^o and D_{150}^o are observed integrated Doppler counts on 400 MHz and 150 MHz, respectively. This correction is either hardwired for some receiver types or applied in preprocessing. The remaining residual ionospheric refraction error can reach up to several percent of the first-order term or up to a few meters in position. Clynch and Altenburg [1979] developed a model for estimating residual ionospheric refraction using the total electron content (TEC) evaluated from a global model; alternatively, Kouba [1981] and Hara and Sato [1982] estimated TEC from the observed two frequency corrections. An improved residual ionospheric refraction model also taking into account the earth magnetic field term is presented by Clynch and Renfro [1982], who also present some initial test results. However, the residual ionospheric refraction modeling should also be used in externally generated orbits in order to prevent small biases for point positionings utilizing the residual ionospheric model and the external orbits.

Tropospheric refraction delays are smaller than those due to the ionosphere, and in the microwave frequency region are frequency independent. The effect of tropospheric refraction is to lengthen the observed ranges to the satellite. A correction for tropospheric refraction, ΔS (to be subtracted from the measured range S^o), is given by

$$\Delta S = \int_0^S (n - 1)\, dS$$

where n is the refraction index. The refractivity N, defined as N = (n - 1) x 10^6, is usually expressed as a sum of dry N_d and wet N_w components which can be evaluated from standard formulae [e.g., E.K. Smith and Weintraub, 1953]:

$$N_d = 77.6\ P/T$$
$$N_w = 3.73 \cdot 10^5\ e/T^2$$

where P is the pressure (in millibars), e is the partial vapor pressure (in millibars) and T is temperature (in degrees Kelvin). There are several tropospheric refraction models based on surface meteorological data [e.g., Hopfield, 1969; Saastamoinen, 1973; Marini, 1972]. Of these, Hopfield's model is most often used for satellite Doppler positioning.

Using the nominal temperature gradient of 6.71°/km and the perfect gas law, Hopfield [1969] derived a convenient quadratic expression for the dry refractivity profile $N_d(h)$ as a function of the height h above the earth's surface:

$$N_d(h) = N_d(0)\ [(h_d - h)/h_d]^4 \qquad h\varepsilon<0;\ h_d>$$

where $N_d(0)$ is the surface refractivity and h_d is the effective height (in meters) of the troposphere above the station given by [Hopfield, 1972]

$$h_d = 40136 + 148.72\ (T - 273.16)$$

where T is temperature in degrees Kelvin.

An analogous expression, though without any theoretical basis, was used by Hopfield [1969] for the wet refractivity $N_w(h)$:

$$N_w(h) = N_w(0)\ [(h_w - h)/h_w]^4 \qquad h\varepsilon\langle 0;\ h_w\rangle$$

with h_w = 11,000 m.

The tropospheric refraction correction is then evaluated from measured surface temperature, pressure, and partial vapor pressure:

$$\Delta S = N_d/h_d^4 \int_0^{h_d} (h_d - h)^4 \frac{ds}{dh}\, dh + N_w/h_w^4 \int_0^{h_w} (h_w - h)^4 \frac{ds}{dh}\, dh$$

Though the above formulae can be integrated in a closed form, neglecting only path bending, serious numerical problems are encountered for high elevation angles. Yionoulis [1970] proposed two series expansions, one for low and one for high elevation angles, which practically eliminate all the numerical problems. Although the above evaluation is not complex, the large quantity of data coupled with other model approximations permits a further simplification. ΔS can be expressed in terms of the zenith corrections ΔZ_d and ΔZ_w:

$$\Delta S = (N_d h_d I_d + N_w h_w I_w)/5 \qquad (9)$$

where

$$I_d = \Delta S_d/\Delta Z_d \text{ and } I_w = \Delta S_w/\Delta Z_w$$

Hopfield [1972] suggested the following approximations:

$$I_d = 1/\sin\ (E^2 + 6.25^{o})^{1/2}$$
$$I_w = 1/\sin\ (E^2 + 2.25^{o})^{1/2}$$

where E is the satellite elevation angle. Such an approximation is within 1% of the corresponding rigorous evaluation for elevations above 5° [Wells, 1974]. Black [1978] developed a similar approximation which is within 0.1% of rigorous formulae for elevations above 3° [Kouba, 1979]. All the above approximations are well below the modeling errors. The dry portion of the refraction is quite accurate, with an rms of about 2% (or 4 cm in the zenith) and an rms of 0.1% (2 mm) once the tropospheric height h_d is adjusted for a particular position and time span [Hopfield, 1972]. However, when the assumption of the perfect gas law (i.e., no horizontal gradients) is violated near weather fronts, the model errors can reach 15%. The wet portion is much more difficult to model (partly because it is difficult to measure humidity accurately). Fortunately, the wet refraction contribution is only about 10% of the total correction. The current limit of wet refraction modeling is about 10-20% or about 3-5 cm in the zenith [see Murray and Marini, 1976]. Neglected path curvature only introduces significant errors (3%) at elevation angles below 3°. To mitigate the refraction modeling errors (as well as multipath), the Doppler data below 5° or 10° elevations are usually not used.

The wet portion of the tropospheric refraction represents a fundamental limitation of Doppler (and any radio based) positioning systems unless the wet refraction delays are directly measured with sufficiently high accuracy [Resch and Claflin, 1979].

Weighting of Doppler Data

The proper assignment of weights for Doppler observations is mandatory for minimum variance solutions as well as realistic error estimates [e.g., Hamilton, 1964].

In the past it was often argued that continuously integrated Doppler (CID) data should be statistically independent, that is have a diagonal weight matrix, which also implies the serial correlation (i.e., ρ = -0.5) for intermittently integrated Doppler data (IID) [e.g., Brown, 1976; Krakiwsky et al., 1972]. We note that for instrumentation with continuous counting, the corresponding CID Doppler counts are obtained by simply summing appropriate IID Doppler counts. However, phase white noise (with σ_p) due to receiver, oscillator, and timing noise, which would be consistent with the above weighting scheme, usually constitutes only a relatively minor portion of the total IID error budget, typically about 5-10 cm for 30 s of counting [Colquitt and Anderle, 1979; Gouldman, 1982]. Most of the error sources are due to instrumentation and atmospheric refraction as discussed above and exhibit white frequency noise which is independent from IID count to count [see Kouba, 1979].

Consequently, the IID data observed with precise oscillators (1 x 10^{-12}/100 s) can be approximated by a diagonal variance-covariance matrix:

$$\sigma_{ii}^2 = \sigma_0^2\, \Delta\tau \qquad \rho_{ij} = 0 \quad i \neq j \qquad (10)$$

where σ_0^2 is the variance of unit time and $\Delta\tau$ is the integration time. The corresponding CID variance-covariance matrix is then nonstationary with a variance-covariance matrix characteristic of the random walk process:

$$\sigma_{ii}^2 = \sigma_0^2\, (\tau_i - \tau_0)$$

and

$$\rho_{ij} = \left(\frac{\tau_i - \tau_o}{\tau_j - \tau_o}\right)^{1/2} \qquad \tau_j > \tau_i > \tau_o \qquad (11)$$

It can be easily confirmed that identical results can be obtained with either CID or IID data treatment, provided that corresponding variance-covariance matrices ((10) or (11)) are taken into account. When a random walk process is subjected to a straight line fit, the residual becomes nearly stationary, and the strong positive correlation diminishes more rapidly than indicated by (11) [Fell, 1980], so that a CID stationary data treatment with an appropriate correlation model as advocated by Brown and Brown [1979] is then still possible.

TABLE 1. Crystal Oscillator Error Contribution Characteristics for Intermittently Integrated Dopper Counts and Crystal Oscillator With Two Sample Allan Variance sigma $\sigma_y = 5 \times 10^{-12}$

Time Interval $\Delta\tau$, s	$\sigma_D(\Delta\tau)$, cm	Correlation Coefficients			
		$\Delta\tau$	$2\Delta\tau$	$3\Delta\tau$	$4\Delta\tau$
5	1.8	0.80	0.69	0.62	0.57
30	9.0	0.70	0.53	0.47	0.43
60	16.6	0.66	0.51	0.43	0.36
120	30.2	0.63	0.44	0.31	0.22

The σ_D for other values of σ_y is obtained by multiplying the tabulated σ_D by a factor $2\sigma_y \times 10^{11}$. The correlation coefficients remain unchanged.

However, when crystal oscillators of lower short-term stability (5×10^{-12}) are used, their error contribution dominates the total error budget, and even IID data have a significant positive correlation persisting over several periods as seen from Table 1. The CID variances then show an error growth nearly proportional to $(\tau_i - \tau_o)^2$ and correlation exceeding those of (11) [Kouba and Orosz, 1982].

The satellite instrumentation introduces a positive correlation between corresponding Doppler counts at different stations observing the same satellite pass. This correlation can be removed either by differencing the corresponding Doppler counts (if it is possible), a method analogous to the very long baseline approach [e.g., Cannon, 1978], or by introducing appropriate correlation between stations [Kouba, 1981], which can also be used for instrumentation with different Doppler gating.

3. Satellite Orbit Determination

The knowledge of satellite positions in space and time is essential for satellite navigation and point positioning. However, if the satellite is used only as a common reference during simultaneous observations from known and unknown stations, this information need not be explicitly available with great accuracy. Either the satellite positions can be treated independently (in so-called geometrical mode), or physical laws governing satellite motion are invoked (dynamical mode) to constrain and strengthen the data reduction. The complexity of such dynamical models depends on the required precision and length of the satellite trajectory. In general, the dynamical models involve either numerical or analytical integrations of the equations of motions [Anderle, 1974a].

$$\ddot{\underline{r}} = \frac{\partial V_E}{\partial \underline{r}} + \frac{\partial V_{SM}}{\partial \underline{r}} + D_g + R_p + T_{SM} \tag{12}$$

where $\ddot{\underline{r}}$ are the second time derivatives of the satellite position vector in an inertial coordinate system, $\partial V_E/\partial \underline{r}$ and $\partial V_{SM}/\partial \underline{r}$ are gravitational and lunisolar body accelerations, D_g and R_p denote air drag and radiation pressure contributions, and T_{SM} are lunisolar tidal effects. For more details, see Anderle [1974a] Kaula [1966], Mueller [1964], and O'Toole [1976].

For computations of a portion of one revolution, the so-called short-arc, all but the first term in (12) can be neglected without any significant effect on orbit shape accuracy. The first term in (12) involves the spherical harmonic expansion of the gravitational potential V_E, and this expansion can be truncated at relatively low degree and order for short-arc orbit computations.

Broadcast Ephemeris

The broadcast ephemerides, which are available (broadcast) in real time in the satellite message, are computed by the U.S. Navy Astronautics Group from Doppler data observed at the four OPNET tracking stations located in California, Minnesota, Maine, and Hawaii. Broadcast ephemerides are fitted to a 30-hour Doppler data span every 12 hours (24 hours during periods of low solar activity), and resulting orbits are then extrapolated for up to 30 hours into the future and finally uploaded into the satellite memory every 12 hours. For uploading, the extrapolated ephemerides (up to 18 hours) are factored into a precessing and osculating Keplerian ellipse (the so-called fixed parameters) with corrections to this ellipse at even minute intervals (the so-called variable parameters). Because of a limited word length the variable parameters are rounded off to the nearest 10 m or 1×10^{-5} degree, which results in positioning errors due to rounding-off errors of up to 6 m.

Even though a comprehensive modeling (equation (12)) is used for broadcast ephemeris computations, including a recent WGS72 gravitational model [Seppelin, 1974] and extrapolated Bureau International de l'Heure (BIH) pole positions, the broadcast ephemeris errors can still approach 100 m [Jenkins and Leroy, 1979], mainly due to uncertainty in orbit predictions. The largest prediction errors are in the along-track direction and are caused by unpredictable changes in the air drag. Even for balanced passes in north-south directions these errors may not average out, particularly in the case of 24-hour computation schedules [Jenkins and Leroy, 1979]. Remaining error sources are due to predicted pole positions, changes in UT1-UTC during the prediction span, initial orbit parameters, etc., all of which can cause position- and time-correlated errors which will not generally average out below 1 or 2 m. The new (NOVA) generation of NNSS satellites are air drag compensated and thus have improved along-track orbit predictions, a stronger signal, and improved clocks [Hoskins, 1979; Eisner and Yionoulis, 1981].

Accuracy of the broadcast ephemeris was investigated by Wells [1974], who also proposed a smoothing of variable parameters to reduce the rounding-off errors with the following base functions: 1, t, sin(2mt), cos(2mt) , where t is time and m is the satellite mean motion. For more recent studies of errors in broadcast orbits, see Arur [1979], Jenkins and Leroy [1979], White et al. [1975], and Ziegler [1979].

Postfitted Precise Ephemeris

Apart from the broadcast ephemeris which is available in real time from the transmitted signal from all operational NNSS satellites, the U.S. Defense Mapping Agency (DMA) regularly computes an ephemeris for at least one NNSS satellite. This so-called precise ephemeris is computed from 48 hour data spans observed by the TRANET tracking network, using different software [O'Toole, 1976] than that used for the broadcast ephemeris and with some differences in the gravity field and program constants [Anderle, 1974a; Jenkins and Leroy, 1979]. Since data are from the worldwide TRANET network [Williams, 1981] and no extrapolations are involved, the precise ephemeris is consistent at the 2-m level [Ziegler, 1979]. The precise ephemeris in the form of earth fixed Cartesian coordinates x, y, z and their time derivatives $\dot{x}$, $\dot{y}$, $\dot{z}$ at 1-min intervals is usually made available by DMA upon request from (qualified) scientific research organizations. As a by-product of the precise ephemeris generation an accurate pole position is computed for every 48-hour data span. This DMA Doppler pole monitoring has proved to be one of the most accurate continuous polar motion determinations to date [Robertson et al., 1979; King, 1981; Guinot, 1980].

Precise ephemerides also imply the coordinate system position, orientation, and scale mainly established by the set of coordinates adopted for the TRANET stations. Precise ephemeris coordinate system NWSC9Z-2 and its predecessor NWL9D were extensively compared to other independent satellite and astronomical systems [e.g., Anderle, 1975, 1980; Hothem, 1979; Kouba, 1978; Grappo, 1979; Lachapelle and Kouba, 1981]. It was shown that the NWL9D (and NWSC9Z-2) requires corrections of about -0.4×10^{-6} in scale, +0.8" in longitude east, and about +4 m in the Z coordinate in order to be consistent with international conventions for scale (the speed of light), pole, and longitude orientations. All but the last correction have been explained [Anderle, 1980]. The 4-m bias in the Z coordinate is still being investigated, and indications are that it may have been caused by a combination of the old value for GM (i.e., the product of the gravitational constant G and the earth mass M) of 398,601 km^3/s^2 instead of a more recent value of 398,600.5 km^3/s^2 and limited TRANET station distribution [Jenkins and Leroy, 1979; Anderle, 1980].

Since 1975 the coordinate system implied by the broadcast ephemeris is compatible with NWSC9Z-2 (or NWL9D) except for small scale correction of $+0.2 \times 10^{-6}$ and a -2 m Z coordinate shift [see Jenkins and Leroy, 1979].

Short-Arc Ephemeris

Short-arc orbits are usually computed within the Doppler reduction software using a rather heavily truncated expansion of the earth's gravitational potential [Brown and Trotter, 1969]. The short-arc computations only make sense when no high resolution (precise) orbits are available, since they represents a considerable additional expense, up to 10 times more than reductions using external (e.g., precise or broadcast), orbits [Nesbo, 1976; Hauck et al., 1979]. However, a short-arc computation which utilizes a least squares polynomial approximation subject to the second time derivative constraints derived from a gravitational model according to (12) does not significantly increase the computation time [Kouba, 1982]. Figure 1 indicates that for NNSS satellites and 5 cm precision in orbit shape, the earth's gravitational field should not be truncated below degree and order 10. Since most recent gravitational models agree quite well at least up to degree and order 10 [Lachapelle, 1975; Khan, 1976; Reigber, 1981; Pospisilova et al., 1982], it should not matter which model is used.

Three different broadcast orbit representations namely, short-arc, smoothed broadcast, and an orbit computed directly from the message are compared to the precise ephemeris in Figure 2 in order to examine orbit shape accuracies of each type of orbit representation. The short-arc computation [Kouba, 1982] used smoothed orbit positions and velocities at -10, 0, and +10 min with respect to the closest approach as initial conditions. The short-arc orbits (generated with a

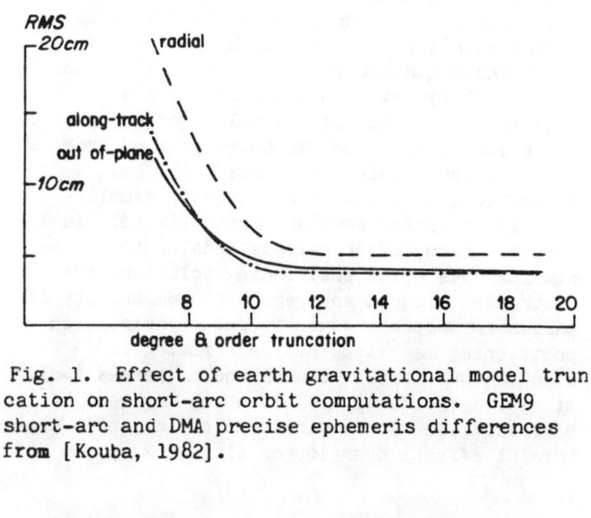

Fig. 1. Effect of earth gravitational model truncation on short-arc orbit computations. GEM9 short-arc and DMA precise ephemeris differences from [Kouba, 1982].

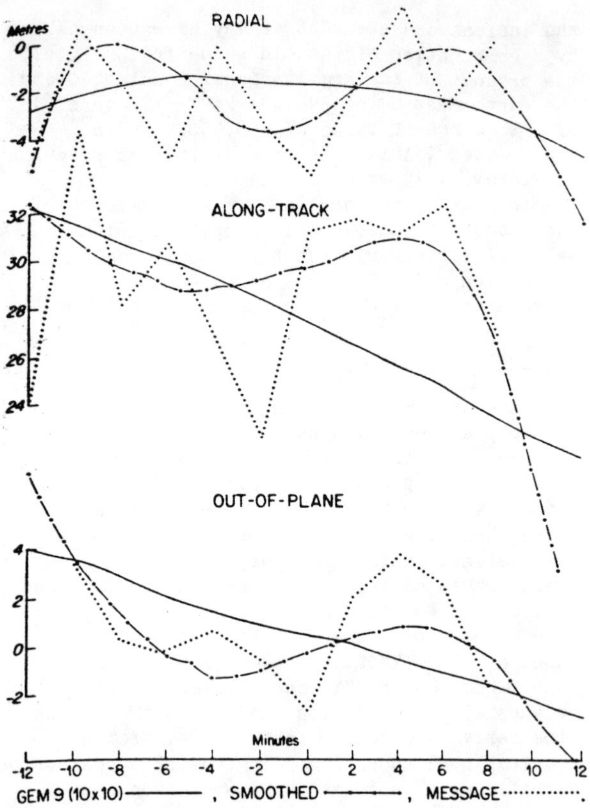

Fig. 2. Comparisons of three types of broadcast orbit representation with DMA precise ephemeris [from Kouba, 1982].

gravitational model truncated at degree and order 10) approximate the precise ephemeris well, except for position biases and small slope, both of which can, to a large extent, be accounted for by changes in the initial orbital elements. The orbit shape errors, in particular the long-wavelength features, can for a small number of passes correlate with some position parameters.

4. Satellite Doppler Positioning

Precise Doppler positioning for a stationary observer (called point positioning) is based on many satellite passes and the assumption that orbital, propagation, and instrumental errors will average out over many passes. This may not be the case for broadcast ephemeris or unfavorable atmospheric and solar conditions if repeatability levels below 1 or 2 m are required. Most of the orbital error and part of the propagation error can be eliminated from relative positioning for short (<250 km) separations when data from several, simultaneously observing stations are utilized. In the simplest case, often called translocation, the simultaneous point positioning solutions are subtracted without any regard to commonality of passes or Doppler counts. Some translocation positioning may also involve trimming to simultaneous Doppler counts and/or passes [Wells et al., 1979; Wells, 1976], which may be beneficial for short distances but may have some harmful effects over longer distances by distorting data balance (east-west and north-south). Logical extensions of translocation are short-arc and semi-short-arc positioning which utilize either locally computed (short-arc) or externally generated orbits (semi-short-arc) in a rigorous least squares adjustment involving data from several stations and many satellite passes.

Doppler Reduction Characteristics

The measured IID counts, corrected for atmospheric propagation delays (section 2), are related to the unknown position (or orbit) parameters through the following observation equations (see (2a),(2b), and (3)):

$$D_i+\varepsilon_i = (S_k - S_j)f_o/c + (t_k - t_j)(f_o - f_s) + d.(\dot{S}_k - \dot{S}_j)f_o/c + R_i + \Delta_i + s_i \quad (13)$$

where S_k, S_j are distances from a ground station with reference frequency f_o and radius vector $\underline{r}_g$ to satellites at $\underline{r}_s$ transmitting frequency f_s; s_i represents instrumental and atmospheric biases which can be modeled either as random biases (uncorrelated between passes) or by time processes reflecting correlation between passes and stations [e.g., Gelb, 1975; Kouba, 1981]; ε_i denotes observational errors, c is the speed of light; d is the mean pass receiver delay; and $\dot{S}_k$, $\dot{S}_j$ are time derivatives of S_k, S_j.

The relativistic correction R_i is given by Boucher [1978] as

$$R_i = -[GM(1/r_g - 1/r_s) - (|\dot{\underline{r}}_s|^2 - |\dot{\underline{r}}_g|^2)/2] .(t_k - t_i).f_o/c^2$$

and the earth's rotation correction [e.g., Kouba and Boal, 1976a] is

$$\Delta_i = (\underline{\omega} \times \underline{r}_g).[\underline{r}_s(t_k) - \underline{r}_s(t_j)]f_o/c^2$$

where ω is the vector of earth rotation. The CID data treatment is analogous to range observations subject only to an initial range unknown S_o and frequency offsets for each station and satellite pass:

$$D_i + \varepsilon_i = (S_i - S_o)f_o/c + (t_i - t_o)(f_o - f_s) + d.(\dot{S}_i - \dot{S}_o)f_o/c + R_i + \Delta_i + s_i \quad (14)$$

The simplest reduction models include at least three position unknowns and for each pass a frequency offset unknown. Geodetic and geodynamic reductions typically utilize many passes (usually a minimum of 30) from a number of simultaneously observing stations, systematic error modeling which may include several bias parameters for each pass (e.g. six orbital unknowns) and for each pass and station (e.g., refraction scaling, time synchronization, frequency, and frequency drift) [e.g., Kouba and Wells, 1976; Brown, 1976; Kouba, 1981]. The solution for position and bias parameters is obtained by a standard least squares analysis [e.g., Hamilton, 1964].

The set of observation equations (13) or (14) may be expressed in a vector form

$$\underline{L} + \underline{E} = \underline{F}(\underline{X})$$

where $\underline{L}$ and $\underline{E}$ denote the vectors of observation and measurement errors, respectively, $\underline{F}(\underline{X})$ is the functional relationship of the right-hand side of the observation equations ((13) and (14)), and the vector $\underline{X}$ consists of unknown position and bias parameters. The above observation equations are linearized around the approximate values $\underline{X}^{\circ}$ of parameters $\underline{X}$ as follows:

$$A \quad + \underline{W} = + \underline{V}$$

where the design matrix $A = \partial\underline{F}/\partial\underline{X}$ and $\underline{W}$ is the misclosure vector (i.e., $\underline{W} = \underline{F}(\underline{X}^{\circ}) - \underline{L}$). The vector of residuals $\underline{V}$ and corrections $\underline{\delta}$ are obtained from minimizing the quadratic sum:

$$\underline{V}^T P \underline{V} + \underline{\delta}^T P_X \underline{\delta} \qquad (15)$$

The weight matrices P and P_X correspond to observations $\underline{L}$ and parameters $\underline{X}$ and are proportional to the inverses of their respective variance-covariance matrices Σ_L^{-1} and Σ_X^{-1}. The solution for $\underline{\delta}$ is then [e.g,. Mikhail, 1976]

$$\underline{\delta} = -(A^T P A + P_X)^{-1} A^T P \underline{W}$$

and finally the adjusted parameters are $\underline{X} = \underline{X}^{o} + \underline{\delta}$. Such solutions usually require several iterations, and each iteration can be done in a batch mode involving all stations and passes [e.g., R.W. Smith et al., 1976] or partitioned in a sequential manner, usually pass by pass, to economize computations and statistical data editing [e.g., Kouba and Boal, 1976a]. Statistical data editing may include normal distribution tests on residuals $\underline{V}$ and the quadratic sum equation (15), usually at or below the 1% significance level. Since gross errors arising from decoding, recording, propagation delays, etc., can hardly be expected to conform to the normal distribution, the statistical data rejections are usually higher than the expected significance level. Usually, the rejections are between 1% and 10%.

For the optimal least squares solution, proper assignments of weight matrices P_X and P are important. The assignment of the matrix P_X is not so crucial as the assignment of P, since the position and frequency unknown parameters do not require any a priori weighting and the remaining bias parameters tend to become deterministic for a number of simultaneously observing stations. In such cases the solution is relatively insensitive to different assignments of P_X.

Proper assignment of the weight matrix P, on the other hand, is mandatory for minimum variance solution [Hamilton, 1964] as well as for realistic error estimates. As discussed in the previous section, the matrix P tends to be diagonal and stationary for IID measurements with precise oscillators when all error sources are considered. On the other hand, the CID treatment may require a fully populated weight matrix P [Kouba and Orosz, 1982; Brown and Brown, 1979].

Software Standardization

Data editing and various error and orbit modeling can potentially influence the final results [e.g., White, 1976; Hothem, 1975]. It is important that different Doppler software give the same results under similar conditions, particularly for applications of a global character [Hothem, 1979; Langley et al., 1979; Kouba, 1978]. Kouba and Hothem [1978] studied compatibility of the U.S. and Canadian Doppler software used by the two national geodetic surveys [R.W. Smith et al., 1976; Kouba and Boal, 1976a] and concluded that both programs agreed at the 20 cm level when comparable editing and other options were used. Subsequent studies identified additional minor differences, and when these were eliminated, the results of point positioning agreed within 10 cm. For this purpose the Special Study Group on Doppler Positioning of the International Association of Geodesy assembled three test data sets which can be used to study Doppler software compatibility and verification [Strange, 1980].

There is no firm agreement on what the best reduction options are, although the practical consensus which has emerged over the years includes (1) - solution for a frequency offset unknown for each pass and station, (2) - solution for the receiver delay, (3) - 3 σ statistical limit for Doppler residual testing, (4) - Hopfield tropospheric refraction model with surface meteorological data input, usually subject to a scaling bias solution with a priori σ of 5-10%, (5) - Doppler data deleted below 5°-10° elevation angles, and (6)- short-arc or semi-short-arc reduction methods for relative positioning of several, simultaneously observing stations.

5. Satellite Doppler Applications

Brown [1970] was one of the first to recognize the potential of Doppler positioning for submeter accuracy, advocating the short-arc method, the reduction method involving short-arc orbit computations and data from several simultaneously observing stations. Anderle [1972] and subsequently the Department of Defense [1972] demonstrated convincingly, using precise ephemeris and independent point positioning, that a 1-m accuracy level was achievable with 50-100 observed passes. From these two approaches, several applications and refinements evolved over the years. In Canada, three organizations (the Bedford Institute of Oceanography, Shell Canada, and the University of New Brunswick) pooled their resources to conduct initial tests in Doppler point positioning and translocation [Hittel et al., 1971; Hittel and Kouba, 1971; Wells et al., 1976].

The first geodetic applications provided a badly needed zero-order geodetic control in Canada at a spacing of 200-500 km (see Figure 3) and supplemented the transcontinental geodimeter traverse in the United States [Bossler, 1978], for which the U.S. National Geodetic Survey of NOAA adhered to precise point positioning using the DOPPLR software [R.W. Smith et al., 1976]. The Geodetic Survey of Canada attempted to use precise ephemeris satellites for absolute positioning as well as data from satellites for which only the broadcast ephemeris was available, in order to strengthen relative

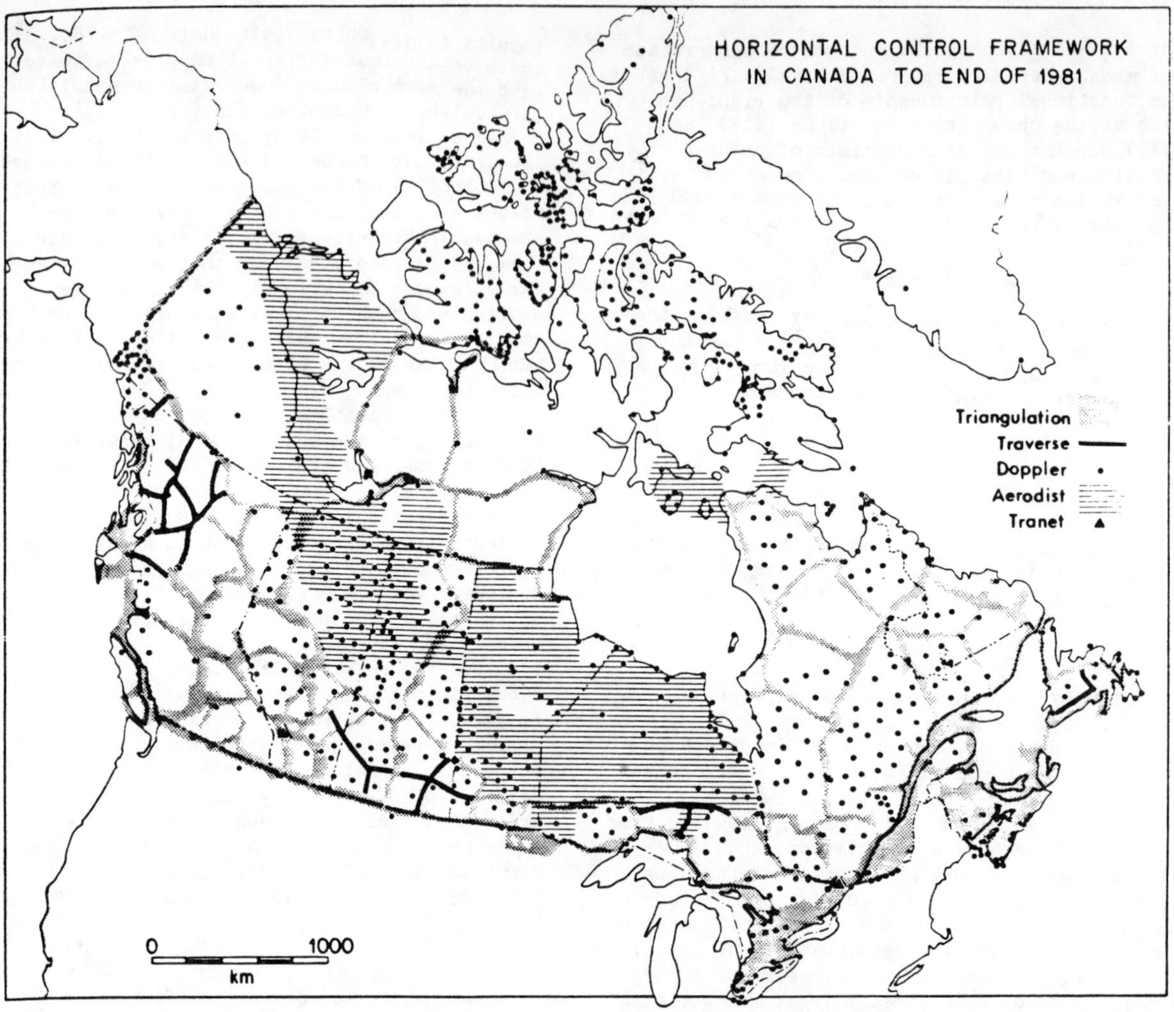

Fig. 3. Horizontal control framework in Canada to the end of 1981, including zero order Doppler network at a spacing of 200-500 km and Doppler densification network at 50- to 100-km spacing.

positioning by using semi-short-arc software [Kouba and Wells, 1976; Kouba and Boal, 1976a]. Similar work was carried out in Europe and Africa [Boucher et al., 1979; Seeger et al., 1979; Rinner and Pesec, 1982; Walker, 1979].

Geodetic Datum

Doppler point positioning utilizing the precise ephemeris (precise point positioning) implies a position and orientation of a geodetic network to much higher accuracy than was possible using classical methods that is, by astronomical and gravity observations. Anderle [1974b] recognized this important aspect of precise Doppler positioning for a geodetic datum realization, that is, positioning and orientation of a reference ellipsoid. Subsequently, precise Doppler positions, scaled and oriented to international conventions, were proposed for a datum realization [Kouba, 1978, 1980; Ashkenazi, 1981]. This new concept of datum realization was tested in several trial adjustments of Canadian networks [Beattie et al., 1978] and is likely to be adopted for the planned readjustment of the North American Datum scheduled for 1983-1984.

Doppler Geodetic Control Densification

With improved Doppler hardware [Blaha, 1979; Icenbice and Loiler, 1979; Stansell, 1979; Humble, 1981] and precise, short-arc and semi-short-arc reduction methods, relative positioning with accuracies below 0.5 m become possible with relatively small data samples of 30-50 passes and 2- to 3-day observing periods (see Figure 4) [Brown, 1976; Kouba and Boal, 1976b; Seeger et al., 1979; Hatch et al., 1979; Usandivaras et al., 1976]. At that accuracy level, Doppler positioning can also be used for densification of geodetic networks at spacings of less than 200 km. In Canada, densification at 50- to 100-km spacing was found to be useful in providing the necessary control for detail surveys by inertial surveying methods (Figure 3) [Babbage, 1977; Carrière et al., 1977].

Doppler Leveling

Doppler positioning is three dimensional, so it also provides a direct determination of the geometrical height h above the reference ellipsoid. When the orthometric heights H (i.e., above the mean sea level) are also known

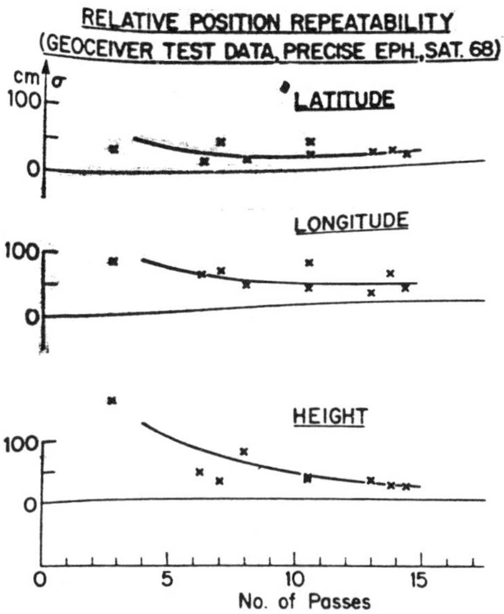

Fig. 4. Typical relative position repeatability for short separations (<250 km) based on short-arc and semi-short-arc methods [after Kouba, 1979].

at some points, usually obtained by geodetic leveling, it is possible to determine geoidal heights:

$$N_o = h - H.$$

Such Doppler-derived geoidal heights are useful for controlling error propagations in continental geoid determinations and also can be used for determinations of a zero order geoid undulation, mean earth ellipsoid and Doppler coordinate shift [e.g., Lachapelle and Kouba, 1981; Grappo, 1979]. Comparisons of Doppler derived geoidal heights with the most accurate geoids are in agreement to well below the 0.5-m level [Seeger et al., 1979]. Conversely, when the geoid is known to a sufficient accuracy, the Doppler ellipsoid heights can be used for orthometric height determinations (H). This method is particularly viable for remote and inaccessible regions.

Geodynamic Applications

As a by-product of the precise ephemeris, since 1972 DMA has solved continuously for the pole positions for every 48 hour span. The Doppler pole positions were forced to coincide (by appropriate orientation of the TRANET station network) with BIH pole positions at 1966.7 [Anderle, 1975]. DMA 2-day pole positions are estimated to be consistent to about 0.5 m and are regularly being included in the final BIH pole coordinates [Oesterwinter, 1979]. The high accuracy of Doppler pole positions was confirmed by a number of independent studies [e.g., King, 1981; Robertson et al., 1979; Guinot, 1980].

Although in principle, given errorless orbits, the solution for variations in the rotation of the earth (UT1 - UTC) is also possible, for polar satellites the UT1-UTC (DUT1) time difference couples almost exactly with the correction $d\Omega$ to the right ascension of the ascending node Ω. In the case of the precise ephemeris the departures from the extrapolated DUT1 are absorbed by $d\Omega$ for every 48-hour span. Anderle and Oesterwinter [1980] analyzed $d\Omega$ corrections in view of the possibility of DUT1 determination with some degree of success, in particular for the high-frequency variations, as low frequencies are corrupted by orbital errors. Research into Doppler DUT1 determination continues [Williams, 1981].

Another potentially useful by-product of precise ephemeris generation may be crustal movement determinations obtained from analyses of long series of Doppler positions of TRANET tracking stations [Malyevac and Anderle, 1979]. Anderle [1968] predicted the possibility of detecting crustal movements at the 5-cm/yr level from a 20-year-long, continuous series of Doppler observations. With some success the feasibility of such studies was already demonstrated by R. J. Anderle (unpublished paper, 1978), Malyevac and Anderle [1979] and Kouba [1981]. The main limitation of such studies is again due to orbital errors, that is, in gravity field and air drag modeling.

Ice Motion Studies

The Doppler positioning is also well suited for continuous ice dynamics monitoring as demonstrated by Wells et al. [1979] and Popelar and Kouba [1982]. Figure 5, from Popelar and Kouba [1982], shows the high degree of resolution for station separations of the three camps in the Lomonosov Ridge Polar Expedition (LOREX) in the spring of 1979 [Weber, 1979]. Here the station motions, up to 20 km/d, between and during passes were modeled as the first-order Gauss Markov processes [see Kouba, 1981].

6. Summary

The original satellite Doppler positioning design accuracy of 100 m (real-time navigation) has improved to the 1-m level for a stationary observer using postfitted precise ephemeris and observation over a period of several days with 30 or more satellite passes. This improvement results from improvements in gravity modeling, hardware, and software. Accuracy in relative positioning below the 0.5-m (1σ) level is now possible with 1- or 2-day observations and at least 30 passes, using either externally generated (e.g., broadcast or precise) ephemeris or short-arc orbits and several simultaneously observing stations.

Potential accuracy of Doppler positioning is estimated to be between 5 and 20 cm (1σ level in relative positioning). The major limitation is the water vapor influence on the radio wave propagation and, for larger separations, residual ionospheric refraction. Doppler hardware design currently limits the achievable accuracies to between 10 and 50 cm. Oscillators with improved stability, improved hardware design, and hardware calibrations would mitigate variations in receiver delay and antenna phase

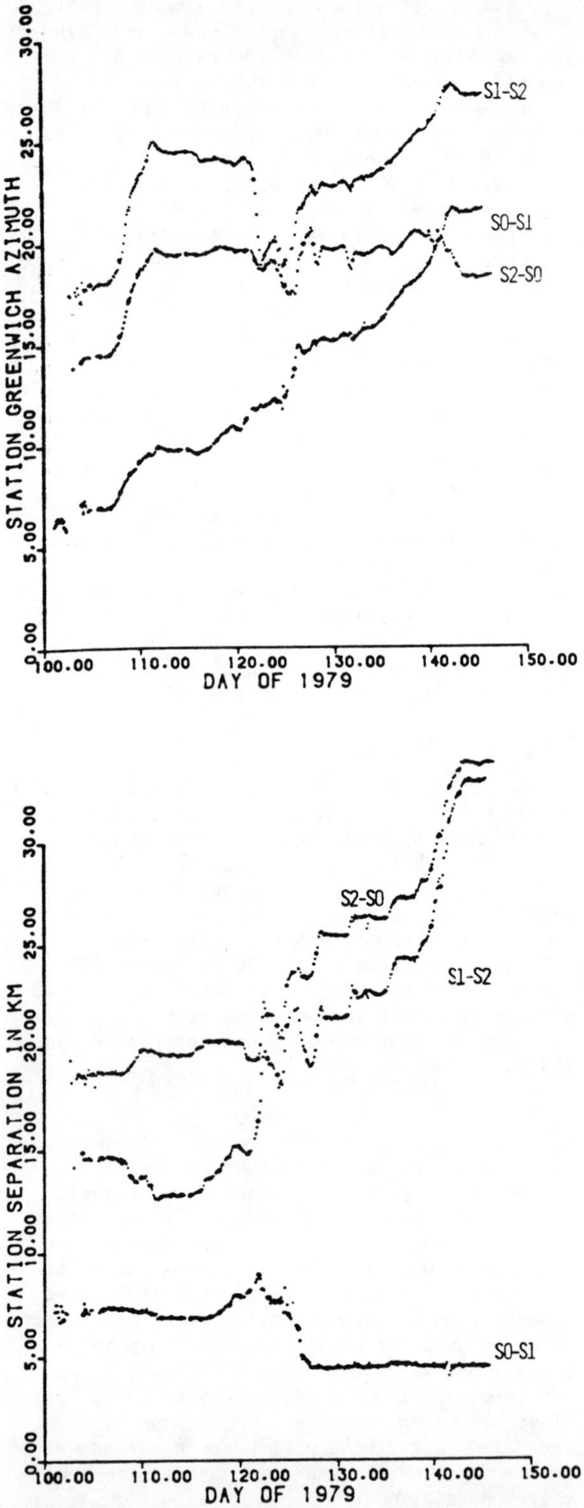

Fig. 5. Relative changes of the LOREX triangle side lengths and orientations (in degrees) [from Popelar and Kouba, 1982], SO is Main Camp; S1, Snowsnake; and S2, Iceman. (Reprinted by permission of Crane, Russak & Co.)

center and should permit Doppler relative positioning at the 5 cm (1σ) level. This would lead to additional geophysical applications, in particular monitoring of local crustal movements.

Acknowledgments. D. E. Wells of the University of New Brunswick and his students Ken Alexander, Harold Janes, and Furaha Lugoe reviewed the manuscript and supplied the author with many useful suggestions and comments. Contribution 1027 of the Earth Physics Branch.

References

Allan, D. W., J. H. Shoaf, and D. Haford, Statistics of time and frequency data analysis, in Time and Frequency: Theory and Fundamentals, NBS Monogr. 140, edited by B. E. Blair, 151-204, National Bureau of Standards, Washington, D.C., 1974.

Anderle, R. J., Accuracy of determination of continental drift from Doppler observations, paper presented at Annual Spring Meeting, AGU, Washington, D. C., April 18-11, 1968.

Anderle, R. J., Accuracy of Doppler determinations of station positions, in Rotation of the Earth, edited by P. Melchior and S. Yumi, 101-103, International Astronomical Union, Marioka, Japan, 1972.

Anderle, R. J., Transformation of terrestrial survey data to Doppler satellite datum, J. Geophys. Res., 79 (35), 5319-5331, 1974a.

Anderle, R. J., Role of artificial earth satellites in redefinition of the North American Datum, Can. Surv., 28(5), 590-597, 1974b.

Anderle, R.J., Long term consistency in positions of sites determined from Doppler satellite observations, Rep. NSWC/DL TR-3433, Nav. Surface Weapons Center, Dahlgren, Vir., April 1975.

Anderle, R. J., Accuracy of mean earth ellipsoid based on Doppler, laser and altimeter observations, Bull. Geod., 54(4), 521-527, 1980.

Anderle, R. J., and C. Oesterwinter, Determination of high frequency variation in earth rotation from Doppler satellite observations, Bull. Geod., 54(4), 544-552, 1980.

Applied Physics Laboratory Staff, Planned improvements in the TRANSIT system, 1975, Navigation Washington, 22, 352-360, 1976.

Arur, M. G., Experiments for improved positioning by means of integrated Doppler satellite observations and NNSS broadcast ephemeris, Proc. Int. Geod. Symp. Satell. Doppler Positioning 2nd, 83-112, 1979.

Ashkenazi, V., Models for controlling national and continental networks, Bull. Geod., 55(1) 49-58, 1981.

Babbage, G., Operation of the inertial survey system: The manager's view, Proc. Int. Symp. Inertial Technol. Surv. Geod. 1st, 312-321, 1977.

Beattie, D. B., J. A. R. Blais, and M. C. Pinch, Test adjustments of Canadian primary horizontal network, Proc. Int. Symp. Probl. Related Redefinition N. Am. Geod. Networks 2nd, 41-54, 1978.

Black, H. D., An easily implemented algorithm for tropospheric range correction, J. Geophys. Res., 83(B4), 1825-1828, 1978.

Black, H. D., The TRANSIT system 1977: Performance, plans and potential, Philos.

Trans. R. Soc. London, Ser. A, 294, 217-236, 1980.

Black, H. D., R. E. Jenkins, and L. L. Pryor, The TRANSIT system, 1975, Tech. Memo. TG1305, Appl. Phys. Lab., Johns Hopkins Univ., Laurel, Md., Dec. 1976.

Blaha, F., Features and capabilities of a new generation CMA-751 satellite Doppler field survey system, Proc. Int. Geod. Symp. Satell. Doppler Positioning 2nd, 535-556, 1979.

Bossler, J. D., Status of the new adjustment in the United States, Proc. Int. Symp. Probl. Related Redefinition N. Am. Geod. Networks 2nd, 31-403, 1978.

Boucher, C., Relativistic correction to satellite Doppler observation, paper presented at Symposium on the Mathematical Structure of Gravity Field, Int. Assoc. of Geod., Lagonissi, Greece, 1978.

Boucher, C., P. Pâquet, and P. Wilson, The second European Doppler observation campaign (EDOC-2): Results and conclusions obtained by EDOC-2 computing centres, Proc. Int. Geod. Symp. Satell. Doppler Positioning 2nd, 819-850, 1979.

Brown, D. C., Near term prospects for positional accuracies of 0.1 to 1.0 m from satellite geodesy, Rep. AFCRL-70-0501, U.S. Air Force Cambridge Res. Lab., Bedford, Mass., 1970.

Brown, D. C., Doppler positioning by the short-arc method, Proc. Int. Geod. Symp. Satell. Doppler Positioning 1st, 97-140, 1976.

Brown, D. C., and J. D. Brown, Short-arc adjustment of 36 Doppler stations, Tech. Rep. 79-015A, Indiaatlantic, Florida, Aug. 1979.

Brown, D. C., and J. E. Trotter, SAGA, a computer program for short-arc geodetic adjustment of satellite observations, Rep. AFCL-69-0080, U.S. Air Force Cambridge Res. Lab., Bedford, Mass., 1969.

Cannon, W. H., The classical analysis of the response of long baseline radio interferometer, Geophys. J. R. Astron. Soc., 53, 503-530, 1978.

Carrière, R. J., J. Kouba and R. C. Penney, Experience with inertial survey system at the Geodetic Survey of Canada, Proc. Int. Symp. Inertial Technol. Surv. Geod. 1st, 146-197, 1977.

Clynch, J. R., and R. A. Altenburg, Ionospheric residual range error model, Proc. Int. Geod. Symp. Satell. Doppler Positioning 2nd, 113-134, 1979.

Clynch, J. R., and B. A. Renfro, Evaluation of ionospheric residual range error model, Proc. Int. Geod. Symp. Satell. Doppler Positioning 3rd, in press, 1982.

Colquitt, E. S., and R. J. Anderle, Effect of oscillator performance on Doppler Geodesy, Proc. Int. Geod. Symp. Satell. Doppler Positioning 2nd, 409-422, 1979.

Department of Defense, Report of the DOD GEOCEIVER test program, DMA Rep. 0001, Def. Mapp. Agency, Washington, D. C., 1972.

Eisner, A., and S. M. Yionoulis, Neutral density variations in 900-1200 km region of upper atmosphere, Ann. Geophys., 37(1), 235-240, 1981.

Fell, P. J., Geodetic positioning using a global positioning system of satellites, Dep. of Geod. Sci. Rep. 299, Ohio State Univ., Columbus, June 1980.

Gelb, A., (Ed.), Applied Optimal Estimation, MIT Press, Cambridge, Mass., 1975.

Gouldman, M. W. S, Oscillator studies, Proc. Int. Geod. Symp. Satell. Doppler Positioning 3rd, in press, 1982.

Grappo, G. A, Determination of the earth's mean equatorial radius and center of gravity from Doppler derived and gravimetric geoid heights, paper presented at 17th General Assembly, Int. Union of Geod. and Gephys., Canberra, Australia, 1979.

Guier, W. H., and G.C. Weiffenbach, Theoretical analysis of Doppler radio signals from earth satellites, Rep. JHU/APL BB-176, Appl. Phys. Lab., Johns Hopkins Univ., Laurel, Md., 1958.

Guier, W. H., and G.C. Weiffenbach, A satellite Doppler navigation system, Proc. Inst. Radio Electron. Eng. Aust., 48, 407-516, 1960.

Guinot, B., Determination of the motion of the pole, and comparison with astrometry, Philos. Trans. R. Soc. London, Ser. A, 294, 119-124, 1980.

Hamilton, W. C., Statistics in Physical Science, Ronald, New York, 1964.

Hara, T., and K. Sato, Electron content distribution and variation around Mizusawa, Japan, Proc. Int. Geod. Symp. Satell. Doppler Positioning 3rd, in press, 1982.

Hatch, R., S. Chamberlain, and J. Moore, MX 1502 Doppler survey software, Proc. Int. Geod. Symp. Satell. Doppler Positioning 2nd, 187-208, 1979.

Hauck, H., K. Herzberger, D. Lelgemann, and W. Schluter, Comparisons of single-point, semi-short arc and short-arc solutions using Doppler data, paper presented at 17th General Assembly, Int. Union of Geod. and Geophys., Canberra, Australia, 1979.

Hittel, A., and J. Kouba, Accuracy tests of doppler satellite positioning in orbital, translocation and simultaneous modes, Proc. Int. Congr. Surv. 17th, 503.1-503.13, 1971.

Hittel, A., J. Kouba, E. J. Krakiwsky, D. E. Wells, and R.M. Eaton, Report on a doppler satellite positioning in Canada, paper presented at the 64th Annual Meeting, Can. Inst. of Surv., Ottawa, Ont., Feb. 3-5, 1971.

Hopfield, H. S., Two-quadratic tropospheric refractivity profile for correcting Satellite Data, J. Geophys. Res., 74 4487-4499, 1969.

Hopfield, H. S, Tropospheric refraction effects on satellite range Measurements, APL Tech. Dig., 11(4), 11-19, 1972.

Hoskins, G. W, The future Navy Navigation Satellite System, Proc. Int. Geod. Symp. Satell. Doppler Positioning 2nd, 21-38, 1979.

Hothem, L. D., Evaluation of precision and error sources associated with Doppler Positioning, paper presented at 16th General Assembly, Int. Union of Geod. and Geophys., Grenoble, France, Aug. 1975.

Hothem, L. D., Determination of accuracy, orientation and scale of satellite Doppler point-positioning coordinates, Proc. Int. Geod. Symp. on Satell. Doppler Positioning 2nd, 609-630, 1979.

Humble, A., The Mini-Ranger satellite survey system Motorola position determining systems, IAG/COSPAR CSTG Bull. 3, 98-105, Int. Assoc.

of Geod., Ohio State Univ., Columbus, July 1981.

Icenbice, P. J., and R. D. Loiler, Major improvements in satellite receiver time recovery for (cm) accuracy, Proc. Int. Geod. Symp. Satell. Doppler Positioning 2nd, 557-573, 1979.

Jenkins, R. E., and C. F. Leroy, "Broadcast" versus "precise" ephemeris - Apples and oranges, Proc. Int. Geod. Symp. Satell. Doppler Positioning 2nd, 39-62, 1979.

Kaula, W. M., Theory of Satellite Geodesy, Blaisdell, Waltham, Mass., 1966.

Kershner, R. B., and R. R. Newton, The TRANSIT system, J. Inst. Navig., 15, 129-144, 1962.

King, R. W., Lunar ranging validation of LAGEOS and Doppler polar motion results, IAG/COSPAR CSTG Bull. 3, 95-96, Int. Assoc. of Geod., Ohio State Univ., Columbus, July 1981.

Khan, M. A., Comparative evaluations of recent global representations of earth's gravity field, Geophys. J. R. Astron. Soc., 46, 435-554, 1976.

Kouba, J., Doppler levelling, Can. Surv., 30(1), 21-31, 1976.

Kouba, J., Datum considerations for test adjustments of Canadian primary horizontal network, Proc. Int. Symp. Probl. Related Redefinition N. Am. Geod. Networks 2nd, 55-62, 1978.

Kouba, J., Improvements in Canadian geodetic Doppler programs, Proc. Int. Geod. Symp. Satell. Doppler Positioning 2nd, 63-82, 1979.

Kouba, J., Geodetic satellite Doppler positioning and application to Canadian test adjustments, Philos. Trans. R. Soc. London, Ser. A, 294, 271-276, 1980.

Kouba, J., Satellite Doppler positioning for geodynamics, Ann. Geophys., 37(1), 205-213, 1981.

Kouba, J., An efficient short-arc orbit computation, submitted to Bull. Geod., 1982.

Kouba, J., and J. D. Boal, Program GEODOP, Publ. 1213E, Surv. and Mapp. Br., Ottawa, Ont., 1976a.

Kouba, J., and J. D. Boal, The Canadian Doppler satellite network, Proc. Int. Geod. Symp. on Satell. Doppler Positioning 1st, 187-206, 1976b.

Kouba, J., and L. D. Hothem, Compatiblity of Canadian and United States Doppler station networks, Proc. Int. Symp. Probl. Related Redefinition N. Am. Geod. Networks 2nd, 195-208, 1978.

Kouba, J., and J. A. Orosz, Evaluation of ELECTRAC receiver and oscillator effects on Doppler data quality at TRANET station 128 Ottawa, Proc. Int. Geod. Symp. on Satell. Doppler Positioning 3rd, in press, 1982.

Kouba, J., and D.E. Wells, Semi-dynamical Doppler satellite positioning, Bull. Geod., 50(1), 27-42, 1976.

Krakiwsky, E. J., D. E. Wells, and B. P. Kirkham, Geodetic control from Doppler satellite observations, Can. Surv., 26(2), 146-162, 1972.

Lachapelle, G., Determination of the geoid using heterogeneous data, Mitt. Geod. Inst. Tech. Univ. Graz, 19, 1975.

Lachapelle, G., and J. Kouba, Relationship between terrestrial and satellite Doppler systems, in Reference Coordinate Systems for Earth Dynamics, edited by E.M. Gaposchkin and B. Kolaczek, 195-203, D. Reidel, Hingham, Mass., 1981.

Langley, R. B., W. H. Cannon, W. T. Petrachenko, and J. Kouba, LBI and satellite Doppler: Baseline comparisons, Proc. Int. Geod. Symp. Satell. Doppler Positioning 2nd, 589-608, 1979.

Malyevac, C., and R. J. Anderle, Determination of plate tectonic motion from Doppler observations of Navy navigation satellites, Proc. Int. Geod. Symp. on Satell. Doppler Positioning 2nd, 695-742, 1979.

Marini, J. W., Correction of satellite tracking data for an arbitrary tropospheric profile, Radio Sci., 7(2), 223-231, 1972.

McClure, T. F., Method of Navigation, U.S. Patent, no 3, 172, 108, filed May 12, 1958. (Issued March 2, 1965).

Mikhail, E. M., Observation and Least Squares, Dun-Donnelley, New York, 1976.

Mueller, I. I., Introduction to Satellite Geodesy, Frederick Ungar, New York, 1964.

Murray, C. W., and J. W. Marini, A comparison of methods using surface meteorological measurements to predict altimeter bias caused by atmospheric water vapor, report Goddard Space Flight Center, Greenbelt, Md., June 1976.

Nesbo, I., Comparison of mathematical models for Doppler satellite positioning,M.Sc. thesis, Dep. of Surv. Eng., Univ. of New Brunswick, Fredericton, Aug. 1976.

Newton, R. R., The United States Doppler geodetic system and its observational accuracy, Philos. Trans. R. Soc. London, Ser. A, 292, 50-66, 1967.

Oesterwinter, C., Polar motion from astrometric, laser and Doppler, Proc. Int. Geod. Symp. Satell. Doppler Positioning 2nd, 751-766, 1979.

O'Toole, J. W., CELEST Computer Program for computing satellite orbits, Rep. NSWC/DL TR 365, Nav. Surface Weapons Center, Dahlgren, Vir., 1976.

Piscane, V. L., Navy Navigation Satellite System status, IAG/COSPAR CSTG Bull. 3, 106-107, Int. Assoc. of Geod., Ohio State Univ., Columbus, July 1981.

Piscane, V. L., B. B. Holland, and H. D. Black, Recent (1973) improvements in Navy Navigation Satellite System, Navigation Washington, 20, 224-229, 1973.

Popelar, J., and J. Kouba, Satellite Doppler determination of differential sea ice motion in the vicinity of the north pole, submitted to Mar. Geod., 1982.

Pospisilova, L., J. Klokocnik, and J. Kostelecky, A comparison of earth models by degree of their harmonic coefficients, Bull. Astron. Inst. Czech., 33, 33-36, 1982.

Reigber. C., Determination of the earth potential and possible improvements, Ann. Geophys., 37(1), 15-23, 1981.

Resch, G. M., and E.S. Claflin, Microwave radiometry as a tool to calibrate tropospheric water-vapor Delay, NASA Conf. Publ.2115, 377-384, 1979.

Rinner, K., and P. Pesec, West-east European Doppler observation campaign, Manuscr. Geod., 7(2), 117-132, 1982.

Robertson, D. S., et al., Polar motion and UT1: Comparison of VLBI, lunar laser, satellite laser, satellite doppler, and conventional astrometric determination, Radio Interferometry Techniques for Geodesy, NASA Conf. Publ. 2115, 33-44, 1979.

Saastamoinen, J., Contribution to the theory of atmospheric refraction, Bull. Geod., (107), 13-34, 1973.

Seeber, G., D., and Egge, Positionsbestimmung mit Satelliten - Doppler verfuren in bewegter Situation, Z. Vermess., 106(3), 123-131, 1981.

Seeger, H., W. Schluter, K. Herzberger, F. Kramer, and G. Soltau. The German-Austrian (German part) and EROS Doppler observation campaigns (DODOC and EROS-DOC), Proc. Int. Geod. Symp. Doppler Satell. Positioning 2nd, 889-924, 1979.

Seppelin, T.O., The Department of Defence World Geodetic Datum 1972, Can. Surv., 28(5), 496-506, 1974.

Smith E.K., Jr., and S. Weintraub, The constants in the equation for atmospheric refraction index of radio frequencies, Proc. Inst. Radio Electron. Eng. Aust., 41, 1035-1037, 1953.

Smith, R. W., G. R. Schwarz, and W. D. Googe, DOPPLR - A point positioning program using integrated doppler satellite observations, Proc. Int. Geod. Symp. Satell. Doppler Positoning 1st, 839-890, 1976.

Stansell, T. A., The many faces of TRANSIT, Navigation Washington, 25, 55-70, 1978.

Stansell, T. A., The 1502 satellite surveyor, Proc. Inter. Geod. Symp. Satell. Doppler Positioning 2nd, 497-534, 1979.

Strange, W. E., International Association of Geodesy, Special Study Group 2.54 -Satellite Radio Techniques (IAG SSG 2.54), IAG/COSPAR CSTG Bull. 1, 63-64, Int. Assoc. of Geod., Ohio State Univ.,Columbus, May 1980.

Usandivaras, J., P. Pâquet, and R. Verbeiren, An ORB Doppler program analysis and its application to European data. Proc. Int. Geod. Symp. Satell. Doppler Positioning 1st, 707-726, 1976.

Walker, J. W., Transformation of Doppler positions to local or national datum, Proc. Int. Geod. Symp. Satell. Doppler Positioning 2nd, 925-946, 1979.

Weber, J. R., The Lomonosov Ridge Experiment: LOREX 79, Eos Trans. AGU, 60(42), 715-721, 1979.

Wells, D. E., Doppler satellite control, Tech. Rep. 29, Dep. of Surv. Eng., Univ. of New Brunswick, Fredericton, Sept. 1974.

Wells, D. E., Concept of satellite Doppler positioning using translocation techniques, Proc. Int. Geod. Symp. Satell. Doppler Positioning 1st, 77-96, 1976.

Wells, D. E., E. J. Krakiwsky, D. B. Thomson, and J. Kouba, Review of Doppler satellite positioning in Canada, Bull. Geod., 50(4), 307-321, 1976.

Wells, D. E., C. R. Gillespie, P. Martin, and A. Thorndike, Error analysis of satellite doppler positioning of AIDJEX manned camps, Proc. Int. Geod. Symp. Satell. Doppler Positioning 2nd, 379-346, 1979.

Williams, O. W., DMA - U.S. Defense Mapping Agency, IAG/COSPAR CSTG Bull. 3, 13-22, Int. Assoc. of Geod., Ohio State Univ., Columbus, July 1981.

White, H. L., D. Huber, and J. Taylor, Comparison between Naval Surface Weapons Center and Navy Astronautic Group ephemeris for GEOCEIVER positioning, Rep. TR-75-001, Def. Mapp. Agency Aerosp. Center, St. Louis, Mo., 1975.

White, H. L., Variation in Doppler positions resulting from differences in computer programs and tropospheric refraction computations, Proc. Int. Geod. Symp. Satell. Doppler Positioning 1st, 551-576, 1976.

Yionoulis, S. M., Algorithm to compute tropospheric refraction effects on range measurements, J. Geophys. Res., 75, 7636-7637, 1970.

Ziegler, R. E., Ephemeris comparisons, Proc. Int. Geod. Symp. Satell. Doppler Positioning 2nd, 135-156, 1979.

(Received July 20, 1982;
accepted September 22, 1982.)

Modern Navigation Systems and their Relation to Timekeeping

WILLIAM J. KLEPCZYNSKI

Invited Paper

***Abstract*—This paper will discuss the interdependence between navigation and Precise Time and Time Interval (PTTI) systems. The modern role that navigation systems play in the field of timekeeping is shown through examples of how Omega, Loran-C, TRANSIT, and the Global Positioning System (GPS) are utilized by the PTTI community.**

Introduction

THE USERS of Precise Time and Time Interval (PTTI) systems are concerned with maintaining and distributing time (epoch) to better than one-millisecond (1-ms) precision and/or accuracy and time interval (frequency) to better than one part in ten to the ninth (1×10^{-9}). They also are interested in synchronizing clocks and intercomparing time scales. Navigation systems, because they are time-ordered systems, have proven to be effective in aiding the PTTI community to reach their goals. Fortunately, the use of navigation systems for PTTI distribution has not interfered with their navigation missions.

The interdependent relationship between PTTI and navigation has been recognized from rather early times. The desire for more precise navigation led to improvements in timekeeping, as witnessed by John Harrison (1693–1776) and the development of his famous clock [1] which sailed on the *HMS Deptford* in 1761. Conversely, in modern times, the development of small rubidium frequency standards (clocks) led to the development of a class of navigation systems which utilized stabilized VLF communication signals for some air navigation problems [2]. The quest for a high-precision, satellite-based, navigation system sent the navigation community to the PTTI community with requests for the development of space-qualified rubidium [3] and cesium [4] frequency standards, and the ongoing development of a space qualified hydrogen-maser frequency standard [5].

The importance of navigation systems to the PTTI community cannot be overestimated. The Bureau International de l'Heure (BIH) presently relies on Loran-C for its formulation of International Atomic Time (TAI) [6]. Omega has been valuable as a time-interval distribution system and has been used as a time-distribution system, under certain conditions [7]. The present NOVA satellites have enhanced TRANSIT's usefulness to the PTTI community [8] for worldwide time distribution. The implementation of the full operational phase of the Global Positioning System (GPS) program is anxiously awaited to see how useful this system will be for international PTTI distribution. Its usefulness for national goals is unquestioned.

The effectiveness of these navigation systems to the PTTI community rests in the fact that they are all referenced to the same time system by DoD Directive [9]. The reference is maintained by a combination of direct control and continuous monitoring. The timing of the navigation signal is controlled by a clock which is kept to the reference time. The signals are continuously monitored to verify that they are on time. The time reference is UTC(USNO), Coordinated Universal Time as determined at the U.S. Naval Observatory (USNO). UTC(USNO) is a time scale which is obtained by averaging 20–25, "selected," commercially available, cesium-beam frequency standards operating independently [10]. All data pertaining to the time scale and the timing of the navigation systems are published regularly in various Time Service Announcements (TSA) are available through the Automated Data Service (ADS) of USNO [11] by use of a terminal and modem (phone number 202/653-1809).

Use of Omega and VLF Transmissions for Timekeeping

Time Interval Determination

In 1957, Pierce [12] described four techniques for frequency comparison at intercontinental distances using VLF transmissions, which are still, in principle, applicable today. In 1962, Echols [13] demonstrated a system which could attain an accuracy of several parts in 10^{-11} in frequency comparison, based on the principles of Pierce. His equipment consisted of a VLF receiver, frequency synthesizer, phase comparator, and a servo-driven phase shifter.

Modern practice for determining time interval from Omega consists in comparing the phase of some received VLF signal with the phase of a signal coming from a local frequency standard. In order to receive one of the Omega frequencies, a special Omega commutator is needed to discern the desired signal from the transmitted Omega sequence. The equipment which does the comparison usually will convert automatically the signal from the local frequency standard to a form that can be compared to the received VLF signal. The result of the comparison usually is recorded on a chart recorder and is scaled so that one division of the ordinate is the equivalent of a microsecond. Any difference in frequency between the local frequency standard and the frequency standard governing the transmissions will be noticed as a displacement of the recorded signal in time. The relative frequency offset between the two standards can be calculated by forming the difference, in microseconds, between a pair of readings and dividing this difference by the interval of time between the two readings.

The accuracy of the frequency offset as determined this way will depend on three factors

Manuscript received July 29, 1983.
The author is with the U.S. Naval Observatory, Washington, DC 20390.

Reprinted from *Proc. IEEE,* vol. 71, no. 10, pp. 1193–1198, Oct. 1983.

a) stability of the frequency source at the transmitter;
b) precision of the phase comparison measurement system;
c) propagation effects affecting the radiated signal.

Santamore [14] gave a summary of the current practices used to control the phase and phase rate of the carrier frequencies transmitted by the Omega stations. Lavanceau [15] shows how to relate Omega system time to UTC. Presently, there are four cesium-beam frequency standards controlling the radiated frequency at each Omega station. Omega system synchronization and frequency control is done by a statistical technique based on both internal (intrasystem) and external monitoring of the transmitted signals. Stability and accuracy of the resulting frequency emissions are estimated to be several parts in 10^{-12}.

The precision attainable with current phase comparison measurement systems should exceed 1 μs and in some instances should approach 0.1 μs. Therefore, assuming no propagation path variations, it should take between 10^{-4} and 10^{-5} (1 day) to achieve a frequency comparison good to a part in 10^{-11}. Obviously, in the limit, one cannot exceed the stability of the frequency source.

One must keep in mind that propagation uncertainties will degrade the results. Winkler [16] discusses that part of propagation delay due to path delay and Reder [17] gives an excellent practical account of various sources of interference and variations, both short-term (diurnal) and long-term (seasonal).

In current practice, the PTTI user usually restricts his use of VLF and Omega for frequency comparison to that part of the day when the received signal has traveled to the user over a path which is completely in daylight. This can present a problem for long east–west paths because of the relatively short "all-daylight" period. However, the use of multiple-frequency techniques offers some interesting possibilities for mitigating the diurnal effects.

Time Determination

Omega also can be used for time distribution in three different ways. The first and simplest method uses the frequency-comparison techniques described earlier. This, in essence, is the same technique practiced by every navigator who rates his clock. Initially, one sets his clock and then determines the difference in rate between the oscillator driving the clock and the frequency as derived from the received Omega signal. Subsequent readings of the clock are then adjusted by the difference in frequency.

The second technique consists of driving the clock from a signal slaved to the received Omega signal. This method also requires that the local clock must first be set by some unspecified technique. It further requires that some means be derived to adjust the received signal for propagation anomalies. If not done correctly, the variations in the received signal will be considered as frequency changes and the clock will be driven erronously for brief periods.

The third technique uses the Omega signals both to set the local clock and to control its frequency. Morgan and Baltzer [18], Chi and Witt [19], and Fey and Looney [20] were among the first to implement the use of dual VLF frequency techniques to derive time signals.

Time is recovered from Omega by the identification of a particular cycle of a carrier signal from the phase difference between two received carrier signals. Perhaps the most successful Omega timing receiver was that reported by Chi and Wardrip [7] which required a 250-Hz frequency difference between the two signals. To achieve a 250-Hz frequency difference, two special, unique frequencies were transmitted experimentally by one of the Omega stations. Cycle identification was achieved with the knowledge of propagation path delay and the implied knowledge of one's position.

Operational results by Chi and Wardrip [7] indicated that an accuracy of 2.5 μs was attainable in daylight hours. This figure degraded to 5–20 μs during evening hours, primarily because of propagation path anomalies. Propagation corrections were entered into the system via thumbwheel switches.

Composite Techniques

In 1968, Pierce [21] suggested a technique known as composite Omega for combining two of the navigation frequencies in order to minimize the diurnal instability and the effects of unusual propagation conditions. Effectively, one tries to determine a propagation path delay which does not vary with time by a linear combination of the 10.2- and 13.6-kHz navigation signals. Papousek and Reder [22] extended Pierce's work and determined an improved set of coefficients for the equation governing the combination.

Brown, Van Allen, and Strohbehn [23]–[25] noted that the day and night group velocities are nearly equal in the 12.0- to 12.5-kHz range. Following Pierce, they theorized that if the propagation link could be made to depend on group delay, in contrast to phase delay, the diurnal shift could be eliminated. Their work has met with only partial success thus far, with the principle difficulties being in the inadequacy of the model to account for the day-to-night transition periods and the inherent nighttime instability of group velocity.

Use of Loran-C for Timekeeping

Introduction

Another navigation system that has played an important role in present-day timekeeping is the Loran-C system. It seems as a primary means of clock synchronization. The Loran-C chains, which use cesium-beam frequency standards to control the frequency of the radiated signals, are maintained by the U.S. Coast Guard [26] so that the navigation pulses are synchronized to UTC(USNO). Once a transmitted navigation pulse is synchronized to some UTC second, there always will be some coincidence with another UTC second at a later time, determined by the Group Repetition Interval (GRI) of the particular chain in question.

The time of arrival of a navigation pulse at some known site will be affected by the distance the signal has traveled from the transmitter, the conductivity of the Earth's surface over which the signal travels, and various instrumental delays. If one can account for these delays, a local clock can then be synchronized to the Loran-C clock. The Coast Guard maintains the synchronization of the Loran-C chains to within 5 μs of UTC(USNO). This can be checked by direct monitoring. However, for those chains which cannot be monitored directly, a remote time scale must be formed for the area of the chain coverage as described by Lukac and Charron [27]. The chain's performance with respect to UTC(USNO) can then be monitored and determined to about 100–200 ns. Guinot [6] points out that Loran-C is the vehicle which allows the broad-based formation of International Atomic Time (TAI). Timekeeping laboratories report to the BIH the differences between individual cesium clocks and a Loran-C chain. After taking into account the differences between the various chains, the BIH calculates TAI through a rather sophisticated algorithm known as ALGOS.

Time from Loran-C

Each station in a Loran-C chain transmits a set of eight navigation pulses (nine from a master station). Each chain repeats the pulses at a specified GRI in order to permit identification of a particular chain and minimize interference from other nearby chains. The first set of pulses, transmitted within a chain, is from the master station of the chain followed by the pulses from the secondary stations, designated *W*, *X*, *Y*, and *Z*. The pulses from the secondaries are delayed by an amount called the Emission Delay in order to assure that, in the area of coverage, the master pulse is received first and that no possibility exists for receiving the pulse groups in the wrong order or misidentifying pulse groups from another chain.

For timing purposes, one must first relate the pulse from a master station of a chain to some epoch. The epoch chosen for all chains is 1 January 1958, 0000 UTC, i.e., it is assumed that all chains emitted their pulses simultaneously at this instant. Future Times-Of-Coincidence (TOC) of a master station's pulse with a UTC second are published by the USNO in TSA, Series 9. Any difference between the actual time of transmission and a UTC second is published in TSA, Series 4.

A special Loran-C Timing Receiver is needed in order to obtain time from Loran-C. The most advanced one currently on the market, is the Austron 2100 Timing Receiver. It is a microprocessor-based, improved version of the highly successful 2000 C Timing Receiver. The operation of the receiver is rather simple. The operator enters in the chain GRI and the station of the chain to be tracked, e.g., master, *W*, *X*, *Y*, or *Z*. The operator also must connect an external reference frequency (1, 5, or 10 MHz) and a 1 PPS to the receiver and enter, into the microprocessor, UTC to the nearest second and the first TOC for the day. The receiver then will generate a 1 PPS called the "fixed 1 PPS" which is synchronized to the received Loran-C pulses. The receiver, utilizing an internal time interval counter, can be programmed to display the difference between the reference 1 PPS and the received "fixed 1 PPS."

The operator then can relate the local reference 1 PPS to UTC(USNO) by taking into account the Total Timing Delay (TTD), which is made up of the antenna coupler delay, receiver delay (calibrated by the manufacturer), cycle correction (normally, for timing, the third cycle of the 100-kHz carrier wave which is 30 μs from the beginning of the pulse), propagation delay (the time for the pulse to travel from the transmitter to the receiver), emission delay, and the USNO correction from TSA, Series 4. The accuracy of the TTD can be checked by a portable clock to about 20–300 ns, depending on the circumstances of the portable clock. Seasonal verification of the TTD is strongly recommended as the propagation delay is known to fluctuate [28]; but, as yet, the variation cannot be modeled.

Upon entering the TTD into the 2100 Loran-C Timing Receiver, a "slewed 1 PPS" will be generated which is then synchronized to UTC(USNO). Once this "synchronization" has occurred, the reference 1 PPS can be removed from the receiver and the output "slewed 1 PPS" will continue to produce UTC(USNO), locked to the Loran-C frequency. This process is limited in accuracy by the ability of the Loran-C chain to keep its drift constant with respect to UTC(USNO). By calling the USNO ADS, using a terminal and modem, it is possible to get the latest values of UTC(USNO)–UTC(Loran-C) and, in some instances, to request that a real-time measurement be made for some chains.

Use of TRANSIT for Timekeeping

Introduction

The TRANSIT System has been fully operational as a navigation system since 1964 [29]. Since that time, several laboratories have employed special receivers to recover precise time from TRANSIT. However, there has not been wide recognition and/or use of the unique capabilities of TRANSIT for dissemination of precise time. Recently, the Navy made a commitment to use TRANSIT as a source of time for some communication systems. The polar orbit configuration makes the service available worldwide, and line-of-sight VHF transmission by TRANSIT eliminates some radio-propagation problems associated with other radio-based time systems. Further, since a Fiducial Time Mark (FTM) is transmitted every 2 min, and this mark is identified in the digital message, it is simple to establish an epoch.

The original navigational requirement for TRANSIT time was for an accuracy of 500 μs. Operational requirements for this precision and accuracy have not changed, but recent developments in receiver technology, utilizing microprocessors, and improvements in spacecraft hardware have increased the usefulness of TRANSIT to the timekeeping community. Time recovery with an accuracy of 25 μs is now available [30] with the current satellite system (3 μs with the Nova spacecraft).

Time via TRANSIT

Each TRANSIT satellite transmits a continuous message on a carrier frequency of 400 MHz. The message contains 6103 bits, organized in 156 words of 39 bits each plus 19 additional bits. Of the 39 bits in each word, the first 36 bits contain data while the last three bits are for parity, telemetry, and clock correction. If the clock correction bit is zero, the divider chain is uninterrupted, but, if it is a one, a pulse in the main divider chain, equivalent to 48 cycles of the 5-MHz master oscillation, is deleted. This delays all operations in the satellite by 9.6 μs, i.e., it lengthens the memory readout by that amount.

A total of 112 time correction bits is available in each message (the other 44 bits are not used for various reasons). This permits a control of about 1 ms in the length of the 2-min interval. The satellite clock is designed to run nominally fast in order to provide positive compensation. Thus frequency offset and epoch provide both positive and negative compensation. When the oscillator is at its nominal 4 999 600 Hz, 56 time corrections per day are required.

Finer control is provided by time control bits in the ephemeral memory words. These control bits are used once when a given ephemeral word is read out at word 50. Since they are used only once in a 12-h memory, these bits have 1/360 the effectiveness of time correction bits in the main memory. With proper use of all control bits it is theoretically possible to hold the satellite clock on time within 9.6 μs over a 12-h memory period.

Under current control procedures, the mean value of clock errors is about ± 30 μs, although individual satellite performance may show a greater error for short periods. Daily values for satellite clock performance, UTC(USNO)–UTC(TRANSIT), are published in the USNO TSA, Series 4.

A TRANSIT time recovery receiver was described by Hunt and Cashion [31]. This receiver is manufactured now by Frequency and Time Systems, Inc. During each satellite pass, the receiver must detect the broadcast time marks and decode the satellite message defining the satellite's position. The received time marks are compared with the local clock, and, after correction for transmission delays, any time difference represents the local clock error. A correction may be applied to the clock to synchronize it with the satellite or the difference can be noted to represent the epoch error.

Use of a microprocessor in the time recovery receiver allows several programmed data operations that reduce statistical errors and thereby improve clock performance. At each 2-min mark, for example, the clock data are edited for slant range and for data fit. This eliminates low elevation data points and noisy data. No single data point is used for a clock correction, but, rather, the mean value of all time marks in a pass determines the clock error for the satellite. Since an individual satellite is likely to have a clock error greater than the mean of all satellites in the constellation, the mean receiver clock error can be reduced by filtering clock corrections over all satellites. Further, if the clock performance of individual satellites is better or worse than the total constellation, this satellite may be selected to control the clock or to be deleted from clock control.

Improvements with the Nova Spacecraft

The Nova spacecraft was one of the several recent improvements to the TRANSIT program [32] that occurred around 1975. Improvements were made in the orbit determination software, such as improved geopotential model (WGS-72), improved reference station coordinates, modeling of solid Earth tides and an improved radiation pressure force computation. Among the hardware improvements are a drag compensation device (DISCOS) and an improved clock system consisting of an incrementally programmable synthesizer which modifies the satellite oscillator frequency (5 MHz) according to the transfer function

$$F_{out} = F_{osc}(1 - 1/B(1 + 1/A))$$

where A and B are control parameters. B exerts coarse control and A is fine control. These two parameters should allow control of the satellite oscillator to 1×10^{-13} and its drift to 1×10^{-13}/day.

With existing TRANSIT timing receivers, one can observe phase corrections for the Nova satellite (designated 30480) being made every 10–30 days. These observed coarse adjustments are at the noise level of the navigation function and a nuisance to the PTTI community. Present monitoring of the Nova satellite indicates a drift rate of about 2 μs/day which holds for the period between phase corrections. The rms scatter of the residuals to a linear regression is about 10 μs, i.e., after removing drift. This is indicative of the present performance of the Nova timing system but is not a real measure of its intrinsic capabilities. Data obtained by the Navy Astronautics Group using special system monitoring equipment are a factor of 10 better than those obtained by the commercial timing receiver.

GPS for Timekeeping

Introduction

The NAVSTAR GPS besides being the primary DoD navigation system for the 1990's and beyond, will be also the primary means for worldwide DoD time distribution of UTC(USNO) and could become the primary means for the international comparison of the various national time scales. It is expected that the fixed user will obtain UTC at the 100-ns level, using the single-frequency Standard Positioning Service (SPS), previously known as the C/A-code [33]. However, through the use of differential GPS, it has been shown that two fixed users can achieve a precision and accuracy of better than 10 ns in synchronizing remotely located clocks through common-view, simultaneous observation of the same satellite. If simultaneity is not met, then the precision of synchronization would quickly degrade to between 25–50 ns [34].

Presently, the USNO provides the time reference, UTC(USNO), for the GPS system. Monitoring of GPS time is done at the U.S. Naval Observatory, using a Stanford Telecommunications, Inc. Prototype Time Transfer Unit (TTU) [35] and an STI 502 TTU [36]. The processed output of the TTU is collected by an HP1000 data acquisition system where it is stored for later analysis and distribution. Normally, up to twenty 5–15-min periods of tracking are done each day. Satellites tracked are selected for common-view visibility experiments with other timekeeping laboratories and for across-the-sky coverage. Tracking schedules are available through the ADS(202/653-1809). All 6-s points obtained during a pass are stored, but only the averaged value for UTC(USNO)–UTC(GPS) is readily available to the ordinary user. Individual clock performance of the cesium clocks in orbit has been monitored and their performance can be considered excellent. Because the system is presently in the development stage, analysis of the data has to be done selectively and carefully.

Time from GPS

Obtaining time from GPS is a three-step process. One must first determine the time kept by the satellite being tracked. Next, one applies corrections to the satellite time to derive UTC(GPS). The reference for UTC(GPS) is an ensemble of three cesium clocks at the Master Control Station at Vandenberg AFB. Then, one applies corrections to UTC(GPS) to obtain UTC(USNO). The necessary corrections are computed from data transmitted within the navigation message [37] which are transmitted from the GPS satellites at two frequencies, 1575.42 MHz (L_1) and 1227.60 MHz (L_2). The TTU's which are now used to monitor time at the USNO, operate only on the L_1 frequency because they use only the C/A-code which is not available on the L_2 frequency. The data are transmitted from the satellites at 50 bits/s. A complete navigation message consists of 1500 bits of information. A subframe of 300 bits is transmitted every 6 s. From Subframe 1 of the navigation message one computes the relationship

$$t = t_{sv} - a_{f0} - a_{f1}(t - t_{oc}) - a_{f2}(t - t_{oc})^2 + (\Delta t_r)$$

where

t	GPS system time
t_{sv}	space vehicle PRN code phase time
a_{f0}, a_{f1}, a_{f2}	polynomial coefficients of Kalman filter determined by MCS
t_{oc}	clock data reference time
Δt_r	relativistic term which user must now calculate $= Fe(A)^{1/2} \sin E_k$
F	constant $= -4.443 \times 10^{-10}$ s/(m)$^{1/2}$.

One next corrects t for several factors, e.g., ionospheric correction, tropospheric correction, dither (if present), etc.

$$t_e = t + \text{corrections}.$$

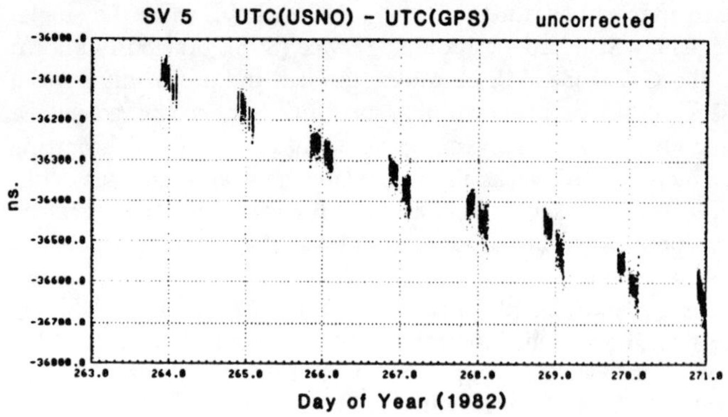

Fig. 1. UTC(USNO)–UTC(GPS), Space Vehicle 5.

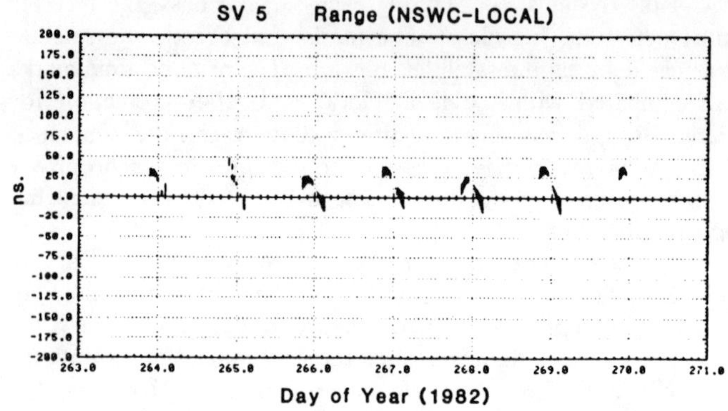

Fig. 2. Difference in range to Space Vehicle 5, obtained by differencing the NSWC batch-fit ephemeris and the transmitted ephemeris from the space vehicle.

One then gets UTC from

$$\text{UTC} = t_e(\text{mod } 86400) - \Delta t_{LS} - A_0 - A_1(t_e - t_{0T})$$

where

Δt_{LS} leap second correction (Subframe 4)

A_0, A_1 coefficients of polynomial (Subframe 4) provided by USNO

t_{0T} reference time for A_0, A_1.

Presently, Δt_{LS} is 3 s, i.e., UTC(USNO) = UTC(GPS) − 3 s. It will increase by 1 s every time a leap second is introduced.

The received time marks (one every 6 s) adjusted by these corrections are now within 100 ns of UTC(USNO). Present steering is such that the rms errors of this adjusted time reference is 25 ns.

The database accumulated for the purpose of monitoring GPS time and evaluating satellite oscillator performance also has proved useful in other studies. Fig. 1 shows the data collected over an eight-day period in 1982. Note the apparent shift in pseudo-range residuals near the midpoint of a pass that occurs over Washington, DC. Similar discontinuities appear in data obtained by different organizations with different TTU's at nearby sites on the east coast of the United States [38]. One cannot attribute these patterns to erratic clock performance. The difference can be traced to minor deficiencies in the transmitted orbit. Fig. 2 shows the difference in range to a satellite obtained by differencing the transmitted ephemeris and a post-fit, batch-processed ephemeris computed by the Naval Surface Weapons Center, Dahlgren, VA. Treating the differences as corrections to be applied to the observed clock data in Fig. 1, the biases disappear (Fig. 3). The scatter in the remaining residuals is a combination of the intrinsic receiver noise; unmodeled tropospheric and ionospheric noise. The peak-to-peak spread is on the order of 70 ns and is indicative of an rms error of about 25 ns.

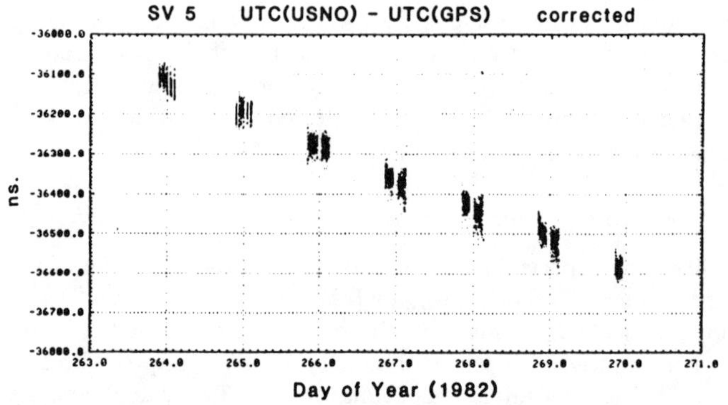

Fig. 3. UTC(USNO)–UTC(GPS), Space Vehicle 5, corrected by range differences shown in Fig. 2.

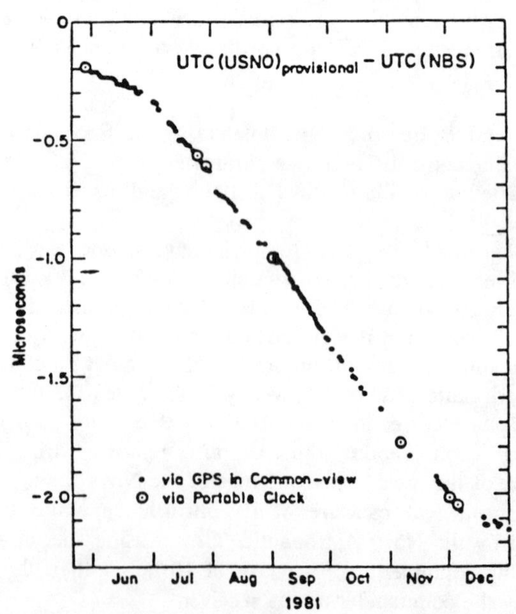

Fig. 4. UTC(USNO)–UTC(NBS) obtained by common-view GPS observations (dots) and portable clock trips (open circles). (Courtesy of David W. Allan, NBS.)

Common-View Time Transfer

Because the current generation of TTU's do not utilize the P-code and because there are specialized users requiring clock synchronization to better than 10 ns, the common-view technique (modified differential GPS) has been utilized by the PTTI community [39]. By making simultaneous observations of the same satellite at two sites, whose coordinates are known, many common errors cancel themselves in this differential mode. User A determines UTC(A)–UTC(GPS), while user B determines UTC(B)–UTC(GPS). Differencing the two relations gives UTC(A)–UTC(B). Errors contained in UTC(GPS), such as ephemeris errors, are common to both A's and B's determination of UTC(GPS) and consequently, for the most part, drop out. Fig. 4 (courtesy of David W. Allan, NBS) shows the comparison of UTC(NBS) with UTC(USNO) obtained by common-view GPS observations. Agreement of the GPS data with portable clock trips is below the 10-ns level.

Discussion

The timekeeping community has effectively utilized different navigation systems for its different needs. The present mixture of navigation systems meets a wide range of PTTI requirements. However, it seems that the future will bring changes to the relation between the timekeeping community and the navigation community. The enhanced performance of GPS will create blessings and problems for the PTTI community. GPS has the capability to provide to the PTTI community the performance of a cesium-beam frequency standard at a fraction of its cost. This ability will allow easy solutions to timing problems. The need for frequency standards may diminish and, if it does, a unique, high-tech manufacturing expertise could be endangered.

Acknowledgment

The material for this article was drawn from the author's notes and past collaborations and discussions with many colleagues, among whom are G. M. R. Winkler (USNO), R. G. Brown (Iowa State University), P. J. Wheeler (USNO), K. Putkovich, D. W. Allan (NBS), and R. Cashion.

References

[1] R. T. Gould, "John Harrison and His Timekeepers," National Maritime Museum, 1958

[2] J. J. Tymczyszyn, "Operational advantages of low cost VLF/Omega navigation equipment," *Navigation*, vol. 19, p. 101, 1972.

[3] D. E. Ringer, J. Gandy, and E. Jechart, "Spaceborne rudidium frequency standard for Navstar-GPS," in *Proc. 7th Annu. Precise Time and Time Interval Applications and Planning Meet.* (X-814-76-45), p. 671, 1975.

[4] J. White, F. Danzy, S. Falncy, A. Frank, and J. Marshall, "NTS-2 cesium beam frequency standard for GPS," in *Proc. 8th Annu. Precise Time and Time Interval Applications and Planning Meet.* (X-814-77-149), p. 637, 1976.

[5] R. L. Easton, "The hydrogen maser program for Navstar-GPS," in *Proc. 8th Annu. Precise Time and Time Interval Applications and Planning Meet.* (X-814-77-149), p. 3, 1976.

[6] B. Guinot, "Review of time scales," in *Proc. 6th Ann. Precise Time and Time Interval Planning Meet.* (X-814-75-117), p. 17, 1974.

[7] A. R. Chi and S. C. Wardrip, "Clock synchronization experiments using Omega transmissions," in *Proc. 5th Annu. NASA and DoD Planning Meet.* (X-814-74-225), p. 769, 1973.

[8] S. M. Spyropoulos and W. J. Klepczynski, "Performance of the NOVA (TRANSIT) for time transfer," in *Proc. IEEE 1982 Position Location and Navigation Symp.*, 1982.

[9] Department of Defense Directive, Number 5160.51, August 31, 1971, "Precise time and time interval standards and calibration facilities for use by Department of Defense components."

[10] D. B. Percival, "The U. S. Naval Observatory clock time scale," *IEEE Trans. Instrum. Meas.*, vol. IM-27, no. 4, pp. 376–385, Dec. 1978.

[11] M. Miranian, "The U. S. Naval Observatory Automated Data Service," presented at the HP1000 Users Group Meet., Sept. 1982.

[12] J. A. Pierce, "Intercontinental frequency comparison by very low frequency radio transmission," *Proc. IRE*, vol. 45, p. 794, June 1957.

[13] J. D. Echols, "Calibrating frequency standards with VLF transmissions," *Electronics*, p. 60, Apr. 27, 1962.

[14] H. J. Santamore, R. N. Schane, and S. F. Donnelly, "Omega synchronization-current operations and future plans," in *Proc. 8th Annu. Precise Time and Time Interval Applications and Planning Meet.* (X-814-77-149), p. 467, 1976.

[15] J. D. Lavanceau, "Omega signals and universal time," presented at the 1st Omega Symp., Tokyo, Japan, June 1975.

[16] G. M. R. Winkler, "Path delay, its variations and some implications for the field use of precise frequency standards," *Proc. IEEE*, vol. 60, p. 522, May 1972.

[17] F. H. Reder, "VLF phase tracking for PTTI application," in *Proc. 3rd Annual DoD PTTI Strategic Planning Meet.*, p. 117, 1971.

[18] A. H. Morgan and O. J. Baltzer, "A VLF timing experiment," *NBS J. Res.*, vol. 68, Nov. 1964.

[19] A. R. Chi and S. N. Witt, "Time synchronization of remote clocks using dual VLF transmissions," in *Proc. 20th Annu. Symp. on Frequency Control*, p. 588, 1966.

[20] L. Fey and C. H. Looney, "A dual frequency VLF timing system," *IEEE Trans. Instrum. Meas.*, vol. IM-15, p. 190, 1966.

[21] J. A. Pierce, "The use of composite signals at very low radio frequencies," Harvard Univ. Tech. Rep. 533, Feb. 1968.

[22] W. Papousek and F. H. Reder, "A modified composite wave technique for Omega," *Navigation*, vol. 20, p. 171, 1973.

[23] R. G. Brown and R. L. Van Allen, "Three frequency difference Omega," *Navigation*, vol. 24, p. 206, 1977.

[24] R. L. Van Allen and R. G. Brown, "An on-line Omega compensation technique utilizing three frequencies," in *Proc. ION Nat. Marine Meet.* (San Diego, CA, 1976).

[25] R. G. Brown, R. L. Van Allen, and K. Strohben, "Calculation of Omega propagation group delay and application to local time standard monitoring," in *Proc. 8th Annu. Precise Time and Time Interval Applications and Planning Meet.*, p. 449, 1976.

[26] W. J. Thrall, "Loran-C, an overview," in *Proc. 13th Annu. Precise Time and Time Interval Applications and Planning Meet.*, NASA Conf. Publ. 2220, p. 121, 1981.

[27] C. F. Lukac and L. G. Charron, "Timing a Loran-C chain," *Navigation*, vol. 29, p. 235, 1982.

[28] C. C. Costain, J. S. Boulanger, H. Daams, D. W. Hanson, R. E. Beehler, A. J. Clements, D. D. Davis, W. J. Klepczynski, L. Veenstra, J. Kaiser, B. Guinot, J. Azoubib, P. Parcelier, G. Freon, and M. Brunet, "Two-way time transfer via geo-stationary satellites NRC/NBS, NRC/USNO, and NBS/USNO via Hermes and NRC/LPTF (France) via Symphonie," in *Proc. 11th Annu. Precise Time and Time Interval Applications and Planning Meet.*, p. 499, 1979.

[29] T. A. Stansell, "Transit, the Navy Navigation Satellite System," *Navigation*, vol. 18, p. 93, 1971.

[30] R. E. Cashion, W. J. Klepczynski, and K. Putkovich, "The use of Transit for time distribution," *Navigation*, vol. 26, p. 63, 1979.

[31] G. A. Hunt and R. E. Cashion, "A Transit satellite timing receiver," in *Proc. 9th Annu. Precise Time and Time Interval Applications and Planning Meet.*, p. 153, 1977.

[32] Staff of the Space Analysis and Computation Group, Applied Physics Laboratory, "Planned improvements in the Transit system (1975)," *Navigation*, vol. 22, p. 352, 1975.

[33] W. J. Klepczynski, Ed., "Global positioning system," *Navigation*, vol. 25, no. 2, 1978.

[34] D. W. Allan and M. A. Weiss, "Accurate time and frequency transfer during common-view of a GPS satellite," in *Proc. 34th Annu. Frequency Control Symp.*, p. 334, 1980.

[35] A. J. Van Dierondonck, Q. O. Hua, J. R. McLean, and A. R. Denz, "Time transfer using NAVSTAR GPS," in *Proc. 13th Annu. Precise Time and Time Interval Applications and Planning Meet.*, p. 389, 1981.

[36] D. L. Hall, J. Handlan, and P. J. Wheeler, "Test results of the STI GPS time transfer receiver," in *Proc. 14th Annu. Precise Time and Time Interval Applications and Planning Meet.*, 1982.

[37] Rockwell International, *ICD-GPS-200, NAVSTAR GPS Space Segment/Navigation User Interface*, 1982.

[38] O. J. Oaks, J. A. Buisson, and S. C. Wardrip, "GPS time transfer receivers for the NASA transportable laser ranging network," in *Proc. 4th Int. Workshop on Laser Ranging Instrumentation*, P. Wilson, ed. (Bonn, Germany, 1982).

[39] D. D. Davis, M. Weiss, A. C. Clements, and D. W. Allan, "Remote synchronization within a few nanoseconds by simultaneous viewing of the 1.575 GHz GPS satellite signals," in *Proc. 37th Annu. Frequency Control Symp.*, 1983.

Part 3
Introduction to Land Navigation

3.1 Ancient Land Navigation

EARLY man rarely wandered more than a night's stay from his camp or village. When he did, he wanted to know "Which way?" and "How far to walk?". His navigation references were local landmarks that he knew intimately from all angles. In some cultures, shepherds regularly migrated with their sheep along traditional routes. These men, adventurous hunters and traders, were probably the first to explore at a distance. At night they easily recognized the polestars, even before Polaris became the present-day North Pole star. In daytime, they measured azimuth by the Sun, offsetting their heading according to the approximate time of day, as do modern hikers. In cloudy weather, they probably did not go beyond landmark distance of the camp or village. In unfamiliar areas, they hired local guides who also served as protectors. In arctic regions, there are tales of men who became lost within hundreds of feet of their huts during whiteout conditions because navigation was impossible.

We can assume that ancient man measured distances in the same way that modern hikers do; he probably measured his walking time by counting paces. As discussed in Part 2.3, pg. 9, length standards were entirely local, originally based on local standard-sized objects or the dimensions of chiefs or priests. In any case, each traveler calibrated his own walking pace.

In 4000 B.C., the two-wheeled cart appeared in Iran, replacing the dragged sledge. This invention certainly increased man's traveling range. A thousand years later, the four-wheeled wagon appeared, harnessed to two animals, allowing the transport of goods over long distances in flat, open country. The chariot appeared in 1700 B.C. in Iran, spurring the breeding of fast horses. Man learned to ride the horse on the steppes of central Asia sometime after 2000 B.C., again increasing his range of travel. Chinese farmers developed a breast harness for horses about A.D. 100 that increased the animal's pulling power without choking it. Stirrups reached Europe about A.D. 600, permitting knights to wear heavy armor. Despite these improvements in transportation, the associated navigation techniques were still based on landmarks, celestial azimuths, and time of travel.

As roads (rarely paved) were cut through desert and forest, man placed markers and signposts to navigate along the one-dimensional space. Mayan and Moghul Indian highways had rest houses at intervals; Roman roads had numbered mileposts. In cities, two-dimensional signposts were erected (and sometimes taken down in times of war, to confuse literate enemies) and early city maps were created. A map of Hellenic Jerusalem survives in the Israel Museum. Maps of the crooked, narrow streets of early cities may not have been distributed outside the government; visiting merchants probably hired guides. The need to ask for directions, as in modern Japanese cities where most streets are neither named nor numbered, ensures some control over strangers.

Most ancient travel was by river and along the coast. Long-distance land travelers were exposed to bandits and often found no roads or provisioning facilities. Off-road overland navigation appears to have started in the Arabian deserts. Spice caravans began to cross the deserts on 2-month trips after the camel was domesticated in 1300 B.C. With clear skies, navigation was probably by the Sun and stars between oases. I suspect that stone markers would not have been useful amidst shifting sands so these travelers must have been the champion land navigators of the ancient world. In 583 B.C., the Persians conquered Babylon and imposed their authority throughout western Asia. In 330 B.C., the Macedonians conquered the Persians and imposed their rule in western Asia. Three hundred years later, the Romans imposed their rule. Each conquest increased the area controlled by a single government, making travel easier by reducing the number of "taxes" to be paid en route. About 150 B.C., a new camel saddle was invented [see (van Beek, 1969)]. The result of central government control and the new saddle was extensive east–west traffic on the "silk route," in short hops between trading centers from Palmyra (in modern Syria) to Dun Huang (in modern China).

The well-known "south-pointing needle" on a chariot, about A.D. 300 in China, was a wooden differential gear, driven by the two wheels. It would have had to be initialized by the charioteer and probably would have drifted rapidly in azimuth as the wheels slipped and bounced on their rough roads. In any case, it was an azimuth reference, not a navigation device.

Land navigation probably evolved little until the astrolabe allowed calculation of the time of day (before A.D. 600) and the magnetic compass measured azimuth reliably, in any weather conditions. Even on a rough road, the compass needle would settle down correctly when the vehicle stopped. There are reports of a floated-needle magnetic compass on camelback in the Mid-Eastern deserts in A.D. 1500.

During the European Renaissance, printed maps became plentiful. Roads were cut through the forests as cities grew, nations were created, and trade prospered. Navigation was along the roads and trails between villages.

American Indians traveled extensively by land and river. The North American nomads probably followed traditional routes. On the Gulf coast, there is evidence of 8th-century land trade routes and a 10th-century coastal trade route, but there is no evidence of portable astronomic instruments for navigation. The American Indians probably navigated entirely by landmark. I doubt that individual Indians traveled far enough to observe the changing elevation of the Sun and stars with latitude; if they did, they did not record it or transmit it orally.

3.2 Railroad Navigation

The purposes of railroad navigation are to prevent collisions, to guide trains through switches, to monitor the location of valuable cargo, and to provide initial conditions for launching missiles.

The first land navigation system was invented for the railroads in the 1870s. Tracks were divided into lengths called "blocks," as long as 200 mi in rural areas (the maximum distance between waterings of a steam locomotive). A train driver received a hard-copy clearance from a trackside dispatcher (he often caught it from a frame while the train was moving) that cleared him into the next block with the intent that he stop at the end of the block if he received no further clearance. Until a downstream dispatcher reported the train exiting the block, no other train would be cleared into it. Dispatchers communicated with each other by telegraph. Hence, navigation was in increments of blocks. The system worked on single tracks (two-way traffic) and on parallel tracks. By this means, the United States and Europe, and later Japan and Russian Asia, were crossed by railroads.

Electric signaling was introduced into metropolitan areas where hundred-meter-long blocks were needed for commuter trains. Rails within a block were insulated from the rails in the adjacent blocks. A 1- to 2-kHz signal was applied between the rails of each block. A train with metal wheels and axles shorted the rails revealing its presence. The short circuit set the lights upstream and, in some commuter lines, set a mechanical arm that would apply the brakes in a train trying to enter an occupied block. Hence, the error of navigation was reduced from a hundred miles to a hundred meters. The block system is still used in older subways and, in more elegant forms, on the most modern rail lines. The Washington and Hong Kong metros, for example, superimpose a computer-dispatch system on the basic block controls. They send speed commands to each train thereby increasing the density of traffic. Still, the block logic will override any attempt by a train to violate the cleared space of another.

For the main-line railroads of North America, a dead-reckoning navigation system is being developed that will radio the position of a train within a block to the dispatcher (see Part 8.3). Old main lines, even in the United States and Europe, still require that the crew stop the train and telephone from trackside. Indeed, a fatal collision occurred in England in 1989 reportedly when a crew stopped a train to report a defective signal by telephone. Most railroads are installing radio telephones. Giles and Simson (1980) have described a UHF radio system that uses trackside 25-W transceivers, each covering ±4 mi of track. Extra transceivers are installed in tunnels. The objectives are to allow new clearances to be radioed to trains and to allow the crew to report local incidents.

3.3 Dead-Reckoning Navigation

The earliest electronic land navigators were expensive and were developed for military services and NASA. Virtually all military land vehicles navigate visually, by landmarks, and steer by magnetic compass. The invention of the tank (1916) placed new demands on navigation because the magnetic compass works poorly in an iron vehicle. Tank commanders have had to climb out and walk away from their vehicle to use their magnetic compass and map. A few military vehicles have been equipped with simple dead-reckoning devices that resolve distance through azimuth and accumulate the eastings and northings. Distance is measured by the rotation of one or more drive wheels or treads, at a typical scale of one pulse per foot of travel. Heading is measured with a free gyroscope or a north-seeking gyrocompass. No attempt is made to compensate for the curvature of the Earth. Hadfield and Leiser (1988), Mueller and Adams (1984), Packard (1985), and Rickords and Yamamoto (1985) have described low-cost gyro-based military navigation systems, most of which can operate off-road. Gyrocompass-odometer systems have exhibited errors less than 2% of cumulative distance traveled on paved roads. The more expensive inertial land navigators are achieving errors less than 0.25% of distance traveled at a cost of $300,000. They are the same as the inertial land surveyors described in Part 2.3 but are operated without periodic ZUPT stops ("zero-velocity updates", Part 2.3) and without retracing routes. Caspary and Konig (1986), Goldin (1985), and Kriegsman and Mahar (1986) described them. In 1989, Germany, France, Egypt, and Israel have production land navigators on military vehicles. The United States is developing a system for its Abrams tank.

In 1971–72, a gyro-odometer navigation system was successfully used in the Lunar Roving Vehicle that explored within five kilometers of the last three lunar bases (Fig. 6-9). It had a rotation counter on each of its four wheels, with voting logic to detect slip and sensor failures. The odometer count was resolved into rectangular coordinates through a 10°/h directional gyroscope that was pendulously leveled. Since the Moon rotates slowly, a north-seeking gyrocompass was impractical. Hence, the gyroscope was realigned periodically with a sundial built into the instrument panel. The gnomon reading was radioed to the ground whereupon the flight controllers calculated the vehicle's heading and radioed it to the Moon. The system achieved an accuracy of 3% of distance traveled. It was not a critical item because the crew could see the base and could follow the rover's tracks home. However, it measured radial distance to the base, so the crew would not drive too far to walk back on their remaining oxygen supply. A lunar rover was carried to the Moon on Apollos 15, 16, and 17. Heffron and LaPiana (1970) (in the Part 6 bibliography) described its navigation system. A full-scale test article is exhibited at the Henry Ford Museum in Dearborn, Michigan.

Inexpensive dead-reckoning navigators measure heading by differential wheel rotation or with a magnetic compass.

Whitcomb (1988) has described the difficulties of locating a magnetic sensor on a steel vehicle such as an automobile or a truck. The vehicle's own magnetic field changes if it is parked for several weeks, if it is driven near direct-current (DC) power lines (e.g., near a DC-powered electric rail line), or if its internal direct-current loads are switched. Whereas a ship's magnetic compasses are compensated (Part 4.4), land-vehicle compasses are usually computer calibrated, the correction being added to the compass reading. By automatically recalibrating on roads of known heading, the vehicle's magnetic deviation can be continuously corrected. Nevertheless, the accuracy of compensated magnetic sensors in a moving vehicle is rarely better than 2°, and is usually much worse. Magnetic variation is stored in the computer, either in tables or as an algorithm.

These heading errors cause errors in the dead-reckoned position that steadily increase. In order to bound the navigation errors in cities, expert system logic is sometimes used, which adjusts the dead-reckoned position to conform to a stored map of the area, knowing the heading of streets and the position of intersections. In hilly cities, pitch profiles could be matched to a map, to improve navigation accuracy (''pitch matching''). The problem is that, in 1989, an accurate vertical sensor that is independent of the automobile's acceleration, is expensive.

The keys to successful performance with low-cost dead-reckoning are:

—low-cost sensors,

—the simplicity of the logic that adjusts the data to the prestored map,

—the convenience and cost of the data base (usually on optic disk, for a city), and

—the method of presenting the routes to the driver. (Voice and pictorial instructions have been tried with various degrees of disturbance to the safe operation of a single-occupant vehicle.)

Specialized dead-reckoning devices are being developed for mobile robots that perform routine tasks in factories, work in hazardous areas (such as nuclear plants), perform dangerous tasks (such as bomb disposal), or explore hostile environments (such as planets). Robots guided by a tape or a stripe on the floor can find their way on predetermined routes of known geometry.

Research is underway at many universities to develop sensors and pattern-recognition algorithms that will create a map of an unknown area, solely from sensor observations, and locate the robot on the map [see (Elfes, 1989)]. For example, planetary robots must follow macronavigation instructions while autonomously avoiding obstacles and selecting a feasible microroute. The detailed steering instructions must be generated on-board because of the 12-min to 40-min round-trip travel time of radio signals to the nearest planets.

3.4 Automotive Navigation

Automotive navigation requirements are taking two forms:

—The driver wants to know the distance and direction to his destination, and perhaps the optimum route. Either radio or dead-reckoning navigation will satisfy the first requirement and part of the second. However, if the calculation of the optimum route is to be based on current traffic conditions, the driver must receive information from outside. In U.S. cities, cellular radio and emergency VHF radio already exist. Western Europe is developing a continentwide cellular radio system.

—A dispatch center wants to know where its vehicles are and what their status is. The center may also want to communicate with the vehicles, two ways. This requirement is typical of a cross-country trucking company, especially one carrying high-value cargo, and of a fleet of emergency vehicles.

Experiments to satisfy both requirements began in the 1960s, using the dead-reckoning techniques described in Part 3.3 and the radio-navigation techniques described in sections 3.5, 3.6, 4.8, and 4.10. In the early 1980s, shippers of high-value cargo sponsored the development of Transit-based and Transit-odometer-magnetic-compass-based systems to deter hijacking. These systems are installed in a few specialized vehicles. Canada and probably the Soviet Union developed multisensor land navigators for travel in their polar regions, as described in McMillan's reprint, pp. 73–81.

In the late 1980s, ultra-low-cost navigators are being developed for automobiles, trucks, emergency vehicles, and rental cars. In open country, they report the vehicle's position and status and may also transmit dispatch instructions. In cities, they locate addresses and may calculate the optimum route. Commuter vehicles can be rerouted depending on observed traffic.

In *open country*, today's land navigators use navigation fixes from GPS (Part 3.6), Loran (Part 4.8), or Transit (Part 4.10). Outside cities, there has been no nationwide mobile communication service anywhere on Earth. Cross-country trucks may transmit their position to a dispatch center by HF radio, via Geostar or Starfix satellite, or experimentally via meteor backscatter. Geostar, for example, transmits the Loran-derived position of several thousand trucks via an L-band transponder carried on a Spacenet satellite. Data arrive at the dispatch center via a telephone gateway. Geostar is described by S. Baker (1989) and by Foley (1988). Another example is the navigation of trucks operated by North American Van Lines, which is experimenting with Loran-derived positions transmitted to its Indianapolis dispatch center. The trucks aim repeated messages at the sky, at VHF. Sensitive receivers at the dispatch centers detect occasional backscatter from randomly occurring meteors. The system, called Transtrack, is said to receive the message within a few minutes. Later generations of open-country land navigators will derive position from the modulation on the nav-comm satellite's carrier. These future systems will conduct two-way communications by satellite, as forecast in Part 8.1.

At least two European groups are planning continentwide nav-comm satellites. In the United States, they are likely to be privately owned; in Europe, they are likely to be publicly owned. Mobile communications in open country are described by Pattan (1987). In 1989, user charges were reported to be \$50/month per vehicle plus less than a dollar per message. Receivers were reported to cost less than \$3500.

In *cities*, radio navigation at low frequency is sometimes used. For example, the reprint by Cortland (1989), pp. 90–100, reports that Detroit is equipping 760 police, fire, and medical vehicles with Loran-C receivers and VHF radios. The

Loran and radio receiver share a common antenna. Five fixed stations, located throughout the city, poll each vehicle every 2- to 10-min according to a priority list. Dispatch instructions are sent to the vehicles on existing VHF communication channels. Cortland reports a position accuracy of 660 ft at 96% of the test points. The dispatch center has a differential Loran receiver whose offsets (assumed to be constant throughout the city) correct the positions of all the vehicles.

At higher navigation frequencies, outages are predicted in cities due to shadowing and reflections. Held and Kricke (1985) predicted that two or more GPS satellites will be visible 30% of the time in streets and 60% of the time at intersections. Frye's measurements (1989) of GPS found outages 2% of the time in open country and 72% of the time in cities. Outages interrupt the ranging data and the ephemeris message (see section 3.6).

Thus, most *city* navigators that are in development depend on dead-reckoning, resolving an odometer through calculated heading. Heading is either measured (e.g., magnetic compass) or inferred from differential wheel rotation. Proximity beacons (''electroposts'') can radio a position update and the street's heading to passing vehicles. For example, in 1989, Nissan was offering a dead-reckoning city navigator using a magnetic compass and differential wheel rotation. The driver updates position manually at known locations. The Nissan system contains digital maps (currently of Tokyo, Osaka, and Nagoya) and a digital road map of Japan on optic disks. The map and the vehicle's position are displayed on a 6-in CRT. Distance and direction (not the optimum route, which is a complex function of one-way streets, protected turns, and traffic conditions) are displayed to 30,000 prestored destinations. I believe that Nissan charges advertisers to be included in the data base.

In 1989, other map-matching navigators were in experimental use in San Mateo, California, and Albuquerque, New Mexico. Dispatcher-controlled navigators were in use in Toronto, Melbourne (Australia), and London (England). Dork (1987) described a General Motors van equipped with many navigation sensors for experiments. Japanese, European, and American engineers have designed and tested low-cost, on-road navigation devices, as described in French's reprint, pp. 101–107.

3.5 Ground-Based Radio Navigation, Aids

The U.S. military sponsored the development, starting in the mid-1970s, of the Joint Tactical Information Distribution System (JTIDS), an L-band radio system in which time slots are allocated to user terminals for communication, identification, and position–location messages. Each terminal carries a precise clock that is periodically updated with respect to the network's master clock. Interterminal messages are timed by these synchronized clocks, allowing each user to calculate his position relative to those of the others by trilateration. If two or more terminals are at surveyed sites, the entire net calculates absolute position. Fried (1978) and Widnall and Kelley (1986) have predicted an accuracy of 30–150 m. Thus far, JTIDS terminals have been airborne or shipborne, but soldier-borne ground terminals are planned. Terminals have been installed in airborne warning aircraft (AWACS) and capital ships, and in F-15, F-16, F-18, and Tornado fighter aircraft. To make jamming difficult, JTIDS ''frequency-hops'' in a predetermined pattern; the transmitters and receivers have synchronized clocks that know when the hops will occur.

The 420- to 450-MHz Position Location Reporting System (PLRS) was developed at the same time for the Army and Marine Corps battlefields. Portable terminals exchange ranging messages and send data to a master terminal that calculates everyone's position within 10 to 50 m. Each PLRS net accommodates as many as 470 users. By 1989, land terminals had been tested in Army and Marine Corps maneuvers; field commanders claimed a 3:1 improvement in productivity due to their instantaneous knowledge of the locations of their own infantry, armor, artillery, and helicopters. Enhanced PLRS (EPLRS) adds the transmission of 1200-bit/sec messages and connects the PLRS net to the JTIDS network via gateways located at command posts. The JTIDS and PLRS navigation systems are complex and expensive, but not easily jammed. They work well in open country, on-road or off-road. The reprint by Kivett and Cook, pp. 82–89, describes PLRS.

3.6 Space-Based Radio Navigation, Aids

The Navstar satellites (part of the Global Positioning System, GPS) have been developed by the U.S. Air Force during the past 20 years. GPS permits worldwide accuracy better than 100-m, using the publicly available coarse-acquisition code (C/A-code, now called ''standard positioning service''). GPS also permits better than 15-m accuracy in military receivers using the precision code (P-code, now called ''precise positioning service''). Sub-meter accuracy is being obtained using either code by recording many hours of observations and by measuring the phase of the carriers, whose wavelength is 20 cm. The U.S. Department of Defense has reserved the option, in the operational satellites, to degrade the accuracy of the ephemeris to 100 m, without notice, if only to test that wartime capability.

Seven experimental satellites and 5 operational satellites were working in 1989, leaving large coverage gaps in space and time. By 1992, at least 21 operational satellites will be in orbit, giving continuous worldwide coverage. Figure 3-1 shows a Navstar during factory assembly. The solar panels and the conical-spiral antenna array that forms the L-band beam are visible. Other antennas are for control and monitoring of the satellite. Navstars are in 12-h orbits so they move rapidly across the sky. Every satellite transmits two L-band signals, L1 at 1575.42 MHz and L2 at 1227.60 MHz, whose carriers are stable to one part in 10^{13}. The carriers from all satellites are coherent but are distinguished from each other by their Doppler shifts and coding. L1 is modulated by three messages:

—the C/A chip-code at 1.023 MHz;

—the P-code at 10.23 MHz, whose 10,230-chip sequence will be kept secret when ''anti-spoof'' operation is required; and

—the ephemeris of the orbit.

The 32 C/A-codes are orthogonal to each other so a matched

Fig. 3-1. GPS spacecraft. Courtesy of Rockwell Corp.

filter in the receiver, knowing the code for each satellite, can lock onto the proper code-train even while all satellites transmit simultaneously on the same frequency. Lock-on occurs in less than 5 min; subsequent carriers are acquired in less than 10 s.

Similarly, the L2 carrier is modulated by two messages: the encrypted P-code and the orbit ephemeris. The P-codes from the twenty-one satellites are orthogonal to each other. The code keys will sometimes be made known, will sometimes be released only to military users, and may be changed daily. Receiving L1 and L2 allows the first-order ionospheric error to be removed by comparing the range differences at both carrier frequencies. The absolute time delay in receiving either P-N code measures range to the satellite, when corrected for the offset of the user's clock (calculated by the user from several satellite signals) and atmospheric propagation delays. Jorgensen (1989) showed that a simple model removes 60% of the ionospheric error, leaving a 15-m uncertainty.

The ephemeris of each satellite is broadcast on L1 and L2 at 50 bits/s with the most significant digits corrupted or encrypted to deny precise positioning service to civil users. Each data bit consists of 20,000 C/A-code chips or 200,000 P-code chips. Each C/A chip consists of approximately 1500 carrier cycles. The format of the satellite's ephemeris has been explained by Cariveau and Therkelsen (1988). Thirty seconds of data are required to receive the ephemerides for standard positioning service; 12.5 min is required for precise positioning service. In the 12.5-min message, each satellite transmits an "almanac" that contains the approximate ephemerides of all satellites. Hence, a newly powered-up receiver can learn the orbits and calculate which satellites to use for optimum results. Green (1989) described the orbital coverage with and without satellite outages.

Within the GPS satellite, the instantaneous orbit is predicted from the ground-uplinked state vector. The prediction algorithm does not use a model of the forces acting on the satellite. Instead, the GPS stores the dimensions of an instantaneous ellipse and the rates of change of the arguments of perigee and the line of nodes. The GPS ground stations uplink these parameters, their rates of change, and the coefficients of a quadratic or third-order power series that describes the satellite's departure from the precessing ellipse. This prediction technique, called "kinematic orbit prediction," has been described by Yunck (1988) in the Part 2 bibliography.

By measuring range to two satellites, a user can calculate its horizontal position for a known height, without a clock correction. By measuring range to three satellites, the user can calculate its horizontal position and clock offset, or its horizontal and vertical positions. By measuring range to four or five satellites, the user can remove ambiguities and solve for clock drift, greatly reducing the position errors. In 1989, receivers were being produced that received the C/A-code from one, two, five, or six (for space geodesy) satellites. Receivers are made for the military-secure precision code (P-code) from one or five satellites. The highest accuracy is achieved by tracking the carrier and counting cycles as range changes. Data are lost whenever the signal is interrupted or when vehicle movement exceeds the bandwidth of the phase-tracking loops ("cycle slip"). The least expensive sets receive the C/A-code on L1 from one satellite at a time. Aggarwal's reprint, pp. 108–119, describes the internal circuitry of a 16-lb receiver that is built in two- and five-channel versions with a variety of outputs. The receiver contains an optimum filter that estimates range and clock errors for a fixed vehicle and speed errors for a moving vehicle. Blank and Rhodes (1988), Euler and Bittner (1984), and Ward (1980) have described other GPS receivers. Ashjaee *et al.* (1988) have described the input–output provisions and operator controls for a survey-quality receiver. At the other extreme, Lennen and Daly (1989) describe an inexpensive one-channel digital receiver. Status monitoring is discussed in Part 8.1, pg. 435.

According to current test reports, moving vehicles carrying five-parallel-channel receivers measure their position with an accuracy of 10–20 m and stationary observers achieve 1 m in real time, while surveyors achieve 10-cm accuracy (Part 2.3). Three-meter navigation can be achieved on a moving vehicle with "differential GPS." Surveyed ground stations are placed in the areas where vehicles are to navigate. The ground station calculates its own position from GPS transmissions and compares it with its known position. The station then broadcasts the offset to vehicles within a radius of a few hundred miles on a mobile communication frequency. The reprint by Teasley *et al.*, pp. 131–138, describes test results for differential GPS. The first one-channel hand-held receivers are already for sale for $3000. A vast literature exists about the GPS constellation, the signals in space, the design of receivers, algorithms for reducing data, and algorithms for mixing GPS with other navigation sensors.

The Soviet Union has a GPS-like satellite system called Glonass, also on L-band. Dale *et al.* describe Glonass in detail in the reprint on pp. 139–146. It differs from GPS in at least the following respects:

—each satellite transmits on its own carrier frequencies, requiring that the receivers be tunable;

—the C/A-code is 511 bits long, versus 1023 bits for GPS;

—the ratio L1/L2 is 0.2% lower in Glonass than in GPS;

and

—the data format differs, certainly for the P-code.

Dale and his colleagues built a Glonass receiver and found 30-m 3-sigma errors in a stationary receiver using the Soviet C/A-code.

The European Space Agency proposed a navigation satellite system whose users receive GPS and Glonass transmissions. It proposed ground stations that monitor the American and Soviet satellites for outages and out-of-tolerance errors, and that broadcast their status and differential corrections. Lenorovitz (1988) has outlined one European proposal and Carnebianca (1988—see the reference list in the Part 4 Introduction) has described another. A joint navigation receiver could monitor itself for satellite outages. The relationship between GPS–Glonass and the next generation of nav-comm satellites is discussed in Part 8.1, pg. 435.

The United States, Canada, the Soviet Union, France, and an ever-increasing number of other countries operate an unusual space-based navigation system that locates emergency beacons that transmit on 121.5, 243, or 406 MHz. Emergency beacons are carried on aircraft and ships and by survey parties. The beacon is activated manually or, on aircraft, automatically upon impact. The system, called Cospas in the Soviet Union and Sarsat in the West, relays beacon transmissions, as received by low-orbit satellites, to the ground where the time history of the Doppler-shift is measured and the location of the beacon is calculated within 20 km. Three U.S. and Canadian satellites carry Sarsat transponders and two Soviet satellites carry Cospas transponders. Any nation may establish ground stations that listen for beacons or emergency voice transmissions on the satellites' downlinks. Upon detecting an emergency transmission, they calculate the location of its transmitter and dispatch search aircraft. More than 1200 people have been rescued worldwide with the aid of this system since the first satellite became operational in 1983. A new user-beacon design may reduce the uncertainty to 5 km and allow reception by synchronous satellites, thus improving coverage. Keightley (1987) has described Sarsat/Cospas. An anonymous article in *Aviation Week* (1988) described the political arrangements among the cooperating nations.

Other transponders, on U.S. weather and environmental satellites, relay signals from animals in the wild and from instrumented buoys in the open ocean. The beacon's position is calculated at a ground listening station.

The future of land navigation is discussed in Part 8.3 in four categories:

—city vehicles;

—open-country trucks, on highways;

—off-road vehicles (mostly military), including those users that require a precise attitude or heading reference; and

—railroads.

References and Bibliography

Aggarwal, A. K. "EURONAV™—A state of the art military GPS receiver," in *Rec. IEEE Position Location and Navigation Symp.*, 1988, pp. 153–164. Reprinted in Part 3, pp. 108–119.

Anon., "Four nations sign 15-year SARSAT pact," *Aviat. Week and Space Tech.*, Oct. 24, 1988. United States, USSR, Canada, and France. U.S. cost $3 million per year, spread among many agencies. Saved 1150 lives so far.

Ashjaee, J. M., *et al.*, "Ashtech XII GPS receiver," in *Rec. IEEE Position Location and Navigation Symp.*, 1988, pp. 426–443. Details of a 12-channel surveying receiver, using L1 only. Descriptions of software, display, and I/O.

Bailey, C. D., "Fixed site comparison of P and C/A code ranging," Johns Hopkins University Rep., Apr. 1982. Thirty-three-day tests show 20-40-m errors for P- and C/A-codes. P-code noise is 0.5 m RMS; C/A-code noise is 5-m RMS.

Baker, S., "Locstar gets ready to go," *Interavia*, pp. 24–28, Jan. 1989. Europeans plan Locstar for truck navigation and low-message-rate communication, under license from Geostar. Also, INMARSAT will offer land-mobile service via PTTs.

Blank, R. W. and W. D. Rhodes, "Field test results prove GPS performance and utility," *IEEE AES Mag.*, pp. 11–20, June 1988. Describes Rockwell Collins GPS set in one-, two-, and five-channel versions. Tested with C/A- and P-codes. Best accuracy is 16-m spherical error probable. Five minutes to converge.

Braff, R., "Ranging and processing mobile satellite," *IEEE Trans. Aerosp. Electron. Syst.*, pp. 14–22, Jan. 1988. Mitre concept of a nav-comm satellite for air traffic in the year 2010. Users measure range to each satellite; satellites poll users. Forecasts 200-m error.

Cariveau, B. K. and K. L. Therkelsen, "Satellite data management in DoD NAVSTAR GPS receivers," in *Rec. IEEE Position Location and Navigation Symp.*, 1988, pp. 134–144. Reprinted in Part 3, pp. 120–130.

Carson, L., *et al.*, "Design and predicted performance of the GPS demonstration receiver for the NASA TOPEX satellite," in *Rec. IEEE Position Location and Navigation Symp.*, 1988, pp. 442–454. Reprinted in Part 6, pp. 353–365.

Caspary, W. F. and R. Konig, "Comparative study of Ferranti, Honeywell, and Litton inertial survey systems," *Bulletin Géodésique*, vol. 60, no. 2, pp. 143–153, 1986. Tests of three systems on a Munich range showing 1-cm accuracy.

Cortland, L. J., "Loran-C vehicle tracking in Detroit's public safety dispatch system," *Navigation*, vol. 36, no. 3, pp. 223–233, Fall 1989. Reprinted in Part 3, pp. 90–100.

Cotton, A. R. and H. B. Calverley, "Hong Kong Mass Transit Railway," *Mech. Eng.*, pp. 37–44, Jan. 1983. Describes signaling system.

Dale, S. A. and P. Daly, "The Soviet Union's Glonass navigation satellites," *IEEE AESS Mag.*, pp. 13–17, May 1987. Describes the constellation and signals in space; C/A- and P-code. Differences from GPS.

Dale, S. A., *et al.*, "Position-fixing using the USSR's GLONASS C/A code," in *Rec. IEEE Position Location and Navigation Symp.*, 1988, pp. 13–20. Reprinted in Part 3, pp. 139–146.

Diederich, P., "The development of civil satellite navigation in Europe," *Navigation*, vol. 36, no. 1, pp. 127–136, Spring 1989. Europeans investigating alternatives to GPS, such as ESA's NAVSAT, German GRANAS, and INMARSAT-based REXSTAR.

Dork, R. A., "Satellite navigation systems for land vehicles," *IEEE AES Mag.*, pp. 2–5, May 1987. Describes GM van with Loran, GPS, flux gate, odometer. Achieves 50–100 m RMS error.

Elfes, A., "Using occupancy grids for mobile robot perception and navigation," *IEEE Computer*, vol. 22, no. 6, pp. 46–57, June 1989. Describes algorithms to calculate probability that robot is in each two- or three-dimensional cell from sensor measurements. Two examples.

Euler, W. C. and M. A. Bittner, "Magnavox user equipment full scale engineering development program for GPS," *Navigation*, pp. 137–148, Jan 1984. Description of 1, 2, 5 channel set to level of LRUs. Concentrates on results of military tests. Acquisition time less than 5 min for first satellite.

Foley, T. M., "Space operations begin using Geostar payload," *Aviat. Week and Space Technology*, pp. 55–56, July 25, 1988. Describes the Geostar system, transmitting Loran fixes, and its truck users. Proposal to launch dedicated satellites with triangulation capability.

French, R. L., "Automobile navigation: Where is it going?", *IEEE AES Mag.*, pp. 6–12, May 1987. Reprinted in Part 3, pp. 101–107.

Fried, W. R., "Principles and simulation of JTIDS relative navigation," *IEEE Trans. Aerosp. Electron. Syst.*, vol. 14, no. 1, pp. 76–84, Jan. 1978. Good system description.

Frye, E. O., "GPS signal availability in land mobile applications," *Navigation*, vol. 36, no. 3, pp. 287–301, Fall 1989. Tested single-channel C/A code receiver. Railroad outage 2%; residential areas, 15%; cities, 72%.

Giles, L. J. and R. E. Simson, "Track-to-train radio," Proc. Communication 80 Conf. (U.K.), 1980. Specifications of UHF radio for ground to train communication. Example of system installed on 65 km of track in England.

Goldin, M., "Vehicle navigation and automated survey equipment," *J. Roy. Inst. Navigation,* Aug. 1985. Good description of U.S. Army land navigation programs.

Green, G. B., *et al.*, "The GPS 21 primary satellite constellation," *Navigation*, vol. 36, no. 1, pp. 9–24, Spring 1989. Describes current orbits and coverages without outages, with one outage, and with two outages.

Hadfield, M. J. and D. E. Leiser, "Field test results on the RLG modular azimuth positioning system," in *Rec. IEEE Position Location and Navigation Symp.,* 1988, pp. 61–72. Brief marketing description including environment. Test data in survey van and howitzer.

Held, V. and K. D. Kricke, "GPS satellite navigation in the urban environment," *J. Roy. Inst. Navigation,* Aug. 1985. Tests in Munich. Predicts 2 + satellites 30% of time in streets, 60% of time at intersections.

Heffron, W. G. and F. LaPiana, "The navigation system of the lunar roving vehicle," NASA CR-11 6268, Dec. 1970. Describes the mission, navigation system, and error analysis.

Hillman, H. D. and O. H. Cerbins, "The Washington metro automatic train supervision system," ASME Paper 73-ICT-79, Sept. 1973. Good description of on-board, trackside, and central controls.

Jeffrey, D. J., "Driver route guidance systems: State of the art," *Int'l Symposium on "Telematics—Transportation and Spatial Development*," Delft Univ. of Technology, April 1988. Describes city traffic system to optimize routes of individual cars. Compares on-board navigation, area broadcast of traffic via Radio Data System, and roadside transmitters. Advantages of two-way communications. Summarizes tests of CARIN, AUTOGUIDE, and LISB.

Jorgensen, P.S., "An assessment of ionospheric effects on the GPS user," *Navigation*, vol. 36, no. 2, pp. 195–204, Summer 1989. Describes a model of ionospheric delay for 1-frequency GPS receiver. The simplest model removes 60% of the propagation errors, resulting in vertical error of 5–15 meters RMS and a horizontal error usually less than 1 meter.

Kalafus, R. M., "GPS integrity channel RTCA working group recommendations," *Navigation*. 36:1 Spring 1989. pg. 25–44. Presents the results of an RTCA committee whose goal is a standard format for worldwide broadcast of GPS integrity, whose messages reach the pilot within 10 s of an outage.

Keightley, R. J., "Simulation of the Canadian GOES signal processor," *IEEE Trans. Aerosp. Electron. Syst.*, pp. 361–370, May 1987. Reviews SARSAT that will use transponder on geostationary GOES-East and GOES-West. Includes 406 MHz.

Kivett, J. A. and R. E. Cook, "Enhancing PLRS with user-to-user data capability," in *Rec. IEEE Position Location and Navigation Symp.*, 1986, pp. 154–161. Reprinted in Part 3, pp. 82–89.

Kivett, J. A. and F. L. Morse, "PLRS engineering development testing—Early results," in *Rec. IEEE Position Location and Navigation Symp.*, 1980, pp. 330–334. How the system works and its then current status.

Krakiwsky, E. J., *et al.*, "Kalman filter for integrating dead reckoning, map matching, and GPS positioning," in *Rec. IEEE Position Location and Navigation Symp.,* 1988, pp. 39-46. Filter either updates magnetic-compass-odometer by GPS or updates at one of 11 preplanned locations in Calgary. Predicts performance with GPS-fix and map-fix.

Kriegsman, B. A. and K. B. Mahar, "Gravity model errors in mobile inertial navigation systems," *AIAA J. Guidance,* vol. 9, May–June 1986. Power-spectral density of gravity along and cross track. Zero-velocity update. Goal is 100 ft/axis.

Lennen, G. R. and P. Daly, "A Navstar GPS C/A code digital receiver," *Navigation*, vol. 36, no. 1, pp. 115–126, Spring 1989. Description of one-channel C/A code receiver by an experienced team at Leeds University. Code and carrier-tracking loops are digital. Test data show range noise less than 15 m.

Lenorovitz, J. M., "ESA offers updated Navsat concept," *Aviat. Week and Space Technology,* pp. 54–55, Jan. 25, 1988. ESA proposal for ground stations to monitor GPS and GLONASS and report when down.

McMillan, J. C., "An integrated land navigation system," in *Rec. IEEE Position Location and Navigation Symp.,* 1988, pp. 221–229. Reprinted in Part 3, pp. 73–81.

Mueller, C. E. and G. Adams, "Laser gyro land navigation system performance predictions and field results," in *Rec. IEEE Position Location and Navigation Symp.,* 1984, pp. 81–90. Decribes Honeywell system.

Murphy, E. E., "All aboard for solid state," *IEEE Spectrum,* pp. 42–45, Dec. 1988. Describes automatic train control and navigation initiatives in North America. Brief mention of dead-reckoning, electropost, and satellite navigation.

Okawa, U. S. and R. Fincher, "PLRS development testing—An update," in *Rec. IEEE Position Location and Navigation Symp.,* 1986, pp. 147–153. How PLRS works; 10- to 50-m results.

Packard, W., "Modular azimuth position system," *J. Royal Institute of Navigation*, Aug. 1985. Survey $\frac{1}{4}$% position, 0.5 mrad elevation for portable howitzer.

Pattan, B., "The advent of land mobile satellite service systems," *IEEE Trans. Aerosp. Electron. Syst.,* vol. 23, no. 5, pp. 691–703, Sept. 1987. Mobile vehicles direct to satellite at L-band.

Petrov, D., *et al.,* "Feasibility study of VLF radio compass for Arctic navigation," *Navigation,* pp. 338–360, Winter 1984–85. Experimental azimuth measurement from VLF stations.

Piggott S., "The beginnings of wheeled transport," *Sci. Amer.*, vol. 240, pp. 82–90, July 1968. Two- and four-wheel carts.

Rickords T. and J. Yamamoto, "A strap-down-configured land navigation system," *J. Royal Institute of Navigation*, vol. 240, pp. 82–90, Aug. 1985. Describes Litton LLN-80.

Teasley, S. P., *et al.,* "Differential GPS navigation," in *Rec. IEEE Position Location and Navigation Symp.,* 1980, pp. 9–16. Reprinted in Part 3, pp. 131–138.

Tsumura, T., *et al.,* "Vehicle position and heading measurement system using corner cube and laser beam," in *Rec. IEEE Position Location and Navigation Symp.,* 1988, pp. 47–53. Experimental transceivers on vehicle scan corner cubes on roadway. Calculate heading and position.

van Beek, G. W., "The rise and fall of Arabia felix," *Sci. Amer.,* vol. 241, pp. 36–46, Dec. 1969. Camel domestication and caravans.

Viets, K. J., "Independent ground monitor coverage of GPS satellites," *IEEE Trans. Aerosp. Electron. Syst.,* vol. 23, no. 5, pp. 678–685, Sept. 1987. Proposal of monitor stations for precision landing with differential GPS.

Ward, P., "Advanced NAVSTAR GPS multiplex receiver," in *Rec. IEEE Position Location and Navigation Symp.,* 1980, pp. 51–58. Receiver multiplexes L1 and L2 and shares multiple satellites. Microprocessor-controlled tuning. Lots of receiver design details.

Weisbin, C. R., *et al.,* "Autonomous mobile robot navigation and learning," *IEEE Computer,* vol. 22, pp. 29–35, June 1989. Laboratory tests of robot for hostile environment. On-board hypercube stores data about laboratory geometry. Future growth to optimum routing and formation of a map of an unknown environment.

Whitcomb, L. A., "Using low-cost magnetic sensors on magnetically hostile land vehicles," in *Rec. IEEE Position Location and Navigation Symp.,* 1988, pp. 34–38. Changes in Earth's field. Stabilized vs. strap-down magnetometers. Compensation.

Widnall, W. S. and J. F. Kelley, "JTIDS relative navigation with measurement sharing: Design and performance," *IEEE Trans. Aerosp. Electron. Syst.,* vol. 22, no. 2, pp. 146–154, Mar. 1986. Proposes sharing raw measurements among JTIDS terminals. Assures stability.

AN INTEGRATED LAND NAVIGATION SYSTEM

J.C. McMillan

DEFENCE RESEARCH ESTABLISHMENT OTTAWA
OTTAWA, ONTARIO, CANADA K1A 0Z4

ABSTRACT

Conditions for land navigation are among the most severe in the Arctic (and Antarctic), where until GPS becomes fully operational, there will be no single system capable of continuously providing the necessary position and heading accuracy. Even when GPS is available, reliability considerations will dictate that a self contained or autonomous backup be included, certainly for the military user. The Defence Research Establishment Ottawa (DREO) has therefore developed a multi-sensor optimally integrated navigation system to satisfy the present operational requirements of the Canadian land forces. This system, called PLANS (Primary Land Arctic Navigation System) was designed and built to be a highly reliable, moderately accurate and moderately priced, nonradiating, automatic navigation system for all-weather off-the-road use. It combines several self contained sensors with two satellite receivers, using an 8 state Kalman filter on a microcomputer, to provide a continuous optimal estimate of position, heading and speed. The self contained sensors consist of a strapdown gyro unit, an odometer pickoff for speed, a magnetic fluxgate sensor and a baroaltimeter. The satellite receivers are a C/A code GPS and dual channel Transit. This paper describes the requirement, the difficulties in meeting it, the design of a multi-sensor integrated system solution, and some test results.

PROBLEM STATEMENT

Perhaps the most stringent requirement for land navigation is imposed by winter operations in the high Arctic. The serious consequences of being lost or failing to find a supply dump in the Arctic make reliability and accuracy vital necessities. Because of weather related poor visibility conditions, it may also be necessary to navigate by instrument very close to the destination in order to actually find it. Most of the Arctic appears completely barren and featureless, rendering the normal map reading skills completely inadequate. Frequent extended periods of low visibility also interferes with visual navigation, and make celestial navigation unreliable.

In many ways Arctic navigation is similar to desert navigation, except that there are several added complications. Proximity to the north geomagnetic pole makes the use of a hand-held magnetic compass futile over much of this area. As shown in Reference [1] random fluctuations in the direction of the field vector will induce heading errors of magnitude inversely proportional to the horizontal field strength. Figure 1 shows how this magnetic heading error magnitude varies with latitude along a selected longitude line (-100° which virtually passes over the magnetic pole). This represents the error of a stationary magnetic sensor, assuming a constant level of field fluctuation, and that the local mean field is known, that the disturbing field from the vehicle is known, and that the accuracy at the equator is 1°.

Of course the usual radio navigation aids do not extend coverage through the Arctic, with the possible exception of Omega [2], which is not sufficiently accurate. The Transit satellite system does provide some position fixing capability, if velocity and height can be continuously provided. Since the Transit satellites are in polar orbits, fixes can be obtained more frequently at higher latitudes, with more than one fix per hour expected above 60°. These are higher elevation fixes however, and their accuracy is degraded, particularly the longitude. The accuracy of a Transit position fix depends upon the accuracy of the velocity provided, but can be quite adequate, especially with a 2 channel receiver (to remove the ionospheric effect) in stationary mode. GPS position and velocity is quite adequate when it is available, even with a C/A code receiver. Until it is fully operational however, the gaps in coverage prevent total reliance on GPS.

This therefore still leaves the requirement to dead reckon (DR) between Transit fixes or through GPS gaps, so that the velocity determination cannot be avoided. Now the speed is easily estimated by an odometer, but the heading presents some difficulty. The first alternative to magnetic direction finding is gyrocompassing which becomes more difficult at higher latitudes λ, where the horizontal earth rate ($\omega\cos\lambda$) decreases quickly as one moves north. Thus the gyrocompassing accuracy of a given sensor varies as the secant of the latitude, as shown in Figure 1. In practice of course for each gyro there is a latitude above which it may not settle at all. For the moderately priced type of unit suitable for land vehicles this is typically given as about 80°. Fortunately, the present location of the geomagnetic pole is such that the land area where

Reprinted from *Rec. IEEE Position Location and Navigation Symp.*, pp. 221-229, 1988.

magnetic heading is not available (using a magnetometer) is not coincident with the land area where gyrocompassing is not reliable, as shown in Figure 2. This figure shows a contour of constant horizontal magnetic field strength, and a constant latitude contour. Of course neither of these regions are sharply delineated, but rather each heading measurement becomes increasingly less accurate moving towards the magnetic or geographic pole. The point at which each measurement becomes useless depends on the particular sensors chosen, how they are used, and the magnetic conditions at the time. The important point to note is that in the extreme northwest, which is the most difficult area, the two methods complement each other.

Figure 1. Magnetic and Gyroscopic Heading Error

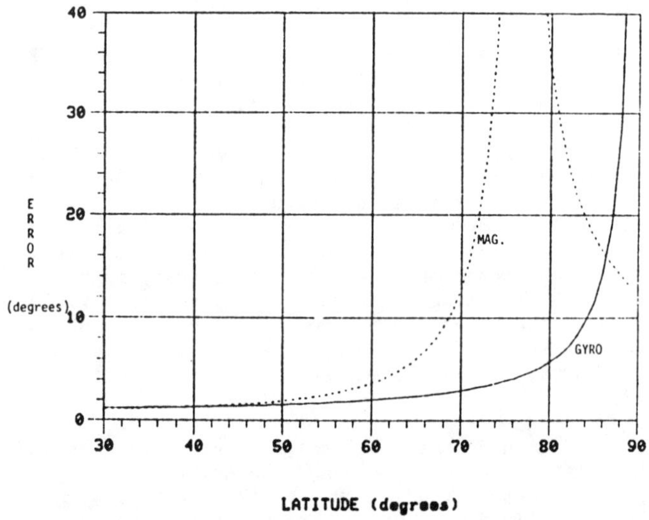

Figure 2. Gyro & Magnetic Heading Problem Areas

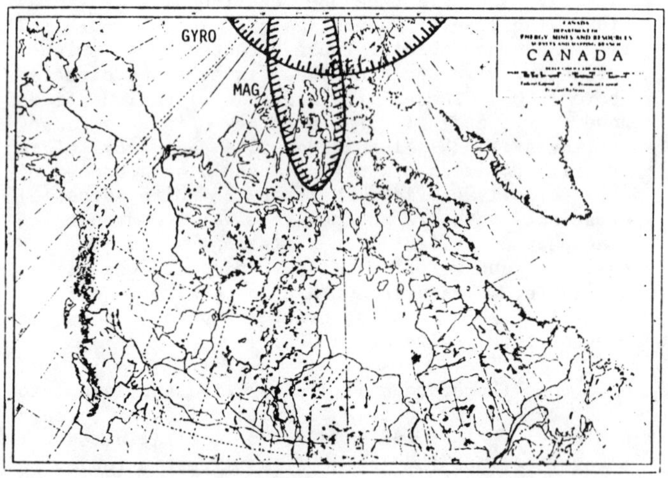

SYSTEM SOLUTION

With the problem as stated above, the obvious solution is to combine one or both of the satellite receivers with a magnetic sensor, a gyrocompass/ directional gyro unit and an odometer pickoff, in an optimally integrated system, using a Kalman filter to properly exploit the strengths of each sensor. After investigating some non-conventional alternatives, such as VLF direction finding and short baseline interferometry using Transit to obtain heading information, it was decided that an integrated system with off-the-shelf components was the preferred solution. Since no such system was available, it was necessary to develop one. Previous experience in the design and development of a Kalman filter based integrated navigation system [3] was influential in this decision, and in the design approach. The major components chosen for PLANS (Primary Land Arctic Navigation System) are shown in Figure 3, in block diagram form.

Figure 3. PLANS Block Diagram

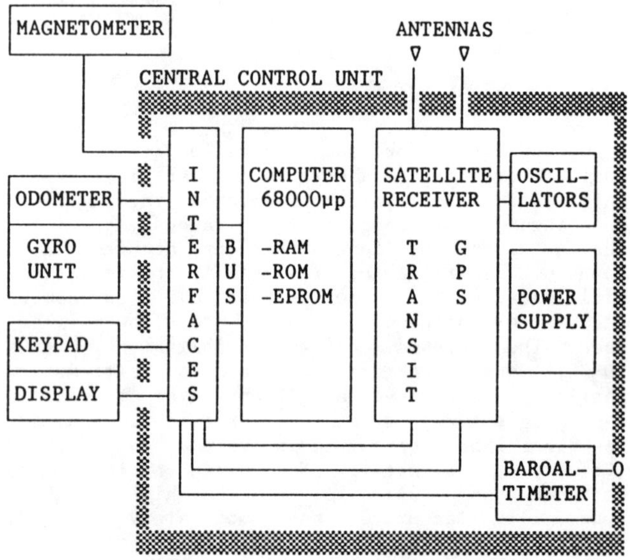

This design is actually a transitional configuration, since PLANS was first built without the GPS receiver, and when GPS is more fully operational the Transit receiver can be eliminated. Since the Transit receiver requires height input, a baroaltimeter has been included, which is augmented with a small terrain elevation database. These are appropriately blended in the Kalman filter. The strapdown gyro unit provides stationary gyrocompassing and thereafter is a directional gyro, providing heading, which can be combined with speed obtained from the vehicle odometer to form the velocity vector. A dead reckoning solution can then be obtained by integrating this velocity. The core of the integrated system is this dead reckoning system. All other sensors can be thought of as aiding sensors.

The magnetometer provides additional heading information, to be combined with the gyro heading through the prefilter and the Kalman filter, as described below. The magnetic measurement of course is compensated to account for the difference between true north and the local magnetic field direction. This requires a geomagnetic field model and a vehicle field model. The geomagnetic model used is the IGRF85 model (International Geomagnetic Reference Field, 1985) [4]. The vehicle field model is a second order

Fourier series representing the vehicles permanent field ("hard iron") and induced field ("soft iron") and is found by using the gyro to calibrate the installed magnetometer while driving the vehicle over the complete range of headings [1].

The gyro unit and magnetometer are not the only sources of heading information available. Of course GPS velocity can be used when available and when the vehicle is moving. It is also possible to obtain heading information from a Transit position fix, in an indirect way, using the known sensitivity of the Transit fix to the error of the velocity fed to the receiver. The Kalman filter, using a detailed model of this relationship, can update its estimate of the velocity error (including heading error) when it processes the Transit position fix. In simulation this technique is quite successful, however it does rely on the Transit receiver providing correct satellite elevation angle and direction of travel information.

SENSOR SELECTION

The gyro unit used in the proof of concept study was a strapdown unit having one tuned rotor gyro (2 axes), one single axis rate gyro and 2 accelerometers. This unit can gyrocompass in about 5 minutes (longer at higher latitudes) if the vehicle is stationary and pointing within about 30° of north. This provides an initial heading with accuracy on the order of one degree secant latitude (1σ), after which the gyro automatically switches to directional gyro mode and the vehicle can be moved.

For the PLANS proof of concept development model, a two axis gimballed magnetic sensor was used, which directly measures the horizontal components of the earth's magnetic field. The manufacturer claims that it will provide magnetic heading accurate to 2° for magnetic dip angle from 0° to 80° which excludes a large portion of northern Canada. Because of this and a potential low temperature problem, a more suitable 3-axis strapdown magnetic sensor unit has been obtained for the next phase of development.

The baroaltimeter sensor chosen was a High Output Pressure Transducer with gage pressure range rated at 0 to 25 psi (0 to 172 kP). The range of interest is roughly 11 to 16 psi corresponding to a barometric height range from about -700 metres to 2,500 metres.

The satellite receiver chosen for the PLANS development model was a commercial unit with a 2-channel Transit receiver and a single channel C/A-code GPS receiver (intended for the marine environment) which provides convenient serial output. The keypad, display and case were discarded, and the necessary electronics (including battery backup) was repackaged with a 68000 based computer system, the interface cards, power supplies and the baroaltimeter. In this way the satellite receiver(s) can be completely controlled by the PLANS processor, using a keypad emulator.

ERROR MODELLING

In order to properly integrate these different sensors using a Kalman filter, all significant stochastic and deterministic errors must be carefully modelled. Deterministic errors, such as magnetic declination, are modelled so that their effects can be directly removed as much as possible, and the remaining errors are modelled as random processes to enable the Kalman filter to estimate them and to calculate the appropriate gain matrix to weight the various measurements. An overview of the modelling process for PLANS is given in [5]. It should be noted that to a large extent this error modelling is generic, with the form of the models depending only on the type of sensors being used. The actual sensor models chosen will of course determine the numerical values of the parameters, enabling the PLANS software to be easily adapted to different qualities of sensors by changing these few numbers. This flexibility allows the cost/performance tradeoff and final sensor selection to be made late in the development program, a necessary measure since PLANS system development began before sensors with the desired qualities were available.

The types of stochastic models used in the PLANS Kalman filter are limited for practical reasons to first order Markov processes and uncorrelated white noise [6].

KALMAN FILTER

It is primarily the errors of the dead reckoning system that the Kalman filter estimates. That is, the north and east position error (in metres), the gyro heading error (in radians), the vehicle odometer scale factor error (dimensionless) and the baroaltimeter height error (metres). In directional gyro mode, the gyro drift rate is also modelled. The Kalman filter also estimates the correlated errors of the aiding sensors, such as the magnetic flux valve heading error (radians) and the error of the digital map height (metres).

The filter estimates these errors by processing all of the available measurements using suitable stochastic and deterministic error models. Reference [6] describes the standard stochastic model types and provides an elementary introduction to Kalman filtering techniques.

State Vector

The PLANS filter employs an eight dimensional state space, defined as follows:

$$\begin{vmatrix} X1 \\ X2 \\ X3 \\ X4 \\ X5 \\ X6 \\ X7 \\ X8 \end{vmatrix} = \begin{vmatrix} \text{true height above geoid - map height} \\ \text{true height - baroaltimeter height} \\ \text{true heading - magnetic heading} \\ \text{gyro drift rate in directional gyro mode} \\ \text{true heading - gyro heading} \\ \text{(true speed - odometer speed)/odom. spd.} \\ (\text{ true } \lambda - \text{ DR } \lambda \text{) in metres} \\ \text{(true L - DR L) in metres} \end{vmatrix} \qquad (1)$$

It should be clarified that these states

represent the Markov, or time correlated portions of the errors. The uncorrelated component of the sensor errors are treated as measurement noise. Also there are actually two state vectors involved, because state X4 is deleted when in gyrocompass mode.

Transition Matrix

DR position error is simply the integral of the DR velocity error. This fact, together with the assumed stochastic error models for the DR velocity errors and the remaining state vector components, is used to propagate the state vector in time, according to the vector differential equation:

$$\dot{X}(t) = F(t)X(t) + W(t) \qquad (2)$$

This is used to propagate X between measurements. To implement this on a digital computer, equation (2) was converted to a discrete difference equation and linearized [1], to obtain the state transition matrix Φ, where:

$$X(t+\Delta t) = \Phi(\Delta t, t)X(t) + W(\Delta t) \qquad (3)$$

The result is

$$\begin{bmatrix} X1 \\ X2 \\ X3 \\ X4 \\ X5 \\ X6 \\ X7 \\ X8 \end{bmatrix}_{t+\Delta t} = \begin{bmatrix} e^{-\Delta t/\tau 1} & & & & & & & \\ & e^{-\Delta t/\tau 2} & & & & & & \\ & & e^{-\Delta t/\tau 3} & & & & & \\ & & & \Box & & & & \\ & & & & & & & \\ & & & & & e^{-\Delta t/\tau 6} & & \\ & & & & \Phi 7,5 & \Phi 7,6 & 1 & 0 \\ & & & & \Phi 8,5 & \Phi 8,6 & 0 & 1 \end{bmatrix} \begin{bmatrix} X1 \\ X2 \\ X3 \\ X4 \\ X5 \\ X6 \\ X7 \\ X8 \end{bmatrix}_{t} + \begin{bmatrix} W1 \\ W2 \\ W3 \\ W4 \\ W5 \\ W6 \\ 0 \\ 0 \end{bmatrix} \qquad (4)$$

where:

$$\begin{aligned} \Phi 7,5 &= -\Delta t\ S \sin(\theta + X5/2) \\ \Phi 7,6 &= \Delta t\ S \cos(\theta + X5) \\ \Phi 8,5 &= \Delta t\ S \cos(\theta + X5/2) \\ \Phi 8,6 &= \Delta t\ S \sin(\theta + X5) \end{aligned} \qquad (5)$$

where Δt is the propagation time interval, S is the dead reckoning speed, θ is the dead reckoning heading, and the submatrix

$$\Box = \begin{cases} \begin{bmatrix} e^{-\Delta t/\tau 4} & 0 \\ \tau 4(1-e^{-\Delta t/\tau 4}) & 1 \end{bmatrix} & \text{in directional gyro mode} \\[2ex] \begin{bmatrix} 0 & 0 \\ 0 & e^{-\Delta t/\tau 5} \end{bmatrix} & \text{in gyrocompass mode} \end{cases} \qquad (6)$$

where:

$$\begin{aligned} \tau 1 &= \text{map height error correlation time} \\ \tau 2 &= \text{baroaltimeter height error correlation time} \\ \tau 3 &= \text{magnetic declination error correlation time} \\ \tau 4 &= \text{gyro drift rate correlation time} \\ \tau 5 &= \text{gyro Markov error correlation time} \\ \tau 6 &= \text{odometer speed factor error correlation time} \end{aligned} \qquad (7)$$

and

$$Wi = \text{driving white noise with covariance } Q_i \text{ (for i = 1 to 6)} \qquad (8)$$

where (see [6]):

$$Qi = Mi(1 - e^{-2\Delta t/\tau i}) \qquad (9)$$

where the Mi are the Markov process steady state rms values (covariances).

Notice from equations (4) and (5) that the state transition equation has not been completely linearized in X, and consequently the transition matrix Φ still exhibits a weak dependence on the state vector X. Therefore an extended Kalman filter is used, rather than a simple linear one. This is necessary because of the potentially large dead reckoning heading error, X5, that could occur in directional gyro mode, and the fact that the DR position error, (X7,X8), varies nonlinearly with this heading error. In fact the linearization is only necessary to propagate the covariance matrix. The state vector components, and in particular X7 and X8, are propagated in a nonlinear manner, and at a higher rate than the covariance.

This nonlinearity in heading and potentially large heading error also led to a closed loop filter design whereby the filtered estimate of velocity and height are fed into the Transit receiver. Periodic resetting of the gyro heading with the filter heading keeps the gyro error X5 bounded. This is important primarily for the purpose of using DR as a backup under certain failure conditions.

Measurement Vector

The odometer "speed" and gyro heading are measured at a 1 Hz rate for dead reckoning, and the digital map is read once a minute to obtain the height. The filter measures the baroaltimeter and the magnetic flux valve once a minute to update its estimate of the state vector X (especially X1, X2 and X3). Transit position fixes are processed whenever they occur to update the filter estimate of X (especially X7 and X8 but also X5 and X6). Whenever the vehicle is in directional gyro mode and is not moving, the gyro drift rate is measured from successive heading measurements, to allow the filter to update the estimate of X4. The GPS position, velocity and height are measured at a 30 second rate, when they are available. The Kalman filter's measurement vector is therefore defined as:

$$
\begin{bmatrix} Z1 \\ Z2 \\ Z3 \\ Z4 \\ Z5 \\ Z6 \\ Z7 \\ Z8 \\ Z9 \\ Z10 \end{bmatrix} = \begin{bmatrix}
\text{map height} & - & \text{barometric height} \\
\text{map height} & - & \text{GPS height} \\
\text{magnetic heading} & - & \text{gyro heading} \\
(\text{gyro heading}(t+\Delta t) & - & \text{gyro heading}(t))/\Delta t \\
\text{Transit latitude} & - & \text{DR latitude} \\
\text{Transit longitude} & - & \text{DR longitude} \\
\text{GPS latitude} & - & \text{DR latitude} \\
\text{GPS longitude} & - & \text{DR longitude} \\
\text{GPS north velocity} & - & \text{DR north velocity} \\
\text{GPS east velocity} & - & \text{DR east velocity}
\end{bmatrix} \tag{10}
$$

As was mentioned, in order to be at all useful at high latitudes, the raw magnetic flux valve measurement must be adjusted by geomagnetic and vehicle field models. The IGRF85 geomagnetic field model [4] predicts all three components of the earth's magnetic field, which allows not only the magnetic declination to be calculated, but also the horizontal field strength and the dip angle. These can be used to give a reasonable indication of the accuracy that can be expected from the magnetic heading measurement.

Measurement Matrix

The relationship between the state vector X and the measurement vector Z must be described in detail for the Kalman filter to properly process the measurements. This description is expressed as a measurement equation:

$$
Z = H X + V \tag{11}
$$

where H is the "measurement matrix" and V is the measurement noise vector. For PLANS the measurement equation is:

$$
Z = \begin{bmatrix}
-1 & 1 & & & & & & \\
-1 & 0 & & & & & & \\
 & & -1 & 0 & 1 & & & \\
 & & 0 & -1 & 0 & & & \\
H51 & 0 & 0 & 0 & H55 & H56 & 1 & 0 \\
H61 & 0 & 0 & 0 & H65 & H66 & 0 & 1 \\
 & & & & & & 1 & 0 \\
 & & & & & & 0 & 1 \\
 & & & & -S\sin\theta & S\cos\theta & & \\
 & & & & S\cos\theta & S\sin\theta & &
\end{bmatrix}
\begin{bmatrix} X1 \\ X2 \\ X3 \\ X4 \\ X5 \\ X6 \\ X7 \\ X8 \end{bmatrix}
+ \begin{bmatrix} V1 \\ V2 \\ V3 \\ V4 \\ V5 \\ V6 \\ V7 \\ V8 \\ V9 \\ V10 \end{bmatrix} \tag{12}
$$

where:

$$
\begin{aligned}
S &= \text{vehicle speed} \\
\theta &= \text{vehicle heading} \\
H51 &= \cos\Psi \tan\sigma \\
H61 &= \sin\Psi \tan\sigma \\
\Psi &= \text{bearing to subpoint} \\
\sigma &= \text{maximum elevation angle}
\end{aligned} \tag{13}
$$

The remaining H components (H55, H56, H65 and H66), which define the sensitivity of the Transit fix to errors in velocity, are described briefly in [1].

The measurement noise vector V is defined as the non-Markov component of the measurement vector Z, given by equation (10). These are assumed to be uncorrelated white noise processes. The covariance matrix R of this noise vector is therefore diagonal of rank 10.

$$
R = E\{VV^T\} = \text{diag}[R1\ R2\ R3\ R4\ R5\ R6\ R7\ R8\ R9\ R10] \tag{14}
$$

Expressions for these components are given in [5].

FILTER IMPLEMENTATION

The filter mechanisation used is Biermans UDU^T [7], which offers both excellent numerical stability (preserving a positive definite symmetric covariance) and reasonably good computational efficiency. Since the state vector is quite small no special measures are required to reduce the processing burden.

Special measures did have to be taken to handle certain practical problems that are expected to arise. One such difficulty is that it may be necessary to start the system in directional gyro mode without an adequate estimate of the initial heading. Because of this possibility it was necessary to implement a "heading adjustment" parameter, to be added to the directional gyro measurements before they are passed to the navigation filter. The value of this parameter will be initially determined by using the magnetic measurement. This value will then be refined at an hourly rate by the Kalman filter itself. This is necessary whenever directional gyro mode is used for any substantial length of time (10 hours or so) because the growing directional gyro heading error will introduce significant nonlinearities into the filter equations, whereas the linearized state transition equation (4) carries the implicit assumption that the gyro heading error X5 is small.

PREFILTER

To be as accurate and reliable as possible, any Kalman filter based integrated system should have a good prefilter to apply any known corrections to the measurements, to remove any spurious data and to detect sensor or subsystem failures.

To maximise the DR accuracy, the gyro must be carefully aligned to the vehicle's direction of motion, and the odometer scale factor must be carefully calibrated. The results of this alignment and calibration are used by the prefilter to offset the gyro measurement and scale the odometer measurements appropriately.

The magnetometer measurement would be useless without the geomagnetic field model, which allows the prefilter to convert from magnetic to geographic or true heading. The prefilter must also apply the magnetic calibration function for the vehicles permanent and induced fields, which is a function of the true and magnetic headings. Even this corrected and calibrated magnetic heading is very noisy. By averaging N magnetic measurements in the prefilter, the uncorrelated portion of this noise can be reduced by $1/\sqrt{N}$. Of course this can only be done over a short time interval over which the correlated error will not change appreciably, and it is the (gyro - magnetic) heading that is averaged, to remove the effect of heading change.

The prefilter also uses various methods to

detect spurious measurements and hard sensor failures, using known physical constraints on the vehicles speed, acceleration, turn rate etc. All measurements are first tested in this simple way, and if they pass this, they then undergo a residual test. The residual test uses the Kalman filter's estimate of what the measurement error should be, based on all previous measurements, along with the filters covariance (degree of uncertainty in the estimate) based on the stochastic error models. This residual test is the most sensitive test, and the last line of defence before the measurement is processed by the filter. These two levels of failure detection significantly improve the accuracy and reliability of the system.

SIMULATIONS

Throughout the PLANS development activity extensive simulations were performed for three important reasons:

1/ system design testing and evaluation
2/ algorithm and software testing and evaluation
3/ system performance prediction

One primary aim of the system design testing was to predict whether or not the Kalman filtering of dynamic Transit position fixes could correct the heading error in the region where both magnetic and gyroscopic errors are large.

Simulations were performed using the full PLANS software package, along with specially created measurement generation routines. Monte-Carlo and Covariance Analysis techniques were employed to predict the PLANS system performance. A very brief summary of some of the basic simulation results is presented here in the form of two Monte-Carlo sets. These were run without GPS since the system will be essentially slaved to GPS when it is available, and before the gyro resets had been implemented. The first set was simulated starting at 65° N 100° W and the second at 75° N 100° W, which is very close to the geomagnetic pole. The first set was run with a constant speed of 10 metres per second, and the second with a periodically varying speed between 0 and 10 metres per second. This was to show the effect of the gyro drift rate measurements (only available in stationary mode), and stationary Transit fixes. In both sets the simulated heading was 45° (northeast), and the duration was 24 hours. The second simulation therefore traverses the area roughly between the magnetic pole and the northern limit of land, providing a worst case scenario.

Each Monte-Carlo set consisted of 20 runs. In each run the deterministic errors were the same and the random errors had the same statistical properties but different values. The results of the 20 runs were combined to produce one representational run. Figures 4 and 5 illustrate the results of the first set, and the numerical results below summarise the basic accuracy results. The solid curve in Figure 4 shows the radial position error and the dashed curve shows the Kalman filter's estimate of its 95% radial position accuracy from the covariance matrix (2σ). Figure 5 shows the PLANS heading error and it's filter estimated covariance (1σ). The overall accuracy results tabulated above are found by taking the time rms of these curves over the 24 hours.

Table 1. SIMULATED ERRORS.

SYSTEM	RMS ERROR POSITION	HEADING
PLANS	282 metres	.67 deg.
Transit	303 metres	NA
DR	21,426 metres	4.5 deg.
Magnetic	NA	2.1 deg.

Figure 4. Simulated Position Error and Covariance.

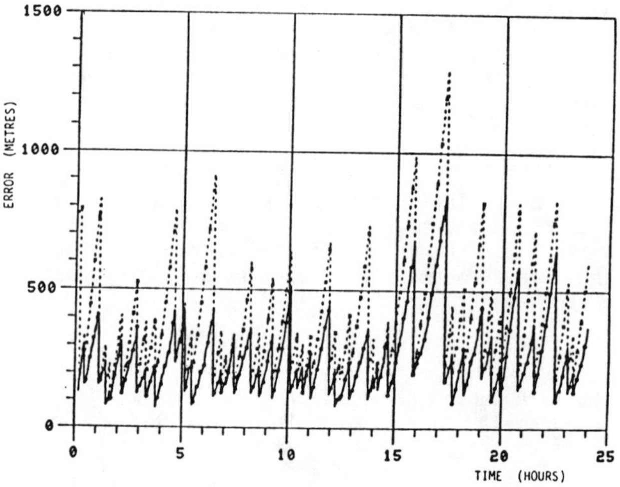

Figure 5. Simulated Heading Error and Covariance.

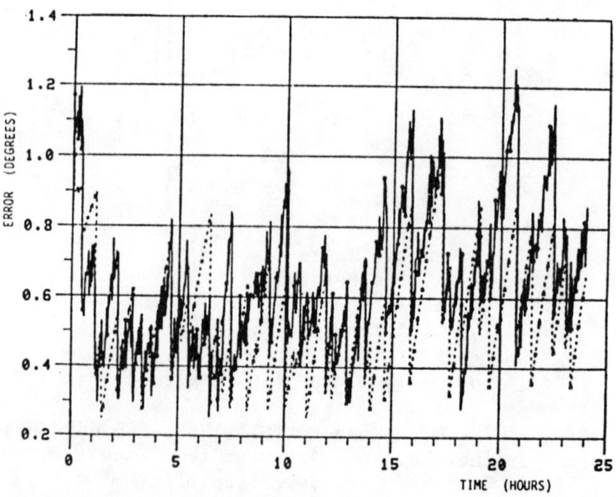

In directional gyro mode, the heading error is expected to increase without bound resulting in an increasingly rapid growth in the DR position error and, more importantly, a potential for unbounded growth in the Transit position error (if the Kalman filter can not adequately estimate and remove the gyro error). Here it can be seen that the filter performed well, keeping the PLANS heading error, and hence the position error, well bounded.

The basic results of the second Monte-Carlo set, with the periodic velocity profile at the higher latitude, northeast from the geomagnetic pole, are summarised in Table 2.

Table 2. SIMULATED HIGH LATITUDE ERRORS.

SYSTEM	RMS ERROR POSITION	HEADING
PLANS	350 metres	.96 deg.
Transit	326 metres	NA
Gyro/Odometer	10,640 metres	6.4 deg.
Magnetic	NA	10.7 deg.

This, along with other simulations, verifies the potential effectiveness of the Transit position fixes as a source of heading information, and demonstrates proper blending of the gyro and magnetic heading.

Sensitivity analysis simulations were also performed, by running various Monte-Carlo sets, each with a significantly increased value being assigned for a major error source. As would be expected the position error was found to be most sensitive to the gyro error.

FIELD TRIALS

Three types of PLANS trials have been conducted to date: local trials conducted on smooth level roads at moderate speeds, mid-latitude trial on a military base over a variety of terrain at maximum speed and an Arctic trial under winter conditions. Naturally the results are very different. The platform for all of these trials was an M577 armoured tracked vehicle.

Local Trials

The local trials were conducted periodically from early 1985 to late 1986 on a course of about 8 kilometres. The primary purpose of these trials was to determine the error model parameters for the various sensors and subsystems, to verify the structure of these models, to verify the simulation results, and gain confidence in the error models. Another goal of these trials was to evaluate different brands of each type of sensor for final sensor selection. By the fall of 1986 these objectives had all been met. The position error never exceeded 100 metres, and typical RMS position error over the course was about 15 metres (due largely to the shortness of the course, and the hard smooth road).

Mid-Latitude Field Trial

The first formal field trial was conducted on a course about 55 kilometres in length, over a one week period in September 1986. The purpose of this trial was to evaluate the system performance and provide a significant database of raw measurements to continue to enhance and refine the system software. Recording all of the raw measurements as well as the reference data allows the trial to be "rerun" back at the lab, providing a powerful tool for further system development and detailed sensor error analysis.

This field trial was designed to generate the largest position error that could be expected for the location (about 46° N, 77° W). This was accomplished by driving at full speed for over half an hour (the expected time between Transit fixes) in a direction generally away from the starting point, and returning along a different path, again at full speed. Each run of one complete circuit of the 55 km course took about 100 minutes.

The reference position information was obtained by placing 27 retro-reflectors at surveyed positions around the course, at approximately 2 km intervals. An infrared emitter/detector was mounted on the vehicle so that as the vehicle drove past each surveyed point a reflection would be detected and recorded. The surveyed position could then be compared to the PLANS position.

The real time PLANS position and heading accuracy results are summarised in Table 3. Although there were 27 reflectors on the course, seldom was a good reflection received from every one, so Table 3 gives the number of data points (good reflections) for each run. The position error growth can be seen in Figure 6 where each reference point equates to about 2 km. Thus the position error was just under 1% of distance travelled, to a maximum of about 300 metres.

The heading errors were found indirectly by comparing the true bearings between surveyed waypoints to the bearings calculating between PLANS position estimates at these waypoints. Since PLANS essentially dead reckons, a constant heading error will simply rotate these bearings.

Table 3: FIELD TRIAL RESULTS.

RUN #	NUMBER OF DATA POINTS	NUMBER OF TRANSIT FIXES	POSITION ERROR RMS (metres)	HEADING ERROR RMS (degrees)
1	27	0	101	.8
2	17	0	339	2.2
3	25	1	135	.7
4	27	1	43	.4
5	23	0	186	1.2
6	16	1	107	.5
7	19	0	66	.5
8	25	1	251	2.4
9	13	0	324	1.9
10	17	1	315	2.1
11	15	0	146	1.3
Total:	224	5	198	1.4

Figure 6. Field Trial Results

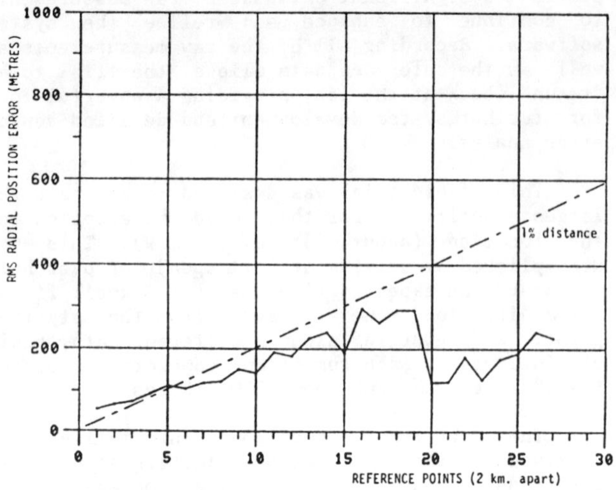

One lesson learned from the simulations and field trials is the importance of the initial gyrocompassing accuracy of the gyro, which effects both the installation alignment and the runtime initialisation. Therefore before the Arctic trial, steps were taken to improve this initial heading accuracy by cycling through the initial gyrocompass phase several times and using the RMS result. In this way N alignments will reduce the uncorrelated component of the error by a factor of $1/\sqrt{N}$. Four cycles would then cut this error in half. At higher latitudes however, this will be offset by the expected increase in gyrocompassing error. For example at 70° latitude this error will be greater than it is at 45° latitude by a factor of:

$$\frac{\cos(45°)}{\cos(70°)} = 2.07$$

Therefore the RMS heading accuracy from Table 3 should also be representative of what can be achieved with PLANS at high latitude, if the multiple gyrocompassing is effective in reducing the initial heading error.

Arctic Trial

An Arctic trial was conducted in February 1987, to test the PLANS equipment under severe weather conditions, at high latitude, over icy terrain. The location selected was Iqaluit, at a latitude of about 64° N. Weather conditions at the time were somewhat worse than normal, with 60 km/hr winds, an ambient temperature of about -40°C and occasional whiteouts. Air transport scheduling for the host vehicle limited the test to three days, one of which was required for preparation. This shortness of time, together with the snow cover, limited the reference data to 4 surveyed points within the town of Iqaluit and one near a radio tower about 5 km away.

The course for this trial consisted of a 3.7 km closed circuit, around the 4 survey points, and a 10.3 km return trip to the outlying radio tower. The surface was mostly hard packed snow on gravel or asphalt, with some drifting. The results from one long run (# 5) and 5 sets of short circuits are summarised in Table 4 and Figure 7.

Table 4. ARCTIC TRIAL RESULTS.

RUN #	NUMBER OF DATA POINTS	NUMBER OF TRANSIT FIXES	DURATION OF RUN (min.)	RMS POSITION ERRORS (metres) WITH TRANSIT	RMS POSITION ERRORS (metres) WITHOUT TRANSIT
1	21	0	143	45	45
2	13	0	83	37	37
3	5	1	52	62	56
4	17	2	132	82	47
5	3	0	65	85	85
6	3	1	70	177	18
Total	62	4	545	71	47

Due to an antenna failure, no GPS data was received on this trial. This led to larger than normal vertical errors, which corrupted the Transit position fixes. In fact for these short runs, given an accurate initial position, the system performed better without Transit than with it, as shown in Table 4. These results without Transit were obtained by rerunning the data through the filter software without the Transit measurements. The small long term error growth seen in Figure 7 (< .5% distance) is due to the cancellation of heading error effects on the return path. The short term growth is a more realistic inter-fix error growth.

Figure 7. Arctic Trial Position Error.

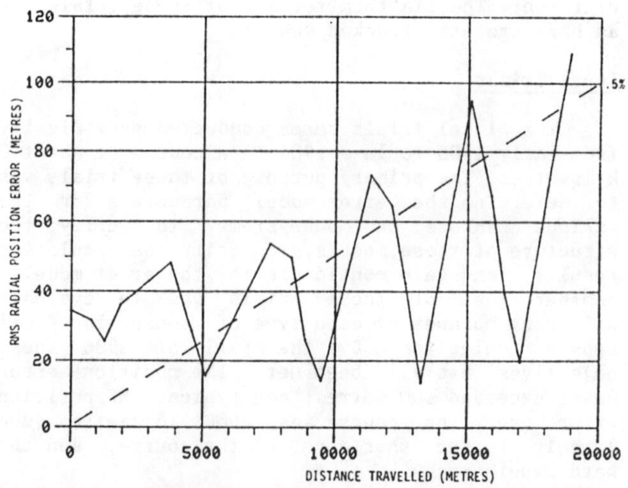

The subsystems that were expected to have difficulty in Arctic conditions, and were therefore of primary interest here, were the gyro, odometer and magnetometer. These trials proved that the integration of these sensors was successful under Arctic conditions.

SUMMARY

An exploratory development model of an optimally integrated system for Arctic land navigation has been designed, implemented and tested. All necessary sensors, processors and displays have been identified, purchased, tested and interfaced to:

1/ Gyrocompass/Directional Gyro Unit
2/ Dual Channel Transit Receiver
3/ C/A Code GPS Receiver
4/ Pickoff for the vehicle odometer
5/ Magnetic Compass Sensor Unit (fluxgate)
6/ High Output Pressure Transducer
7/ Motorola 68000 microprocessor
8/ Custom LCD displays

The necessary algorithms and data bases have been acquired or developed to correct the deterministic errors of these sensor measurements, including a full geomagnetic field model algorithm and a digital Arctic elevation map data base. A detailed functional relationship has been derived which explicitly describes the dynamic Transit position fix errors in such a way as to allow heading information to be extracted from the Transit position fixes.

A two stage prefilter, and an 8 state extended Kalman filter have been designed and implemented to optimally integrate all valid sensor information. Extensive simulations, road trials, field trials and an Arctic trial have been conducted to test the PLANS system design with positive results.

In summary an automatic, all weather vehicle navigation system has been developed to the exploratory development model stage, capable of providing the accuracy and reliability necessary for the Arctic application. Funding has been allocated for the manufacture of one ADM and several production models. The contract should be in place by late 1988, which will transfer the PLANS technology to industry.

FUTURE DIRECTIONS

An extended high Arctic trial is planned, to evaluate the very high latitude and cold temperature performance of PLANS.

As an integrated system PLANS has been designed with sufficient flexibility to allow new sensors to replace old with a minimum of reconfiguration. The PLANS sensor suite has already changed several times since work began, as better sensors have become available. The existing sensor suite is still not entirely satisfactory and some possible replacements are presently being considered. There are also several new sensors, expected to be available in the near future which are good candidates for integration into PLANS. A moderately priced RLG or DTG AHRS would be very desirable. Part of the emerging technology is a GPS receiver that also measures heading and attitude, using several antennae, which could also be very useful.

REFERENCES

[1] McMillan, J.C., "Design of an Optimally Integrated Primary Land Arctic Navigation System, Vol. I System Design", DREO Report 946, 1986.

[2] Gupta, R.R., Donnelly, S.F., Creamer, P.M., and Sayer, S., "Omega Signal Coverage Prediction Diagrams for 10.2 kHz", Document prepared by The Analytic Sciences Corporation for the U.S. Department of Transportation, U.S. Coast Guard, October 1980.

[3] McMillan,J.C., "MINS-B II: A Marine Integrated Navigation System", Proceedings of the Position Location and Navigation Symposium, PLANS 88, Dec. 1988.

[4] Malin, S.R.C., and Barraclough, D.R., Computers and Geosciences, Vol. 7 No. 4, 1981, pp. 401-405.

[5] McMillan,J.C., "An Integrated System For Land Navigation", NATO AGARDograph GCP/AG.314 on: "Analysis, Design & Synthesis Methods for G&C Systems", 1988.

[6] Gelb, A., Editor, The Analytic Sciences Corporation, Applied Optimal Estimation, Cambridge, Massachusetts, The M.I.T. Press, 1974.

[7] Bierman, G.J., Factorisation Methods for Discrete Sequential Estimation, New York, Academic Press, 1977.

ENHANCING PLRS WITH USER-TO-USER DATA CAPABILITY*

J. A. Kivett and R. E. Cook

Hughes Aircraft Company
Communications Systems Division
Ground Systems Group
Fullerton, California

ABSTRACT

The Enhanced Position Location Reporting System (EPLRS) maintains all of the basic PLRS capabilities while greatly increasing the user-to-user data capability. The EPLRS utilizes the proven PLRS control network for monitoring and controlling large communities of user terminals including the positioning, position reporting, navigation aid, cryptographic key distribution, and status reporting functions. In addition, the control network is utilized for distributing communications circuit assignments and monitoring user-to-user communications performances.

Both duplex (point-to-point) and group addressed (broadcast) types of service are available via the same user terminal. Each terminal can support many user circuits (needlines) simultaneously with a composite (receive plus transmit) information rate in excess of three kilobits per second. The primary user interface is via the PLRS/JTIDS Hybrid Interface (PJHI), which uses CCITT X.25 protocols, but an FSK interface is also available for backward compatibility with existing systems such as TACFIRE.

The EPLRS concepts have been proven through live testing using prototype terminals, including interfacing with six military host systems. In addition, extensive computer modeling and large scale user community simulation have been used to confirm extension of performance to full scale operation with up to one thousand user terminals in a division area. Limited production of over two hundred terminals is underway, with development and operational testing scheduled for early in 1988.

INTRODUCTION

The Enhanced Position Location Reporting System (EPLRS) extends the original PLRS concept to include user-to-user data communications. The capabilities of the USMC/USA Position Location and Reporting System (PLRS) uses advanced Time and Frequency Division Multiple Access Spread Spectrum technology. PLRS is in the process of being put into production and EPLRS is finishing Engineering Development.

The EPLRS system will support the Army's near term requirements for data distribution, position location, navigation, and identification in a battlefield environment. EPLRS, under a five phase development plan, will go into Developmental Testing/Operational Testing evaluation in 1988 and subsequently evolve into the objective Army Data Distribution System.

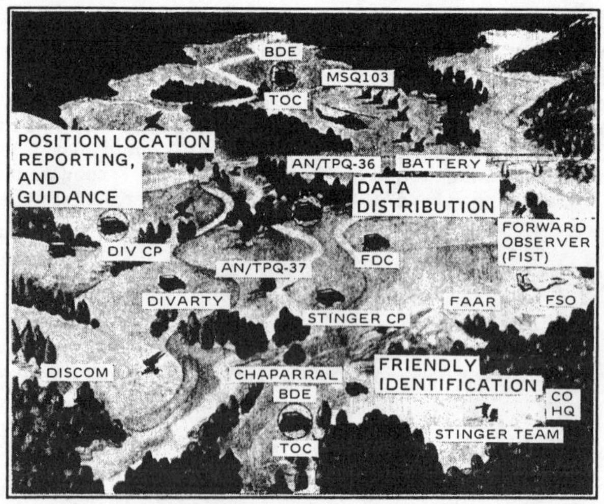

61153-3

Figure 1. EPLRS Maximizes Battlefield Effectiveness.

PLRS/EPLRS interoperability allows for mutual support and coordinated operations between Basic PLRS and EPLRS equipped units. Coordination of resource allocation, Areas of Responsibility, and the lateral transfer of Command and Control information are all accommodated by the architecture.

DEPLOYMENT

<u>EPLRS System Elements</u> - EPLRS supports all five general mission areas in the Army (i.e., Air Defense Artillery, Fire Support , Intelligence/ Electronic Warfare, Combat Service Support , and Maneuver Control. EPLRS contains two primary

*This work was supported by the U.S. Army under contract DAAB07-82-C-J096, administered by U.S. Army Communications-Electronics Command, Fort Monmouth,N.J.

Reprinted from *Rec. IEEE Position Location and Navigation Symp.*, pp. 154–161, 1986.

equipment elements: Enhanced PLRS Terminals (PTs) and Network Control Stations (NCSs). The PTs are assigned to Division and Corps units that participate in data communications, identification, and position location/navigation. Between 500 and 820 PTs are assigned to each Division. In addition approximately 970 PTs and 6 NCSs are assigned to the Heavy Corps. Each separate Brigade and Armored Cavalry Regiment is assigned one NCS and from 170 to 290 PTs. Additionally approximately 970 PTs and 6 NCSs will be deployed in a Heavy Corps rear area.

Area Coverage - A key consideration in planning for Ultra High Frequency radio communication is the line of sight propagation characteristics inherent in this band. The PT operates in the 420-450 MHz range. Modeling has shown that percent connectivity on the order of 10% is adequate. A deployment of PTs providing good area coverage is essential to take full advantage of the EPLRS system's integral relay feature. The adaptive automatic relay capability of the network allows the support of long range circuits. The proper employment of PTs deployed as dedicated relays is important to planning for and utilization of the system. Otherwise, too sparse a distribution of PTs in hilly or heavily foliated terrain may result in increased percentage of unsatisfied circuits.

Gateways - Data communications requirements which extend beyond the boundary of a Division sized area is supported by gateways. Two PTs connected back-to-back by a cable between their data ports form a gateway. One PT is in one NCS community and one PT is in the other. Information is passed between the two PTs at baseband so that each of the gateway PTs can maintain timing and cryptographic synchronization within its own community. Data communications between a source PT in one community and a destination PT in another is accomplished by setting up a circuit between the source and the gateway PT. A similar procedure takes place in the other community. Data transverses circuits within the two different communities and across the gateway, providing the functional intercommunity circuit.

SYSTEM ARCHITECTURE

The NCS establishes control circuits between PTs and the NCS via the control network. This control network supports the position location, navigation aid, friendly unit identification, and over the air rekey functions. Basic PLRS and EPLRS terminals operate compatibly using the control network, but user-to-user communications are supported solely by EPLRS Terminals (PTs). All PTs are capable of being assigned as relays in both the control and communications networks. PTs can simultaneously support control, duplex communications and group communications assignments as indicated in Figure 2.

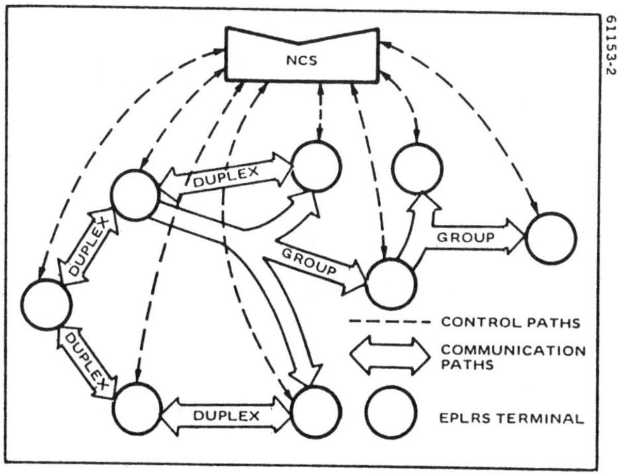

Figure 2. Control and Communications Networks in a Single NCS Community.

Host-to-Host Data Cmmunications - Host-to-Host data communications are provided via the EPLRS communications network. The communications network consists of both duplex and group addressed circuits. Duplex circuits are used to support requirements between two users requiring end to end acknowledgment control. A maximum of four PTs can be assigned to relay on a duplex circuit. Group addressed circuits are used to support requirements between multiple users not requiring acknowledgment control. Only one PT may source information on a group addressed circuit. A maximum of three PTs can be assigned as relays on a group addressed circuit.

Data Communication Message Flow - Communications messages are input to the system via the PLRS/JTIDS Hybrid Interface (PJHI). This is an X.25 type interface where data flows in packets containing up to 128 bytes (1024 bits) of user information. Destination packet addressing information is integral to the header of each data packet in the form of an 8 bit Logical Channel Number. Each host message is mapped into PJHI packets with a corresponding Logical Channel Number by the Host so that the subsequent transmission of the message can take place over the proper circuit. The PTs convert PJHI packets into a form for EPLRS transmission called Transmission Units. Each Transmission Unit contains 80 information bits. Transmission Units are transported through the communications network via NCS assigned virtual circuits. The PTs are responsible for acknowledging and sequencing Transmission Units and circuit monitoring. The NCS assigns PTs to participate on communication circuits as sources, destinations, or relays. The Transmission Units are sent between source and destination PTs through relays, where the adaptive circuit routing is controlled by an NCS. Transmission Units are converted back to PJHI packets at the destination PT.

All control network traffic is nodal to the extent that the message routing takes place through the NCS as opposed to the PT-to-PT traffic (via relays) that occurs on the communication network.

From a message flow standpoint, the control network provides an "order-wire" capability that can be used to establish and maintain the communication network. In addition to the order-wire, the control network also supports NCS services. These services provide the PT user with a wide range of position location, navigation, and identification information. The communication network continues to function in the absence of NCS control for as long as the PT deployment locations remain relatively static and a steady state environment exists (i.e., no severe jamming attack occurs).

EPLRS Circuits - An EPLRS circuit is defined by the following parameters: (1) the source and destination Military ID and associated Logical Channel Numbers, (2) circuit priority, (3) circuit rate, and (4) duplex or group addressed circuit.

The Logical Channel Numbers are used in place of addresses. Since PTs are capable of operating with multiple Logical Channels simultaneously, the assignmnt of Logical Channel Numbers is important so the host messages may be routed over the correct virtual circuits.

Four circuit priorities are accommodated. Circuit priority is a key software consideration in the circuit activation process. In case of heavy system loading, requests to activate low priority circuits receive delayed responses, or in extreme cases the lowest priority circuits are not implemented.

The rate selected for a circuit dictates how large a block of timeslot resource will be assigned to satisfy the requirement. This rate is determined based on both the throughput and response time requirements. The Acknowledgment control determines whether a duplex or group addressed circuit is used.

Relay - The relay PTs programmed by the NCS are responsible for relaying Transmission Units between source and destination PTs. The relays can not read or alter the Transmission Unit data content, but they do perform error correction decoding and encoding, and validation of received Transmission Units.

EPLRS communication circuit (duplex and group addressed) timeslot assignments are structured to allow for multiple receive opportunities. The source, destination, and relay PTs have specific times to receive defined by recurrence period, start frame, and timeslot index.

Duplex Circuits - Source and destination PTs transmit on duplex circuits in a periodic manner. They transmit once per Recurrence Period. On duplex circuits, the timeslots are arranged such that for each source transmission there is a destination transmission allowing for balanced communications with acknowledgment between the two PTs.

Each duplex circuit structure has at least two paths. Path 0 is used for the first transmission in the epoch and every other transmission after that, and path 1 is used for the second transmission in the epoch and every other transmission after that.

Group Addressed Circuits - As in the duplex circuit structure, the group addressed circuit structure is periodic with respect to the Recurrence Period, and the timeslots are defined by start frame and timeslot. These parameters allow a source to know exactly when to transmit. Because the reception of messages on group addressed circuits are not acknowledged, the relays need only transmit in one direction. The timeslots are arranged so that the source can continuously transmit (up to 1200 bps). The relays and destinations on a group addressed circuit are not programmed for a specific path or level. Therefore, it is not known what path(s) will be used to relay the messages to the destinations. This approach is commonly known as path diversity broadcast.

Second Source Receive - Both duplex and group addressed circuits use the second source receive technique. This technique assigns a set of timeslots in which relays and the destination PT "listens" for the source PT's transmission and all other upstream relay transmissions. This allows PTs to receive transmissions early and increases the probability of receiving transmissions. For example, the fourth-level PT could conceivably receive a message during the transmission from the source or any subsequent relay of the message.

The PTs provide direct user-to-user communication capability with integral relay as well as the ability to perform Time-of-Arrival measurements between units. The communication links are configurable by commands from the NCS to allow the hosts to exchange digital data.

ENHANCED PLRS USER UNIT (EPUU)

The PT consists of an Enhanced PLRS User Unit (EPUU), a User Readout device, and an appropriate installation kit for ground, surface vehicle, or airborne use. The EPUU can be configured to support Hosts via either the PJHI, or a Frequency Shift Keying (FSK) Interface. The EPUU is designed to meet the environmental requirements of ground and airborne installations. The User Readout device serves as a control panel for the PT, and for handheld use with ground based installations. The data access capabilities of the User Readout is limited to short message exchanges via the NCS.

The PT performs user-to-user communications and generates ranging information for position location/navigation. The NCS can assign any PT as a relay of opportunity. This extends the communications range and keeps communication links operable during jamming or changes in user location.

Data Flow - The key data flow steps in the EPUU are shown in Figure 3. After striping the user specific protocol and replacing it with the EPLRS network protocol, each Transmission Unit is encrypted with a user variable, then a (104,94) Cyclic Redundancy Check error detection code is applied. After this, the Transmission Unit is transmission encrypted, split into 26 4-bit blocks. and a (7,4) Hamming error correction code is applied to each block. The encoded blocks are interleaved to protect against burst errors. Finally, each channel bit is modulated by a 5 Mchip-/sec Pseudo-Noise code for spread spectrum protection and the transmission unit is frequency hopped to provide increased antijam protection. At the receiving terminal the process is reversed.

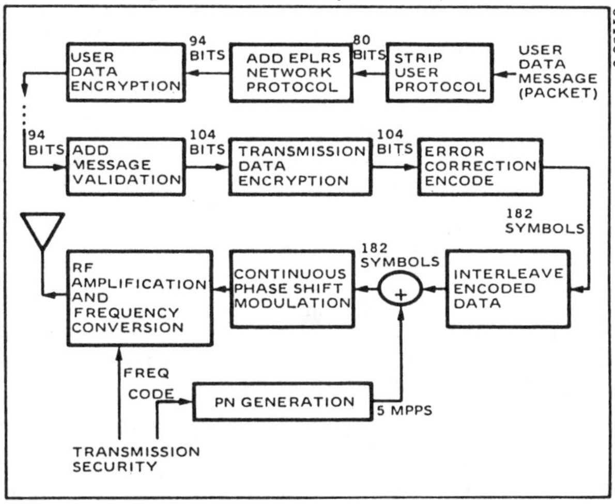

Figure 3. PT Data Transmission Process.

The EPUU is a multifunction radio that generates and processes EPLRS messages. Centralized control of the EPUU by a microprocessor within the Message Processor allows partitioning of processing into logical EPUU functions as shown in Figure 4.

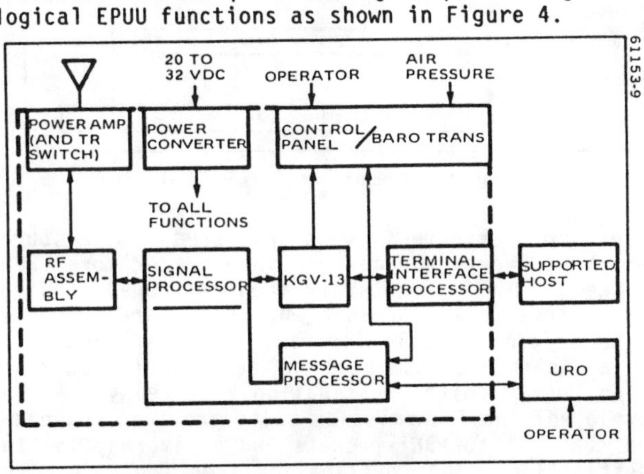

Figure 4. Major PT Functions.

<u>RF/IF Function</u> - The EPLRS RF/IF function performs frequency conversion, amplification, and filtering of the transmitted and received signals. During receive, this function performs an A/D conversion of the incoming signals using a 2-bit adaptive A/D converter. During transmit, the digital output of the Signal Processor is used to generate the Continuous Phase Shift Modulation.

<u>Signal Procesor Function</u> - The Signal Processor function performs preamble detection/ generation, interleaving/deinterleaving and error correction/ error detection encoding/decoding, Pseudo-noise generation, data correlation, and Time-of-Arrival measurement.

<u>Secure Data Unit Function</u> - The Secure Data Unit function is performed by the KGV-13 which encrypts/decrypts transmitted and received signals, and provides message validation.

<u>Message Processor Function</u> - The message processor contains a CMOS microprocessor that is the central controller of the EPUU. This function controls all processes done within the EPUU. The message processor, additionally, generates, and decodes link messages.

<u>Terminal Interface Function</u> - The Terminal Interface Processor provides the interface to the PJHI link or the Frequency Shift Keying interface. The PJHI has as its objective the formulation of a standard interface to which all tactical data systems can design. This interface defines packet generation for data distribution in a rigorous and logical manner through the first three levels of the International Standards Organization (ISO) model. The Frequency Shift Keying interface is capable of supporting a full duplex channel between the Frequency Shift Keying tactical data system and the PTs.

NET CONTROL STATION

The NCS is a tactically packaged computer facility providing centralized control and monitoring of a EPLRS community. The NCS performs dynamic network management of all PTs under its control as indicated in Figure 5. The NCS is the central control point of the EPLRS network, allowing a technical control operator to maintain a real time overview of the network within his responsibility.

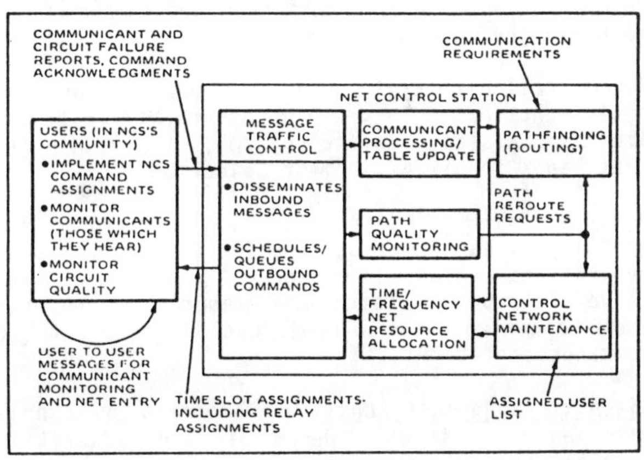

Figure 5. Major NCS Circuit Management Functions.

The NCS software operates in realtime. The NCS's operational software provides centralized network management and automatic distribution of position, navigation, and identification information.

The terminals accept and implement NCS issued commands and report status and communicant data (i.e., who can hear whom). These reports are essential for accomplishing the automatic adaptive routing capability of the EPLRS system.

NCS Operator - The principal human interface to the system is at the NCS. The NCS operator, as the communications technical controller, is responsible for network initialization and monitoring ongoing community performance. As PTs enter the network, the NCS operator monitors the performance of the network using the graphic display. Normal operation requires only minimal operator intervention. His work is intended to be primarily exception processing, taking actions to help the automated network management.

Network Management - The Network Management software functions automatically provide for dynamic network monitoring and resource assignment. In the battlefield evironment, the operating radio links, or connectivity, available in the network can be very limited and rapidly changing. There are two design features of the EPLRS system that enable it to provide continuous service in such an environment. First is the high antijam capabilities of the spread spectrum EPLRS radios, without which there would be very little connectivity in a jamming environment. Second (and just as important in the dynamic environment) is the automatic network management function which controls the EPLRS network configurations in real time. The network management function within the NCS performs dynamic, realtime communications routing and timeslot assignment to the PTs and monitors and controls the subsequent network performance.

The communication network management functions are distributed to both PTs and NCS. The PTs communication network management role is to implement the commands issued by the NCS, detect circuit failures, and control the flow of data packets into and out of the network. The NCS uses the control network to deliver the commands used to build and maintain the communication network.

Communication network management consists of three basic functions. They can be generically categorized as: Message Traffic Control, Circuit Management, and Circuit Assignment Selection.

Message Traffic Control Function - The main responsibilities of Message Traffic Control for communication network management are to receive/distribute status and measurement reports, and to control the acknowledged delivery of the commands sent to PTs.

Circuit Management Function - The main responsibilities of Circuit Management are prioritized queuing of circuit requests and maintaining circuit status.

Circuit Assignment Selection Function - The Circuit Assignment Selection finds paths and resources that satisfy the circuit requirements. The paths are found using a minimum cost path finding algorithm. Circuit assignment or reassignment requests are processed on a priority basis. Requests with equal circuit priorities are selected on a first-in, first-out basis. Successful attempts at finding circuit assignments results in generating Communication Circuit Assignment commands for the PTs selected to participate on the circuit.

Crypto Management Function - An additional NCS function is the generation and over the air distribution of cryptovariables to all PTs. The EPLRS security architecture provides the system with both SECRET and CONFIDENTIAL cryptographic data encryption.

HOST INTERFACES

EPLRS implements a Virtual Circuit approach to satisfy host-to-host data communication requirements. A wide range of information rates and response times are provided in support of both duplex and group addressed requirements.

The EPLRS system uses the PJHI (X.25 type) protocol, in which hosts input data into the communication network on Logical Channel Numbers in the form of packets via the standard PJHI. The data delivered through the network is transparent to the host. X.25 is an International Telephone and Telegaph Consultive Committee (CCIT) standard three-layer protocol and is virtual circuit oriented. PTs can participate on multiple simultaneous Virtual Circuits using different Logical Channel Numbers.

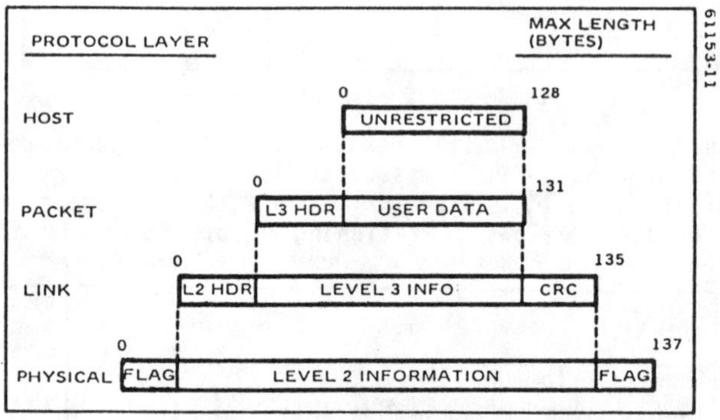

Figure 6. Information Units in PJHI.

User to user data flow is initiated by the source which converts the message into one or more PJHI packets. The packets are then delivered to the PT's Interface Processor where they are converted into a form suitable for radio transmission. The data is then transmitted in the form of transmission units to the destination PT. Relays may be employed. The Transmission Units are then converted back into PJHI packets and delivered to the destination. The destination then sequences the PJHI packets into a message identical to the one that the source sent to its PT.

PJHI utilizes packets to organize the efficient exchange of information across the interface. Messages which are going to be transferred on the PJHI are first broken into segments, called data packets. The maximum size of a PJHI data packet is 128 bytes. These data packets are transferred across the PJHI independently allowing messages to several destinations to be time multiplexed This independent transfer of packets increases efficiency of information flow.

The PJHI is a functionally layered set of communications protocols implementing the bottom three layers of the International Standards Organization Open System Interconnection reference model. The three layers implemented by X.25 and PJHI are the physical layer or level 1, the link layer or level 2 and the packet layer or level 3.

Physical Layer - The PJHI physical layer provides the media to transfer bits from one physical device to another. The physical media in the PJHI is either NRZ digital baseband or conditioned diphase.

The NRZ digital baseband interface may be utilized for devices that are located within 5 meters of the PT. This interface consists of six signal pairs, two pairs each of data, clock, and control. This interface is similar in design and function to the Electronic Industries Association standard RS-449/422.

The conditioned diphase interface may be utilized to remote the PJHI devices up to 1000 meters. This interface requires only two signal pairs and a shield.

Link Layer - The PJHI link layer provides the functional and procedural means to maintain a connection to another PJHI device while detecting and possibly correcting errors which may occur in the physical layer. The PJHI performs the X.25 link level protocol by utilizing the Link Access Procedure B and the High Level Data Link Control procedure. The unit of information transferred at the link layer is called a frame, PJHI supports three general classes of frames; Information, Supervisory, and Unnumbered. The function of the information frame is to transfer sequentially numbered frames containing information for the packet layer. Information frames are variable in size depending upon the amount of information they contain. The maximum length of an Information frame is 131 bytes. The function of the Supervisory frames is to control supervisory link functions such as acknowledging, requesting retransmission, or temporarily suspending transmission of Information frames. Unnumbered frames perform the function of supplementary control of the link such as setting up, disconnecting, or stating status of the link.

Packet Layer - The PJHI packet layer defines the protocol that routes packets from one host system to another. To efficiently control packets to several hosts, the PJHI utilizes the concept of a logical channel. A logical channel is a virtual connection from one host system to another. When a logical channel is routed through a network, such as EPLRS, the Logical Channel Number has only local significance.

In the PJHI packet layer, the transmission of data is controlled separately for each logical channel in each direction and is based upon acknowledgments from the receiving PJHI. To perform this flow control protocol, the PJHI sequences packets and defines a window protocol. Data packets are sequenced by attaching to each data packet a sequentially numbered tag. Data packets can be transferred across the PJHI when the data packet is within the window.

If a data packet is outside the window, then the packet will be rejected. The receiver can convey authorization for additional data packets by incrementing a packet receive sequence number. In this way the receiving PJHI packet layer can control, on a logical channel basis, the flow of incoming data packets.

The PJHI packet layer also performs error control for recoverable errors and error diagnosis for unrecoverable errors at the packet layer. The packet layer performs error control by resetting the affected logical channel and starting again.

WAVEFORM

The EPLRS network resource is organized into a Time and Frequency Division Multiple Access structure. Each member of a community is assigned one or more timeslots in which it can transmit a burst while other PTs receive. To accomplish this, each PT must possess a clock that is synchronized to the clock of the NCS and other PTs.

Basic Architecture - The development of EPLRS is based on synchronous Time Division Multiple Access and spread spectrum technology. Integral cryptographic security, error detection and correction coding, and frequency hopping are employed to provide robust communications in a hostile Radio Electronic Combat environment. EPLRS contributes significant performance advantages including rapid response times and effective data throughout, Communications Security, resistance to Electronic Countermeasures /Electronic Support Measures, integral automatic relay, transmissions for ranging measurements, and freedom from voice/data contention.

Resource Definition - The network utilizes the resources of time, frequency, and code to multiplex the many operations necessary for system operation. The structured use of time allows a convenient and efficient method for gathering Time-of-Arrival data, and for managing the multiple relay levels within the network. The use of the frequency resource provides additional antijam protection and allows for non-interfering, coordinated operation.

The Time Resource - Each PT is commanded by the NCS to perform specific transmissions and receptions at specific times. The time division structure simultaneously accommodates both minimum and maximum network access requirements. For control, these range from manpack PTs requiring update rates approximately once every minute to high performance fixed-wing aircraft with desired updates 30 times per minute. For communications, these typically range from field artillery circuits needing two source transmissions per second to maneuver control circuits needing one source transmission per 32 seconds.

Epoch - The epoch is the longest time division which is significant to the network. The longest period for which a PT can be programmed to repeat its assignments is once every 256 frames or once per 64 second epoch. The programmed cyclic transmissions and receptions do not change from epoch to epoch unless the NCS alters the PT's assignments.

Frame - The frame contains 128 timeslots and has a period of 1/4 second. The significance of the frame is that it represents the shortest period at which a PT can be programmed to participate in the Control Network. On communication circuits, PTs can be assigned up to four transmissions within a frame on a single circuit.

Timeslot - The fundamental time division is the timeslot. See Figure 7. The timeslot length is 1.95 milliseconds. The burst transmission accounts for 800 microseconds and 600 microseconds is allocated to RF propagation delay. The remaining time is required for processing overhead such as message encoding, validation, and guard time.

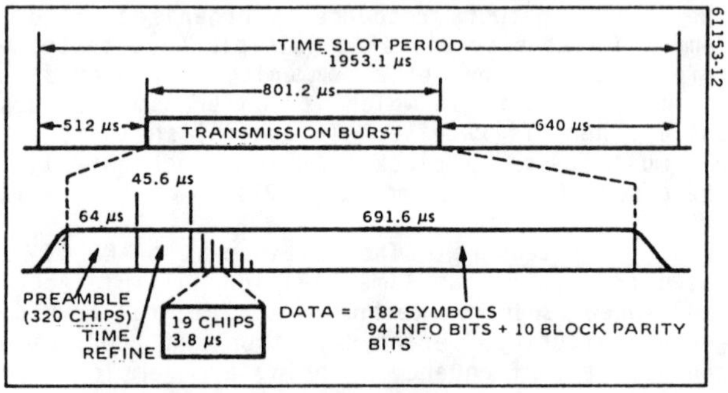

Figure 7. PLRS Timeslot Signal Structure.

The Frequency Resource - Each transmission occurs on a particular frequency channel. When operating in the hop mode, each channel is hopped across the 420 to 450 MHz band. Frequency hopping provides the network with additional Antijam performance. There are eight channels in the EPLRS, thereby increasing the capacity of the network by allowing simultaneous transmissions to occur with a low probability of mutual interference.

The Code Resource - In addition to the time and frequency resources, EPLRS uses a Pseudo-Noise code resource. This code resource provides different codes for each circuit in the communication network. These codes are used to determine the spread spectrum pattern used. In this way, a third dimension is added to the network capacity.

SYSTEM CAPACITIES

The EPLRS system capacities fall into four major categories: network, NCS, circuit, and PT capacities.

Network Capacity - An EPLRS community consisting of 250 PTs has a nominal capacity of 60 Kbps in support of duplex circuits, plus a nominal capacity of over 100 Kbps in support of group addressed circuits.

NCS Capacities - The unit libraries at the NCS define the participating PTs and their communication requirements. A maximum of 1000 PTs can reside in the unit library. A nominal NCS community consists of 250 PTs and 375 circuits. An NCS is capable of supporting a maximum community size of 460 PTs. The Division wide library is capable of storing up to 1500 circuits.

Circuit Capacities - The EPLRS circuit capacities more than satisfy the highest rate EPLRS users as defined in the circuit requirements found in the Battlefield Communication Review Data Distribution database. The limiting factor on circuit capacity is the ability to realistically manage the circuit resources with adaptive relaying in a large network.

Duplex circuits are capable of supporting acknowledged data rates of up to 600 bps. Group addressed circuits are capable of supporting non-acknowledged data rates up to 1200 bps.

Terminal Capacities - The capacities of the PTs in the EPLRS system are limited by the number of circuit assignments they can store and the timeslot allocation. A PT can support a maximum of 60 circuit assignments (30 source/destination, 30 relay). The maximum total duplex and group addressed circuit rates assignable to a PT are 900 bps and 2400 bps respectively.

SYSTEM PERFORMANCE

The EPLRS system is required to build to support eighty five percent of required circuits within the desired time durations. These time durations account for up to ten percent mobility of the total PTs. For this discussion, mobility refers to the fraction of PTs in motion at each instant, not the fraction capable of motion. The normal EPLRS community size refers to a Brigade Area of Operations containing at least 40 and up to 250 simultaneously active PTs under control of a single NCS. The initial build of both the control and communication networks for the normal community size can be achieved within 17 minutes under Electronic Countermeasures threat conditions.

Under Electronic Countermeasures conditions, the EPLRS system continues to provide communications and position location services for PTs retaining control and communications connectivity. Ninety percent of the PTs which experience an interruption in service due to Electronic Countermeasures conditions are detected and reported to the NCS within 120 seconds of the onset of the interruption. The EPLRS system then automatically repairs the communications and control paths of these PTs.

For all circuits above 150 bits per second, the Network Management function allocates the resources of the timeslot architecture and available relay PTs to achieve a one second response time for short messages. The major component of the response time is access delay, which is based on the circuit's recurrence period. The other components of median response time include source encryption, relay delay, destination decryption, signal processing and miscellaneous data transfer delays (e.g., the PJHI data transfer).

FUTURE GROWTH

VHSIC Insertion - A key planned improvement to EPLRS is the VHSIC insertion program. The primary objective of VHSIC insertion into the PT is to increase the data throughput and antijam performance.

The programability of the Very High Speed Integrated Circuit (VHSIC) chip set permits its adaptation to a wide range of signal processing parameters in real time. The benefits of this capability to EPLRS are achieved by the three new waveform modes for the data communication circuits, thereby providing adaptive increases in traffic capacity or antijam margin in response to change in tactical demands. See Figure 8. In many situations and deployments, a large number of the communication circuits will not experience the worst-case jamming threat. These circuits can employ the maximum throughput waveform mode. Conversely, high priority messages on critical circuits under severe jamming can use the maximum ECCM waveform mode for greater protection.

The spread spectrum burst transmission waveforms used in the VHSIC EPUU allow data transfer between units in the network at rates from 80 to 240 bits per transmission unit. In addition to spread spectrum processing gain, increased antijam protection is added to each waveform by the improved error correction coding.

The mode 0 waveform is the standard EPLRS waveform using an 800 microsecond burst to transmit 80 bits of data. The burst mode is used to communicate with Non-VHSIC EPUUs or PLRS user units lacking the waveform capabilities.

Modes 1, 2, and 3 use a transmission burst of about 1000 microsecond and transmit 80, 160, and 240 bits respectively. This improves antijam protection or improves throughput over mode 0, depending on mode used. The 1000 microsecond burst has a shorter usable range than the 800 microseconds burst because the extra 200 microseconds is taken out of the propagation guard band while maintaining the overall length of the timeslot. However the higher speed signal processing partially compensates for this loss.

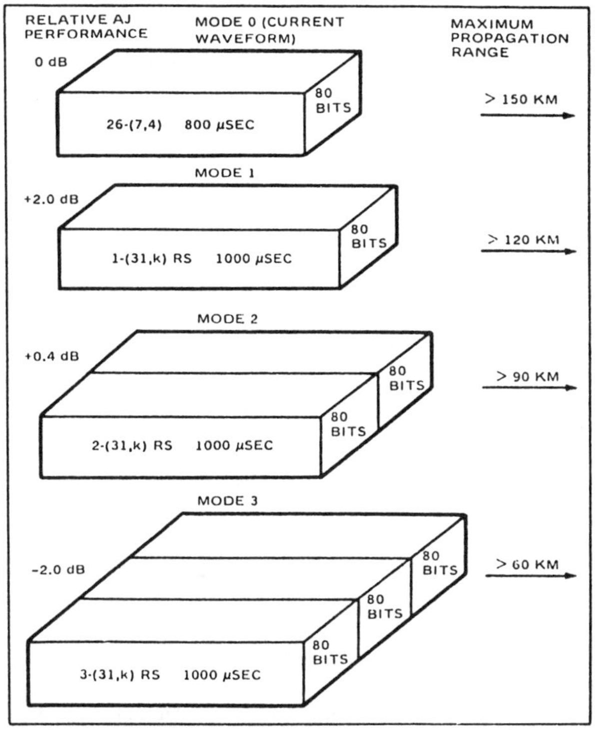

Figure 8. Performance Comparison of VHSIC Waveforms.

Loran-C Vehicle Tracking in Detroit's Public Safety Dispatch System

LAURENCE J. CORTLAND

Il Morrow Inc., Salem, Oregon
Received January 1989
Revised June 1989

ABSTRACT

In 1928, the city of Detroit had the distinction of being the first city in the world to introduce police radios and two-way radio police dispatch. Sixty years later, another world first by the City of Detroit Police Department was announced by Mayor Coleman Young: the introduction of emergency vehicle tracking integrated with the "Enhanced 911" emergency dispatch system. This system is the most ambitious undertaking of its kind and incorporates many innovative features, including automatic tracking of the largest fleet of Loran-C equipped vehicles. A complete floor of the Detroit Police Headquarters has been renovated to provide the most advanced facility for public safety dispatch. This paper presents concepts and some details of the Automatic Vehicle Location System developed to complete the elements of the Detroit Public Safety Dispatch System.

INTRODUCTION

The overall objective of the Detroit Public Safety Dispatch System is to improve the quality of service to Detroit citizens by minimizing the average response time of field resources to calls for service. To that end, on 20 January 1988, the city of Detroit announced selection of the II Morrow Inc./Advanced Control Technology team to develop a customized Automatic Vehicle Location System (AVLS). This system was to interface with the city's "Enhanced 911" emergency telephone system, the Unisys Computer Aided Dispatch (CAD) System, and the Electrocom Mobile Data Terminal (MDT) System, all of which had earlier been awarded.

The AVLS would consist of about 760 police, fire, and Emergency Medical Services (EMS) vehicles that serve the city of Detroit. The position of each vehicle would be determined by a mobile Loran-C transponder and transmitted back to the Dispatch Center. There the data would be processed through Micro Vax computers and displayed on one or more of the proposed six police, two fire, and two EMS interactive graphic dispatch workstations. The city-developed Geobase file would provide graphic maps linked to addresses so that incident reports would be identified on the map displays and in the CAD incident file for automatic selection of the required vehicles nearest to the incident. This information would be presented to the dispatcher for final review and assigning of response unit(s). The incident information could then be automatically downloaded to the assigned police unit through the MDT system.

Reprinted with permission from *Navigation*, vol. 36, no. 3, pp. 223–233, Fall 1989.

All of this action would be taken without transmission of voice communications from either the vehicle or the dispatcher, thus providing rapid response with security and privacy of transmission for the added safety of public safety personnel.

AUTOMATIC VEHICLE LOCATION SYSTEM DESCRIPTION

The AVLS consists of the following:

1) Great Lakes Loran-C chain coverage of the Detroit area
2) Loran-C transponders and VHF transceivers in each of 760 vehicles to be tracked
3) A fully redundant two-channel dedicated data radio network that covers the entire Detroit city area
4) A radio/telephone interface with fully redundant channel selector/polling units
5) A Dispatch Center fully redundant computer system serving 24 interactive workstations with data I/O capabilities to both the CAD and MDT systems
6) Data log recording equipment

An overall block diagram of the AVLS components as they interconnect with each other and to the MDT and CAD systems is shown in Figure 1. Figure 2 shows the basic block diagram of the mobile equipment that was installed in Detroit, and Figure 3 shows the basic block diagram for the fixed-end equipment.

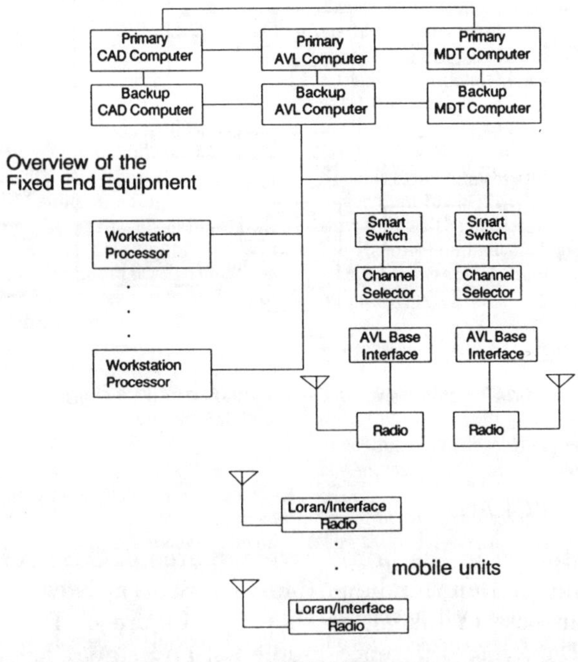

Fig. 1—Block Diagram of Major System Components

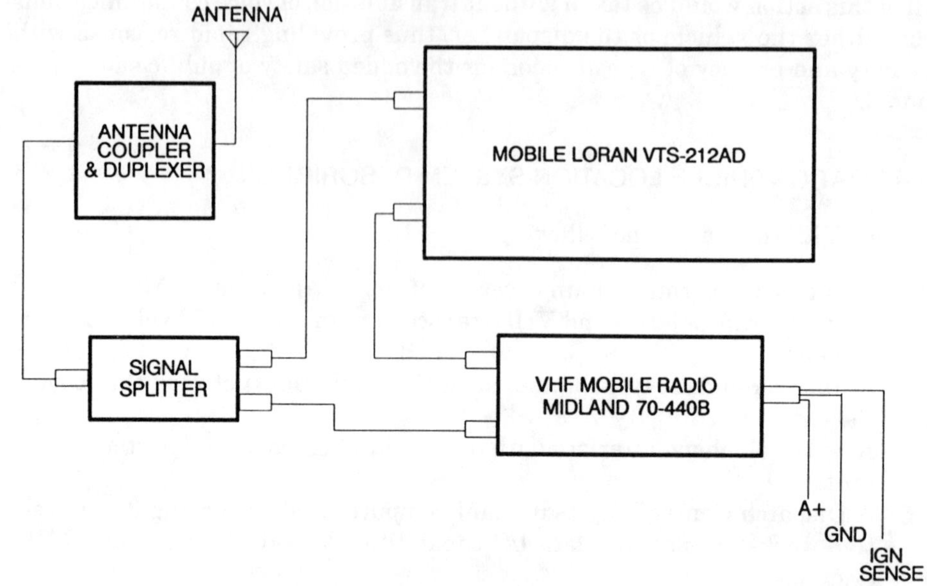

Fig. 2—Detroit AVLS Mobile Block Diagram

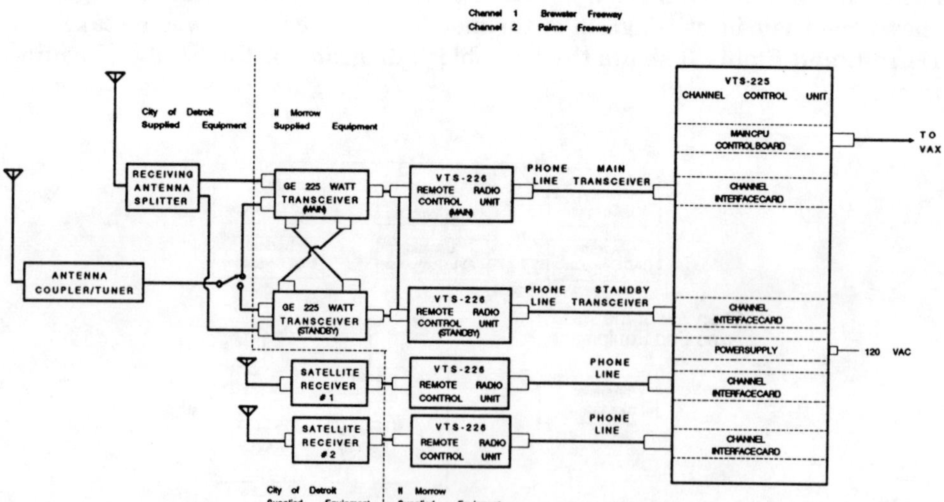

Fig. 3—Detroit AVLS Fixed-End Basic Block Diagram

DETROIT AREA LORAN-C COVERAGE

Detroit is situated in the prime coverage area of the Great Lakes Chain (8970 GRI) triad of Dana, Indiana (Master), Seneca, New York (Xray), and Baudette, Minnesota (Yankee), as shown in Figure 4. The calculated time differences (TDs), time difference gradients, and signal level estimates are shown in Table 1. The crossing angles of the secondaries were plotted and are

Fig. 4—Great Lakes Loran-C Chain Coverage Area

Table 1—Loran-C Performance Data Calculations for the Detroit Area

Station	Time Difference (TD)	Gradient	Signal Strength
Dana, IN	NA	NA	78.0 dB/1μV/m
Seneca, NY	31376.50 μs	508.5 ft/μs	63.7
Baudette, MN	49984.25	780.8	51.4
Crossing Angle = 65.9 deg		dRMS = 98.4 ft	

shown in Figure 5. This data indicates excellent Loran-C signal coverage for the position accuracy of the AVLS.

VEHICLE POSITION ACCURACY

The radial position accuracy requirement was specified as 660 ft over 95 percent of the city for 90 percent of the time. Preliminary plots within the Detroit area were taken to demonstrate the feasibility of attaining these accuracies. See Figures 6 and 7. These plots were taken with a Mariner Plotter and Loran system installed in a vehicle that was driven on the streets as shown. To test the position accuracy, the city area of Detroit was divided up into a grid of squares about 0.5 mi on a side. Readings were taken at intersections as near to the center of each square as possible. The names of these crossroads were recorded, along with an average of 10 samples of the Loran data. This data was then compared to the tabulated geodetic coordinates taken from the Geobase for these known locations, and 96.2 percent of the readings were found to be within 660 ft of the known location. Table 2 is a summary of the radial accuracy by percentage within the specified number of feet. Figure 8 shows the outline of the city limits of Detroit and the locations where the data was gathered.

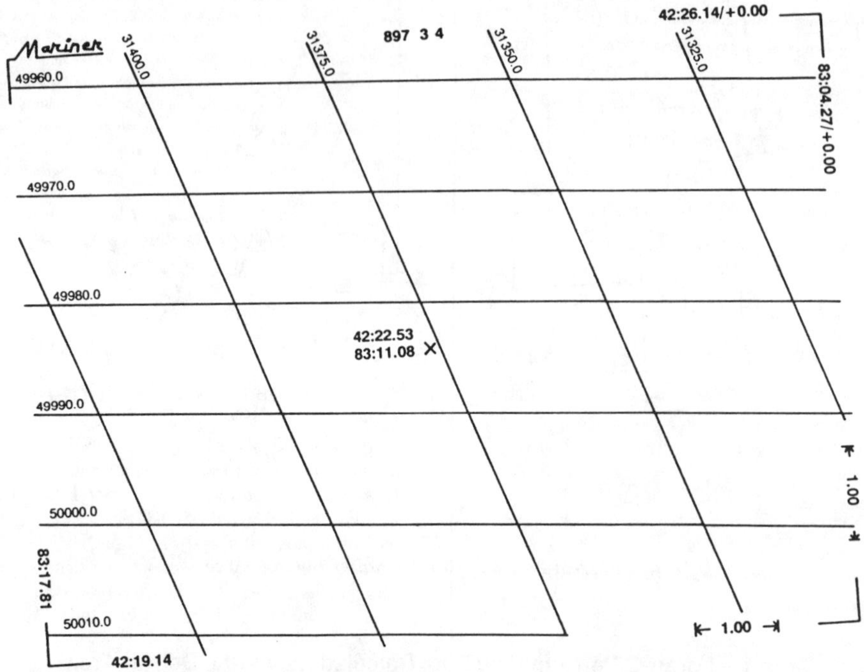

Fig. 5—Loran-C Crossing Angle Grid for Seneca and Baudette Secondaries

VHF RADIO NETWORK

A separate dedicated VHF radio network was established for the AVLS. Two existing police frequencies were assigned, and new GE Master II 225 W transmitters with complete standby transmitters replaced the older equipment at each site. Each transmitter pair operates on one of the two frequencies.

Strategically located through out the city are five supplemental receivers, three on one frequency and two on the other. For every response to a poll, the Vehicle Tracking System (VTS)-225 Channel Control Unit receives up to three good responses and compares them for agreement to ensure valid data. Spreading the supplemental receivers throughout the city ensures that responses are received from the lower-powered mobile radios no matter where the vehicle is located. Each transmitted poll request is also monitored by all supplemental receivers and correlated with the intended transmission data to determine whether the transmitter is operating properly. If a response is not received for five consecutive polls, the VTS-225 redirects the data to the standby transmitter, and the AVLS VAX computer flags the dispatcher that there is a problem with the first transmitter.

To properly support the requirements of the City of Detroit, the radio network provides the following: (1) redundancy of operation to allow continued operation in the event of a component failure, (2) complete and reliable radio coverage of the entire Detroit area, and (3) optimum use of the two dedicated frequencies to obtain the fastest possible polling times.

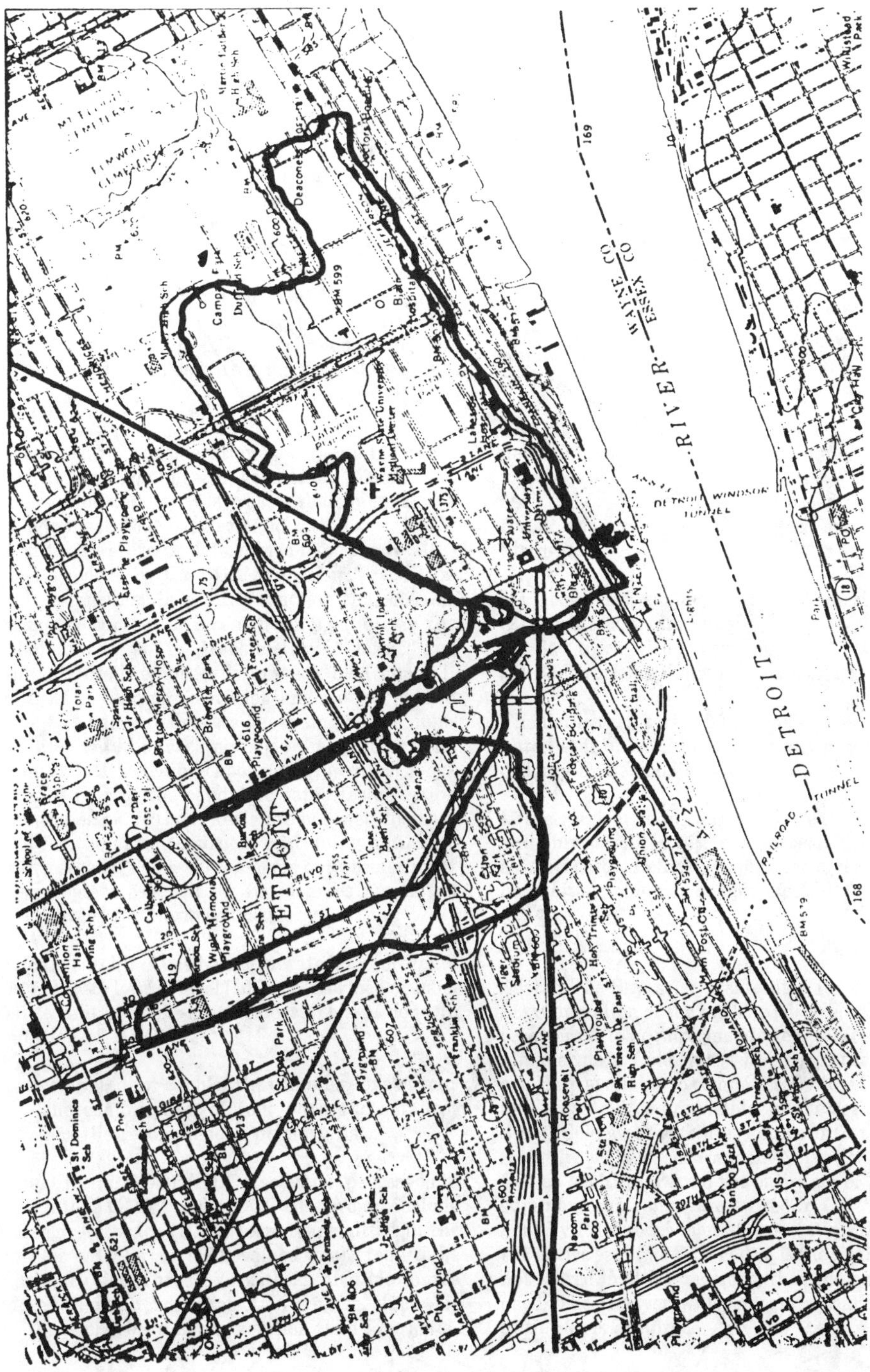

Fig. 6—Recorded Track of Loran-C Position Data

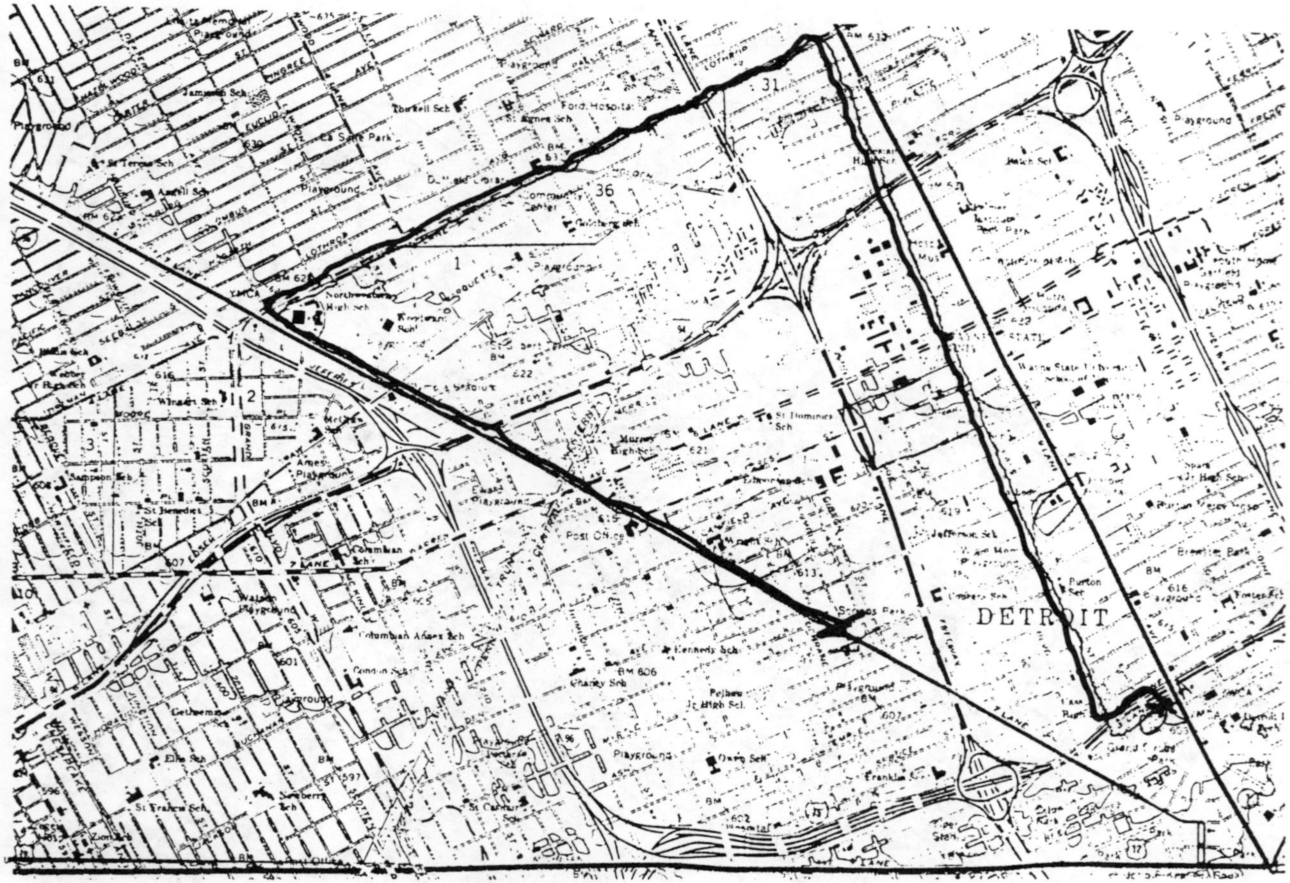

Fig. 7—Additional Track of Loran-C Position Data

Table 2—Summary of Position Error Test Data by Percentage of Samples

Average Latitude Error	209.2 ft
Average Longitude Error	153.4 ft
Summary of Radial Errors in Position (Percentage of Sample within Distance Listed Below)	
(ft)	(%)
100	17.9
200	50.8
300	71.6
400	83.1
500	91.7
600	95.9
660	96.2

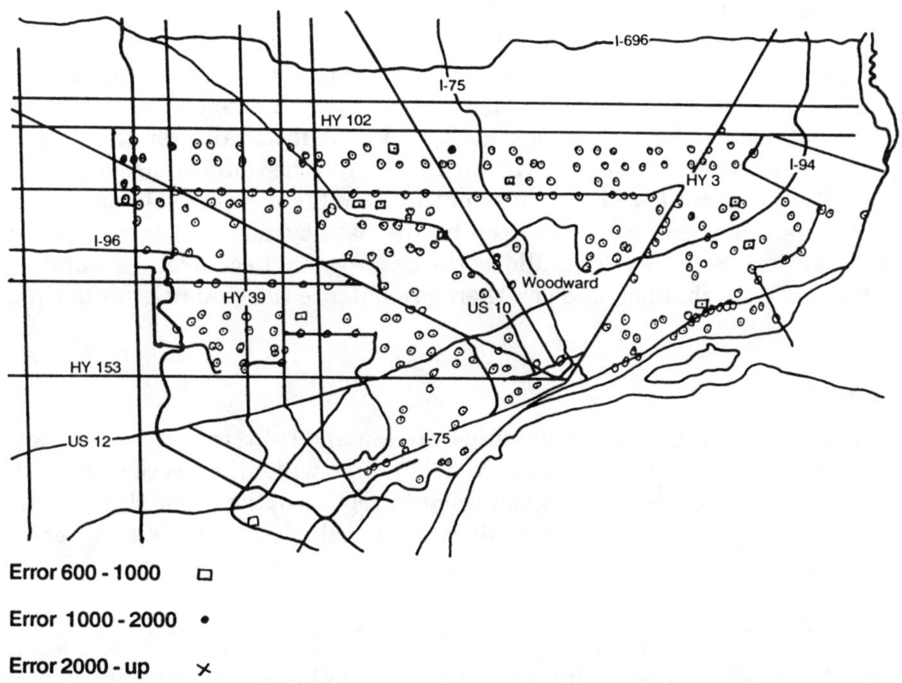

Fig. 8—Map of Position Error Samples

POLLING TIMES

There are seven polling lists. Vehicles on the highest-priority list are polled as often as possible while service is still maintained for vehicles on the lower-priority lists. The former list usually includes vehicles that are in an emergency condition and therefore need the most current position possible. There are three polling rates for the remaining polling lists, designed for polling each vehicle every 2, 5, or 10 min under normal loading of vehicles per list. The last two polling rates can be either locked in or dynamically selected by the status of the vehicle (whether the ignition is turned off, and whether the radio and Loran have been turned off). This allows for faster polling of active vehicles. The

AVLS computer keeps track of the timing for the servicing of the lists for each of the two frequencies and can reassign vehicles between the frequencies to maintain a balance in the use of the channels.

VEHICLE EQUIPMENT

Loran Transponder

The VTS-212AD contains the Loran sensor and the radio control board. The Loran sensor continuously monitors the Loran signals and converts the time differences to latitude and longitude position data. The Loran has been modified specifically to address the problems of land-based operations, such as the noise and frequency interference associated with power lines. The radio control board constantly monitors the radio for poll requests from the base unit. When it detects a poll addressed to its four-digit identification code, it responds by initiating a push to talk, and inserts the frequency shift keying (FSK) signal that contains the latitude and longitude position, along with the status of the vehicle's ignition, onto the radio microphone input. It also controls the power to the radio such that at a preprogrammed time interval after the vehicle's ignition has been turned off, it will turn itself and the radio off as well.

The radio control board also has control over the channel selection within the mobile radio, and when directed by the base system will force the radio onto the other frequency. This allows the base system to control the number of vehicles on each channel, and can therefore balance the channels for optimum usage.

Duplexed Antenna

A single 21 inch trunk mount whip antenna and splitter system was developed for this program. This duplexed antenna will allow reception of the 100 kHz Loran-C signals and transmission/reception of the 150 MHz band radio data. Thus only one antenna was added to the vehicle instead of two—one for the radio and one for the Loran-C receiver.

Mobile Radios

The dedicated data radios integrated into the AVLS mobile units are Midland 70 440B VHS 40 W mobile radios.

DISPATCH CENTER EQUIPMENT

VTS-225 Channel Control Unit

The VTS-225 channel control unit does the polling management and directs communication between the mobile equipment and the computer. It holds in memory a list of vehicles to poll and sorts them by priority. It can have up to 16 channel control boards that direct the polls up the telephone line to the transmitter and listen to the telephone line for responses from the mobile units. In the Detroit configuration, there are four channel select control boards installed in one of the VTS-225s and five channel select control boards installed in the other. In both cases, two channel boards are used to control the main and standby transmitters, and the remaining boards are interconnected to the supplemental receivers.

VTS-226 Remote Radio Control Unit

The VTS-226 receives poll requests on the telephone line, and directs the FSK signal onto the radio microphone input while it initiates a push to talk. It listens to the radio for responses from the mobile units and passes these responses on to the telephone line for detection at the VTS-225. The unit also monitors the radio for a busy condition; if it senses that the channel is busy, it waits for a clear channel before it allows transmission of the poll request. This allows for transmission of the Morse code equivalent of the call letters for the channels, which occurs every 30 min.

Differential Loran-C Receivers

A pair of II Morrow 611 Loran-C monitor receivers mounted in a special AVLS rack at the Dispatch Center supply location information to the computers. This information is used for calculating differential corrections to update all vehicle location data used and displayed within the Detroit Public Safety Dispatch System. This automatically accounts for season- and weather-related changes to the propagation times of the Loran-C signals.

AUTOMATIC VEHICLE LOCATION COMPUTER

Computer System

The AVLS computer system is made up of two DEC Micro VAX II computers. This was done for redundancy to ensure continued operation should a hardware failure occur. In the event one of the major systems (CAD, MDT, or AVLS) fails, system communications are routed so that the remaining two systems can continue to operate. The computers include communications subsystems for CAD, MDT, and workstation interfacing. AVLS polling is monitored and controlled through the computers. Also, Geobase data is reduced and directed to its proper destination.

Dispatch Workstations

The dispatch workstations were designed to combine CAD, MDT, and AVLS information and dispatch it through the use of windows and mouse-driven selection criteria. The menu levels were set up to minimize the number of keystrokes needed to access often-used functions. High-resolution graphics were installed for high-quality maps. State-of-the-art graphics boards were used for quick drawing of the maps when the dispatcher wants to zoom in or out to smaller or larger areas. Vehicles are shown on the maps by icons that also show their status by color and shape. The shapes quickly display to the dispatcher those vehicles which are currently unassigned.

SUMMARY

Shipment of the first vehicle Loran radio units began early in June. Factory acceptance tests of the Automatic Vehicle Location System (AVLS) were completed in September 1988, and the system was shipped and installed in the Dispatch Center at Police Headquarters in Detroit during October 1988. Since then, system integration of the AVLS computer with the Computer Aided Dispatch (CAD) and Mobile Data Terminal (MDT) systems has been accomplished, along with the operational and accuracy testing of the fleet tracking. Parallel testing of the new automated system as compared against the manual method of operation was completed in late June 1989. With the completion of this testing, the Detroit Public Safety Dispatch System once again made history by leading this nation with the most sophisticated automatic dispatch center in the world.

ACKNOWLEDGMENT

The author wishes to thank the II Morrow and ACT design teams for the innovative and dedicated work that has produced a major advance in the systems approach to vehicle tracking and dispatching. Thanks are also extended to Lyle Gibby for his assistance in the preparation of this paper.

Based on a paper presented at The Institute of Navigation National Technical Meeting, San Mateo, CA, January 1989.

AUTOMOBILE NAVIGATION: WHERE IS IT GOING?

Robert L. French
Consultant
Fort Worth, TX 76107

ABSTRACT

Automobile navigation systems based on dead reckoning, map matching, satellite positioning and other navigation technologies are under active development. "Map intelligent" systems achieve high relative accuracy by matching dead-reckoned paths with road geometry encoded in a digital map data base that may also serve other functions such as vehicle routing and geocoding. Satellite-based navigation systems achieve high absolute accuracy but require dead reckoning augmentation because of signal aberrations in the automotive environment. Future systems will probably include multiple navigation technologies. Issues influencing the design of future systems include safety concerns regarding the driver interface, and the future availability of comprehensive map data bases, real-time traffic data, and mobile data communication links necessary for on-board generation of optimum routes.

INTRODUCTION

The capabilities and functions of automobile navigation systems in the 1990s will be shaped by important issues encountered in the 1980s.

- Choosing Technology Paths
- Integrating Overall Systems
- Resolving Driver Interfaces
- Providing Map Data Bases
- Coordinating Mobile Communications

This paper reviews each of these issues, and discusses possible scenarios for the future based upon current directions here and abroad. It is a companion piece to an earlier paper which traces the history of automobile navigation and describes contemporary navigation technologies [1].

Based on a presentation at PLANS '86.

Vehicular navigation systems, digital maps, and mobile communications equipment are interrelated product areas that have large potential markets because of their wide applicability to consumer as well as business vehicles. In the United States, for example, there are approximately 150,000,000 automobiles in use, and an additional 10,000,000 are sold each year. U.S. business fleets include 3,400,000 automobiles, 1,600,000 light trucks and vans, 600,000 heavy trucks, and approximately 500,000 buses.

Digital maps and mobile data communications combine synergistically with vehicular navigation to multiply its usefulness and effectiveness, and to enhance the potential market for both consumer and commercial applications. Table I lists system capabilities that may be realized by vehicle navigation alone, by navigation in conjunction with stored digital maps, and in combination with both digital maps and data communications. Clearly, the direction and pace of vehicular navigation system development and deployment will be strongly influenced by the availability of comprehensive digital map data bases and supporting data communications.

Table I. Vehicular System Capabilities

SYSTEM FUNCTIONAL CAPABILITIES	NAVI-GATION ONLY	WITH MAP DATA	WITH DATA COMM.
Location Information			
In-Vehicle	X	X	X
Central Dispatch			X
Route Guidance			
Historical		X	X
Dynamic			X

Reprinted from *IEEE AES Mag.*, pp. 6–12, May 1987.

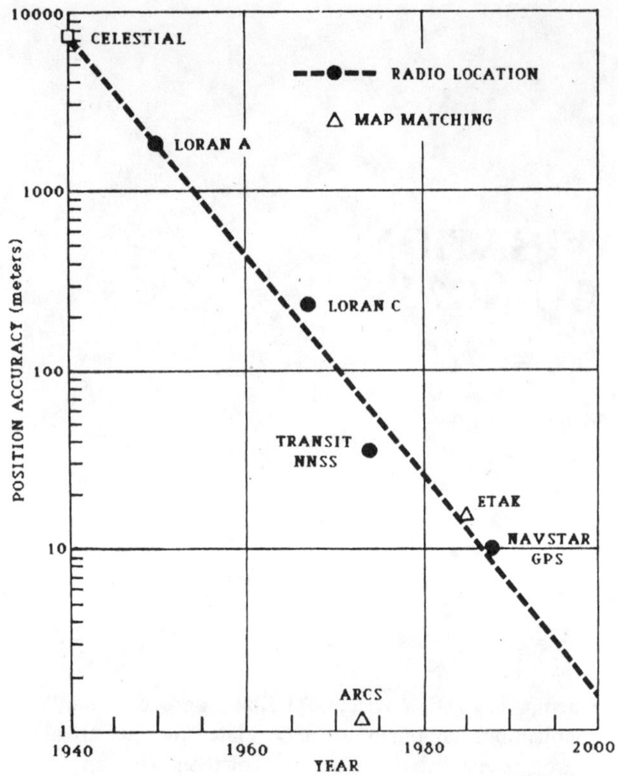

Figure 1. Improvements in Position Accuracy

TECHNOLOGY PATHS

The effectiveness of navigation technology is traditionally rated in terms of position accuracy. Figure 1 shows the improvements in accuracy that have been achieved in recent decades with multi-purpose radio-location technologies as reported by Luse and Malla [2]. The accuracy of celestial navigation in 1940 [2], and the accuracy of recent map matching technologies reported by French and Lang [3] and by Honey and Zavoli [4] are included for comparison.

Radio Location

The exponential improvement in radio-location accuracy with time is striking, but it should be noted that the target date of 1988 for Navstar Global Positioning System (GPS) availability will not be met as a result of the presently diminished U.S. capacity for launching satellites. Loran is now gaining some popularity as a means of tracking land vehicle location from a central location [5,6], but its modest accuracy, along with the persisting lack of mid-continent coverage, limits its usefulness for automotive navigation.

Although the absolute accuracy of Navstar GPS has great appeal for future automotive navigation, it is not the panacea inferred by reports following Chrysler's introduction of the CLASS concept at the 1984 World Fair in New Orleans [7] and the investigation of satellite navigation by other automobile manufacturers [8]. GPS signal disruption by buildings, bridges, foliage, etc., may produce positioning discontinuities when GPS receivers are operated in automobiles. Auxiliary dead reckoning subsystems are thus required for effective automotive navigation using GPS.

Other future satellite systems may also offer vehicle navigation as well as mobile data communication services [9]. In addition, there is potential for new radio-trilateration approaches using subcarrier signals from cellular radio transmitters [10].

Dead Reckoning

Dead reckoning is the process of determining a vehicle's location relative to an initial position by integrating measured increments and directions of travel. Most dead reckoning technologies used in automobile navigation systems were developed long before the automobile itself [1]. These include the odometer, the differential odometer, and the magnetic compass. The gyro compass, developed in 1906, has seen only limited use in automobile systems, and inertial systems are presently expensive and ill-suited for the harsh automotive environment.

Dead reckoning accuracy is difficult to quantize because it continuously decreases with time and/or the distance a vehicle is driven. For example, the vehicular odograph, a WW-II military navigation system based on the odometer and compass, accumulated 1 mile of error per 50 to 150 miles driven [11]. Even the most precise dead reckoning navigation systems require periodic reinitialization.

Map Matching

Artificial intelligence concepts may be applied to match dead-reckoned vehicle paths with road maps which are digitized and stored in computer memory. With map matching, sensed mathematical features of the vehicle path are continuously associated with those encoded in a map data base, just as a driver associates observed landmarks and road features with those depicted on a paper map to recognize position. Thus a vehicle's dead-reckoned location may be automatically reinitialized at every turn to prevent accumulation of dead reckoning error.

The first vehicular application of map matching technology was in the Automatic Route Control System (ARCS) which used a differential odometer for dead reckoning [3]. ARCS had an average location accuracy of 1.15 meters (see Figure 1) relative to predetermined routes that were initially digitized by direct field measurements using the ARCS equipment. Since the digital route map data were from replicable field measurements, the 1.15-meter location accuracy is more a measure of the effectiveness of the map matching algorithm rather than the accuracy of the dead reckoning technique or of the map data being matched.

The more recent Etak map-matching system uses a solid state flux-gate compass as well as the differential odometer to dead reckon paths for matching with digitized maps compiled from Census Bureau GBF/DIME files, U.S. Geological Survey topographical maps, and aerial photographs [4]. As shown in Figure 1, the Etak system exhibits an accuracy of approximately 16 meters relative to the digitized maps. The Etak map data base is within 16 meters of ground truth, and has a relative accuracy of 5 meters on each map according to Dial [12].

Interestingly, neither the source nor the magnitude of dead reckoning and map data errors need be overriding concerns when map matching is used. As long as the street and road connectivity is accurately defined, the map matching process identifies position relative to the road network as visually perceived by the vehicle driver. The information actually needed by the driver may thus be presented in a customary frame of reference. After all, vehicle location coordinates are of little value to a driver unless used in conjunction with a map [13].

Even relatively crude dead reckoning is sufficiently robust to support map matching when operating on a defined road network. However, good dead-reckoning accuracy is required to achieve reinitialization through map matching upon returning to the road network after off-road operation such as in a parking lot.

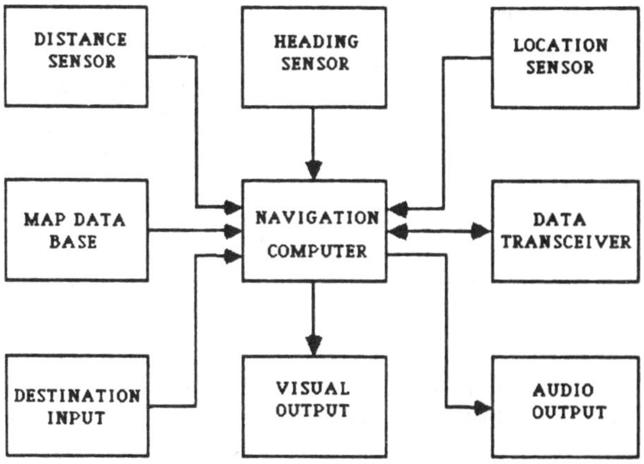

Figure 2. Future System Configuration

Proximity Beacon

This approach uses strategically located short-range transmitters, and the very reception of their location-coded signals infers the receiving vehicle's instantaneous location. Several variations of the proximity approach, some involving two-way communications with equipped vehicles, have been investigated for interactive route guidance [14]. Typically, the driver enters a destination code on the vehicle panel for automatic transmission to a roadside unit as the vehicle approaches an instrumented intersection. The roadside unit, which may be networked with a traffic management system, analyzes the destination code and instantly transmits route instructions for display on the vehicle's panel. Alternatively, the roadside unit may only transmit its location to the vehicle where an on-board computer, using stored road network data, generates the route instructions from the identified location.

The proximity beacon approach, inactive in the U.S. since Congress mandated termination of the ERGS project in 1970 [15], has been the subject of further development and testing in Japan [16] and West Germany [17] where the task of instrumenting the road networks is less formidable. For example, the ALI-SCOUT proximity beacon system, which uses dead reckoning and map matching techniques between beacons which download updated map and traffic data, is nearing testing in West Berlin [18].

FUTURE INTEGRATED SYSTEMS

What configurations will prevail in automobile navigation systems of the future? Patterns already set in the design of other automotive systems, as well as computer and entertainment systems, suggest commonality and modularity as likely system characteristics.

Figure 2 is one concept for the type of general purpose systems likely to appear in the next 15 years. It represents a composite of state-of-the-art developments, provides the modularity necessary for adaptation to specific applications, and leaves room for vendor differentiation.

Distance and heading sensing will be included even if precise location sensing, such as Navstar GPS, is available. Distance and heading sensing may be accomplished by differential odometry using input signals from existing antilock braking systems, or in combination with software compensated flux-gate magnetic compasses. The fibre optics gyroscope [19] also shows potential as an inexpensive and rugged means for accurately sensing heading changes.

Digitized maps are as essential to the automotive navigation systems of the future as paper maps and charts were to the professional maritime navigators of the past. Large capacity storage is required for useful amounts of map data. Cooke estimates that 60 percent of the U.S. population lives on about one million streets represented in the Census Bureau's GBF/DIME files [20]. Assuming 120 to 150 bytes per street, simple extrapolation allowing for rural areas suggests that a nationwide digital map would fit on one 500 MByte CD-ROM disk.

Map matching systems based upon dead reckoning alone can become confused if digital maps are not current, or from extensive driving off the defined road network. Thus absolute location sensors, such as satellite positioning or proximity beacons, will be required if occasional manual reinitialization is to be avoided.

Data communications will also be a feature of future systems in countries that integrate traffic management with in-vehicle route guidance to enhance the benefits. One-way communication from the infrastructure to the vehicle is the most useful link. However, additional system benefits would be provided by vehicle-to-infrastructure data communications. This could provide destination information to central traffic management systems for planning optimal traffic flow or, as in the case of the ALI-SCOUT [18], eliminate the need for traffic sensors by reporting recent travel experience to the central traffic management system.

The lower three blocks of Figure 2 comprise the driver interface, an important and controversial topic which is discussed separately.

DRIVER INTERFACES

The approach to the interface between an on-board navigation system and the vehicle operator must take into account ergonomics and safety considerations as well as functional requirements. Most systems proposed or developed to date use detailed map displays or some combination of symbolic graphics, alphanumeric messages, and audio signals.

There is considerable information available to automotive systems design engineers on display and control surface location, display brightness, contrast, color, font type, character density, analog vs digital representation, and many other signal properties have been well defined as a result of many years of intensive aerospace research. Good sources of data are also available that specifically relate to automobile and traffic control device designs. However, considerably less information is available on when to present information to motorists; priorities for presentation of different types of information; temporal, spatial, and modal redundancy; and resolution of apparent conflicts in information derived from outside and inside the vehicle [21].

Insights on the driver interface issue may be gained by reviewing approaches that have been tried or proposed here and abroad.

United States

Some of the most serious research on driver interfaces was in connection with the ERGS project of the late 1960's [22]. The ERGS vehicle unit back-illuminated various combinations of arrows and words on a guidance panel. Tests were also conducted with "heads up" displays which projected simple combinations of arrows and words on the windshield so that the driver did not have to remove his eyes from the road [23].

Safety concerns about automotive use of CRT displays and elaborate control panels more typical of aircraft are exemplified by those of Zwahlen [24]. He points out the potential for visual overload leading to impairment of lateral steering control.

In the early 1970s, ARCS, an automatic route control system developed for operation over programmed newspaper routes, used prerecorded audio route guidance instructions during extensive operational tests [3]. This approach worked well, but tape recorded audio instructions were awkward to prepare and control, and systhesized speech was not yet available. Therefore, an improved version used a plasma display panel to give route guidance in the form of shaped arrows along with street names, etc. [25]. Subsequent research reported by Streeter [26] established that drivers who listened to directions drove to destinations in fewer miles, less time, and with about 70 percent fewer errors than drivers using customized route maps.

The next generation of systems development in the United States used CRT displays for the driver interface. These include satellite-based navigation systems demonstrated by Ford [27], General Motors [28], and Chrysler [7], all of which displayed detailed map images. The Etak Navigator, the only advanced automotive navigation system actually on the market in the U.S., carries the trend a step further by displaying a map that rotates to match vehicle heading [4].

Europe

European automobile navigation system designs seldom use elaborate visual displays for the driver interface. The principal exception is the Philips CARIN system [29] which includes a color CRT map display for showing vehicle location relative to the surroundings. CARIN includes synthetic voice for conveying instructions to the driver when operating in the route guidance mode.

The earliest example of a route guidance system incorporating automatic route generation using on-board digital maps was "Micropilot" developed in England in 1981 [30]. This system used an audio interface in the form of digitized voice — not speech synthesis — with a vocabulary of twenty six words.

West German designs invariably use simplified visual displays, sometimes in combination with audio messages, for conveying route instructions to the driver. Virtually all West German systems to date use some combination of short visual messages, symbolic graphics, and/or voice.

One example is the route guidance system described by Haeussermann in a 1984 paper [31]. When on highway networks, this system uses a 2-line LCD display with 16 characters per line to give the next route point and remaining distance. When on city streets, an alternate LCD display with a pointer indicates the direction of the destination, and shows numerals to indicate the remaining distance.

The first generation EVA system developed by Bosch-Blaupunkt also uses simplified graphics to convey route instructions [32]. The original prototype includes differential odometer, map data base with map matching, and route search software to generate explicit route guidance instructions. The main display includes a vertical LCD panel for graphics and a small horizontal LCD strip for character display. Voice capability is included, and is used to confirm destination entries. Turns at complicated intersections, lane changes, etc., are specified to the driver in the form of simplified diagrams which show lane boundries and use arrows to indicate the path to be taken.

Bosch-Blaupunkt also provides the interface equipment for the ALI-SCOUT system, a joint project of the West German Government, Siemens, Volkswagen, Blaupunkt and others. ALI-SCOUT is also a route guidance system, but rather than being autonomous, it depends upon the reception of area road network data and recommended route data broadcast from strategically located IR beacons [18]. Driving directions are presented much the same way as in EVA, but with an additional feature similar to the "Wolfsburg wave" [33]. The Wolfsburg wave is essentially a bar graph that, in this application, gives a "count down" to the exact point where the vehicle is to turn, thus clearly delineating among closely spaced turns.

Destination input, as well as system control, for the next generation of ALI-SCOUT (which will be subjected to large scale user tests in West Berlin starting next year) will be via a hand held wireless remote control unit with shift keys for alphanumeric information. Thus initializing the system for a trip will be much like remotely programming a VCR, and may be done by anyone in the automobile.

Japan

Current driver interface approaches for Japanese automobile navigation systems seem to align more with those of the U.S. than with Europe. However, Japan's first major step toward route guidance was the CACS (Comprehensive Automobile Traffic Control System) project [16] of the 1970s which was patterned after the U.S. ERGS (Electronic

Route Guidance System) project of the late 1960s [22]. Like ERGS, CACS used simple combinations of direction indicators and descriptors to indicate routes to be taken.

Autonomous navigation systems started appearing in Japan in the early 1980s. The Nissan Driver Guide System [34] displayed information in a simplified graphical form "in order to make the bearings and distance to the destination easily viewable while driving." Directional arrows showed the direction to the destination, and a bar graph indicated the fractional distance remaining.

Appearing about the same time was the Honda "Electro Gyro-Cator" navigation system [35] which used a CRT display to show a plot of the vehicle's path. Location could be established by using a transparent map overlay.

Subsequent systems shown in Japan incorporate color CRT map displays. These include systems displayed by several automobile manufacturers at the October 1985 Tokyo Auto Show [36]. Another example of current directions is a Loran-based Nissan delivery truck system which also uses a color CRT map display [37].

A new integrated systems approach currently under joint development by Japanese industry and government groups potentially includes an on-board computer, CRT display, compact disk with road network data, dead reckoning sensors, roadside electronic signposts for location confirmation as equipped cars pass within range, and map matching to augment dead reckoning between signposts [36].

MAP DATA BASES

"Not since cartographer Gerardus Mercator started mapping the world as round instead of flat in the 16th century has the science of map making been as stirred up." Thus observed the *Wall Street Journal* in an April 18, 1985 article on the impact of digital maps on cartography. Indeed, the introduction of the computer to map making is being hailed as a revolution that might be comparable to the development of the printing press. The market for computer mapping systems is expected to grow from $250 million to $1 billion during the last half of this decade [38].

Irrespective of the progress being made in preparing and using maps in digital form, a major barrier to widespread implementation of vehicular navigation systems will be the lack of comprehensive map data bases to support advanced navigation functions such as route guidance. Developing and maintaining the large amounts of map data required are costly tasks that are difficult to justify commercially until enough map data is available for realistic feedback on the demand [13].

The most extensive body of vector-encoded map data is the GBF/DIME (Geographic Base File/Dual Independent Map Encoding) System developed by U.S. Census Bureau for virtually all Standard Metropolitan Statistical Areas (SMSA). GBF/DIME includes street names and address ranges between nodes, as well as a variety of information not relevant to vehicle navigation. This map data base has numerous errors and represents the street network as of 1977 [20]. Nonetheless, it is a highly useful resource as it provides an established map encoding approach which is applicable to digital maps for use with vehicular navigation systems.

In particular, graph theory is used as a conceptual framework for mathematically defining the location of all intersections and the interconnecting streets or roads, thus reducing conventional road map information to computer files. Road geometry is modeled in terms of nodes and segments with nodes representing the location of intersections, turns, etc., and connecting segments representing increments of road.

In order to discuss map data requirements for vehicular navigation systems it is necessary to consider the functions to be performed [13]. The driver presumably needs a user friendly road map for trip planning, vehicle location, orientation, etc. Digitized map data is also needed for map matching regardless of what location technique is used. A third use of digital map data is for automatic routing, an especially demanding application which requires that the map data base include information on traffic directions, turn restrictions and other ephemeral data. Other digital map uses relating to vehicle navigation include geocoding (what is the location of a specific address?) and location directory (where is the nearest garage?).

Standards for digital map data bases are not yet available although the newly organized Automotive Navigational Aids Subcommittee of the Society of Automotive Engineers (SAE) has formed a task group to address the need for map data base standards. Nonetheless, one automobile navigation systems company, Etak, Inc., has undertaken development of a proprietary digital map data base for the entire U.S. [4]. The Etak map data base will support navigation systems which display vehicle location, destination, etc., relative to the road network, as does Etak's own system and the version being developed under license by General Motors for introduction in 1989. Systems which automatically generate an appropriate route to an input destination will require a map data base containing additional attributes as described above.

MOBILE DATA COMMUNICATIONS

Major potential roles for data communications in future automobile navigation systems will be to provide current updates (road additions, closures, detours, etc.) for on-board map data bases, and to provide real-time information on traffic conditions for systems that include on-board route generation. Mobile data communications offer additional advantages for commercial vehicles using navigation systems, including centralized vehicle location monitoring and dispatch control.

Mobile cellular radio offers great potential for two-way data communications with vehicles equipped with navigation systems. As the public adopts the idea of the "mobile office", there will be considerable demand for cellular radio for the transmission of data as well as voice. The necessary modems are already beginning to appear on the market [39].

For one-way data communications with vehicle navigation units, an alternative area broadcast approach such as the Radio Data System (RDS) proposed by the European Broadcasting Union [40] may be advantageous. This enables digital information to be superimposed on the normal broadcasts of FM radio stations by means of sub-carrier modulation.

Although appropriate technologies may be available for communicating updated map and traffic data to vehicular systems, there are numerous institutional barriers to the implementation of supporting data communications. Standards are not only required for the frequency, modulation techniques, communications mode, error detection and correction techniques, etc., but the data content of transmissions should follow a standard structure. Some progress has been made on these issues in vehicular navigation in Europe [40], but efforts to address these requirements in the U.S. have not yet been initiated.

A critical prerequisite for dynamic route guidance is coordination among traffic authorities at the local, regional, state and national levels for the uniform collection and dissemination of traffic information for communication to vehicular systems. Without initiative on the part of the federal government, there does not appear to be a viable mechanism for bringing about the necessary coordination in the U.S.

CONCLUSIONS

The various technologies required for future automobile navigation and route guidance systems that will relieve drivers from the tedium of planning routes and finding the way over them have already been developed, or are within reach. The main developments yet to come are in the information and institutional areas. In particular, large private sector investments will be required for the development and maintenance of comprehensive digital map data bases, and coordination among public sector organizations will be required for collecting, standardizing, and communicating real-time information on traffic and road conditions.

REFERENCES

1. R. L. French, "Historical Overview of Automobile Navigation Technology," *Proceedings of the 36th IEEE Vehicular Technology Conference,* 350-358, Dallas, Texas, May 20-22, 1986.

2. James D. Luse and Rajendra Malla, "Geodesy from ASTROLABE to GPS — A Navigator's View," *Navigation,* Vol. 32, No. 2, 101-113 (Summer 1986).

3. R. L. French and G. M. Lang, "Automatic Route Control System," *IEEE Transactions on Vehicular Technology,* Vol. VT-22, 36-41 (1973).

4. S. K. Honey and W. B. Zavoli, "A Novel Approach to Automotive Navigation and Map Display," *RIN Conference Proceedings "Land Navigation and Location for Mobile Applications,"* York, England (1985).

5. R. V. Janc, "Consideration of the Various Error Sources in a Practical Automatic Vehicle Location System, " *34th IEEE Vehicular Technology Conference Record,* 277-284 (1984).

6. R. Bronson, W. Sears and L. Cortland, "II Morrow's Loran C Based Vehicle Tracking System," *RIN Conference Proceedings "Land Navigation and Location for Mobile Applications,"* York, England (1985).

7. M. Lemonick, "Now: Driving by Satellite," *Science Digest, 92,* 34 (1984).

8. D. Gable, "Automobile Navigation: Science.Fiction Moves Closer to Reality," *Electronic Engineering Times, 296,* D14, D18 (1984).

9. Roy E. Anderson, "Mobile Satellite Services for Transportation Industries," presented at the 1986 Annual Meeting of the Transportation Research Board (1986).

10. Lewis Sabounghi, "Applications to Goods Movements," presented at the 1986 Annual Meeting of the Transportation Research Board (1986).

11. J. D. Faustman, "Automatic Map Tracer for Land Navigation," *Electronics,* Vol. 17, No. 11, 94-99 (1944).

12. Robert B. Dial, "Etak's Nationwide Road Network Database: Its Attributes and Applications in Urban Transportation Planning," presented at the 1986 Annual Meeting of the Transportation Research Board (1986).

13. Gary M. Andrew, "Map Database Requirements and Outlook," presented at the 1986 Annual Meeting of the Transportation Research Board (1986).

14. R. L. Gordon, B. Soicher and E. Donaldson, "Study of the Feasibility and Design Configuration for In-Vehicle Route Guidance," Sperry Corp. Report No. FHWA/RD-81/056 (1981).

15. Dan A. Rosen, "Applications to Traffic Management," presented at the 1986 Annual Meeting of the Transportation Research Board (1986).

16. H. Fujii, "The CACS Project and Now: Dynamic Route Guidance as the Final Target," presented at the 1986 Annual Meeting of the Transportation Research Board (1986).

17. Peter Braegas, "Function, Equipment, and Field Testing of a Route Guidance and Information System for Drivers (ALI)," *IEEE Transactions on Vehicular Technology,* Vol. VT-29, 216-225 (1980).

18. R. von Tomkewitsch, "ALI-SCOUT — A Universal Guidance and Information System for Road Traffic," *Proceedings of the IEE International Conference on Road Traffic Control,* 22-25, London, England, April 15-18, 1986.

19. Y. Yoon Kim and Herbert John Shaw, "Fiber-Optic Gyroscopes," *IEEE Spectrum,* Vol. 23, No. 3, 54-60 (1986).

20. D. F. Cooke, "Vehicle Navigation Appliances," *Auto-Carto 7 Proceedings,* 108-115 (1985).

21. Burton W. Stephens, "Overview: Human Factors Design Considerations," presented at the Annual Meeting of the Transportation Research Board (1986).

22. D. A. Rosen, F. J. Mammano and R. Favout, "An Electronic Route Guidance System for Highway Vehicles," *IEEE Transactions on Vehicular Technology,* Vol. VT-19, 143-152 (1970).

23. R. W. Benzinger and E. Bell, "Experimental Route Guidance Heads Up Display Res?earch," Highway Research Board, Research Record No. 265, 62-70 (1968).

24. Helmut T. Zwahlen, "Competition Between Searching for In-Vehicle Informantion Displays and Vehicular Lateral Control Performance," presented at the 1986 Annual Meeting of the Transportation Research Board (1986).

25. R. L. French, "On-Board Vehicle Route Instructions via Plasma Display Panel," *SID International Symposium Digest of Technical Papers,* Vol. 5, 146-147 (1974).

26. Lynn A. Streeter, "Interface Considerations in the Design of an Electronic Navigator," presented at 1985 International Symposium on Computer Assisted Cartography (Auto-Carto 7), Washington, DC, March 11-14, 1985.

27. M. W. Jarvis and R. C. Berry, "Cathode-Ray Tube Information Center with Automotive Navigation," SAE Technical Paper Series, No. 840313 (1984).

28. R. Dork, "Automotive Navigation System," *Detroit Engineer & Michigan Engineering,* Vol. 44, No. 8, 11-13 (1983).

29. M. L. G. Thoone and R. M. A. M. Bruekers, "Application of the Compact Disc in Car Information and Navigation Systems," SAE Technical Paper Series, No. 840156 (1984).

30. H. J. Wootton and A. C. Brett, "Route Information Systems — Signposts of the Future," *Proceedings of the 2nd IEE International Conference on Road Traffic Control,* 15-17, London, England, April 15-18, 1986.

31. P. Haeussermann, "On-Board Computer System for Navigation, Orientation and Route Optimization," SAE Technical Paper Series, No. 840485 (1984).

32. Otmar Pilsak, "EVA — An Electronic Traffic Pilot for Motorists," SAE Technical Paper Series, No. 860346 (1986).

33. W. Zindahl, "Guidelines and Some Developments for a New Modular Driver Information System," *Proceedings of the 34th IEEE Vehicular Technology Conference,* 178-182, Pittsburgh, Pennsylvania, May 21-23, 1984.

34. K. Mitamura, S. Chujo and T. Senoo, "The Driver Guide System, " SAE Technical Paper Series, No. 830660 (1983).

35. K. Tagami, T. Takahashi and F. Takahashi, "Electro Gyro-cator, New Inertial Navigation System for Use in Automobiles," SAE Technical Paper Series, No. 830659 (1983).

36. M. Shibata, "Navigation Technology Status and Directions in Japan," presented at the Annual Meeting of the Transportation Research Board (1986).

37. "Vehicle Navigation System Takes Care of Business," *High Technology,* Vol. 5, No. 8, 7 (1985).

38. "Grady: $1 Billion Mapping Market by 1990," *Computer Graphics Today,* Vol. 3, No. 2, 4, 9, 23 (1986).

39. E. Ryerson Case, "Overview of Automobile Navigation Systems and Potential Applications," presented at Institute of Transportation Engineers District 6 & District 7 Annual Meeting, Vancouver, BC, July 27-31, 1986.

40. K. Russam and E. J. Jeffery, "Route Guidance and Driver Information Systems — An Overview," *Proceedings of the 2nd IEE International Conference on Road Traffic Control,* 1-5, London, England, April 15-18, 1986. ■

EURONAV^{TM} - A STATE OF THE ART MILITARY GPS RECEIVER

DR. ANIL K. AGGARWAL

MAGNAVOX ADVANCED PRODUCTS AND SYSTEMS COMPANY

ABSTRACT

Magnavox has developed two and five channel EURONAV^{TM} GPS receivers for military applications. Both types of receivers have a wide range of features in addition to providing the basic GPS navigation functions. These receivers provide high performance, selective availability/anti-spoofing, high anti-jam GPS capabilities whose design has evolved over two decades of Magnavox experience in designing and manufacturing military GPS user equipment. By the use of advanced technology components, Magnavox has been able to reduce the size of a five channel GPS receiver from the GPS Phase II qualified 3/4 ATR long to 3/8 ATR short, with a corresponding weight and power reduction. This was accomplished without loss of performance capabilities and was designed to meet the performance requirements of the SS-US-200 specification for medium and high dynamic aircraft.

EURONAV^{TM} receivers have been built with a multitude of interfaces to support integration with a multiplicity of host vehicles. Specifically, these receivers provide the following interfaces: precise-time, time interval interface to output highly accurate timing data; Havequick; MIL-STD-1553B; dual channel low or high speed ARINC 429 for use in interfacing INS and digital flight instruments; RS-422 serial for control display or as an expansion port; a second RS-422 serial for instrumentation (ICD-GPS-204A); and an altimeter (AAU-32) input usable for GPS aiding. Additional interface with KYK-13 is provided to support full denial of accuracy and anti-spoofing. The standard features of the EURONAV^{TM} receivers include a security module, low power time source for self-initialization of the receiver, battery backed non-volatile RAM, waypoint storage, ARINC or circular connectors and configuration discretes with selectable switches for easy adaptation to many platforms without hardware or software modifications.

1.0 INTRODUCTION

A family of compact, lightweight, militarized navigation receiver processors, (RXP) has been designed by Magnavox for use as a part of the user system segment of the NAVSTAR Global Positioning System (GPS). They provide long-range accurate navigation for ships, aircraft and ground operation. RXP sets may be employed as a navigation sensor for aircraft (A-RXP) or at sea (N-RXP). The RXP is produced in 2-, and 5-channel versions. In the 2-channel Sets, the GPS satellites are tracked sequentially. The 5-channel Sets can track up to 5 GPS satellites simultaneously, allowing for greater dynamics by the Host Vehicle.

The 2-channel Sequential EURONAV^{TM} Receiver (A-RXP-2) and the 5-channel Continuous Tracking EURONAV^{TM} Receiver (A-RXP-5) with high anti-jam capability for medium to high dynamic performance aircraft have evolved from twenty years of Magnavox experience in designing and manufacturing military GPS user equipments. The EURONAV^{TM} system design evolved from the GPS Phase II architecture, algorithm and field test experience with a new hardware and software design implementation. To this end the EURONAV^{TM} retains the proven hard limiting - phase sampling digital receiver design and the software has both unaided and aided (loosely or tightly coupled) navigation modes fully compliant with the U.S. military user equipment specifications.

The A-RXP-2/5 receiver processors are designed and built through the use of common modules. These modules include RF/IF/synthesizer correlator, reference oscillator, baseband, CPU/memory, Input/Output Processor (IOP) and power supply. Figure 1 is a block diagram representation of the A-RXP-2 and A-RXP-5 receivers. Except for the number of channels employed in the two receivers, the hardware for these receivers is identical. This common module design approach is important since it lowers hardware design and development cost and at the same time improves user maintainability.

To allow for future evolution, either in cost reduction or performance improvement, the overall system and chassis design has been implemented in such a manner as to allow insertion of new technology parts. This forward going technology insertion approach will minimize technology obsolescence and provide the highest possible reliability while maintaining form-fit-function with earlier units.

First article tests in the plant as described later in this paper and field tests in France and Germany have shown remarkable improvement in performance from Phase II GPS Sets with regards to repeatability, reliability, jamming, navigation and automaticity in operation.

Reprinted from *Rec. IEEE Position Location and Navigation Symp.*, pp. 153-164, 1988.

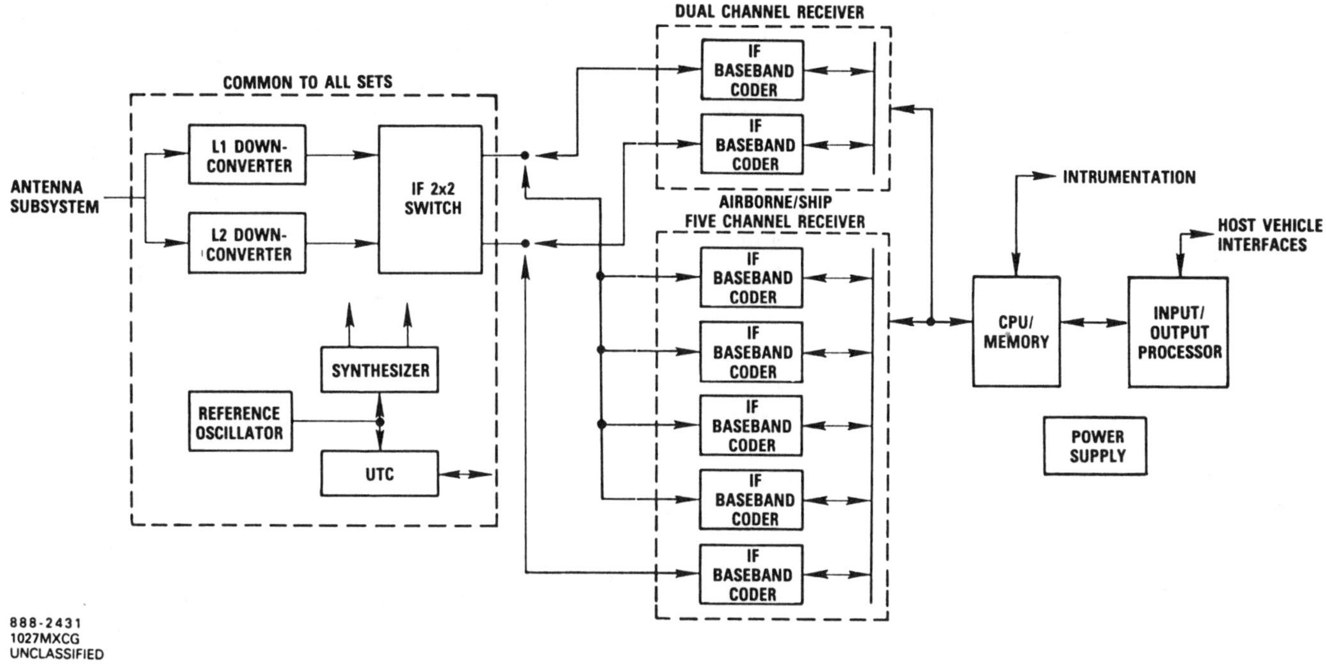

Figure 1. EURONAV™ Set Architecture

2.0 PRODUCT SPECIFICATION

A-RXP-2/5

As a compact and lightweight navigation aid, the A-RXP is designed for installation in an aircraft. The EURONAV™ receivers provide full capability for military missions; namely, dual frequencies, P-code, access to Selective Availability (SA) and Anti-Spoofing (AS) for those users with appropriate U.S. Government approval, jamming resistance and Class 2 MIL-E-5400 environmental specification. Table 1 describes the characteristics of the EURONAV™ family of GPS receivers.

CDU

The Control Display Unit (CDU) is an ATI standard size box with a CRT primary display, LED auxiliary displays, and a control keyboard on the front panel. Its weight is less than 3.5 kg.

Antenna/PAU

The fixed radiation pattern atenna (FRPA-4) is used in conjunction with an optional RF preamplifier unit (PAU) for the reception of GPS satellite signals. It receives both L1 and L2 signals and is designed for aircraft installation with a low physical profile. The associated PAU consists of a low noise L-band amplifier with a maximum gain of 20 dB. The PAU is adjustable for different lengths of antenna cables by selection of one or more of the PAU's internal attenuators. The PAU filters and amplifies L1 and L2 signals received from the antenna and feeds them to the A-RXP via coaxial cable. This coaxial cable is also used to provide the PAU with the required power from the A-RXP.

The EURONAV™ User Equipment is shown in Figure 2.

3.0 SYSTEM DESCRIPTION

EURONAV™ design is based on the Magnavox Phase II Full Scale Engineering Development Program of the US-DOD for the GPS User Equipment. The major changes in the equipment design from Phase II are: changing the receiver frequency plan to match the Phase III standard GPS Receiver 3A, changing the SOS-20 (based on UYK-20 architecture) computer to Intel 80286/80287, changing the baseband microprocessors (5 in all) to a single Intel 80386 processor, and shifting to 'C' language from the assembly level language for bit manipulation. Following is the description of the updated design for the receiver, signal processing and navigation filter.

Receiver Design

RF signal processing techniques similar to Phase II hardware are used from the antenna/preamp to IF (3 IF levels: 34 F, 12.25 F, .25 F, F= 5.115 MHz). The code correlation is done after the second IF at 12.25 F followed by a post correlation bandpass filter (BPF) with a bandwidth of 30 KHz to accommodate signal doppler in an unaided receiver application. The BPF signal output, converted to the last IF at 1/4 F (1.28 MHz) is passed through a phase sampler, accumulated and multiplied by stored sine and cosine tables to form the I and Q signal for the tracking loops. The I, Q samples are processed in the receiver software to form

Table I. EURONAV™ Receiver Processor Characteristics

	A-RXP-2/A-RXP-5
Application	Airborne
Environment	MIL-E-5400 Class 2
Temperature	-55°C to +71°C
Size	3.6" x 7.6" x 12.7"
Weight	16 pounds
Power	44 watts (5-ch), 40 watts (2-ch)
Maximum Velocity	1200 m/sec.
Maximum Acceleration	118 m/sec.^{2} (5-ch) 20 m/sec.^{2} (2-ch)
Maximum Jerk	180 m/sec.^{3} (5-ch) 20 m/sec.^{3} (2-ch)
Position Accuracy (SEP)	16m
Velocity Accuracy (1 sigma per axis)	0.1 m/sec.
Time Accuracy (1 sigma)	100 nsec.
Time-to-First Fix	120 sec. (5-ch), 360 sec. (2-ch)
Reaction Time	7 min. (5-ch) 9 min. (2-ch)
J/S Capability	53 dB
EMP Protection	Yes
Kalman Filter	11 state
BIT	95% Fault Detection 95% SRU ISO (with IMTS)
MTBF	≥2100 hrs. (5-ch) ≥5050 hrs. (2-ch)
1553 Interface	USAF Standard Protocol
ARINC INS	0,1,2
Baro Altimeter	Input
STANAG 4156	Option
KYK-13	Yes
Inst. Port	Yes
PTTI	Yes
Atomic Clock	1 PPS Input
CDU Input	0,1
HAVEQUICK	Output
SA/AS Capability	Unclassified Set when keyed

code and carrier tracking functions to develop a signal-to-noise power ratio estimate (C/No) and to perform pseudorange, and deltarange measurement processing.

Within each receiver channel (used N times in an N-channel receiver) the signal is despread, filtered, hard limited, and later processed by an

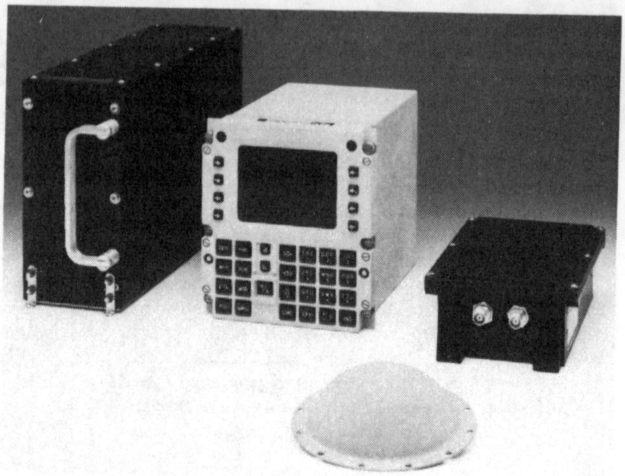

1087-2792
UNCLASSIFIED

Figure 2.

Intel 80386 microcomputer. The Intel 80386 performs the usual baseband processes such as code synchronization, code tracking, AFC discrimination, and Costas detection. Use of the on-time (t-dithered) phase of the feedback code allows closing both the code and carrier loop through the same IF signal path. The result is a size, power and weight savings over a system that utilizes separate code and carrier channels. An expanded block diagram representation of the baseband processing functions as implemented in the Intel 80386 is shown in Figure 3.

The carrier tracking function is implemented as an AFC/Costas loop. The AFC tracking loop is second order for the EURONAV receivers. The loop noise bandwidth is selectable as a function of receiver mode and C/No. The AFC reduces carrier frequency uncertainty to allow Costas acquisition which is required for data demodulation and precision deltarange measurements. The Costas loop is implemented as a second/third order filter depending upon the vehicle dynamics requirements.

The 5-channel high dynamics receiver set (A-RXP-5) consists of 5 channels. Each of these channels tracks one satellite's code and carrier. Four of the five channels track four visible satellite with optimal constellation whereas the fifth channel is used for iono measurements and alternate satellite tracking or ephemeris refresh. This design approach reduces the hardware and computer processing requirements.

Navigation Filter Design

The navigation parameters are estimated and controlled by a state covariance filter, commonly known as the linearized Kalman Filter with state error correction after each successful receiver measurement cycle. The filter is implemented by the upper triangular diagonal (U-D) factorization of the covariance matrix. The U-D Kalman filter states will be those which estimate the errors in navigation states. Three different environments are presented by land vehicles, shipboard platforms

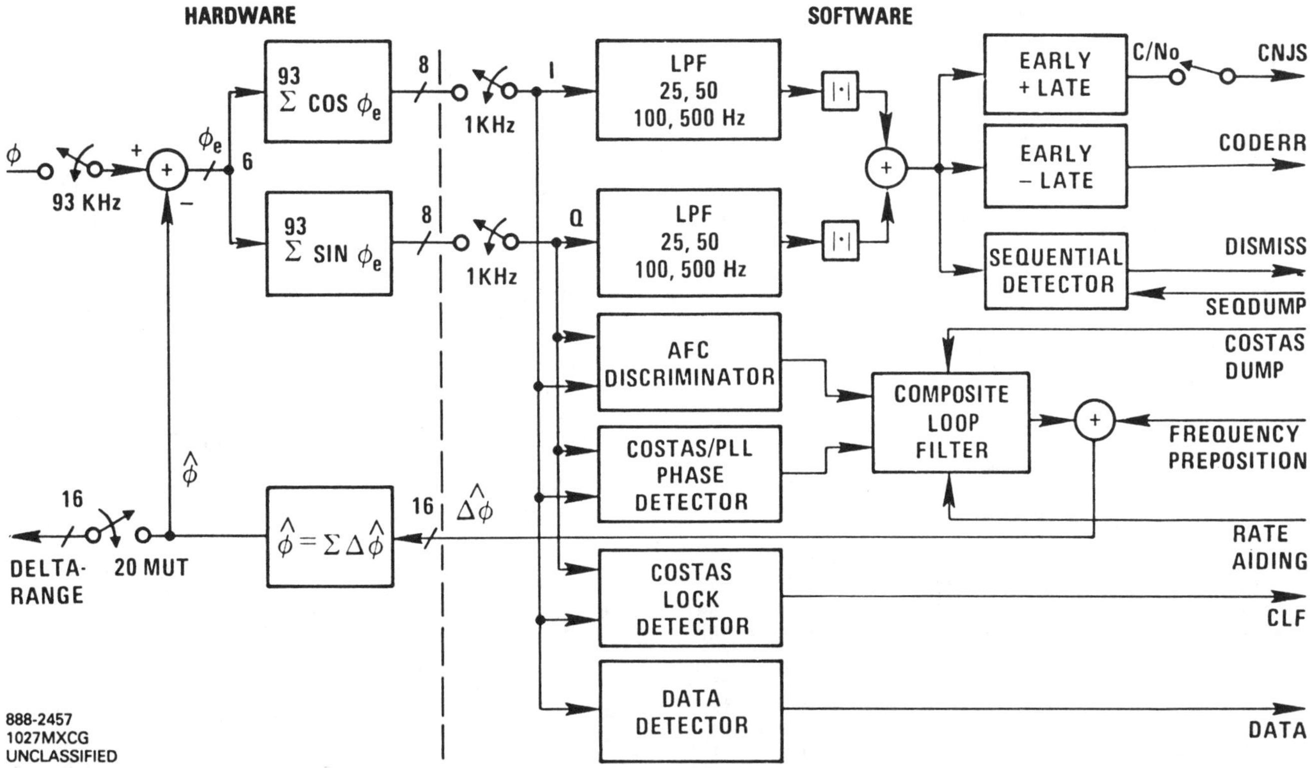

Figure 3. Block Diagram of Receiver Baseband Processing Functions

and low, medium, or high dynamics airborne platforms. These configurations are listed in Table II.

5-State Configuration - This configuration is used for stationary systems. The estimation state vector consists of clock parameters and 3-position components.

8 State Configuration - This configuration is used for low to medium dynamic, land, naval, and airborne platforms. The estimation state vector is augmented by 3-velocity components.

11 State Configuration - This configuration is used for medium to high dynamic airborne platforms. The estimation state vector is augmented by 3-acceleration components.

Another feature of the filter design is the coupling mode of the GPS navigation mechanization with the host vehicle aiding sensors. There are two such options: loosely or tightly coupled. Loosely coupled is when the navigation states are updated only from the GPS receiver measurements. Tightly coupled is when the navigation state propagation is made using external Host Vehicle sensor inputs and corrections to these states are made by GPS receiver measurements. The tightly coupled basic navigation state propagation may be either velocity coupled (if velocity is the Host Vehicle sensor output) or acceleration coupled (if acceleration or delta-V is the Host Vehicle sensor output). The preferred method is to have the basic navigation state propagation be tightly coupled.

In a loosely coupled configuration, the GPS Set uses the GPS navigation solution to calibrate all participating host vehicle sensors by employing a simple low-pass filtering scheme. If fewer than four satellites are available or if, instead, the system covariance matrix exceeds a predetermined caution-threshold, then the previously calibrated host vehicle sensor data is to be used only as a

Table II. Kalman Filter Error States

High Dynamic INS-Aided	High Dynamic Unaided	Medium/Low Dynamic Aided or Unaided	Stationary
Position (3D) Velocity (3D) GPS Clock (2D) Misalignment (3D)	Position (3D) Velocity (3D) GPS Clock (2D) Acceleration (3D)	Position (3D) Velocity (3D) GPS Clock (2D)	Position (3D) GPS Clock (2D)

measurement for the Kalman filter and further sensor calibration ceases until such time as all conditions again warrant this calibration procedure to resume.

Quite a bit of thought and design effort was spent in modeling the effect of adaptive process noise and measurement noise. The satellite selection algorithm, now includes the effect of jamming and shading. The J/S indicator is used along with GDOP to select a new satellite in the constellation. The criterion for including a replacement for a shaded satellite is that it provides the maximum (180°) separation between the line-of-sight vectors to the satellite being replaced and to the one being included.

4.0 HARDWARE DESCRIPTION

RF/IF/Synthesizer Module

The RF/IF/Synthesizer module accepts the GPS L-band signals, downconverts these signals, mixes these signals with a generated code, and provides a hard limited signal to the baseband module. Its subfunctions are described below.

RF amplifier and First IF: Integral to the front end RF is the signal reception electronics and associated pre-selection filters for signal processing from the system antenna or remote antenna electronics. Specifically, the L-band signals are amplified using low noise amplifiers, modulated with a slow code, split into L1 (1575.4 MHz) and L2 (1227.6 MHz) frequencies, amplified separately and filtered with a 30 Mhz bandpass filter. The RF amplifier gain is sufficient to operate without a preamplifier if located close to the antenna. For long cable runs, a preamplifier is necessary. Both the L1 and L2 RF signals are separately downconverted to the first IF frequency of 34F where the letter F represents multiples of the frequency 5.115 MHz.

Second IF: The second downconverter section converts the amplified 34F (IF) signal to 12.25F intermediate frequency signal. This mixer also modulates the frequency with another slow code, which limits the adverse effects of leakage in the correlator mixer.

Two-by-Two Switch: The 12.25F signals are fed to the [2x2] switch. The switch is controlled by the navigation computer located on the CPU/memory module. The switch routes the L1 path signal and L2 path signal independently to the upper set of one (for A-RXP-2) to four (for A-RXP-5) IF channels, A through D, or to the lower one, E as shown in Figure 4. In the A-RXP-2, channels A and E track independently different satellites on the same frequency (L1 or L2) or the same satellite on two different frequencies. In the A-RXP-5, channels A through D usually track four satellites, while channel E searches for a fifth satellite or tracks one of the same four satellites, but on the second frequency of the GPS signal (L2 if the other four are tracking on L1).

PN Correlation and Third IF: The IF channels are all alike. Each accepts a 12.25F signal (62.66 MHz), and amplifies it. The first mixer removes slow code 1 (S1) and the pseudonoise (PN) code of the satellite of interest. The S1 and PN codes for the mixer are both contained in one signal, which originates in another printed wiring board called the baseband board. The first mixer simultaneously spreads jammer signals and signals of other satellites. The second mixer removes slow code 2 (S2), and spreads any spurious signals induced in the preceding circuits. At this point, the desired signal consists of the carrier, biphase modulated with data from one satellite at 50 bits per second; hence, its bandwidth is only 100 Hz. The bandpass filter shown (labeled BPF) removes wideband noise and the signals which have been spread by the two mixers, including jamming signals and spurious signals induced in earlier amplifier stages. The output signal from the bandpass filter is just the desired signal. The third mixer in the IF channel converts the signal frequency to 0.25F (1.29 MHz). The output amplifier limits the signal, and delivers it to the demodulator circuits, which are on the baseband circuit board.

Frequency Synthesizer: The frequency synthesizer creates four signals, fixed in Frequency at 274F (1401.5 MHz), 21.75F (111.25 MHz), 12F (61.4 MHz), and 4F (20.5 MHz). The first three are used for frequency conversion in the RF/IF section, and the fourth is used for reference by the baseband board. All four are locked to the reference signal of frequency 2F (10.23 Mhz), which originates in a temperature controlled crystal oscillator.

Baseband/Receiver Processor

The baseband module accepts the hard limited signal from the RF/IF/synthesizer correlator module and provides the processing capabilities to generate and control the code sequences, perform signal acquisition, signal track and make pseudorange and deltarange measurements.

Baseband Processor: The baseband processor as shown in Figure 5 is based on Intel 80386 supported by 64Kx16 bits of EPROM program memory, and 32Kx32 bits of RAM memory. On power up, the 96K bytes of program is transferred using a block move instruction from EPROM to RAM within 30 milliseconds. During execution, 80386 runs at 12 MHz with zero wait states in conjunction with 70 nsec. memory speed to service the 5-channel baseband processing at 1 KHz.

Baseband Coder: The main function of this custom chip are to generate P, C/A and slow code sequences for code correlation, code control to slew the coders to a code position, code phase generation by generating doppler corrected 2F clock, and carrier/phase digital sampling. This chip also interface with the Auxiliary Output Chip (AOC) for Y code generation.

Security Subsystem - The present U.S. Government policy is to deny unauthorized users the full capability and accuracy of the GPS navigation

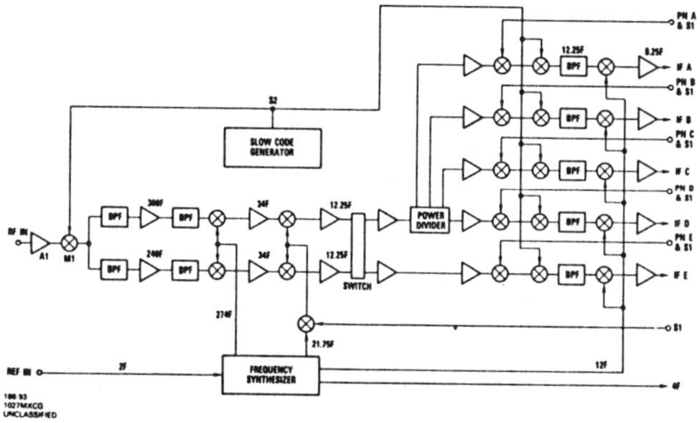

Figure 4. 5-Channel RF/IF Correlator Module Block Diagram

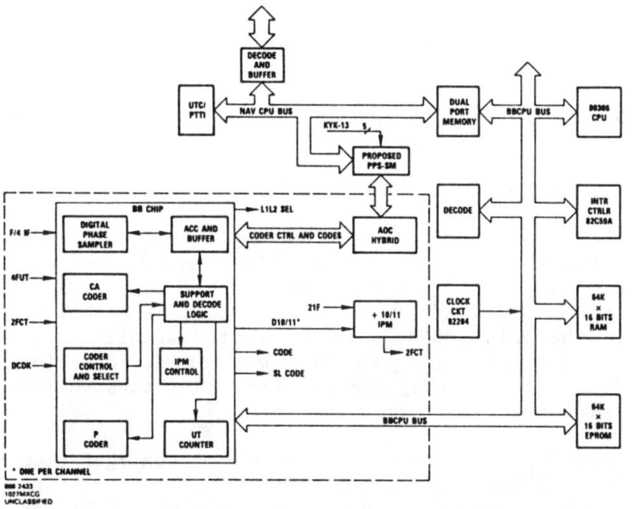

Figure 5. A-RXP-2/5 Baseband Module Block Diagram

system. A two step process known as Selective Access/Anti-Spoofing (SA/AS) to degrade the available navigation accuracy of the Set and to deny the P-code capability. To overcome these limitations an authorized user must have access to the crypto variables (keys) and a means of loading them into the GPS set by means of an authorized COMSEC device. Magnavox has implemented an agreement with the National Security Agency (NSA) to design and produce a tamper-proof module which houses the keys in a secure hardware subsystem. This subsystem, called the Precise Positioning System Security Module (PPS/SM) allows the A-RXP to be unclassified when keyed.

UTC/PTTI: A third custom chip provides User Time Clock (UTC) control and Precise-Time-Time-Interval logic. The UTC helps in making pseudorange and deltarange measurements at a specific time epoch. It also services the 10 msec, 100 msec and 1 sec time frames for software tasks execution. PTTI hardware function accepts as well as transmits 1 PPS and 1 PPM BCD time and timing fault discrete.

CPU/Memory

The main navigation processor design is based upon Intel 80286 microprocessor and a numerical coprocessor Intel 80287 as shown in Figure 6. The CPU is supported by 512K bytes of EPROM program memory, 64K bytes static RAM, and 16K bytes of battery powered critical RAM. The processor provides the processing capability to support functions required for receiver control and navigation processing. The processor also controls the system clock and timing registers.

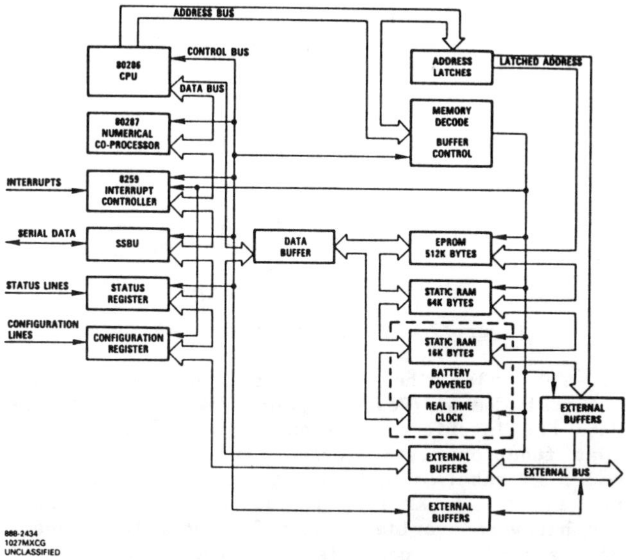

Figure 6. CPU/Memory Block Diagram

The main processor connects to the receiver Baseband Processor (Intel 80386) and Input/Output Processor (Intel 8086) over the system bus. Aiding data from I/O processor is used to produce the GPS position, velocity and time when coupled with the output of the Kalman filter. The main processor also contains the hardware/software to support the ICD-GPS-204 serial test instrumentation interface.

Input/Output (IOP) Processor

The I/O processor passes data between the Navigation Computer Program and the Host Vehicle. Inputs from the Host Vehicle include initialization data, aiding data, and moding commands. Outputs to the host vehicle include position, velocity, time as determined by GPS measurements and/or the aiding data. The I/O Processor contains logic to time-tag and format the data from the Host Vehicle as well as the formatting and control processing for the keyboard and display of a man/machine interface device. This processor design is based on 80C86 microprocessor with 256K bytes of program memory, 64K bytes of static RAM and 64K bytes of static RAM for dual port access as shown in Figure 7. The IOP board supports the following interfaces:

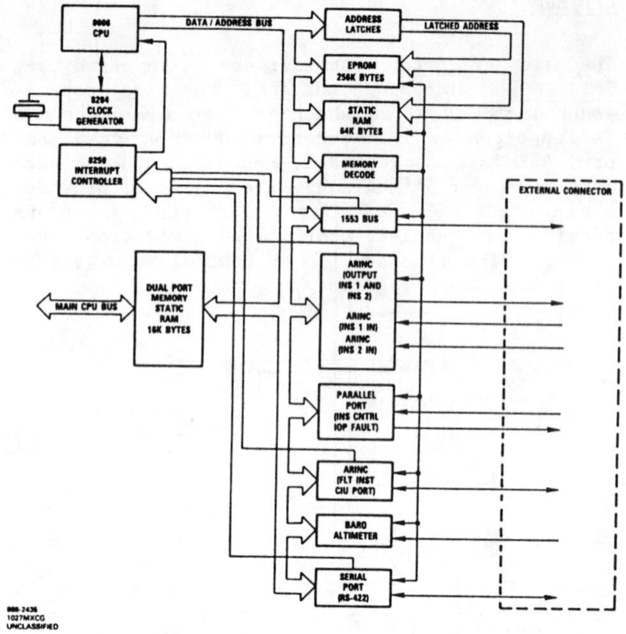

Figure 7. IOP Module Block Diagram

MIL-STD-1553: The A-RXP contains the aircraft internal Command/Response time division multiplex data bus interface electronics, a serial data bus that functions asynchronously in Command/Response mode. Transmission occurs in a half-duplex manner and the A-RXP responds as a remote terminal. The hardware provides dual redundant data channels and associated controls for the MIL-STD-1553B interface and can provide either direct or stub coupling.

ARINC 429: The A-RXP contains a dual ARINC interface which conforms to the dual ARINC 429 electrical requirements as described in ICD-GPS-70. The dual interface is under software control and message protocols have been established by the Interface Control Document with the ARINC interface providing a full-duplex, two-wire, dedicated two way data transfer system.

EIA RS-422 Interface: For communication with a Control/Display Unit or for an instrumentation interface, the A-RXP has an RS-422 serial data interface which conforms to the ASCII standard for data interchange. The basic control for the RS-422 interface is a Universal Asynchronous Receiver/Transmitter (UART).

Chassis Design

The complete A-RXP system is housed in an ARINC 3/8 ATR short chassis. The assembly consists of nine main components: the chassis, the I/O connector, (circular or rack and panel), the oscillator, the power supply, and five unique electronic modules, providing a producible, low-cost, and easily maintainable assembly. The chassis is of an aluminum alloy material to provide reduced equipment weight and cost, ease of fabrication, and excellent thermal conductivity. Center partitions are used to increase chassis strength and to provide a mounting interface for internal components. A removable top cover furnishes access to the oscillator, power supply, and electronic modules. External connections are made with either the high-density front panel circular connectors qualified to MIL-C-38999, or the rear-mounted rack and panel connector per ARINC 404.

Oscillator

The GPS oscillator is an extremely accurate and stable element required to ensure the precise position accuracy of the GPS navigation solution. To achieve the GPS performance characteristics and to be consistent with reliable yet low cost receivers, Magnavox has specified, tested, and evaluated oscillators for use in GPS for over 12 years. The A-RXP oscillator was developed specifically for GPS user equipment.

Because of the criticality of the oscillator and the difficulty of replacement with assurance of accurate performance, the oscillator is considered a part of the chassis and is only replaceable during depot level (third or fourth line) maintenance.

5.0 SOFTWARE DESIGN DESCRIPTION

The software has been evolved by Magnavox over many years of GPS user equipment development, and has been thoroughly tested during previous programs. For the EURONAV™ program, the software is hosted in the Intel™ 80X86 family computer architecture. The EURONAV™ program has resolved deficiencies from previous programs and has added substantial enhancements and improvements.

The A-RXP has three Computer Program Configuration Items (CPCI) as shown in Figure 8. The Receiver Computer Program (RCP), which is responsible for receiver baseband hardware control and signal processing, executes in an 80386 processor with the "C" language as its environment. The RCP communicates with the Navigation Computer Program (NAVCP) which performs the navigation solution, system control and receiver moding. The NAVCP is coded in a structured version of FORTRAN-77 known as Meltran-86 and executes in an 80286/80287 processor. The NAVCP communicates with the Input/Output Computer Program (IOPCP) which performs the control of the interface protocols and related data formatting. Except for the Control Display Processing software, the IOPCP software is coded in Meltran-86. The CDU software is programmed in C language. The IOPCP runs in an 80C86 processor. The brief description of the three programs is given below.

Receiver Computer Program (RCP)

The RCP implements receiver signal processing functions that are required for acquisition and tracking of GPS signals, data demodulation, and performing pseudorange and deltarange measurements. RCP consists of two major sub-functions:

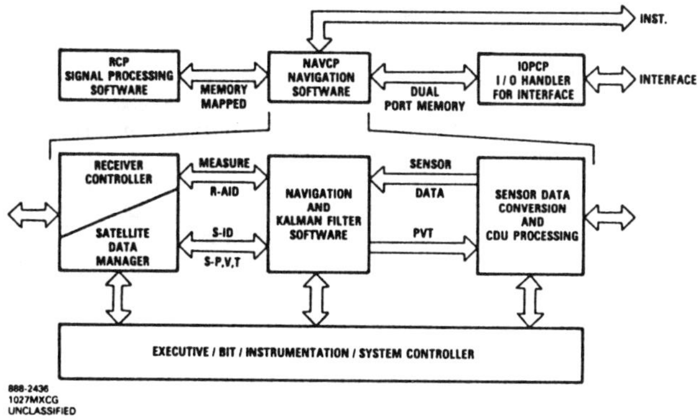

Figure 8. EURONAV Computer Programs Overall Block Diagram

dwell control and receiver signal processing. The dwell controller generates commands like enable Costas track, select bandwidth, etc. to control the following receiver signal processing operations.

- Channel-time establishment (coder slews, bit sync, sample sync)
- Code-search control
- Code tracking
- Code-sync detection
- AFC/Costas carrier tracking
- Costas-lock detection
- Data demodulation
- Pseudorange and deltarange measurements
- C/No estimation
- Jamming-level estimation

RCP utilizes in-phase (I) and quadrature-phase (Q) samples from the hardware correlators to implement these signal processing operations. Samples are read after each one-millisecond, user-time interrupt. The one millisecond sampling intervals are synchronized to 20-millisecond, channel-time databit transitions by RCP. RCP also controls the carrier phase and PN-coder phase by outputting $\Delta\phi$, Δt, and coder-slew commands to the hardware.

Navigation Computer Program (NAVCP)

The NAVCP consists of five major functions: Receiver Controller, Satellite Data Manager, Navigation and Kalman Filter processing, Sensor Data/ CDU processing and Executive software/BIT/Instrumentation processing. This top level partitioning is shown in Figure 8. These functions are described below in brief.

Receiver Controller: The receiver controller processing acts as an interface between the receiver processing software contained in the baseband processor and the other software modules. Receiver controller contains the control logic for the commanding of the different dwells performed by receiver processing.

Satellite Data Manager: The satellite selection module determines the optimal constellation of four satellites for initial acquisition, keeps this constellation current by periodic review processing, and selects alternate satellites when one or more satellites from the optimal four are unavailable due to vehicle dynamics and/or jamming or satellite setting.

The satellite data gathering function will collect satellite data words demodulated by the Receiver Processing module for the building of ephemeris and almanac, clock correction terms, and other satellite data.

The satellite position computation modules expands almanac data, computes ephemeris derived terms, computes satellite position/velocity/acceleration from either almanac or ephemeris data, and computes relativistic corrections to satellite clock parameters.

Navigation and Kalman Filter Processing: The navigation controller module determines the best source of set initialization (i.e., position, velocity, and time) and performs coordination among the other navigation functions.

The navigation input data conditioning accomplishes preprocessing of host sensor data and interpolation/extrapolation of host vehicle aiding data as required to reconcile host vehicle and GPS Set data rates.

The navigation inner loop initializes and maintains the fundamental GPS navigation functions as well as the sensor only dead reckoning parameters associated with modes of operation when GPS signals are unavailable.

The GPS Kalman filter processing provides error estimates to the navigation inner loop module using receiver and other measurement data. The Kalman filter responds to environment and host vehicle configurations.

The receiver aiding/pre-positioning estimates the code phase and carrier frequency (doppler) for the selected GPS satellites. When the GPS set is aided with an inertial navigation system or any other velocity sensor, this function will provide aiding to the receiver tracking loops by generating line-of-sight range, range rate, and range acceleration with respect to the GPS satellite.

Sensor Data and CDU Processing: The Host Vehicle I/O processing provides an interface to host vehicle systems. The structure of this module allows easy addition of future I/O protocols as they are defined. The Host Vehicle I/O currently supports ARINC and MIL-STD-1553B and baro interfaces.

The A-RXP may have a Control/Display Unit (CDU) to provide a man machine interface. The man/machine interface software receives commands and data from the CDU. The commands and data are deformatted and processed and may consist of mode commands or data parameters such as initialization time or waypoint coordinates. When requested by operator, the man/ machine interface software will retrieve or compute data for display (e.g., current position, distance to waypoints, etc.).

Executive and Monitor Processing: The Executive provides the scheduling and enablement of system tasks based on time and priority in the multi-tasking environment.

The system controller module controls the orderly transition of system modes of operation.

BIT/AFI provides both an off-line determination of Set readiness (BIT) as well as an on-line continuous monitoring of Set health (AFI).

The instrumentation module provides a test interface between the GPS software and the outside world via the instrumentation link defined in ICD-GPS-204A. This module controls the formatting and transmitting of instrumented and the reception and deformatting of received commands.

Input/Output Computer Program (IOPCP)

IOP computer program is an autonomous software package which performs the initialization of the hardware interfaces e.g., ARINC, Baro, MIL-STD-1553, RS-422, etc., commands self-test and selects/controls any subset of these interfaces depending upon the System configuration discretes. During tracking mode, the software receives host sensor input blocks, performs parity and checksum verification, data rate adjustment, scaling (if necessary) and time tagging. It also transmits GPS position, velocity, time, nav-covariances, channel status and other related data to the host vehicle via the selected interfaces.

6.0 TEST PROGRAM AND TEST RESULTS

The A-RXP-2/5 testing is being jointly performed between Magnavox and team members along with their Ministry of Defense (DOD) representatives. These tests are comprised of computer program Preliminary Qualification Tests (PQT), Receiver Performance Tests and Environmental Qualification Tests.

Performance testing occurs in two distinctly different environments. One, under controlled conditions, occurs in-house with special test equipment which simulates some or all of the operational environment and allows evaluation of performance from data gathered by engineers/specialists. The second occurs in an operational environment and relies heavily on the pilot's notes of aircraft performance when aided by the equipment. The A-RXP-2 has undergone both levels of testing. The in-house testing is described below for the A-RXP-2. The 5-channel A-RXP-5 is currently under qualification and its results will be presented during the conference. The operational environment testing will be described in the next section.

The formal performance tests were conducted at the Magnavox Torrance facility using the Global Positioning System Simulator shown in Figure 9 and using the Block-1 GPS satellite with L-band antenna and preamplifier installed on the premises of the Magnavox facility. The GPS simulator is a full military qualified system which provides a precise emulation of the operational GPS environment including: simulation of NAVSTAR GPS satellite constellation dynamics, simulation of user vehicle dynamics, generation of GPS RF signals which mirror the dynamic satellite user geometry, simulation of Host Vehicle equipment such as inertial navigation systems, emulation of host vehicle interfaces, and simulation of scenarios with special effects, (such as jamming and signal blockage). The GPS Simulator is ideal for testing and evaluation of the GPS Receivers, because it facilitates the performance characterization through its capabilities of data reduction and mission replay aids.

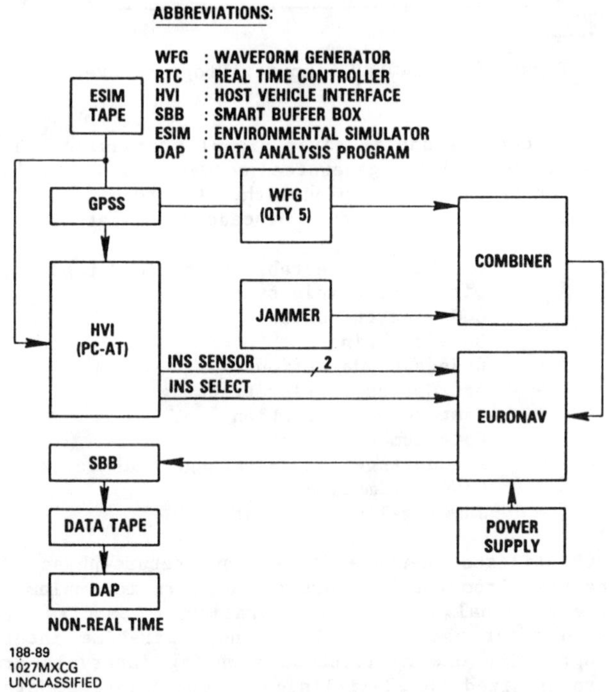

Figure 9. GPS Simulator

Eight groups of tests were conducted. They were: Signal acquisition, Tracking accuracy, Ionospheric measurements, Recovery, C/No, Position and Velocity Accuracy, Jamming, and Dynamic Threshold.

Table III shows the eight groups of tests, any special conditions relevant to the test and summarizes the test results against the specification.

Navigation Accuracy. During normal Costas tracking of four satellites when GDOP was less than or equal to 3.6, the navigation position accuracy was:

i) 10m spherical error probability SEP in stationary case

ii) 16m SEP in standalone (unaided) GPS under constant velocity profile

iii) 16m SEP in aided GPS with INS. However, the position error grew as a function of acceleration and data latency between INS and GPS.

Table III. A-RXP-2 Performance Test Results

	Test Parameters	Results/Spec.
Signal Acquisition		
Low Power	L1-C/A, -130 dBm 10 m/s^2 20 m/s^3	31.7/78 sec.
High Power	L1-C/A, -120 dBm 10 m/s^2 20 m/s^3	32.6/78 sec.
Search the Sky	L1-C/A with Dynamics	13/No min. spec.
TTFF - No Jamming		3.36/4 min.
- Jamming	24 dB J/S	3.42/4 min.
Reaction	24 dB J/S	7.5/9 min.
Tracking Accuracy	L1-P, -133 dBm 900 m/sec., J/S=40	1.45/1.5m PR .018/.02 m/sec. DR
	L1-C/A, -130 dBm 20 m/s^2, 40 m/s^3, J/S=30	100% Lock/90% Lock
	L2-C/A, -133 dBm 900 m/s, J/S=30	1.07/1.5 m PR .02/.02 m/sec. DR
Ionospheric Correction	L1, L2-P, -133 dBm 900 M/S, J/S=40	1.0/2.2m
Data Recovery	L1, L2-P, -133 dBm 900 M/S, J/S=40	0/10^{-5} Bit Error Rate
Sequential Lock	L1-C/A and P, L2-P at 2G	100% Lock/90% Lock
C/No Measurements	L1-P, -133 dBm 900 M/S	1/1 dB-Hz
Position, Velocity Accuracy		
INS AIDED	Vel = 150 m/s	15/16M, .4/.2 m/sec.
BARO AIDED	Acceleration = 20 m/s^2	15/16M, .4/.1 m/sec.
UNAIDED (GOOD SATS)	Altitude = 3000m	15/16M, .5/.1 m/sec.
UNAIDED (DEGRADED SATS)	J/S = 40 dB	Navigated/No Spec.
UNAIDED (MEMORY MODE)	GDOP = 3.6	Navigated/No spec.
STATIONARY	Angular change during	7/16M
IMU ALIGNMENT	Gs = 540 degrees	15/16M, .2/.1 m/sec.
ALTITUDE HOLD		15/16M, .2/.1 m/sec.
Dynamic Thresh of Performance		150 m/sec. at 2.4 G
Jamming		
CODE TRACK	Dynamics conditions as	53/53 dB
AFC TRACK	specified above	46/46 dB
COSTAS TRACK		40/40 dB
SIGNAL REACQUISITION AT 53 dB		26/30 sec.

The plots of these tests are shown in Figures 10 through 15.

Jamming Performance. The Set was able to acquire GPS signals on C/A code when L1 was jammed at J/S = 24 dB Hz. The time to first fix was still under four minutes.

The Set was able to track GPS signals on C/A code at J/S = 30 dB Hz and P-code at J/S = 40 dB Hz. During the availability of INS aiding, Set was code tracking GPS signals at J/S 53 dB Hz.

When the Set lost GPS signals due to masking, the signals were recovered within 10 seconds on signal visibility and when the jamming was equal to or less than 40 dB Hz. The signal reacquisition average measured time increased to 26 seconds when the J/S had increased to 53 dB/Hz.

Instrumentation - The EURONAV Set outputted data for post-test analysis from which the navigation performance plots were created. This instrumentation design and protocol were matched to the standards set by U.S. DOD on GPS programs.

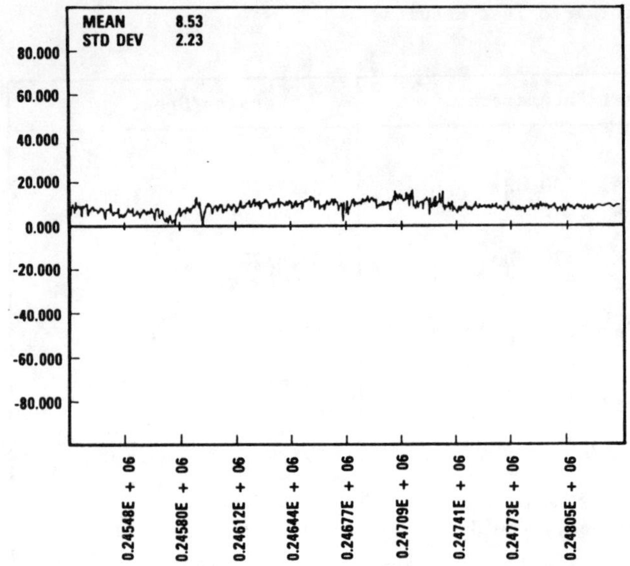

188-80
1027MXCG
UNCLASSIFIED

Figure 10. East Position Error

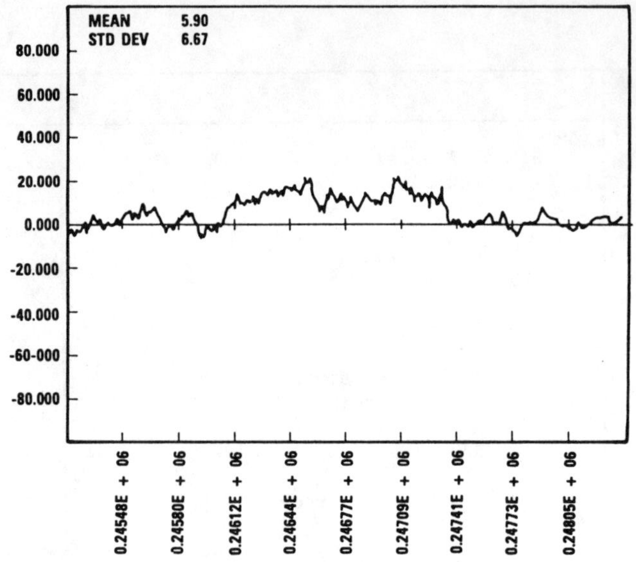

188-82
1027MXCG
UNCLASSIFIED

Figure 12. Vertical Position Error

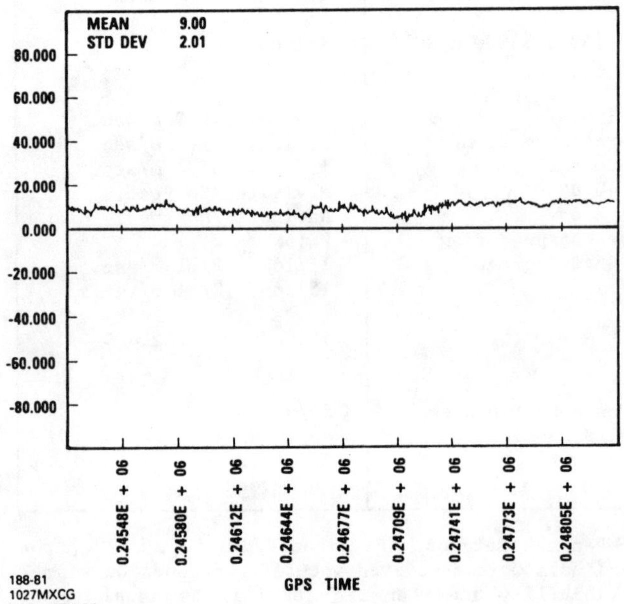

188-81
1027MXCG
UNCLASSIFIED

Figure 11. North Position Error

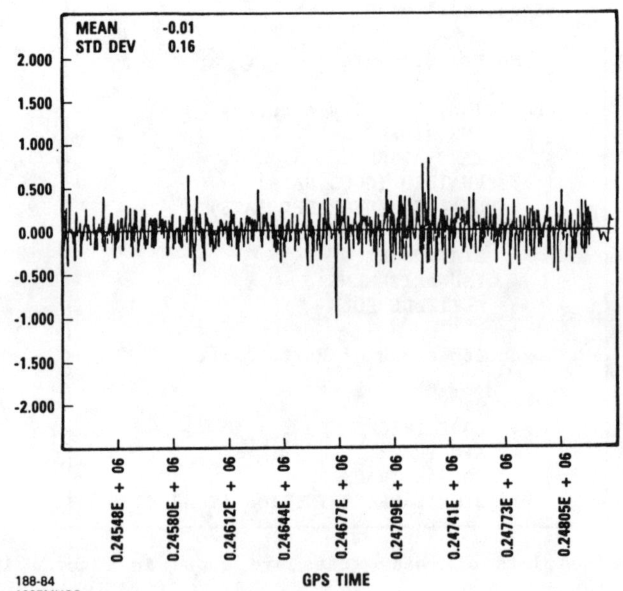

188-84
1027MXCG
UNCLASSIFIED

Figure 13. East Velocity Error

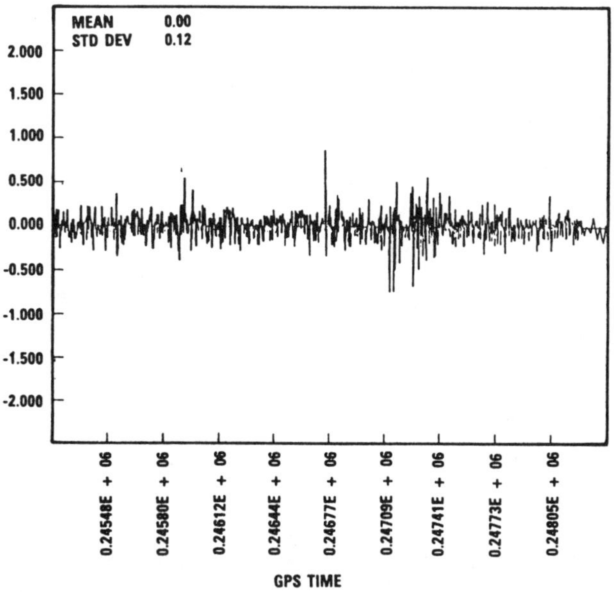

188-85
1027MXCG
UNCLASSIFIED

Figure 14. North Velocity Error

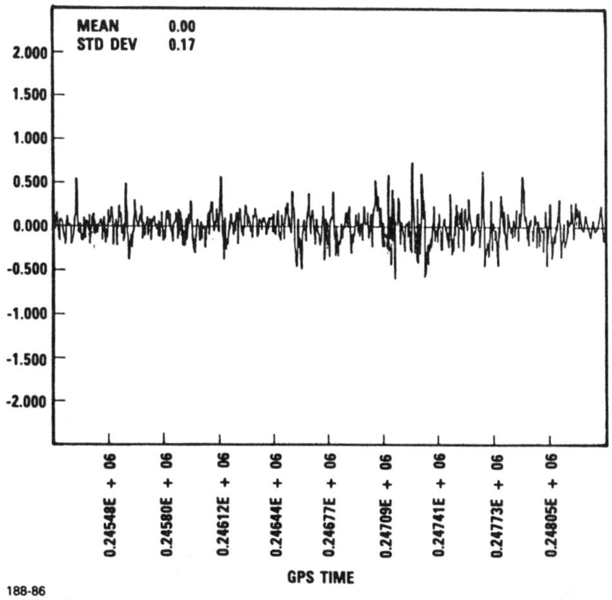

188-86
1027MXCG
UNCLASSIFIED

Figure 15. Vertical Velocity Error

Miscellaneous Performance Observation - The Set has 22% excess throughput capacity for future growth. The throughput was measured under the worst case scenario when the Set was integrated with INS, and the GPS Set was tracking few satellites and searching for few others along with steering data output to a waypoint.

The Set has battery backed memory which stores critical data to sustain close-air-support missions, e.g., ephemeris, navigation solutions, waypoints, etc.

The Set also includes a feature of turn-key system. Due to critical memory and a wrist-watch quality real-time clock, the Set will automatically go into navigation without operator action.

Aircraft Testing. The A-RXP-2 has undergone several successful flight tests conducted by the German and French Team Members with their respective Government's witness. Two tests during which the GPS flew unaided were conducted on the Caravelle aircraft. Ten other tests, with the GPS operating in a loosely coupled mode with the GPS-INS aboard the aircraft are occurring through May of 1988. In these tests the GPS performance is checked using the Trident system as the prime reference and Trident data coupled with GPS-INS data used as a hybrid reference for post-test evaluation. Thirty further flights are also underway in Germany during which the GPS is installed in a Transall C-160 and first aided by an altimeter than later INS-aided. A-RXP-5 testing is also planned for September 1988.

SATELLITE DATA MANAGEMENT IN DoD NAVSTAR GPS RECEIVERS

B.K. Cariveau, K.L. Therkelsen
Collins Government Avionics Division, Rockwell International

ABSTRACT

This paper describes the software implementation of the Satellite Management function in the DoD NAVSTAR GPS (Global Positioning System) receivers.

The software implementation addresses three functional requirements:

(1) Data base management of satellite almanac, ephemeris, and deterministic corrections data.
(2) Computation of precise satellite position and velocity for use by Navigation software.
(3) Using satellite and receiver position data to periodically calculate the constellation of four satellites with optimum geometry for Navigation.

The implementation description details how these functions are accomplished, including the interfaces with the receiver satellite tracking functions necessary to acquire the 50-Hz downlink data. Its goal is to illustrate the solutions that were designed into the NAVSTAR User Equipment software to deal with the obstacles presented in creating the efficient, effective, and timely data management capability required by a real time Navigation system.

INTRODUCTION

The implementation of the satellite management functions in DoD receivers has been divided among three functions collectively named Satellite Manager (SM). The three functions are Satellite-Data-Base-Manager (SDBM), SV-Position-Velocity-Acceleration (SVPVA), and Select-Satellites (SS). The source code for each of these functions is common for the family of one, two and five channel DoD receivers.[1-3] Figure 1 shows the functional relationships these functions have with each other and other software functions in the receiver.

The major interfaces Satellite Manager has are with Receiver Manager (RM) and Receiver Processing (RP) which perform the task of acquiring and tracking satellites, and collecting subframe data; with Navigation (NAV) which needs satellite position, velocity, and acceleration data to compute the navigation solution; with the Classified (CL) function which applies denial-of-accuracy corrections to the processed satellite ephemeris data; with Interface Manager (IM) and Data Loader (DL) functions which can provide external sources of satellite almanac, health, and ephemeris data; and with the CDU manager which displays age of data and visible satellite information to the operator (figure 1). Additional details on these and other software functions in the receiver are available in.[1-3]

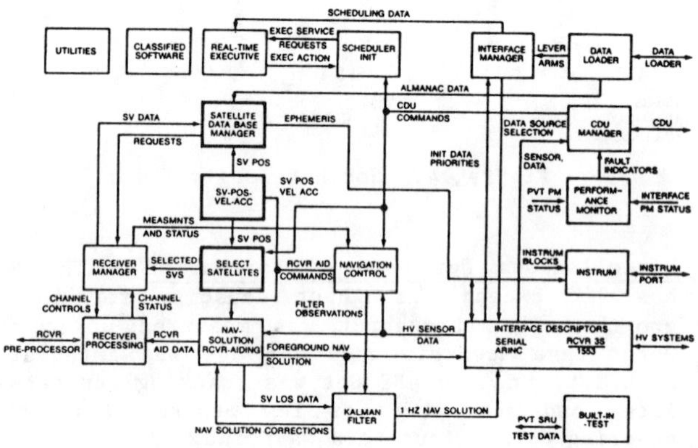

Figure 1. Receiver, Functional Block Diagram

The goal of software commonality for the SM functions in the one, two, and five channel receivers, balanced with processor throughput and memory requirements for each of the receiver types, was the primary consideration driving the design. There was also a need to be able to transition from Block I to Block II satellite data formats early in the Phase II development. Written in JOVIAL J73/I, table 1 illustrates the current memory requirements of each SM function.

Table 1. Memory Usage by Function.

	EPROM*	RAM*	C/NV RAM*
SDBM	9461	4556	1023
SS	1574	76	N/A
SVPVA	1584	969	N/A
TOTALS	12619	5601	1023
* 16 bit words			

This paper will provide a functional description and general requirements discussion, followed by a section showing the resulting implementation at the software module level, for each of the SM functions. The salient data and control structures used and interfaces with other functions will be included, along with some of the rationale behind the implementation.

Reprinted from *Rec. IEEE Position Location and Navigation Symp.*, pp. 134–144, 1988.

SATELLITE MANAGER

SATELLITE-DATA-BASE-MANAGER

FUNCTIONAL DESCRIPTION AND REQUIREMENTS

Many of the Requirements of SDBM are derived from ICD-GPS-200.[4] SDBM has four major requirements to fulfill, which map directly to four of its subfunctions. These are (1) Satellite data base initialization, (2) Make data collection requests, (3) Processing of the satellite data collected, and (4) Providing satellite deterministic corrections. The Satellite Data Base itself will first be discussed, followed by each of the four subfunctions.

The Satellite Data Base

The satellite data base is divided into two major segments split between dynamic RAM and Critical/Nonvolatile (C/NV) memory. Nonvolatile memory is described as "Critical" due to its capability to survive an EMP (Electromagnetic Pulse) from a nuclear event with all data intact. The dynamic RAM data base contains the collected satellite data in a scaled form usable by other functions in real time, while packed portions of the same satellite data are stored in C/NV memory for subsequent power on initializations. Data organized in C/NV memory is driven by two main considerations. The first is the requirement to meet both short-term and long-term power outages. The second is the size limitation imposed by a 6-month memory battery life requirement, coupled with the cost of C/NV memory.

The SDBM function has a generalized indirect table indexing scheme for processing, storage, and retrieval of data in the data base. It allows efficient movement of data by pointer allocation or pointer swap and provides a basic structure that is common to the seven types of data that are maintained on a Satellite Vehicle (SV) basis. The major elements of each segment of the data base and their capacities are listed in table 2.

The Table Size entries listed in table 2 include the slots necessary for temporary storage of data during processing. The number of temporary pointers, indirect indices, and number of table pointers are also listed. Every table pointer will contain one of the indirect indices that point to a table entry if data is present, or a null value if data is not present or is invalid. All external functions use the table pointers to access the respective data. The 32 dimension pointer tables are indexed with the Satellite Vehicle Identifier (SVID), in the range of 1 to 32, that is of interest. The temporary pointers will contain indirect indices that identify the current table entries that are being used as temporary slots for processing data. Indirect indices are obtained from a stack (not listed in table 2) of available indirect indices maintained for each data type. Data is "released" to the data base by setting the table pointer to the value of the temporary pointer where the data was processed. If the table pointer is null, new temporary pointers are assigned from the stack of available indirect indices. If the table pointer is not null, the table pointer and temporary pointer are simply swapped. If data is deleted, the indirect index is put back on the stack of available indirect indices, and the table pointer is set to a null value.

Note that the Midcourse-Ephemeris is a packed version (data bits only, no parity bits) of the ephemeris data that can be transmitted and received from an external "Midcourse Receiver" via a 1553 bus,[4] and transmitted via the RS-422 Instrumentation Port. This data is also used as the source of the C/NV ephemeris data sets that are maintained for the four satellites currently being used for Navigation.

The C/NV ionospheric data is maintained for the four satellites currently being used for navigation. Data for only four satellites is stored in C/NV memory because this is the minimum necessary to meet short-term power outage satellite reacquisition requirements. Storage room is available in C/NV memory to store raw downlink data almanac for as many as 31 satellites. Almanac data is stored in C/NV memory in order to meet longer term (> 7 seconds) and routine (operator controlled) power outage requirements. Due to memory constraints and requirements limitations on the number of satellites in the space segment, the system is designed to process and store only 24 of the C/NV memory almanacs in a form that is readily utilized by real time software. Therefore, no more than 24 satellites will be recognized by a DoD GPS receiver. Receiver hardware biases relative to a reference channel are maintained for up to four hardware channels in C/NV memory.

Other data maintained in the data base includes both data that is internal to the SDBM function and global items for use by other functions. The internal data includes data collection status flags that are used in communicating short term collection status information between the Process Satellite Data and the Make Data Collection Requests functions. Internal and external absolute hardware path delays for the L1 and L2 frequencies are combined and stored by the Satellite Data Base Initialization function, and are used in the Providing Deterministic Corrections routines. The global items include C/NV memory validity bits, and current constellation and health status summary words. In the C/NV portion of the data base, each major table entry has an associated data validity bit that must be set in order for data to be considered valid. The constellation status summary word contains a summary of the healthy and unhealthy satellites that are usable or that may potentially become usable. The Health Status summary simply contains a summary of all currently healthy satellites. The summary words are provided in two thirty-two bit words, with one bit for each satellite in each word. Each bit is set using the most timely of the three sources of health information available from the downlink data stream.

Table 2. Major Elements/Capacities of the Satellite Data Base.

Element	Table Size		Temporary Pointers		Indirect Table Index		Table Pointers	
	RAM	C/NV	RAM	C/NV	RAM	C/NV	RAM	C/NV
Ephemeris	17	5	5	1	17	5	32	32
Almanac-Offset	13	0	1	0	13	0	32	0
Midcourse-Ephemeris	17	0	5	0	*	*	*	*
Partial Subframe One	13	0	1	0	13	0	32	0
Almanac	25	33	1	1	25	33	32	32
Message	2	0	1	1	2	0	1	1
UTC Parameters	2	1	1	1	2	2	1	1
Health-Sum/Config-Codes/A-S Status**	2	2	1	1	2	2	1	1
Classified**	2	0	1	0	2	0	1	0
Ionospheric-Corrections	13	5	1	1	13	5	32	32
Receiver Cal Data	0	4	0	0	0	0	0	0

* Uses Ephemeris indices and pointers
** 2 each

Satellite Data Base Initialization

Satellite data base initialization uses data from Critical/Nonvolatile (C/NV) memory, the Data Loader (DL), 1553 bus Interface Descriptor (1553 ID) inputs, and/or default values. The DL function can load satellite almanac and health, for up to 24 satellites, directly into C/NV memory while the receiver is in the INIT mode.[1] Also while in the INIT mode, the 1553 Interface Descriptor function can load current satellite ephemeris (received from another GPS receiver via 1553 bus) for up to 12 visible satellites. Data base initialization is a separate "task" that executes once on the transition from the INIT to NAV operating modes.[1,2] This task executes immediately prior to the SS and SVPVA initialization that also occurs on the transition to NAV mode.

Satellite Data Base Initialization uses the Process Satellite Data function as a subfunction when initializing satellite downlink portions of the data base. Satellite Data Base Initialization acts as a "driver" for the Process Satellite Data function. Therefore, the set of modules used to process and release the satellite downlink data when it is collected from the satellite is the same set that is used to process and release it at initialization.

Appropriate portions of the data base are cleared and all stacks, table pointers, temporary pointers, counters, collection status flags, and global constellation and health status words are set to initialization values. The C/NV pointers and validity bits are checked for range and consistency. Available, valid downlink data from 1553 inputs (if supplied) and C/NV memory are first unpacked into a data structure that is equivalent to the data structure used when the data was first collected and sent to SDBM. The unpacked data is then processed and "released" to the data base using the Process Satellite Data function. Collection status flags, constellation, and health status words are cleared or set appropriately as data is processed.

Ionospheric data stored in C/NV memory is checked for senescence and released to the RAM data base. Since the relative channel to channel delays require recalibration after every mode transition (e.g. INIT to NAV), the C/NV receiver calibration data is invalidated. The absolute hardware path delays for internal receiver circuitry and external cable and antenna electronics are combined for the L1 and L2 frequencies.

Make Data Collection Requests

Data transmitted by the satellites is formatted as a 1500 bit long frame made up of 5 subframes, each subframe being 300 bits long. Subframes 4 and 5 are subcommutated 25 times each; therefore, a complete data message requires the transmission of 25 full frames. Data is transmitted at 50 Hz, resulting in a transmission time of 30 seconds per frame, with the full 25 frames taking 12.5 minutes. Each 6-second subframe consists of ten thirty-bit data words, with each data word containing 24 data bits and 6 parity bits. Table 3 provides a brief definition of each class of subframe data collected and its position in the bit stream. Table 4 provides a summary of the frequency of collection of the various subframe classes.

Table 3. Functional Data Description and Location In Bit Stream.

Ephemeris	- A set of parameters used to calculate SV position, and SV clock error. Transmitted in subframes 1, 2, and 3, in all frames. Each SV transmits only its own ephemeris.
Partial Subframe 1	- The first three words of subframe 1, which contains the ephemeris health status bits. Subframe 1 transmission, all frames.
Almanac	- A reduced precision subset of the SV ephemeris and clock parameters. Transmitted in subframe 5, frames 1-24, for SVs 1-24, and in subframe 4, frames 2-5 and 7-10 for SVs 25-32. Each SV transmits all SVs almanac. Each SVs almanac also contains the SVs health.
Message	- A set of 22 ASCII characters used to contain Control Segment special messages. Transmitted in subframe 4, frame 17.
UTC	- A set of parameters to convert GPS time to UTC time. Transmitted in subframe 4, frame 18.
Health Summary	- A summary of all satellites health status, configuration codes, and Anti-Spoofing status. Transmitted in subframes 4 and 5, frame 25.
Classified	- Data that allows authorized users to overcome denial of accuracy effects imposed on unauthorized users. Transmitted in various subframes.

Navigation performance requirements in sequential receivers dictate that data collection must be minimized, since the continuous tracking required lengthens the time interval between visits to the other satellites in the sequential 1-Hz tracking cycle.

Table 4. Data Types and Frequency of Collection.

Data Type	Receiver Tracking Strategy			
	Initial Acquisition			Normal Track
	Cold Start	Normal Start	Hot Start	
Almanac	Yes	No	No	Daily
Ephemeris	Yes	Yes	No*	Hourly
Part Subf 1**	No	No	No	1 SV every 2 minutes
Almanac Offset	No	No	No	Aperiodic***
Message	Yes	No	No	Hourly
UTC	Yes	No	No	Hourly
Health Summary	Yes	No	No	Hourly
Classified	Yes	Yes	No	On Demand by CL
Ionospheric Corr	Yes	Yes	No	Every 30 Seconds****
Receiver Calibrate	No	No	No	Every 3 Minutes****

* May be requested if not available.

** Frequency of collection is greater in five channel receiver

*** See Table 3 definition.

**** Performed by RM with no request from SDBM

Data is normally collected on a subframe basis, with each subframe taking 6 to 8 seconds of continuous dwell time to collect (up to 2 seconds of synchronization overhead). The exception to the normal data collect case is the Partial Subframe 1 collection. Partial Subframe 1 consists of the first three words of subframe 1, and requires 2 to 4 seconds of dwell time to collect. The reason for collecting this data is that this portion of subframe 1 contains the satellites most current health status bits and must be monitored more frequently, since, otherwise, the collection of the ephemeris and health summary subframes is performed only on an hourly basis. Only a partial subframe is collected to minimize the dwell time required to obtain the health data.

As indicated in table 4, data collection requests are subdivided into two main categories depending on whether the Receiver is in the Initial Acquisition mode of satellite acquisition or in the Normal Track navigation mode (see figure 2). Initial acquisition requests are further divided into three subcategories depending on the Initial Acquisition Start type the receiver is performing; these are Cold Start Requests, Normal Start Requests, or Hot Start Requests.

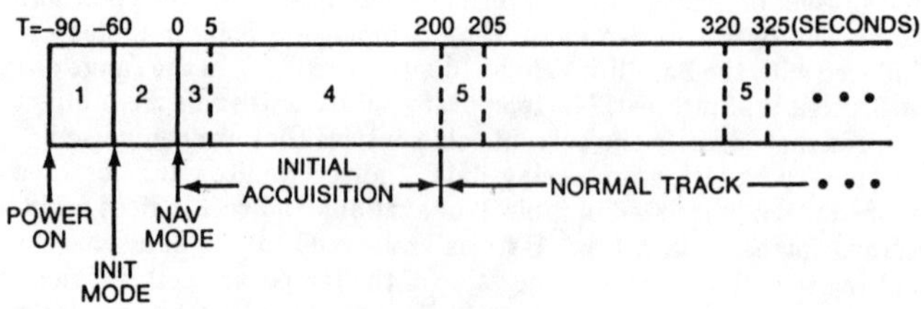

1. INITIAL TEST (AUTOMATIC EXECUTION)
2. OPERATOR INITIALIZATION (OPTIONAL)
3. INITIAL GDOP/HDOP SATELLITE SELECTION
4. INITIAL SATELLITE ACQUISITION AND DATA COLLECTION (4 SVs)
5. NORMAL TRACK SATELLITE SELECT EXECUTION (EVERY 2 MINUTES)
   NOTE: FIRST TIME EXECUTION ALSO GENERATES INITIAL SV-POS-VEL INTERPOLATION COEFFICIENTS.

Figure 2. Nominal GPS Time Line

Since the Make Data Collection Request was designed as a common software function for all receiver types (1, 2, and 5 channel), data collection requests are all performed in an identical fashion, independent of a given receiver's efficiency in executing the requests.

Data collection requests are prioritized and issued to the Receiver Manager (RM) function. The satellite data base is checked for elements that are old or not present. Subframes needed from the upcoming frame are all requested 5 seconds prior to frame start, every 30 seconds, for up to 5 different SVs. Requests are placed in a collection request buffer that consists of 5 frame requests, each of which contain 5 subframe requests. The subframe requests each consist of an SVID, Priority, and Special Processing Requirement. The highest priority requests are always contained in frame one of the request buffer, next highest priority in frame two, and so forth. Therefore, the single channel RM function need only scan the first frame of the request buffer for the highest priority requests, and similarly for two and five channel RM functions. The priority is also used by RM as a guide in performing data collection when there is a conflict with other tasks it must perform. The relative priorities of the collection requests are listed in table 5. The Special Processing Requirement is used to identify partial subframe one requests, ephemeris requests using almanac offset, or partial subframe one using almanac offset.

Table 5. Summary of Relative Priorities of Collection Requests.

Data Type	Previously Collected	Not Present
SDBM Requests:		
Health Summary	120	120
Classified	120	120
Partial Subframe 1		
Navigation SVs	76	76
Ephemeris	71	75
Almanac Offset		
Healthy SVs	74	74
Almanac	64	72
Partial Subframe 1		
Alternate SVs	62	62
Message	61	69
UTC	61	69
Almanac Offset		
Unhealthy SVs	60	60
Partial Subframe 1		
Unhealthy SVs	58	58
RM Initiated:		
SV Recovery	125	125
Receiver Calibrate	90	90
SV Switch	85	85
Iono Measurement	40	40
SV Expansion	40	40

RM will command the RP function to collect data via a "data collect enable flag" in the RM-RP interface. RP will demodulate the 50-Hz downlink data from the signal and notify RM that the end of a subframe has been reached via an "end of subframe flag" in the RP-RM interface. The RP function performs the parity checking on all subframe words, and provides a subframe parity summary status in the RP-SD interface. RP continues to collect data until the "data collect enable flag" is turned off by RM.[(1)] Upon receipt of the "end of subframe flag", if the subframe is one that SDBM requested, RM will notify SDBM the subframe was collected, via procedure call to the Process Satellite Data function of SDBM. RM provides a pointer to the subframe data in the RP-SD interface. Figure 3 illustrates the satellite downlink data path from the SV to the Satellite Manager Function.

While in Initial Acquisition, RM may also make collection requests on demand via a procedure call, for satellites that are acquired between the normal 30 second data collection request epochs. The on demand requests are required in Initial Acquisition to help minimize Time To First Fix (TTFF).

Almanac Offset measurements will be requested in Normal Track, for healthy satellites that do not have ephemeris in the data base,

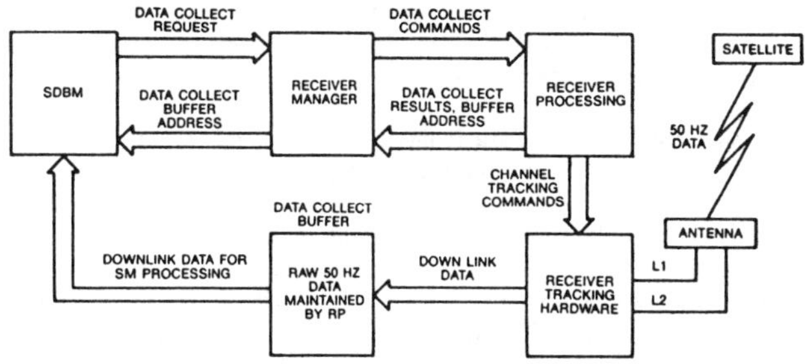

Figure 3. Downlink Data Path From SV To Satellite Manager

prior to the collection of ephemeris data. The almanac offset measurement is made to obtain an accurate delta value between a satellite position as defined by its almanac data and as actually determined by tracking the satellite. During subsequent normal track operation use of the offset data allows the Receiver Management and Navigation software to accurately command a direct P-code track without first acquiring ephemeris data. The Almanac Offset requests are inserted into a buffer separate but similar to the data collection request buffer. Almanac Offset measurements may be requested for up to 5 different SVs, with each request consisting of an SVID and Priority. RM will command acquisition of the satellite in response to the request. Once the satellite is acquired, RM notifies SDBM the acquisition was successful. SDBM will then request ephemeris subframes for the satellite, setting the Special Processing Requirement of the subframe requests to a value indicating that RM is to use the Almanac Offset measurement in subsequent reacquisitions to collect any remaining ephemeris subframes. RM also has the option of proceeding to collect ephemeris subframes following the acquisition of the satellite, without notifying SDBM of the acquisition. In this case RM makes one attempt at collecting subframes 1, 2, and 3, prior to informing SDBM of the successful acquisition. In the event any of the subframes were missed, SDBM will follow up by requesting the missed subframes, setting the Special Processing Requirement accordingly.

Almanac Offset measurements will also be requested in Normal Track for unhealthy visible satellites, but only when the number of visible healthy satellites with ephemeris is less than five. These measurements are performed prior to the collection of Partial Subframe 1 from the unhealthy satellites, to attempt to detect any unhealthy satellites that have transitioned to a healthy status. The request mechanism is the same as the almanac offset measurement requests for healthy SVs, but with a different Priority and setting for the Special Processing Requirement flag. The priority of the requests is lower than for healthy satellites, and the Partial subframe 1 request that follows the successful acquisition will have the Special Processing Requirement set to a value indicating that RM is to collect only Partial subframe 1, using the almanac offset if necessary to reacquire the satellite.

Processing of the Satellite Data Collected

The processing of each subframe consists of extracting subframe data fields and scaling, units conversion, temporary storage, validity checking of the data, various housekeeping, packing for storage in C/NV memory, and finally release of the data to the satellite database.

Much of the processing is performed with table driven algorithms. The table driven approach was chosen as an attempt at obtaining a more generalized algorithm for processing the seven types of subframes in the downlink data stream. The tables provide a detailed template of the parameters in each of the subframes, and define most of the operations that must be performed on them. When the transition from Block I satellite data format to Block II format was done, the change necessary was to substitute Block II table definitions in place of Block I, with minor modifications of the algorithms performing the processing.

Upon receipt of the data from RM/RP, the parity summary flag is checked, and a temporary table entry for the data is assigned. The data item(s) in each word of the subframe being processed are first extracted, or "parsed" to a local table of extended float point data type, applying any necessary scale factors in the assignment. Parsing is a table driven algorithm using a table that specifies the first bit, number of bits, and scale factor for each of the items in the subframe. Processing continues on the parsed list with an "adjust" process that concatenates those variables that were parsed from a noncontiguous bit stream (such as the AF0 term of the almanac[4], corrects for any sign errors created in the parse step, and performs any necessary units conversions such as semicircles to radians. A "store" process is then performed on the adjusted floating point list. Two predefined tables are used to define the target data type and a base address of each item as it appears in its respective table in the data base. The offset of the temporary is calculated and added to the base address of the item and a conversion from extended floating point to the defined type is performed in the assignment to the temporary table item.

Additional validity checks are performed on the data, constellation and health status words are updated if health information is in the subframe, and collection status flags are updated. The data is then released to the data base. The release step includes pointer manipulation as described in the Satellite Data Base Section. For ephemeris data, the Classified function will perform selective availability processing, if the set is authorized, on the data while it resides in the temporary table entry.

If the data is to be stored in C/NV memory, the raw subframe is packed, retaining only the 24 data bits of each subframe word, moved to a temporary in C/NV memory, then released. Ephemeris subframes are packed into a temporary in the Midcourse Ephemeris table, and Midcourse-Ephemeris is released at the same time as ephemeris, since the Midcourse-Ephemeris data uses the same pointer set as ephemeris.

The Process Satellite data function has no cognizance of what data was requested by the Make Data Collection Request function. It does set collection status flags indicating it has processed a particular subframe of data. The Make Data Collection Request function has the responsibility of checking the data base to see if all data needed is present and current.

Providing Deterministic Corrections

Deterministic corrections are the known delays and biases that can be predicted or measured that are inherent in the pseudorange and deltarange measurements.[1] The Deterministic corrections consist of ionospheric delay, tropospheric delay, SV clock delay, receiver hardware channel absolute and relative delays, and delays introduced by Selective Availability effects. The Deterministic corrections are provided to the Navigation function for SVs that are to be tracked in the next 1-Hz cycle.

The ionospheric differential delay estimate is the output of a single-state recursive variable-gain filter, maintained for each visible satellite. Iono differential delay measurements are taken in initial acquisition and subsequently in Normal Track at a nominal rate of every 30 seconds for the four navigation satellites being tracked. The steady-state gain of the filter is approximately 0.26. This filter design, executed at the 30-second rate, allows the receiver to maintain accurate iono differential delay estimates even during intervals of maximum rate of ionospheric delay change.

Ionospheric differential delay measurements are formed by tracking the same satellite on L1 and L2 frequencies, and taking the difference between the pseudoranges, accounting for any relative channel delays. The RM function is responsible for scheduling the periodic iono measurements. RM passes pseudoranges and deltaranges to SDBM, which forms the measurement and processes it in the filter. Multichannel receivers simultaneously track the same satellite on L1 and L2. The single channel receiver forms the measurement by sequentially tracking on L1 and L2, and time coordinating the pseudoranges using the deltarange. The initial filter output and associated variance are "released" to the satellite data base using the same methods described earlier for other types of data. Subsequent measurement updates are made to the released data. The measurement variance is updated at a 1-Hz rate, as is the maintenance of the C/NV iono data for the four navigation satellites.

The iono filter output is converted to an absolute delay at L1 or L2 when used in the Deterministic correction. In the absence of filter data for a satellite, an average of other satellites current filter estimates is used as a default value.

Tropospheric delay estimates are obtained from a simple model that is a function of user altitude and elevation angle to the satellite. The model is based on the Central Radio Propagation Laboratory Reference Refractivity Atmosphere - 1958 (CRPL - 1958)[5].

Satellite clock delay estimates are computed using the clock correction terms transmitted in the 50-Hz downlink data (ephemeris subframe 1 and almanac subframes) and the algorithm specified in ICD-GPS-200[4] (multichannel receivers only).

Relative internal hardware channel delays are measured by simultaneously tracking the same satellite on the same code type and frequency on different hardware channels. The pseudoranges are passed to SDBM, which filters the difference in the pseudoranges in a simple fixed gain filter. The RM function is responsible for scheduling receiver calibration for each non-reference channel every three minutes. This time interval was chosen to allow accurate tracking of internal delay changes driven primarily by temperature fluctuations experienced by the receiver. The filter outputs are maintained by SDBM for up to four reference channels, and are used on a 1-Hz basis in the deterministic corrections. Absolute receiver hardware delays are constants for L1 and L2 and are divided between internal and external delays. The internal delays are ROM constants and are the mean delays measured on a number of receivers. The external delay is the delay for external cable and antenna electronics, and can be entered on the Control Display Unit (CDU), or can be received from host vehicle interfaces. The internal and external delays are combined during Satellite Data Base Initialization for use in Providing Deterministic Corrections.

Selective Availability effects require incorporation of additional clock correction delay estimates provided to SDBM by the Classified software function to correct for SV clock perturbations.

Implementation

Given the information in the previous section, the partitioning of SDBM into functional modules is fairly straightforward. Figures 4 through 6 show the partitioning of SDBM into modules.

Figure 4 shows the Make Data Collection Request function. The SI'MONITOR process is one of the main 1-Hz tasks in the receiver. It calls the highest level data request module SDDATAREQ, which determines the tracking mode the receiver is in and calls the appropriate subroutine for Cold Start, Normal or Hot Start, or Normal Track. As noted earlier this software executes 5 seconds prior to the start of every 30-second frame boundary. As can be seen in the figure, all the lower level request modules insert collection requests into

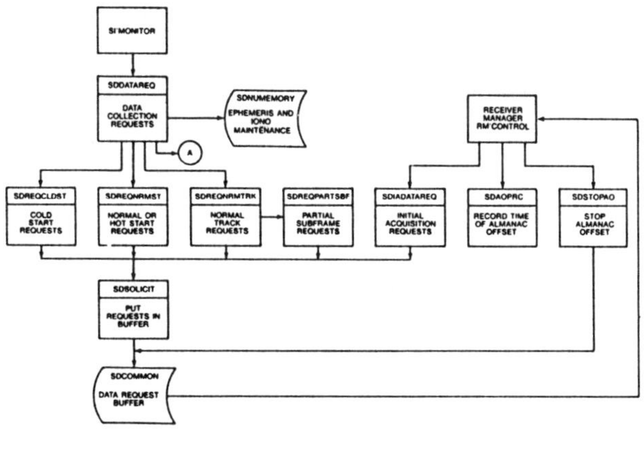

Figure 4. Make Data Collection Request

the collection request buffer using a common utility, SDSOLICIT. Note that SDDATAREQ is the 1-Hz function that maintains the C/NV Ionospheric and ephemeris data for the four navigation satellites currently being tracked. The off page connector A from figure 4 to figure 5 shows the 1-Hz update of the Iono filter variances.

RM is also shown in figure 4 with the procedure interface for Initial Acquisition collection requests, and two service routines for almanac offset processing. Recall that RM will call the Initial Acquisition request module for SVs it acquires between frame boundaries. The two service routines for almanac offset processing consist of SDAOPRC, which is called to inform SDBM that acquisition of a satellite has been successful, and SDSTOPAO, which is used by RM to inform SDBM to stop making almanac offset requests on a satellite for a period of time (normally because acquisition of the satellite as been attempted and failed).

Figure 5 shows the Satellite Data Base Initialization, Process Satellite Data, and the Provide Deterministic Corrections functions.

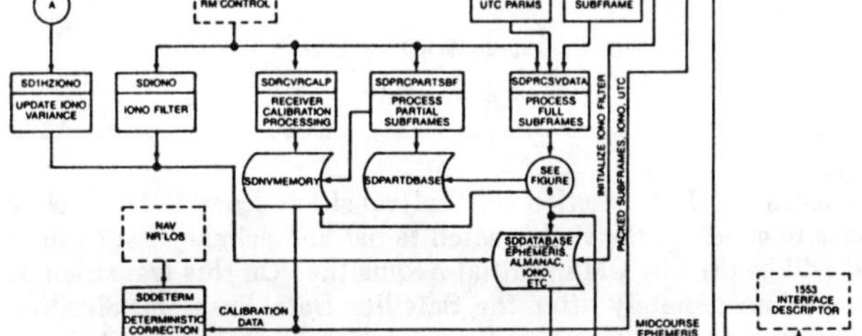

Figure 5. Deterministic Corrections and Process Data

The highest level module for Satellite Data Base Initialization is SDDBINIT, which executes as a task, scheduled by the SI'MONITOR function, on the transition from INIT to NAV modes. Inputs shown are from the C/NV memory area SDNVMEMORY, and from the Midcourse-Ephemeris table in SDINEPHEM. After unpacking the subframes with SDUNPACK, the remaining processing is performed by the highest level module in the Process Satellite Data function, SDPRCSVDATA. Initialization of UTC parameters is broken out in a separate module, since this portion of the initialization is also performed immediately at power on by the Scheduler Init function.

SDPRCSVDAT CALLS AND SUBFRAME PROCESSES

SDPRCSVDAT CALLS	EPHEMERIS SUBFRAMES 1, 2 AND 3	ALMANAC	UTC	MESSAGE SPECIAL MESSAGE	HEALTH SV'S 1-24	HEALTH SV'S 25 –32, AS –CC
SDWHICHTEMP	X					
SDPARSE	X	X	X	X	X	X
SDADJUST	X	X	X	X	X	X
SDSTORE	X	X	X	X	X	X
SDDATAPACK	X	X	X		X	X
SDUPDHLTST	X					
SDUPALMH		X				
NR'ESTGPSWK	X					
SDNVHLTUDATE	X				X	X
INSTRUM	X	X	.	X	X	X
SDNVUPDATE		X	X		X	X
SDRELEASE* ** (TO C/NV RAM)		X	X		X	X
SDRELEASE* (TO RAM)	X	X	X	X	X	X
SDUPDHLTSUML					X	
SDUPDHLTSUMU						X
SDUPALMCST					X	X

*SAME ROUTINE LISTED TWICE TO MAINTAIN GENERAL CALLING SEQUENCE.
**EPHEMERIS DATA IS MAINTAINED IN $C/_{NV}$ MEMORY BY SDDATAREQ AT A 1 HZ RATE. NO MESSAGE SUBFRAME IS SAVED IN C/NV MEMORY.

Figure 6. SDPRCSVDAT Procedure Calls

The highest level module for Provide Deterministic Corrections is SDDETERM, which is called by the Navigation function at a 1-Hz rate. Navigation provides the list of SVIDs and their elevation angles via parameters, and receives their deterministic corrections, correction rates, and variances in return. SDDETERM calls submodules to obtain components of the deterministic correction. They are SDTROPCORR, SDCLCORR, and SDIONOCORR. SDTROPCORR implements the tropo model. SDCLCORR provides the SV clock delay algorithm, and SDIONOCORR scales the iono filter output by L1 or L2 scale factors. Receiver calibration data and absolute path delays are taken directly out of the C/NV memory area, SDNVMEMORY, by SDDETERM. Supporting the deterministic corrections are the iono filter and receiver calibration filter, SDIONO, and SDRCVRCALP, respectively, which are both called by the RM function when the respective pseudoranges have been collected for each of these types of measurements. SD1HZIONO is called at 1 Hz out of the 1-Hz Make Data Collection Requests function, SDDATAREQ, to update the variances of the iono filters.

The Highest level modules for the Process Satellite Data function are SDPRCSVDATA and SDPRCPARTSBF. These are called by RM upon receipt of full subframes and partial subframes respectively. Figure 6 shows the list of modules SDPRCSVDATA calls when processing a particular subframe of data. Table 6 provides a description of the function of each of the modules listed in figure 6.

Table 6. SDBM Module Functional Descriptions.

MODULE	FUNCTIONAL DESCRIPTION
SDWHICHTEMP	used to find which of the 5 ephemeris temporary slots to use to process a particular SVIDs ephemeris subframes.
SDPARSE	extracts data fields from a subframe applying necessary scale factors.
SDADJUST	concatenates variables, corrects sign errors created in SDPARSE, and performs units conversions.
SDSTORE	stores parsed and adjusted data in the temporary slot assigned for the data.
SDDATAPACK	packs the raw subframe data, retaining only the 24 data bits of each of the 10 words of a subframe.
SDUPDHLTST	updates the constellation status and health status words given any 6 bit health pattern for a SVID.
SDUPALMH	updates the constellation status and health status words given the 8 bit health pattern transmitted in the satellite almanac.

Table 6. SDBM Module Functional Descriptions (Continued).

MODULE	FUNCTIONAL DESCRIPTION
NR'ESTGPSWK	provides the GPS week number transmitted in the ephemeris to the Navigation function for GPS time maintenance.
SDNVHLTUDATE	updates an individual SVIDs health in C/NV memory.
INSTRUM	called to output test data over the Instrumentation Port.
SDNVUPDATE	moves packed subframes to C/NV memory temporary slots.
SDRELEASE	the module used to "release" all types of data to the data base.
SDUPDHLTSUML	updates the constellation status and health status words given the 6 bit health patterns for SVIDs 25 through 32 that are contained in the health summary subframe 5.
SDUPDHLTSUMU	updates the constellation status and health status words given the 6 bit health patterns for SVIDs 1 through 24 that are contained in the health summary subframe 4.
SDUPALMCST	updates the almanac collection status words when an almanac is processed.
SDPRCPARTSBF	called by RM to process all of the partial subframes that are collected. This routine also uses SDNVHLTUDATE, SDUPDHLTSTAT, and SDRELEASE in processing of partial subframe one data.

SATELLITE MANAGER SELECT-SATELLITES FUNCTIONAL DESCRIPTION AND REQUIREMENTS

The purpose of the Select-Satellites function is to select the satellites whose pseudorange and deltarange measurements will be processed to derive position, velocity, and time. Select-Satellites consists of three types of selection processes: GDOP/HDOP-Select, Quick-Select, and Search-The-Sky. Designed for a 24-satellite constellation, the Select Satellites functions all have a capacity for 12 visible satellites. The normal selection method, GDOP/HDOP select will be discussed first, followed by the other two selection methods.

GDOP/HDOP-Select

GDOP/HDOP-Select produces a list of satellites currently visible to the user and selects a set of four primary satellites that will be used for navigation. The four primary navigation satellites are chosen on the basis of their relative geometry, from a list of candidate satellites that have been screened from the visible satellites list. Geometric Dilution Of Precision (GDOP) or Horizontal Dilution Of Precision (HDOP), which provide the measure of the relative geometry, are computed for each combination of four candidate satellites. The combination of four candidates with the smallest GDOP or HDOP will be the set of primaries recommended to the RM function for acquisition. These primary satellites will normally be acquired in Initial Acquisition with either the Normal or Hot Start acquisition strategy. Subsequent updates to recompute the set of four primary satellites are done on a routine basis while in Normal Track (see figure 2).

The HDOP selection is the default selection criteria on five channel shipborn sets (RCVR3S) only, with the operator having the option of specifying GDOP as the selection criteria via a Control Display Unit (CDU). The software remains common for all receiver types, but with the GDOP selection as the default for the other receivers. Table 7 lists the equations implemented in the GDOP/HDOP selection process.

Almanac or ephemeris data must be available to the SVPVA function to obtain satellite position, and a reasonably good user position and time estimate must be available from the Navigation function for the visible and primary lists to be accurate. The RM function bases its calculation of receiver start type, Cold, Normal or Hot, on the contents of the primary list. If no primary satellites are recommended, the receiver will enter Cold start operation.

The GDOP/HDOP Select function executes as an autonomous "task"[1] that is scheduled for a single execution (see figure 2, #3) by

Table 7. Equations for GDOP/HDOP Selection Process.

A. Generation of Visible Satellites List equations. The equations required are those that compute the elevation angle of satellites, as measured in locally level coordinates.

(1) Extrapolate current TFS to approximate midpoint of the reselection interval obtaining T.

$$T = TFS - 60$$

Where TFS is frequency synthesizer time in seconds.

(2) Use navigation function to compute user position ($\bar{P}_E$) and unit vertical vector ($\overline{UP}$) at time T

(3) Compute position of satellite ($\bar{S}_E$) at GPS time associated with T via Satellite Position Velocity Acceleration Function

$$\bar{S}_E = (x_k, y_k, z_k)$$

(4) Compute the line of sight vector ($\bar{L}_E$), unit line of sight vector ($\overline{LU}_E$), and elevation angle (E) for each satellite.

$$\bar{L}_E = \bar{S}_E - \bar{P}_E$$

$$\overline{LU}_E = \frac{\bar{L}_E}{|\bar{L}_E|}$$

$$E = \arcsin(\overline{LU}_E \cdot \overline{UP})$$

B. Equations for the formation of GDOP and HDOP.

(1) Convert user position from ECEF ($\bar{P}_E$) to locally level coordinates ($\bar{P}_T$).

(2) Use user position ($\bar{P}_T$) to form a coordinate transformation matrix (C_{ET}):

$$C_{ET} = \begin{bmatrix} -\cos^{\alpha}\sin^{\lambda} - \sin^{\alpha}\sin^{\phi}\cos^{\lambda} & \cos^{\alpha}\cos^{\lambda} - \sin^{\alpha}\sin^{\phi}\sin^{\lambda} & \sin^{\alpha}\cos^{\phi} \\ +\sin^{\alpha}\sin^{\lambda} - \cos^{\alpha}\sin^{\phi}\cos^{\lambda} & -\sin^{\alpha}\cos^{\lambda} - \cos^{\alpha}\sin^{\phi}\sin^{\lambda} & \cos^{\alpha}\cos^{\phi} \\ \cos^{\phi}\cos^{\lambda} & \cos^{\phi}\sin^{\lambda} & \sin^{\phi} \end{bmatrix}$$

(3) Transform the unit line sight vector from ECEF ($\overline{LU}_E$) to locally level coordinates ($\overline{LU}_T$)

$$\overline{LU}_T = C_{ET}\overline{LU}_E$$

(4) Form the matrix G from four unit line of sight vectors $\overline{LU}_{Ti}$, where $i = 1, 2, 3, 4$.

$$G = \begin{bmatrix} LU_{1X} & LU_{1Y} & LU_{1Z} & 1 \\ LU_{2X} & LU_{2Y} & LU_{2Z} & 1 \\ LU_{3X} & LU_{3Y} & LU_{3Z} & 1 \\ LU_{4X} & LU_{4Y} & LU_{4Z} & 1 \end{bmatrix}$$

(5) Take the inverse of G and multiply the result by its transpose to obtain the solution matrix H.

$$H = G^{-1}G^{-T}$$

(6) Compute GDOP from the trace of H.

$$\text{GDOP} = (H_{11} + H_{22} + H_{33} + H_{44})^{1/2}$$

(7) Compute HDOP from H (Applicable to RCVR3S only).

$$\text{HDOP} = (H_{11} + H_{22})^{1/2}$$

the Scheduler Init function on the transition from INIT to NAV modes to generate the visible satellite list and primary list (if any) that will be used by RM in Initial Acquisition. On this transition, it executes immediately after the Satellite Data Base Initialization task mentioned earlier. At the completion of Initial Acquisition, upon transition to Normal Track, it is then rescheduled for cyclic execution by RM, with the cyclic execution time being once every two minutes. During the cyclic execution in Normal Track, the task executes as a "background task" that is interruptable by higher priority tasks in the system.[1] Operation as a "background task" every two minutes allows for enough processing time to generate and compute GDOP and HDOP for every combination of four candidate satellites.

The final organization of the visible satellites list contains both the healthy and unhealthy satellites, with the healthy group appearing first in the list, and each group being organized by descending elevation angle order. Unhealthy satellites are included in the visible satellites list since the SDBM function will request periodic partial subframe one data collections from these satellites when the number of visible healthy satellites with ephemeris data is less than five. Satellite visibility is computed using a position estimate obtained from the Navigation function at a time one minute in the future. Satellite position is obtained using the SVPVA function. The initial satellite visibility computations are all performed in Earth Centered Earth Fixed (ECEF) coordinates, with later translation of candidate satellites data to locally level coordinates for GDOP/HDOP calculations. Inclusion in the visible satellite list requires that the satellite have an elevation angle greater than or equal to zero degree, and also that the constellation status word (as set by SDBM) indicates it is a potentially trackable satellite. The satellite health status word is checked to identify healthy satellites.

In order to be considered as a candidate satellite for inclusion in the recommended primary selection, satellites must have their satellite positions calculated with the right class of data necessary to support the tracking mode the receiver is in. While in Initial acquisition, either almanac or ephemeris can be used for the satellite position calculation. While in Normal Track, ephemeris data is required since pseudorange and deltarange measurements are only processed by the Navigation function for satellites that have ephemeris data present in the receiver. Ephemeris data is not required in Initial Acquisition since it will be requested and collected if it is not present or if it is old. Whether or not ephemeris is required is determined by the RM function setting an "ephemeris required" flag for the Select Satellites function.

Additional criteria required for consideration as a candidate satellite are that the satellite have good health, good track history, and that it must have an elevation angle that is greater than or equal to a "sliding" elevation angle mask that is selected at each execution of the task. The health of the satellites is determined from the Health-Status summary word provided by the SDBM function. Track History is maintained and determined by the RM function and contains information concerning the current or recent tracking status of the satellite. A "track summary" flag of the track history is checked and indicates whether the satellite is empirically untrackable (though it may have good health). The "sliding" elevation angle mask is then assigned values of ten, five and zero degrees depending on the number of candidate satellites and the tracking mode the receiver is in. The elevation angle mask is selected by choosing the greatest mask such that a minimum of four of the candidate satellites elevation angles meet or exceed the mask. If the criteria are not met, the next lowest mask is tested for the same condition, continuing until zero degrees is reached if necessary. The tracking mode determines what the starting elevation angle mask will be; ten degrees for Initial Acquisition, and five degrees for Normal Track.

Once the list of candidate satellites has been generated, a coordinate transformation from ECEF to locally level (LL) is performed on each of the user to satellite unit line of sight vectors. This is done to facilitate the computation of HDOP.

GDOP and HDOP are computed for each of the combinations of four candidate satellites. The combination of four with the smallest GDOP or HDOP becomes the recommended primary list.

A planned improvement in the GDOP/HDOP selection process is to calculate a "noise weighted" GDOP and HDOP that would include the satellites User Range Accuracy (URA) term transmitted in the downlink data. Satellites with larger URA values would tend to be excluded from the primary list. Three SVs with an altitude source could also be considered as an alternate selection criteria, since this combination may at times provide a better solution than four satellites.

Quick-Select

The Quick Select function is used by RM to select replacement satellites for those recommended primary satellites that cannot be acquired or have lost track. During Initial Acquisition, RM will call this function for replacement SVs for any of the initial recommended primary SVs that cannot be acquired. In Normal Track, RM will call this function for replacement of primary navigation satellites that have lost lock. Quick Select returns SVIDs that are not currently Navigation satellites, that are the highest in the sky with the best available track history, and that have the data support necessary for the tracking mode the receiver is in. The underlying assumption is that the satellites with greatest elevation angles will be the ones that have the highest probability of being acquired in most cases, and the next execution of the GDOP/HDOP Select function will provide a new constellation with the best geometry (excluding satellites with bad track history). Selection of alternates based on geometry was considered too costly in terms of processor memory and throughput, and added complexity.

An SVID is chosen from the list of alternate satellites organized by elevation angle, data support, and track history. Returned with the SVID is a list pass number indicating the pass number being made through the current alternate list. A new alternate list is built whenever a reselection-initialization parameter is set by the RM function to command the Quick Select function to build a new list. The pass number is incremented each time the list is exhausted and a new pass is begun through the list.

The alternate list is built by first copying the visible satellites list, deleting unhealthy satellites and any satellites that are navigation satellites as indicated in the RM track history. Next, each satellite is assigned one of four priorities based on the data support available for it, and the status of its track summary flag in the RM track history:

(1) Good track summary with ephemeris
(2) Good track with almanac
(3) Bad track with ephemeris
(4) Bad track with almanac or no data.

The list is then sorted by ascending priority and the resulting alternate list is ordered in four priority groups with each priority group ordered by descending elevation angle.

If the receiver is in Normal Track then ephemeris data is required and only satellites with ephemeris data support will be passed back as alternate satellites. Replacement SVs in Initial Acquisition can have either almanac or ephemeris data support.

Search-The-Sky

Search-The-Sky selects satellites for acquisition during the Cold Start operation of the set. The purpose of Cold Start in DoD standard receivers is to acquire one of the satellites to collect current satellite almanac data. Cold start will occur if there is no almanac or ephemeris data available to compute satellite visibility, or if data is present and no satellites are visible, or the set operator commands a Cold Start. Cold start satellites returned to the RM function are chosen from the set of Pseudorandom Noise (PRN) codes (or SVIDs) identified as healthy, or from the entire set of 32 PRN codes. If the health summary subframes are stored in C/NV memory, then the list of healthy PRN codes is used, exhausting this list twice prior to using the entire set of 32 PRN codes. If the health summary subframes are not stored in C/NV memory, then the entire set of 32 PRN codes is used. The receiver will perform a search for each SVID returned until one is acquired.

Implementation

Figure 7 shows the partitioning of the Select Satellites function into modules. The GDOP/HDOP function is implemented in three routines labeled SSGDOPSELECT, SSSCREENSATS, and SSSCREEN-CANS. SSGDOPSELECT obtains the user position and time and calls SSSCREENSATS to generate the visible and candidate satellite lists, and the operator commanded lists. SSSCREENCANS is called to screen the candidate satellites according to the sliding elevation angle mask. SSCONVE2L performs the coordinate transformations on the remaining candidate satellites. SSGDOPSELECT continues processing with the logic for combination generation, and the GDOP/HDOP calculations.

The Quick Select and Search-The-Sky functions are shown in figure 7 as the subroutines SSQUIKSELECT and SSSEARCHSKY, called on demand by the RM function. The module SSSSINIT is called by the SI'MONITOR function prior to the scheduling of SSGDOPSELECT on the transition from INIT to NAV modes.

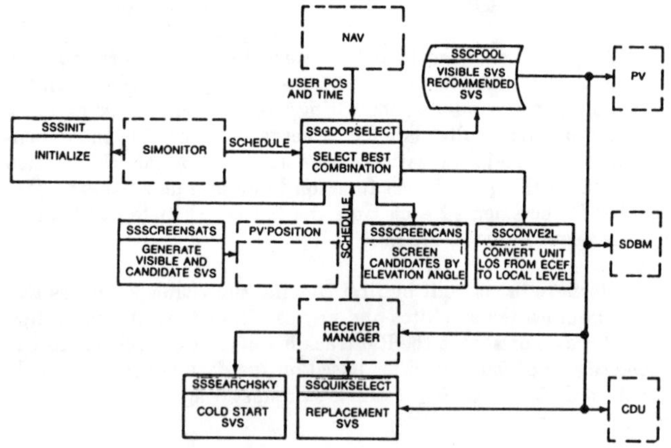

Figure 7. Select Satellites

SATELLITE MANAGER

SATELLITE-POSITION-VELOCITY-ACCELERATION

FUNCTIONAL DESCRIPTION AND REQUIREMENTS

The Satellite Position Velocity Acceleration (SVPVA) function is divided in two functions which map readily into the requirements it must fulfill in the DoD receiver. The first function is SV-Pos-Kepler, which computes satellite position directly from the Keplerian like almanac or ephemeris parameters transmitted in the satellite down-link data stream. The second function is SV-Pos-Vel-Acc, which computes satellite position, velocity, and acceleration using an efficient three-point parabolic interpolation scheme. SV-Pos-Kepler is used by the Select Satellites function for satellite visibility computations, and by SV-Pos-Vel-Acc to obtain the three interpolation points. SV-Pos-Vel-Acc is used by the Navigation function to obtain satellite position, velocity, and acceleration for use in satellite line-of-sight computations. Satellite position, velocity, and acceleration calculations are all performed in the WGS-84 Earth Centered Earth Fixed (ECEF) reference frame. SV-Pos-Kepler will be discussed first followed by the SV-Pos-Vel-Acc function.

SV-Pos-Kepler

The SV-Pos-Kepler function implements the equations found in ICD-GPS-200[(4)] for computing satellite position. These equations are listed in table 8. The algorithm can be called on demand to produce ECEF satellite position at the time specified by the calling routine. Inputs to the function are an SVID, the GPS time of the desired satellite position, and a data source flag indicating whether to use almanac or ephemeris data in the computations. Either ephemeris or almanac can be specified as the data source or an "available" specification can be used. When "available" is specified the routine will use almanac or ephemeris as the source but with ephemeris being the preferred source. The almanac or ephemeris data used is copied from the satellite data base maintained by SDBM. Outputs of the function are ECEF satellite position, and a flag indicating the data source used, or error condition if no position was computed.

Table 8. Equations for SV-Pos-Kepler.

$\mu = 3.986005 \times 10^{14}$ meters^3 sec^{-2}
WGS 84 value of the earth's universal gravitational parameter
(Mass of earth's atmosphere included)

$\dot{\Omega}_e = 7.2921151467 \times 10^{-5}$ rad sec^{-1}
WGS 84 value of the earth's rotation rate

$A = (\sqrt{A})^2$ semi-major axis

$n_o = \sqrt{\frac{\mu}{A^3}}$ computed mean motion

IF (Ephemeris Data)
BEGIN
$t_k = t - t_{oe}$ "time from reference epoch"
IF $t_k < -302400$ THEN $t_k = t_k + 604800$ "end of week crossover correction"
IF $t_k > 302400$ THEN $t_k = t_k - 604800$ "end of week crossover correction"
END

IF (Almanac Data)
$t_k = (GPSWEEK - GPSREFWK)604800 + (t - t_{oa})$ "time from reference epoch"

$n = n_o + \Delta n$ "corrected mean motion"

$M_k = M_o + nt_k$ "mean anomaly"

$E_k = M_k + e \sin M_k$ "initial guess eccentric anomaly"

FOR J:1 by 1 while ≤ 2 "iterate twice for final value"
$E_k = (e(\sin E_k - E_k \cos E_k) + M_k)/(1 - e \cos E_k)$

$\cos v_k = (\cos E_k - e)/(1 - e \cos E_k)$ "cosine of true anomaly"

$\sin v_k = \sqrt{1 - e^2} \sin E_k/(1 - e \cos E_k)$ "line of true anomaly"

$\phi_k = v_k + \omega$ "argument of latitude"

IF (Almanac Data)
BEGIN
$C_{us} = C_{uc} = C_{rc} = C_{rs} = 0$
$C_{ic} = C_{is} = IDOT = 0$
$i_o = 0.3$ semicircles
END

$\delta u_k = C_{us} \sin 2\phi_k + C_{uc} \cos 2\phi_k$
"argument of latitude correction (second harmonic perturbation)"

$\delta r_k = C_{rc} \cos 2\phi_k + C_{rs} \sin 2\phi_k$
"radius correction (second harmonic perturbation)"

$\delta i_k = C_{ic} \cos 2\phi_k + C_{is} \sin 2\phi_k$
"correction to inclination (second harmonic perturbation)"

$u_k = \phi_k + \delta u_k$ "corrected argument of latitude"
$r_k = A(1 - e \cos E_k) + \delta$ "corrected radius"
$i_k = i_o + \delta_{ik} + \text{IDOT} \times t_k$ "corrected inclination"
$x'_k = r_k \cos u_k$ "position in orbital plane"
$y'_k = r_k \sin u_k$ "position in orbital plane"
$\Omega_k = \Omega_o + (\dot{\Omega} - \dot{\Omega}_e)t_k - \dot{\Omega}_e t_d$ "corrected longitude of ascending node"
$x_k = x'_k \cos \Omega_k - y'_k \cos i_k \sin \Omega_k$ "earth fixed coordinate"
$y_k = x'_k \sin \Omega_k + y'_k \cos i_k \cos \Omega_k$ "earth fixed coordinate"
$z_k = y'_k \sin i_k$ "earth fixed coordinate"

Where:

AODE	age of data (ephemeris)
C_{ic}	amplitude of the cosine harmonic correction term to the angle of inclination
C_{is}	amplitude of the sine harmonic correction term to the angle of inclination
C_{rc}	amplitude of the cosine harmonic correction term to the orbit radius
C_{rs}	amplitude of the sine harmonic correction term to the orbit radius
C_{uc}	amplitude of the cosine harmonic correction term to the argument of latitude
C_{us}	amplitude of the sine harmonic correction term to the argument of latitude
Δn	mean motion difference from computed value
e	eccentricity

Table 8. Equations for SV-Pos-Kepler (Continued).

GPSREFWK	reference week number of the almanac t_{oa}
GPSWEEK	GPS week number (0 to 1023)
IDOT	rate of inclination angle
i_o	inclination angle at reference time
M_o	mean anomaly at reference time
ω	argument of perigee
$\dot{\Omega}$	rate of right ascension
Ω_o	right ascension at reference time
t	GPS time of week
t_d	t_{oe} if ephemeris data, t_{oa} if almanac data
t_{oa}	reference time almanac
t_{oe}	reference time ephemeris
$\sqrt{A}$	square root of the semi-major axis

SV-Pos-Vel-Acc

The SV-Pos-Vel-Acc function is responsible for providing the Navigation function satellite position, velocity, and acceleration for up to 5 different SVIDs on a 1-Hz basis. To minimize processor throughput loading, interpolated values are used that provide the satellite position, velocity, and acceleration using quadratic, linear, and constant vector functions of time. The interpolation points are three satellite positions, thirty seconds apart, computed with the SV-Pos-Kepler function. A parabolic curve is used to fit the points on the orbit plane. A circular curve fit would provide a smaller interpolation error, but would require the calculation of a sine and cosine each time an interpolation value was needed. A test computer program has shown that the parabolic curve fit used with the thirty-second time interval has a maximum error of only 0.12 meters.

To provide the interpolated satellite data SV-Pos-Vel-Acc is divided into two subfunctions. The first is responsible for initializing and maintaining the necessary interpolation coefficients. The second is responsible for computing the satellite position, velocity, and acceleration using the interpolation coefficients.

Initial values of the interpolation coefficients are derived by an autonomous "task" that is scheduled for a single execution by the Scheduler Init function on the transition from INIT to NAV operating modes (see figure 2). The interpolation initialization task executes immediately following Satellite Data Base Initialization and the GDOP/HDOP tasks described earlier. Interpolation initialization generates interpolation coefficients for all visible satellites based on satellite almanac or ephemeris data. A copy of the visible satellites generated by GDOP/HDOP select is made. The SV-Pos-Kepler function is called three times for each visible satellite to obtain satellite positions at times t_1, t_2, and t_3 (Δt = 30 seconds). Interpolation coefficients are then computed and stored in an interpolation coefficient data base. The equations necessary for computing the interpolation coefficients are shown in table 9.

Table 9. Equations for SV-Pos-Vel-Acc.

A. Equations for computing Interpolation Coefficients Given the following:

$P1(x_1, y_1, z_1)$ = Satellite ECEF position at GPS time t_1

$P2(x_2, y_2, z_2)$ = Satellite ECEF position at GPS time t_2

$P3(x_3, y_3, z_3)$ = Satellite ECEF position at GPS time t_3

E_1 = eccentric anomaly associated with $P1$ at t_1

E_2 = eccentric anomaly associated with $P2$ at t_2

E_3 = eccentric anomaly associated with $P3$ at t_3

$t_1 = t_2 - \Delta t$

t_2 = midpoint of time interval for which interpolation shall occur

$t_3 = t_2 + \Delta t$

Δt = 30 seconds

Table 9. Equations for SV-Pos-Vel-Acc (Continued).

The interpolation coefficients shall be defined as:

$$a_x = (-x_1 + 2x_2 - x_3)/(-2(\Delta t)^2)$$

$$b_x = (x_1 - x_3)/(-2\Delta t)$$

$$c_x = x_2$$

The coefficients $a_y, b_y, c_y, a_z, b_z, c_z$ are identical to the above 'x' coefficients with 'x' replaced by 'y' and 'z'.

The eccentric anomaly coefficients shall be defined as:

$$a_E = (-E_1 + 2E_2 - E_3)/(-2(\Delta t)^2)$$

$$b_E = (E_1 - E_3)/(-2\Delta t)$$

$$c_E = E_2$$

B. Equations for computing satellite position, velocity, and acceleration Given the following:

interpolation coefficients: $a_x, a_y, a_z, b_x, b_y, b_z, c_x, c_y, c_z, a_E, b_E, c_E$

t is the time at which the interpolated position is required

t_2 is the midpoint of the interpolation interval t_1, t_2, t_3

The interpolated position of the satellite vehicle ($PI(x_I, y_I, z_I)$) shall be:

$$x_I(t) = a_x(t - t_2)^2 + b_x(t - t_2) + c_x$$

$$y_I(t) = a_y(t - t_2)^2 + b_y(t - t_2) + c_y$$

$$z_I(t) = a_z(t - t_2)^2 + b_z(t - t_2) + c_z$$

This interpolation scheme shall produce acceptable values for satellite position for any GPS time (t_x) such that $t_1 \leq t_x \leq t_3$. Satellite velocity and acceleration shall be the first and second derivatives of the equation of motion x_I, y_I, z_I. Satellite velocity:

$$v_x(t) = 2a_x(t - t_2) + b_x$$

$$v_y(t) = 2a_y(t - t_2) + b_y$$

$$v_z(t) = 2a_z(t - t_2) + b_z$$

Satellite acceleration:

$$a_x(t) = 2a_x$$

$$a_y(t) = 2a_y$$

$$a_z(t) = 2a_z$$

Satellite Position Variance:

```
    δ²SV = 10.0 × 10^6 Meters² default variance
    IF (ephemeris data was used)
        IF (Fit Interval indicates 4 hour fit)
            IF ( t2 lies within 4 hour fit interval)
                δ²SV = 0.0
            ELSE
                δ²SV =(function of time outside fit)
            ENDIF
        ELSE
            IF ( t lies with 6 hour fit interval)
                δ²SV = 0.0
            ELSE
                δ²SV =(function of time outside fit)
            ENDIF
        ENDIF
    ENDIF
```

The interpolated eccentric anomaly shall be:

$$E(t) = a_E(t - t_2)^2 + b_E(t - t_2) + c_E$$

Note: $E(t)$ is used by SDBM to calculate SV clock corrections.

In order to ensure the accuracy of the satellite position, velocity, and acceleration data for the Navigation function, GPS time must lie between t_1 and t_3, the Δt between the points must be 30 seconds. Additionally, the data used to compute the satellite positions at each of the interpolation points must be from the same set of almanac or ephemeris parameters.

Maintenance of the interpolation data base is performed as a 1-Hz process. Maintenance of the interpolation coefficients includes three types of updates to the interpolation data base. The first type of update is the normal update that begins every thirty seconds and only requires one new satellite position computation for each visible

SV. In this case one visible SV is updated per second once the update begins, with the update taking N seconds to complete for N visible satellites. The update will begin when GPS time is greater that t_2. The satellite positions at the old t_2 and t_3 become the positions at the new t_1 and t_2, and a new satellite position is obtained at the new t_3 from SV-Pos-Kepler. Interpolation coefficients are then recomputed with the satellite positions at the new t_1, t_2, and t_3. The normal update process also deletes interpolation data for any satellites that are no longer visible. The second type of update occurs when there is a large correction made to the receivers estimated GPS time, which may occur in the acquisition sequence of the first satellite. If time t_2 differs from GPS time by more that 30 seconds, then interpolation coefficients must be recomputed using three entirely new satellite positions from SV-Pos-Kepler for the new t_1, t_2, and t_3. This time adjustment type of update is also performed at the rate of one SV per second. The third type of update occurs when new ephemeris data is collected for the satellites. New ephemeris updates also require three new satellite position calculations, but are performed for up to 5 satellites per second to ensure completion of the updates as quickly as possible. All updates to the interpolation data occur in the 1-Hz sequence prior to the execution of the Navigation function line-of-sight calculations.

The process of computing satellite position, velocity, and acceleration from the interpolation coefficients is straightforward. The equations are also listed in table 9. An error condition is returned if no data exists in the interpolation data base, or if the time specified for the data is outside the interval (t_1, t_3).

Note: E(t) is used by SDBM to calculate SV clock corrections.

<u>Implementation</u>

Figure 8 shows the partitioning of the Satellite Position Velocity Acceleration function into modules. The SV-Pos-Kepler function is implemented in the module labeled PVPOSITION. The SV-Pos-Vel-Acc function is implemented in the four modules labeled PVINITIAL, PVONEHERTZ, PVCALCCOEF, and PVPVAINTRP. PVINITIAL is the initialization "task" scheduled for execution by the Scheduler Init module SI'MONITOR on the transition from INIT to NAV modes. PVINITIAL uses PVPOSITION to obtain satellite positions, and calls PVCALCCOEF with three satellite positions to calculate and store the interpolation coefficients. PVCALCCOEF also stores the satellite positions for times t_1 and t_2 for possible use in the normal 30 second update. PVONEHERTZ performs the maintenance of the interpolation coefficients at a 1-Hz rate, also using PVPOSITION and PVCALCCOEF to calculate satellite positions and for calculation and storage of interpolation coefficients and satellite positions. PVPVAINTRP is the routine called by the Navigation function to obtain interpolated values for satellite position, velocity, and acceleration.

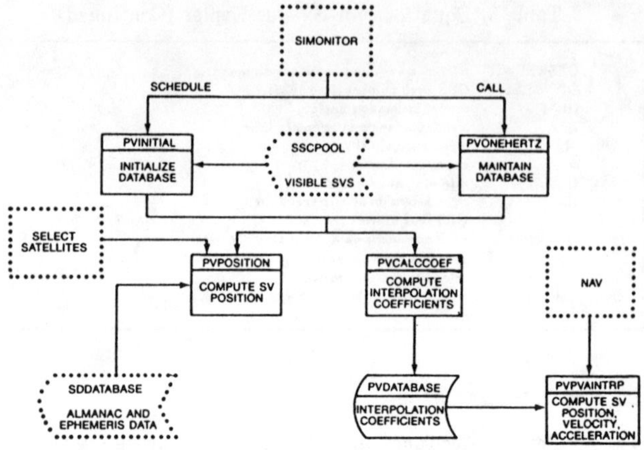

Figure 8. Satellite Position, Velocity, Acceleration Function

SUMMARY

The satellite data management implementation described in this paper fulfills all the requirements of the DoD for NAVSTAR GPS receivers. As a testament to its commonality, efficiency, and the flexibility of the interfaces between Satellite Manager and the rest of the system, it should be noted that this design, with minor tailoring as necessary, is also utilized in GPS commercial products software as well as in other Government sponsored programs for unique GPS applications.

REFERENCES

1. Bartholomew, et al: *Software Architecture of the Family of DoD Standard GPS Receivers*

2. Vacherlon, et al: *GPS Phase III Multichannel User Equipment*

3. Krishnamurti, et al: *The Design and Performance of GPS Phase II User Equipment Navigation Software* NAVIGATION: Journal of the Institute of Navigation, Vol. 32, No 3, Fall 1985.

4. ICD-GPS-200, Revision B, Interface Control Document, NAVSTAR GPS Space Segment/Navigation User Interfaces, November 30 1987.

5. Bean, et al: *Radio Meteorology* United States Department of Commerce, National Bureau of Standards (1966).

DIFFERENTIAL GPS NAVIGATION

Stewart P. Teasley, Wayne M. Hoover, and Charles R. Johnson

Texas Instruments Incorporated

ABSTRACT

The concept of a differential Global Positioning System (GPS) is introduced. A description of the required components of a generalized differential system is presented. The specific implementation using the Texas Instruments Phase I High Dynamic User Equipment (HDUE) and the available ground-based equipment at the Yuma Proving Ground (YPG), Yuma, Arizona, is described and representative test results are presented. A discussion of possible future implementations with next-generation GPS user equipment sets is included.

INTRODUCTION

Users of radionavigation aids such as Loran C, OMEGA, and TRANSIT have developed differential techniques when high accuracies are required. Differential navigation means either that one user set is navigating relative to another user set via a data link or that the user has been at the destination at some prior time and is navigating relative to coordinates measured at that point. This technique provides increased accuracy because bias errors caused by transmitter imperfections, ionosphere, etc., are cancelled out since they are common at both the user location and the destination. These common errors, which are usually the largest component of the total error, are subtracted out by the differential technique. Therefore, the differential technique provides a remarkable improvement in accuracy.

For example, at points along the Gulf Coast, the absolute accuracy of Loran C is worse than 1,000 feet. However, Loran C can be used in a differential mode to return to port with an accuracy as good as 50 feet. Therefore, a differential navigation capability is a logical application for the GPS. The gains that can be made in accuracy are summarized in Table 1. When the user equivalent range error is multiplied by a position dilution of precision (PDOP) of 3, an accuracy of 12 meters results. It should be noted that this estimate of accuracy is for a point solution to the GPS ranging equations and can be improved with a navigation filter. The navigation filter will be particularly effective in reducing these errors because they are mostly random in character. Since random errors in most applications can be reduced at least to a factor of 1/3 by filtering, it is not unreasonable to expect the differential errors to be reduced to less than 4 meters (Table 1).

Normally, a paper investigating the use of a new navigation technique would provide many simulations that prove the point but no real test data to furnish final proof. The equipment available from Phase I testing at YPG, however, was ideal to test differential GPS. The differential technique was implemented and tested at YPG, and the results are discussed in this paper.

SYSTEM DESIGN CONCEPTS

Differential GPS represents a cost-effective, straightforward method to significantly improve the accuracy of GPS user sets. The focal point of the differential system is a GPS receiver (differential sensor) that operates from a surveyed antenna. The true values of this receiver's navigation fix are compared against the measured values and the differences become the differential corrections. These corrections may be transmitted to area user sets in real time (Figure 1), or they may be recorded for postmission use. In either case, the users apply the corrections to their navigation data to produce position fixes that are free of GPS-related biases. Since the rate of change of the differential terms is rather slow, the data entry into the user set does not need to be sophisticated and conceivably could be done by hand through a control display unit (CDU). Some significant issues relevant to the design of a differential GPS (Figure 2) are discussed in the following paragraphs.

Differential Sensor

The differential sensor may be a standard GPS receiver capable of tracking all visible satellites through a surveyed antenna. Requirements for simultaneous versus sequential tracking are dictated primarily by the accuracy requirement in the final correction terms. Since user dynamics are nil in this receiver, a set that produces a few low-noise measurements is ideal. The set applies all normal corrections to the measurements (e.g., group delays, tropospheric delay, ephemeris clock corrections) but should retain the option to inhibit ionospheric corrections. From the user set side, group delays and tropospheric delays are uncorrelated with respect to the differential sensor and need to be calculated by the user sets in the normal manner. The ephemeris clock corrections are applied by both the user sets and the differential sensor. Since the estimation of ionospheric delays is imprecise and requires an L2-compatible receiver in the user set, these delays may be inhibited in both the user sets and the differential sensor. In this way, the true ionospheric delays become part of the differential correction terms.

Clock Corrections Versus Navigation Solution Corrections

A basic design issue is the nature of the corrections themselves. The most general approach is to implement the correction terms in the form of static satellite clock corrections. That is, the differential terms become a correction at the

Reprinted from *Rec. IEEE Position Location and Navigation Symp.*, pp. 9–16, 1980.

TABLE 1. GPS ERROR BUDGET

Error Source	Absolute (meters) Bias	Absolute (meters) Random	Absolute (meters) Total	Differential (meters) Bias	Differential (meters) Random	Differential (meters) Total
Clock and navigation subsystem stability	0	2.7	2.7	0	2.7	2.7
Predictability of satellite perturbations	1.0	0	1.0	0	0	0
Other	0	0.866	0.866	0	0,866	0.866
Ephemeris and clock prediction	2.5	0	2.5	0	0	0
Ionospheric delay compensation	2.3	0	2.3	0	0	0
Tropospheric delay compensation	0	2.0	2.0	0	2.0	2.0
Receiver noise and resolution	0	1.5	1.5	0	1.5	1.5
Multipath	0	1.2	1.2	0	1.2	1.2
1σ system user equivalent range error (UERE)	3.54	3.97	5.3	0	3.97	3.97

Absolute:

Point solution accuracy = PDOP X ranging accuracy = 3 X 5.3 ≅ 16 meters

Filtered solution accuracy =

$$\sqrt{(3 \times 3.54)^2 + \left(\frac{3 \times 3.97}{3}\right)^2} \cong 11 \text{ meters}$$

Differential:

Point solution accuracy = PDOP X ranging accuracy = 3 X 3.97 ≅ 12 meters

Filtered solution accuracy =

$$\frac{3 \times 3.97}{3} \cong 4 \text{ meters}$$

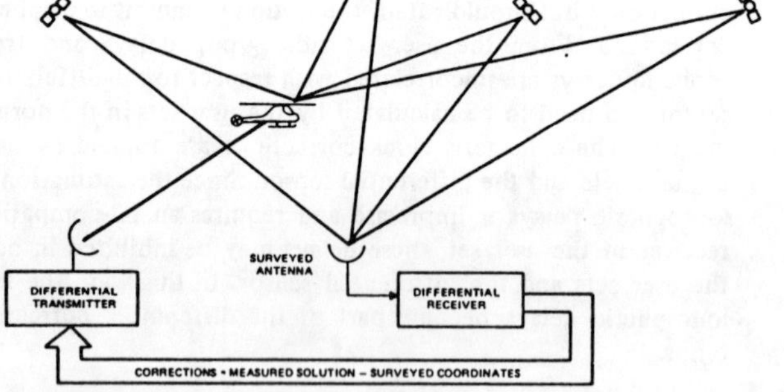

Figure 1. Differential GPS Operation

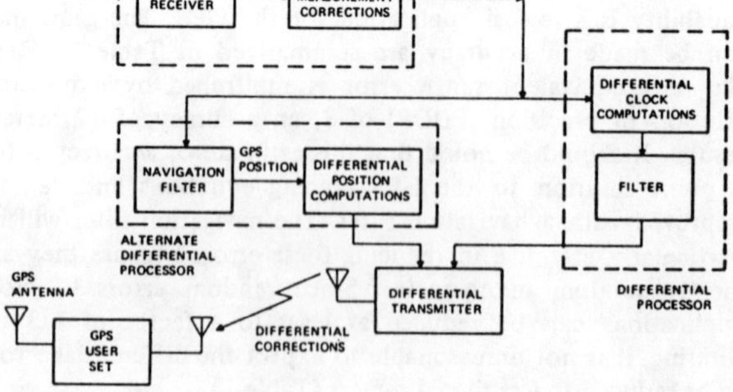

Figure 2. Differential GPS Block Diagram

pseudorange measurement level similar to the a_0 clock correction term in the GPS ephemeris data. This approach requires that the differential sensor track all visible satellites and pass the associated data to the user sets. The user sets are then free to choose their own satellite group (constellation) and use the corresponding corrections. Many standard GPS receivers, however, do not provide the filtering at the measurement level that would be necessary to generate smoothed differential correction terms. An alternate approach is to produce a navigation solution with the measurements made from a four-satellite constellation. The resulting solution is compared with the surveyed antenna location and the differential corrections are generated as an [X,Y,Z] vector in a convenient coordinate system. This technique has the disadvantage that the user and the differential source must operate with the same constellation. Smoothed navigation data, however, may be more readily available in an ordinary GPS receiver than smoothed measurement data. For this reason, this technique may be more workable if a system is to be built around existing equipment designs.

Real-Time Versus Off-Line Corrections

Two features must be incorporated into the design to accommodate real-time requirements:

(1) Some form of data link between the differential processor and the user sets wishing to incorporate the corrections.

(2) Real-time generation of the differential terms themselves.

For a real-time system, the major design considerations are the method of data entry into the user set and the data rate needed to support accurate navigation. This rate is determined primarily by the uncompensated clock error rate associated with the satellites. While this behavior is difficult to predict, an update rate of once per 5 minutes would appear to be compatible with the satellite characteristics seen to date. (The initial field tests were run with an update rate of 30 seconds.) The data entry system involves a communication link between the differential processor and the user sets. The link might be as simple as an open broadcast of the corrections to all users in the area, or it could be more sophisticated in the form of a secure digital communications link. The field tests used a GPS ground transmitter with a modified 50-Hz data frame to transmit to a differential user that was capable of tracking the transmitter's signal.

An off-line system would require recording of both the user set solution and the differential corrections. These files would then be merged as an off-line operation, producing a corrected user solution.

Examples of user scenarios that would require real-time differential corrections are precision landing approaches, rendezvous, and any type of precision navigation. Scenarios that are compatible with off-line or postmission application of the corrections are land or sea surveys, nonreal-time precision navigation, and instrumentation system augmentation.

IMPLEMENTATION WITH TEXAS INSTRUMENTS HDUE

A real-time differential GPS implementation using the Texas Instruments built High Dynamic User Equipment and the available support equipment at the Yuma Proving Ground has been verified and tested. The Inverted Range Control Center (IRCC) at YPG contains a multichannel receiver that provides continuous tracking of all GPS satellites being used by the various user sets on the range. One of the purposes of the IRCC is to provide "ground truth" data for postmission analysis. This information is used as a standard for comparison against user GPS solutions to allocate user errors to the user itself and to the GPS satellites. The IRCC also provides real-time control over four ground-based GPS transmitters used as primary navigation sources before four GPS satellites were available. These ground transmitters contain sufficient spare words in their transmitted data blocks to allow the differential correction terms to be added to these blocks. The ground-based portion of the system was built around the IRCC and one of the ground transmitters (GTs) as follows:

(1) Since the GPS antenna at the IRCC is surveyed, the differential corrections are calculated by determining the apparent error in the navigation solution obtained at the IRCC. These corrections are in the form of position solution corrections rather than satellite clock corrections.

(2) The option is retained at the IRCC to inhibit the incorporation of ionospheric delay compensation into its measurements. In any case, the receiver passes to the differential processor the value of the ionospheric delays that it is currently incorporating for each satellite.

(3) The differential processor receives the three items necessary to support the differential system and incorporates these terms into the 50-Hz data frame of a GT at YPG. In this manner, the terms have an update rate of 30 seconds. The three items in the data stream are:

    Four satellite identifier codes

    Differential corrections in the earth-centered, earth-fixed coordinate system

    Ionospheric delay compensation incorporated at the IRCC for each satellite.

The HDUE (Figure 3) contains a five-channel receiver that provides up to five simultaneous pseudorange and range-rate measurements at a 320-millisecond rate. It contains an 11-state Kalman filter for navigation purposes that provides user state propagation and estimation at the same 320-millisecond rate. A control display unit (CDU) is used to display navigation data to the operator and to receive commands and other navigation requests. Three SBP 9900 microprocessors provide the processing environment in which the HDUE real-time software resides. Figure 4 shows the HDUE installed in its test pallet along with the associated instrumentation system for installation in the test vehicle. In this configuration, the HDUE has consistently provided excellent navigation results during the Phase I test program at YPG.

Figure 3. HDUE System

Figure 4. HDUE in Test Pallet

The HDUE is ideally suited for the differential application for several reasons:

(1) Its fifth receiver channel, normally used for L2 ionospheric measurements, can be used directly to receive the differential corrections transmitted by the GT.

(2) Its computer memory is random access and is thus a flexible test bed.

(3) Its constellation management allows a fifth GPS source to be selected for tracking without requiring its use in the navigation solution.

(4) The HDUE controls a pilot steering display (PSD) that provides horizontal and vertical steering commands to aid in precision approaches.

(5) It is configured with full data recording to support detailed postmission analysis.

The HDUE software was expanded to include the differential capability in addition to all its normal functions. The primary additions to the baseline HDUE capabilities are in the following areas:

(1) An interface established through the CDU allows the operator to select the differential mode and enter a ground transmitter number as the differential source. This action causes the HDUE to acquire and track this GT on its fifth channel. The satellite management function then inhibits the GT measurements from being incorporated into the navigation solution.

(2) The data block processing function has been modified to recover the differential terms added to spare areas in the GT 50-Hz data frame.

(3) All ionospheric measurement corrections are read from the differential data frame rather than being calculated from L2 measurements.

(4) The navigation solution is corrected based on the received differential correction terms.

(5) The corrected solution is then used for all normal HDUE navigation activities such as absolute position determination, relative navigation to a selected waypoint, and control of the steering display to maintain a desired approach.

FIELD TEST RESULTS

Period of Testing

The essential baseline software modifications to incorporate the differential GPS mode as well as the PSD control capability were installed in the HDUE early in December 1979. System shakedown flights in a UH-1H aircraft began on 13 December 1979 and, following year-end closedown at YPG, continued until 8 January 1980. These missions permitted the checkout of the added functions in the HDUE and the modifications in the IRCC real-time processing, including the updated IRCC-to-GT-to-HDUE links. In addition, substantial emphasis was placed on "tuning" the unique PSD drive algorithm in the HDUE based on the experienced degree of accommodation of the aircraft flight characteristics and pilot reaction.

The differential GPS data collection missions were concluded on 16 January 1980. The data obtained during these last flights are addressed in this paper.

Description of Tests

All flight tests were conducted in the UH-1H at the YPG range. Typical missions consisted of repeated flights over a prescribed closed course followed by simulated landing approaches.

The closed course was a box pattern established by selecting in the HDUE four three-dimensional waypoints representing the box corners. The relative navigation feature of the HDUE was used to drive the PSD horizontal and vertical needles so that the pilot could steer to, and then follow, selected flight vectors to each successive waypoint, all with the precision afforded by the differential technique. Figure 5 is a plot of one circuit around the pattern counterclockwise, illustrating the overshoot at each corner waypoint crossing and the return to the selected vector directed to the next waypoint. Landing approaches were handled in the same manner, but in this case the selected waypoint was the desired touchdown point in three

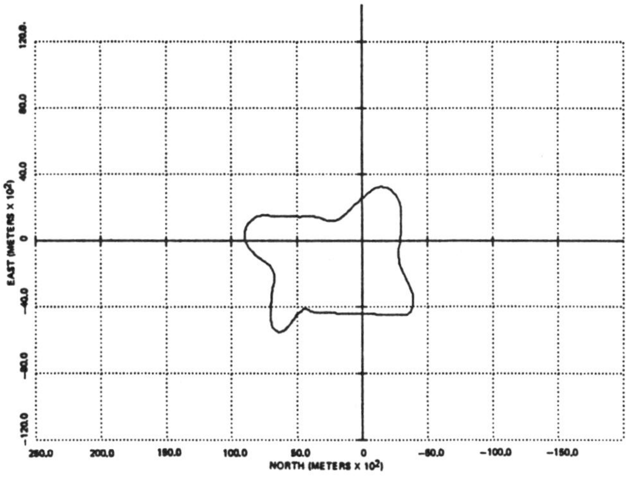

Figure 5. Position Plot

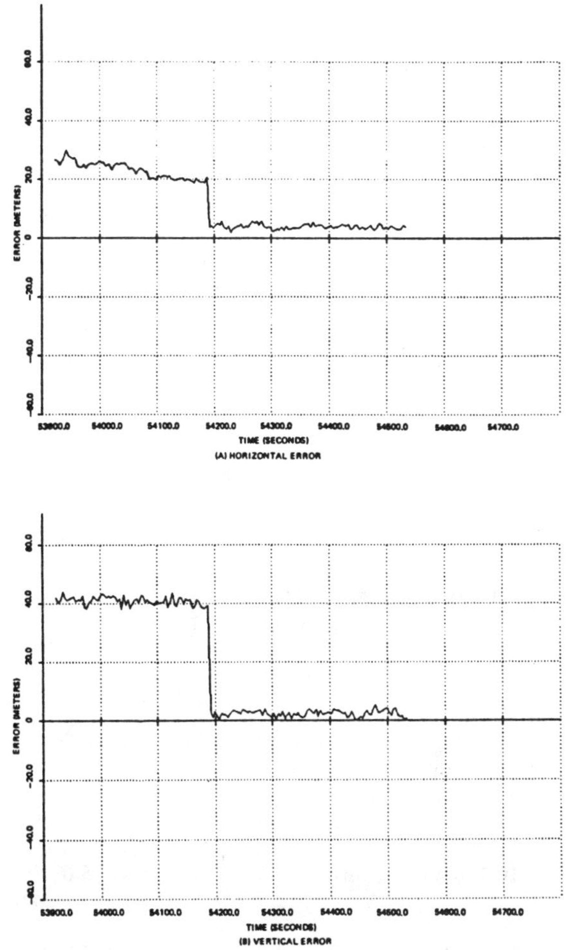

Figure 6. Horizontal and Vertical Errors

dimensions, and the selected flight vector was the desired approach heading and slope.

Exceptionally good results were obtained with this mechanization and the pilots adapted to it well. Results of the differential GPS aspects of the test, however, concern only the difference between the differential HDUE-derived position of the aircraft and its range instrumentation-derived position. The PSD is mentioned as the means of guiding the data collecting flights, and its observed performance precision with the pilot interface is not a part of this paper.

Representative Test Results

A clear illustration of the effect produced by the differential technique is shown in Figure 6, which is a segment of the 11 January 1980 mission data. When operating in the normal nondifferential mode, the HDUE provided position estimates that had horizontal errors greater than 20 meters and vertical errors near 40 meters. When the HDUE was placed in the differential mode, these errors immediately reduced to the neighborhood of 5 meters in this example.

The large nondifferential errors were a result of satellite errors, and as shown in Figures 7 (vertical error) and 8 (east component of horizontal error), a major feature of the technique is the compensation for large satellite errors when the user equipment is operating in the differential mode. The immediate inference is that precision navigation is possible even with constellations containing well-aged ephemeris data.

A tabulation of representative differential performance during four different missions is shown in Table 2. Values shown in the table were obtained from the analysis of a 10-minute segment from each mission. This time span was a limitation imposed by the capacity of the data reduction system used, and represents approximately two-thirds of the circuit around the box course. GDOP during the times the data were taken ranged between approximately 3.0 and 3.6.

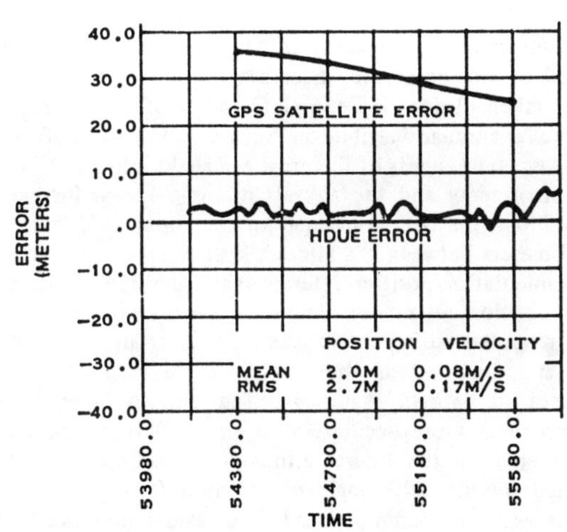

Figure 7. Vertical Error Versus Time

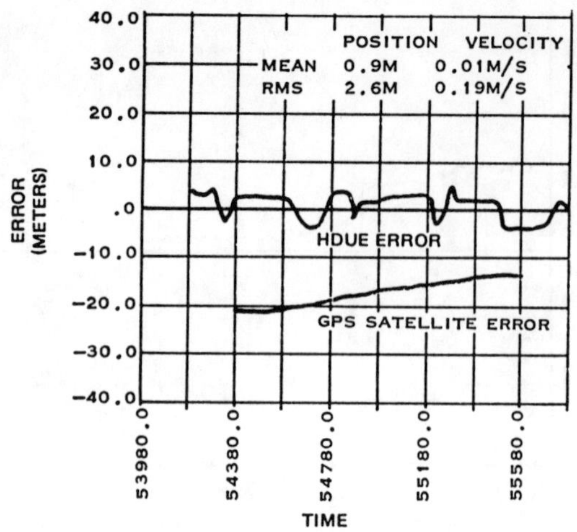

Figure 8. East Error Versus Time

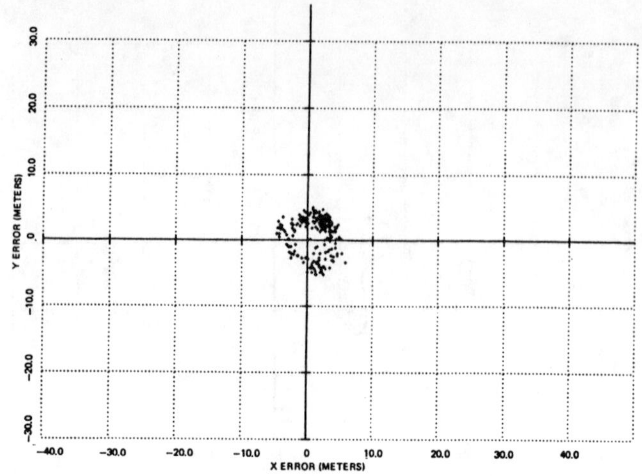

Figure 9. X Error Versus Y Error

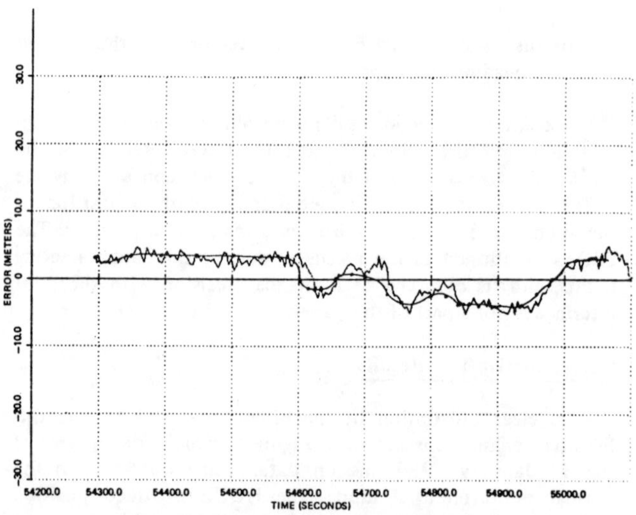

Figure 10. North Error Versus Time

TABLE 2. REPRESENTATIVE DIFFERENTIAL PERFORMANCE

		Error (meters)				
Date		Mean	RMS	Std Dev	CEP	PE
9 Jan 1980	Horiz	4.2	4.7	2.1	3.8	
	Vert	−5.2	5.5	1.9		5.2
10 Jan 1980	Horiz	5.0	5.3	1.8	5.0	
	Vert	3.9	4.5	2.2		4.0
11 Jan 1980	Horiz	3.8	3.8	0.8	3.7	
	Vert	1.4	2.0	1.5		1.5
16 Jan 1980	Horiz	3.6	3.8	1.5	3.4	
	Vert	2.1	2.6	1.5		2.1

An anomaly appearing in the analysis of all data is illustrated by the horizontal error scattergram of Figure 9. These data, taken during one complete circuit of the course, exhibit a peculiar annular distribution. Subsequent analysis of the north and east components of the error revealed a correlation between the error sense and the aircraft heading. It was further determined that the data contained an uncompensated lever arm of 3.29 meters between the aircraft laser retroreflector (the range instrumentation position reference) and the HDUE antenna (the user position reference). Based on direction-of-flight information recorded during the mission, this lever arm was calculated and superimposed as the smooth curve in the north error component data in Figure 10. The smooth curve, therefore, represents a more accurate zero error reference for the measured error in the figure, although an additional component created by the drift angle of the aircraft to compensate for winds was not accommodated. The results are indeed remarkable, and indicate a similar treatment of the data contained in Table 2 would produce a more accurate (and improved) performance measurement.

Additional Flights

Continued operation at YPG using the HDUE in the differential mode–in both the UH-1H and C-141 aircraft–have produced essentially the same results. Its demonstrated precision and consistency have allowed the technique to now be used as a tool in developing other GPS applications that, to be feasible, require this precision. Nevertheless, much more is to be learned in further characterization of the technique.

FUTURE APPLICATIONS

Area Navigation and Precision Approach

A global positioning system using the GPS C/A code for enroute navigation and the differential GPS C/A code for area navigation and precision approach has many advantages for the commercial and civil aviation community. An overview of this type of system is shown in Figure 11. The ground equipment will consist of a GPS receiver to calculate range corrections and a transmitter for transferring the range corrections to the aircraft. The position of the ground station is determined precisely during installation using the GPS set in a surveying mode with long-term averaging. During normal operation, the ground-based GPS set tracks all visible satellites and solves for position and time bias. These data then provide combined ionospheric, satellite delay, clock, and range corrections for each visible satellite. In addition, airport-related data, such as approach heading and touchdown coordinates, could be transmitted for use by the airborne equipment. The airborne equipment includes a C/A GPS receiver and an antenna, as well as the data-link receiver and antenna. The data-link receiver provides the GPS range corrections to the GPS receiver where they are applied to the pseudoranges before the navigation computations are made. To reduce the cost to the user, a Texas Instruments multiplexed GPS receiver could be used with a GPS pseudolite transmitter on the ground to replace the data-link equipment. The multiplexed receiver samples satellite signals at a 5- or 10-millisecond rate with one hardware channel but maintains four tracking loops in software continuously. One of these 5- or 10-millisecond time slots every 20 milliseconds could be used for the data-link function. Obviously, the military services can use the same technique but with even more accuracy since they will have access to the P-code. P-code accuracies from the data gathered at YPG are close to those required for category I landings.

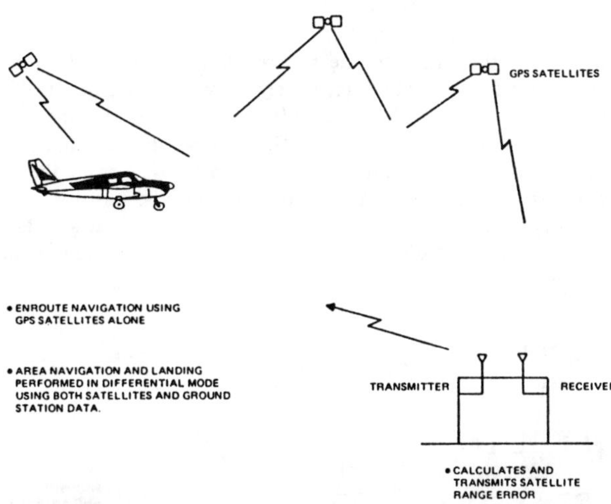

Figure 11. Low-Cost GPS–Commercial and Civil Air

Close Air Support

The close air support application of differential GPS is similar to the landing system just discussed except the object is to drop weapons on the target that threatens ground troops. Figure 12 illustrates the close air support differential system. The troops on the ground use their laser ranger and attitude sensor to determine range, bearing, and declination of the target relative to themselves. These data are relayed to the attack aircraft along with the GPS-derived position of the troops. The attack aircraft uses this information to determine the

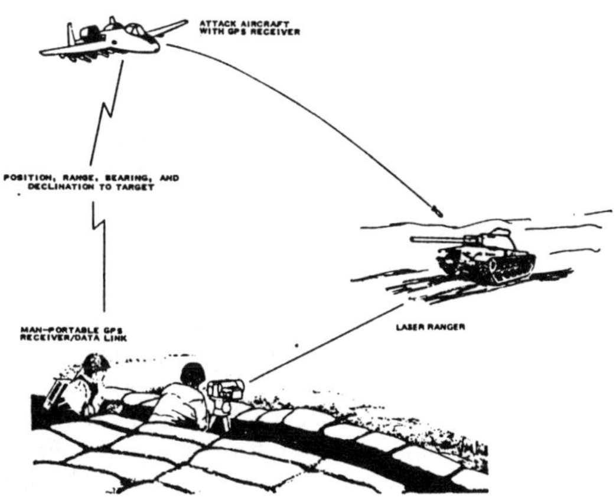

Figure 12. Close Air Support Application of Differential GPS

location of the target and to destroy it using differential GPS (using the same satellites as the ground troops).

Ship Navigation in Narrow Waterways

Shipping in close waterways, such as the St. Lawrence Seaway, is brought to a halt by bad weather at great cost to the world economy. The English Channel is traversed at great peril. Dangers of environmental damage caused by oil spills from collisions in narrow crowded waterways increase daily. Figure 13 illustrates the use of differential GPS for navigation in narrow waterways. The ground antenna must be elevated to provide the coverage required. The GPS data allow each ship to remain in an assigned lane and thereby avoid traffic going the other way. Other navigation data, such as location of anchored ships, can be reported on the data link as well. A more elaborate, but perhaps necessary, system would be for each ship to transmit its differential GPS position to the ground

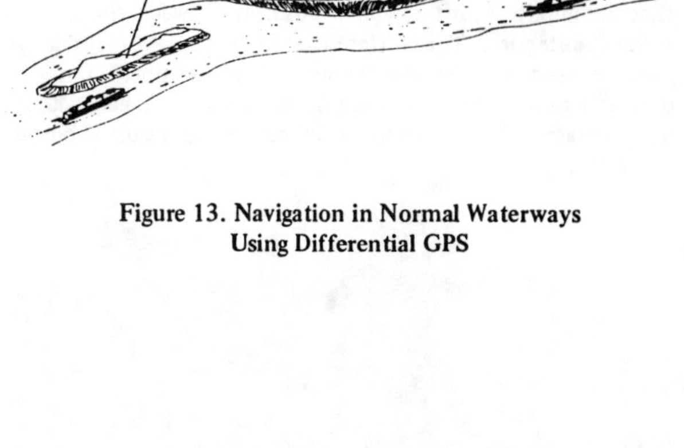

Figure 13. Navigation in Normal Waterways Using Differential GPS

station, which would then determine a map of all traffic and retransmit it to all ships for their use.

Station Keeping

Because the extremely high GPS real-time accuracy for a wide range of dynamics, it is suggested that a differential system would make an excellent station-keeping system, as illustrated in Figure 14. Note that all users transmit present position, time, and velocity to all other users. Each user in the network then makes station-keeping calculations using these data. All users navigate using the same satellite constellation; therefore, their relative station-keeping calculations result in differential navigation with the attendant ultra-high accuracies needed for the operation. GPS time could be used effectively in this application to control a time-division-multiplexed data link in which each vehicle is assigned a separate time slot. Examples of station-keeping applications include cargo aircraft, hydrofoils, tanks, and refueling aircraft.

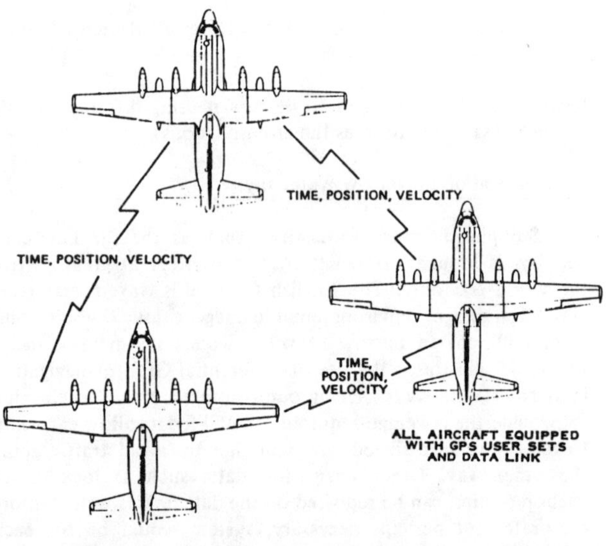

Figure 14. Station Keeping Using Differential GPS

Artillery Surveying

Not only do artillery positions have to be surveyed, but their pointing angles have to be determined also. It is suggested that a system of two antennas separated by more than 100 meters with a multiplexed receiver can be used to obtain extremely high angular data relative to north, as shown in Figure 15. The multiplexed receiver will multiplex rapidly between the two antennas. Two navigation solutions will be maintained from the information measured: position of the first antenna and difference in position between the first and second antennas. The differences in antenna positions will then be used to calculate the angle between north and a line drawn between the two antenna locations. It is believed that an accuracy of less than 1 milliradian can result from this technique.

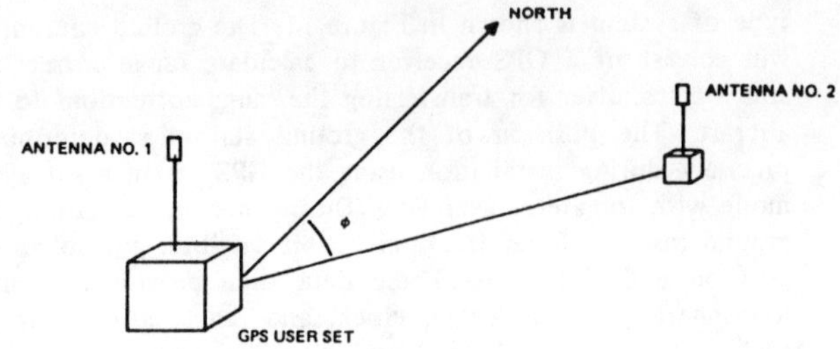

Figure 15. Artillery Survey Using Differential GPS

Seismic Geophysical Exploration

New 3D seismic data-processing algorithms require a high degree of accuracy to find deep pockets of oil. The near-shore areas have already been explored to a great extent and the geophysical industry is looking toward deeper water farther offshore. Unfortunately, current navaid coverage is restricted to the near-shore case. For these reasons, the worldwide coverage of precision differential GPS navigation is required to find new energy sources to help make the United States more energy-independent. The application of differential GPS to seismic exploration is illustrated in Figure 16. No real-time data link is required between the shore-based GPS set and the shipboard set since the seismic data are recorded and processed at a later time. In this application, the shore-based GPS set must record the GPS data that are later time correlated with the seismic data and shipboard GPS data and used in a differential way to determine the position of the ship. Another differential GPS application also illustrated in Figure 16 is that both the ship and the sources and sensors in the towed array must be located accurately. This is accomplished by placing an extra GPS antenna/preamplifier in the tail buoy at the end of the array. The GPS set would then multiplex between the two antennas in the same manner as in the previous artillery survey example to determine the position of the tail buoy relative to the ship. A model of the towed array motion is used to determine location of each of the sources and sensors in the array.

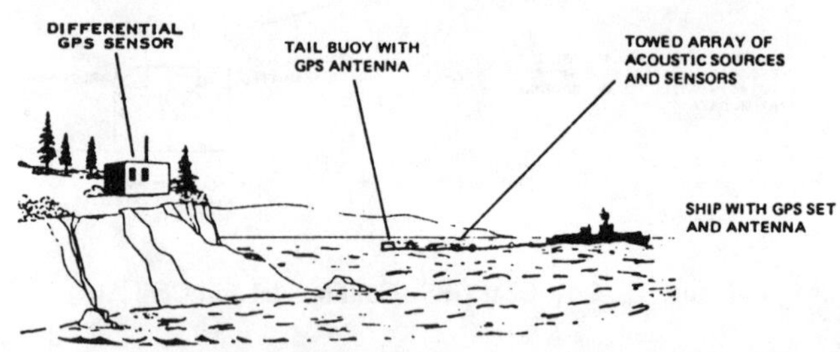

Figure 16. Seismic Exploration Using Differential GPS

ACKNOWLEDGEMENTS

The authors are particularly grateful to USAF Major James Bybee and to Mr. James Robbins of the Electronics Division of General Dynamics Corporation, whose support of the program at the Yuma Proving Ground has been invaluable.

POSITION-FIXING USING THE USSR'S GLONASS C/A CODE

S A DALE, I D KITCHING and P DALY

Department of Electrical and Electronic Engineering
University of Leeds, Leeds LS2 9JT, United Kingdom

ABSTRACT

Details of the USSR's global satellite navigation system GLONASS currently available allow the system to be used to carry out position-fixing and timing measurements. The results can be compared with those achievable with Navstar GPS and conclusion drawn as to the possibility of employing a common navigation system. The answer to this question will only be reached following exhaustive testing of the two systems in a variety of satellite and receiver modes and configurations. In this paper we report on first attempts to carry out basic position-fixing and timing measurements with a single channel, sequencing digital receiver in a C/A code phase-measurement mode. The number of satellites available in the present pre-operational phase heavily restricts the time and satellite configurations which can be tested. However the results encourage us to propose a range of experiments in the future at several laboratories aimed at evaluating the two systems and eventual integration.

INTRODUCTION

Global satellite navigation systems have been under development by the United States and the Soviet Union since the 1970's and initially planned to become operational towards the end of the 1980's. The US NAVSTAR GPS system saw its first launch in 1978; the USSR's GLONASS system was inaugurated 4 years later. Since then both systems have been developing towards an operational state now expected around 1991-1992 following a variety of problems mostly connected with launch failures. The payload carried by both NAVSTAR GPS and GLONASS includes a military (P-code) and a civil (C/A) component; only the latter is of concern to us here. Prior to May 1988, only NAVSTAR GPS was registered with international bodies such as ICAO and Inmarsat as a candidate for future navigation systems. In May 1988 at the ICAO meeting in Montreal, the USSR released details (1) of the GLONASS C/A system sufficient to provide the user with a position-fixing and timing capability quoted at 100m and 1 μsec. An account of the GLONASS system based on measurements and interpretation at the University of Leeds during the previous 18 months was made available (2) at the same time. The latter paper includes information not available in the Soviet release, such as an age-of-data (AOD) parameter referring to ephemeris and clock corrections and also an account of the method by which parity checks on incoming data are actually performed. The release of technical data by the USSR to international bodies such as ICAO gives greater credibility to the prospects (3) of a joint NAVSTAR GPS/ GLONASS satellite navigation system being adopted for international use.

In November 1986 details were published (4) concerning the radio-frequency structure of GLONASS signals. Later publications on the same general theme referred to aspects of GLONASS orbits relevant to navigation (5) and interpretation of almanac and ephemeris data in the transmitted message (6). Details of the transmitted space vehicle timing and frequency offsets and their relation to GLONASS system time were given later (2). An additional parameter in the message giving the difference between GLONASS system time and a second reference was also described. This second reference is quoted (1) as Moscow time and it is not yet clear how this latter time is related to USSR coordinated time UTC(SU). For the purposes of global time comparison, the GLONASS system is certain to be linked to UTC(SU) and thence to international coordinated time (UTC).

In this paper preliminary results are reported of the use of four GLONASS satellites for the purposes of extracting position and time references at our location. The receiver used to produce the fixes was a digital, sequencing unit with a single stage of down-conversion and followed by 1-bit quantisation. Code acquisition was accomplished using I-Q channels measuring code phase.

GLONASS SYSTEM

Current status

TABLE 1 presents the international identifiers, Cosmos numbers and GLONASS numbers of all known launches since the first one in 1982; all triple launches have taken place into one of two orbital planes (referred to in the TABLE as planes A and B) separated by 120 degrees.
It should be noted that, as the time approaches when GLONASS is anticipated to become operational (1991-92), the interval between successive launches has been reduced to 94 days (interval between launch 11 and launch 12). The USSR is clearly now capable of four launches per year carrying a total of 12 GLONASS satellites. At this rate (which could almost certainly be accelerated), the entire GLONASS operational system could be installed "ab initio" in 2 years.

Reprinted from *Rec. IEEE Position Location and Navigation Symp.*, pp. 13–20, 1988.

TABLE 1

TRIPLE GLONASS LAUNCHES

STATUS - end July 1988

LAUNCH	INTERNATIONAL DESIGNATOR	COSMOS	GLONASS	PLANE
1	1982-100A,D,E	1413 1414 1415	1 2 3	B
2	1983- 84A,B,C	1490 1491 1492	4 5 6	B
3	1983-127A,B,C	1519 1520 1521	7 8 9	A
4	1984- 47A,B,C	1554 1555 1556	10 11 12	A
5	1984- 95A,B,C	1593 1594 1595	13 14 15	B
6	1985- 37A,B,C	1650 1651 1652	16 17 18	B
7	1985-118A,B,C	1710 1711 1712	19 20 21	A
8	1986- 71A,B,C	1778 1779 1780	22 23 24	B
9	1987- 36A,B,C	1838 1839 1840	25 26 27	A *
10	1987- 79A,B,C	1883 1884 1885	28 29 30	A
11	1988- 9A,B,C	1917 1918 1919	31 32 33	B *
12	1988- 43A,B,C	1946 1947 1948	34 35 36	B

* failed to achieve final orbit

GLONASS orbital plan

According to the available evidence, it is clear that the GLONASS satellite navigation system foresees an operational configuration of 24 satellites with 8 satellites in each of 3 orbital planes separated in Right Ascension of the ascending node (RAAN) of 120 degrees. Taking plane A with RAAN around 73 degrees on day 250, 1987 as reference, planes B and C have RAAN's of +120 and +240 degrees respectively. The situation is shown in FIGURE 1 where the separation in argument of latitude or orbital phase in the plane, is a multiple of 45 degrees. However, as the FIGURE shows, there is a displacement of +30 degrees for satellites in planes B with reference to plane A. The convention used assumes satellite direction of travel is such as to cause a decrease in phase with motion.

Relative positions of satellites remain very stable over long periods because they have very much the same, small rates of change of RAAN amounting to about -0.03 degrees/day for near-circular GLONASS orbits. All satellites have the same nominal orbital period of 675.73 minutes with longitude change of 169.41 degrees W each orbit. This orbit produces a ground-track repeat every 17 orbits lasting 8 whole days less 32.56 minutes. This diurnal offset of ΔT = 4.07 minutes from a full 24-hour day coincides with that of NAVSTAR GPS and is very nearly the difference between a solar and sidereal day (3.93 minutes). This implies that each complete day less ΔT minutes a satellite performs 17/8 orbits or 2 whole revolutions plus an additional 1/8 revolution, equivalent to 45 degrees. It

GLONASS ORBITAL PLAN

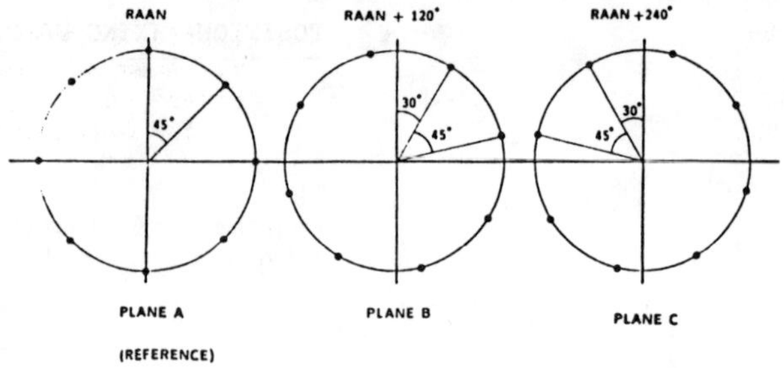

FIGURE 1

follows that two satellites in the same plane but separated by 45 degrees in orbital phase, appear at precisely the same position on successive days less ΔT minutes. During that interval, the earth has rotated very nearly 360 degrees with the result that the ground-based observer sees both satellites at the same pointing azimuth and elevation. Over a ground track repeat interval of 8 days then, all satellites in the same plane with separation of 45 degrees appear in turn at the same position at intervals of 1 day less ΔT minutes. When discussing the commonality of NAVSTAR GPS and GLONASS orbits from the point of view of combined operation, geometrical factors which determine the effective PDOP need only be investigated over a single day. This arises because, in the operational stage, orbital positions occupied by

satellites on any one day are again occupied on successive days, not (in the case of GLONASS) by the same satellites but by others a multiple of 45 degrees away.

Current occupation of orbital slots

As of the end of July 1988, no GLONASS satellite had yet been launched into plane C. Using the satellite Cosmos 1884 (GLONASS 29) as a convenient reference at zero phase in Plane A, the current situation is shown in FIGURE 2.
Only six positions are occupied at present as shown - in slots at -45 degrees, 0 degrees, and +45 degrees in plane A (GLONASS 30, 29 and 28) and at slots -60 degrees, -15 degrees and +30 degrees in plane B (GLONASS 35, 34 and 36). The three older satellites in plane B (GLONASS 22, 23 and 24) were replaced following the launch of May 21, 1988 as will be described in the next section.

The current constellation provides a full 4-satellite navigation facility for several hours per day on average. It should be stressed that only those satellites which currently transmit a full navigation message are included in this constellation. Occasionally older satellites transmit either plain carrier, or carrier plus code or carrier plus code plus timing information but not on a regular, reliable basis.

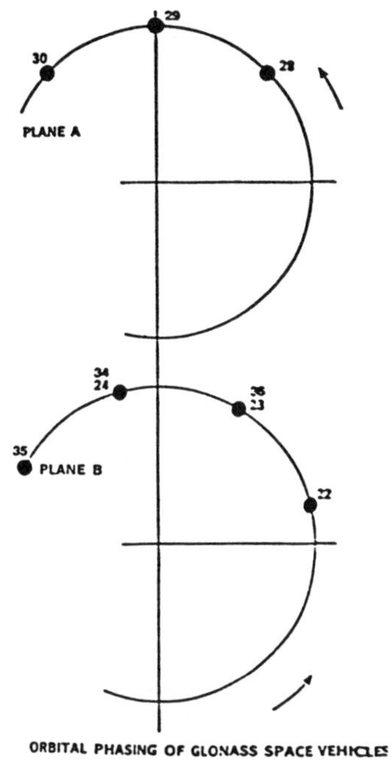

FIGURE 2

Procedure following GLONASS launch

We have already discussed the timing of GLONASS launches and the rate at which satellites can be replaced, now reduced in the most recent launch to 94 days. It is important now to discuss how long it takes following launch for satellites to enter into active service. Apart from the two launch failures, the procedure adopted after launch is standard and illustrated by the events which took place after the 21st May 1988.

On Saturday, 21 May 1988 at 1759 GMT, three GLONASS satellites (GLONASS 34,35 and 36) were launched from Tyuratam. The NASA orbital bulletin for 1988-43A on day 142 (21 May) gives rev 0 at epoch 142.68 with RA 183 degrees, eccentricity 0.002 and period 676.41 mins. The satellites were placed in plane B and clearly intended to replace one or more of the ailing satellites in the same plane (GLONASS 22, 23 and 24). All three satellites were very close together in position drifting slowly past the position of GLONASS 24.

The next manoeuvre began around 25 May and sent satellites 1988- 43B and 43C in opposite directions. GLONASS 35 was moved ahead of GLONASS 24 at the rate of -6.72 minutes per orbit and GLONASS fell behind by +4.80 minutes per orbit. Given the 45 degree spacing in orbital plane now standard in the system, GLONASS 35 was expected to move to a position 45 degrees ahead of GLONASS 24 so that the manoeuvre would be complete by 31 May. Similarly GLONASS 36 was expected to be in position 45 degrees behind GLONASS 24 by 3 June. Later data showed GLONASS 34 moving in the same direction as GLONASS 36 at a much slower rate of +1.13 minutes/orbit probably returning to its initial designated position coincident with GLONASS 24.

As of 1 June no transmissions from any of the new satellites had been observed in Leeds although monitoring was confined to examining spectrum as one of the two satellites with known orbits (GLONASS 35 and 36) passed through the main beam of our 3-m antenna. On Saturday, 4 June at 1110, GLONASS 34 was received with GLONASS 24 in the main beam of the 3-m dish transmitting on channel 12 (1608.75 MHz). On the same evening at 2000 both GLONASS 36 and 23 were received (again in the main dish), the new satellite using channel 24 (1615.5 MHz). This is the first case of channel re-use ever observed with GLONASS - see previous use of channel 24 by GLONASS 8, Cosmos 1520, 1983-127B. By searching the other as yet unused channels, GLONASS 35 was discovered using channel 23 (1614.9375 MHz) on the same day. All three new satellites were transmitting time only on the first line of the data message. The interval between launch and full payload switch-on was in this case 2 weeks, the normal period.

On the 9th June, GLONASS 36 was observed transmitting ephemeris briefly. Orbits were finalised by day 157 (5 June) with the three new satellites separated by 45 degrees in plane B as anticipated; GLONASS 36 occupies a position coincident with GLONASS 23, GLONASS 34 is coincident with GLONASS 24 and GLONASS 35 is in an as-yet unoccupied position 45 degrees in advance of GLONASS 34. It seemed more than likely that GLONASS 22 and 23 would be retired very shortly. In fact all three of the satellites in this grouping were removed from active service shortly afterwards. All three new satellites were transmitting full ephemeris and entered into almanacs by 15 June. We can conclude that following launch, it takes around 3 weeks to bring newly-launched satellites into service.

SIGNAL TRANSMISSION

Communications from GLONASS

For the purposes of allowing the user to compute his own position, navigation satellites transmit details of their own positions and a time reference. The transmission carrier frequencies chosen (3) for the GLONASS satellite navigation system lies in L-band, around 1250 MHz (L_2) and 1600 MHz (L_1). Only the L_1 frequency carries the civil C/A code. Radio-frequency carriers used by GLONASS are channelised within the bands 1240-1260 MHz and 1597-1617 MHz, the channel spacing being 7/16 or 0.4375 MHz at the lower frequencies and 9/16 or 0.5625 MHz at the higher frequencies. The carrier frequencies themselves are also multiples of channel spacing and the number of planned channels is 24. Recently launched GLONASS L_1 transmit carrier frequencies are presented in TABLE 2 (corresponding frequencies at L_2 are in the ratio 7/9).

TABLE 2

TRANSMIT FREQUENCIES AND CHANNEL NUMBER (CHN)

SAT_ID	COSMOS	GLONASS	CHN	L1 (MHz)
1987- 79A	1883	28	14	1609.8750
1987- 79B	1884	29	21	1613.8125
1987- 79C	1885	30	5	1604.8125
1988- 43A	1946	34	12	1608.7500
1988- 43B	1947	35	23	1614.9375
1988- 43C	1948	36	24	1615.5000

In all, only 5 channels of the 24 possible have not been observed during the time scale of measurements (roughly 1986-88) and one might reasonably assume that 3 of these channels were allocated to satellites in one of the failed launches and the remaining 2 channels to satellites in stable orbit which had already ceased transmitting before our observations started at the beginning of 1986.

It is also significant that the previous practice of launching three satellites with only two reaching stable orbit and transmitting has now been terminated. In all three of the last successful launches, the three satellites have been placed in stable orbit and have transmitted a full navigation message.

It was reported earlier (4) that certain of the GLONASS satellites (in particular Cosmos 1710 and 1711) were transmitting sub-carriers at the P-code nulls. This practice, deliberate or accidental, has also been discontinued in all of the last three successful launches.

Information transmission

For the purposes of allowing the user to compute his own position, navigation satellites transmit details of their own positions and a time reference. Each satellite sends data at low-speed from which its own position at any reference time may be calculated. This data is sent at a 50-baud rate and superimposed on a pseudo-random noise (PRN) code which is, in fact, periodic and very much longer than a single data bit. The GLONASS low-precision code has length 511 bits as compared to NAVSTAR's 1023 bits for its equivalent code. A code sequence lasts only 1 msec so that each data bit occupies 20 entire code sequences, the code itself or its inverse being sent depending on whether the data bit is a '0' or a '1'. In this manner, the information spectrum is spread across a wide bandwidth.

The code rate can be seen from the figures already given to be 511 kbits/s and 1023 kbits/s for the civil GLONASS and NAVSTAR codes respectively. In order to transmit the encoded data, a binary phase-shift keyed (BPSK) modulation technique is employed, the first nulls in the transmitted spectrum at +/- the bit rate. Hence, bandwidths for the GLONASS transmission can be taken at 1 MHz and 10 MHz for the civil and military codes respectively. These figures compare with 2 MHz and 20 MHz for NAVSTAR's equivalent bandwidths.

Ephemeris in the data message

The GLONASS data message is formatted in frames of 3000 bits whose duration is 30 seconds. Details of the transmitted message are available in the literature (1), (2). The ephemeris is transmitted as a set of position, velocity and acceleration coordinates in a Cartesian earth-centred, earth-fixed (ECEF) coordinate system. A new ephemeris is available every half-hour, the ephemeris being valid at the following quarter-hour. It is intended that the ephemeris should provide a satellite's coordinates at any time in the interval +/- 15 minutes around the time of validity. In this manner, the user should have a new ephemeris every half-hour with enough data to compute the satellite's coordinates until the next half-hour is reached. The process is then repeated. Occasionally, in practice, ephemerides are only provided every hour and valid on the half-hour. In this case, of course, the accuracy of the computed position and velocity at the ends of the time interval of integration (+/- 30 minutes) will have degraded.

Accuracy of satellite position and velocity

In TABLE 3, the ephemeris coordinates transmitted are shown together with their units, ranges, resolution and number of bits.

The range of these parameters is naturally such as

TABLE 3

SATELLITE EPHEMERIS

	Bits	Units	Range	Resolution	Comment
Time	7	1/4 hour	2^7	1	MST
Position	27	km	$\pm 2^{15}$	2^{-11}	
Velocity	24	km/s	$\pm 2^3$	2^{-20}	
Acceleration	5	km/s/s	$\pm 2^{-26}$	2^{-30}	correction

to accommodate a typical GLONASS orbit with maximum position coordinate of the order of 25510 km and maximum velocity of the order of 4 km/s. The time reference MST stands for Moscow Standard Time which is equal to GMT + 3 hours. The acceleration coordinate must be a correction term as the actual acceleration of the space vehicle is several orders of magnitude larger than the range of the parameter given. Examining the orders of magnitude of the various gravitational terms acting on an earth-satellite (7), we observe that the potential consists of gravitational forces experienced due to a spherical centre of attraction plus higher-order tesseral harmonics of decreasing magnitude. It is easily shown that only the first of these harmonics is important for the purposes of satellite predictions over +/- 15 minutes to the resolution of the given coordinates (0.5m in position and 1 mm/s in velocity). Terms higher than the first order in the geopotential field are smaller by an order of magnitude or more than the acceleration correction term quoted in the TABLE. We reasonably conclude that this acceleration correction term arises from external forces - lunar and solar forces - and the order of magnitude of the term bears out this assumption. It should be pointed out that in reference (2), the authors were unable to decode the acceleration term and in fact attributed in error bit 25 to the velocity coordinate.

Cartesian coordinate systems

The use of ephemeris over half-hourly intervals can proceed either in the given ECEF coordinate system or in a star-fixed (inertial) frame. It is probably simpler to adopt the latter procedure since in ECEF coordinates, the gravitational terms are complicated by rotational (Coriolis) forces. Transformation to the star-fixed frame is simple given the time angle, γ, known as Greenwich Sidereal Time (GST) which is easily computed (8). Thereafter position and velocity in the new coordinate system are found as follows:-

$$X_S = X_E \cos\gamma + Y_E \sin\gamma \qquad (1)$$

$$Y_S = - X_E \sin\gamma + Y_E \cos\gamma \qquad (2)$$

$$\dot{X}_S = - \dot{X}_E \cos\gamma + \dot{Y}_E \sin\gamma + \omega_E Y_S \qquad (3)$$

$$\dot{Y}_S = - \dot{X}_E \sin\gamma + \dot{Y}_E \cos\gamma - \omega_E X_S \qquad (4)$$

where ω_E is the earth rotation rate. The Z-coordinates along the earth's axis are of course unchanged by this transformation. Following the transformation, integration of the force equation in the inertial frame given an initial position and velocity is a standard problem in computational mathematics solved by the use of a fourth-order Runge-Kutta integration. The terms neglected are of fifth order in the derivatives and interval of integration. The latter can be shown to be around 120 seconds if the aim is to provide the necessary accuracy in prediction but prudence would dictate an integration time of 60 seconds. Interpolation between the integration points is then quite simple. There is bound to be some uncertainty in the integration of the differential equations as one is unaware of the precise orbital model being used; it is also not clear what values are to be used for the earth's rotation rate and the geocentric gravitational constant GM as various values are quoted in the literature. In practice, it is found that position and velocity predicts with accuracy of 1m in position and 2mm/s are feasible, perfectly adequate for position-fixing purposes with expected accuracy of 20m or so.

Pseudo-ranging and time offsets

In principle, once satellites' positions and velocities are known, it is possible to measure pseudo-range to four satellites and then to compute one's unknown position (X_o, Y_o, Z_o) and time offset (t_o) from the equations:-

$$(X_o - X_{si})^2 + (Y_o - Y_{si})^2 + (Z_o - Z_{si})^2 = (\rho_{si} - ct_o)^2$$

where the subscripts s and i (1,2,3 and 4) refer to the four satellites and ρ_{si} is the pseudo-range from the observer to each satellite. There are two major unknowns however which affect the above picture. The first question refers to the coordinate reference system used in GLONASS. If the reference system is either the WGS-72 or WGS-84 ellipsoid employed by NAVSTAR GPS, then one would expect the same position-fix as with NAVSTAR GPS. Otherwise, on the assumption that the reference system used is merely a translation of WGS-72 or WGS-84, then we naturally expect a corresponding translated position. However, the situation could easily be even more complicated by involving not only a translation of coordinates but also rotations in all three axes. We make the assumption, at this stage, that only translation is involved and leave the question of rotations to be resolved by simultaneous measurements of several well-space observers at a later stage.

In addition to the problem of reference coordinates, there is also the question of GLONASS time reference. An offset in time between GLONASS time and the user's time appears in the solution as t_o and not in the computed position. In many applications however it is important to produce a time strobe referred to coordinated time. It is also obvious that in any combined application of global navigation systems, reference time must be coordinated.

Time coordinate reference systems

For many years the measurement of time has been coordinated at the Bureau de l'Heure in Paris. Inputs from various national laboratories running atomic standards are used to compute Universal Coordinated Time (UTC) and the differences between UTC the various national standards are published regularly. For example, UTC(USNO) refers to the atomic master clock at the US Naval Observatory and UTC(SU) refers to the hydrogen atomic clock in the USSR (Moscow). It is well known that, in satellite navigation systems such as NAVSTAR GPS and GLONASS, that the master control stations use an intermediate time reference for individual space vehicles. This time is known as NAVSTAR GPS (or GLONASS) system time and is transmitted as part of the normal data message. Individual space vehicle time offsets are expressed in terms of system time. From the above, it is clear that a common reference time is UTC itself. However the situation is clouded by the fact that, although UTC(USNO) is referred to UTC directly by satellite link, UTC(SU) is at present referred by a Loran C link, resulting in an uncertainty of UTC(SU) of the order of 1μs. Naturally this does not imply an uncertainty in the USSR atomic standards, only an uncertainty in the method of reference itself. The relationship therefore between UTC(USNO) and UTC(SU) remains unresolved at an accuracy better than 1μs, which corresponds to a range error of 300m.

In what follows, it therefore remains a major preoccupation not only to provide accurate position-fixes using the GLONASS C/A code but also to refer GLONASS system time and hence UTC(SU) to a common reference. The transmitted satellite clocks offsets (1) refer to space vehicle time offset with respect to system time (a0) and system time offset with respect to Moscow time (A0). It is not clear at this stage whether Moscow time is the same as

UTC(SU) or related to it in a known way. It would seem likely that the ultimate time standard to which the operational system would be referred would be UTC(SU) itself.

RECEPTION OF GLONASS SIGNALS

Receiver design

The receive chain is designed to allow operation with either GLONASS or Navstar GPS for the purposes of (i) locking to individual satellites (ii) monitoring of incoming data (iii) making epoch timing measurements with respect to a common reference (either a Rubidium clock or a Navstar GPS receiver time strobe output) and (iv) to produce position fixes using the GLONASS code phase only.

A block diagram of the receiver is reproduced in FIGURE 3 showing all the major features of operation. The digital receiver begins as usual with an antenna with hemispherical coverage followed by low-noise preamplifier and a coaxial cable connection to the laboratory. One-stage down-conversion is employed to an i.f frequency of 21.25 MHz, so that image frequencies can easily be removed. The local oscillator is locked (as are all other sources) to the same Rubidium atomic frequency standard. Following amplification, the signal is 1-bit quantised at a sampling rate of 5 MHz. The signal is then transferred to baseband producing I- and Q-channels which are both correlated using early, late and punctual versions of the local code, separated by 1/2 of a code chip. Code generation is programmable and codes may then be down-loaded to RAM so that current and future codes can be accommodated. The correlation outputs are summed over 1ms and dumped into software at the local epoch. The software uses comparisons of early and late channels to yield the code phase correction which drives the loop into lock. The code phase resolution is 1/16 of a chip which is low for accurate code phase measurements but resolution is increased in software by the simple expedient of signal averaging.

Control of the receiver including the local oscillators is achieved with an IBM PC/AT and the IEEE bus. The same microcomputer is used for all numerical computation and is interrupted every 1ms (at local epoch) by the hardware; software written in 80286 assembler code performs all signal processing functions. Navigational and other software is written in Microsoft 'C' and occupies around 15,000 lines. Currently the receiver operates in either a Navstar GPS or GLONASS mode. Apart from the utilities provided the main tasks performed are navigation, time measurement/time transfer and system status monitor.

Position-fixing with GLONASS C/A code

Following system checkout with GLONASS satellites over a period of months which involved making constant, epoch-timing measurements with all available GLONASS satellites, a series of position-fixing tests were carried out using 4-satellite configurations so as to produce a 3-dimensional position and a receiver timing error referenced for convenience to GLONASS system time roughly once every 6 seconds.

The results produced in FIGURE 4 and FIGURE 5 are typical when all satellites are flagged healthy and known to be functioning normally. In FIGURE 4, a scatterplot is shown of latitude and longitude over a period of roughly 1/2 hour with GLONASS 35, 28, 34 and 30 (two satellites each in orbital planes separated by 120 degrees) with a PDOP of 2.5 - 2.8 during the measurement. A circle of radius 30m is drawn with centre at the measured Navstar GPS coordinates for comparison. Also given are the position coordinates and their 1-sigma values.

For the purposes of comparison, FIGURE 5 shows the same situation with GLONASS 35 replaced by GLONASS 36, nearly 2 hours later in the same day where the PDOP was 3.8 - 8.0. An increase in the spread of the second set of results can be attributed to the higher PDOPs during the period of measurement.
These two sets of data show agreement in position well within the 100-m accuracy quoted (1) by the USSR delegates in Montreal.

Those familiar with position-fixing test with Navstar GPS code phase during the current pre-operational phase will not be surprised that the results presented above are not reproducible day-in, day-out on a reliable basis. It can be observed that average position may change by 20-30m depending on configuration, time of day and satellite health. This is the order of magnitude of error one would expect when ignoring propagation errors. However measurements do indicate a level of performance comparable with that of Navstar GPS when both systems operate in the same mode (C/A code phase) and with the same type of receiver. It is also well known that the C/A code on Navstar GPS allows very much higher accuracy if carrier phase tracking is employed. It is not yet clear whether GLONASS offers the same property.

Conclusions

Preliminary results have been produced showing the kind of accuracy and precision one might expect with GLONASS satellites using a simple, digital, sequencing C/A code receiver measuring code phase only. The figures presented lead one to conclude that GLONASS is capable, when operated with C/A code phase in the same manner as NAVSTAR GPS, of the same level of performance. In terms of position-fixing, we observe a position fixing capability using code phase only and signal averaging, well within the quoted accuracy of 100m, probably of the order of 10-30m. With a pre-operational status and without direct access to the control segment, one must exercise extreme caution in expressing a view of the operational capabilities of the GLONASS system.

Without the assistance of at least two additional ground stations to corroborate measurement data, it is extremely difficult for the single observer to be certain that he has eliminated all possible sources of systematic error. With this proviso in mind, it is still possible to observe anomalies in the system which give rise to concern. For example, fixes indicating a drifting position, symptomatic of a satellite clock error have been

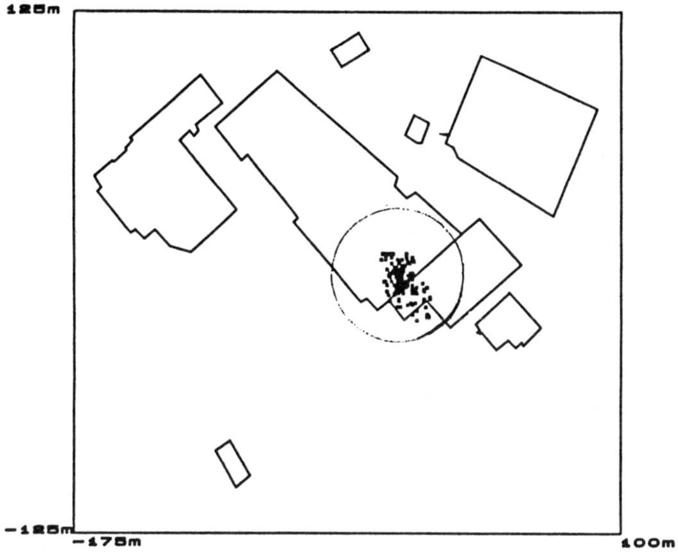

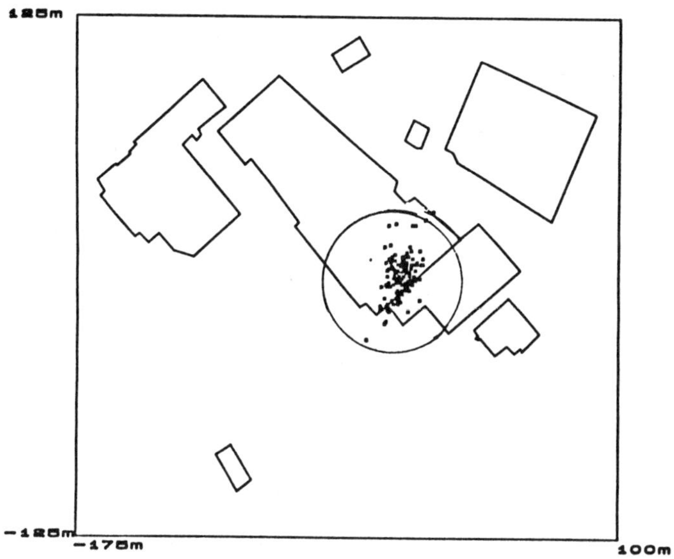

Coordinate	Average	1-sigma
X(m)	3773,064	7.4
Y(m)	-102,455	5.4
Z(m)	5124,372	8.0
Lat (°)	53.809363	0.000067
Long (°)	-1.555442	0.000084
Height(m)	161.2	7.9

Position fix with GLONASS 35, 28, 34 and 30.
Data for 10 August 1988 at 10:11 GMT.

FIGURE 4

Coordinate	Average	1-sigma
X(m)	3773,061	10.5
Y(m)	-102,455	5.3
Z(m)	5124,376	19.0
Lat (°)	53.809402	0.000087
Long (°)	-1.555453	0.000082
Height(m)	162.8	19.5

Position fix with GLONASS 36, 28, 34 and 30.
Data for 10 August 1988 at 13:30 GMT.

FIGURE 5

observed even when all satellites are flagged healthy. Clock error in one of the satellites has been confirmed by time transfer measurement on these occasions.

In the future, it is hoped to compare results with other observers and to pursue measurement accuracy by refined techniques such as integrated Doppler and carrier phase measurements. In parallel a programme of work is underway to integrate the user segments of GLONASS and Navstar GPS giving position-fixing and time transfer using satellites of both systems.

References

1 T G Anodina
Working paper
"Global positioning system GLONASS"
Special committee on

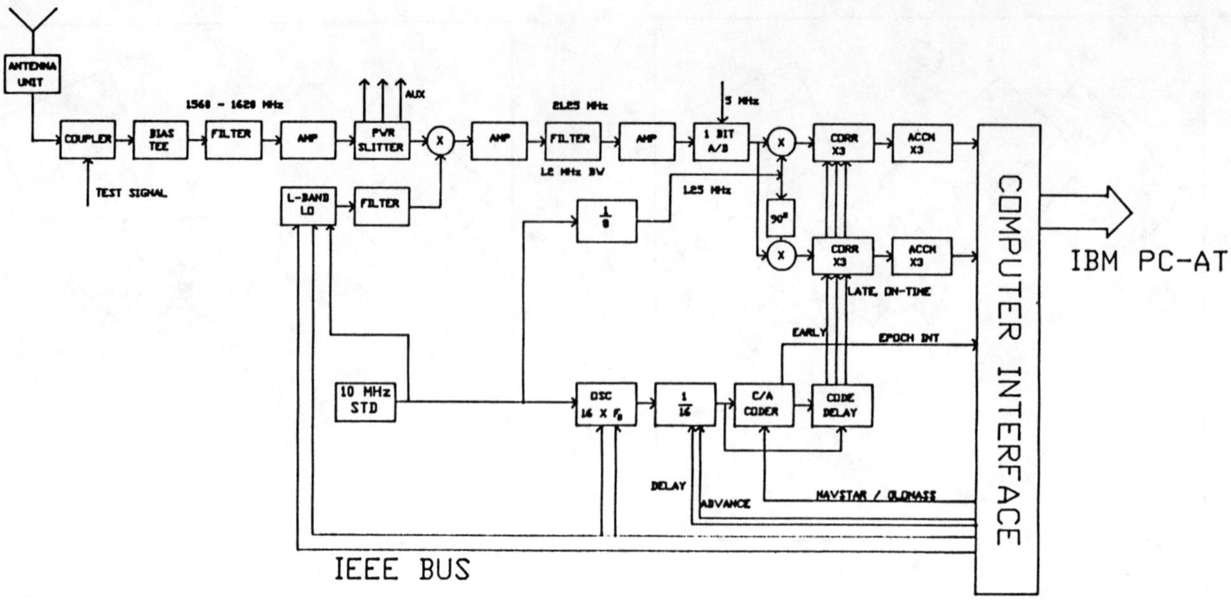

RECEIVER BLOCK DIAGRAM

FIG 3

future air navigation systems (FANS/4)
International Civil Aviation Organisation
(ICAO) Montreal, 2-20 May, 1988.

2 S A Dale, I D Kitching and P Daly
"Understanding signals from
GLONASS navigation satellites"
Int. J.Sat. Comms. (Wiley)
Vol.6, 1988 (to appear)

3 I D Kitching, S A Dale and P Daly
"Potential for a unified GLONASS/NAVSTAR
civil navigation system"
Fourth IEE International Conference
on Satellite Systems for Mobile
Communications and Navigation, London
17-19 October 1988

4 S A Dale and P Daly
"Recent observations on the Soviet Union's
GLONASS navigation satellites"
IEEE Plans'86, Las Vegas,
4-8 November, 1986, pp 20-25.

5 P Daly
"Aspects of the Soviet Union's GLONASS
satellite navigation system"
Journal of Navigation, Vol. 41 No.2
May 1988, pp 186-198

6 S A Dale and P Daly
"Developments in interpretation of the
GLONASS navigation satellite data structure"
IEEE Naecon'88 symposium, Dayton
23-27 May 1988, pp 292-297.

7 A E Roy
"Orbital motion"
Adam Hilger (2nd ed), 1982, p 169

8 E J Durwen
"Determination of sun interference periods
for geostationary satellite communication
links"
Space Communication & Broadcasting
(North-Holland)
Vol.5 No.3, 1987, pp 183-195

Part 4 Introduction to Ship Navigation

4.1 Ancient Voyages

THE earliest boats were probably used for fishing in rivers and lakes, for crossing rivers, and for travel to rich alluvial islands where food could be found and birds hunted. The boats were poled or rowed in protected waters. They were made of:

(a) rafts of small logs or bundles of reeds lashed together (in the Mediterranean, the ends were eventually tapered); or

(b) animal skins stretched over thin wood frames;

(c) a half-cylinder of tree bark, sewn at the ends and kept open by wood stretchers; or

(d) large logs, dug out by fire and primitive scrapers. In later years, planks (''wales'') were lashed to the top of the dug-out log to raise the sides; still later, the log was softened by water and spread with sticks (''thwarts'') to give a wide-beamed, flat-bottom boat.

In historical times, Peruvian Indians and Gulf Arabs made reed boats, Australian aborigines and North American Indians made bark boats, the Gaelic people made animal-hide boats (called ''coracles'' or ''curraghs'') and South American Indians made dugouts. The existence of Irish curraghs was reported in 600 B.C by Mediterranean visitors; they are sometimes made even today. Riverboats had to be light enough to be carried around rapids by one or two men.

Navigation was entirely visual by landmark. The navigator/captain sat close to the water and knew the details of the area intimately, even the color and odor of the water, the type of floating debris that could be seen (flotsam), local birdlife, and the nature of the bottom. The only instrument was a sounding rod used to prevent grounding and to judge the distance from shore.

As boat technology matured, men ventured out to sea to fish and to reach visible islands. Dugouts and reed boats were occasionally used at sea but the major seagoing civilizations were those who had large trees close to shore and who learned to build rib-and-plank boats (extensions of built-up dugouts or rafts). Ship historians say that butted planks, sewn together, dominated the Indian Ocean, while overlapping planks dominated the Mediterranean. Pre–Bronze-Age planks were lashed or doweled. After the invention of metals (3000 B.C.), planks could be nailed, but nails corroded and were, at first, more expensive than thongs or dowels. The invention of metal tools reduced the effort needed to build boats. Hence, boats became larger and more numerous. Western Hemisphere shipwrights did not have metal tools until the arrival of the Europeans. Yet, the Stone-Age Indians of the California coast reached the Channel Islands. 15 nm offshore, 10,000 years ago [see (Miller, 1988)]. The islands are clearly visible from the mainland and were probably reached by hungry hunters in reed boats. Later, Chumash Indians built dugout canoes which they used in protected waters. They apparently learned to sew planks to the dugouts, raising the sides for safer and regular ocean trips 2,000 years ago. They did not use internal stiffening frames, as did the canoe builders of Europe and Asia.

I suppose that every populated river, lake, and seacoast had some fishing and trade by boat by 3000 B.C. The first recorded rock drawings of boats are from 7000 B.C. An Egyptian painting of 3500 B.C. shows a reed sailboat. With the invention of sails, long trips were possible, following the coastline for fishing, trading, and raiding. Ship historians say that square-rigged sails (better for sailing with the wind) were invented in the Mediterranean, whereas triangular lateen sails (better for tacking upwind) were invented in the Orient. The Egyptians are sometimes credited with the invention of decking. On Egyptian boats, a long prow (jutting forward and out of the water) served as a platform for a man with a sounding pole to test the bottom ahead of the submerged part of the boat.

Many ancient tales tell of coasting ships that beached at night so the crew could sleep and gather provisions, sometimes from a hostile population. Folktales tell of villages that lit fires on offshore shoals to lure coasting ships to their doom so the villagers could plunder the wreckage. Travel was hazardous in those days; land travel was subject to bandits and rough terrain, while coasting was subject to pirates and shoals. Therefore, merchant mariners from distant lands probably sailed just over the horizon so they were not visible from shore. They required the cooperation of the local population for safe passage and nightly stops. Seagoing powers established beach stops, probably a day's sail apart, where their boats could find a friendly reception. The modern city of Caesarea, Israel, is built on a Phoenician beach stop. A trading boat or military vessel paid less tax at a friendly beach stop and was more likely to depart safely the next day. As for navigation, celestial azimuth measurements were made on clear days and nights, but ships did not deliberately venture far from shore.

Early man may have reached some islands (e.g., England and Japan) during the Riss glaciation, 150,000 years ago. Man certainly reached them during the next glaciation (Würm), 15,000 years ago. Stone-Age people filled all of the world's continents (except Antarctica), proving that they knew how to build rafts and boats that could cross wide rivers, tens of thousands of years ago. As sea levels rose again after 10,000 B.C., Japan, the British Isles, Malta, the Canary Islands, and the islands in northwestern North America continued to be populated by people on daylong or overnight voyages in boats. ''Populated'' is the term for the first to arrive; ''invaded'' is the term for the later ones. Crete, 100 mi from Rhodes and accessible from the Turkish mainland in 30-mi hops, was populated by 7000 B.C. In the Western Hemisphere, there is evidence of coastal trade between Maine and Labrador, 1000 mi apart, in 2000 B.C. Goods probably moved in short coastal segments, each tribe trading from one neighbor to another.

There are Egyptian hieroglyphs, from 2000 to 1500 B.C., of ships sailing along the coasts of the Red Sea. Babylonians, at the same time, were trading (probably by ship) for spices with the people on the Arabian peninsula. From 1500 to 500 B.C., Semitic Phoenicians from the eastern Mediterranean dominated trade westward, at least as far as the Mycenaean kingdom of Greece and southward into the Red Sea. They had regular routes along the coasts and across the Mediterranean from Tunis to Italy, among the Aegean islands where they could island hop, and probably from Crete to Egypt. The salvage, in 1986, of a 50-ft merchantman of 1400 B.C. yielded tin, copper, ebony, ivory, and weapons from many Mediterranean cultures. The Bible mentions trading fleets in 900 B.C. that connected ancient Israel with the Red Sea and East Africa. Homer (700 B.C.) told of Odysseus steering by the Great Bear (Big Dipper). He also told of a direct route south from Greece to Africa, probably via the islands. Deliberate multiday voyages out of sight of land had to await larger ships, better navigation instruments, and anti-scurvy remedies.

Coastal navigation was by dead-reckoning, sounding the bottom, observing swells and weather, and sighting landmarks. During storms, captains were forced to land (if in a friendly area) or go several miles out to sea, with the risk of never returning.

The oldest known navigation aid was the Greek lighthouse at Sigeum, described by a poet in 660 B.C. It may have been part of a temple dedicated to sea gods. Greek and Persian oared galleys of this period used no special navigational techniques; they traveled long distances under sail, having to stop often because of the small space for stores (they were probably accompanied by supply ships). In battle, they left their sails ashore and rowed with the objective of ramming the enemy ship (with a copper-sheathed prow) or of boarding it for hand-to-hand combat. Navigational books of the time did not describe instruments; they were guides to the shorelines and hospitality (or hostility) of the local inhabitants. For example, the *Periplus of Scylax* was a Greek guide to the Mediterranean, published about 500 B.C.

In 320 B.C., Greek-speaking Alexandrians established colonies in the Persian Gulf as transshipment points for war elephants (what Hourani called "the military tank of its day"). By A.D. 100 Egyptian and Greek ships were sailing the direct, safe (from pirates) route to India, across the open Indian Ocean. They probably sailed southward to the latitude of their destination, then east to India (see Part 4.3). Yet, the New Testament tells of sailors at the same time in the Mediterranean who gave up hope of survival when the Sun and stars were not visible for a few days.

Great migrations populated the Southeast Asian islands until historical times. Arabs, Chinese, Indians, Indonesians, Malays, Persians, and others traveled the Indian Ocean. Persian and Arab traders from Bahrain (then Dilmun) may have traveled to India and Africa by coasting in wooden boats as early as 2000 B.C. By A.D. 100 they too had apparently developed large enough boats and good enough navigation technology to use the safer, faster routes directly across the Indian Ocean. They understood how to measure latitude from the elevation of the polestar on clear nights, how to follow a heading that led to the destination's latitude but windward, and how to sail downwind at constant latitude. Arab trade routes extended southwest to modern Kenya. Starting in A.D. 950, Swahili traders coasted in East African waters to collect ivory, gold, rock crystal, and slaves to be exchanged with Arabs north of Zanzibar for ultimate delivery to Europe, China, Egypt, and Baghdad. Recent excavations at Shanga, on the northeast coast of Africa, found a trading post from A.D. 1000. The African boats were 30-ft long and built of lashed planks in Indonesian fashion. The east Africans traded along the coast with Arabs in the north but apparently were unable to reach Madagascar, only 120 nm of open ocean offshore via the Comoro Islands. Indonesians apparently colonized Madagascar just before A.D. 500.

I am astonished that trade existed at all. Ancient traders must have been clever negotiators to pursuade their customers that there was a long-term advantage in buying merchandise rather than stealing it. I suppose a navy backed up their trading activities with some threat of reprisals for piracy and theft. Long-term trade patterns were often negotiated as an exchange of "gifts" between rulers, the merchants being merely emissaries. Woe unto the merchant who arrived after his customers altered their foreign policy. The idea of exchanging gifts among rulers, rather then trading, was still prevalent among the Chinese in the 1600s. The existence of trade surely required a stable economic order throughout the area. Only the hungriest or most adventurous traders explored beyond the politically secure region. To underscore this risk, I report that the legal limit on interest under the Byzantine emperor, Justinian, was 8% except for loans on ships, for which the legal limit was 12%.

The Vikings and Celts populated the previously deserted Faroes and Shetland Islands, 200 mi from shore, by A.D. 700. In good weather, land is never out of sight when traveling to those islands from Scotland or Norway. From the Faroes, the Irish sailed to Iceland by A.D. 870. The first Irish explorers observed geese migrating and thus suspected the existence of offshore islands. Monks, seeking isolation, followed the birds northward in the spring and southward in the fall. One hundred years later, the Vikings reached Iceland from Norway in wooden boats built of overlapping planks on wood frames with side-steering oars. They traveled by way of the Shetlands so land was almost always in sight in good weather. The Celts and Vikings traveled without magnetic compasses, probably in the summer. Both carried birds (said to be ravens in the Norse sagas) that, when released, flew toward land. In India of 1500 B.C., trading ships carried shore-seeking birds as did Noah, in Mesopotamia, somewhat earlier. Bird cages must have been a fixture of ancient ships and of the supply stores in port.

The southwest Pacific was initially populated by Australoid people (akin to the Australian aborigines) during the last ice age, 40,000 to 10,000 years ago, when sea level was lower, when Indonesia, Borneo, and Malaysia constituted a single peninsula, when many islands around New Guinea were contiguous, and when the distances between islands in the western Pacific were much shorter than they are today.

In India, an Indo-European people called Dravidians spread southward by land after 2000 B.C., displacing the natives in the north. Neither the Australoids nor the Dravidians could reach Diego Garcia, 900 nm offshore via the Maldive Islands, though the Australoids reached the Andaman Islands (now 200 mi from the Burmese mainland). The first recorded Chinese ocean voyages occurred about A.D. 500. A Chinese expedition to the mouth of the Euphrates in A.D. 760 is recorded. Korean "ironclad" ships repelled a Japanese invasion fleet in A.D. 1592.

By A.D. 1100, Arab traders from India, West Africa, and Arabia coexisted with Chinese traders in the Indian Ocean. Chinese exploration reached its peak from 1405 to 1433, when voyages reached Borneo, the Philippines, and East Africa. After 1433, Arab competition and an insular Chinese bureaucracy ended ocean voyages by Chinese ships. I do not know of any Chinese, Indian, or Arab sea-charts that survived.

Sometime after 4000 B.C., an Asian people, called Austronesians, spread through the western Pacific in boats, killing or intermarrying with the indigenous population and becoming the modern Melanesians. On the Asian mainland, the Australoids were almost entirely exterminated. Their remnants exist in Taiwan, Laos, India, the Andaman Islands, and the Philippines. In 1642, European visitors to Tasmania found that the Australoid aborigines, having arrived during the ice age, were unable to cross to the Australian mainland, a maximum overwater leg of 50 nm via islands. However, the Austronesians crossed 100 mi of ocean outside of the line of sight, long before the Christian era. In their primitive boats, probably constructed of bark or skins, they took their chances at capsizing or being lost in fog.

Later Asian populations (of the Lapita culture) invaded the Micronesian Islands about 1500 B.C., as described in the reference by Bellwood (1980). They traveled in 35-ft canoes made of a dug-out log with sewn planks that extended the sides upward. Sometimes repeated waves of people from Southeast Asia and overpopulation of the islands caused endless migrations and wars. The most recent wave departed Micronesia about A.D. 400 and colonized the widely spaced islands of the eastern Pacific to become the modern Polynesians. Traveling in double-hulled catamarans, they discovered volcanic islands (e.g., Hawaii, New Zealand, Tahiti) that were never connected to mainlands and hence were uninhabited since their emergence from the Pacific. Their boat technology eventually allowed them to build 70-ft-long double canoes, with which they settled the remotest Pacific islands from A.D. 500 to A.D. 1000, as far as Easter Island and New Zealand. There is a large body of literature on Polynesian migration, some realistic [see (Sharp, 1963) and (Lewis, 1978), for example] and some fanciful.

The voyages of the Micronesians and Polynesians have been romanticized and great accuracy has been attributed to them. In fact, clans set sail when their food supplies ran out or when they were exiled. When Captain Cook arrived in Polynesia, the natives thought he had been exiled from a famine-stricken country and wondered why he had not brought women! It is not surprising that the Polynesians were hostile; they were accustomed to killing strangers who landed on their island. Even in historical times, clans migrating to populated islands had to fight their way ashore. The traditions of the Pacific islanders tell that after clan battles, the losers were forced into exile. The losing clan put their provisions in their catamarans (equipped with one or two steering oars, not rudders) and set sail in the hope of reaching islands known to be inhabited by friendly clans, or in the hope of finding either uninhabited new land or inhabited islands they could muscle in on. Such an invasion, about A.D. 1600, probably ended the era of statue-building on Easter Island. The crews of large canoes on long voyages were reported to have arrived "starved, thirsty, and sunburned" and to have refrained from travel for several months after their ordeal.

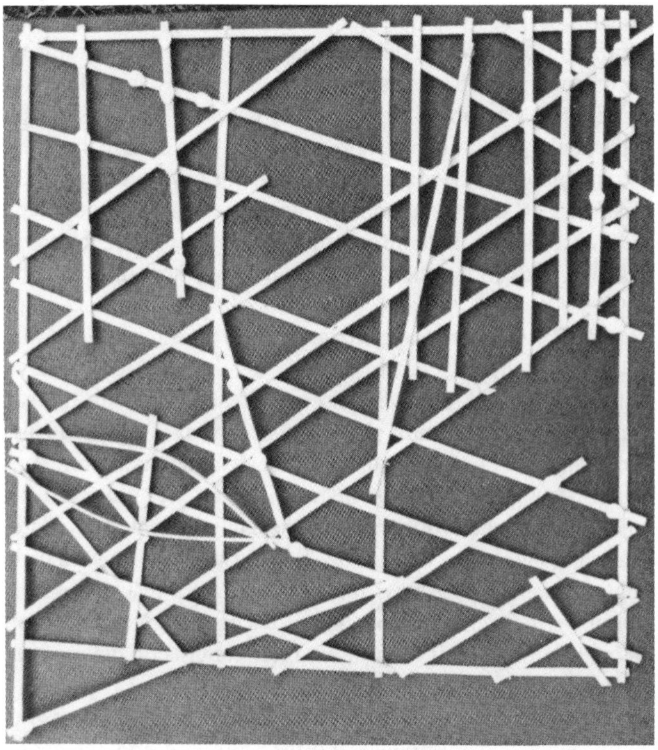

Fig. 4-1. Micronesian stick map. Photographed by M. Kayton from the collection of Robert Sharples.

These Pacific island fishermen memorized sailing directions between island pairs. They memorized guide stars, wave conditions, descriptions of intervening reefs and islands, the birds likely to be seen near each island, and magic formulas that included secret names for islands and interisland routes. They measured distance by counting verses of slow, rhythmic chants. Supposed evidence of the existence of a seaborne Polynesian water clock was disproved 60 years ago.

In the Marshall Islands, maps of wave patterns were used as a training aid to apprentices. Figure 4-1 depicts a 500-nm square of the Marshall Islands, photographed with north approximately on top. The left uppermost island is Bikini; the right uppermost island is Rongerik. The lines are diffraction patterns of long ocean waves as they pass around the islands in a particular prevailing current and wind. I am certain that these fragile maps of sticks and shells could not have been carried to sea in a wet, open boat and that, in any case, they would have been useless on a thousand-mile trip. They were certainly teaching aids in the men's lodges; they are now produced for tourists.

The Polynesians traveled far enough north–south so they understood the concept of latitude, at the level of medieval Europeans and Arabs who associated each port with the star that is overhead at midnight. The Micronesians traveled east–west and seem not to have made latitude measurements. Micronesian mens' lodges, in historic times, had observatories under the thatched roof on which the elders mounted simulated stars and taught the appearance of the sky at various times of night, at various seasons, for each voyage; a feat of memory common to preliterate people. We can imagine navigators being briefed in their lodges, with star patterns on the roof and stick maps of waves on the floor, prior to a long trip.

In the central Pacific, the polestars are near the horizon, often concealed in clouds, and the stars rise and set nearly vertically. Hence the navigators could hold a heading by memorizing the positions of a succession of stars. The Micronesians divided the horizon into 32 unequally spaced "compass" points corresponding to the rising and setting of 15 stars. Each would be used for a few hours as a heading reference, with alternates available in case of overcast. The daytime azimuth reference was the Sun, which was calibrated against the stars visible at sunrise and sunset (they had no systematic knowledge of the Sun's seasonal motion). Thus, Micronesian navigation was based on memorized azimuths between island pairs. An unpredicted lateral current or a storm would put the boat off course, where the navigator could not recalculate his position. He would then set sail at a heading he thought would lead toward a long island chain, hoping to intercept it before he and his crew died. If the lost boat reached an unfriendly shore, the crew was likely to have been enslaved or sacrificed.

East–west trips could be made more easily than north–south trips since azimuth could not be corrected on a north–south trip due to the inability to measure longitude. An error of two Western compass points will give a 300-mi lateral error on a thousand-mile north–south trip. There are records, from historic times, of appalling rates of losses of ships sailing for islands only 300 mi away. The European colonizers at first required elaborate safety equipment for these trips, then forbade long trips entirely. Early in the 20th century, the young islanders sailed anyway, even when steamer service was provided, much as American boys today joyride in automobiles.

Attributing precise long-distance navigational accuracy to such people is probably another romantic story about noble savages. I suggest that the mechanism of expansion eastward consisted of the following consecutive stages:

1. The accidental discovery of an island by a fishing boat blown off course, or by a desperate, exiled family. The most exacting accuracy of navigation cannot help the first boatloads of immigrants, until someone returns to the ancestral homeland to report the discovery and can find it again. Lots of nonsense has been written about precise navigation toward a previously unknown island.

2. Taking possession of the island as the secret property of a clan, as a source of food, and as a destination for excess population.

3. A return voyage to seek friendly settlers that will diversify the gene pool. A round-trip return voyage was probable for distances less than 200 nm.

4. Invasion by competitive clans; hence further diversification of the gene pool. The probability of invasion also diminished with distance from populated islands.

It seems to me that this is close to what happened in the peopling of offshore islands everywhere on Earth.

4.2 Azimuth

The concept of azimuth arose in parallel among early astronomers and sailors. Babylonian astronomers divided the horizon circle into 360 parts according to the motion of the Sun (Part 2.1, pg. 5). Mediterranean sailors divided the horizon according to the prevailing winds. The Greeks and Romans named four winds and subdivided the horizon circle into 8 parts ("points," halving the winds) or 12 points (associating each wind with two companions). Eight-point roses were still used in Europe in 1250 and 1295. After the invention of the magnetic compass, Italian sailors of the late Middle Ages divided the 8 Roman points again to get 16 compass points, because more accurate steering was then possible. Chaucer (1391) knew of the division of the horizon circle into 16 parts. By the 1500s, Dutch sailors from Antwerp and Bruges (which was then on the seacoast) divided the horizon circle into 32 points ("rhumbs"). The 16- and 32-point circles became the standards for sailors during the great age of European exploration.

Arabs and Pacific Islanders divided the horizon into 32 unequally spaced points, identified with the rising and setting azimuths of 15 stars, as noted in section 4.1. A star rising in the eastern sky had a corresponding star setting in the western sky, 180° away. Such a system did not permit quantitative navigation to develop. As explained on page 149, Pacific Island navigation was based on memorized headings relative to those stars. Because stars rise and set vertically near the equator, star headings were usable for hours without a time correction for the rotation of the Earth.

Pigafetta, in 1519, proposed using the 360° circle for azimuth. I had the pleasure of seeing a 360° circle surrounding a 32-point compass rose in an original 1704 manuscript. The 360° horizon, following astronomical practice, did not routinely go to sea until the 19th and 20th centuries, when mathematical techniques were adopted by navigators. Trigonometric tables were always published in astronomical degrees, not maritime compass points, but no celestial measurements of azimuth angles were made at sea (except to calibrate the magnetic compass).

The ability of the lodestone (naturally magnetized magnetite) to attract iron was known to the ancient Greeks

(600 B.C.) and the Chinese (by A.D. 100 since they entered the Iron Age later). The English term ''lode-stone'' or ''lode-star'' came from the old English ''lad,'' meaning ''to lead.'' I suspect that some Greek and Roman temples may have been aligned to magnetic north because temples constructed at successive dates near each other are sometimes closely aligned, within a few degrees of each other and of north. Perhaps someone will analyze the history of magnetic variation at those sites, during the period of construction of each temple. If my guess is correct, it would prove that Greek and Roman diviners attributed magical properties to magnetic north, as did Chinese diviners.

The Chinese used lodestones for fortune-telling in a board game played by the person whose fortune was being told. The south-finding property of the lodestone may have been used for fortune-telling in China after A.D. 400. Prof. Needham has said (1970) that in A.D. 800, the Chinese learned to magnetize a thin needle with a lodestone (the Greeks knew how in 600 B.C.) and made their fortune-telling boards more elaborate. (Many Chinese still use lodestones to locate auspicious grave sites and building sites in 1989.) Needham has reported that, by A.D. 1080, they used a floating magnetized needle on a boat, a ''magician'' being part of the crew to remagnetize it periodically from a lodestone. It thus appears that Chinese sailors ''invented'' the magnetic compass. Why did it take a thousand years for magnetic technology to diffuse from grave-diviners and temple priests to ship captains, whether in Europe or in China? Though the Chinese fortune-telling magicians worked at the royal courts where the lowly ship captains did not see their instruments, I would expect some word to have leaked out.

By the first century, some sailors in the Indian Ocean, Pacific Islands, and North Atlantic were attempting direct voyages in order to avoid pirates and tax collectors (often indistinguishable) along the coasts. Despite the crews' ability to measure latitude, direct trips were exceptionally risky since, under an overcast, only wind and waves could offer azimuth cues. A big technological breakthrough came just after A.D. 1100, when there were reports of magnetic compasses in China, Persia, and northern Europe. The magnetic compass greatly reduced the risk of direct routings. In the 1200s, handwritten Chinese and European tables listed magnetic bearings from principal ports. Nevertheless, two-dimensional position-fixing (navigation) was still impossible at sea.

The magnetic compass, mysteriously attracted by Polaris, was accorded supernatural status in several cultures. Captains carried a lodestone with which they periodically remagnetized the needle in a religious ritual. Even calibration of the compass by sighting with respect to Polaris was sometimes done ritually. Permanently magnetized compass needles were not used until the early 1700s. I read an original navigation text of 1661 in which the author, Thomas Philipot, speculates for several pages about the kind of material and magnetizing process needed to make a true-north-pointing compass. His alchemist-like attempts should caution us to understand cause and effect before inventing new systems.

With the magnetic compass floating on its cork or under its compass-card, northern-European ships grew in size to the 100-ton 12th-century ''cog'' cargo ship which was the first Western ship (perhaps the first anywhere) to have a rudder hinged to a sternpost, in place of a steering oar. By the 15th century, ships had grown to 400 t and could carry provisions for month-long trips. The galleon and caravel were rib-and-plank ships on which a European ''castle'' was mounted as living quarters. By the time of Columbus, these ships were making regular voyages to the Canary Islands, the Azores, and West Africa, crossing 800 nm of open ocean.

At the same time (1200–1450), the Chinese were building flat-bottom, compartmented rib-and-plank sampans of several hundred tons which traveled the Indian Ocean and to islands of Southeast Asia. Needham's estimate of 1500 t for Chinese ships is probably exaggerated since he also reported that the Chinese considered the Portuguese ships (400 t) to be larger than their own. The Chinese navigators of this period probably used a primitive magnetic compass to supplement celestial sightings. In the 1400s, government-subsidized voyages ended in China. At the same time, the Chinese began to change from side-steering oars to a hinged, center-mounted rudder. As these changes occurred, their navy was being dismantled for political reasons and trade was becoming localized, even discouraged. That situation remained until the forcible opening of trade by the West in the 17th century.

In 1569, Mercator published his world map in a mathematically exact projection. Many maps had a rectangular grid of parallels and meridians. His innovation was to distort the north–south scale so that it was everywhere equal to the east–west scale, which expands as the secant of latitude. The resulting log-tangent spacing of the parallels of latitude is familiar to all of us. The fact that a constant-true-heading course mapped as a straight line made his chart popular among mariners for hundreds of years. It was perfectly suited to steering by magnetic compass, especially after variation (the horizontal angle between true north and magnetic north) was identified in 1580. The Mercator map and the invention of printing rendered obsolete the hand-drawn Portolan charts (Part 2.2). Astronomers and educated mariners thereafter understood that the magnetic variation is different at different points on Earth and changes slowly with time as well. Magnetic variation began to appear on maps and tables.

After metal ships came into use in the 1870s, an elaborate procedure was gradually developed to compensate for the compass ''deviation'' produced by the ship's ''hard iron'' (permanent magnetism) and ''soft iron.'' Deviation was measured relative to the stars, with the ship's pelorus, (a telescope that sights stars, usually Polaris, relative to the compass) at many headings. On the basis of these data, as many as nine iron rods of calculated length, position, orientation, and permeability were placed in the wooden compass binnacle. By this procedure, 40° compass deviations were reduced to less than 3°. The compensation procedure had to be repeated when iron cargo was loaded or unloaded, when magnetic latitude changed appreciably, or when the ship was docked at a fixed orientation for several weeks. In modern vehicles, the deviation is calibrated by computer, fitting a power series in azimuth to the measured error (Part 3.3). Calibration is far simpler than compensation and results in less error. The gyrocompass and the advent of metal ships are described in section 4.7. The gyrocompass has been the primary heading reference on iron ships during most of the 20th century.

4.3 Latitude

The period A.D. 1350–1750 was the age of measuring latitude at sea. At the beginning, the marine astrolabe (a simple weight and sightline, calibrated in degrees) measured the Sun's peak elevation at noon or a star's peak elevation at midnight. The navigator then calculated latitude using his tables (probably kept secret by maritime guilds). The calculation required no trigonometric tables. Latitude was found by subtracting the Sun's declination (from a table) from its measured noon elevation and subtracting the difference from 90°. The elevation measurements were with respect to a plumb bob vertical that oscillated violently on the small ships of the day.

In this period, tables were published in "Arabic" numbers, Roman numerals having vanished in tables by A.D. 1600. The name, "Arabic," was taken from Central Asian nomads, writing in Arabic script, who copied the Indian practice of writing a single symbol for each number from 0 to 9 in decimal-place order. They transmitted those symbols to Europe. (Modern Arabic numerals are different from European, American, or Japanese numerals).

Prince Henry's academy at Sagres promoted the technique of "constant latitude sailing" in which a ship would sail to the desired latitude, well upwind of its destination, then sail downwind to landfall. Before Henry's time, Spanish and Portuguese sailors used Portolan charts (Part 2.3), which did not show latitude or longitude.

By 1328, the first references appear to the cross-staff, which replaced the marine astrolabe (some books credit its invention to Prince Henry's academy). The cross-staff was a stick that was held vertically and moved away from the navigator's eye until it just filled the space between the pole star and the visible horizon. A second, horizontal stick measured the distance of the vertical stick from the eye, hence latitude. Early cross-staffs were marked with the names of various ports. Numerical values were not used nor did the sailors think in terms of a spherical Earth. They thought of latitude as an attribute of each port, associated with the distance of the stick from the eye (elevation angle of the stars). They knew which way to steer (north or south) to make the elevation angle increase or decrease.

Cross-staffs were in widespread use by 1400. A back-staff version allowed the navigator to sight the sun without looking at it. With a backstaff, the navigator looked away from the Sun and observed the shadow of a stick. The cross-staff and its descendants referenced the star and Sun elevation angles to the horizon, thus eliminating those errors in the plumb bob vertical that affected the readings of the astrolabe.

Indian-Ocean sailors of the 1300s (Arabs, Indonesians, and Indians) used a cross-staff–like instrument (called the "kamal") that consisted of a stick at the end of a knotted string. The knots represented the latitudes of important ports. The stick was held in the navigator's outstretched hand, its bottom appearing to touch the horizon and its top appearing to touch the polestars. The string was held in the teeth and pulled taut, so the kamal could be operated by one hand. The knot nearest the navigator's mouth represented latitude. Another form of the kamal had a fixed-length string with a choice of sticks, one length for each port. As did the sailors who used the cross-staff, the kamal-using sailors thought of latitude as a nonnumerical attribute of each port, associated with the length of the stick or the position of the knot. These Southern Hemisphere sailors faced the added complexity of not having a bright polestar.

In 1475, the first tables of the declination of the sun were published in Spain; this leads me to believe that they already existed within maritime guilds. The invention of movable-type printing presses at the same time made it difficult to keep the tables secret. In 1508, the Spanish business school in Seville, the Casa de Contratación, was teaching navigation, a subject that was more useful than writing business plans or merging unproductive companies.

Columbus, on his first voyage to America, measured the latitude of Cuba on land with his quadrant, using the wrong pole star. He was 11° too high. He later corrected it in his logbook but still estimated 26° north latitude instead of the correct 21° N., a north–south error of 300 nm. Magellan's five ships carried seven marine astrolabes and twenty-one quadrants.

By 1650, a navigator in a calm sea without overcast could measure his latitude from the elevation angle of the Sun or stars and could calculate the approximate time of day with an astrolabe and sandglass. In overcast weather, the navigator could dead-reckon by speed and magnetic compass (see section 4.5). Even in the early 1700s, Dutch and English navigators were using old Spanish and Portuguese tables of the Sun's declination.

When the telescope appeared in Europe after 1600, European inventors combined it with the cross-staff to form a quadrant. The quadrant evolved into the octant in the late 1700s. The navigator rotated a mirror in order to place the image of a heavenly body on the visual horizon. A scale on the mirror measured elevation angle directly. It evolved into the sextant that was in widespread use until today, when it is being rapidly displaced by navigation satellite receivers (section 4.10, pg. 157).

4.4 Longitude and Time

Part 2.5 described the ancient methods of measuring time. The errors were so large that longitude could not be measured astronomically; it was estimated from east–west distance traveled. Hipparchus and Ptolemy knew that star measurements made from different places on Earth during a lunar eclipse would be simultaneous, but few, if any, longitude measurements were made that way until the great age of European exploration. Ptolemy showed the length of the Mediterreanean as 62° of longitude instead of the actual 42°. This 50% error persisted on maps until the 18th century. In the 1520s, the Flemish Professor Frisius again showed how to calculate the longitude difference between two points on Earth if time could be measured. Knowing the time, the navigator could measure a star's elevation, calculate its position in the sky at the known time, then calculate longitude from a spherical triangle.

Early Northern Hemisphere sailors estimated time at sea by

memorizing the position of the pointer stars of the Big Dipper (also called the Great Bear) at midnight during the various seasons. The "nocturnal" was first described in 1272 and improved in 1510 as a memory aid. The navigator aligned its arm with the pointer stars of the Dipper and read time from a scale. Prior to the 18th century, clocks (mechanical, water, or sand) could not maintain adequate accuracy for navigation at sea. Huygens' watch of 1675 (with an oscillating balance-wheel) kept time no better than 1 min/day at sea (10-mi longitude error per day at mid-latitude), even when mounted in gimbals.

When a European ship landed at a strange location, the navigator set up an astronomical station and measured latitude and longitude. A fine painting of Capt. Cook (in the Vancouver Maritime Museum) ashore in British Columbia (1778) clearly shows such a station under a tent. The ship's navigator was often asked to measure the local time of a celestial event (e.g., an occultation of a star or planet by the Moon) whose time an astronomer would measure at his home observatory. For example, Amerigo Vespucci, on one of his voyages to North America, landed to observe an eclipse. He measured time with a sandglass, after the celestially observed local noon.

On Columbus' second voyage, he measured the time of a lunar eclipse on Santo Domingo and compared it to the predicted time of occurrence at the Nuremberg Observatory (from his almanac). He calculated that Santo Domingo was 102° 30′ west of Nuremberg, too far west by 23°. This means his clock was 1.5 h slow relative to Santo Domingo time, even though he probably measured the time carefully with an astrolabe or nocturnal. His prediction of this eclipse terrified the Indians into giving his men free food.

By these means, the longitudinal accuracy of maps gradually improved. Ships meeting at sea compared longitudes; usually the estimate of the last ship to leave port was accepted by both. Thus was the longitude of each new landfall discovered. Errors were enormous; the Portuguese Alvaro Mendaña de Neyra, in 1567, was 60° in error (4000 mi) in his measurement of the longitude of islands in the Solomons. He was unable to find them again on his second expedition 27 years later.

During the period 1598–1714, Spain, Portugal, and Venice (later Holland, France, and England) offered rich prizes for a method of "finding the longitude" at sea that would be accurate to 30–60 min of arc (20 to 40 nm in mid-latitudes). Astronomers favored sighting the orbit of the Earth's Moon or observing the eclipses of the moons of Jupiter. Either observation would measure time well enough to calculate longitude but the first was computationally complex while the second required a high-power stabilized telescope. The method of Jovian eclipses was widely used at fixed stations by French-trained astronomers throughout Europe beginning 1680 (Part 2.3). However, ships' navigators observed the magnetic variation (angle between the compass and the polestar, sighted with a telescope called a "pelorus") and, from their charts, got a rough measure of longitude (perhaps 300 nm in error). That method was used until 1800.

The solution to measuring longitude, which was to last until radio broadcasting of time in 1923, was found by the Englishmen, John Harrison and his son, in the 1760s. They improved the technology of mechanical clocks, building one that could keep time to one part in 100,000 on a pitching, rolling ship, after compensation for a constant clock-drift rate. It used unlubricated counterrotating torsion pendulums in place of the swinging pendulum that was used in land clocks. The presence of a seaborne clock forced navigators to do spherical trigonometry at sea. Latitude calculations from a noon fix or meridian-transit of a star required only the subtraction of tabulated declination from the measured elevation. However, with an on-board clock it was possible to calculate latitude and longitude any time of night by sighting several stars, not necessarily on the meridian. Ship-borne trigonometry was done with logarithms, invented by Napier in 1614 (he also invented the decimal point). Hence, national observatories redoubled their efforts to produce easy tables for mariners to use. The Greenwich Observatory issued its first *Nautical Almanac* (still being produced) in 1767. Each national observatory served as the "zero longitude" for its country's ships. It was only in 1884 that an international conference accepted England's observatory, Greenwich, as zero longitude (because England was the dominant sea power of the time) and established the 24 worldwide time zones.

As late as 1878, Belgian railroad time differed from French railroad time by 28 min, Swiss time differed from French time by 26 min, and German time differed by 25 min. Standard time zones eliminated the use of local time on land and at sea.

Time is also discussed in 2.6.

4.5 Speed and Distance

The distance dimension of dead-reckoning at sea was measured by estimating speed and elapsed time. At first, speed was estimated by the captain, whose store of knowledge must have included the distance traveled by his ship in one sandglass interval under various conditions of wind, loading, sail deployment, and sea state. Calibration must have been difficult. A 2-kt current would produce a 50-nm error in a day, 25–30% of distance traveled. Hence, dead-reckoning by the day's sail must have been in error by 50%. When sandglasses came into use for measuring time intervals (perhaps during the Roman era, but certainly by the 13th century when inexpensive glass was being produced), a chip was thrown into the sea from the bow and the time for it to pass to the stern was measured. By 1550, throwing a chip had evolved into throwing a "log" overboard and counting knots on an attached rope unwinding from a reel. In 1688, a towed, vaned rotor (called a "patent log") was invented in England.

Modern ships infer speed from Pitot or electromagnetic measurements. The Pitot tube or "pitometer" measures the dynamic pressure of the onrushing water. The "electromagnetic log" measures the speed of the conductive seawater passing through an electric field between two conductors in a tubular "stinger" or "sword." Both the Pitot tube and electromagnetic sword are subject to fouling and must be calibrated upon installation because the speed of the local water flow is usually faster than the speed of the ship. The reprint by Griswold, pp. 179–187, describes traditional speed-measuring sensors. None of them can measure the transverse speed of the ship (''leeway'') which is significant when sailing in a crosswind, especially with a shallow keel.

On today's ships, an acoustic sonar can measure true ground speed as long as bottom reflections can be received; else, it measures speed relative to low-lying layers of water. With three narrow, noncoplanar beams looking downward, a Doppler sonar can measure two components of the ship's velocity, as described in the reprint by Wapner *et al.*, pp. 188–193.

In 1989, only a few Doppler sonars were installed in ships. Commercial sonar logs for surface ships formed two beams, to correct for pitch angle. They solved only for the forward component (in the plane of the beams) of the ship's velocity. A principal advantage of a sonar log is that it need not be calibrated; it measures ground speed and is insensitive to errors in mounting. However, sonars for submarines require pressure sealing. Sonars emit acoustic energy that betrays the presence of the ship. Hence, submarines look for other means of navigation (section 4.9).

The distance made good is calculated from these measurements of speed and heading. Cryer (1988) described an analog dead-reckoning computer carried in 16th-century ships. The ship's distance and heading were marked with pegs in a perforated wooden board called a ''traverse table'' as the ship tacked in the wind. At the end of each 4-hr watch, the average eastings and northings were measured on the board and transferred to the chart. The navigator thus determined the latitude traveled and, from tables of the cosine of latitude, he calculated the longitude traveled during the watch. The major dead-reckoning errors were in measuring ship's speed, estimating the velocity of water currents, estimating side-slip under sail (''leeway''), and extrapolating time with the half-hour sandglass (used by the Portuguese at least since 1500) from a noon fix or with a sandglass or nocturnal from a midnight fix.

4.6 Almanacs

Almanacs of the late 1400s contained sailing directions. The first known navigation handbook, published in Portugal in 1481, contained tables of the offset of Polaris from the true pole, tables of the declination of the Sun versus time of the year, tables of the latitudes of various ports, and a procedure for calculating latitude from the noon sighting of the Sun. If a ship captain had been asked to make astronomic observations, he would also have carried tables predicting eclipses.

By 1800, time-of-day was measurable; a telescopic sextant, tables of the heavenly bodies, and log-trig tables were on-board ship. Dead-reckoning was perfected with log and magnetic compass during overcast days. A spate of books appeared in all maritime countries instructing ships' officers on how to navigate. The British publisher, Mount and Page, was sustained by such books for more than 150 years (its predecessor began in 1650 by translating Dutch manuals into English). The most famous American text was Nathaniel Bowditch's *The American Practical Navigator* (1802). Flinders' work on correcting compass deviations with peripheral hard-iron and soft-iron rods followed shortly as did Sumner's line of position (1837), a major advance in algorithms for hand calculation. Sumner's method calculated the elevation of a heavenly body at the assumed position of the ship, then used the difference between the estimated elevation and an actual measurement to calculate the corrections to the assumed fix graphically. Bowditch's and Sumner's work made possible the clipper-ship trading voyages from New England to the Orient in the mid-1800s. The same iterative technique is widely used for many kinds of calculations today, including navigation fixes in modern digital computers. The computational tedium to ''reduce'' a celestial fix has been eliminated by the advent of portable microcomputers, just as celestial navigation is disappearing in favor of satellite navigation (section 4.10).

In the next century, almanac data will probably be broadcast in real time to ships at sea from digital data bases. The mariner will get his information immediately while the expense and delay of hard-copy distribution will be eliminated.

4.7 Gyrocompass

When the sun or stars were not visible, ships often spent days dead-reckoning with log and magnetic compass. The magnetic compass was subject to many errors; caused, for example, by the rolling, pitching, and turning of the ship, and by the ship's own iron.

The use of iron aboard ship is at least as old as the Romans, who used iron spikes in construction. Two thousand years later, substantial quantities of iron were introduced in steam engines and in the hull. Cast-iron steam engines were mounted on wooden river boats in France (1783), the United States (1787), and Scotland (1787). Beginning in 1810, passenger and freight traffic in the western world grew prodigiously in wooden river boats, driven by steam-powered paddle wheels. Magnetic compasses, remotely mounted from the engine, would have worked but were probably not used in river service. As iron and steel became less expensive, large amounts were used in hulls. Some historians say that the Korean ''turtle ships,'' that repelled a Japanese invasion in 1592, were iron-sheathed, iron-roofed wooden warships. I am skeptical about that claim because guns of the time were not powerful enough to penetrate wooden planking; hence iron would have been unnecessary, heavy, expensive, and destabilizing in roll. The first certain naval use of iron-cladding on wooden hulls was in the French warship, *Gloire*, 1860, that was powered by sail and steam. Construction of the *Gloire* started a naval arms race that led to World War I.

However, merchant ships were ahead of the navies in the use of iron. The first all-iron hulls, of wrought iron, were on river passenger boats in England (1821), to reduce the risk of fire caused by sparks from steam engines. In 1846, the *Great Britain*, an iron-hulled liner, was making regular trips to Australia under sail, with an auxiliary steam engine. Seagoing iron ships carried magnetic compasses that were compensated for permanent magnetism (''hard'' iron) and magnetism induced by the Earth's field (''soft'' iron), as explained in Part 4.2.

As iron ships came to dominate new construction at the end of the 19th century, there was demand for a nonmagnetic compass. The Anschütz-Kämpfe Company, in Hamburg,

Fig. 4-2. World War I marine gyrocompass, with repeaters. Courtesy of the Sperry Gyroscope Company.

Germany, produced a gyrocompass about 1911 that consisted of a horizontally spinning wheel (the electric motor had been recently invented) in a floating gimbal that was weighted to remain approximately vertical. In the steady state, the axis of the gyro-wheel settled out in the meridian plane, tilted upward by a fraction of a degree, thereby allowing the gravity torque produced by the weight to precess the gyro at the Earth's rotation rate. Max Schüler, who worked for Anschütz, calculated that by correctly selecting the offset between the pivot and center of gravity of the ball, the gyro would be insensitive to the ship's horizontal acceleration. The "Schüler-tuned" compass behaved like a north-seeking gyro mounted on a pendulum whose plumb bob was at the center of the Earth and whose period was 84 min. A complex theory was created during the 1910s and 1920s to correct and damp the gyrocompass in all three axes. Ingenious proprietary hydraulic and mechanical devices were invented and have been described by Rawlings (1944). The licensing of the Anschütz design to other companies was described by Schaafhausen (1962). Just after 1911, Sperry (United States; Fig. 4-2) and Brown (United Kingdom) repackaged the gyrocompass for their own national navies, just in time for World War I. The gyrocompass gradually became the primary heading reference of all large ships. These ships also carried magnetic compasses to monitor the gyrocompass and to substitute for it in case of failure. In 1980, miniature and marine gyrocompasses were available, using inertial navigation gyros. Because of their damping errors, modern marine gyrocompasses are usually tuned to a natural period closer to 100 min than to the Schüler period of 84.4 min.

4.8 Radio Navigation

During World War I, radio messages were sent to and received by ships at sea and the first 375-kHz radio beacons were installed along the American coast. These electronic lighthouses sent an identification signal, in Morse code, that allowed mariners a thousand miles from shore to find the direction toward the radio station by looking for the null in a rotating loop antenna. By taking cross-bearings to two or more stations, the ships could establish their position at sea in all weather, without celestial measurements.

In 1922, the U.S. Navy invented sonar, which started the long process of replacing sounding lines in shallow water, finding obstacles and submarines underwater, and mapping the ocean floor. Beginning in the 1960s, the United States established 36 networks of passive sonar sensors on the seafloor, along the American coastline, and under key ocean passes (such as the Strait of Gibraltar and the Greenland–Iceland–England gap). These networks allow shore stations to identify and track ships and submarines. The passive arrays are probably augmented by sound sources that can illuminate quiet vessels.

In 1935, the French ocean liner, Normandie, had VHF radar for navigation near shore and for collision avoidance. In 1939, the German Navy installed a VHF radar (approximately 125 MHz) on the battleship Graf Spee with the added intention of finding targets. The United States put radar on every capital ship during World War II.

Loran-A was invented during World War II. A receiver measured the difference in the times of arrival of 2-MHz pulses sent from pairs of shore stations. The lines of constant time difference were distorted hyperboloids of revolution whose intersections with the reference ellipsoid were printed on a maritime chart. The navigator observed a CRT in the receiver to measure the time differences from two pairs of stations and plotted time on a chart. By observing at least two station pairs, the ship's position was located at the intersection of two hyperbolas. A few World War II-era Loran-A stations still operate in China and Japan, used by local fishermen. The last occidental Loran-A chain was decommissioned before 1980.

Fig. 4-3. Loran-C transmitting antenna at Kure, Hawaii. Courtesy of the U.S. Coast Guard.

A regional navaid, Decca, has been widely used in the waters around the United Kingdom since World War II, as described in the reference by Powell (1985). It is a 100-kHz hyperbolic system whose chains were set up by a private company and sold to the British Lighthouse Authority which still operates Decca.

The U.S. Coast Guard, NATO, Japan, and a few other countries maintain more than 20 Loran-C navigation chains in busy shipping lanes throughout the world. Loran-C is a 100-kHz phase-difference navigation aid that is used by 400,000 ships. It is little used by military and air-carrier aircraft, though general aviation use is increasing slowly. Figure 4-3 shows a Loran transmitting antenna. The master ground station has an atomic clock synchronized within 2.5 μs of U.S. Naval Observatory time, as described by Sakahara (1988).

A portable version of Loran-C, called Loran-D, has been used by the U.S. military, for example by the rapid deployment forces. The traditional hyperbolic fix requires reception of a master and at least two secondary (''slave'') stations in a chain of three to five stations, whose coverage extends about 1000 nm. The basic measurement is the phase difference between the ground wave from a master and the ground waves from at least two slaves. Lines of constant phase difference are hyperbola-like curves on the reference ellipsoid, whose foci are at the transmitting stations. Another way to use Loran is to carry an atomic clock on the vehicle and measure one-way range to any two stations, thus fixing the vehicle's position at the intersection of two circles. In one-way ranging, the receiver must know the absolute time at which pulses are emitted from each station. One-way range is called ''pseudorange'' until it is corrected for propagation speed. It is much more subject to timing errors than is hyperbolic fixing. A third pair of stations (for hyperbolic fixes) or a third station (for circular fixes) will resolve ambiguities.

Loran receivers include computers that store the coordinates of stations and their signal properties worldwide. These receivers automatically select stations and calculate vehicle position with: (a) a simple formula for the propagation speed of Loran ground waves over seawater, or (b) a more complex formula that includes ''secondary phase factors'' to account for propagation speed over land and ice. The reprint by Woodman *et al.*, pp. 221–228, shows how to calculate the field strength and propagation delay as a function of the conductivity and permittivity of the ground.

An accuracy better than 0.25 nm CEP is achievable within 1000 nm of the stations. At one time, the slaves transmitted when they received the master's signal. Soon, when all Loran master stations will time their emissions relative to an atomic clock, the distinction between chains will weaken. Fixes will therefore be possible, using stations in different chains ("cross-chaining"), though with degraded accuracy. The reprint by Shapiro, pp. 211–220, describes Loran-C in detail, at the waveform level.

All Loran stations have cesium clocks. Master stations close to the U.S. Naval Observatory are synchronized by radio. Distant masters are synchronized by comparing their atomic clocks to portable atomic clocks that are shipped among stations and the naval observatory. After years of observation, the stations' clocks have been calibrated within 10 μs. As described in the reprint by Klepczynski, pp. 58–63, Loran masters will soon be synchronized more cheaply with reference to a GPS receiver at each master site, to an accuracy of 0.1 μs of UTC. Loran secondary stations are presently synchronized by observing the phase of the signals received at monitor stations within each chain. Secondary stations, too, may soon have GPS receivers, thus improving the accuracy of one-way range fixes.

Monitor stations could be "differential-Loran" stations that transmit their measured position offsets to ships in the area. I know of no U.S. government-operated differential-Loran stations, though they are predicted to reduce absolute position-fix errors from 0.25 nm to 50 ft. Receivers that read-out in latitude–longitude can be purchased for as little as $500. Loran timing receivers that measure absolute time within 10 μs, after compensation for propagation speed over land and ice, can also be purchased. Timing receivers identify the first three cycles of the carrier (before the sky wave arrives) and compare their known shapes to the return. Current improvements in Loran were described by Fuentes (1986). Loran for aircraft navigation is discussed in Part 5, pg. 232.

The Russian Chaika ("sea gull") is identical to Loran-C except in the fine structure of its pulse shape. Experiments have shown that Loran and Chaika sets can receive each others' signals, as shown in the reference by Westling (1988). A joint Loran-C–Chaika chain is being established in the Bering Sea.

4.9 Inertial Navigation

Dead-reckoning was greatly improved in 1955 when the first Ship's Inertial Navigation System (SINS) was tested on the U.S. Navy's Compass Island. The heading and level references were gyroscopic, whereas distance was calculated by measuring the horizontal components of acceleration precisely, and double-integrating them in a computer. Position could be kept accurately for several hours between radio fixes. Today, the Mark 2 SINS (two or three per ship, for redundancy) is used on all U.S. nuclear submarines and naval capital ships. It not only navigates the ship, it initializes missile position, velocity, and azimuth. It serves as a vertical and azimuth reference for gunfire. Aircraft carriers use SINS to initialize the position, velocity, and attitude of their aircraft before launch. The reprint by McKelvie and Galt, pp. 194–206, describes the evolution of SINS through its various versions.

By offsetting the tilt of the SINS platform to account for the local deflection of the vertical, navigation accuracy can be improved to tenths of a mile per hour between fixes.

Experiments have been done with gravity gradiometers, with the intention of matching observed gradients with stored gradients as a navigation aid [see (Wells and Breakwell, 1982)]. Because gradiometers do not emit acoustic energy (as does a sonar), they are attractive for military vessels. The obstacles to a successful gradiometer are:

—it is difficult to build a precise instrument,

—whether direct reading or differencing to nearby accelerometer,

—the normal to the ellipsoid must be found by calculation and precisely tracked, and

—the variable mass of the ship contaminates the measurement.

The reference by Molny (1980) has described a gradiometer.

4.10 Satellite Navigation

The most important recent advance in ship navigation came with the launch of the first U.S. Navy Transit satellite in 1964. Transit has grown to a constellation of seven satellites called the "Navy Navigation Satellite System." By measuring the history of the Doppler shifts of a 150-MHz tone and a 400-MHz tone emitted by a satellite at 1 to 5 Watts during a 10-min pass, the ship's computer can calculate its position within hundreds of meters using any of several algorithms. Receiving two frequencies allows the calculation of a first-order atmospheric correction, reducing position errors from ± 90 to ± 30 m RMS on a stationary vehicle. The reprint by Danchik, pp. 207–210, briefly describes Transit; the reference by Stansell (1978) describes it in more detail.

Position is calculated by an on-board algorithm. A typical algorithm iterates like Sumner's line-of-position calculation (Section 4.6). It assumes a position and calculates the accumulated Doppler shifts (Doppler count) in 23-s increments throughout the pass. It compares them to the measured counts, from which it computes the along-orbit and cross-orbit corrections to the assumed position. Solutions are obtained in two or three iterations, even when the initial error is 300 km, per Stansell's chapter in the reference by Laurila (1976) in the Introduction to Part 2. The usual time between fixes is an hour. A 3-h gap occurs every 3 days in mid-latitudes. Fewer gaps occur at sea, where the satellite's signals can be used at elevations as low as 5°, than on land, where elevation must be greater than 15°. Maximum elevation for a usable fix is about 70°.

The ship's velocity (north and east components) must be measured and included in the calculation. The sensitivity to velocity errors is 0.1 to 1 nm/kt depending on the orbit geometry and the ship's heading. This is the largest error source on a moving vehicle. Because of it, a Doppler sonar dead-reckoning system is required if Transit position errors are to be less than 0.5 nm. Another source of error is the measurement of the ship's height above the reference ellipsoid. The horizontal error is usually two or three times the altitude uncertainty. This error is not a problem for ships at sea but it is important for ships on inland waterways and for land surveys.

Transit receivers have been used for surveys with reported errors of approximately 1 m in the length of the measured arc, by observing the same pass from both ends of the arc. Transit disseminates absolute time within 30 μs.

Three of the satellites have orbit stationkeeping motors. The orbits of the other four drift relative to the Earth and relative to each other. Four Navy-operated monitor stations receive the satellites' signals and transmit them to a command center at Pt. Mugu, California, which calculates the orbits and uplinks the orbit elements once or twice per day. The orbit elements are broadcast from each satellite every 2 min. Transit geodesy has been based on its own measurements and its own datum. It is now converting to the WGS-84 datum (pg. 11).

In 1989, 75,000 to 100,000 Transit receivers (costing as little as $2000) were in service around the world. The Soviet Union operates a Transit-like system called Cicada.

The American Global Positioning System (Navstar-GPS) and its Soviet counterpart, Glonass, are already being used by civil and military ships. The reference by Noe *et al.* describes a marine GPS receiver and its test results.

The two Marecs satellites in orbit in 1989 permit ships in the Atlantic, Pacific, and Indian oceans to communicate with shore stations (and to national telephone networks) and with each other on VHF. The satellites are owned by INMARSAT, an intergovernment consortium. They are not used for navigation. However, privately owned navigation satellites exist. Geostar relays the user's Loran-derived position to a control center using a leased transponder on a commercial communication satellite. In later versions, the Geostar satellites will be tracked and their range to the user will be measured. Geostar is selling its services in the continental United States and to oil drillers offshore. Its technology has been licensed to Europe, for Geostar (pg. 67).

Other satellites are being proposed that will offer communication and navigation services over oceans and transcontinentally. Examples are ARINC's Avsat, Starfix, satellites to be launched by the American Mobile Satellite Company, and proposed European and Japanese systems. The International Civil Aviation Organization has proposed a worldwide mobile nav–comm system, for aircraft and ships. However it comes about, ocean navigation coverage beyond that offered by the low-capacity Marecs is likely to be a monopoly, perhaps also owned by a consortium of air and ship traffic control agencies. Part 8.1 projects the future of navigation and communication services on land, sea, and air.

4.11 Integrated Navigation Systems

Multisensor ship navigators are the most accurate, most reliable, and most expensive ways to measure position and velocity. A marine gyrocompass/electromagnetic-log navigator was designed in the 1970s for U.S. destroyers [see (Nash, 1976)]. It maintains 1-mi position and 1° heading accuracy when updated hourly by Transit satellite. Deep-submergence vehicles also use gyrocompass/electromagnetic-log dead-reckoning, updated from bottom-mounted acoustic transponders placed in the vicinity of the submarine cable or oil well on which they are working. Navigation errors of 2 to 10 m have been reported, as decribed in the references by Meathe *et al.* (1978) and O'Brien and Bojarsky (1978).

Inertial navigators (section 4.9) are the major means of dead-reckoning on naval capital ships and submarines. They are updated periodically from satellite fixes, by Omega and Loran, from sonar mapping of the bottom, or from Tercom-like matching of the measured sea-bottom profile with a stored profile, p. 236.

The reprint by McMillan, pp. 161–170, describes the standard Canadian naval navigator, which uses a gyrocompass, a magnetic compass, Doppler sonar, Omega, Transit, Loran, and GPS. Dunham and Kriegsman (1988) have described a similar American system in development. These multisensor systems achieve 100- to 200-m accuracy at sea with all sensors working, degrading to 1000 m with a few outages. In the historical perspective of this book, these accuracies are astounding for a ship in the open sea.

With the advent of supertankers in the 1970s, micronavigation of a ship in narrow channels (called "conning") became an important problem which was addressed by building the first electronic simulators of commercial ships to train pilots. The major conning aids are the captain's eyes, his radar, and his sonar depth indicator.

References and Bibliography

Anderson, R. and R. C. Anderson, *The Sailing Ship*. New York: W. Norton, 1963, 212 pages. Evolution of European sailing ship from ancient Egypt to the early 20th century.

Atkinson, J., *Epitome of Navigation*. London: R. Mount, 1704. Compass rose with 32 points and 360° scale on the periphery on p. 65.

Bass, G. F., "Oldest known shipwreck," *National Geographic,* Dec. 87. Ulu Burun, Turkey excavation of a 50-ft-long, 1400 B.C. Phoenician boat, on the sea bottom.

Bellwood, P., *The Polynesians*. London: Thames and Hudson, 1978, 180 pgs. Mostly on large migration patterns. Little on canoes and navigation.

Bellwood, P. S., "The peopling of the Pacific," *Sci. Amer.*, vol. 252, pp. 174–186, Nov. 1980. Sequence of settlement from *Homo erectus* in Southest Asia, 300,000 B.P., to Polynesians in New Zealand, A.D. 800. Archaeological and linguistic evidence.

Bowditch, N., *The American Practical Navigator*. Wash., DC: U.S. Navy Hydrographic Office, issued regularly since 1802. Excellent manual on ship navigation.

Brøgger, A. W. and H. Shetelig, *The Viking Ships: Their Ancestry and Evolution*. Oslo: Dreyers Forlag, 1951, 251 pgs. Authoritative account from paleolithic Norwegian boats to plank boats of first millenium A.D., to Viking ships of A.D. 600–900, and Saga-era boats A.D. 800–1300, that made voyages to Iceland, Greenland, America.

Carnebianca, C., "Regional to global satellite-based navigation system," in *Rec. IEEE Position Location and Navigation Symp.*, 1988, pp. 25–33. Another proposal for a European navigation satellite constellation that would not depend on the United States and the Soviet Union.

Clautice, W. G., "Submarine navigation," *Navigation,* pp. 34–39, Spring 1978. Too brief. Loran-C towed antennas.

Coast Guard, *Loran-C Users' Handbook* (U.S. Coast Guard Publ. COMDTINST M-16562.3), 1980. Semitechnical treatment for boat owners.

Coast Guard, *OMEGA Global Radionavigation, a Guide for Users* (U.S. Coast Guard Publ. COMDTINST P-16566.2), 1983. Detailed treatment with numbers and signals in space.

Cryer, J. P., "Navigation on Nuestra Senora de Atocha 1622," *Astrolabe,* vol. 4:3, pp. 21–32, Summer 1988. Review of navigation techniques used on a Spanish galleon and discussion of the instruments found in the salvaged vessel.

Culver, C., "The future of Loran in the GPS age," *American Sailor*, vol. 9, no. 11, pp. 3A–4A, Nov. 1989. Predicts that cheapest GPS receivers will always be three times as expensive as the cheapest Loran receivers. Small boats do not need worldwide coverage.

Danchik, R. J., "Navy navigation satellite system status," in *Rec. IEEE Position Location and Navigation Symp.,* 1988, pp. 21–24. Reprinted in Part 4, pp. 207–210.

Dunham, S. J. and J. Kriegsman, "Improved ship's navigation accuracy through precision data reconstruction," in *Rec. IEEE Position Location and Navigation Symp.,* 1988, pp. 509–513. Naval Air Development Center description of integrated sensors (inertial, Loran, Doppler log, GPS, and Transit); 23-state filter for postcruise calculation of position of bathymetric survey ship.

Dunlap, G. D., "Major developments in marine navigation during the last 25 years," *Navigation*, vol. 18, Spring 1971. Loran-A, radar, gyrocompass, beacons, sextant, chronometer, celestial tables, Mercator charts, Decca, Transit, Loran-C, Omega.

Enge, P. K., and J. R. McCullough, "Aiding GPS with calibrated Loran-C," *Navigation*, vol. 35, no. 4, pp. 469–482, Winter 1988–89. Describes a hybrid receiver that combines Loran and GPS lines of position. GPS calibrates Loran propagation delays. Would reduce Loran errors in half. Can serve as an integrity monitor for GPS.

Foley, V. and W. Soedel, "Ancient oared warships," *Sci. Amer.,* vol. 253, pp. 148–163, Apr. 1981. Good analysis of Greek warships.

Fuentes, F., "Loran-C in the 21st Century," *IEEE Position Location and Navigation Conf.*, 1986, pp. 277–79. U.S. Loran stations replacing transmitters, adding monitors for remote operation. USSR, European, mid-East, Chinese expansions.

Gilson, J., ed., *Symposium on Sharp's Theory of Accidental Voyages,* Wellington, NZ: Polynesian Society, 1963. Polynesian migrations. Men's houses have planetariums. Data on percentage of arrivals for recorded mass voyages.

Gladwin, T., *East Is a Big Bird: Navigation and Logic on Puluwat Atoll.* Cambridge, MA: Harvard University Press, 1970. Anthropological study of Micronesian islanders. Description of their boats and star compass. Memorizing cues about sea-lanes.

Greenhill, B., *Archaeology of the Boat.* Middletown, CT: Wesleyan University Press, 1976. Superb history of ancient boat technology in Europe and Asia.

Griswold, L. W., "Underwater logs," *Navigation,* vol. 15, no. 2, pp. 127–135, Summer 1968. Reprinted in Part 4, pp. 179–187.

Hapgood, C. H., *Map of the Ancient Sea Kings.* New York, NY: Dutton, 1966. Turkish map of 1513 by Piri ibn Haji Memmed also called Piri Re'is. Proposes Phoenician origin of "normal Portolans."

Hay, D. and J. Hay, *No Star at the Pole.* London: Knight, 1972. Entirely on pre-electronic ships. Practical problems of ancient sailors.

Henry, T. D., "Acoustic transponder navigation," in *Rec. IEEE Position Location and Navigation Symp.,* 1978, pp. 237–244. Reprinted in Part 4, pp. 171–178.

Hine, A., *Magnetic Compasses and Magnetometers.* Toronto: University of Toronto Press, 1968. Definitive treatment of magnetic needles and flux gate. Detailed treatment of compass compensation.

Hobbs, R. R., *Marine Navigation 2, Celestial and Electronic.* Annapolis, MD: Naval Institute Press, pp. 86–93, 1981. How-to-do-it book for naval officers.

Horton, M., "The Swahili corridor," *Sci. Amer.,* vol. 259, pp. 86–93, Sept. 1987. Boat commerce from East Africa to Zanzibar.

Hourani, G. F., *Arab Seafaring in the Indian Ocean.* Princeton, NJ: Princeton University Press, 1951. Good survey.

Howse, D. and M. Sanderson, *The Sea Chart, an Historical Survey Based on the Collections in the National Maritime Museum.* New York, NY: McGraw-Hill, 1973. Ptolemy's *Geographia* (table of the latitude-longitude of 8000 places). Roman highway maps, European maps. Marinus of Tyre (A.D. 100) early maps.

Hughes, R., *The Fatal Shore.* New York, NY: A. Knopf, 1986. History of early Australia. References to aborigine canoes and Alvaro de Mendaña's 1568 expedition.

Lewis, David, *The Voyaging Stars: Pacific Island Navigators.* Sydney: Collins, 1978. By a Polynesian M.D. Austronesian migrations. Accidental voyages. Lots of material on boat construction, with sewn planks. Handling. Cues from zenith star, steering star, birds, clouds, water, and wind.

MacKenzie, K. V., "Early history of deep submergence," *Navigation,* vol. 17:1, pp. 3–12, Jan. 1970. Twenty-foot accuracy to retrieve points, 6–8 h of operation. Acoustic obstacle avoidance and target detection.

McGrail, S., *The Ship—Rafts, Boats, and Ships.* London: H. M. Stationers Office, 1981. Good text on ancient and modern boat construction. Use of boats by ancient people. The first of ten books on the history of ships.

McKelvie, B. and H. Galt, Jr., "The evolution of the ship's inertial navigation system for the fleet ballistic missile program," *Navigation,* vol. 25, no. 3, pp. 310–322, Fall 1978. Reprinted in Part 4, pp. 194–206.

McMillan, J. C., "MINS-B II: A marine integrated navigation system," in *Rec. IEEE Position Location and Navigation Symp.,* 1988, pp. 499–508. Reprinted in Part 4, pp. 161–170.

Meathe, W. F., *et al.*, "SCARAB acoustic location," in *Rec. IEEE Position Location and Navigation Symp.,* 1978, pp. 251–254. SCARAB submersible has an acoustic phased array.

Miller, B. M., *Chumash—A Picture of their World.* Los Osos, Ca.: Sand River Press, 1988, 144 pgs. Indian remains found on the island of Santa Rosa from 10,000 B.P. The Chumash used reed, dugout, and planked canoes when the Spanish arrived. Canoes were built by a guild of fishermen and traders in secret shops.

Mollat du Jourdin, M. and M. de la Roncière, *Sea Charts of the Early Explorers: 13–17th Century.* London: Thames and Hudson, 1984. (Translated from *Les Portulan (sic), cartes marines du 13–17 siècles.*) Portolan construction. Maps are luxury decorations for kings.

Molny, M. J., "Gravity gradiometry in shipboard environment," in *Rec. IEEE Position Location and Navigation Symp.*, 1980, pp. 59–63. Bell gradiometer. Error sources.

Morison, S. E., *Christopher Columbus, Mariner.* Mentor Books, 1942. Account of his voyages to the Americas.

Nambiar, O. K., *Our Seafaring in the Indian Ocean.* Bangalore, India: Jeevan Publishers, 192 pp., 1975. Indian and Chinese voyages from 3000 B.C.

Nard, G. and R. Gounon, "SERCEL high-precision real-time differential GPS; principles and results," in *Rec. IEEE Position Location and Navigation Symp.,* 1988, pp. 528–535. Transmits GPS corrections on 1.6–2.5 MHz. Gets 0.9 to 0.95 probability of reception at 600-km range; 1- to 2-m errors observed within 200 km.

Nash, J. M., "An integrated marine navigation system," in *Rec. IEEE Position Location and Navigation Symp.,* 1976, pp. 177–183. For DD-963 destroyer. Gyrocompass, EM log, Omega, Transit, growth to GPS. No test data.

Needham, J., *Clerks and Craftsmen in China and the West.* Cambridge: Cambridge University Press, 1970. Collected lectures on sea voyages, ship construction (rudder, sails), and magnetic compass.

Noe, P. S., *et al.,* "The C/A code GPS receiver at sea," in *Rec. IEEE Position Location and Navigation Symp.,* 1980, pp. 48–50. Tests by Texas A&M University of one-channel, C/A-code GPS receiver at sea, versus Loran. All fixes agreed within 200 m.

O'Brien, M. D. and S. M. Bojarsky, "Deep submergence rescue vehicle navigation system," in *Rec. IEEE Position Location and Navigation Symp.,* 1978, pp. 219–224. Qualitative description.

Ott, L. E., "STARFIX—Commercial satellite positioning," in *Rec. IEEE Position Location and Navigation Symp.,* 1988, pp. 8–12. Ten U.S. tracking stations measure satellite position and calculate user positions; 5-m accuracy.

The Portolan Chart of Angellino De Dalorto, 1325. Reproduced in 1929 by the Royal Geographic Society from the maps of Prince Corsini, Florence. Good plates, with a short introduction.

Powell, C., "Early history of the DECCA navigation system," *J. Inst. Electronic and Radio Engineers (UK),* pp. 203–209, June 1985. U.S. invention sold in England after World War II.

Rawlings, A. L., *The Theory of the Gyroscopic Compass and Its Deviations.* New York, NY: MacMillan, 1944. The best on marine gyrocompasses.

Ruane, M. F., *et al.*, "Test experiences using marine radiobeacons for DGPS communications," in *Rec. IEEE Position Location and Navigation Symp.,* 1988, pp. 303–309. Adds side tones to marine beacons at 285–325 kHz to send 50-Hz differential GPS data.

Sakahara, M.D., "100 ns synchronization and the users of Loran-C," in *Rec. IEEE Position Location and Navigation Symp.,* 1988, pp. 518–522. USNO observes masters and sends corrections to adjust frequency. Proposes GPS receiver at each master, latter at secondaries.

Schaafhausen, F. W., *The History of the Firm, C. Plath, 1862–1962.* Hamburg: Verlag Hanseatische Merkur, 1962. 138 pp. Built 50,000 handheld sextants, as Germany industrialized. Bought license to build Anschütz gyrocompass, 1911, for merchant marine. Expanded into aircraft instruments, 1930. Appendix describes the history of the sextant.

Schwarz, K. P. *et al.*, "Marine positioning with GPS-aided inertial navigation systems," *Navigation,* vol. 31, pp. 38–45. Jan. 1984. Tested GPS with Ferranti inertial land surveyor, 1981. Results at sea show 15-m error after smoothing.

Sentman, O. L., "Navy navigation satellite system (TRANSIT)," *IEEE AES Mag.,* pp. 25–26, July 1987. Control center at Pt. Mugu and 4 tracking stations. Update Oscar twice per day, Nova once per day. No more launches planned after 1989; system to be phased out in 1994.

Shapiro, L. D., "Time synchronization from Loran-C," *IEEE Spectrum,* pp. 46–55, Aug. 1968. Reprinted in Part 4, pp. 211–220.

Sharp, A., *Ancient Voyages in Polynesia.* Berkeley: U.C. Press, 1963. Many examples supporting accidental voyages to eastern Pacific islands. Successful colonization of uninhabited islands occurred when women happened to be aboard. Details about New Zealand.

Smith, D. E. and J. Ginsburg, "From numbers to numerals," in *The World of Mathematics,* J. R. Newman, Ed. New York, NY: Simon & Schuster, 1956, pp. 442–464. Roman and Arabic numerals.

Sridharan, K., *A Maritime History of India,* 2nd ed. New Delhi: Ministry of Information and Broadcasting, 1982. Ship voyages by Indian princes to Ceylon from 3000 B.C. to A.D. 1000.

Stansell, T. A., *The TRANSIT Navigation Satellite System.* Torrance, Ca.: Magnavox Corp., 1978. System description and thorough error model. Also in Laurila book on geodesy. (See Part 2 bibliography.)

Taylor, E. G. R., *The Haven-Finding Art.* London: Hollis and Carter, 1956. The best history of pre-electronic ship navigation.

Taylor, E. G. R., *The Geometrical Seaman.* London: Hollis and Carter, 1962. By the same author as above; equally good, but shorter.

Turner E. E., *et al.,* "The Raytheon acoustic Doppler navigator," *Navigation,* vol. 13, pp. 210–221, Autumn 1966. Raytheon Doppler sonar.

Wapner, M., *et al.*, "Doppler sonar velocity log for attack submarines," in *Rec. IEEE Position Location and Navigation Symp.,* 1980, pp. 146–151. Reprinted in Part 4, pp. 188–193.

Weiffenbach, G. C., "The genesis of Transit," *IEEE AES Mag.,* pp. 474–482, July 1986. History of Transit navigation, including early experiments to define system characteristics.

Wells, E. M. and J. V. Breakwell, "Real-time gravity gradiometer utilization to improve INS accuracy," *AIAA J. Guidance and Control,* pp. 113–117, Mar. 1982. Two horizontal-axis gradiometers improve estimate of velocity error. Assumes unknown horizontal gravity field has 20-nm correlation length. Steady-state Weiner filter.

Westling, G. R., "Joint Soviet-American Loran operations, Bering Sea," in *Rec. IEEE Position Location and Navigation Symp.,* 1988, pp. 523–527. Soviet Chaika versus U.S. Loran-C. Plans new chain with two Russian stations and one U.S. station.

Woodman, K. F., *et al.,* "Methodology for the determination of the cover and secondary phase correction plots for a Loran-C chain," in *Rec. IEEE Position Location and Navigation Symp.,* 1988, pp. 207–214. Reprinted in Part 4, pp. 221–228.

Worcester, G. R. G., *Sail and Sweep in China.* London: H.M. Stationer's Office, 1966. History of Chinese boats from 2000 B.C. to A.D. 1500.

MINS-B II: A MARINE INTEGRATED NAVIGATION SYSTEM

J.C. McMillan

DEFENCE RESEARCH ESTABLISHMENT OTTAWA
OTTAWA, ONTARIO, CANADA K1A 0Z4

ABSTRACT

A low cost navigation system integrator has been developed for the Canadian Navy. This system, AN\SYN-501 (MINS-B II), employs a 17 state, dynamically reconfigurable extended Kalman filter, along with extensive error detection and correction algorithms. Designed to improve the navigation accuracy, efficiency and reliability on board major ocean going vessels, this system optimally integrates most standard marine navigation sensors and systems: GPS, Transit, Omega, Loran-C, Gyrocompass, Speedlog and operator entered sextant measurements or position fixes. The MINS is designed to be very flexible, with microprocessor controlled interfaces to collect sensor input from a wide variety of brands and models of these systems and sensors. The integration software has comprehensive failure mode capability allowing operation with any viable subset of these inputs. Physically MINS consists of three units: a central electronics unit, a control display unit and a data entry unit.

Over the last few years MINS has undergone extensive sea trial evaluation (by both the Canadian and U.S. Navies) as well as environmental testing, and is going into service throughout the Canadian Navy. This paper briefly describes the salient points of the MINS integration scheme and presents some recent seatrial results.

INTRODUCTION

Design Objectives

A common situation on major ocean going vessels is the presence of many different navigation sensors, operating independently and requiring varying degrees of manual intervention. The MINS development objective was to improve the operational efficiency, reliability and accuracy by automating all necessary calculations, optimally integrating all sensor information and providing a central unit for digital output, visual display and recording (if desired) of all navigation data.

Development History

The MINS concept and system design were developed at DREO over the period 1980-1988. The initial design and simulation study is summarised in reference [1], at which time only the gyrocompass, speedlog and Omega were considered for integration. In 1981 Loran-C was added, still at the simulation level. This design, along with some of the analysis that was used for its development, is described in reference [2].

A laboratory development model was then assembled, using a general purpose computer. This lab model was tested at sea in 1982, providing the first valuable data. This led to the development of error detection methods for handling spurious speedlog data, Omega lane slips, Loran cycle selection errors etc. Reference [3] describes some of this work.

In 1983 Transit position fix processing capability was added to MINS, creating MINS-B, as described in reference [4] which presents the 1983 sea trial results. The sensors for this and subsequent trials were:

1-Sperry Mk 23 Mod C-3 gyrocompass
2-Sperry SRD-301 Doppler speedlog
3-Magnavox MX1105 Transit-Omega (GPS added later)
4-Internav LC204 Loran-C receiver

In 1983 an experimental development model of MINS-B was built using a Motorola 68000 based microcomputer with a keyboard, display and the necessary interfaces, power supplies etc. In 1984 this MINS-B XDM, was tested on both the west and east coasts. The major verification sea trial was in October 1984 as reported in reference [5].

In 1985 the U.S. Navy acquired an upgraded MINS-B XDM for evaluation. The U.S. Naval Air Development Centre (NADC) conducted sea trials (for NAVSEA) with this unit in 1985, 86 and 87 on the USS Reasoner in the Pacific and on the Vanguard in the Atlantic. In 1986 DREO incorporated GPS position processing capability into the MINS filter and added a water speed error state to the DR error model to allow filter reconfiguration for an optimal speedlog failure mode. In 1986 and 1987 DREO added software enhancements to improve the operability, the accuracy and the failure modes.

In 1986 a MINS-B II ADM was built. This had significantly improved hardware, with a 68020 processor, a 68881 co-processor, more compact interface hardware and a separate control/display unit. In 1987 production began to equip all major Canadian Naval vessels with a MINS-B II unit.

Reprinted from *Rec. IEEE Position Location and Navigation Symp.*, pp. 499–508, 1988.

MINS DESIGN OUTLINE

Figure 1 shows the major components of MINS-B II in block diagram form, along with the sensor input that it can integrate.

Figure 1. MINS Block Diagram.

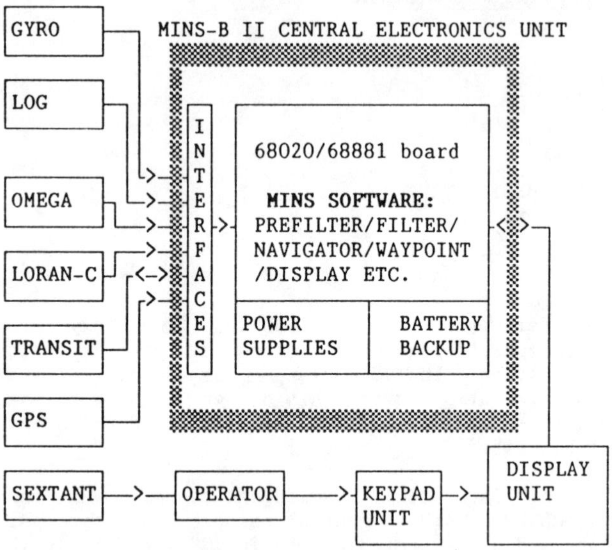

Figure 2 is a photo of the preproduction unit of MINS-B II, showing the three units that comprise MINS: the plasma display unit, the keypad unit and the central electronics unit.

Figure 2. MINS-B II Photo.

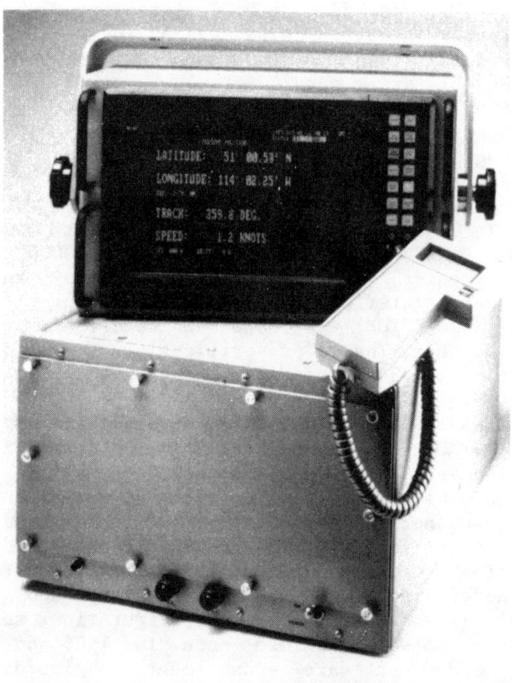

The display unit has dedicated keys for calling up each of the three standard pages of information (raw data, subsystem position/velocity, optimal position/velocity) and for initiating each of the eight operator interaction functions (system initialisation, position fix entry, waypoint voyage planning etc.). All interactive functions are menu driven with suitable prompts from the display for ease of use. Figure 2 also shows the central electronics unit, which houses the sensor interfaces, processor/ memory board, power supplies and battery backup.

The MINS processor is a Motorola 68020/68881 (32 bit machine) with about one half megabyte of memory. The MINS software consists of 27 tasks, written in "C". Figure 3 shows the major data flow into and out of the MIN$ filter.

Figure 3. MINS Data Flow Diagram.

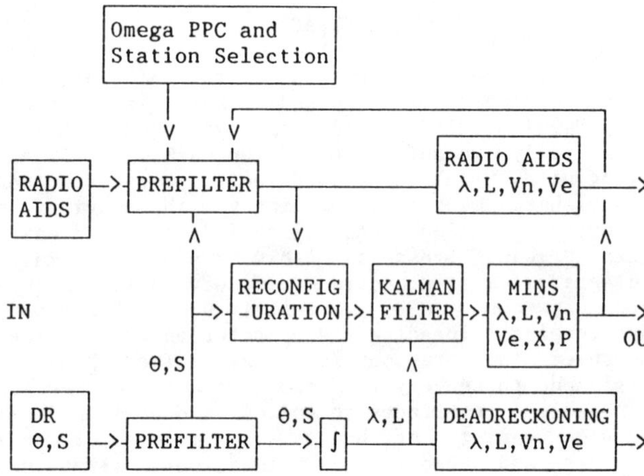

The dead reckoning (DR) data is collected from the gyrocompass and speedlog at a 4 second rate and this DR velocity is integrated to give a continuous position estimate. MINS uses this smooth DR position as a base for the high frequency component of the ship's track. The radio aid data is collected at a 20 second rate and is used to remove the bias and drift of the DR data. The optimal position and velocity output is the DR position and velocity, corrected by the filter estimate of the DR position and velocity error. The filter's state covariance matrix will provide a statistical estimate of how accurate this filtered position and velocity are. This design ensures that MINS will continue to provide optimal filtered position and velocity output under a broad range of sensor failure conditions.

ERROR MODELLING

In order to design an optimally integrated system using the sensors and subsystems displayed in figure 1, it is first necessary to carefully model the errors of each system as stochastic processes, and to model the dynamics of these errors.

Dead Reckoning Errors

The DR position (λ,L) is determined by

integrating the DR velocity measurements (S, θ) starting from some known initial position:

$$\lambda(t+\Delta t) = \lambda(t) + \frac{\Delta t\ S\ \cos\theta}{R_n} \quad (1)$$

$$L(t+\Delta t) = L(t) + \frac{\Delta t\ S\ \sin\theta}{R_e \cos\lambda} \quad (2)$$

where R_n and R_e are the north and east radii of curvature of the WGS84 ellipsoid. If numerical error is avoided, the only significant errors here will be the DR velocity component errors. Since the speedlog measures water speed, the ocean current is a dominant source of error, followed by speedlog sensor error and gyrocompass heading error. With the gyros and speedlogs typically used on major vessels, the contribution of the heading error is relatively insignificant. Since the heading error is also nonlinear it was decided on the basis of numerical simulations that the benefit of including it explicitly is marginal and that it could be absorbed into the ocean current error model.

On a large scale, the ocean current is quite random, correlated both in time and space. It will be bounded and continuous and it's north and east components will be nominally zero mean (over the earth). Therefore each component can be modelled as a standard first order Markov process, defined by its correlation time and its steady state variance (as described in reference [6]).

The speedlog is assumed to be water-tracking (rather than bottom-tracking). In this case its error is primarily from calibration error, water turbulence, the pitching, rolling, sideslipping etc. motion of the ship's hull, and from sensor noise. If pulse counts are used (rather than synchro or BCD), then there may also be significant quantization error. In any case the speed error is randomly varying, nominally zero mean and bounded. This can therefore also be modelled as a first order Markov process.

Omega Errors

Of all the sensors used in MINS, Omega has the largest and most difficult errors to model. The Omega signals effectively propagate in the waveguide formed by the earth's surface and the D-region of the ionosphere. Therefore it is the conditions of these two surfaces which primarily effect Omega phase lag errors. The height of the ionosphere is primarily effected by solar illumination, which gives a pronounced diurnal component to the Omega phase lag, and the anisotropy of the ionosphere is effected by the geomagnetic field. The earth's surface conductivity along the signal path also has a significant effect on the phase lags. To a large extent these factors can be modelled and compensated for, as described in references [7] and [8]. These are known as the "Phase Propagation Corrections", or PPCs.

Of course these and other factors, such as meteorological and geomagnetic, cannot be completely modelled, so significant propagation errors remain, as illustrated in figure 4, which shows the remaining error in a PPC corrected phase difference measurement between two Omega stations in lanes, over 4 days. Except for receiver loss-of-lock error, which is handled by the prefilter, the Omega phase error is bounded. Except for lane slips, also handled by the prefilter, this error is continuous. Finally this error is also zero mean, and can therefore be modelled as first order Markov processes. This error has both a significant "bias" that is very slowly changing (but still zero mean in the long term) and a more rapidly changing component. Therefore the Omega phase error is modelled in MINS as the sum of two Markov processes: one with a very long correlation time and another with a relatively short one. This is actually a reduced state model, since significant diurnal and other periodic errors remain (which would require 2 states each). However, as shown in reference [3], the reduced state model described here is actually more effective.

Figure 4. Omega LOP Error.

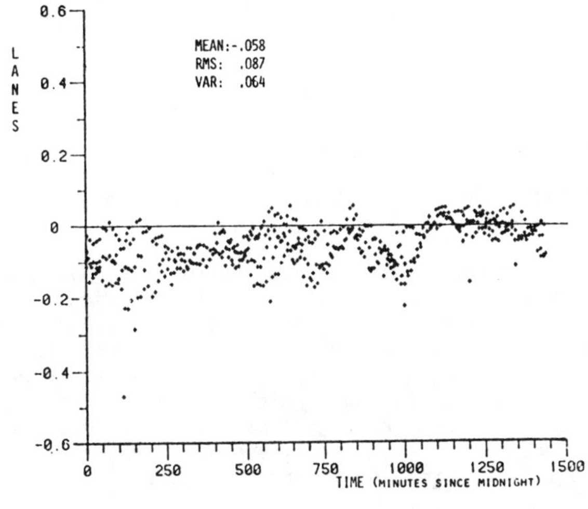

Loran-C Errors

Loran-C operates under the same general principle as Omega, but is higher frequency, shorter range, and pulse modulated rather than continuous. The time delay between arrival of pulses from different Loran-C stations (master-slave pairs) provides range difference information, or equivalently a hyperbolic line of position, similar to Omega. The Loran-C errors are also primarily phase lag errors. Corrections for these errors known as SPF, (Secondary Phase Factor) and ASF (Additional Secondary Factor) are also modelled and calculated to improve the accuracy of the time delay measurement, as described in reference [9]. Figure 5 shows typical errors in an ASF and SPF corrected Loran-C time delay measurement, converted to distance.

Loran-C also has a discontinuous error component known as cycle selection error, which is handled by the prefilter. Therefore the filter will see a continuous, bounded and zero-mean error, which is modelled as a first order Markov process.

Figure 5. Loran-C Time Delay Error.

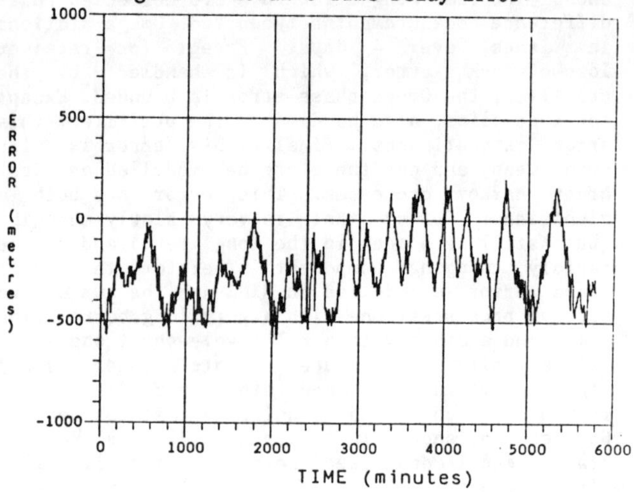

Transit Fix Errors

Transit is a satellite based navigation system, described in reference [10]. Transit satellites are in circular polar orbits, transmitting continuously at two very stable frequencies. The receiver obtains a position fix by measuring the Doppler shifts due to the relative motion. To do this the receiver must be provided velocity and height information. Any error in this velocity and height input will lead to an error in the Transit position fix.

The exact relationship between the velocity and height error input and the position error output is described in [11]. Figure 6 illustrates typical Transit/DR position errors, taken from the November 1987 sea trial. Aside from the known effect of velocity and height input errors, the Transit errors are largely random, zero mean, and independent of each other, and are treated by the MINS filter purely as measurement noise.

Figure 6. Transit/DR Position Errors.

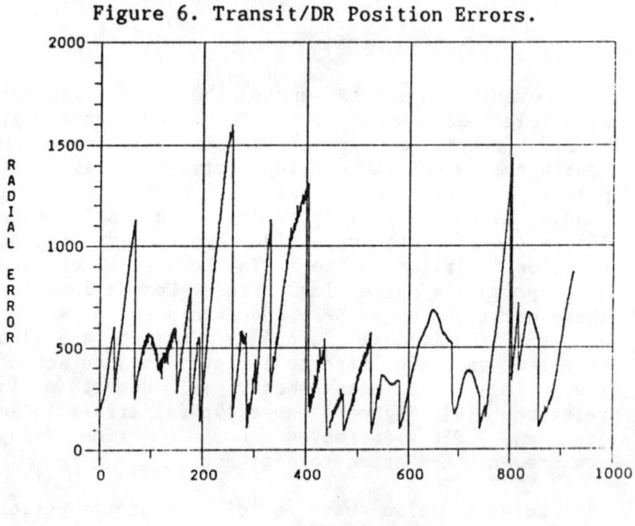

GPS Errors

The GPS satellite constellation currently provides about 8 hours per day of useful two dimensional coverage in the test area used for MINS evaluation. GPS satellites are in much higher orbits than Transit, broadcasting at higher frequencies, and pseudo ranging rather than Doppler fixing is used. These and other factors make GPS a much more accurate system. Reference [12] gives a good description of the GPS system and equipment.

GPS position error, even from a C/A code receiver, is generally significantly smaller than that from any other sensors used by MINS. Typical C/A code GPS position errors are shown in Figure 7, taken from the November 1987 sea trial. These position measurements have zero mean, bounded errors, which are nominally continuous. Because they are not observable however, these errors are treated by the filter simply as measurement noise. The expected magnitude of this error depends primarily on the geometric dilution of precision, which is generally provided by the receiver.

Figure 7. GPS Position Errors.

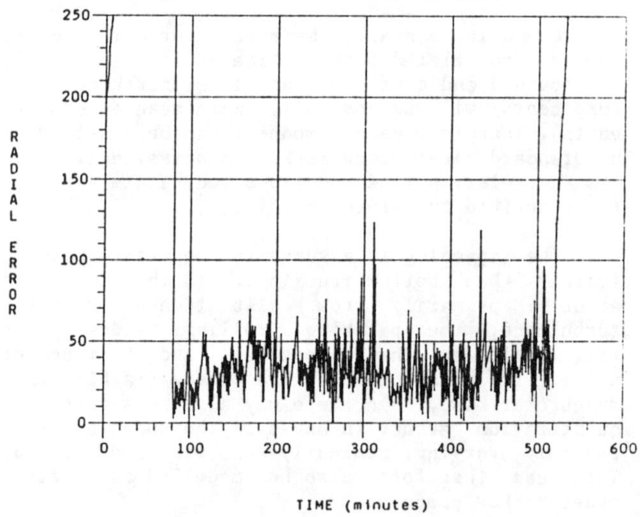

MINS KALMAN FILTER

State Vector

The MINS state vector is dynamically reconfigured depending on the number of Omega stations being used and whether or not the speedlog measurements are available. It has up to 17 error states to model the time correlated components of the position and velocity errors of the dead reckoning (DR) system, the phase errors from up to five Omega stations and the time delay errors from two Loran-C master-slave pairs. The state vector X is arranged as shown below in (3). These states are all modelled as stationary first order Markov processes, with the exception of the first two, which are obtained by integrating the speed and ocean current states.

Various other state space models were tested at DREO before finalising on this design, and the many MINS sea trials, from 1982 to 1988, have allowed refinement and verification of the error model parameters used in MINS.

$$
\begin{bmatrix} X1\\ X2\\ X3\\ X4\\ X5\\ \\ X6\\ X7\\ X8\\ X9\\ X10\\ X11\\ X12\\ X13\\ X14\\ X15\\ \\ X16\\ X17 \end{bmatrix}
\equiv
\begin{bmatrix} d\lambda\\ dL\\ dS\\ NOC\\ EOC\\ \\ \Omega B1\\ \Omega M1\\ \Omega B2\\ \Omega M2\\ \Omega B3\\ \Omega M3\\ \Omega B4\\ \Omega M4\\ \Omega B5\\ \Omega M5\\ \\ LM1\\ LM2 \end{bmatrix}
\begin{matrix} \text{DR North Position (Latitude) error}\\ \text{DR East Position (Longitude) error}\\ \text{Water Speed error}\\ \text{North Ocean Current}\\ \text{East Ocean Current}\\ \\ \text{Omega Station1 Bias error}\\ \text{Omega Station1 Markov error}\\ \text{Omega Station2 Bias error}\\ \text{Omega Station2 Markov error}\\ \text{Omega Station3 Bias error}\\ \text{Omega Station3 Markov error}\\ \text{Omega Station4 Bias error}\\ \text{Omega Station4 Markov error}\\ \text{Omega Station5 Bias error}\\ \text{Omega Station5 Markov error}\\ \\ \text{Loran-C LOP1 Markov error}\\ \text{Loran-C LOP2 Markov error} \end{matrix}
\qquad (3)
$$

The MINS optimal position and velocity are defined as the DR position and velocity, corrected by the Kalman filter estimates of the DR error states X1 to X5. Thus the filter is run open loop on the DR, allowing the pure DR solution to always be available for failure conditions. The slight nonlinearity that this causes in the Kalman filter measurement equation is dealt with by using an extended filter formulation.

Propagation Matrix

The continuous differential form of the MINS state propagation model can be shown explicitly as:

$$\dot{X} = F\,X + W \qquad (4)$$

where

F =

$$
\begin{bmatrix}
\begin{matrix} 0 & 0 & \cos\theta & 1 & 0\\ 0 & 0 & \sin\theta & 0 & 1\\ 0 & 0 & -\beta s & 0 & 0\\ 0 & 0 & 0 & -\beta c & 0\\ 0 & 0 & 0 & 0 & -\beta c \end{matrix} & & \\
& \begin{matrix}
-\beta b & 0 & 0 & 0 & 0 & 0 & 0 & 0 & 0 & 0\\
0 & -\beta m & 0 & 0 & 0 & 0 & 0 & 0 & 0 & 0\\
0 & 0 & -\beta b & 0 & 0 & 0 & 0 & 0 & 0 & 0\\
0 & 0 & 0 & -\beta m & 0 & 0 & 0 & 0 & 0 & 0\\
0 & 0 & 0 & 0 & -\beta b & 0 & 0 & 0 & 0 & 0\\
0 & 0 & 0 & 0 & 0 & -\beta m & 0 & 0 & 0 & 0\\
0 & 0 & 0 & 0 & 0 & 0 & -\beta b & 0 & 0 & 0\\
0 & 0 & 0 & 0 & 0 & 0 & 0 & -\beta m & 0 & 0\\
0 & 0 & 0 & 0 & 0 & 0 & 0 & 0 & -\beta b & 0\\
0 & 0 & 0 & 0 & 0 & 0 & 0 & 0 & 0 & -\beta m
\end{matrix} & \\
& & \begin{matrix} -\beta L & 0\\ 0 & -\beta L \end{matrix}
\end{bmatrix}
\qquad (5)
$$

where θ is the ship's heading and βi (i=s,c,b,m,L) are the inverses of the error state correlation times τi

$$\beta i = 1/\tau i \qquad i=s,c,b,m,L \qquad (6)$$

The magnitude of the driving noise (needed to propagate the covariance matrix) is described by the power spectral density matrix of W. This is a simple diagonal matrix with elements

$$Q(\tau) = \text{diag}[\ 0\ 0\ qs\ qc\ qc\ \ qb\ qm\ qb\ qm\ qb\ qm\ \ qb\ qm\ qb\ qm\ \ qL\ qL\ ] \qquad (7)$$

where

$$qi = 2pi/\tau i \qquad i=s,c,b,m,L \qquad (8)$$

Evidently this system is driven by 15 Markov processes, (5 different types), each of which is stochastically described by it's correlation time τi and it's steady state covariance pi.

For implementation on a digital computer the discrete state transition model is required, which is of the form:

$$X(t+\Delta t) = \Phi(\Delta t,t)\,X(t) + W(\Delta t) \qquad (9)$$

where the discrete noise vector $W(\Delta t)$ is not the same as the continuous noise process W of equation (4). The discrete driving noise covariance matrix $Q(\Delta t)$ is

$$Q(\Delta t) \equiv \int_t^{t+\Delta t} \Phi(\Delta t,\tau)Q(\tau)\Phi^T(\Delta t,\tau)\,d\tau \qquad (10)$$

which is evaluated in reference [13].

The transition matrix Φ is required by the MINS filter to extrapolate both the state vector estimate and its covariance matrix according to the standard Kalman filter equations. In the stationary case (where **F** is independent of time), or for small Δt, Φ is related to the system dynamics matrix **F** as follows:

$$\Phi(\Delta t,t) \simeq e^{F(t)\Delta t} \qquad (11)$$

The only nondiagonal block of the transition matrix Φ is for the DR states. Exponentiating the DR block of $F\Delta t$ can be done exactly, using eigenvectors as in reference [1], which yields:

$$
\Phi_{DR} = \begin{pmatrix}
1 & 0 & A\cos\theta & B & 0\\
0 & 1 & A\sin\theta & 0 & B\\
0 & 0 & e^{-\Delta t/\tau_s} & 0 & 0\\
0 & 0 & 0 & e^{-\Delta t/\tau_c} & 0\\
0 & 0 & 0 & 0 & e^{-\Delta t/\tau_c}
\end{pmatrix} \qquad (12)
$$

where

$$A = \tau_s\left(1 - e^{-\Delta t/\tau_s}\right) \qquad B = \tau_c\left(1 - e^{-\Delta t/\tau_c}\right) \qquad (13)$$

The dependence of this Φ_{DR} matrix on the ship's heading θ makes this a non-stationary system. The variable components of Φ_{DR} are however strictly bounded, and the only consequence is that the propagation period Δt should be kept small enough that $\sin\theta$ and $\cos\theta$ do not change significantly over a single propagation period.

The Omega and Loran-C blocks are found by exponentiating the corresponding diagonal blocks of Φ, which yields simply diagonal elements:

$$\phi_i = e^{-\Delta t/\tau_i} \qquad i=b,m,L \tag{14}$$

If the speedlog fails however, the speed error state is replaced by a full speed state which is propagated as a random walk (by replacing $-1/\tau_s$ with 0 in the **F** matrix) so that in the absence of measurements it will not decay as a Markov process would. However the corresponding covariance is propagated as a Markov covariance so that it is not unbounded, as a random walk covariance is. This hybrid state conforms to the physical reality, and is easily implemented.

Measurement Vector

Generally the measurement vector Z is some function of the state vector:

$$Z = H(X) + V \tag{15}$$

with some additive noise V. Although the implementation method used in MINS de-correlates the measurements and processes them individually as scalars, for the purpose of discussion it can be arranged in a vector as follows:

$$Z = \begin{bmatrix} Z\Omega1 & \text{Measured - Predicted Omega LOP1} \\ Z\Omega2 & \text{Measured - Predicted Omega LOP2} \\ Z\Omega3 & \text{Measured - Predicted Omega LOP3} \\ Z\Omega4 & \text{Measured - Predicted Omega LOP4} \\ & \\ ZL1 & \text{Measured - Predicted Loran-C LOP1} \\ ZL2 & \text{Measured - Predicted Loran-C LOP2} \\ & \\ ZT\lambda & \text{Measured - Predicted Transit } \lambda \\ ZTL & \text{Measured - Predicted Transit L} \\ & \\ ZGPS\lambda & \text{Measured - Predicted GPS } \lambda \\ ZGPSL & \text{Measured - Predicted GPS L} \\ ZGVN & \text{Measured - Predicted GPS VN} \\ ZGVE & \text{Measured - Predicted GPS VE} \\ & \\ ZO\lambda & \text{Measured - Predicted Operator Fix } \lambda \\ ZOL & \text{Measured - Predicted Operator Fix L} \\ & \\ ZC\xi & \text{Measured - Pred. Celestial Altitude} \end{bmatrix} \tag{16}$$

where "predicted" means calculated from the DR position and velocity.

The filter actually processes the measurement innovation, or residual ν, which is the difference between Z and it's expected value:

$$\nu \equiv Z - E\{Z|\hat{\zeta}\} \tag{17}$$

where ζ is the space of all previous measurements. If equation (15) is slightly nonlinear, as with MINS, then an "extended" Kalman filter should be used. This primarily amounts to using the filter estimate of position and velocity instead of the DR values for the "predictions" in (16).

Measurement Matrix

Although the innovations used in the state vector update equation are formed in their full nonlinear form, as shown in (17), a linearized relationship is needed to update the covariance matrix and to form the filter gain matrix. This is described by the measurement matrix **H**, where

$$Z = HX + V \tag{18}$$

From (18) we see that the inner product of the ith row of **H** with the state vector estimate X is the expected measurement component Z_i. Since the individual Omega station errors are modelled while the lines of position (LOPs) formed from pairs of stations are measured, the form of the measurement matrix will depend on the LOP selection. The general form of the measurement matrix can be given as follows:

$$H = \left[\begin{array}{ccccccccccccccccc}
\multicolumn{2}{c}{\boxed{H_{\Omega1}}} & 0 & 0 & 0 & \multicolumn{10}{c}{\boxed{H_{\Omega2}}} & 0 & 0 \\
 & & 0 & 0 & 0 & & & & & & & & & & & 0 & 0 \\
 & & 0 & 0 & 0 & & & & & & & & & & & 0 & 0 \\
 & & 0 & 0 & 0 & & & & & & & & & & & 0 & 0 \\
\cos\Psi s1 - \cos\Psi m & \sin\Psi s1 - \sin\Psi m & 0 & 0 & 0 & 0 & 0 & 0 & 0 & 0 & 0 & 0 & 0 & 0 & 0 & 1 & 0 \\
\cos\Psi s2 - \cos\Psi m & \sin\Psi s2 - \sin\Psi m & 0 & 0 & 0 & 0 & 0 & 0 & 0 & 0 & 0 & 0 & 0 & 0 & 0 & 0 & 1 \\
1 & 0 & F\cos\theta + G\sin\theta & F & G & 0 & 0 & 0 & 0 & 0 & 0 & 0 & 0 & 0 & 0 & 0 & 0 \\
0 & 1 & J\cos\theta + K\sin\theta & J & K & 0 & 0 & 0 & 0 & 0 & 0 & 0 & 0 & 0 & 0 & 0 & 0 \\
1 & 0 & 0 & 0 & 0 & 0 & 0 & 0 & 0 & 0 & 0 & 0 & 0 & 0 & 0 & 0 & 0 \\
0 & 1 & 0 & 0 & 0 & 0 & 0 & 0 & 0 & 0 & 0 & 0 & 0 & 0 & 0 & 0 & 0 \\
0 & 0 & \cos\theta & 1 & 0 & 0 & 0 & 0 & 0 & 0 & 0 & 0 & 0 & 0 & 0 & 0 & 0 \\
0 & 0 & \sin\theta & 0 & 1 & 0 & 0 & 0 & 0 & 0 & 0 & 0 & 0 & 0 & 0 & 0 & 0 \\
1 & 0 & 0 & 0 & 0 & 0 & 0 & 0 & 0 & 0 & 0 & 0 & 0 & 0 & 0 & 0 & 0 \\
0 & 1 & 0 & 0 & 0 & 0 & 0 & 0 & 0 & 0 & 0 & 0 & 0 & 0 & 0 & 0 & 0 \\
R\cos\gamma & R\sin\gamma & 0 & 0 & 0 & 0 & 0 & 0 & 0 & 0 & 0 & 0 & 0 & 0 & 0 & 0 & 0
\end{array}\right] \tag{19}$$

Here the rows of **H** are blocked off as the measurement vector of equation (16) (corresponding to different subsystems) and the columns are blocked off as the state vector of equation (3) (corresponding to subsystems whose errors are being estimated: DR, Omega and Loran-C).

In equation (19) the Ψi for i=s1,s2,m are the bearing angles (clockwise from north) from the ship to the appropriate Loran-C slave and master stations. The angle γ is the predicted azimuth of the celestial object whose sextant altitude ξ is used, and R is the earth's radius of curvature. The values of F,G,J and K are the sensitivities of the Transit position fix to the error in the velocity components fed into the Transit receiver, as calculated by a detailed mathematical model.

Sample Omega submatrices can be given for a particular station selection: where LOPi is formed from station i minus station i+1, for i=1,2,3,4):

$$H_{\Omega1} = \begin{bmatrix} \cos\Psi2-\cos\Psi1 & \sin\Psi2-\sin\Psi1 \\ \cos\Psi3-\cos\Psi2 & \sin\Psi3-\sin\Psi2 \\ \cos\Psi4-\cos\Psi3 & \sin\Psi4-\sin\Psi3 \\ \cos\Psi5-\cos\Psi4 & \sin\Psi5-\sin\Psi4 \end{bmatrix} \tag{20}$$

and

$$H_{\Omega 2} = \begin{bmatrix} 1 & 1 & -1 & -1 & 0 & 0 & 0 & 0 & 0 & 0 \\ 0 & 0 & 1 & 1 & -1 & -1 & 0 & 0 & 0 & 0 \\ 0 & 0 & 0 & 0 & 1 & 1 & -1 & -1 & 0 & 0 \\ 0 & 0 & 0 & 0 & 0 & 0 & 1 & 1 & -1 & -1 \end{bmatrix} \quad (21)$$

Here the Ψi are the bearing angles from the ship to Omega station i. The form of $H_{\Omega 1}$ is derived in reference [1] and is simply the geometric relationship between a line of position error and the resulting north and east error (similar to the Loran-C block of (19)).

FILTER IMPLEMENTATION

The details of the MINS filter implementation are important because they determine the stability and accuracy of the filter, as well as the computational efficiency, which in turn can determine the maximum rate at which measurements can be processed. In particular the propagation of the covariance matrix is the most critical factor determining computational efficiency.

Covariance Propagation Method

Prior to July 1987, the UDU^T formulation was used to update and propagate the state vector and covariance matrix, as described in reference [14] on pages 76-81 and 124-129. This is a very accurate and stable formulation which is quite efficient for updating the filter with the measurements, but is not very efficient for propagating the covariance matrix between measurements. The operation count for this propagation routine (modified weighted Gram-Schmidt) varies as the cube of the rank of the state vector n. When n was 15, before an additional Omega LOP was added to MINS in 1987, this was already the limiting factor in determining the filter update rate.

Fortunately Bierman had already approached this problem (page 147 of [14]) and building on the work of Agee and Turner [15], Bierman developed an efficient method which can take advantage of certain special forms of Φ and Q. The MINS Φ matrix can be blocked off to satisfy the requirement exactly, but not the discrete Q matrix. However the continuous Q for MINS does meet the requirement, suggesting that for sufficiently small Δt the discrete model should be an adequate approximation. In reference [13] this was shown to be true.

ERROR DETECTION

A crucial aspect of MINS is the extensive error detection that is performed on the sensor measurements by the prefilter and the filter to detect different types of failure. The following briefly lists some of these tests and the action(s) taken upon failure.

Gyrocompass

The gyro measurement is rejected if (1) the gyro data is not available, or (2) the gyro heading fails the continuity test.

The gyrocompass is declared to be a Failed Sensor if (1) the operator has disabled it, or (2) the gyro measurement has been rejected continuously for > 30 minutes.

If a gyro measurement is rejected, then the last accepted value is used. If the gyrocompass is declared to be a failed sensor then the heading is obtained from the best available alternate source (GPS, Loran-C or Omega).

Speedlog

The speedlog measurement is rejected if (1) the speedlog data is not available, (2) the speedlog measurement exceeds the ship's maximum speed, or (3) the continuity test fails.

The speedlog is declared to be a Failed Sensor if (1) the operator has disabled it, or (2) the speedlog measurement has been rejected continuously for >5 minutes.

If a speedlog measurement is rejected, then the last accepted value is used. If the speedlog sensor has failed then the filter is reconfigured to optimally estimate speed from the aiding sensor measurements.

Omega

The Omega LOP measurement is rejected if (1) the Omega data is not available, (2) it's prefilter reasonableness test fails (3) it's Kalman filter residual test fails (4) it's residual average test fails, or (5) it's residual slope test fails.

The purpose of the residual slope test is to detect loss of lock. This would otherwise cause the residual to ramp, and because of the automatic lane selection this ramp would periodically cross zero, be accepted, and try to drag the filter off.

The Omega LOP is declared Failed if (1) the operator has disabled one of it's stations, (2) it's PPC is not available, or (3) the ship is in a region of likely modal interference or "wrong-way-path" for one of it's stations.

Condition 2 only occurs at initialisation, because the PPC task takes several minutes to calculate each PPC. Condition 3 is unlikely to ever occur because the stations are automatically selected every 30 minutes with these conditions being avoided, as described in reference [16].

Loran-C

The Loran-C measurement is rejected if (1) the data is not available, (2) the ship is too close to a baseline extension, or (3) the filter residual test fails.

A Loran-C time delay is declared to be Failed if the operator has disabled it. Also a warning is issued when a Loran-C cycle selection error is detected. Since this error is automatically corrected, it is not necessary to reject the measurement. It is however an indication of poor signal shape, likely due to skywave interference, and a change of station selection should be considered.

Transit

The Transit measurement is rejected if (1) the data is not available, (2) time since the fix mark exceeds 20 min. or (3) one component fails the filter residual test.

The Transit receiver is declared failed if the operator has disabled it. Since the north and east components are not independent, if either one fails then neither are processed. The time limit is to safeguard against the possible association of a position fix with the incorrect fix mark.

GPS

The GPS position measurement is rejected if (1) the data is not available, (2) one component fails the filter residual test.

The GPS velocity measurements are rejected if (1) the data is not available, (2) one component fails the filter residual test, or (3) the GPS velocity exceeds the ship's maximum.

The GPS receiver is declared to be a failed sensor if the operator has disabled it. As with Transit the north and east position components are not independent. The velocity components are treated independently of position but are similarly linked to each other.

Operator Entered Celestial or Position Fix

An operator entered fix is never rejected by MINS. Before the fix is processed however, the distance from the entered fix to the MINS position is displayed to the operator, who is required to either abort the fix or confirm it's acceptability.

MEASUREMENT CORRECTION

Some of the measurement corrections applied by MINS are listed here:

(1) Omega phase propagation corrections (PPCs).
(2) Omega **lane counts** are automatically corrected.
(3) At every good Transit fix the Omega LOPs are tested for **lane slips** based on the Transit fix position.
(4) Loran-C phase corrections (SPF and ASF).
(5) Loran-C **cycle selection errors** are detected and corrected primarily by a continuity test, but also at each Transit fix.
(6) The effect of velocity errors on transit position fixes is calculated. This is a function of the satellite direction of travel, which unfortunately some Transit receivers do not correctly supply (including the MX1105). If this direction of travel is available then the measurement can be compensated, otherwise the magnitude of the velocity error induced position error can be calculated and used in the measurement noise covariance matrix.
(7) The usual refraction and height of eye corrections are automatically applied to any operator entered sextant measurement.

ADAPTIVE SCHEMES

Some of the adaptive schemes used by MINS are listed here:

(1) The filter adapting scheme for speedlog failure has been described above. In this case the state transition model is redefined, (Φ and Q) and the state vector estimate X and it's covariance matrix P are adjusted.

(2) Speedlog re-acquisition is handled in a way very similar to speedlog failure, allowing the filter to restore itself to its error state formulation.

(3) In the "full speed state" mode, where the ship's speed is modelled as a random walk, a manoeuvre detection scheme is in place to adapt to sudden changes in speed that are likely to be associated with a manoeuvre. After any sudden substantial manoeuvre the speed estimate covariance $P_{3,3}$ is boosted to increase the measurement gain, thereby helping the speed state X3 react to any change in speed.

(4) When a good Transit fix fails the residual test and Loran-C is being used, then the Loran-C cycle slip test is performed with this Transit fix. If this indicates that a cycle slip has gone undetected then the Loran state vector elements are reinitialised, the position error covariance elements are boosted and the Transit fix is accepted.

(5) When Omega stations are reselected, the state vector and covariance matrix are rearranged so as to preserve the error estimates and covariances of any stations that have not been dropped.

SAMPLE RESULTS

In the course of developing and improving MINS, formal sea trials were periodically conducted to verify the accuracy and operability of the system. The 1983 and 84 sea trial results are presented in reference [5], and are summarised in table 1, along with 1987 and preliminary 1988 sea trial results. Table 1 shows general agreement between simulation and seatrials. The apparent exceptions are due to Loran cycle selection errors in 83 and very poor Omega geometry in 1984.

Table 1. Summary of MINS & Subsystem Accuracy.

	68% RADIAL POSITION ERROR IN METRES				
	1982 SIMULATE	1983 SEA TRIAL	1984 SEA TRIAL	1987 SEA TRIAL	1988 SEA TRIAL
MINS	370	254	380	138	112
Loran-C	770	3445	335	204	185
Transit/DR	1700	1533	1575	571	1217
Omega	2060	2209	50K+	8641	8895
DR	8780	6775	7540	5672	34143
GPS/DR	NA	NA	NA	668	2666

It should also be noted that the DR error depends largely on the time between resets, which was very long in 1988. Each of these trials embody about a week of data, with precise reference data provided by shore based transponders (Syledis or Maxiran). The exception is the 1987 trial (the first to use GPS), for which only about 16 hours of Syledis reference data was available.

This 1987 data was also rerun through the MINS software in the lab with various subsystems disabled to evaluate failure mode performance, as summarised in Table 2. In almost all cases, this shows MINS to be substantially better than the best input provided to it. The two exceptions (DR/Loran/Ω and DR/GPS/Ω) are due to the effect of large Omega errors.

Table 2. 1987 Sea Trial Reruns (vs. Syledis).

68%	RMS	95%	SENSORS USED
MINS RERUNS:			
66	86	184	GYRO/Log/Loran/GPS/Transit/
71	91	187	GYRO/Log/Loran/GPS/Transit/Ω
128	120	256	GYRO/Log/Loran/GPS/ /
130	132	252	GYRO/Log/Loran/ /Transit/
144	142	261	GYRO/Log/Loran/ /Transit/Ω
167	152	323	GYRO/Log/Loran/GPS/ /Ω
207	206	330	GYRO/Log/Loran/ / /
218	252	506	GYRO/Log/Loran/ / /Ω
247	356	752	GYRO/Log/ /GPS/Transit/
371	427	849	GYRO/Log/ /GPS/Transit/Ω
544	553	1051	GYRO/Log/ / /Transit/
233	619	1712	GYRO/Log/ /GPS/ /
554	621	1257	GYRO/Log/ / /Transit/Ω
1353	1243	2354	GYRO/Log/ /GPS/ /Ω
3278	3402	7193	GYRO/Log/ / / /Ω
5672	4915	6411	GYRO/Log/ / / /
102	103	188	GYRO/ /Loran/GPS/Transit/
120	133	291	GYRO/ /Loran/GPS/Transit/Ω
180	161	338	GYRO/ /Loran/GPS/ /Ω
195	192	348	GYRO/ /Loran/ /Transit/Ω
1250	1324	2897	GYRO/ / /GPS/Transit/Ω
MINS REALTIME:			
138	125	211	GYRO/Log/Loran/GPS/Transit/
SUBSYSTEMS:			
5672	4928	6411	DR
8641	70411	>100K	Omega
571	608	1205	Transit/DR
668	1127	1973	GPS/DR
204	198	325	Loran-C

Representative radial error time histories are shown in Figures 8, 9 and 10, from the reruns with all sensors, all but Loran, and all but the speedlog respectively. These figures also show the EEP (estimated error probable) which is the 95% radial position accuracy estimate derived from the Kalman filter covariance matrix. It is clear from these figures that the MINS EEP does provide a very good statistical indication of MINS accuracy under diverse conditions. Besides confirming the accuracy of this EEP, which is very useful to the navigator, this also furnishes strong evidence that the filter error models are realistic. (Agreement between filter residual and residual variance also provides evidence of MINS error model fidelity.)

Figure 8. MINS Error & EEP With All Sensors.

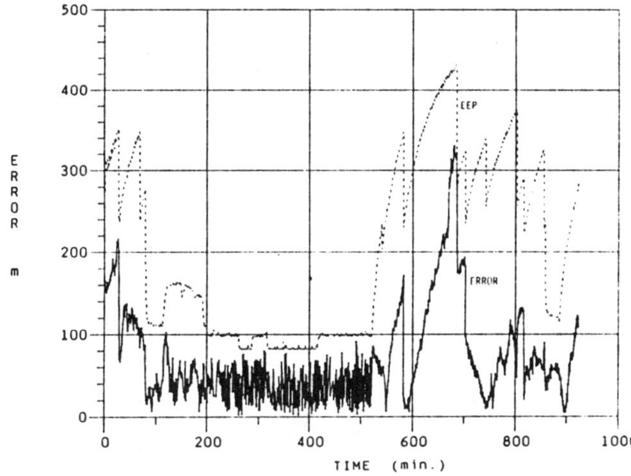

Figure 9. MINS Error & EEP Without Loran.

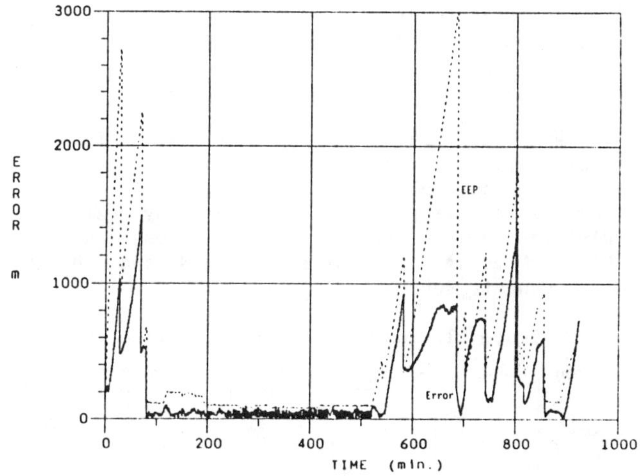

Figure 10. MINS Error & EEP Without Speedlog.

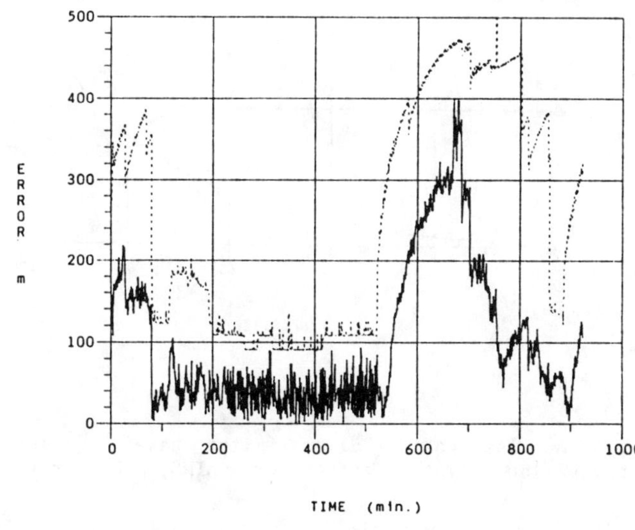

The full 1987 data set (20,000 samples over 110 hours) was run through the MINS system in the lab with Loran disabled, so that Loran could be used as reference. This allowed the "mid-ocean" performance of MINS to be evaluated with a larger data set. Where possible the Syledis and GPS data were used to verify the Loran accuracy. Table 3 summarises these results. Since Loran-C is used as reference there is about 200 metres uncertainty in these results. It should be mentioned that during these reruns the DR position was reinitialised every 16 hours, and therefore did not drift as much as it otherwise would have. Most of the GPS/DR error is due to DR error when GPS is not available.

Table 3. 1987 Sea Trial Reruns (vs. Loran).

68%	RMS	95%	SENSORS USED
MINS RERUNS:			
762	894	1836	GYRO/Log/ /Transit/
849	1070	2166	GYRO/Log/ /Transit/Ω
512	817	1691	GYRO/Log/GPS/Transit/Ω
SUBSYSTEMS:			
7910	7343	13891	DR
5214	20975	34748	Omega
1827	3363	9034	GPS/DR
1008	1138	2441	Transit/DR

With this large data set the MINS performance is statistically superior to that of any subsystem, as is expected of an optimally integrated system, and has been repeatedly demonstrated by seatrials. This can be seen more graphically from Figure 11, which shows all percentiles of MINS and the two most accurate subsystems.

Figure 11. Position Error Percentiles.

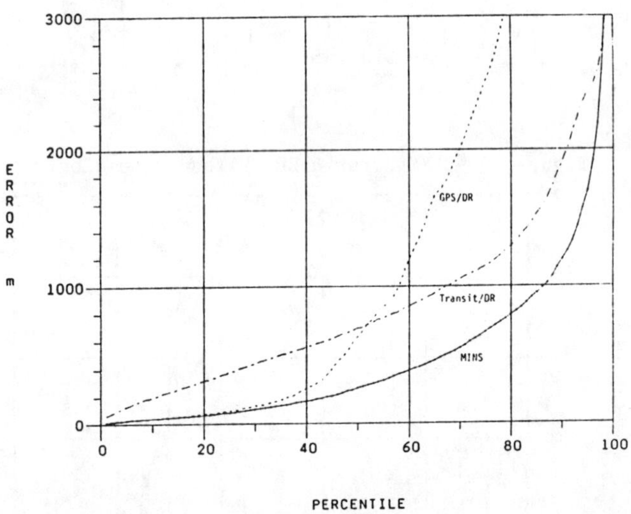

CONCLUSION

A low cost, Kalman filter based, marine navigation system integrator, MINS, has been developed and proven to significantly improve the navigational accuracy, reliability and efficiency onboard major vessels. Work has now begun to expand and reconfigure MINS for optimal integration on dual-inertial equipped ships, with full sensor level and system level failure detection, isolation and reconfiguration (FDIR).

REFERENCES

[1] J.C. McMillan, "A Kalman Filter for Marine Navigation", MPhil thesis, U. of Waterloo, 1980.

[2] D.F. Liang & J.C. McMillan, "Development Analysis of a Marine Integrated Navigation System", Proceeding of IEEE PLANS 82, December 1982.

[3] D.F. Liang, J.C. McMillan et el, "Low Cost Integrated Marine Navigation System", Navigation, Journal of the Institute of Navigation, Vol. 30, No.4, Winter 1983-84.

[4] D.F. Liang, J.C. McMillan & C.A. Maskell, "Design and Test Evaluation of a Marine Integrated Navigation System", Proceedings of the Institute of Navigation National Technical Meeting, San Diego, January 1984.

[5] C.A. Maskell, "System Analysis and Sea-Trial Results of the DREO Marine Integrated Navigation System (MINS)", DREO Report No. 943, Feb. 1986.

[6] Applied Optimal Estimation, A. Gelb, ed., The Analytic Sciences Corporation, M.I.T. Press, 1974.

[7] P.B. Morris and R.R Gupta, "Omega Navigation Signal Characteristics", Navigation, Journal of the Institute of Navigation, Vol. 33, No. 3, Fall 1986.

[8] P.B. Morris and M.Y. Cha, "Omega Propagation Corrections: Background and Computational Algorithm", NTIS Report No. ONSOD-01-74, 1974.

[9] D.H. Gray, "The Preparation of Loran-C Lattices for Canadian Charts", The Canadian Surveyor, Vol. 34, No. 3, September 1980.

[10] T.A. Stansell, "The Transit Navigation Satellite System- Status, Theory, Performance, Applications", Magnavox Government and Industrial Electronics Company, Report No. R-5933, 1978.

[11] J.C. McMillan, "Design of an Optimally Integrated Primary Land Navigation System", DREO Report 946, Sept. 1986.

[12] GPS SPECIAL ISSUE, Navigation, Journal of the Institute of Navigation, Vol. 25, No.2, 1978.

[13] J.C. McMillan, "Technical Description of the DREO Marine Integrated Navigation System (MINS)", DREO Report, 1988.

[14] G.J.Bierman, Factorisation Methods for Discrete Sequential Estimation, Academic Press, 1977.

[15] W.S. Agee, R.H. Turner, "Triangular Decomposition of a Positive Definite Matrix Plus a Symmetric Dyad With Application to Kalman Filters", White Sands Missile Range Tech. Report No. 38,1972.

[16] J.C. McMillan, "Automatic Omega Station and LOP Selection", DREO Report 976, December 1987.

ACOUSTIC TRANSPONDER NAVIGATION

T. D. Henry

Western Electric

A B S T R A C T

The increasing interest in deep sea explorations has brought acoustic transponders into the forefront as an important navigation tool. Besides surface navigation, they are used to provide three-dimensional position information for subsurface vehicles and devices.

This paper addresses surface navigation using a pulsed system that operates on baselines which frequently approach four nautical miles. Some of the topics include deployment techniques and the navigation and calibration considerations that makes possible surface navigational accuracies of 2 to 5 meters.

INTRODUCTION

For a number of years Western Electric has been funded by the Navy to provide various hydrographic, oceanographic and acoustic surveys in numerous parts of the world. To provide these surveys has required varying degrees of precision navigation. Historically there have been such electronic systems as SHORAN, LORAC, DECCA, LAMDA DECCA and LORAN A - to only name a few. In more recent years, higher precision and longer range systems have evolved that are currently being used, such as: LORAN C, the Navy Transit Satellite System, acoustic transponders and integrated systems such as IDWNS (Integrated Deep Water Navigation System.) The purpose of this paper, however, will be to describe the applied usage of acoustic transponders as we at Western Electric currently use them in the on-going Navy program.

Acoustic transponders can provide highly accurate navigation fixes over a relatively small area. They often fill a gap where other existing navigation facilities fail in providing coverage or accuracy. In addition to surface navigation, they can provide three-dimensional fixes in the cases involving towed vehicles or other devices located somewhere between the surface and ocean floor. This paper, however, will primarily address their usage in surface navigation.

Transponders are frequently deployed in groups of three or more to provide the area of coverage required. Usable slant ranges tend to fall between 3 and 7 nautical miles depending on local bathymetry and weather conditions. The surface (horizontal) ranges, however, tend to average 3.5 nautical miles with a maximum of 5 nautical miles depending on depth.

Recent advances in the technologies of both transponders and the shipboard receiving equipment has raised the usable ranges as well as reducing the frequency of false returns and erroneous transponder interrogations. This improves working in areas of high acoustic noise, such as found in moderate to heavy seas. The most serious consideration to transponder navigation in previous years was local weather (sea state) conditions in which higher than normal components of acoustic noise exist. Along with the hardware advances, software and improved calibration techniques have reduced the positional uncertainty to where a given fix position is accurate in the 2 to 5 meter range, relative to the transponder net. If a satellite receiver is available, the Geodetic position of the transponder net can be determined within 25 to 40 meters after twenty or more carefully edited satellite passes.

OPERATION

The use of bottom mounted transponders generally consists of making various range measurements. An interrogation pulse from the ship (for the case of surface navigation) initiates the timing sequence including a shipboard timer or clock. When the interrogation pulse reaches the transponder, the transponder then determines if the incoming signal is a legitimate interrogation and, if so, answers with an acoustic pulse of its own. Once this signal is received by the shipboard equipment a similar validity check is made on the incoming signal before the timers are stopped. The nature of the validity checks vary from manufacturer to manufacturer. The interrogation and reply paths are not straight lines, but will usually be curved owing to refraction along the ray path. In most cases, the refraction distortion is minimal and often ignored to reduce computer storage requirements and computational difficulties. Additionally, the local velocity information is frequently sparce and may not be sufficient to perform more elaborate ray tracing. If the foregoing limitations do not apply, it would be worthwhile to model an appropriate velocity structure; particularly in areas where transponder depths vary considerably,

Reprinted from *Rec. IEEE Position Location and Navigation Symp.*, pp. 237-244, 1978.

and thus reduce localized grid distortions. Slant range can be represented by:

$$R = \tfrac{1}{2}V(T-t'\pm\delta t)$$

Where:

- R - distance along the interrogation path
- V - an average velocity in the water column above the transponder
- T - time reading of the clock or counters
- t´ - correction term representing hardware delays
- δt - a computed correction term owing to ships' movement between interrogation and reply

Since the quantity in parenthesis represents a two-way travel time, the whole expression is multiplied by one-half.

The shipboard equipment used by Western Electric has one millisecond resolution. The time base is derived from a one(1) Kilohertz source stabilized by a Cesium Beam Frequency standard. Hardware time measurement errors should not exceed one(1) millisecond.

TRANSPONDER POSITIONING

As with many other endeavors, successful transponder operation requires careful preparation. Certainly any hardware to be deployed in the ocean must be thoroughly scrutinized and every reasonable performance check made prior to deployment. Equally important is the choice of transponder locations. Generally, they are deployed in uniform geometric configurations to cover the area of interest. Distance (baseline) between transponders is normally selected between 2½ to 3½ nautical miles. This arrangement usually works satisfactorily on flat, or nearly flat ocean bottoms.

In areas having strong bathymetric features, or other forms of strong relief, special precautions must be taken. In these areas it is best to carefully study any existing bathymetric contour charts to avoid line-of-sight acoustic blocking. If this cannot be avoided, it may be necessary to deploy additional transponders to provide suitable navigation coverage when one or more transponders is in a shadow zone resulting from localized bottom topography. Depth recorders should be monitored continuously during deployment periods so that last minute decisions can be made as to the suitability of a proposed transponder site.

Whether the bottom is rough or flat, it is recommended that transponder baselines be held to a maximum of 3½ nautical miles. The only exception to this is the situation where the area of interest is small and can be entirely incorporated within the geometric figure formed by the transponders.

NAVIGATION FIXES

The most basic form of transponder navigation is that of using two transponders. For those who do not have computer aided automatic plotting capability, one of the best alternatives is to create plotting sheets having lines-of-position that represent slant ranges projected onto the surface or X-Y plane. These take the form of concentric circles. Figure 1 represents two transponders, A and B, and each circle represents a particular two-way travel time. The time annotation of the circles are round numbers with the requirement that they yield nearly evenly spaced circles. For hand-drawn systems of circles, it is easiest to assume a rectilinear coordinate system where the X-Y axis are in the surface plane, depth being the Z coordinate. Computerized drawing of these "Range Rings" may be programmed to take earths' curvature effects into consideration. These are usually very small over an area of a dozen or so square miles.

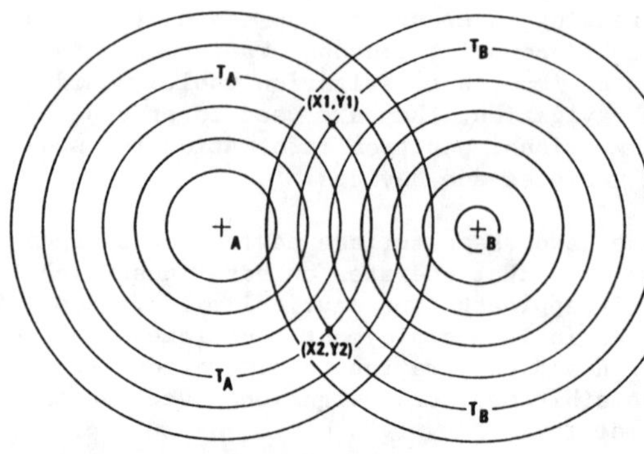

Figure 1

The radii of these circles is calculated from Pythagorean relationships such that:

$$R_a = \{SR_a^2 - Z_a^2\}^{\frac{1}{2}} \text{ and } R_b = \{SR_b^2 - Z_b^2\}^{\frac{1}{2}}$$

where

- R_a, R_b - respective distances (radii) from the locations of transponders A and B
- SR_a, SR_b - respective slant ranges for two-way travel times T_a and T_b
- Z_a, Z_b - respective depths of transponders A and B

where

$$SR = \tfrac{1}{2}VT$$

and

- V - an average velocity for the water column of a transponder
- T - two-way travel time

Obviously, the X, Y, Z coordinates of the transponders must be known (or assumed) prior to navigation or the drawing of range rings. The transponder calibration techniques will be discussed in a later section.

Referring again to Figure 1, it can be seen that for a given set of two-way travel times, two possible fix solutions result. Usually it is possible to determine which side of the baseline one is on by maneuvering the ship and observing the plotted heading versus the ships' compass heading.

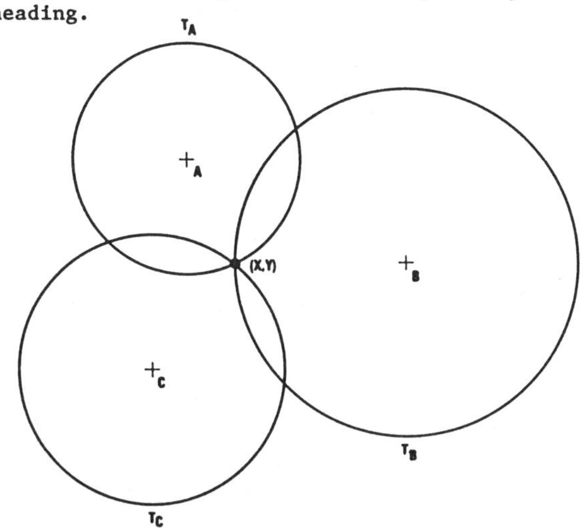

Figure 2

A better method, however, is to have a third transponder. Figure 2 shows how a unique position can be determined from a set of two-way travel times: T_a, T_b and T_c. Unfortunately, the usual case is for cumulative errors to produce not one solution, but three (in the case of three transponders) with the resultant spherical triangle of uncertainty. A computerized solution should pick a point that is best in the Least Squared Error sense. Graphical solutions must be satisfied with the human eye or "apparent average."

COMPUTER NAVIGATION

Computers have many advantages over the manual techniques of transponder navigation discussed in the previous section. Hand plotting more than two transponders quickly becomes tedious and subject to errors. Computers can make decisions and calculations quickly as to a "best fix," provide real-time error analysis of fix computations and maintain records for post analysis. A number of computational techniques are possible, but only one will be presented here.

ITERATIVE METHOD

The equations for two circles can be written as

$$f_1(X,Y) = (X-X_1)^2+(Y-Y_1)^2-R_1{}^2 = 0 \quad (1)$$

$$f_2(X,Y) = (X-X_2)^2+(Y-Y_2)^2-R_2{}^2 = 0 \quad (2)$$

Forming a truncated Taylor series for functions f_1 (X,Y) and f_2 (X,Y) suitable for an iterative scheme

$$f_1(X_i,Y_i)+\frac{\partial f_1}{\partial X}(X_{i+1}-X_i)+\frac{\partial f_1}{\partial Y}(Y_{i+1}-Y_i) = 0 \quad (3)$$

$$f_2(X_i,Y_i)+\frac{\partial f_2}{\partial X}(X_{i+1}-X_i)+\frac{\partial f_2}{\partial Y}(Y_{i+1}-Y_i) = 0 \quad (4)$$

To simplify notation let:

$$X' = X_{i+1}-X_i \text{ and } Y' = Y_{i+1}-Y_i$$

substituting in (3) and (4)

$$f_1(X_i,Y_i)+\frac{\partial f_1}{\partial X} X'+\frac{\partial f_1}{\partial Y} Y' = 0 \quad (5)$$

$$f_2(X_i,Y_i)+\frac{\partial f_2}{\partial X} X'+\frac{\partial f_2}{\partial Y} Y' = 0 \quad (6)$$

or,

$$\frac{\partial f_1}{\partial X} X'+\frac{\partial f_1}{\partial Y} Y' = -f_1(X_i,Y_i) \quad (7)$$

$$\frac{\partial f_2}{\partial X} X'+\frac{\partial f_2}{\partial Y} Y' = -f_2(X_i,Y_i) \quad (8)$$

Equations (7) and (8) are now in a form suitable for solving by determinants and the Newton-Raphson formulation becomes

$$X' = \frac{\begin{vmatrix} -f_1(X_i,Y_i) & \frac{\partial f_1}{\partial Y} \\ -f_2(X_i,Y_i) & \frac{\partial f_2}{\partial Y} \end{vmatrix}}{J} \qquad Y' = \frac{\begin{vmatrix} \frac{\partial f_1}{\partial X} & -f_1(X_i,Y_i) \\ \frac{\partial f_2}{\partial X} & -f_2(X_i,Y_i) \end{vmatrix}}{J}$$

Where J is the Jacobian of the system of simultaneous equations and is equal to:

$$J = \begin{vmatrix} \frac{\partial f_1}{\partial X} & \frac{\partial f_1}{\partial Y} \\ \frac{\partial f_2}{\partial X} & \frac{\partial f_2}{\partial Y} \end{vmatrix} \neq 0$$

Care must be taken to see that the Jacobian is not zero (or near zero) before attempting to calculate X′ and Y′.

Returning to the expressions for X′ and Y′

$$X' = X_{i+1}-X_i \text{ and } Y' = Y_{i+1}-Y_i$$

The iterative equations become:

$$X_{i+1} = X_i+X' \quad (9)$$

$$Y_{i+1} = Y_i+Y' \quad (10)$$

The order of computation is such that initial values are assigned to X_i and Y_i. X' and Y' are then computed and this establishes X_{i+1} and Y_{i+1}. When X' and Y' become equal to, or less than some remainder term (R_m), the iteration is stopped and the solution becomes the present value of X_{i+1} and Y_{i+1}. Figure 3 indicates the software flow diagram.

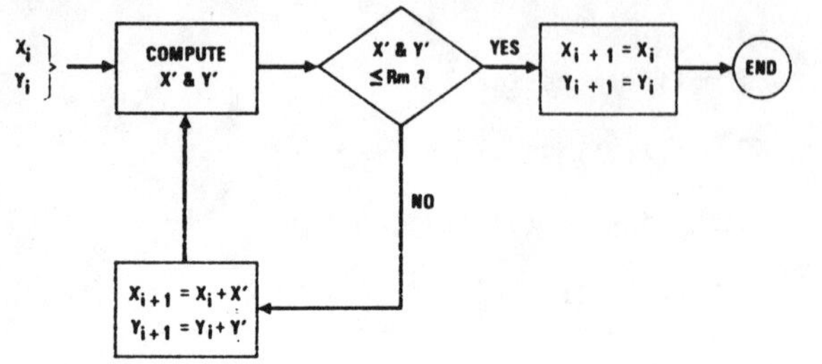

Figure 3

TRANSPONDER CALIBRATION

Before examining procedures to locate transponders, it is appropriate to discuss some velocity considerations. It is important to have access to recent data for a given area in which velocity is plotted versus depth; or in a tabular format of specific depths with corresponding sound velocities. This makes it possible to construct an average vertical velocity in the water column and has proven to be a convenient and accurate technique where more sophisticated results employing refraction correction are not available.

Once the transponders have been deployed and a choice made for sound velocity representation, one can begin determining where the transponders are relative to one another.

The usual procedure is to deploy two to four units initially, using the best existing navigation control. This is important to minimize misplacement and skew (orientation) errors.

The ships' coordinates at drop time are usually used as the "guess" positions for the beginning of the calibration. A choice is made as to which transponder one has the most confidence in as to position, and it is used as a fixed reference in the X-Y plane. Another transponder is also chosen so that the baseline joining it and the reference transponder form an angular reference. The purpose of this is to establish an orientation between two transponders, and all subsequent transponders, so that one can navigate a ship in a manner that the computed course will agree with the true course. If this is not true, then the transponder net is said to have a SKEW error and the X-Y positions of all transponders must be adjusted to eliminate this problem. The adjustment or net rotation around a reference point is usually made after the preliminary calibration of the first three or four transponders. More will be said about this later.

Basically, there are two procedures in current use. Procedure I requires the most time but yields highly accurate results. Procedure II also gives good results, but the time savings over Procedure I in large transponder nets of four or more is considerable.

PROCEDURE I

Referring to figure 4, the ship is driven in a "clover leaf" pattern while interrogating a transponder. The purpose of this is to find an absolute minimum two-way travel time. Theoretically, this occurs when the ship is directly over the top of the unit.. Often the ships' position is not accurately known, but by turning the ship each time a minimum is reached one can converge on a reasonably accurate absolute minimum.

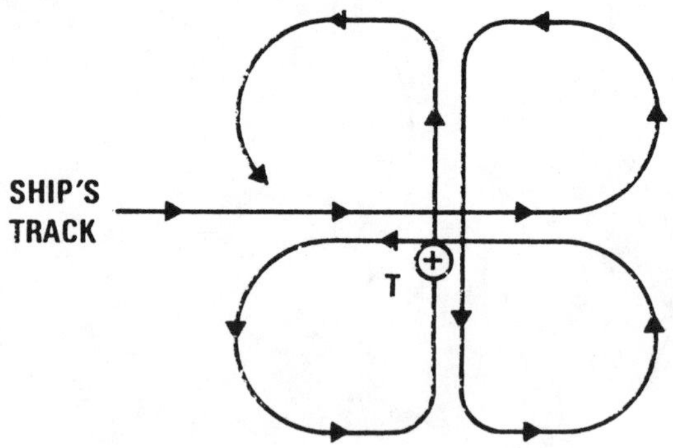

Figure 4

Once the minimum two-way travel time (T) has been determined, it is a simple matter to determine the depth and corresponding sound velocity.

Sound velocity tends to vary linearly with depth in deep water. From the general equation of a straight line, the linear relationship between depth and velocity can be used as follows:

$$d - d_1 = \frac{d_2 - d_1}{v_2 - v_1}(v - v_1) \tag{11}$$

where

d - depth
v - average velocity

and

$$d_2 > d > d_1$$
$$v_2 > v > v_1 \tag{12}$$

let

$$m = \frac{d_2 - d_1}{v_2 - v_1}$$

Knowing that

$$v = \frac{d}{T} \tag{13}$$

Where

$$T = \tfrac{1}{2} T_{min}$$

substituting

$$d-d_1 = m\left(\frac{d}{T} - v_1\right) \qquad (14)$$

rearranging

$$d = \frac{T(d_1 - mv_1)}{T-m} \qquad (15)$$

If in determining the depth and velocity terms of (13) and (15), either or both of the solutions violate the inequalities of (12), then new boundaries must be selected and (13) and (15) recalculated.

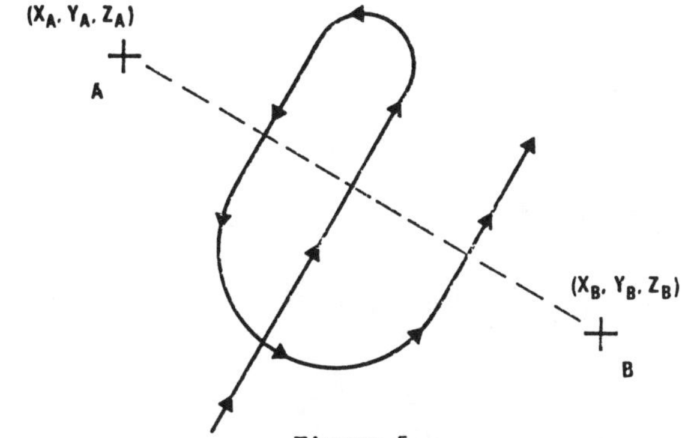

Figure 5

The next step is to determine the baseline lengths as they would be if projected onto the surface or X-Y plane. One method to do this is shown in Figure 5. The ship is maneuvered in such a manner that the baseline is crossed approximately midway and at a 90 degree angle. The transponders are interrogated about once every ten seconds and the ship is driven at approximately 5 Knots. When two-way travel times from both transponders have reached minimums, the baseline has been crossed. The minimums will occur simultaneously if the ships' track is normal to the baseline. The baseline length can then be calculated:

$$B_{AB} = \{(\tfrac{1}{2}V_A T_A)^2 - d_A^2\}^{\frac{1}{2}} + \{(\tfrac{1}{2}V_B T_B)^2 - d_B^2\}^{\frac{1}{2}} \qquad (16)$$

where

- V_A, V_B - average velocities for transponders A and B
- T_A, T_B - Two-way travel time minimums
- d_A, d_B - depths of transponders A and B

If the crossing(s) of the baseline was not normal to the baseline, the two-way travel time minimums will occur at different times. It is then necessary to make several calculations of B_{AB} between the two minimums of T_A and T_B until a minimum B_{AB} is determined.

Figure 6 depicts a typical situation involving three transponders. In this case, transponder A was chosen as a reference, and to simplify the calculations, the origin in the X-Y plane is moved to this position. The positions of A, B and C are the initial drop or "guess" positions. The B´ and C´ positions are the new, or calibrated positions derived geometrically from the three baselines. Note that B was moved to a new position, B´, but the angular orientation (θ) between A and B was not changed.

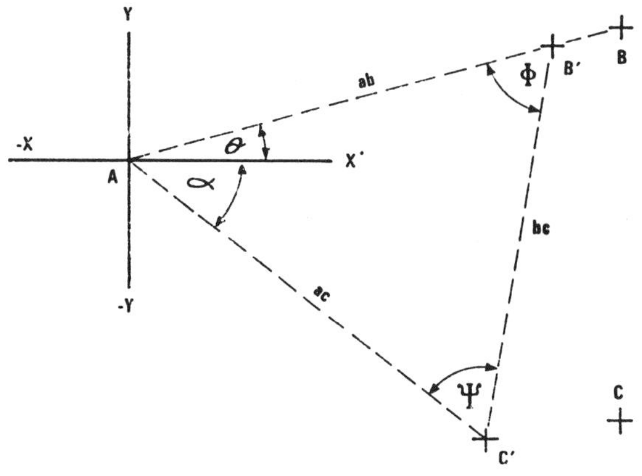

Figure 6

The procedure that has just been described can be extended to any number of transponders. However, for each additional transponder, a minimum of two additional baselines must be determined. If there were six (6) transponders, then nine (9) baselines and six (6) depth determinations would be required. It is easily seen that these procedures become very tedious and time consuming, particularly if sea state and wind conditions are affecting the ships' maneuvering. Another method will now be described.

PROCEDURE II

The basic difference between this technique and Procedure I is that a more sophisticated software algorithm is sought to solve for the transponder positions. A net of two transponders is established as in Procedure I, and these are used to provide navigation fix positions during the calibration of subsequent transponders. The actual working equations will not be developed here, but a generalized technique will be discussed.

Figure 7 depicts a starting point for the solution of the problem. Transponders A and B are assumed to be known relative to one another and it remains to determine the coordinates of C. The drop position is used for the initial position of C.

If the ship was driven to a point designated as P, it would be possible to fix this position as previously described using transponders A and B. At the same time a two-way travel time is available from C. The computed distance from the observation point P, to transponder C, becomes:

$$d_p = \{(X_p - X_c)^2 + (Y_p - Y_c)^2 + Z_c^2\}^{\frac{1}{2}} \qquad (17)$$

where

X_p, Y_p - ships' coordinates in X-Y plane

X_c, Y_c, Z_c - transponder (C) initial coordinates

Forming a residual (R_p) for transponder C at point P

$$R_p = d_p - VT_p \qquad (18)$$

where

V - initial average vertical velocity for transponder C

T_p - one-way travel time from transponder C

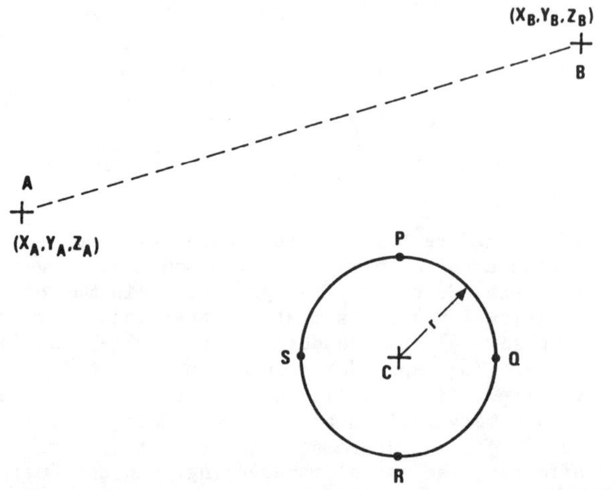

Figure 7

If the above sequence is repeated at other observation points such as Q, R, S as shown in Figure 7, then other remainder terms could be computed in a similar fashion to form R_Q, R_R and R_S. In practice, usually ten to twenty such observations are made although it is not necessary to move the ship to all these points. The ship is maneuvered to three or four positions, approximately equidistant and uniformly spaced around the un-calibrated transponder. Three to five observations are made at each location. The distance r is chosen as a workable range greater than the depth of transponder C.

From all the residuals produced at the various observation points it is possible to define an error function F such that:

$$F = \sum_{i=1}^{N} R_i^2 \qquad (19)$$

where

$$R_i = d_i - VT_i \qquad (20)$$

Obviously, if the transponder position (X_c, Y_c, Z_c) was exactly known, then F is equal to zero. The problem then becomes one in which to minimize F and thereby locate transponder C.

It would seem that some sort of technique is needed in which the gradient information is biased in some optimum fashion. Such a technique is available and is called the method of Steepest Descent. Without proof this method will be quickly summarized.

if

$$A\overline{X} = \overline{C} \qquad (21)$$

describes a system of equations that one desires to solve in an iterative fashion, then a residual at step K can be defined:

$$\overline{r}_K = \overline{C} - A\overline{X}_K \qquad (22)$$

Then it can be shown that the iterative sequence allowing rapid convergence is:

$$\overline{X}_{K+1} = \overline{X}_K + \lambda_K (\tfrac{1}{2}) \nabla_X F(\overline{X}_K) \qquad (23)$$

where

$F(\overline{X}_K)$ - squared risk function, at step K, such as defined in (19)

∇_X - gradient operator

λ_K - scale factor

The real trick here is choosing a scale factor that is optimal. There are a number of methods each having computational advantages. One example of a good choice is:

$$\lambda_K = \frac{\overline{r}_{K-1}^T \, \overline{r}_{K-1}}{\overline{r}_{K-1}^T A \overline{r}_{K-1}} \qquad (24)$$

Several calibration procedures have been outlined in this section and each has certain advantages in particular situations. Experience has proven that no one technique is suitable for all transponder nets because of the numerous environmental situations encountered.

After completing the relative positioning of transponders it remains to ensure that the heading or course, as computed from transponder fixes, agrees with other available information. In many of the ocean areas it is possible to maintain simultaneous plots from other navigation sources such as Loran C. If the two courses are not parallel, then it is necessary to rotate the entire transponder net until the two courses agree. In practice, one transponder is usually held fixed and the others pivoted around it. If, for example, the transponder net indicated a course of 050 degrees and a corresponding Loran C plot (visually smoothed) indicated a course of 045 degrees, then it would be necessary to rotate the net 5 degrees counterclockwise.

Figure 8 indicates the required procedure.

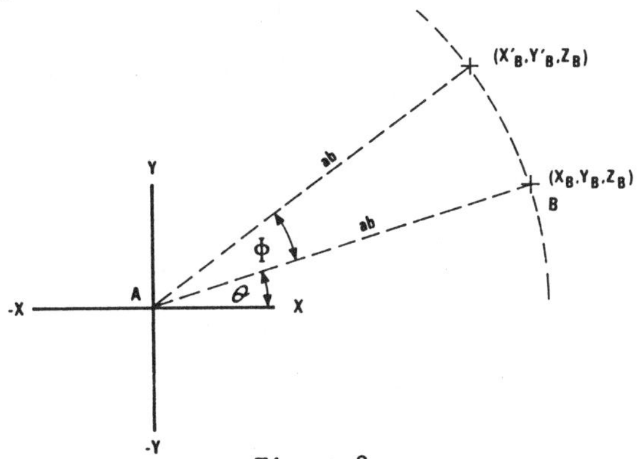

Figure 8

The new coordinates of transponder B become:

$$X'_B = (ab)\cos(\psi) \quad (25)$$

$$Y'_B = (ab)\sin(\psi) \quad (26)$$

for counterclockwise rotation

$$\psi = \theta + \phi \quad (27)$$

and clockwise rotation

$$\psi = \theta - \phi \quad (28)$$

where

ϕ - angle of rotation
θ - reference angle, A to B

It should be noted that the sign convention of a standard Cartesian coordinate system must be observed for the sin and cos terms. The other transponders, if they exist, are rotated in the same fashion.

The last topic to be discussed under calibration is that of Geodetic positioning of the transponder net. In mid-ocean areas this is best accomplished with satellite fixes.

The subject matter of determining quality satellite fixes is too extensive for this paper, but generally, twenty to thirty "good" satellite fixes will ensure transponder net geodetic accuracy in the 25 to 40 meter range. By correlating in time the satellite fixes versus those of the transponder net, average offsets can be computed.

Definitions:

ΔX_i = signed difference between the satellite fix and transponder position in X direction at some time T_i.

ΔY_i = signed difference between the satellite fix and transponder position in Y direction at some time T_i.

Then, the average values of ΔX and ΔY can be expressed as:

$$\overline{\Delta X} = \frac{1}{M}\sum_{i=1}^{M} \Delta X_i \quad (29)$$

$$\overline{\Delta Y} = \frac{1}{M}\sum_{i=1}^{M} \Delta Y_i \quad (30)$$

M - number of satellite passes

The correction terms $\overline{\Delta X}$ and $\overline{\Delta Y}$ are then applied to all transponder positions.

Fortunately, most operations at sea can continue as the satellite data is being accumulated. The final net correction is the last step in the calibration process.

SUMMARY

The purpose of this paper is to provide a general overview of how Western Electric utilizes bottom mounted acoustic transponders for navigation in accomplishing numerous missions for the United States Navy. The subject matter is generalized, but an attempt has been made to discuss a number of subjects and some of the important findings, along with developed techniques that are the results of years of experience in deploying many hundreds of transponders.

There are still significant portions of the oceans not covered by precision navigation systems such as Loran C. Even satellite fixes can be hours apart depending on latitude and other factors. It is frequently desirable to have a highly accurate navigation system that is not subject to temporal pertubations; and, if the areas of interest do not exceed a dozen or so square miles, acoustic transponders become a very attractive and viable alternative to other more complex and expensive systems. Under most conditions a net of three or four transponders will provide excellent coverage in an area of 15 to 20 square miles.

Generally speaking, the keys to successful and accurate transponder navigation and calibration seem to lie in two areas. The first, is to ensure that reasonably accurate sound velocity data is available in the area(s) of interest. This becomes increasingly important when the transponder depths differ significantly. The second, is a careful scrutiny of the data being used, whether this is done manually or by appropriately constructed computer software. There are many areas where errors can enter the system. A detailed error analysis is appropriate but would be too lengthy for this paper. Many good papers exist on this subject such as in reference five(5).

A few possibilities regarding computational techniques were brought forth to illustrate methods of navigation and calibration. Mathematical algorithms are highly subjective and no attempt is made to include all possibilities. Those that were mentioned work however, and most of the computations are simple enough to be performed on hand-held calculators. The situation is even better if the calculators are programmable. When time, navigation accuracy and data logging factors become important the assistance of a computerized operation becomes an invaluable necessity.

REFERENCES

1. Deutsch, Ralph, "System Analysis Techniques", Englewood Cliffs, N. J., Prentice-Hall, Inc., 1969.
2. McCalla, Thomas Richard, "Introduction to Numerical Methods and FORTRAN Programming", New York, John Wiley and Sons, Inc., 1967.
3. Stark, Peter A., "Introduction to Numerical Methods", New York, The MacMillan Company, 1971.
4. Henry, Thomas D., "Theoretical, Practical and Limitational Aspects of Transponder Operations and Software", Western Electric Company, Inc., technical memorandum, April, 1973.
5. Cestone, J. A.; Cyr, R. J.; Roesler, G.; E. St. George, Jr., "Latest Highlights in Acoustic Underwater Navigation", paper presented to the International Navigation Congress in August, 1976, Boston, Massachusetts.

Underwater Logs

LYMAN W. GRISWOLD

Abstract

THIS PAPER REVIEWS the principles of operation of the major types of ship's water speed sensors in use in underwater log systems since 1950 and discusses in more detail the operation of the electromagnetic underwater log widely used by western navies. The capabilities and limitations of ship's water speed measurement are discussed briefly in order to provide proper perspective to the review of sensor characteristics.

Introduction

In 1938 Dr. Hoppe, in the Transactions of the Institute of Engineers and Shipbuilders,[1]* reviewed the techniques of ship's water speed measurement in common use before that time and provided some useful comments on their applications. Most recently published literature has been nearly devoid of discussion of the subject of ship's speed measurement principles. The recent literature of navigation makes only cursory mention of ship's water speed measuring equipment and the problem of water speed measurement in general, in spite of the fact that substantial development of underwater log systems has taken place since 1938. Paradoxically, almost everyone agrees that accurate ships speed measurement is essential but there seems to be some lack of understanding as to what constitutes accurate water speed, or what it is used for, or how to achieve it.

* Superscripts refer to references listed at the end of the paper.

Mr. Griswold is with the Naval Ship Research and Development Center, Annapolis, Md. He presented this paper at The Institute's National Marine Navigation Meeting at the U. S. Naval Academy on October 11, 1967.

The classical question asked by junior officers and log system maintenance personnel onboard ship is, "What good is water speed anyway?" The candid answer that, "It is better than no speed at all," is seldom satisfying. Although logs can perform important functions in the areas of fire control, ship control, and ship status monitoring, it is the navigation functions which are usually of primary interest. Historically, ship's logs have provided one of two primary inputs to the ship's dead reckoning system, the other being the compass. Modern technology has added two new requirements: The ships inertial navigation system (SINS) needs an accurate ships speed input for damping the Schuler oscillations in the vertical indicating loop of the system. Doppler types of satellite navigation systems also require accurate ship's speed inputs to give optimum fix accuracy. In short the requirement continues for ship's speed as an essential input to the ship's dead reckoning system, and the requirements of SINS and satellite navigation systems lend added emphasis to that need. This paper is devoted to reviewing the principles employed since 1950 in meeting these needs and to providing a perspective on the application of present systems.

Principles of water speed sensors

In order to review the various principles employed for water speed measurement it is useful to classify them in some logical fashion. Table 1 lists the various phenomena employed in ship's sensing according to the basic quantity sensed, i.e., (distance, velocity, or acceleration). Of the phenomena listed, the inertial principles have received adequate coverage in the literature and will not be discussed further here.

Impeller logs have been used for many years as

Reprinted with permission from *Navigation*, vol. 15, no. 2, pp. 127–135, Summer 1968.

ship's speed sensors, both as towed logs and hull installed logs. Fig. 1 illustrates the principle of the impeller log used in a hull installation and shows the characteristic curve of output frequency (f) versus speed (V) to be typically linear except at very low speeds—typically less than 10% of the maximum design speed. The slightly downward trend of the curve at low speed is the result of bearing drag on the otherwise freely turning impeller.

Table 1—Classification of Ships Logs

Quantity Sensed	Phenomenon Employed
Distance	Rotational Measure Linear Measure Vortex Shedding
Velocity	Force (Resistance) Pressure (Pitot) Doppler (Acoustic) Doppler (Electromagnetic) Electromagnetic Induction
Acceleration	Inertia

Interest in impeller logs has revived in recent years primarily as a result of the technical advances in the design and fabrication of impellers, bearings, and electronic instrumentation and the upsurgence of the small boat business. A properly designed impeller sensor can give a very precise and linear response over a wide speed range for a cost of the order of a few hundred dollars. It is, however, susceptible to wear and various forms of fouling present in the ocean environment and is best suited to intermittent-use applications. Over the past ten years the United States Navy has developed and used towed impeller logs on an experimental basis for ship's speed trials.

One such device, the Knotmeter shown in Fig. 2, was developed at the David Taylor Model Basin. The significant feature of the Knotmeter is the incorporation of negative-life wings to depress the sensor below the undesirable influences of the flow in the ship's wake. Today's impeller logs, including the Knotmeter, customarily employ a magnetic-induction-type of pulse frequency generator in conjunction with the impeller so that no physical contact, other than bearing surface, is required between the rotor

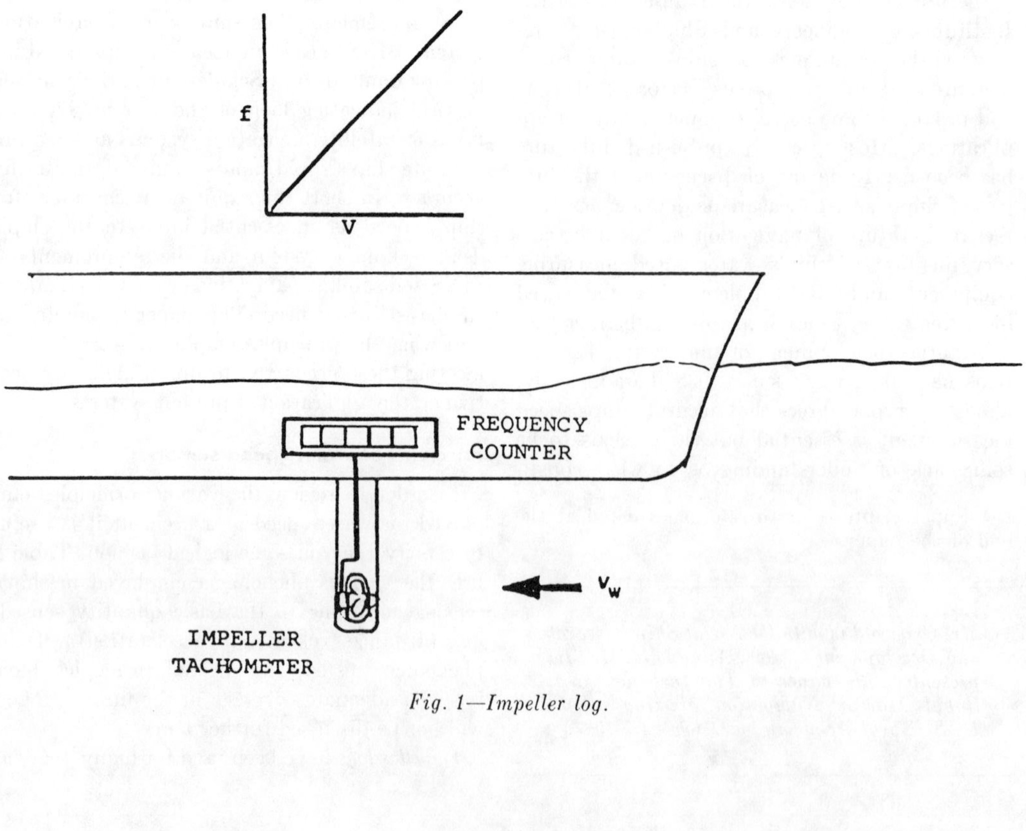

Fig. 1—Impeller log.

and the body of the device. This also permits the use of simple and accurate frequency measuring instrumentation.

The *acoustic log* is a linear measure type of system— the modern equivalent to the Dutchman's log or chip log used by seamen for four centuries. Fig. 3 illustrates the principle of the acoustic log which is based on measuring the difference (Δt) between the times required for simultaneously-transmitted acoustic pulses to traverse a fixed path length, upstream and downstream, in a fluid flow. By suitable electronic processing, the time difference may be used to compute the water speed and distance made good. Systems of this nature are presently used to measure the water velocity across the missile decks of Polaris submarines. A similar system has been proposed and partially developed as a ship's log. By appropriate design, such systems can be made self-compensating for variations of the sound velocity in the water and can also provide a measure of that velocity. Cost figures are not presently available for log systems of this type, but a reasonable estimate would range between $10,000 and $20,000 for a military version.

The *vortex shedding* behind poorly streamlined bodies in a fluid flow has been applied by Dr. Hisamitsu Shiba[2] to measuring ship's speed during trials. The Shiba log principle is shown in Fig. 4. The square cylinder in the flow stream sheds vortices over a wide range of Reynolds Number at a rate proportional to flow velocity. If a suitable detector is placed behind the cylinder, the vortex shedding may be converted to electrical impulses and measured by conventional pulse counting techniques to produce either distance or speed indications. The practical limitations of this simple device are its threshold speed, which is governed by the choice of strut size, and cavitation problems, which appear to place an upper limit on the speed. To the author's knowledge, the only application of this technique to logs has been the towed log development by Dr. Shiba who applied it to several merchant ship trials.

Fig. 2—DTMB knotmeter.

The *force log* (resistance log) is perhaps the oldest of the velocity sensing logs still used today. The principle of the force log is shown symbolically in Fig. 5. The drag force on the strut is proportional to the square of the vessel's speed and may be transmitted mechanically, hydraulically, or electrically to a dial or gauge calibrated in knots. In most applications today the drag strut is hull-mounted and uses a hydraulic transmitter and pressure gauge to obtain remote indication. The chief disadvantage of these relatively simple logs is their square law dependence which results in insensitivity at low speeds. The typical cost of this type of system for small boats is between $100 and $200.

The well known *differential pressure or pitometer log* is still used extensively today in naval vessels and pleasure craft throughout the world. The pitometer log principle, exemplified by Fig. 6, is dependent on the same square law relationship as the force log but is in many cases instrumented by sophisticated methods which produce high accuracy indications except at low speeds. Costs of the differential pressure type of system range from a few dollars for a hand device to $10,000 for a complex militarized version.

One of the major operating difficulties encountered with the pitometer type of log is well illustrated by a quotation from Navy maintenance instructions for these systems. "The rodmeter may become clogged with jellyfish, mud, seaweed, or other foreign matter. Consequently, the pressure difference will not be normal and the mechanism will not register correctly."

The basic principles and literature of the pitot tube were reviewed by Folsom[3] in a paper pub-

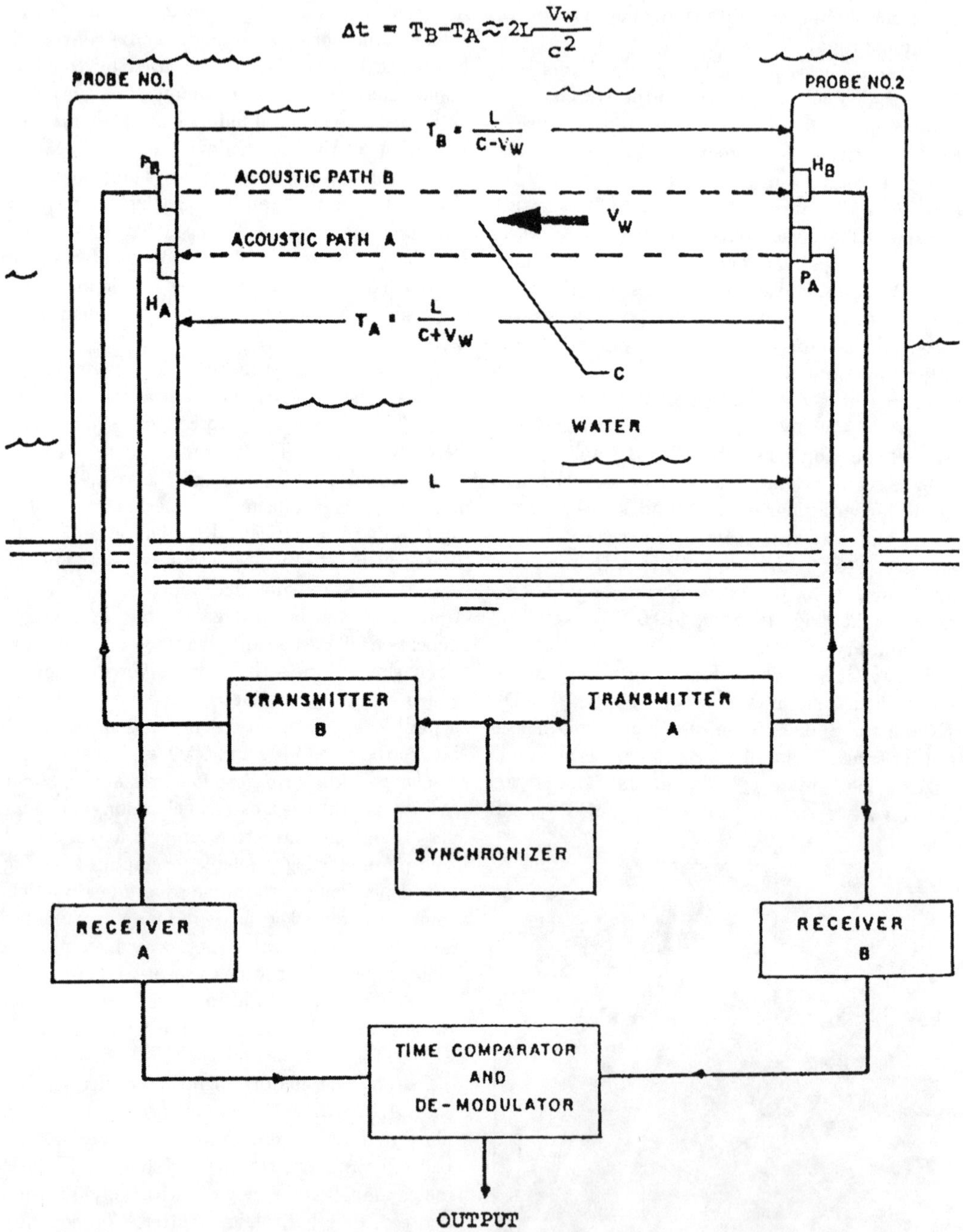

Fig. 3—Time-difference acoustic log.

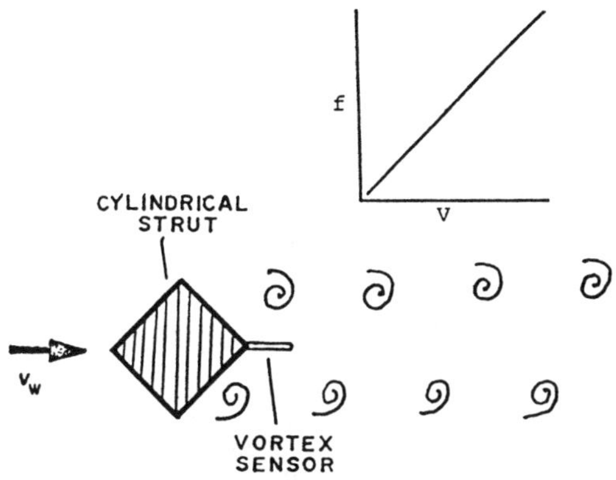

Fig. 4—Vortex (Shiba) log.

lished in the Transactions of the ASME, that paper includes an extensive bibliography of the subject, which may be of value to the interested researcher.

Doppler acoustic techniques have long been employed for fluid flow measurement and for estimating sonar target behavior. Aircraft Doppler radar navigators have been in use for nearly two decades now. In recent years the familiar techniques of Doppler radar navigation have begun to be applied to underwater logs using ultrasonic transducers in a Janus configuration. Fig. 7 shows application of this principle to a surface ship installation. The equation expresses,

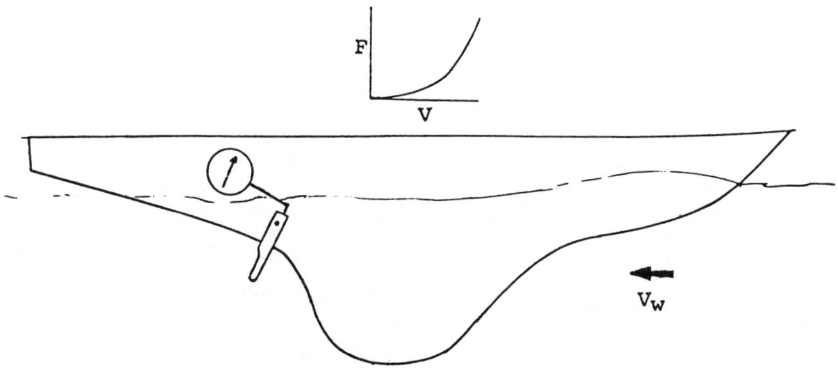

Fig. 5—Force (resistance) log.

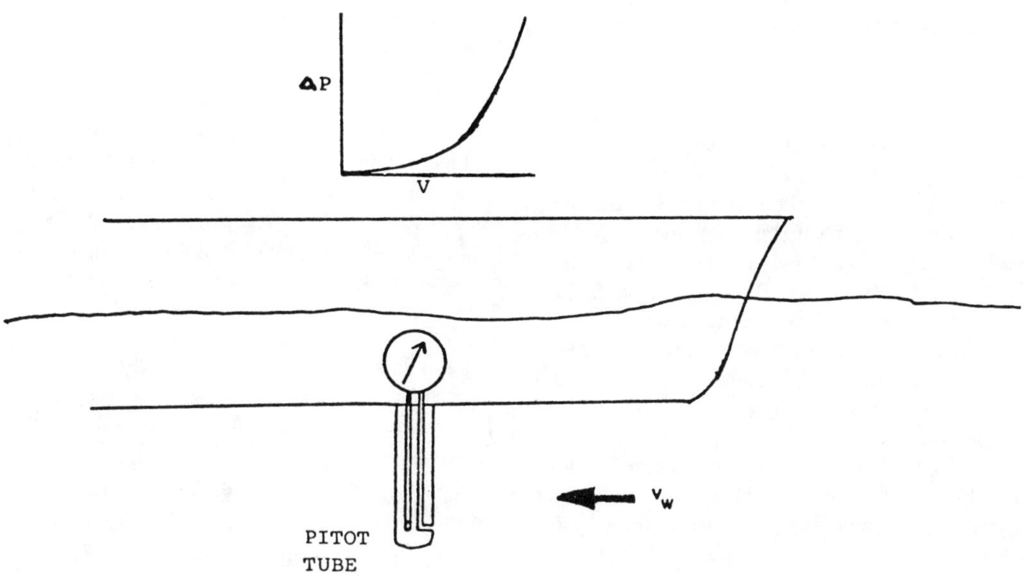

Fig. 6—Pressure difference (Pitometer) log

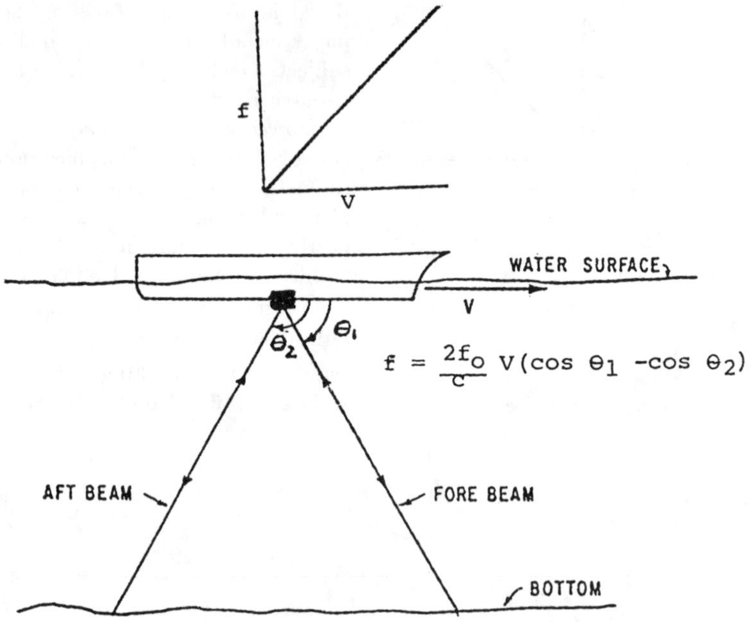

Fig. 7—Doppler acoustic log.

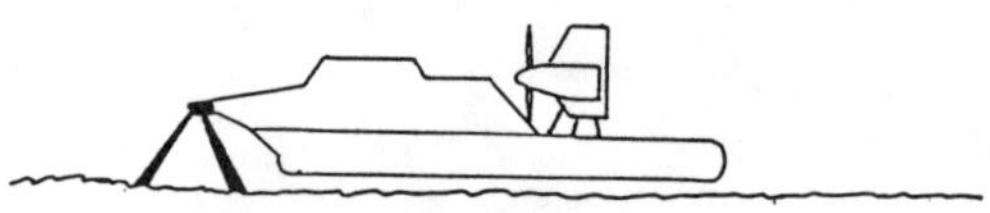

Fig. 8—Doppler radar log.

to a good approximation, the relationship between the Doppler difference frequency (Δf) and the *c-w* transmitted frequency (*fo*) in water having sound velocity (c). Here, as with time-difference acoustic techniques, clever design can make the system self-compensating for variations in *c*, and a high degree of accuracy can be maintained, provided that the motion of the platform is not excessive or can be compensated for. As with aircraft Doppler navigators, a second pair of beams is customarily added, in spacial quadrature to the first pair, to permit drift measurement. If two pairs of athwartships beams are provided a Doppler compass can be devised although the author is not aware of any present application of this technique.

Although the designed mode of operation is to use bottom reflection to provide track and ground speed information in shallow water, a properly designed system will sense water velocity from volume reverberation when bottom contact is lost. Problems of ship motion, bubble sweep-down, depth limitations, and sound transmission variability all must be considered when applying Doppler Acoustic systems. To date the most fruitful potential application seems to be for ocean bottom navigation of deep submersibles.

Although they are not truly "underwater" logs, Doppler radar systems have been applied in recent years, on an experimental basis, to measuring the speed of surface vessels and surface-skimmer-type craft as shown in Fig. 8. Technical problems which arise in adapting aircraft Doppler navigators can be overcome but the basic limitations of the air-water interface appear to limit the application of this method to very high-speed craft. Errors resulting from surface water motion typically reach one or two knots and can be as high as seven knots. Except at the higher speeds, errors of this magnitude are generally unacceptable in a navigation system. Another application of electromagnetic Doppler principles is the *laser log*. Although only in the laboratory experimental stages at the present time, the concept shown in Fig. 9 is an underwater log application of electromagnetic Doppler techniques. The same basic principles apply as with Dop-

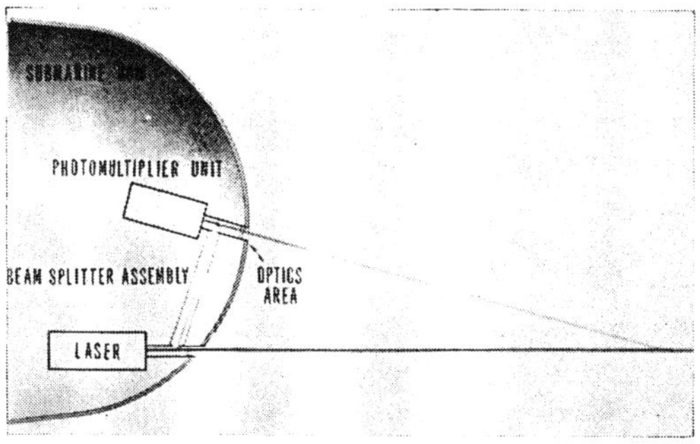

Fig. 9—Doppler laser log.

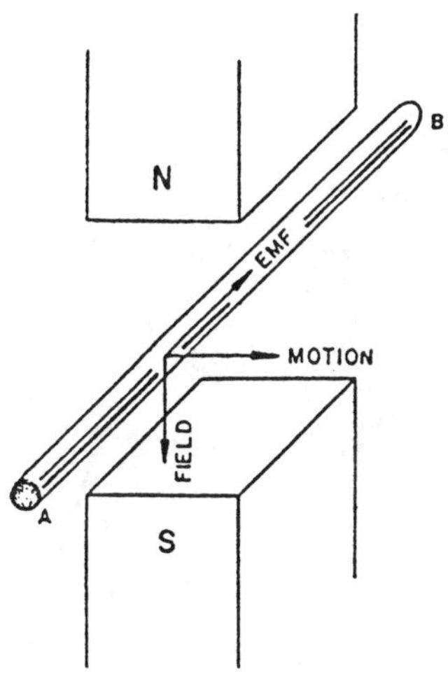

Fig. 10—Electromagnetic induction generator.

pler radar, but in the concept shown here laser light backscattered from particles in the water is mixed with directly transmitted light from the same *c-w* source. The difference frequency is proportional to the water speed relative to the point observed on the beam.

One of the most widely used naval log systems in the world today is the *electromagnetic* or "EM" *log*. Approximately one thousand EM logs are in service in the United States Navy today and additional ones in the British and Canadian Navies.

The basis of the EM log is the principle of electromagnetic induction first applied (unsuccessfully) to water flow measurement by Faraday in 1831, first patented in the form of a ship's speed sensor in 1917, by Smith and Slepian,[5] and finally successfully applied by the United States Navy from 1950 to 1955 in the development of a practical ship's log. Fig. 10 illustrates the principle of electromagnetic induction which is familiar to all as the basis for an electric generator. This principle is applied to the electromagnetic speed sensor in the manner shown in Fig. 11 where the conductor is the seawater surrounding the solenoid housed in the plastic boot attached to the end of a strut projecting from the ship's hull. As the conducting water cuts the magnetic field surrounding the solenoid, the voltage generated in the water is picked up by the electrodes which are embedded in the surface of the boot. By Faraday's Law this voltage is directly proportional to the water flow velocity (V_w) relative to the magnetic field (B).

$$\phi \sim V_w \times B$$

If B is constant the voltage versus speed relationship is linear, as shown graphically in the inset of Fig. 11. The detailed mathematics of the principle as applied to pipe flow meters are well documented by Cushing[6] and others and will not be discussed further here. The mathematics of the particular sensor configuration used in electromagnetic log systems have not been adequately worked out and new sensor designs must be semi-empirically derived.

In order to accurately measure the precise speed signal generated by the sensor, the electromagnetic log system employs a nullbalance type of electronic voltmeter. Systems currently in Naval service employ an electromechanical servo and dial indicator system for this purpose and include a roller and disc integrater for determining distance travelled. Fig. 12 is a functional block diagram of one of these systems.

System characteristics include:

- sensor output signal—325 microvolts/knot
- instrument speed accuracy—0.05 knot
- sensitivity to speed changes—0.01 knot
- instrument distance accuracy—better than 1% of the water distance travelled
- capability to compensate for non-linearities in hull flow characteristics.
- freedom from effects resulting from variations of water characteristics

At least one commercial version of the EM log system is available in an all electronic package with a digital output and slightly less precision.

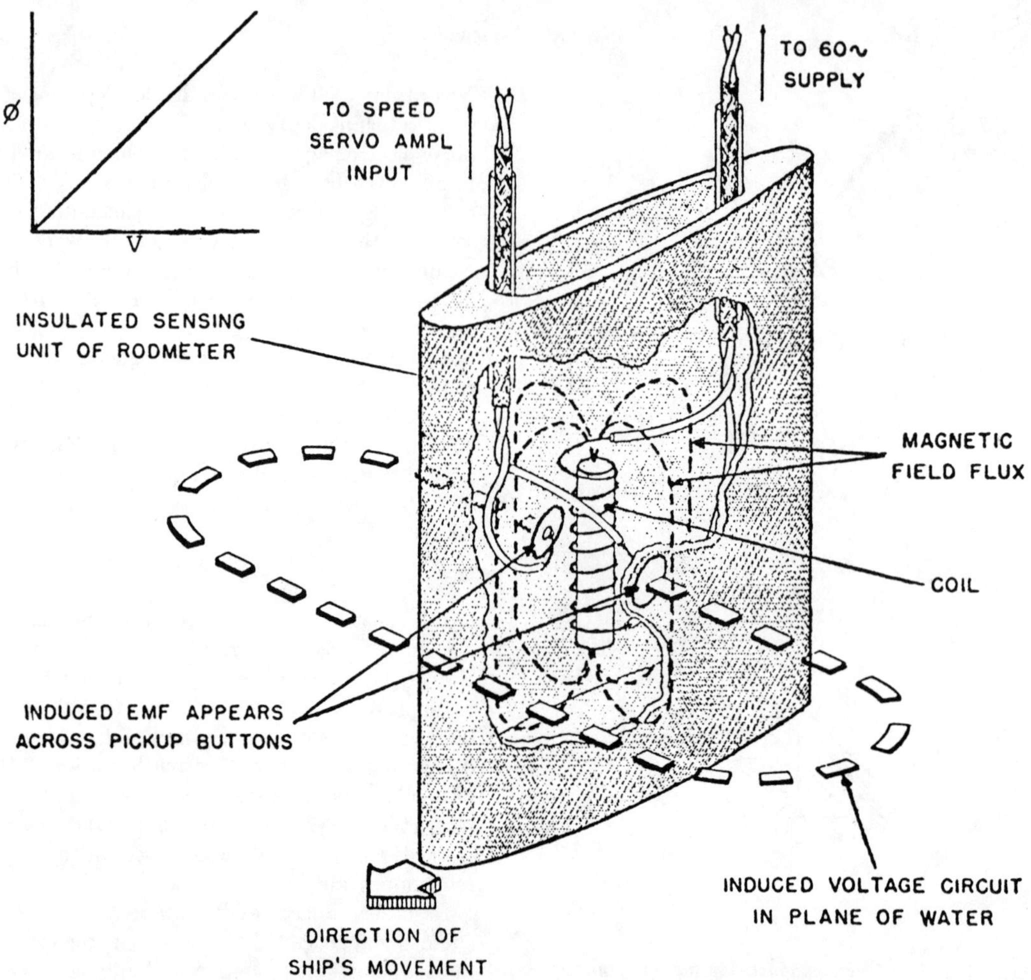

Fig. 11—Electromagnetic water speed sensor.

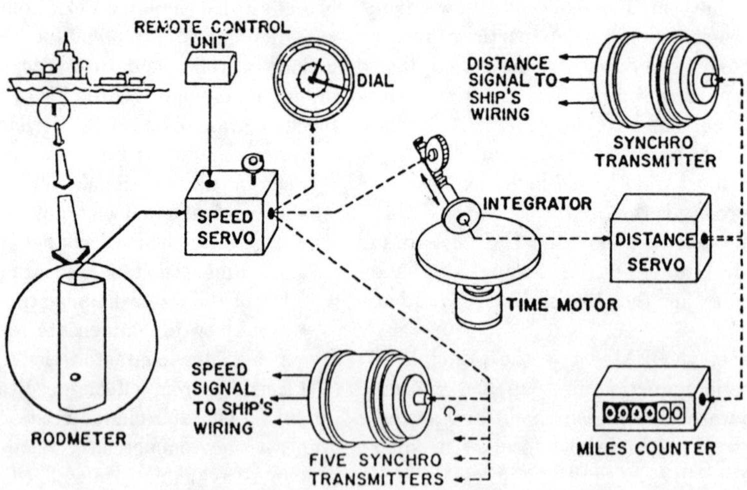

Fig. 12 Electromagnetic log functional diagram.

The chief advantages of the electromagnetic log are its good sensitivity, linear output from zero speed up to full design speed of 40 knots, long term calibration stability, and resistance to fouling elements present in the ocean. Its chief disadvantage is its cost (about $10,000 per system), but present and future developments are expected to contribute to a substantial reduction in this figure.

Perspective

Although the electromagnetic log offers some significant advantages over previously operational log systems, there are several important points about log systems in general which do not seem to be as generally appreciated as they might be:

- The total log system includes the ship's hull and appendages.
- Most speed sensors can only sense *water speed at the sensor* which is within the hydrodynamic influence of the ship's hull.
- Each different hull configuration or log installation sees different local velocities because the velocity distributions and potential flow regions are different. Consequently each system (hull and sensor) must be calibrated.
- Ship's motions affect the relationship between the ship's speed through the water and the speed seen by the sensor.
- Ocean currents affecting navigation are of a highly variable and statistical nature.
- For purposes of position determination, a 0.5 degree compass error for 100 miles is equivalent to a 0.2 knot log system error during five hours.

In view of the above points, one may well ask if it is possible for water speed sensing systems to meet navigational needs. The answer is a qualified yes. Available underwater log systems can give accuracies commensurate with present water speed measurement needs but navigation systems—dead reckoning, SINS, and Doppler satellite—may eventually demand ground speed inputs.

One significant span in the bridge between water speed and ground would be a better capability to measure and predict ocean currents for operational purposes. There are other spans in that bridge. If and when they can all be linked together, underwater logs can provide more useful navigational information.

References

1. Hoppe, H., "Ship's Speed Meters," *Transactions of the Institute of Engineers and Shipbuilders*, Vol. 81, Paper 988; April 12, 1938.
2. Shiba, H., *A New Type of Speed Meter; Application of the Karman Vortex Phenomenon to the Design of a Speed Meter*, Technical Research Laboratory, Japan; 1963.
3. Folsom, R. G., "Review of the Pitot Tube," *Transactions of the ASME;* October 1956.
4. Faraday, M., "Bakerian Lecture," *Diary;* January 12, 1832.
5. Smith and Slepian, *Electromagnetic Ship's Log*, U. S. Patent No. 1,249,530; 1917.
6. Cushing, V., "Induction Flowmeter," *The Review of Scientific Instruments*, Vol. 29, No. 8; August 1958.

DOPPLER SONAR VELOCITY LOG FOR ATTACK SUBMARINES

Martin Wapner Neal Barnett* Eugene Grudzien*

Naval Sea Systems Command, Washington, DC
*Naval Air Development Center, Warminster, PA

ABSTRACT

The Naval Sea Systems Command has a program to develop a Doppler Sonar Velocity Log (DSVL) as a direct replacement for the electromagnetic log aboard SSN attack class submarines. The DSVL is to provide three-axis velocity and is intended primarily as a sensor of velocity relative to water, though a limited absolute velocity capability will be available in shallow water.

A DSVL engineering development model is being fabricated and will undergo technical evaluation aboard an operational submarine. The DSVL circuitry will consist largely of standard electronic modules (SEM) with fully implemented built-in test equipment (BITE) for fault location. The program includes a comprehensive reliability and maintainability program and integrated logistics system (ILS) program.

It is anticipated that the DSVL concept will also be applicable to fleet ballistic missile (FBM) submarines and all conventional Naval surface ships that currently use the electromagnetic log.

INTRODUCTION

The Naval Sea Systems Command with technical assistance from the Naval Air Development Center is conducting a program to develop a Doppler Sonar Velocity Log (DSVL) for nuclear attack class submarines to replace the electromagnetic log (EM Log). The objectives of this engineering development program are (1) to provide the fleet with a high accuracy velocity sensor to improve the effectiveness of tactical forces, and (2) to provide a system with much better performance capabilities than the present EM Log. Full scale development of a DSVL was authorized by Navy Decision Coordinating Paper (NDCP) S0253-CC (1).

General performance capabilities of the DSVL to satisfy program requirements include:

- o High accuracy
- o Measurement of forward and athwartship components of velocity under both surfaced and submerged conditions.
- o Continuity of accuracy during maneuvers (no degradation due to hydrodynamic effects)
- o Capability of electronic calibration without external references; long term stability (no need for constant recalibration)
- o Sensor mounting flush with outer hull; minimum chance of sensor damage and fouling
- o Simplicity of design for ease of operation and maintenance
- o High reliability and maintainability
- o Applicable to all submarines and conventional surface craft which require a speed sensor
- o No compromise to submarine security
- o Direct interchangeability with EM log

The feasibility of utilizing a doppler sonar for velocity measurement on Navy ships was established during the exploratory development phase of this program. Extensive tests were performed using commercial doppler sonars (Ametek-Straza 2015A and Sperry Marine Systems SRD-301). Included were tests in a tow tank (2), on a fresh water lake (3), on the U.S.S. Finback (SSN 670) a nuclear attack submarine (4), and on the U.S.S. Compass Island (AG 153), a large Navy surface vessel. Results of each of these tests and a summary of the entire NAVSEA exploratory development program are contained in reference (5). Each of these tests indicated a high velocity accuracy capability, with no degradation during maneuvers. In addition, the tests aboard the AG-153 indicated improvement in the performance of the ship's inertial navigation system (SINS) with doppler sonar damping of the SIN's vertical loops versus EM Log damping. Dead reckoning accuracy was also improved by approximately 75% when using doppler derived velocity over that of the EM Log. The submarine tests were performed while submerged in the Sargasso Sea and demonstrated doppler sonar capability in very clean waters with low scatterer densities. No evidence of long-term drift or major system failures were observed during 2800 hours of operation.

This extensive testing of commercial doppler sonars demonstrated sufficient gains over the EM log as a ship speed sensor to warrant further development for eventual Fleet applications. It also showed that with certain improvements (better electronics design, transducer design tailored for submarine use, interface compatibility with user equipments, etc.) a highly reliable military system could be produced with current technology. Accordingly, the program was advanced to its current engineering development phase. Subsequent to issuance of an RFP by NAVSEA, and an extensive competitive technical and cost evaluation of the offerer's proposals, Sperry Marine Systems division of Sperry-Rand Corporation was selected

Reprinted from *Rec. IEEE Position Location and Navigation Symp.*, pp. 146–151, 1980.

to produce the engineering development model Doppler Sonar Velocity Log (DSVL).

DOPPLER SONAR VS. EM LOG

The EM Log, the current speed sensor on most naval ships, includes a sensor and an electronics unit (Indicator/Transmitter). The sensor consists of a rodmeter, or sword, that protrudes from the ship's hull. Within the sword is a coil of wire that sets up a magnetic field in the nearby water. Motion of the ship through this magnetic field induces currents in the conductive sea water. As a result of these electrical currents, a difference in potential appears between two conductive buttons on the otherwise nonconductive sword. The magnitude of this difference in potential is an indication of ownship's speed. This effect is physically explained by Faraday's Law.

The EM Log is subject to several deficiencies which limit its measurement accuracy and operational utility. The fundamental limitation is common to all speed logs that sense ownship's motion relative to a water mass near the ship's hull. The speed measurement is subject to error because the water motion in the vicinity of the rodmeter is governed by the hydrodynamic viscous and potential flow fields generated by the ship's profile and movement.

Investigations (6) have been conducted on the relationship between the deviation from the free-stream flow velocity and the distance from the hull. The studies (made on prolate spheroids as approximations to submarine hulls) indicate that at midship's, the velocity of the water at the surface of the hull differs from the free-flow velocity by 1% to 2%. Even at a distance of two hull diameters, the potential flow velocity differs from free-stream by 0.5% to 1%. Similar effects due to potential flow and appendage wakes (from sonar domes, etc.) apply to surface ships as well. Compensation for these hydrodynamic effects requires a complex calibration procedure since the flow characteristics at the sensor vary with sensor location, hull shape, speed, depth, maneuvers and sea water conductivity.

Other limitations of the EM Logs currently used by the U.S. Navy include a velocity measurement capability along the ship's longitudinal axis only (no athwartship velocity measurement) and speed measurement relative to water only, with no capability for measurement relative to ground, or absolute velocity measurement. In addition the EM Log sword, protruding from the hull, is subject to breakage and vulnerable to marine fouling.

A doppler sonar operates by projecting an acoustic signal of known frequency into the water and measuring the frequency of the scattered return signal. The frequency shift, caused by the doppler effect, is directly proportional to ship's speed. Because the return signal is from scatterers at some distance from the ship's hull, the doppler sonar velocity measurements are minimally affected by the ship's hydrodynamic potential flow field. Accuracy is primarily a function of the signal processing techniques in the receiver. In shallow waters, where the bottom is within range of the acoustic signal, speed relative to ground, or absolute velocity, can be measured. By utilizing three acoustic beams, the DSVL can have a 3-axis capability, measuring the ship's longitudinal, athwartship and vertical velocity components.

DSVL HARDWARE DESCRIPTION (PRELIMINARY)

The engineering development model Doppler Sonar Velocity Log is being designed in accordance with MIL-E-16400 and the NAVSEA Engineering Development Requirements document (7). The DSVL is a 3-unit system consisting of a transducer, a transceiver and an electronics control unit (ECU) (See Figure 1). Figure 2 is a simplified block diagram of the system. The requirements on the transducer, which will emit three narrow acoustic beams (see Figure 3), are that it be flush mounted to the hull and that it fit into the currently existing attack submarine EM log rodmeter foundation, including both pressure hull and ballast tank installations. This requirement limits the transducer to a diameter of six inches. To obtain three narrow acoustic beams, the transducer will utilize a polystyrene lens to focus the emitted energy of three small crystals placed within the transducer housing (see Figure 4). The space between the lens and the crystals is filled with RTV-602 rubber, which has very low losses at the operating frequency. This frequency (which is classified) was chosen on the basis of a trade-off study between signal-to-noise ratio at remote water-track ranges and considerations of the detectability of the DSVL transmissions and submarine security requirements. The acoustic signals will be transmitted as pulsed CW into the water.

The speed of sound in the RTV-602 rubber and in the outside sea water enter into the velocity determination as calibration constants. Since the speed of sound in any medium is a function of both temperature and pressure, a pressure sensor and two thermolinear components will be placed within the DSVL transducer to monitor the rubber and water temperatures and pressure. The transducer housing will be made of 70-30 Cupro-Nickel alloy which has good anti-fouling properties.

A single cable, having a maximum length of 40 meters, will run from the transducer to the transceiver mounted within the pressure hull. The transceiver will contain the front-end circuitry including transmit frequency oscillator, output power amplifiers, receiver preamp and amplifier, zero crossing circuits and threshold comparison circuitry for each of the three channels. The zero crossing detection circuit allows measurement of the frequency of the return signal. The threshold circuitry sets a comparison level determined by continuous, automatic monitoring of the received noise level to indicate when the signal-to-noise ratio is high enough to ensure good data on the zero-crossings line. The zero crossing data and threshold level are sent across an interface to the Electronics Control Unit. The

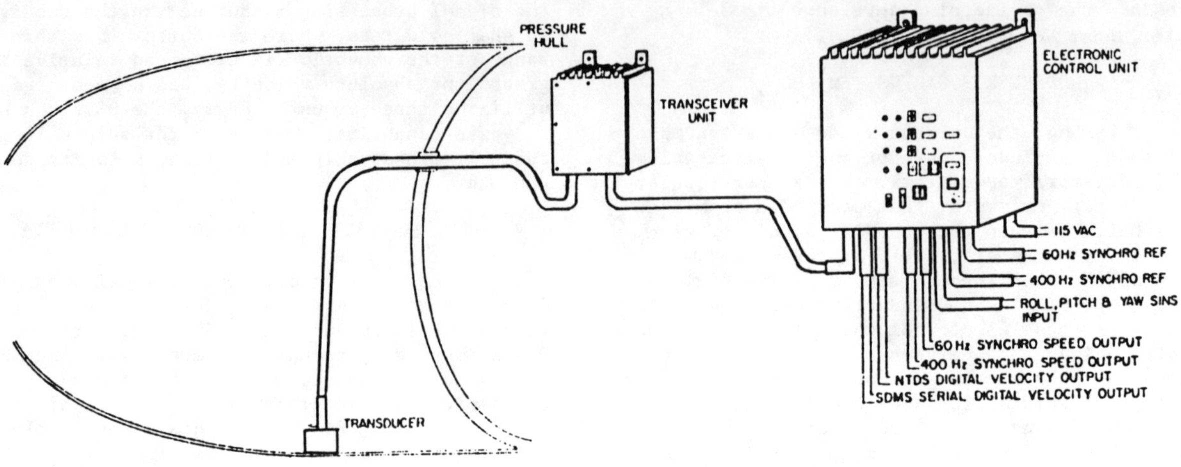

Figure 1. Doppler Sonar Velocity Log System Configuration

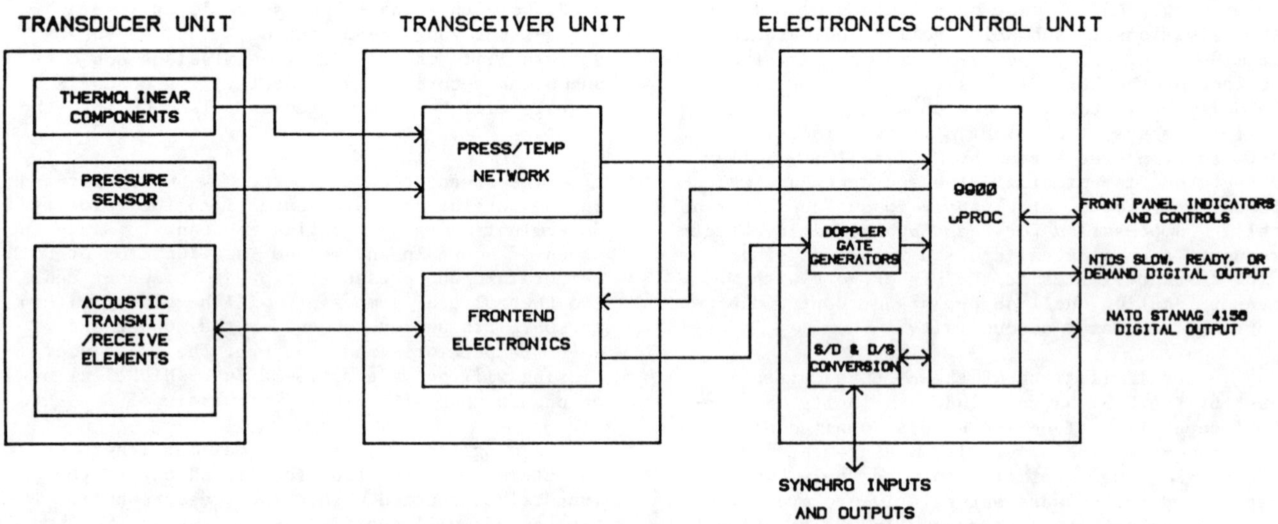

Figure 2. Doppler Sonar Velocity Log Block Diagram

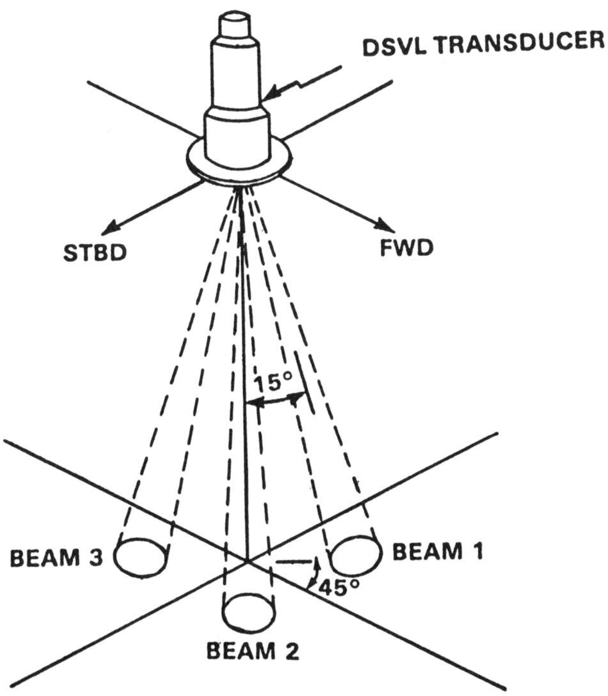

Figure 3. DSVL Acoustic Beam Configuration

outputs of the transducer pressure and temperature sensors are processed in the transceiver and also sent to the ECU, after analog-to-digital conversion.

The ECU sends several control signals across the interface to the transceiver. Modulation signals are sent to the transmitters to produce the pulsed CW transmissions. Automatic gain control enable and Time-Varying-Gain (TVG) signals are sent to the receivers for detection and tracking of bottom returns. For equal ranges, a return signal off the bottom is much stronger than a water mass return. Therefore, by having the TVG low at the start of reception, a signal which exceeds threshold must, of necessity, be from the bottom. The TVG characteristic is subsequently matched to the time varying bottom return signal strength and maintains the amplified level constant at a few db below post-amplifier saturation.

During transmit, the receivers are blanked. The ECU sends unblank signals to the receivers at appropriate times to either ensure detection of the bottom track signal, or ensure reception of the water track signal from a water mass sufficiently beyond the potential flow region of the submarine.

Within the ECU, the threshold level from the transceiver enables doppler gate generators which begin counting the zero crossings of the received signals in each of the three channels to generate speed counts. A transmit gate generator counts zero crossings of the transmitted signal to form a reference count. The threshold signal also goes to circuitry for bottom detection and ranging and for loss of track. The outputs of this circuitry plus the speed and reference counts are inputted to a 9900 microprocessor.

The microprocessor is the heart of the DSVL system, performing all computations, controlling all system timing and generating all real time functions and control signals. Using the temperature and pressure data from the transducer, it processes the difference between the speed and reference counts into raw velocities along the three acoustic beam axes, transforms the velocity components along the ship's longitudinal, athwartship and normal axes, and then, with attitude inputs from the SINS, into horizontal and local vertical components. It then digitally filters the raw data to produce smoothed velocity estimates.

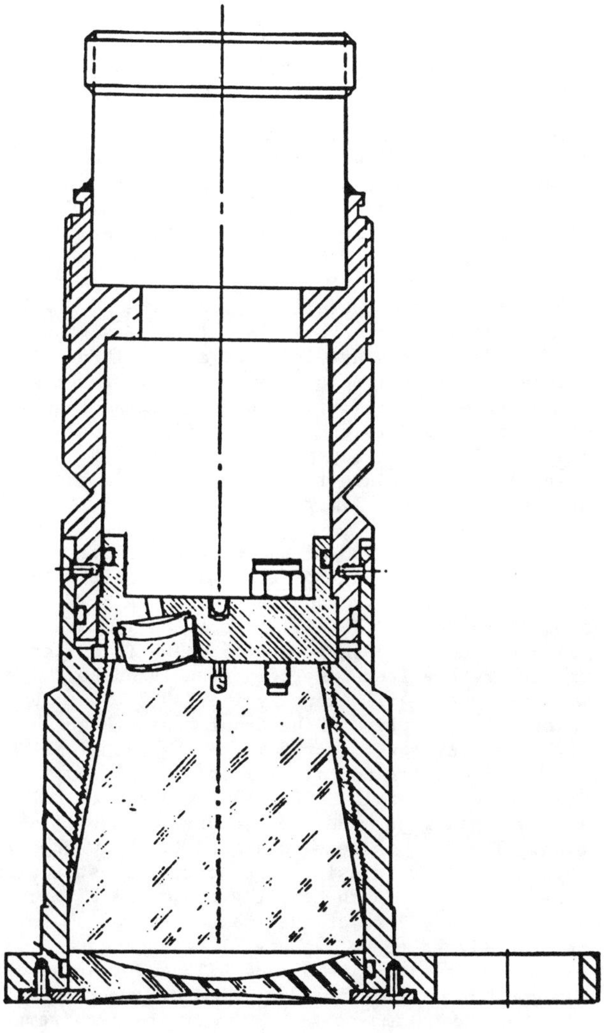

Figure 4. DSVL Transducer (Preliminary)

The DSVL is being designed so that an operator can manually select either water or bottom track modes of operation. Since there are cases when the selected mode is not possible to achieve, a front panel display will indicate the actual mode selected by the microprocessor.

DSVL output data will be both in synchro and digital form. The fore/aft velocity will be distributed to user equipments in synchro format in a manner identical to current EM Log outputs so that no immediate modification will be required to user equipment. In addition a 16 bit digital parallel output channel will be available with the capability to provide all component velocities in either of the following formats: NTDS Slow, Demand Digital and Ready Digital. The DSVL will also have a NATO STANAG 4156 digital interface. This interface is compatible with the Navy's Shipboard Data Multiplexing System (SDMS) which is expected to be installed aboard future ships and submarines. The digital data will also contain status bits indicating system mode and operational status.

The DSVL front panel will be located on the ECU (figure 5). Component velocities, total velocity magnitude, operational mode, on-line monitor functions and a BITE fault localization code will be displayed on the front panel. The front panel will also contain a power on switch, a battle short switch, a test select switch, and switches for manually inputting velocities for test purposes and for distribution to user equipments as required.

While in the operational mode, the DSVL will periodically cycle to a self-test mode, during which test signals are injected at the three channel receiver inputs to exercise the receiver and processing circuitry. A preprogrammed sequence of tests is used to check the validity of signal characteristics at key points, the process culminating in a comparison test of speed along each axis versus stored limits.

A major function of the microprocessor will be system performance monitoring and fault detection and isolation through an extensive BITE capability. The BITE will detect and isolate failures down to a fault group, which is a list of replaceable items averaging ten in number. The microprocessor will identify the fault group, by three digit display on the front panel (Figure 5). The operator then refers to the technical manual which will contain a list of fault numbers. Under each number is a list of replaceable items in the fault group and a procedure for substitution. The DSVL is being designed to allow organizational repair by using a sequential substitution scheme of items in the fault group. The BITE can also detect failure of the transducer, which is being designed to be replaceable at-sea by a diver. This extensive use of BITE and the use of Standard Electronic Modules (SEM) is expected to keep the mean-time-to-repair below the requirement of 0.5 hours and simplify future logistic support. In addition, the DSVL has a reliability goal of 4000 hours.

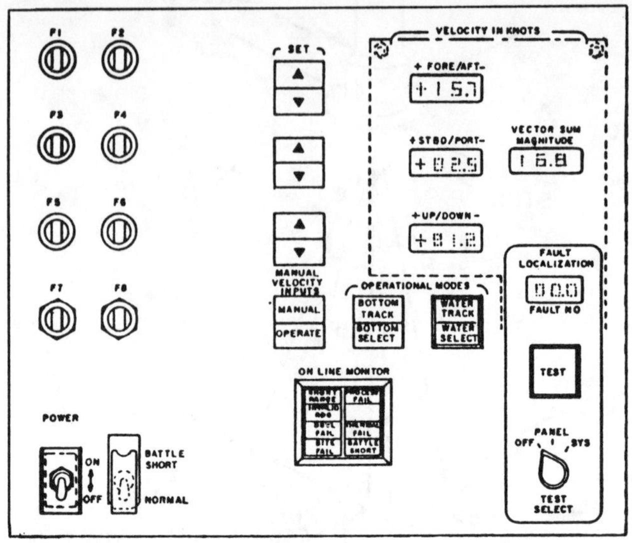

Figure 5. DSVL Front Panel (Preliminary)

Four enginering development models (EDM's) are to be delivered under this program. These EDM's will undergo extensive first article testing including reliability and maintainability demonstrations to insure compliance with the engineering development requirements.

FUTURE OF THE PROGRAM

Subsequent to delivery, one of these EDM's will undergo TECHEVAL and OPEVAL aboard a nuclear attack submarine during FY 82-83. Following Approval for Service Use (ASU), it is expected that a DSVL production contract would permit installation of the DSVL on all classes of nuclear attack submarines. Initial production contract award is expected in FY 1985.

It is further expected that the DSVL will be applicable to fleet ballistic missile (FBM) submarines and the majority of surface combatants which now employ the EM Log.

SUMMARY

The current speed reference (EM Log), used by submarines and conventional surface ships, has accuracy and operational deficiencies. Extensive testing of commercial doppler-sonars demonstrated that a high accuracy, high reliability military doppler sonar, could be produced with current technology. A doppler sonar designed to meet MIL-E-16400 is presently in engineering development and will be subsequently evaluated aboard an operational nuclear attack submarine.

REFERENCES

(1) "Doppler Log (U)" Navy Decision Coordinating Paper (NDCP) NDCP-S0253-CC, 19 January 1978 (C)

(2) "Doppler Sonar Velocity Log Program - Phase I Test Report (U)", NSSNF Project 96-2; Progress Report No. 1, June 1972 (C)

(3) "Doppler Sonar Velocity Log Program - Phase II Test Report (U)", NSSNF Project 96-2; Progress Report No. 2, November 1972 (C)

(4) "Report of Near Term Doppler Log Tests (U)", NUSC Technical Document 4402, 7 May 1973 (C) letter PM13:JBH:bss, Ser C106PM-13 of 12 July 1973.

(5) "Doppler Sonar Velocity Log Exploratory Development Program - Executive Summary 1970-1974 (U)", Naval Air Development Center Report No. NADC-74208-60, 1 October 1974 (C)

(6) Laster, D. R., "Ship's Water Speed Measurement," NSRDC Report No. 421/66, 6 July 1967

(7) Engineering Development Requirements for Doppler Sonar Velocity Log, April 1978, Naval Ship Engineering Center, (C)

The Evolution of the Ship's Inertial Navigation System for the Fleet Ballistic Missile Program

B. McKELVIE and H. GALT, Jr.

INTRODUCTION

On 15 November 1960, the nuclear powered submarine, USS GEORGE WASHINGTON, carrying 16 Polaris A-1 1200-nautical mile missiles, departed Charleston, South Carolina on the first 60-day operational patrol of this nation's Fleet Ballistic Missile submarines.

On board the submarine was an inertial navigator, the MK 2 MOD 0 Ships Inertial Navigation System, known by the acronym, SINS. The SINS was self-contained. It was the heart of a new navigation subsystem that provided precise navigational data without the need for continuous external position references or reliance on the standard dead reckoning techniques of plotting ship's position. The SINS enabled the USS GEORGE WASHINGTON to take advantage of its nuclear propulsion and remain submerged for extended periods of time, minimizing its probability of detection, while remaining in a constant state of readiness to fire its Polaris missiles. The nation's newest strategic deterrent force, the Fleet Ballistic Missile and the FBM submarine, were on station.

To have a ballistic missile hit its target thousands of miles away, the location of the launch point as well as the target must be precisely known. In this case, the launch point is a moving platform, the FBM submarine (Fig. 1). The task of the SINS is to provide the precise instantaneous location of that moving launch point to the missile fire control subsystem which then uses that information, together with the pre-set target locations, to solve the ballistic equations. The SINS must also provide continuous ship's attitude and instantaneous velocity data so that the missile's guidance system can be rapidly and accurately erected and aligned just prior to missile launch.

Eight key parameters are provided by the SINS (Fig. 2). SINS latitude and longitude outputs are used to compute the missile's instantaneous azimuth angle and range angle to target. SINS roll and pitch outputs are differentiated by fire control to obtain pitch and roll rates required for moment arm correction of ship velocity. Ship's heading information is required by fire control to train the missile to the proper bearing for the selected target. SINS velocity is used to correct the computed missile initial velocity to account for ship's motion.

CHANGING REQUIREMENTS

Today, 17 years later, the U.S. Navy has 41 operational FBM submarines, each carrying 16 Polaris A-3 or Poseidon C-3 missiles and a complement of MK 2 SINS. The MK 2 SINS will also be installed in the Navy's newest class of missile-carrying submarines, the Trident.

Advances in missile capability (range, MIRV) and the predicted capability of the oppositions ASW forces have been the derivation of the FBM submarine's changing require-

Messrs. McKelvie and Galt, Jr., are with Autonetics Marine Systems Division, Electronic Systems Group, Rockwell International, 3370 Miraloma Avenue, P. O. Box 4921, Anaheim, CA 92803. They presented this paper at the Thirty-third Annual Meeting of the Institute at Costa Mesa, CA on June 22, 1978.

Reprinted with permission from *Navigation,* vol. 25, no. 3, pp. 310-322, Fall 1978.

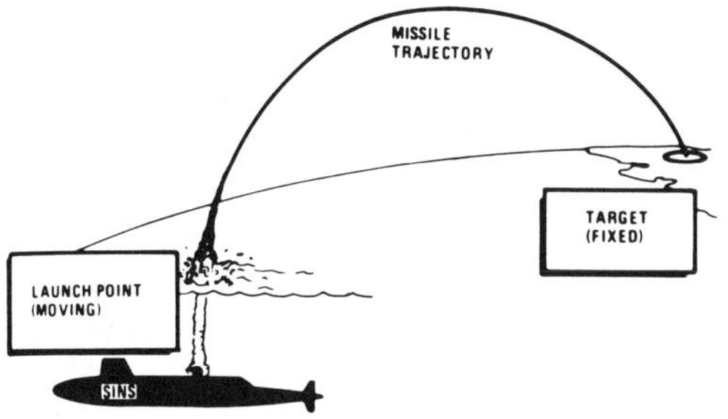

Fig. 1 – Moving Launch Point.

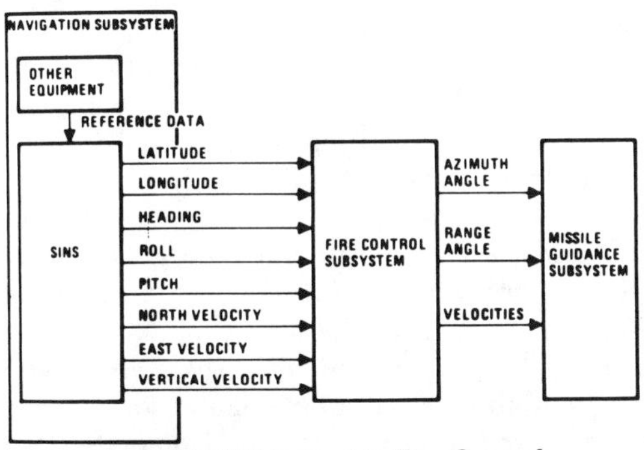

Fig. 2 – SINS Output to Fire Control.

ments for increased survivability, more accurate performance and operational flexibility. This paper traces the SINS evolution in response to these changing requirements.

Ballistic missile error at the target, normally characterized as CEP, is directly affected by accuracy of the launch point information. In the case of the FBM submarines this information is supplied by the SINS. Reductions in missile CEP can require corresponding improvements in the accuracy of the SINS output data.

Submarine survivability affects SINS requirements in two ways. First, the increase in probability of survival is approximately proportional to the square of the increase in missile range. The submarine has a larger area of the oceans in which it can operate while remaining within range of its designated targets. The contribution of SINS errors to the missile target error increases with missile range. Therefore, as missile range increases, SINS accuracy must improve if the missile CEP is to remain constant for the longer ranges.

Second, the SINS outputs, like any inertial navigator, have errors that increase with time and must be periodically corrected or reset. This subjects the submarine to potential detection since the information needed for the "reset" is derived from external position data generally provided by the submarine raising a periscope or antenna. Extending the time interval between the need for a SINS reset reduces the opportunity for detection and increases survivability. Therefore, the time interval of SINS in-specification performance becomes a key parameter influencing the submarine's probability of survival.

The SINS performance requirements are specified classically in terms of the "time RMS accuracy" for a specified reset interval (Fig. 3). Flexibility exists wherein a specific SINS can provide high accuracy for short reset intervals or less accuracy over longer reset intervals (see Fig. 4). The SINS evolution may best be described as an improvement in "effectiveness"

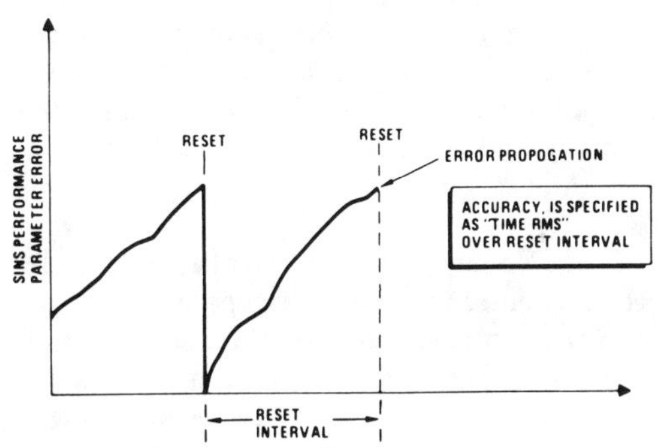

Fig. 3 – SINS Performance.

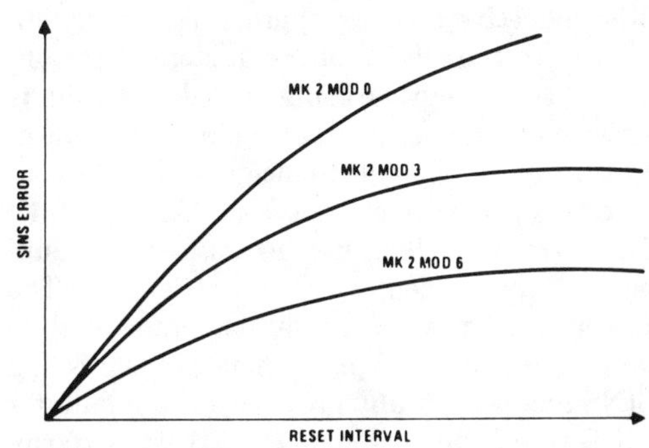

Fig. 4 – SINS Effectiveness, A Function of Accuracy and Reset Interval.

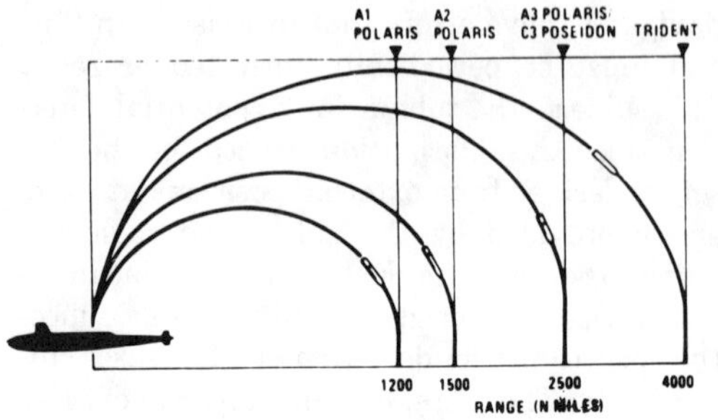

Fig. 5—Advances in Missile Range.

where "effectiveness" is defined in terms of a combination of "accuracy" requirements and "reset interval" requirements.

Fig. 5 illustrates the historically changing requirements for the FBM weapon systems beginning with the Polaris A-1 missile and continuing through the Trident C-4 missile. Each evolutionary step has increased the missile's range. Concurrently, missile accuracy requirements have been modified. The net effect has been increasingly stringent performance requirements on the SINS with the introduction of each new missile—tighten/relax the accuracy and/or extend/shorten the extrapolation interval—i.e., improve the SINS "effectiveness".

The SINS has evolved to meet these changing requirements. In some cases, operating experience was sufficient to meet the need. In other cases, special development was required including the introduction of new gyroscopes as well as modification of the basic system design.

Fig. 6, tracing this evolution through the various SINS generations, illustrates that SINS effectiveness has approached an asymptote. With the advent of the Trident Program, a new technology is being introduced that is expected to provide a quantum improvement in effectiveness. That technology is the Electrostatically Supported Gyroscope-ESG (Ref 1). This gyro is used on the Electrostatically Supported Gyro Monitor-ESGM (Ref 2). The ESGM, which is an inertial navigator, will be used to monitor the performance of the MK 2 SINS and periodically provide position data for SINS resets. In so doing, the ESGM performance is in effect transferred to the MK 2 SINS.

THE FIRST SINS

The development of the first SINS, the MK 2 MOD 0, was accomplished during the late 1950's; a period of technological development in the industry that may be best described as revolutionary. Nuclear power for propulsion, solid fuel for rocket motors, inertial guidance for missiles and submarines were new engineering achievements. The transistor was replacing the vacuum tube, and digital computers, with their programming flexibility, were making analog computers obsolete. Large scale digital computers were becoming available to the design engineer and analyst to permit him to solve the more complex problems associated with the emerging technology. "System Engineering" and "Reliability Engineering" were new disciplines.

The MK 2 SINS is the direct outgrowth of a program initiated in 1954 by the Underseas Warfare Branch of the Office of Naval Research. This program called for the preliminary design of an inertial autonavigator for submarine use, and for design, fabrication and testing of a high-precision gyroscope using an autolubricated gas spin bearing.

During the early design and development period of the SINS, at-sea tests were performed on board the USS COMPASS ISLAND with a modified N6A Inertial Navigator obtained from the Air Force's Navaho Cruise Missile Program. The tests demonstrated the capability of at-sea operation of an inertial navigator for long periods of time. This capability was dra-

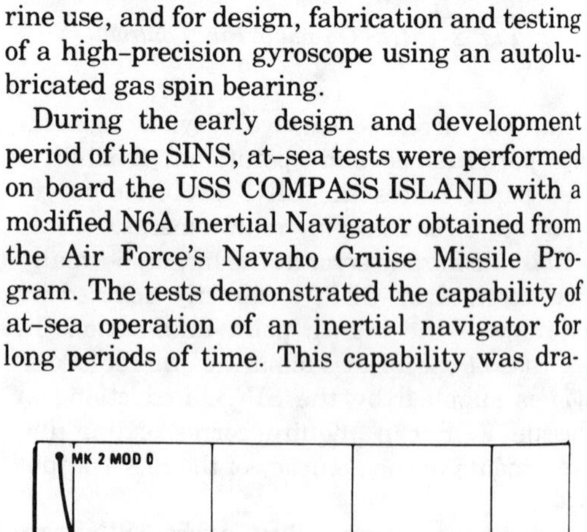

Fig. 6—SINS Evolution.

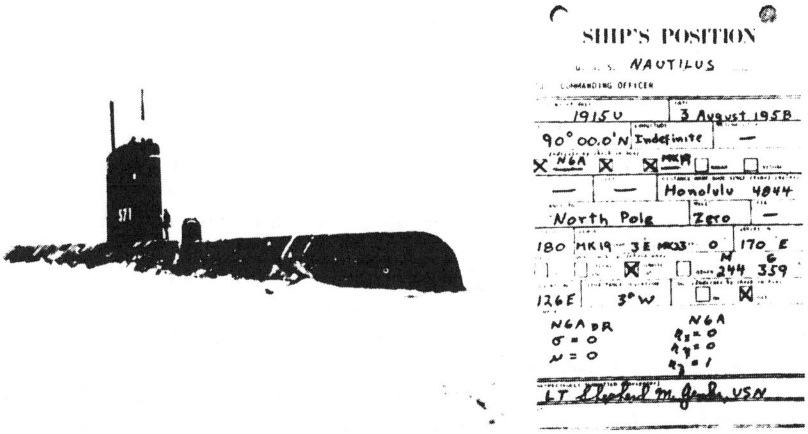

Fig. 7 – N6A Inertial Navigator Guides USS Nautilus to North Pole.

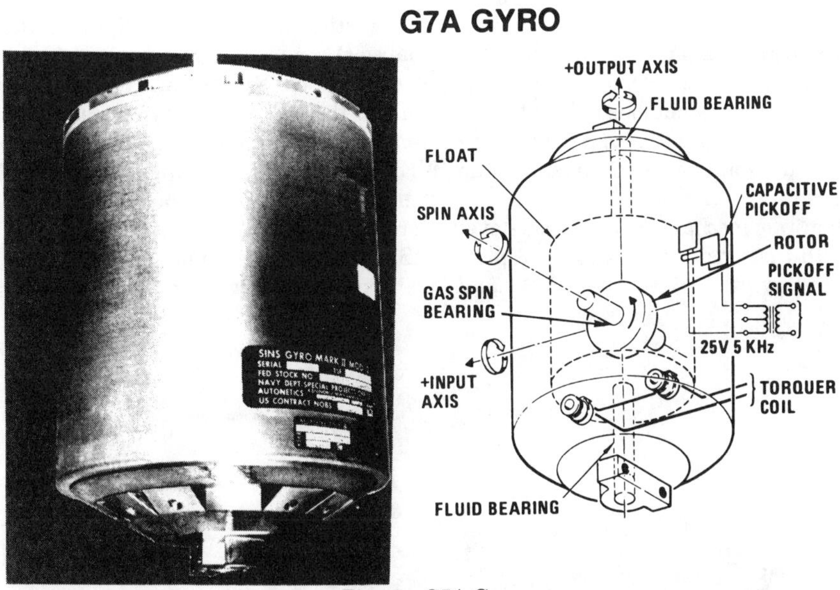

Fig. 8 – G7A Gyro.

matically underscored in 1958 when the USS Nautilus, navigating with an N6A, made its historic submerged transit of the Arctic Ocean. The submarine's precise location at the north pole was determined by the N6A. A photocopy of the handwritten log, recording that major milestone in the history of navigation, is shown in Fig. 7.

The most revolutionary aspect of the first SINS was the conversion of the experimental concept of a gas spin bearing into an operational gyro, the G7A. The associated reduction in unwanted torques, a major source of gyro drift rates, resulted in an order of magnitude improvement in gyro performance over previous ball bearing gyros and provided a gyro that would operate reliably for long periods of time.

The G7A gyro is a single degree of freedom, floated gyro, whose rotor consists of an inertial wheel cemented to a cylindrical hub, supported by an autolubricated gas-spin bearing. An external view of the gyro is shown in Fig. 8.

The float assembly, which houses the rotor

and spin motor is filled with a mixture of helium and hydrogen gases that act as the spin bearing lubricant and support. This assembly is enclosed in an outer sealed housing. The space between the float and outer housing is filled with a dense fluid, fluorolube. Proper choice of the fluid and gyro operating temperature causes the float to be almost perfectly buoyant. This reduces the load on the hydrostatic output axis. Precision torquing of the gyro is provided by a torquer coil mounted on the float assembly.

Three, single-integrating accelerometers (called velocity meters) are used on the system. The velocity meters sense acceleration by means of a pendulous element. The deflection of the element generates a servo signal that drives a dc motor to develop a restoring force opposite to the acceleration. This force is provided by rotating magnets that induce eddy currents in the pendulous element. The magnetic field returns the pendulous element to its null position. The angle through which the motor shaft must rotate is proportional to the sensed velocity.

The system was mechanized as a local-level platform with the sensitive axes of the 3 platform-mounted gyros oriented N-S, E-W, and vertical. The platform is shown in Fig. 9. The local-level platform provided the simplest set of mechanization equations, reduced accelerometer scale factor requirements, and provided ship's roll, pitch, and heading directly from the platform's precision angular readout devices (inductosyns) without the need for complex coordinate transformations. This basic mechanization has remained unchanged through all modifications of the MK 2 SINS. Local level is maintained by computing the earth rate for the submarine's current position plus submarine motion and providing precision torquing to the gyros. The development of precision gyro torquing was a significant achievement and a critical factor in selection of the local-level mechanization.

A digital computer containing a general purpose section and a DDA (digital differential analyzer) section, is programmed to solve the mechanization equations. The computer, designated the VERDAN, was already in production for an Air Force inertial navigator. The VERDAN has a rotating, single-sided disk memory limiting its computational speed to 100 iterations/second.

Significant at the time (1958) was the use of transistors which were coming into widespread

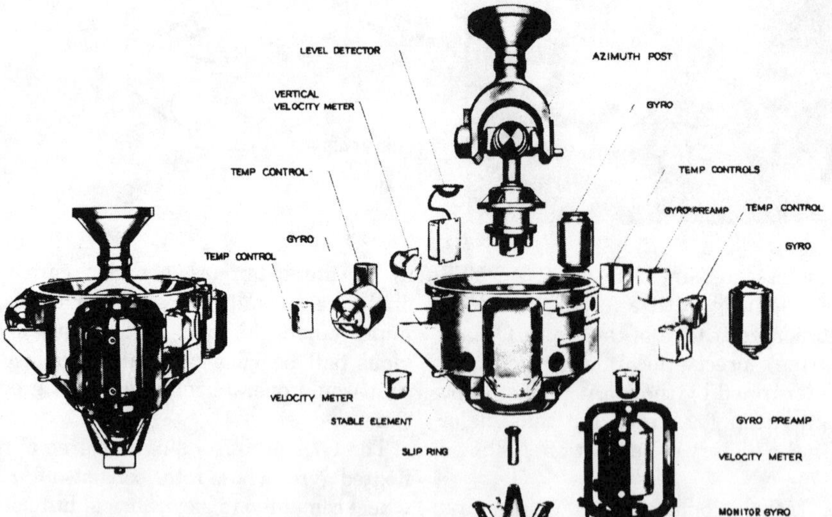

Fig. 9—Stable Platform.

Fig. 10 – MK2 MOD 0 SINS.

application. The SINS was one of the earliest transistorized electronic equipments introduced into Navy inventory.

A picture of the MK 2 MOD 0 SINS is shown in Fig. 10. The basic elements include the stable platform, enclosed in an environmental housing and referred to as the Binnacle which is suspended from a bedplate in the submarine, and a Navigation Console containing the controls, Verdan computer, servo electronics, and precision torquing electronics.

Three SINS were installed in each of the early FBM submarines to ensure the availability of SINS outputs, critical to missile launch and accuracy, during the entire patrol period, generally greater than 60 days.

THE MK 2 SINS EVOLUTION

The evolution of the SINS began as the USS GEORGE WASHINGTON pulled away from the pier at Charleston, South Carolina in November 1960. The MK 2 MOD 0 SINS began accumulating months and years of experience under operational conditions. A total of 56 MK 2 MOD 0 SINS was produced with 3 SINS installed in each of the FBM submarines that carried the 1200 mile Polaris A-1 missile.

Equipment modifications were developed as a product of this experience. Many of these were implemented via the Navy's SPALT program (SPALT refers to Special Projects Alteration). A number of proposed modifications were accumulated and implemented as a block change during the first shipyard overhaul of the submarines. At that time, the SINS were off-loaded and returned to the factory for refurbishment, modification and retest and then returned to the shipyard.

A major element of the block change was the design of a new environmental housing, or binnacle, for the stable platform. The new binnacle contained less temperature gradients and air flow variations – sources of error in the gyro and, ultimately, system performance.

Today, after 15 years experience, the original MK 2 MOD 0 SINS, redesignated the MK 2 MOD 4 after the block change modifications and improvements, and through more efficient use of reference information to reset and update the system outputs, is capable of supporting the performance requirements of the Polaris A-3 missile whose 2500-mile range is over twice that of the original A-1 missile.

However, in 1962 when the planning was underway to add the A-3 missile to the fleet, the improved accuracy required of the SINS could not be achieved by a simple overhaul and refurbishment of the MK 2 MOD 0 SINS. Instead, a new SINS was developed, the MK 2 MOD 2.

The key to this new SINS was a new gyro. This gyro, the G7B, is still a floated, single degree of freedom gas-spin bearing gyro with a hydrostatic output axis. But, experience gained with the G7A gyro indicated that gyro performance could be significantly improved by making certain changes in the gyro's design and selection of materials. By redesigning the induction motor and the rotor, the gyro's angular momentum was increased by a factor of 2.7, reducing the effect of external torques acting on the float and thereby producing lower random drift rates.

The G7A gyro was constructed almost entirely of aluminum, with the exception of the electromagnetic components. The G7B was constructed with a beryllium float assembly. The lighter weight of the beryllium metal allowed mass to be added to the rotor itself, thereby increasing the angular momentum, while maintaining proper float buoyancy. The inherent stability of beryllium also contributed to improved gyro performance. Additional magnetic shielding was included and the design was made more symmetrical to reduce sensitivities to temperature gradients and airflow.

The stable platform environmental enclosure

was also designed to minimize variations in airflow and temperature gradients and so reduce the driving functions operating on the gyro's sensitivity to these sources.

With only two years of operational experience with the MK 2 MOD 0 SINS, it was already apparent that additional computer capacity was required. A new computer, the MARDAN, was designed. This computer used many of the same circuit boards and logic boards in common with the VERDAN, but had a new double-sided disk memory giving it increased capacity (twice the DDA section and four times the GP capacity of VERDAN).

The MK 2 MOD 2 SINS were installed in the 10 627-class submarines.

Perhaps the most significant evolutionary improvement to the SINS was the addition of a fourth gyro, contained in its own gimbal, mounted on the stable platform (Fig. 11). This change was provided for in the MK 2 MOD 2 SINS design, but was not implemented until a comprehensive test and evaluation program was conducted that verified the technique. The change was made in-line in production and retrofit kits were provided to modify, in the submarine, the delivered MOD 2 systems to the MOD 3 configuration.

This gimbaled fourth gyro is referred to as the monitor gyro. The monitor gyro improves the SINS performance by affording a rapid and self-contained method of calibrating the level controlling gyro drift rates.

The monitor gyro is also a G7B, identical to the other three gyros mounted on the platform. The gyro is mounted on a separate gimbal

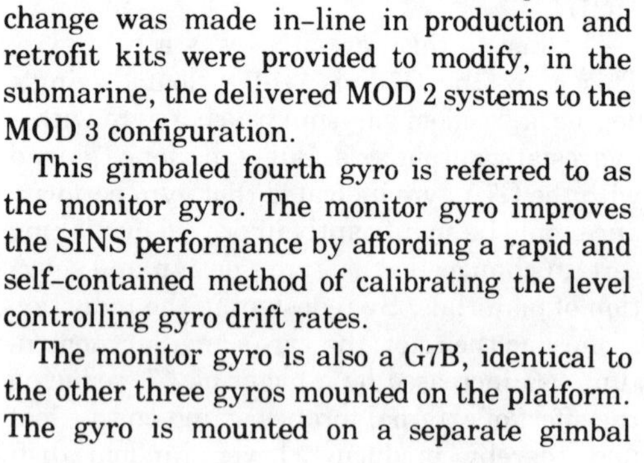

Fig. 11 – Monitor Gyro.

which can be rotated in azimuth, but is otherwise fixed to the SINS stable platform. By precise positioning of the monitor gyro about its output axis, it is possible to align its input axis along the input axes of the north-pointing and east-pointing or level controlling gyros and 180 deg. opposite their input axes.

The monitor gyro is operated in a cage mode where the pickoff signal is used to drive a torquing loop closed about the monitor gyro. The monitor gyro "caging rates," recorded while in the four orientations, are used to calculate the required bias corrections for compensating the drift rates of the two level controlling gyros.

Of particular importance was the monitor gyro's ability to compensate for the drift rate of the east-pointing gyro. Uncompensated east-gyro drift rate is directly proportional to and is a major source of SINS heading error. This error could not be corrected by the normal reset procedures using position information. The early Polaris submarines contained a Type 11 star-tracking periscope that did provide periodic heading reference data used to correct the SINS heading output. However the Type 11's operation was limited to clear skies, minimum sea-states and required the submarine to stay at periscope depth for extended periods while star acquisition and tracking were underway.

The MK 2 MOD 3 SINS (Fig. 12), including the field modifications of the MK 2 MOD 2's, were installed in 27 of the submarines (the 608, 627 and 640 classes). With the heading performance improvement resulting from the monitor gyro, and the overall improvement in system reliability, the Navy elected to install two SINS per submarine vice the original three SINS installed on the earlier classes.

The introduction of the Poseidon C-3 missile, with its 2500-mile range and MIRV capability imposed a new set of performance requirements on the SINS. Specific attention was placed on improved heading accuracy and the MK 2 MOD 3 SINS was modified to a MOD 6 configuration.

A redesigned heading inductosyn package was incorporated into the heading readout loop to reduce the transmission error.

Provisions were incorporated in the SINS to add a mode of operation to better define the

Fig. 12 – MK2 MOD 3 SINS.

stable platform's pitch axis with respect to the submarine's optical reference unit and permit calibration of the east-pointing gyro's input axis misalignment with respect to the pitch axis. This mode of operation is called Pitch Axis Perturbation/Pitch Axis Definition (PAP/PAD) and is discussed later in this paper.

To further improve performance, a gyro selection process was initiated (Fig. 13). G7B gyros were screened during factory selloff tests. These tests included the final drift test, self induced vibration test, and output axis hysteresis test. If the gyro met specified performance criteria for these tests, it was designated a C-3 gyro for use on the MK 2 MOD 6 SINS. If the gyro exceeded the C-3 criteria, but met a less stringent set of criteria, it was designated an A-3 gyro for use on the SINS installed in the submarines still carrying Polaris A-3 missiles.

The MK 2 MOD 6 SINS are currently installed in the 31 submarines carrying the Poseidon C-3 missiles.

THE SOFTWARE EVOLUTION

The improvement with time in system performance, as illustrated by Fig. 6, is not totally the product of changes and advances in hardware design, described above as the SINS evolution. Concurrent with this evolution has been an equivalent evolution in what may be termed the system software. System software includes the procedures and computer programs used to calibrate, reset and operate the SINS.

Early calibration and reset techniques seem primitive and inefficient when compared to present day techniques employing statistical estimating processes and over 80 coefficients in the state vector to describe the system, predict performance and compensate for system errors.

The error model for the MK 2 MOD 0 SINS in 1960 assumed a single, steady-state drift rate for each of the three gyroscopes. The present day error model for the two ESGs installed in the ESGM system has 65 coefficients. These include g- and g^2-sensitivities, input axis misalignments as well as the steady-state drift rates.

CALIBRATION

The performance of an inertial navigation system is dependent on the accuracy in initially determining and then calibrating the systems parameters. These parameters include the inertial instrument's scale factors, biases, alignment of the sensitive axes to the reference coordinate system, g-sensitivities, etc. After initial calibration any variations in these parameters represent system error sources. To prevent the SINS data outputs – position, velocity and attitude – from exceeding the desired performance level, the error sources must be compensated by either a new calibration or by a system reset using reference data.

The SINS has three levels of calibration – factory, dockside (or tender side) and at-sea (Fig. 14). The evolutionary process for calibration has included the development of accurate dockside and at-sea techniques, replacing the factory derived parameters and the dependence on those parameters remaining stable over exceptionally long time periods.

The dockside calibration technique for the MK 2 MOD 0 SINS consisted of a determination of velocity meter biases by rotating one SINS about its azimuth axis while being slaved in its

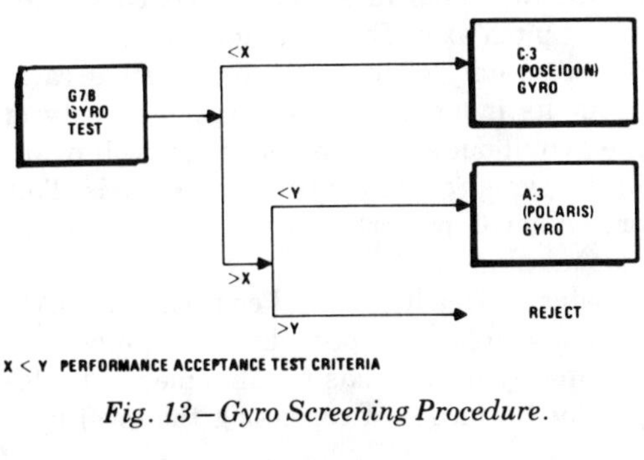

Fig. 13 – Gyro Screening Procedure.

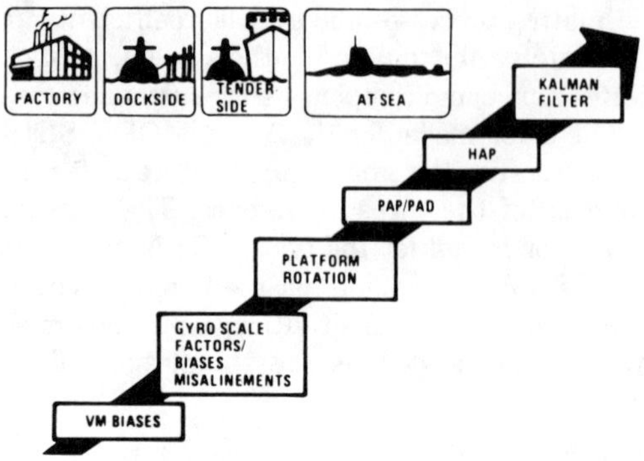

Fig. 14—Evolution of Calibration.

roll and pitch axes to the second SINS then using a rather primitive pulse-counting technique in the gyro torquing loops to determine the initial values of the north-pointing and vertical gyro biases.

The first major step in the evolution of SINS calibration was the development of a technique that collected gyro torquing pulse data with the stable platform's north-pointing, or x-axis, oriented at each of the four cardinal headings 0, 90, 180 and 270 deg. Using the collected data, and solving simultaneous equations permitted the separation of gyro scale factor from bias, thus obtaining dockside, the scale factors for all three gyros.

Developments in heading calibration included Pitch Axis Perturbation/Pitch Axis Definition techniques (PAP/PAD) for the Poseidon MK 2 MOD 6 SINS. The PAD technique optically measures the physical alignment between the SINS pitch gimbal axis and the ship's Optical Reference Unit athwartship's axis. A special mirror was added to the SINS platform for this purpose.

After PAD, PAP is performed using a controlled rotational rate applied about the platform's pitch axis. This technique makes use of the sensitivity of the gyro to rotational rates about its input axis. Any perpendicularity of the gyro input axis to the pitch axis will result in the gyro precessing about the roll axis. This precession is proportional to the gyro's input axis misalignment.

Later, a Heading Axis Perturbation (HAP) technique was developed which determined the monitor gyro train axis misalignments. It also provided a means of correcting the level gyro axis misalignments, velocity meter biases, and the vertical gyro scale factor.

RESET

SINS data outputs may develop significant errors over a period of time. To ensure that the SINS data remains within the desired accuracy requirements, it is necessary to periodically correct the outputs. To minimize the growth of future errors, it is necessary, to the extent possible, to compensate the SINS error sources at the same time the data outputs are corrected. The process of computing and implementing corrections to the SINS outputs and compensations for the SINS error sources is called "reset."

Within the Navigation Subsystem on the submarine there are various fix-taking equipments available (Fig. 15). These equipments provide the position information (and reference heading via the Type 11 Periscope) for the periodic resets. Like the SINS the fix-taking equipment has also undergone an evolution. This evolution has followed three paths: (1) improvement in the hardware (e.g., Loran A replaced with Loran C); (2) addition of new hardware, such as the Transit Satellite and the ESGM; and (3) improvement in the processing of the fix data from these equipments.

Today, the information provided by the fix-taking equipment is position data only. The Type 11 star-tracking periscope was installed in the early classes of submarines. This periscope provided own-ships-heading information that was used to correct the heading output of the SINS. With improvements in SINS performance via the hardware and software evo-

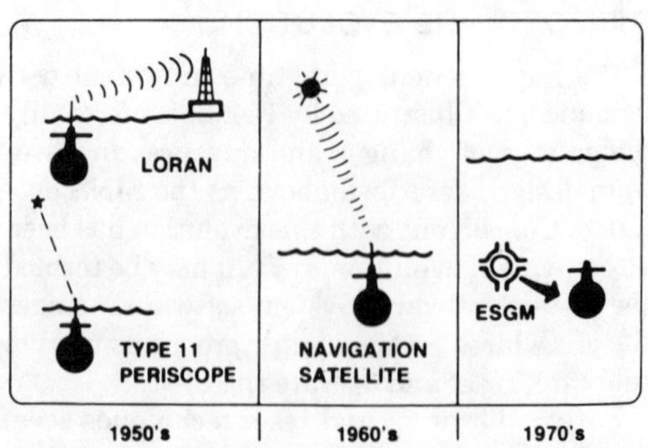

Fig. 15—Navigation Subsystem Fix Taking Equipment.

lution, the SINS heading accuracy surpassed that of the Type 11 and it has since been removed from all submarines.

The first reset technique was called a 3-fix reset. This technique used the data from three position fixes, separated by approximately four hours, together with the equations for the SINS latitude, longitude and heading error propagation, to develop a set of simultaneous equations that could be solved for the SINS output data corrections and error-source compensation. The error-source compensation was limited to the north-pointing and azimuth gyro bias and a correction of the SINS heading output (limited to the drift rate of the east-pointing gyro). The first resets were computed by hand or with the aid of a simple "circle diagram" graph and assumed the heading error correction was made at the instant of the third position fix. This did not allow for the time to compute the reset and the heading correction was, therefore, always out-of-date. The first step in the evolution was to modify the reset equations to permit a "delayed" reset, acknowledging the computational time delay, for the heading correction. This delay also provided the operator an opportunity to validate the accuracy of the fix information before using it in the reset equations.

An extrapolated, three-fix reset was the next evolutionary change. This reset technique permitted the computation of a complete reset with the addition of only one new fix by revalidating the two previously used fixes.

The next step was least-squares fitting of redundant fix data. Three fix values of latitude and two of longitude are all that are required for a complete reset. Additional data are redundant, but can be used to compensate the effect of random error in the fix measurements, thereby improving the accuracy of the reset. The redundant data can also include a large number of position fixes taken at close intervals.

Least-squares fitting was further modified to include age-weighting of the fix data wherein the more recent fixes were weighted the most in the computations. The age-weighting factor is a decaying exponential. Age-weighting helped account for the fact that gyro drift rates change with time and are not constant error sources as had to be assumed up to that time.

Further improvements in reset accuracy were achieved with the use of Kalman Filtering. The Kalman filter takes into account the average error in the optimum estimate of the system, the average error in the fix measurement, and the effect of gyro noise on the SINS outputs. This was one of the first applications of the Kalman technique to operational equipment.

PROCEDURES

The SINS operating procedures have also undergone an evolutionary development which has contributed to the SINS performance improvement.

Through an information feedback process from equipment operators, navigators, Navy and contractor test personnel, analysts and engineers, operational data are evaluated and procedures revised or updated.

Procedures that have been developed over the years include:

1. Monitoring all position reference sources and optimizing the selection process for determining which source to use for a reset.
2. Monitoring of performance trends to determine or forecast when resets are required to maintain system accuracy requirements.
3. Development of criteria and procedures for various operating modes including when to use free-inertial rather than the damped-inertial mode.
4. Applying criteria to determine if and when recalibration is required.
5. Comparing the data outputs of each SINS as a means of early detection of system error growth or malfunction.

THE ELECTROSTATICALLY SUPPORTED GYRO MONITOR – ESGM

An objective of the Navy's newest Fleet Ballistic Missile Weapon System, the Trident, is to reduce the submarine's probability of detection. This objective is to be accomplished by: (1) increasing the submarine's operating area and by (2) significantly extending the time interval between resets (derived from external position fixes) needed to maintain the SINS output data accuracy.

The first objective will be achieved with the Trident C-4 missile which has a range of over 4000 miles. The poseidon C-3 missile range is 2500 miles.

The second objective will be achieved by the addition of the Electrostatically Supported Gyro Monitor (ESGM) to the equipments in the submarine's navigation center. Design modifications are also being made to the SINS to reduce the probability of the ship's detection through unwanted EMI and noise radiation. (The modified SINS has been designated the MK 2 MOD 7.) However, it was judged that a single degree of freedom, floated gas-spin bearing gyroscope such as the G7B could not provide the performance improvement needed for the objective extension of the SINS reset interval.

A new gyro technology was selected – the Electrostatically Supported Gyro (ESG). The ESG is a two degree-of-freedom gyro whose spherical rotor is spinning while suspended in an electrostatic field inside a cavity maintained at a near-perfect vacuum. Such a gyroscope represents the best approximation to a theoretically perfect gyro, free of the classical bearing friction and many of the random torques that produce unwanted drift rates. Preliminary test data of the ESG indicated a quantum improvement in performance with a system using ESGs over one using conventional SINS gyroscopes. However, the program was still in the development state when the Navy needed to make a decision regarding the Trident submarines complement of navigation equipments. To minimize risk they elected to retain two MK 2 MOD 7 SINS and add one system with ESGs, designated the ESGM. The role of the ESGM is to act as a source of position fix data, internal to the submarine, that can be used to periodically reset the SINS, thereby imparting its performance capability to the SINS and in so doing, achieve the objective extension in the time interval between resets using external position fix information (Fig. 16).

The ESGM has been designed with the consideration of adapting it as an inertial navigator with the same outputs as the MK 2 SINS. After sufficient operational experience is gained, it may replace the MK 2 SINS.

The development of the ESG did not occur overnight (see Fig. 17). Initial studies began at

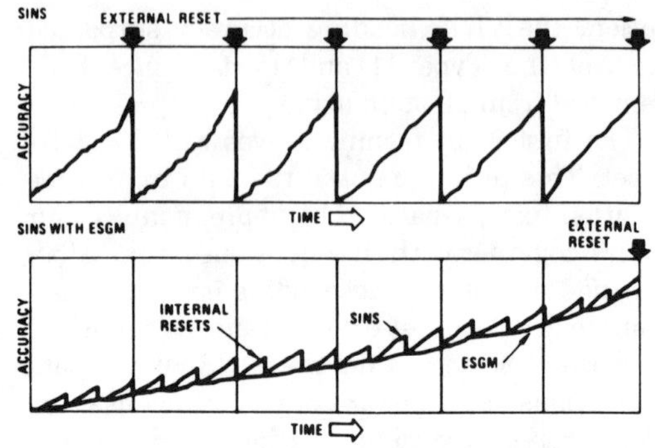

Fig. 16 – Reset Interval Extended with ESGM.

Autonetics in 1959, at the same time the first SINS, the MK 2 MOD 0, was being built. In 1970, a gimbaled ESG was tested that demonstrated the feasibility of using the design for accurate long-term navigation.

Fig. 18 is a photograph of the ESG, designated the G11A. The gyro contains a 1-centimeter diameter solid beryllium rotor, spinning at 216,000 rpm (3600 Hz) in a vacuum and supported by a servoed electrostatic field. The gap between the rotor and cavity is on the order of a few hundred microinches. Sphericity of the rotor at operating speed must be maintained within a few microinches to minimize g- and g^2-sensitivity.

The gyro electronics are contained within the gyro case and are multipurpose in that they provide the suspension servo forces, pickoff signal, and speed control. The gyro has two degrees-of-freedom and makes use of a deliberate radial mass unbalance in the rotor to generate the pickoff signals. The gyro is also periodically case reversed 180 deg about its spin axis to effect a cancellation of case fixed drift rates.

The mechanization of the ESGM system is a departure from the MK 2 SINS mechanization. The SINS is mechanized as a local-level system. To avoid torquing the ESGs (a source of error), the ESGM is a space-stabilized system. One gyro is oriented with its spin axis polar and the other gyro is oriented with its spin axis nominally in the equatorial plane.

Mounted to the same platform member as the gyros are three electromagnetic accelerometers (EMAs). The EMA (Fig. 19) is conceptually an extremely reliable instrument since it has only one moving part, a proof mass. Because of the space-stable platform mechanization of the ESGM, the EMAs tumble with respect to the local gravity vector. This imposes stringent requirements on the EMA's scale factor linearity, long term scale factor stability, and long term bias stability. The EMA has met these stringent requirements and is now in production.

In addition to the gyro and the accelerometer, the ESGM represents several other changes from the basic design used on the SINS.

The space stable platform requires coordinate transformations from the platform coordinates to the submarine's roll, pitch, and heading coordinates. The SINS local-level mechanization provided these outputs directly. For the ESGM, the transformation is accomplished in the ESGM dedicated computer as part of the navigation program mechanization. A dual speed, solid-state A/D converter was developed

1959 — INITIAL STUDIES BEGUN

1961 — FIRST ROTOR SUSPENDED

1963 — FIRST BERYLLIA CAVITIES (1⅝" DIA)

1965 — SIX-PLATE, HOLLOW-ROTOR GYRO TESTED

1966 — TWELVE PLATE, SOLID-ROTOR GYRO TESTED

1968 — FIRST 1 CM ESG TESTED

1970 — FIRST GIMBALED 1 CM ESG TESTED

1972 — FIRST G11A MARINE ESG TESTED

1974 — ESGM PRE PRODUCTION CONTRACT AWARD

Fig. 17—History of ESG Development.

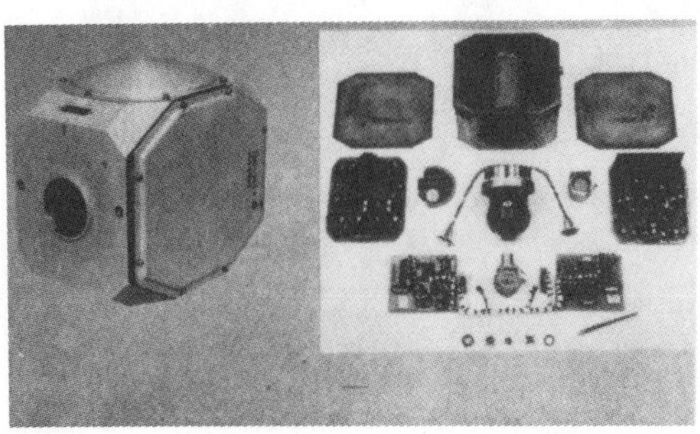

Fig. 18—G11A Gyro (ESG).

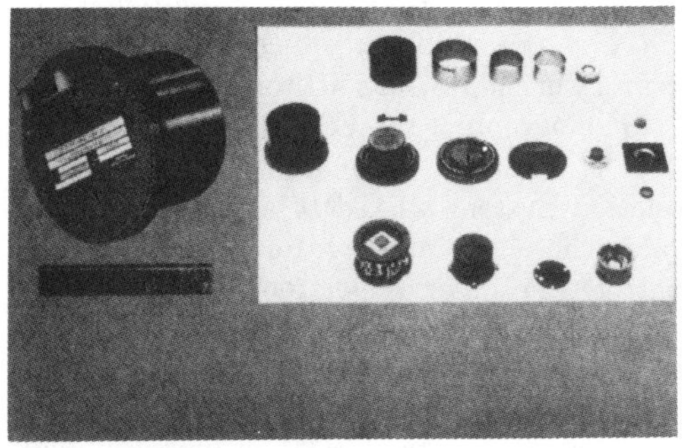

Fig. 19—A188 Electromagnetic Accelerometer.

Fig. 20—Electrostatically Supported Gyro Monitor (ESGM).

which is able to track, with the necessary precision, gimbal rates in excess of 50 degrees per second.

Thermal sensitivities of the inertial instruments required a binnacle temperature control system using a platform shroud to maintain a fixed flow of air across the inertial instruments regardless of ship's heading or attitude. A successful design has been demonstrated, with temperature control of the internal parts of the gyro to better than 0.01°F.

Separated from the cavity walls by a few hundred microinches, and supported only by the electrostatic forces, the gyro rotor is susceptible to catastrophic damage if, while spinning, it should contact the cavity wall due to shock or vibration. A rectilinear shock isolation system was designed for the stable platform to protect the ESG's from such damaging inputs. A rectilinear system avoids introducing rotational motion into the platform which can affect the accuracy of the ESGM's attitude outputs.

A photograph of the ESGM is shown in Fig. 20. The binnacle is mounted in the Shock Isolation System affixed to a submarine bedplate. The electronics control console contains control panels, servo electronics and interface buffers. The electronics are designed to conform to the Navy's Standard Electronic Module (SEM) program.

Calibration of the ESGM is more complex than the SINS. This is due in part to the higher accuracy required of the system and the resultant extension of the reset interval. The longer reset interval means the submarine may operate over a wider latitude between resets. Certain system/gyro parameters are latitude sensitive. Previously, the errors derived from these parameters could be ignored since the potential latitude differential was less because the reset interval was shorter. In the ESGM, however, these sensitivities need to be calibrated. This requirement, coupled with the extensive modeling of the gyro resulted in a total of over 80 coefficients in the state vector defining the ESGM system.

To calibrate all of these terms, and separate out those that are latitude sensitive, a calibration procedure has been developed that involves positioning the stable platform in four orientations while collecting data.

In addition to the number of terms, the ESGM calibration is complicated by the accuracy required. Extension of the reset interval requires that terms that contribute to growing latitude and longitude error be calibrated to an accuracy in direct proportion to the length of the reset interval. In the ESGM, many terms must be calibrated to an order of magnitude better than SINS.

Four preproduction ESGM systems have been manufactured and are currently undergoing test and evaluation. One of these is installed on board the USS COMPASS ISLAND, the same Navy Ship on which the N6A system was installed almost 20 years ago to initially demonstrate the feasibility of inertial navigation at–sea.

The Navy has authorized the manufacture of the first 19 production ESGM systems.

The initial phase of the ESGM evolution has begun.

REFERENCES

1. M. W. McMurran and J. A. Ling, "Development and Testing of a Precise Marine Electrostatic Gyroscope," presented to The Institute of Navigation National Meeting, West Point, N.Y., 27–29 June 1972.
2. O. L. Elton and J. P. Moore, "Marine ESG Navigation as a Capability for the Present," presented to The Institute of Navigation National Marine Meeting, U.S. Naval Academy, Annapolis, Md., 19 October 1972.

NAVY NAVIGATION SATELLITE SYSTEM STATUS

Robert J. Danchik
The Johns Hopkins University Applied Physics Laboratory
Laurel, Maryland 20707

Presented at
Position Location & Navigation Symposium, 1988
Orlando, Florida
November 29 - December 2, 1988

The Navy Navigation Satellite System (TRANSIT) provides global system for navigation, surveying, and time distribution.

TRANSIT is widely used by the U. S. Navy, by commercial users, and surveyors as a worldwide, highly reliable, precise all-weather navigation/survey system. Currently, there are 11 spacecraft in the constellation. Three of these spacecraft are of the newer NOVA design, and seven are of the older OSCAR design. Three NOVA and four OSCAR Spacecraft are operational. Four OSCAR Spacecraft are being stored in orbit. We will review the current status and future of the TRANSIT System.

SYSTEM OVERVIEW

The TRANSIT System was invented in 1957 by the late F. T. McClure while employed at The Johns Hopkins University Applied Physics Laboratory (APL). The APL has played a central role in the development of TRANSIT and continues today as the Navy's Technical Agent. GE ASTRO has produced the production spacecraft. The Naval Space Command's Navy Astronautics Group maintains and operates the spacecraft in orbit.

The TRANSIT System was developed to provide periodic navigation fix updates to the Fleet Ballistic Missile Weapon System and continues to be managed by Strategic Systems Programs. NASA, Langley, Va., provides the procurement and management for the Scout Launch Vehicle shown in this viewgraph. The Scout is produced by the LTV Missiles and Electronics Group Missiles Division. The U. S. Air Force Space Division provides the Range and Support activity necessary to provide successful launches out of the Vandenberg Air Force Base.

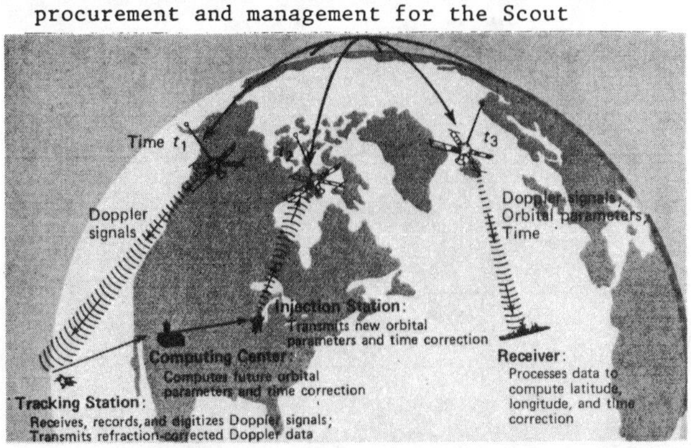

Figure 1

Figure 2

The Doppler Navigation System concept is shown in Figure 1. One satellite provides full system precision. Navigation fix availability improves as the number of satellites increase.

The Scout launch vehicle (Figure 2) has been a very reliable vehicle for the TRANSIT Program. It has successfully launched every operational satellite attempted. We had only one failure to launch an experimental satellite in the mid 1960's. The Scout is 84 of 88 since it became operational. The last launch failure was in December 1975.

Reprinted from *Rec. IEEE Position Location and Navigation Symp.*, pp. 21-24, 1988.

CONSTELLATION

Three launches are planned for GFY 1988. With this plan, sufficient spacecraft should be in the constellation to provide adequate navigation fix availability thru 1996. At this time, two Scout launches have successfully added 1 NOVA and 2 OSCARs to the constellation of spacecraft. The orbit configuration as viewed from a position above the poles is shown in Figure 3. Figure 4 shows the artist concept of the NOVA in the foreground and the OSCAR in the background. These spacecraft will look the same to a user except for the higher signal level that is transmitted by NOVA. There is an OSCAR launch configuration that is called the Stacked-OSCARS-on-Scout (SOOS). Summary characteristics for the spacecraft are shown in Figure 5.

The orbit configuration we showed earlier shows the constellation as it exists today. Because the OSCAR Spacecraft have a long lifetime (14 years demonstrated in orbit) and the orbit planes of the OSCAR Spacecraft precess relative to each other, this constellation view will change with time. Figure 6 shows how the orbit planes precess relative to each other. This precession is due to the launch accuracy of the Scout and is not a program concern. NAVASTROGRU will maintain spacecraft in an operational mode as necessary to provide good earth coverage. The current constellation status is shown in Figure 7.

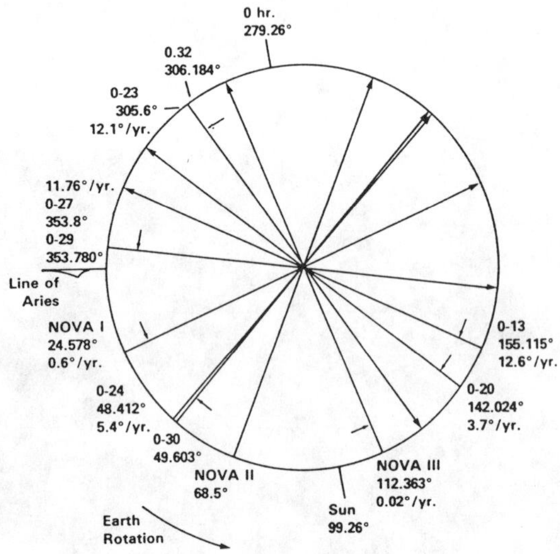

Figure 3

Figure 4

CHARACTERISTIC	OSCAR	SOOS	NOVA
WEIGHT (POUNDS)	130	280	370
EPHEMERIS	16 HOURS	16 HOURS	8 DAYS
POWER OUTPUT/POLARIZATION			
400 MEGAHERTZ	2 WATTS/RHC	2 WATTS/LHC	5 WATTS/LHC
150 MEGAHERTA	1 WATT/LHC	1 WATT/LHC	3 WATTS/LHC
GRAVITY GRADIENT STABILIZED	YES	YES	YES
STATION SEEKING/ORBIT ADJUST	NO	NO	YES
STATION KEEPING/ORBIT MAINTENANCE	NO	NO	YES
DRAG COMPENSATION	NO	NO	YES
ATTITUDE CONTROL	1 AXIS	1 AXIS	3 AXIS
OPERATIONAL FREQUENCY OFFSET (PARTS PER MILLION)	≈−80	≈−80	−84.48
MAINTENANCE FREQUENCY OFFSET	NO	YES	YES
PROGRAMMABLE COMPUTER	NO	NO	YES

RHC = RIGHT-HAND CIRCULAR POLARIZATION
LHC = LEFT-HAND CIRCULAR POLARIZATION

Figure 5

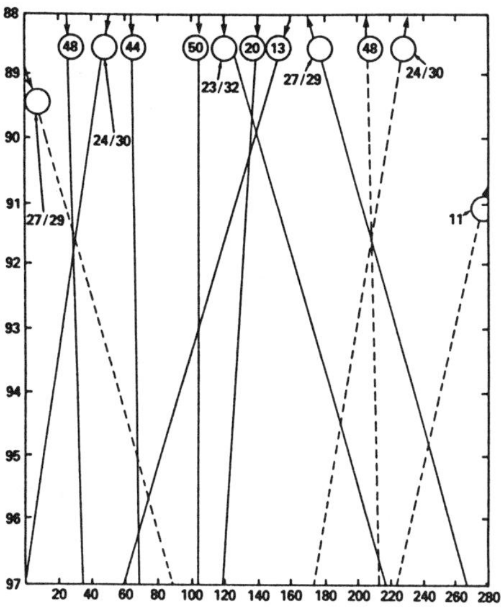

Figure 6

GROUND SYSTEM

The system is operated and maintained by the NAVASTROGRU division of the Naval Space Command.

Headquarters is located at Pt. Mugu, California. Spacecraft tracking and command sites are located in Maine, Minnesota, California, and Hawaii.

Like the space element, the nature of the ground system provides extensive redundancy. NAVASTROGRU personnel provide an efficient and cost effective operation to the U. S. Navy. Satellite availability is better than 99.9%. A designed operational satellite is rarely out of service.

USERS

The number of system users continues to expand, particularly at the low cost end of the spectrum, where single channel receivers are sold at prices beginning at about $1,000.

I am not familiar with how the single channel receivers are implemented, however, it should be noted that users with single channel receivers may see larger than usual navigation errors during periods of large solar storms. This is independent of the spacecraft position error that NAVASTROGRU will be managing, although the source is the same solar activity. This error is caused by an apparent change in the distance between the Satellite and the User caused by the disturbance to the ionosphere. In order to reduce this error, the spacecraft broadcasts the average daily S number which is a measure of the solar activity. An algorithm to provide an error correction was published during the last peak solar cycle.

Two frequency receivers should not see an increase in the navigated position error because the two frequencies provide for good ionospheric error correction. Figure 8 shows the nominal coverage the system will provide to users over the life of the program.

Over the past five years, industry has been meeting the challenge for relatively inexpensive receivers; there has been a significant increase in the number of pleasure craft using the system.

GEODETIC REFERENCE SYSTEM

Since the initial geodetic work in the early 1960's, the TRANSIT System has constrained the tracking station coordinates to be consistent whenever a new geodetic model was introduced.

In 1989, there is a plan to change to the WGS-84 reference system. When this change occurs, precision users will see a shift in longitude position. These changes will be on the order of 20 meters at the equator and will become less with the geometric sine of the latitude as you proceed toward the

NNSS CONSTELLATION STATUS

(JULY 1988)

SATELLITE		TIME IN SERVICE	STATUS
0-13		21 YR 2 MO	OPERATIONAL
0-20		14 YR 4 MO	OPERATIONAL
0-24	SOOS-1	3 YR 4 MO	OPERATIONAL
0-30	SOOS-1	3 YR 4 MO	STORED IN ORBIT
0-27	SOOS-2	10 MO	STORED IN ORBIT
0-29	SOOS-2	10 MO	OPERATIONAL
0-23	SOOS-3	3 MO	STORED IN ORBIT
0-32	SOOS-3	3 MO	STORED IN ORBIT
N-48		7 YR 1 MO	OPERATIONAL
N-50		3 YR 10 MO	OPERATIONAL
N-49		1 MO	OPERATIONAL
0-25	SOOS-4	–	AUGUST LAUNCH
0-31	SOOS-4	–	AUGUST LAUNCH
0-11		10 YR 8 MO	INACTIVE

Figure 7

poles. Discussion of this topic is beyond the scope of this paper. Those interested should refer to two papers presented at the Fourth International Geodetic Symposium on Satellite Positioning held at The University of Texas at Austin, April 28 - May 2, 1986. The first paper is "World Geodetic System 1984" by B. Louis Decker, Defense Mapping Agency, Aerospace Center, St. Louis, Missouri. The second paper, "Testing of the World Geodetic System - 1984 In Precise Orbit Determination and Point Positioning" by James P. Cunningham, Carol Malyevac, and Robert Whitsell of the Naval Surface Weapons Center, Dahlgren, Va., and Stephen Malp of the Defense Mapping Agency (HTC), Washington, D. C.

Another reference is report number DMA TR 8350.2 prepared by the DMA WGS-84 Development Committee titled "Department of Defense World Geodetic System 1984 - Its Definition and Relationships with Local Geodetic Systems". It is available through the Defense Mapping Agency.

The TRANSIT System continues to be the reference system for ship navigation and surveyors both on land and in the broad ocean area, and for time dissemination.

FUTURE PLANS

The U. S. Navy plans to continue to operate the system through 1996.

The plan is to operate the NOVA and OSCAR Spacecraft to provide uniform global coverage during the remainder of the program with up to seven operational spacecraft consistent with satellite on orbit life and orbit dispersion.

SUMMARY

This is a slide (Figure 9) showing the history of the system from conception to the end of the program in the late 1990's.

The TRANSIT System is healthy, and the plans are to continue to provide high accuracy outputs and good global earth coverage from the spacecraft.

NNSS COVERAGE SUMMARY

Number of Satellites	Latitude	Average Wait Time	Number of Gaps (2-4 hrs)	Number of Gaps (4-6 hrs)	Number of Gaps (>8 hrs)	Percentage (%) (>8 hrs)
1	0	8.23	0	0	49	56
	45	6.20	8	30	59	50
2	0	4.50	28	32	15	9
	45	3.10	60	57	3	1
3	0	3.02	88	22	6	3
	45	2.17	110	12	0	0
4	0	2.31	87	20	4	1
	45	1.72	79	14	0	0
5	0	1.95	82	23	1	0
	45	1.47	79	13	1	0
6	0	1.66	81	20	1	0
	45	1.25	74	6	1	0
7	0	1.41	88	15	0	0
	45	1.08	56	5	0	0

NOTE:
1: Days 020-050, 1988 - 024
2: Days 020-050, 1988 - 024, 013
3: Days 020-050, 1988 - 024, 013, NIII
4: Days 020-050, 1988 - 024, 013, NIII, NI
5: Days 020-050, 1988 - 024, 013, NIII, NI, 020
6: Days 020-050, 1988 - 024, 013, NIII, NI, 020, 011
7: Days 020-050, 1988 - 024, 013, NIII, NI, 020, 011, 030
8: 15°-75° EL
9: 5 min Interference

Figure 8

THE LIFE AND TIMES OF TRANSIT (USN PERSPECTIVE)

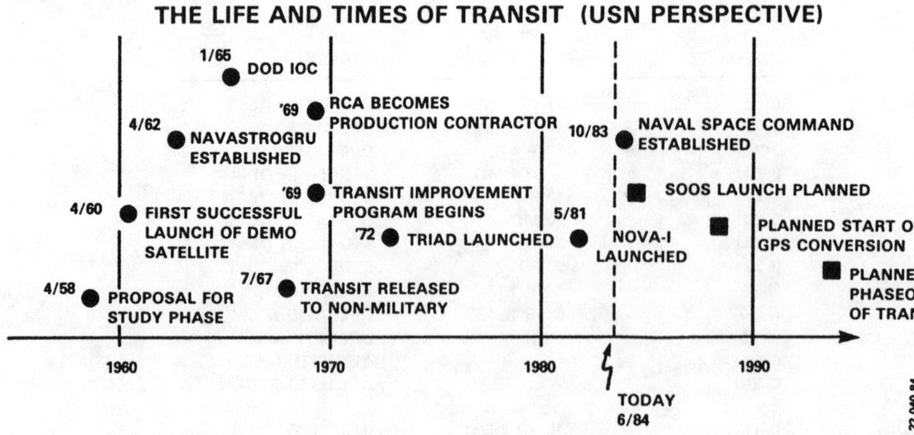

Figure 9

Time synchronization from Loran-C

The capabilities of Loran-C for time synchronization are opening new vistas relating to the conception and implementation of communications and measurement systems

L. Dennis Shapiro *Aerospace Research, Inc.*

Loran-C navigational transmissions, at a frequency of 100 kHz, can be used to provide clock synchronization over long distances to accuracies within ±1 μs. Results to date represent improvements as great as three magnitudes over other operational long-range radio systems. This article discusses the timing capabilities of Loran-C and illustrates the techniques that are used for the derivation of time and frequency.

Loran-C, a navigation system, was first recognized as a means of microsecond-time dissemination by the National Bureau of Standards.[1,2] The U.S. Naval Observatory (USNO) has conducted Loran-C experiments[3-5] since 1961, and is currently engaged in a cooperative program with the U.S. Coast Guard to provide time services with Loran-C over much of the world.

At the present time, synchronization to within a microsecond of the USNO master clock may be accomplished over much of the continental United States, Puerto Rico, the Hawaiian Islands, and the Northwest Pacific (including Japan). In August 1968, time services will be extended to additional Loran-C networks providing coverage over the North Atlantic and Northwestern Europe.

I. Basic group repetition rates

Identification	Pulse Groups per Second
H	33 1/3
L	25
S	20
SH	16 2/3
SL	12 1/2
SS	10

II. Loran-C group periods for basic and specific rates, microseconds

Specific Rate	Basic Rate SS	SL	SH	S	L	H
0	100 000	80 000	60 000	50 000	40 000	30 000
1	99 900	79 900	59 900	49 900	39 900	29 900
2	99 800	79 800	59 800	49 800	39 800	29 800
3	99 700	79 700	59 700	49 700	39 700	29 700
4	99 600	79 600	59 600	49 600	39 600	29 600
5	99 500	79 500	59 500	49 500	39 500	29 500
6	99 400	79 400	59 400	49 400	39 400	29 400
7	99 300	79 300	59 300	49 300	39 300	29 300

The loran system

Loran (LOng RAnge Navigation)[6] was developed during World War II, at the M.I.T. Radiation Laboratory, to provide ships and aircraft with a means of precise navigation. Basically, a loran chain consists of a master and two or more slave stations. A pulse transmitted from the master station is received via ground wave by a slave station—which, in turn, transmits its pulse at a fixed time later. This fixed time, known as the coding delay, is kept constant by monitor stations that "steer" the chain. Its value is selected to ensure that signals in a chain will always arrive at a receiver in sequence, and not overlap. Coding delay is not to be confused with emission delay, which will be discussed later.

The time difference between reception of a master–slave pulse pair determines a hyperbolic line; the intersection of two such lines gives position. Standard loran, also known as Loran-A, operates in the 2-MHz band. Each station transmits pulses of 45-μs duration, which occupy a bandwidth of 75 kHz. Envelope match is the basis for time-difference measurement. This is normally accomplished by causing the master and slave envelopes

Reprinted from *IEEE Spectrum,* pp. 46–55, Aug. 1968.

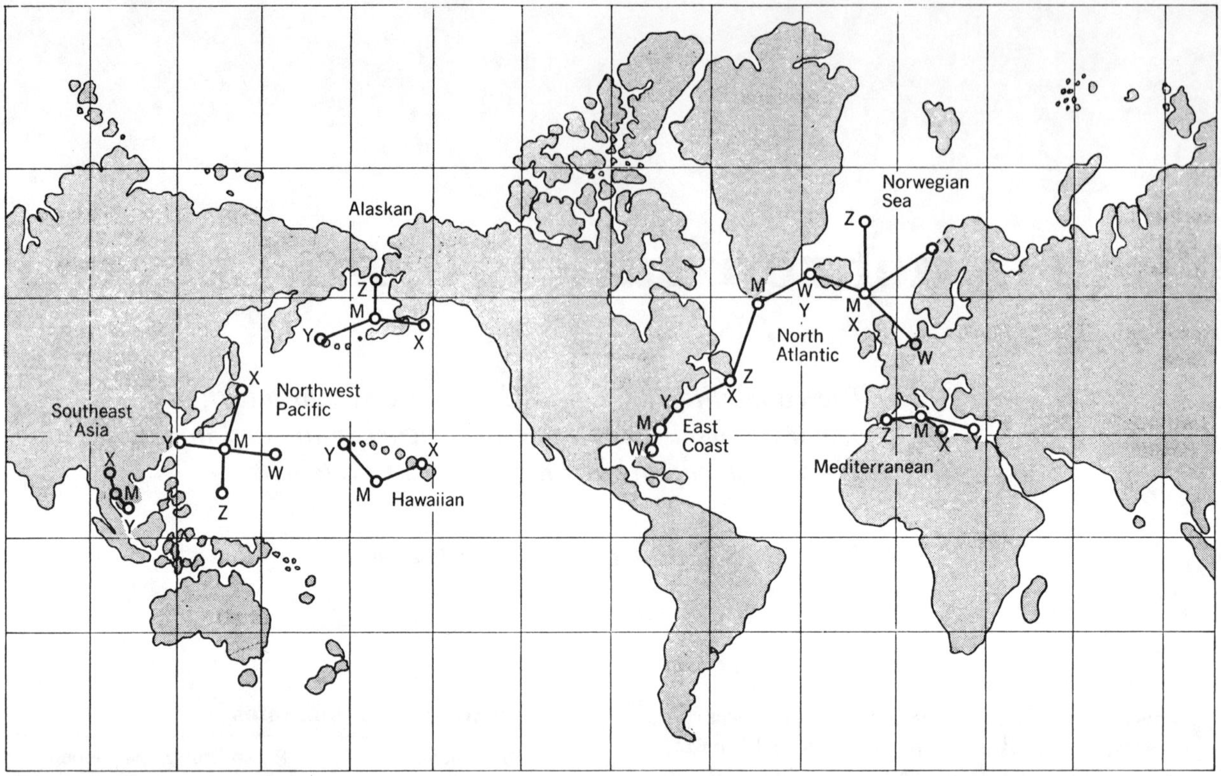

FIGURE 1. Active Loran-C stations.

to be superimposed by adjustment of calibrated delay controls while visually monitoring the waveforms on an oscilloscope.

Loran-C

Loran-C[7,8] was developed to extend loran coverage with fewer stations. Use of a 100-kHz carrier frequency allows greater range due to lower ground-wave attenuation and, consequently, permits use of longer station-to-station baselines. High precision is obtained by both envelope matching of the pulse and phase matching of the cycles within the pulse. The pulse length is approximately 300 μs, and the channel frequency allocation is from 90 to 110 kHz. All Loran-C stations operate at 100 kHz. Adjacent chains employ different repetition rates for identification. The particular rates are selected to be commensurate with the baselines and to provide minimum cross-rate interference. Basic pulse group repetition rates allocated for Loran-C appear in Table I. Coupled with these are offsets in 100-μs period increments that yield a group of so-called specific rates. A complete listing of allocated Loran-C repetition periods is shown in Table II. A Loran-C station may be shared by adjacent chains, in which case the station is pulsed at two rates—one of them taking precedence during instances of overlap.

At the present time, 30 Loran-C stations are in operation. The chains are composed of two triplets, three groups of four, and three of five stations each, as shown in Fig. 1. Table III lists the locations of the stations, the repetition rates, and other pertinent information. Loran-C signals are usable for timing to a microsecond on ground-wave reception to a distance of about 1500 km landward, or 3000 km seaward, and on sky-wave reception at reduced accuracy to about 10 000 km.

The high accuracies obtained with Loran-C result from the ability to separate the ground wave from the sky wave at the receiver and to identify particular ground-wave cycles within the pulse. Loran-C transmitters and their antennas are specially designed to produce a signal

III. Loran-C station data

Chain	Rate	Period, μs	Master[a]	Slaves[a,b] W	X	Y	Z
East Coast	SS-0	100 000	Cape Fear, N.C. ("T" slave sometimes broadcasts from Wildwood, N.J., at ED 84028.7 μs)	Jupiter, Fla. 13 695.5	Cape Race, Newfoundland (4 MW) 36 389.6	Nantucket Island, Mass. 52 542.5	Dana, Ind. 68 564.2
Central Pacific (Hawaiian)	SH-4	59 600	Johnston Island		Upolo Point, Hawaii 15 971.8	Kure Island, Hawaii 35 252.4	
Mediterranean	SL-4	79 600	Simeri Chichi (Catanzaro), Italy		Matratin, Libya 14 107.6	Targabarun, Turkey 32 273.3	Estartit, Spain 50 999.7
North Atlantic	SL-7	79 300	Angissoq, Greenland (500 kW)	Sandur, Iceland (4 MW) 15 068.2	Ejde, Faeroe Islands 27 803.9		Cape Race, Newfoundland (4 MW) 48 212.3
Northern Pacific (Alaskan)	SL-2	79 800	St. Paul, Pribiloff Islands		Sitkinak, Alaska 14 284.4	Attu, Aleutian Islands 31 875.3	Port Clarence, Alaska (1850 kW) 53 069.1
Norwegian Sea	SL-3	79 700	Ejde, Faeroe Islands	Sylt, Germany 30 065.6	Bo, Norway 15 048.1	Sandur, Iceland 48 944.5	Jan Mayen, Norway 63 216.3
Northwest Pacific	SS-3	99 700	Iwo Jima (4 MW)	Marcus Island (4 MW) 15 283.3	Hokkaido, Japan 36 685.2	Gesashi, Okinawa 59 463.0	Yap Island (4 MW) 80 746.5
Southeast Asia	S-3	49 700	Sattahip, Thailand		Lampang, Thailand 13 182.8 (computed)	Con Son Island, South Vietnam 29 522.1 (computed)	

[a] Power 250–400 kW, except where noted.
[b] Emission delay given with reference to master. Information current but subject to change.

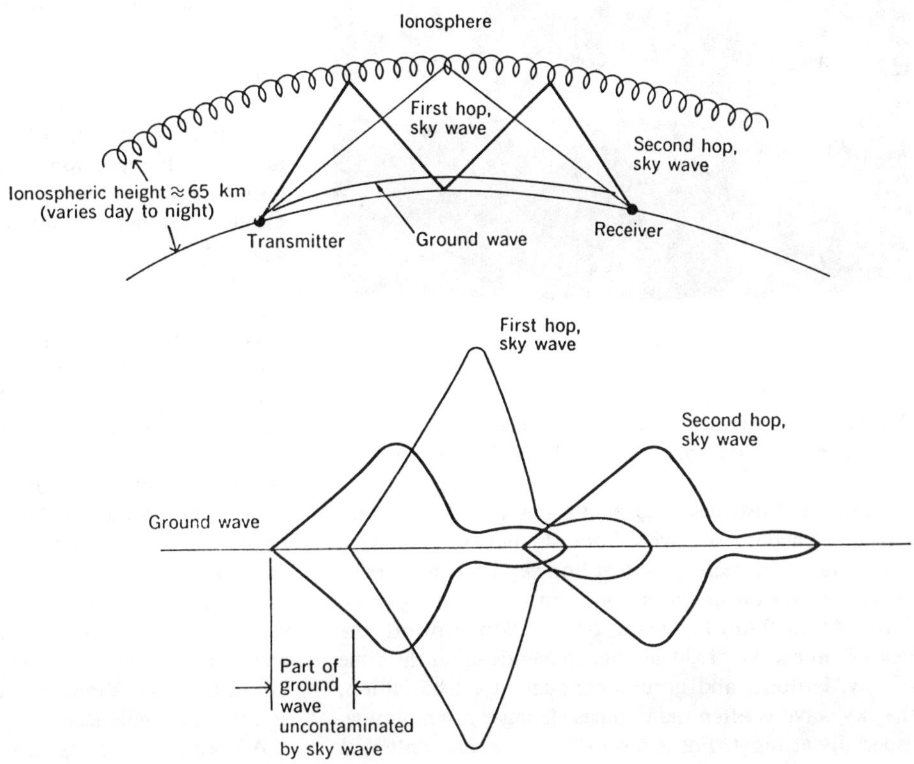

FIGURE 2. Typical ground-wave/sky-wave pattern. As range increases, the sky wave will arrive closer to the beginning of the ground wave. Because of the curvature of the earth, the minimum delay at maximum range reaches an asymptote of about 30 μs. The composite of all the waves may vary in envelope according to the current ionospheric conditions (see Fig. 6).

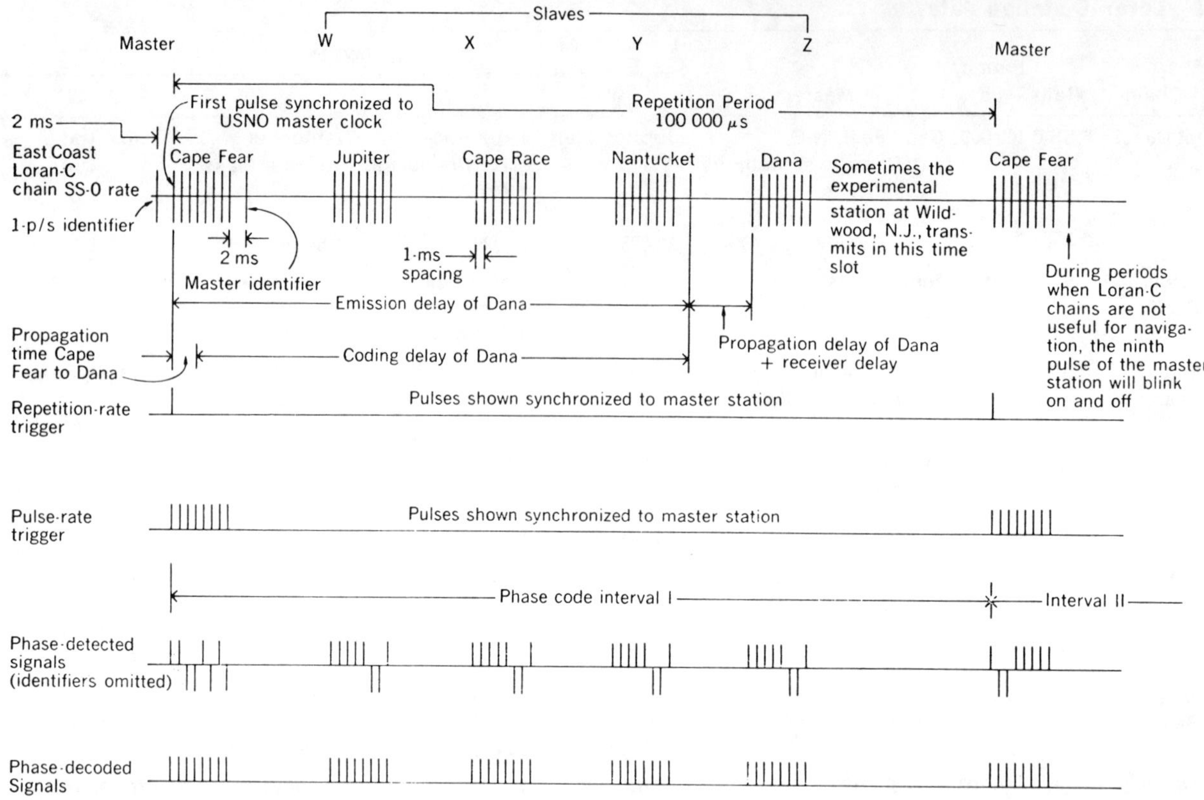

FIGURE 3. Loran-C format and waveforms.

FIGURE 4. Nighttime oscillogram of East Coast Loran-C chain, as taken from Boston. From left to right are the master station at Cape Fear, N.C., and the slaves at Jupiter, Fla.; Cape Race, Nfld.; Nantucket, Mass.; and Dana, Ind. Cape Fear is shown twice to illustrate the presence of the 1-p/s identifier occurring only at the left, positioned 2 ms prior to the first pulse of the master group.

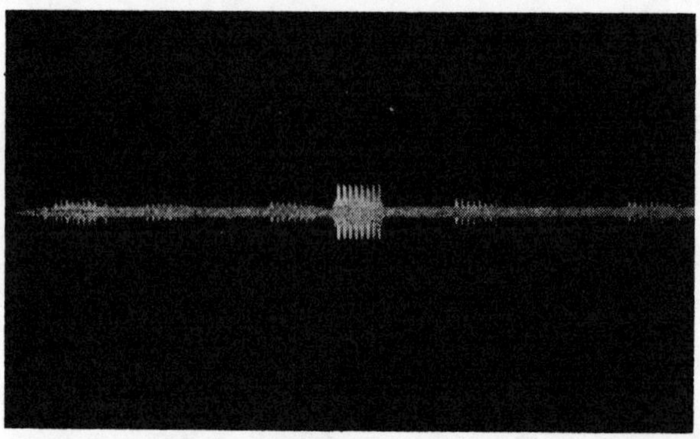

that rises rapidly, achieving an amplitude of 50 percent peak within the first three cycles (30 μs) of the transmitted pulse.

Figure 2 illustrates a ground-wave/sky-wave pattern typically encountered with Loran-C pulses. It is seen that as range increases, the first-hop sky wave will arrive at the receiver progressively closer in time to the ground wave. At 1500 km landward, or 3000 km seaward, the delay is in excess of 30 μs[7] and depends upon the time of day, latitude, and ground conductivity. In addition, the sky wave is often many times stronger in amplitude, especially at night. For a precise measurement, only that

IV. Loran-C phase code

	Master	Slave
Pulse	1 2 3 4 5 6 7 8	1 2 3 4 5 6 7 8
Interval I	+ + − − + − + −	+ + + + + − − +
Interval II	+ − − + + + + +	+ − + − + + − −

+ represents 0° phase; − represents 180° phase. Intervals I and II refer to alternate repetition periods. Slaves always transmit the same interval as the master during any repetition period. Reference to Fig. 3 will help in understanding the phase-code format.

part of the ground wave is used that is not contaminated by the sky-wave signal. Because of this, Loran-C stations transmit a burst of eight pulses spaced 1 ms apart during each repetition period to maximize the uncontaminated ground-wave information at the receiver. The master station transmits a ninth pulse, 2 ms after the eighth, for identification. Figure 3 is a representation of a Loran-C format using the East Coast chain as the model. Additional waveforms are shown to provide an aid to the understanding of Loran-C techniques. Figure 4 shows an oscillogram of the East Coast chain as received in Boston.

It is not unusual, especially at night, for multihop sky waves to arrive at the receiver delayed a millisecond from the ground wave, thereby causing interference to the second through eighth pulses. To minimize this contamination and to reduce the effects of coherent noise, the Loran-C system employs phase coding of alternate pulses and groups. Table IV summarizes the phase code in use on all the Loran-C chains. An examination of Fig. 3 in conjunction with Table IV will help to illustrate the phase-code format. Figure 5 is an oscillogram of a Loran-C signal showing the eight pulses superimposed and the same pulses decoded.

A Loran-C period is initiated by the transmission from

the master station. At certain times later, the slaves transmit their bursts of eight pulses. These times, called emission delays, are known to about 0.1 μs.* The emission delay is equal to the propagation time from the master to a slave, plus the coding delay, as shown in Fig. 3. At the end of the repetition period, a new cycle begins with the transmission of the master station. It is planned that master stations transmit a 1-pulse-per-second identifier. On the East Coast chain, the Cape Fear station transmits a pulse 2 ms previous to the first pulse of the groups synchronized to the USNO seconds' tick.

Ground-wave arrival time is consistent to within 0.1 μs. Beyond ground-wave range, both navigation and timing can be achieved, but with reduced accuracy. Sky-wave consistency can be of the order of a few microseconds when measured according to a prescribed procedure.[9,10] Using sky waves most effectively involves selecting the portion of the sky-wave pulse that is least contaminated by ground-wave or other sky-wave modes. Reference to Fig. 6 will show that this point may not be at the beginning of the pulse.

* The values of the emission delays given in Table III are derived from the data given on Loran-C navigation charts, published by the U.S. Naval Oceanographic Office. Values for the Southeast Asia chain, which are preliminary, were received by the author in a personal communication with the U.S. Coast Guard.

Time derivation

Derivation of time for Loran-C signals involves the precise determination of the time of arrival of the signals at the receiver. From Fig. 5, it is seen that the Loran-C pulse is composed of cycles of the 100-kHz carrier. Time synchronization depends on identifying a particular cycle and then comparing a crossover within the cycle with the reference clock. In order to facilitate cycle identification, the leading edge of the envelope is carefully controlled at the transmitter to conform to the expression

$$f(t) = (kt)^2 e^{-2(kt-1)} \qquad \text{where } k = \frac{1}{72.5 \times 10^{-6}}$$

so that the relative amplitudes of the first few cycles are identical for every transmitter. The envelope reaches a maximum at the positive peak of the seventh cycle, at 72.5 μs (see Fig. 5).

There have been numerous techniques developed for automatic cycle identification. One that we have found to be particularly powerful involves sampling of the relative amplitudes of the first six half-cycles of the pulse (that portion free from sky-wave contamination), and applying these to a matched filter circuit, which will indicate that the cycles selected are, in fact, the ones of interest. The matched filter presents a neutral indication when sampling the proper sequence, and yields negative or positive output readings for early or late sampling, respectively.

One might question the reason for employing special devices such as matched filters for cycle selection because the cycles forming the pulse shown in Fig. 5 are well defined, and one may visually count cycles back from peak, or up from the beginning. Figure 5 shows a simulated pulse uncontaminated by sky waves; however,

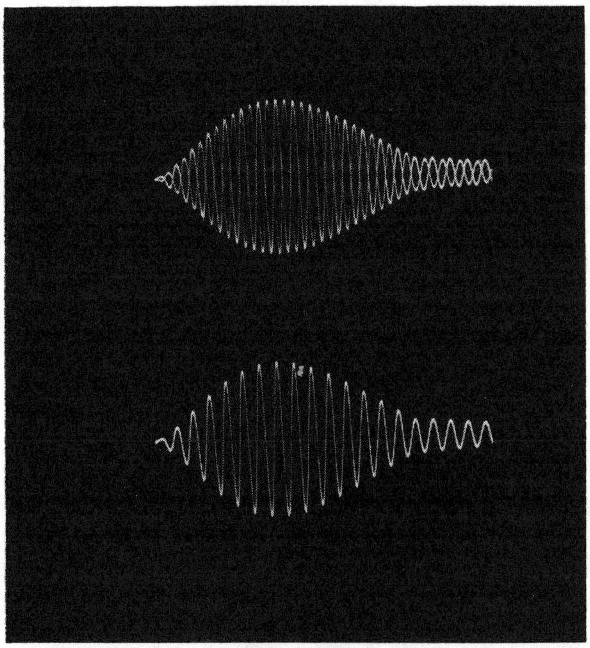

FIGURE 5. Oscillogram of a Loran-C signal showing eight pulses overlaid (top) and the same pulses decoded (bottom). (From Loran-C simulator.)

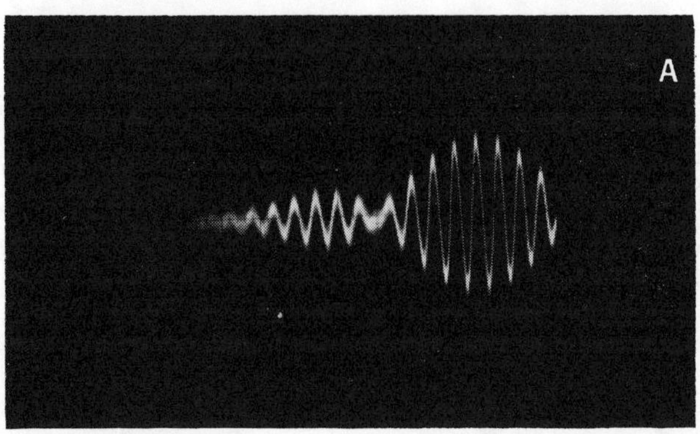

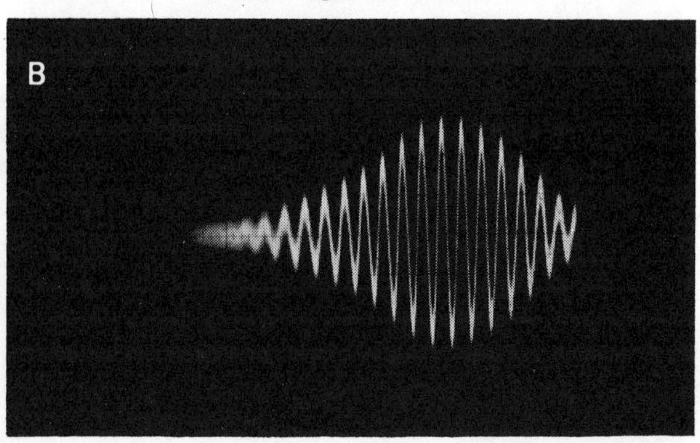

FIGURE 6. Oscillograms of a Loran-C signal, illustrating sky-wave effects. A—Sky wave arriving in phase opposition to the ground-wave component. B—In-phase arrival. It is evident that visual cycle selection is complicated by the sky-wave signal. Oscillograms are of the Cape Fear signal as received in Boston at night.

more typical waveforms are illustrated in Fig. 6. It is evident that selection of the ground-wave peak can be uncertain. Counting from the beginning of the pulse requires that the beginning be well defined. This will be discussed in the next section. In general, the precision of clock synchronization using Loran-C by visual cycle selection is about $\pm 20\ \mu s$.

Cycle identification

The ability to derive time to accuracies of better than about $\pm 20\ \mu s$ when Loran-C pulses are employed depends upon the operator's ability to properly identify a particular cycle. Much has been said and written regarding the cycle-selection problem although little information is available that fully describes the proper technique required.

Historically, Loran-C measurements have been made by observing what appears to be the first three cycles of the pulse, which are assumed to be free of sky-wave contamination. However, when a band-limited receiver is used, observation of these three cycles as being the first three cycles will happen only until the gain of the receiver is increased, or the input level increased; earlier cycles will then be observed, adding complications to this cycle-selection problem.

These earlier cycles, or "precursors," are the low-level output of the receiver that commences upon application of the input pulse. After a delay of about 25 μs, there occurs a sharp upswing of the output envelope function conforming to the pulse output that is normally used for measurement purposes. Consequently, sampling of the received Loran-C pulse during the first three cycles must be redefined as "the first three cycles of significance," which are those occurring approximately 25 μs following the application of the pulse to the receiver. Cycle-selector circuits should be designed to match to these first three cycles of significance and, consequently, will indicate that the receiver is correctly positioned when set for these cycles. This technique is valid as long as the sky-wave amplitude at the receiver is somewhat less than 40 dB in excess of the ground-wave amplitude—at which point the precursor from the sky-wave signal will be of the same order of magnitude as the sampled portion of the ground-wave signal. The receivers used for Loran-C operation should be designed for the minimum precursor for a particular bandwidth.

It is important that the delay through the receiver be known precisely. Normally, this will be specified by the manufacturer to within a microsecond. For absolute measurements requiring greater precision, it is recommended that the exact delay be determined by use of a Loran-C simulator.

Timing receivers

There are three basic types of receiving equipment used for frequency and time synchronization to Loran-C. They are automatic receivers, phase-tracking receivers, and visual receivers. Each will be discussed and comments made about their relative application. The configurations shown are representative of the general complexity required in each case. Various commercial units differ in detail.

The automatic receiver shown in Fig. 7 takes its reference input from a local oscillator at 1 MHz. Internal digital circuitry is switched to select the repetition rate of the station to be observed and to provide the proper master- or slave-phase decoding function. An oscilloscope, which is required for initial adjustment, takes its vertical input from the decoded RF output and its trigger from the pulse rate output. While monitoring the oscilloscope, the trigger pulses are slewed to coincide with the beginning of the Loran-C pulse. After setting the sample

FIGURE 7. Simplified block diagram of an automatic receiver. The oscilloscope used for initial adjustment normally takes its vertical input from the decoded RF output and its trigger from the pulse-rate output.

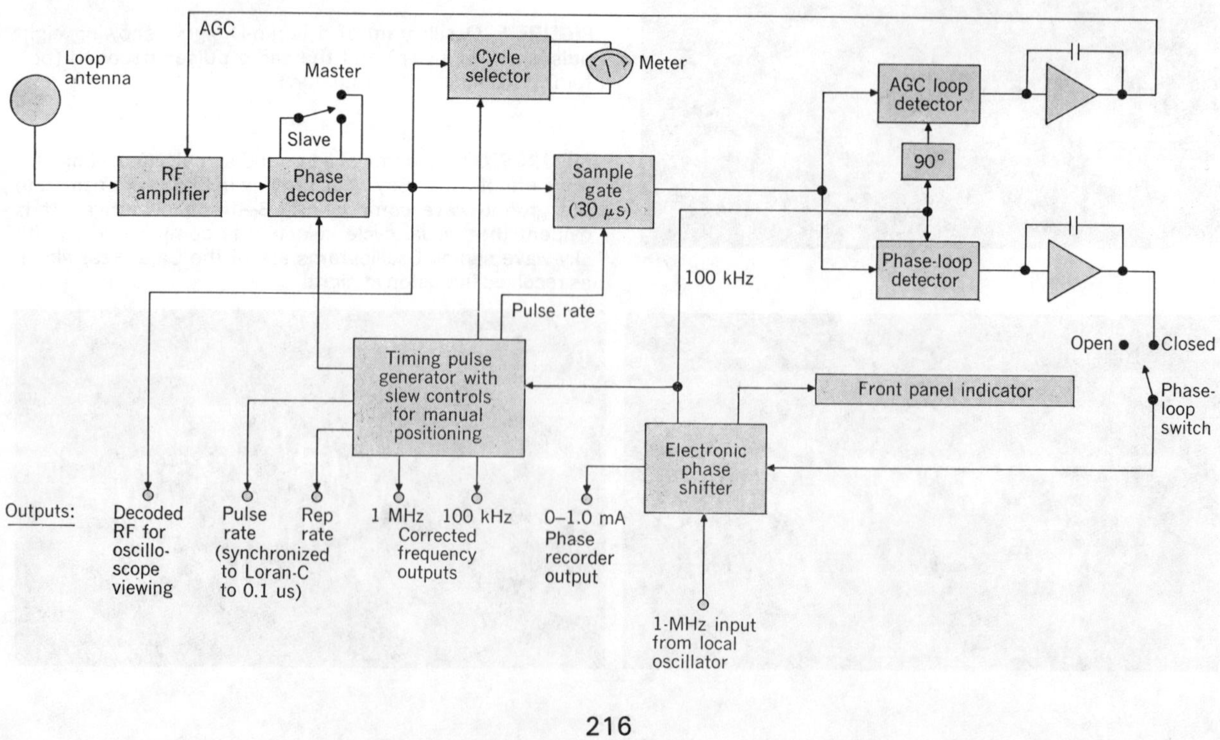

gate, the phase-tracking servo loop is activated. The receiver will automatically adjust the timing of this gate by means of the electronic phase shifter to bracket the three cycles to within 0.1 μs. When the phase loop has stabilized, the cycles that have been selected are sampled by the cycle selector, and output is provided to a front panel meter to indicate whether the proper cycles of the pulse have been selected and, if not, whether the sample gate should be advanced or retarded in 10-μs (one-cycle) steps.

An automatic receiver will indicate when the proper cycle has been selected and will develop a repetition-rate pulse that is synchronized to the time of arrival of the Loran-C signal to within 0.1 μs, after receiver delay correction. A digital counter, calibrated in tenths of microseconds, indicates the phase-error accumulation between the local oscillator and the Loran-C signal; terminals are normally provided for connection to a pen recorder for recording oscillator drift on a continuous basis. "Corrected" frequency outputs, usually available at 1 MHz and 100 kHz, are locked to the phase of the Loran-C signal. This "corrected" frequency possesses the long-term stability of the Loran-C station and can be used to drive clocks directly or to provide steering to local oscillators.

A phase-tracking receiver operates with respect to Loran-C in a similar fashion as a standard very-low-frequency phase comparator to VLF signals. Its phase record of ground-wave signals does not possess diurnal shift, and consequently comparisons can be made faster and with greater precision. An oscilloscope must be used as described previously to position the sample gate. The unit is similar to an automatic receiver in most respects except that it does not contain cycle-selection circuitry. Time reference may be obtained from a Loran-C phase-tracking receiver in the same way as will be described for the visual receiver.

A visual receiver is basically a 100-kHz RF amplifier designed to provide amplification of the Loran-C signal with a maximum degree of fidelity, while at the same time rejecting out-of-band noise and interference. In general, it is used with an external repetition-rate generator whose output pulses trigger an oscilloscope. These pulses must be capable of being moved in the time domain (slewed) so that the Loran-C signal can be positioned appropriately. With simple repetition rates such as those used with the East Coast chain, a 1-p/s (pulse-per-second) trigger may be used to view the Loran-C signal. With other chains it is necessary to use somewhat more sophisticated techniques, which will be discussed later.

Use of both automatic and phase-tracking receivers is most effective within ground-wave range of the Loran-C station. Beyond the ground-wave range, phase-tracking and cycle-selection circuitry become less useful because of ionospheric variations, and more reliance is made on visual techniques for positioning the local clock pulses with respect to the Loran-C station. Both whip and loop antennas are commonly used with Loran-C. For timing purposes (fixed station), a loop is more convenient in that it may be electrostatically shielded and positioned to null local noise centers.

Loran-C time services

With the cooperation of the U.S. Coast Guard, the Naval Observatory has taken steps to synchronize Loran-C transmissions to the observatory's master clock. Cesium-beam oscillators are being installed at the Loran-C master stations to provide a basic stability to the system of the order of a microsecond per day, which is further improved with steering. Since the master clock is on UTC (Universal Time, Coordinated), all frequencies derived from the Loran-C transmissions will be on this scale, which for 1968 is offset by -300 parts in 10^{10} from Atomic Standard Frequency.

Transmissions of Loran-C master stations are steered by the observatory to be synchronous with the UTC second marker pulse, with reference to an epoch. The reference epoch has been established as 0000 hours UT, January 1, 1958, at which time it is assumed that the first pulse of the master station group of each Loran-C chain so synchronized was transmitted. Times of coincidence (TOCs) between this pulse and the UTC second will occur periodically, at which times clocks can be set to the Loran-C transmissions in a straightforward manner. TOC tables for the various Loran-C chains are published by the Naval Observatory. Values of T for currently operating chains, and their synchronization status, are presented in Table V.

Monitor stations have been set up by the observatory in the Far East and Hawaii to provide the steering information. Stations will be added in other areas in the future. Reference to the master clock is made by a combination of traveling-clock, VLF phase-comparison, and satellite techniques. Because of the nature of the Loran-C system, slave transmissions are automatically synchronized to master transmissions to about 0.1 μs and, consequently, may also be used for time and frequency reference.

In addition to the standard navigational transmissions, a number of the Loran-C master stations transmit 1-p/s identifiers synchronized to UTC. They are transmitted without phase code (+ polarity) and on time, except on the East Coast chain, where the Cape Fear 1-p/s identifier is transmitted 2 ms early, since the first pulse of the group is always on time. At present, only Cape Fear and Johnston Island (Central Pacific chain) transmit identifiers. It is planned that other chains will follow this pattern as they become synchronized later this year.

Presently, the observatory is controlling the time of transmissions from the East Coast and the Central Pacific Loran-C chains,[11] from the Northwest Pacific chain,[12] and from the North Atlantic and Norwegian

V. Coincidence interval T and synchronization status for current Loran-C chains

Chain	T, seconds	Synchronization Status*
East Coast	1	operational now
Central Pacific (Hawaiian)	149	operational now
Mediterranean	199	indefinite
North Atlantic	793	operational now
Northern Pacific (Alaskan)	399	indefinite
Norwegian Sea	797	operational now
Northwest Pacific	997	operational now
Southeast Asia	497	indefinite

*Source: U.S. Naval Observatory. Information subject to change.

Sea chains.[13] Phase values published by the USNO of observations with reference to the beginning of the Loran-C pulse from each master station provide corrections to about 1 μs relative to the master clock.

Synchronizing a clock to Loran-C

Synchronization of a clock to Loran-C transmissions is a two-step procedure. External means must be provided for initially setting the clock to within one half of a repetition period to resolve the ambiguity inherent in a pulsed system. For the currently synchronized chains, the half-period interval varies from 30 to 50 ms. Resolution to this degree is obtained from standard HF or VLF time transmissions.

After the period ambiguity is resolved, the clock can be synchronized to Loran-C. This is accomplished by comparing the time of arrival of the Loran-C pulse with the seconds' tick of the clock at a TOC. For the East Coast chain, a TOC occurs every second and, thus, a measurement can be made at any second. Other chains have TOCs only once every *T* seconds and, consequently, a device such as an epoch monitor (Fig. 8) must be used to facilitate the synchronization procedure.

Precise synchronization to the USNO master clock requires a knowledge of propagation delay for the particular transmitter–receiver geometry. This delay can be calculated for ground-wave conditions to within about a microsecond over seawater and a few microseconds over land, and to within about ±20 μs under sky-wave conditions. Measurement of the delay by a portable precision clock will provide a ground-wave synchronization accuracy limited principally by the stability of that clock. By use of certain techniques,[9,10] a sky-wave synchronization accuracy of a few microseconds can be achieved over many paths. The USNO will supply computed propagation delays, on request, to qualified users.

The mechanics of station clock synchronization to Loran-C transmissions may vary according to the type of receiver being used, and the timing precision required. Reference to the configurations shown in Figs. 9 through 12 will help to understand the various procedures. Automatic receivers provide a repetition-rate pulse train that is synchronized to the time of arrival of the Loran-C signal to within 0.1 μs. Use is made of a time-interval counter whose count is initiated by a TOC pulse selected from the station clock by an epoch monitor and terminated by the repetition-rate pulse from the receiver. Once the clock has been set by external reference to within half of a repetition period, the counter reading, accurate to 0.1 μs, will equal the algebraic sum of the propagation delay, receiver delay, emission delay, and the clock error. Since any variations in ground-wave propagation delay, receiver delay, and emission delay are less than 0.1 μs, variations in the counter reading over a period of time will be a direct indication of the local clock drift as referenced to the Loran-C transmissions.

Visual receivers do not provide a pulse output for direct measurements. The operator must visually adjust the timing of pulses from a local Loran-C repetition-rate generator so that they coincide with the beginning of the received Loran-C pulses. Once this has been adjusted, clock synchronization is performed in the same way as if the automatic receiver were used. Accuracy is dependent on the ability of the operator to adjust the local pulses to the beginning of the Loran-C signal as viewed on an oscilloscope and is dependent on the signal-to-noise ratio at the receiver. The degree of accuracy can be enhanced by integration using photographic time exposure.

If proper cycle selection is accomplished visually, synchronization can be achieved to within about a microsecond. Phase-tracking receivers will provide a pulse output that is automatically positioned to bracket three cycles of the Loran-C signal to within 0.1 μs. Since the receiver does not possess automatic cycle selection,

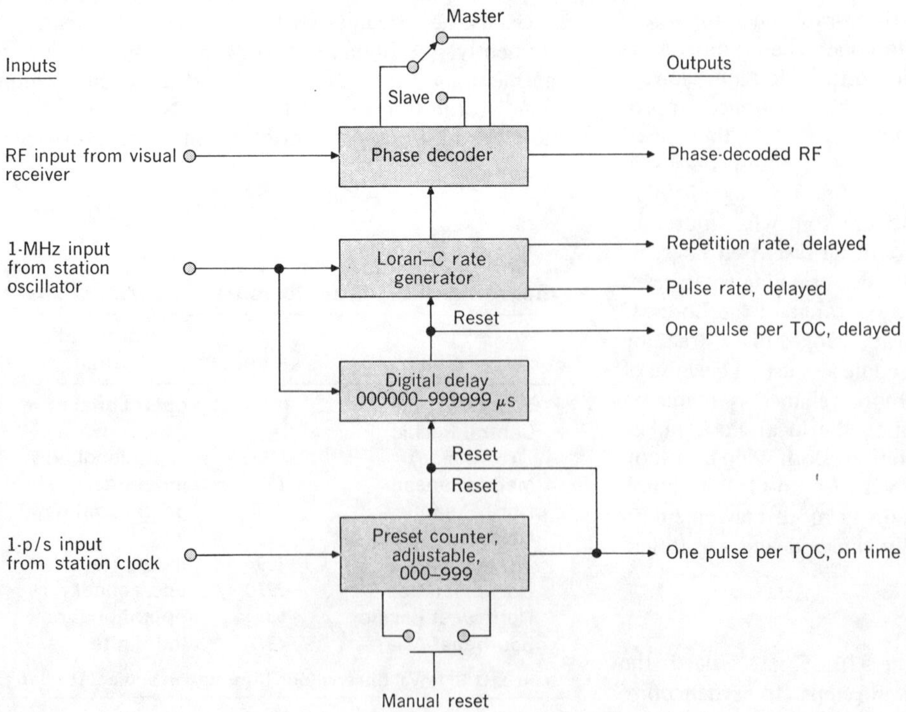

FIGURE 8. Simplified block diagram of an epoch monitor system. Taking a 1-p/s input from the station clock, the unit produces an output each TOC after being manually reset at a TOC. This output is delayed an integral number of microseconds equal to the total of the propagation and system delays and used to reset a rate generator. The rate generator output should then be coincidental with time of arrival of Loran-C pulses.

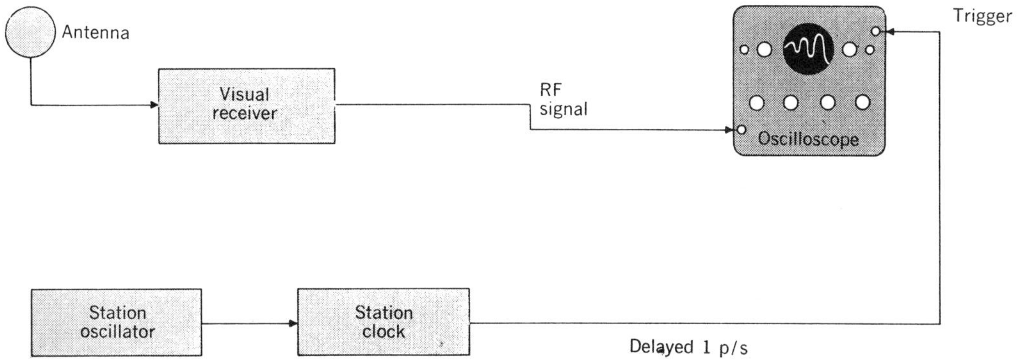

FIGURE 9. Simple visual timing system. This configuration can be used within a few hundred kilometers of a master station transmitting 1-p/s identifiers or within that range of any station on the East Coast chain. The useful range may be expanded by means of photographic integration. Accuracies to ±20 μs for ground waves can be achieved, depending on range and viewing technique. A 1-p/s delay from the station clock is used for positioning the Loran-C signal. A clock not possessing the calibrated delay feature may be used if the accuracy of delays obtained from reading the oscilloscope calibration is sufficient.

FIGURE 10. Visual timing system, recommended for use in any sky-wave situation and for ground waves in which automatic-receiver accuracies are not required. In addition to the selection of TOC pulses, an epoch monitor system as shown in Fig. 8 produces a pulse rate (groups of eight trigger pulses, spaced 1 ms apart, at the repetition rate) that is reset to the delayed 1 p/s, and also phase decodes the RF signal. The pulse-rate output may be further delayed in 1-μs steps for repositioning of the viewed pulse in the event the station clock is off. Accuracies are basically limited by the ability to cycle-select (±20 μs for ground waves) and may be improved by the use of photographic integration techniques.

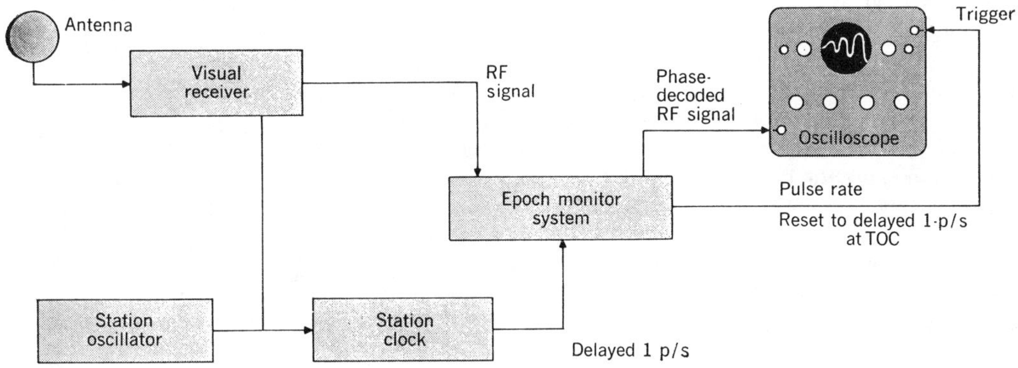

FIGURE 11. Phase-tracking system, which provides a record of the local-oscillator phase drift with respect to Loran-C. Recommended for use within the ground-wave range of a Loran-C station, it provides a record free of diurnal shift that can be used to compare the station oscillator to the cesium reference at the Loran-C station to within one or two parts in 10^{12} in a 24-hour period. Additional outputs are provided for timing purposes, to be used in the same manner as with an automatic receiver. Accuracies will be dependent upon the ability to cycle-select visually and, in general, will be within the range of ±20 μs.

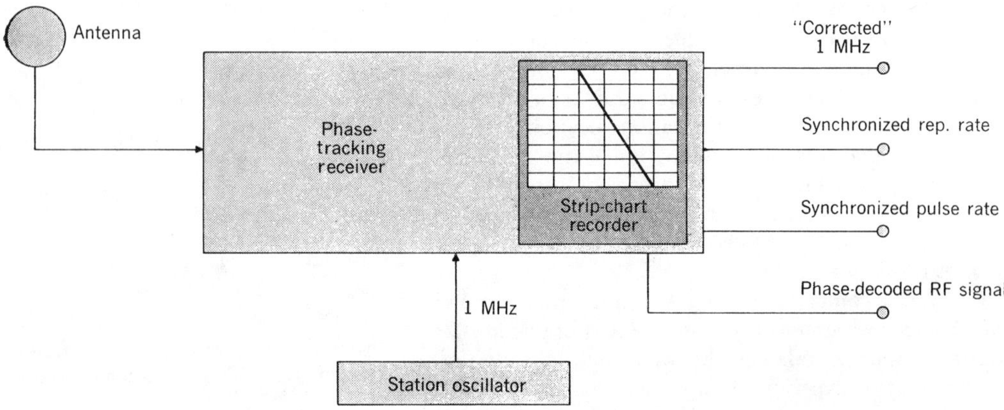

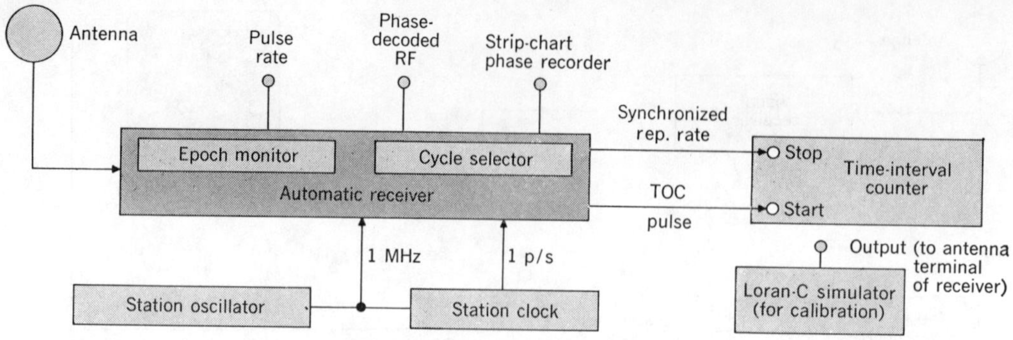

FIGURE 12. Automatic receiver system. Depending on the particular chain to be monitored, various automatic-receiver configurations may be used. When within range of the East Coast chain, an epoch monitor is unnecessary since a TOC occurs every second. The 1-p/s counter is then started by the clock. The configuration shown is for the general case, in which odd-repetition-rate signals are received. The time-interval counter may be replaced by an epoch monitor system as described in Fig. 8, and synchronization can be determined by means of oscilloscopic time comparison of the receiver and epoch monitor system repetition-rate pulses. An automatic receiver has cycle-selection capability, and accuracies to 0.1 μs are readily attainable. A Loran-C simulator is useful for providing system calibration.

synchronization must be accomplished visually. If cycle selection is uncertain, the synchronization accuracy is consequently reduced by the amount of the uncertainty.

Of particular use with visual receivers is an epoch monitor system such as shown in Fig. 8. In addition to selecting TOC pulses automatically, it contains a Loran-C repetition-rate generator that is reset by a delayed TOC pulse for use in viewing the Loran-C signal, thereby correctly positioning the Loran-C for optimum viewing. Radio-frequency input from the visual receiver is phase decoded in either the master or slave format for oscilloscope viewing, using the delayed pulse rate output as the trigger.

The techniques of clock synchronization to Loran-C can be substantially simplified when used with a station that has a simple repetition rate such as in the East Coast chain, or when located within range of a master station of any chain that transmits 1-p/s identifiers. In these cases, the oscilloscope can be triggered by a delayed 1 p/s from the station clock, and the Loran-C pulse from a visual receiver positioned by adjusting the delay. Because the one-per-second sweeps are difficult to view with precision, it is usually necessary to use photographic time exposure for accuracies better than about ± 30 μs.

Conclusions

It has been shown how Loran-C signals can be used for clock synchronization over long distances to accuracies as fine as 0.1 μs. Low-frequency propagation is reliable and definite schedules for station personnel can be maintained. The complexity (and cost) of a Loran-C timing system will vary with station location and accuracy requirements. For many applications, simple visual equipment may be used to provide a substantial improvement in synchronization capability, compared with other techniques, without imposing complex procedures on station personnel.

Frequency comparison using Loran-C ground waves can provide measurements to within a part or two in 10^{12} over a 24-hour period. The absence of diurnal shift suggests numerous experiments in which oscillators distant from one another can be slaved to a common stable reference.

The author wishes to acknowledge the many helpful comments and suggestions given by Dr. G. M. R. Winkler and Mr. J.-D. Lavanceau of the U.S. Naval Observatory, and Dr. L. R. Schissler and Mr. V. Skov of Aerospace Research, Inc.

REFERENCES

1. Doherty, R. H., Hefley, G., and Linfield, R. F., "Timing and space navigation with an existing ground-based system," Rept. 6721, National Bureau of Standards, Oct. 1960.

2. Doherty, R. H., Hefley, G., and Linfield, R. F., "Timing potentials of Loran-C," *Proc. IRE*, vol. 49, pp. 1659–1673, Nov. 1961.

3. Markowitz, W., "Time measurement techniques in the microsecond region," *Engineer's Digest*, published by the U.S. Coast Guard, vol. 135, pp. 9–18, July–Aug. 1962.

4. Markowitz, W., "Time determination and distribution: current developments and problems," *Proc. Internat'l Conf. on Chronometry*, Lausanne, Switzerland, June 1964.

5. Winkler, G. M. R., "Recent improvements in the U.S. Naval Observatory timekeeping and time distribution operations," presented at the 22nd Annual Frequency Control Symposium, U.S. Army Electronics Command, Fort Monmouth, N.J., Apr. 23, 1968.

6. Pierce, J. A., McKenzie, A. A., and Woodward, R. H., *Loran*. New York: McGraw-Hill, 1948.

7. Henry, W. O., "Some developments in loran," *J. Geophys. Res.*, vol. 65, pp. 506–513, Feb. 1960.

8. "Engineering evaluation of the Loran-C navigation system," Jansky and Bailey, Inc., Washington, D.C., Sept. 1959.

9. Shapiro, L. D., "Use of Loran-C for timing and frequency comparison," presented at the 20th Annual Frequency Control Symposium, U.S. Army Electronics Command, Fort Monmouth, N.J., Apr. 21, 1966.

10. Shapiro, L. D., "Loran-C skywave delay measurements," presented at the Conference on Precision Electromagnetic Measurements, National Bureau of Standards, Boulder, Colo., June 27, 1968.

11. "Time service letter," U.S. Naval Observatory, Washington, D.C., Oct. 19, 1967.

12. "Time service announcement, series 9," no. 3, U.S. Naval Observatory, Apr. 1, 1968.

13. "Time service announcement, series 9," no. 6, U.S. Naval Observatory, June 4, 1968.

BIBLIOGRAPHY

Shapiro, L. D., "Loran-C timing," *Frequency*, Mar.–Apr. 1965.

Shapiro, L. D., "Loran-C timing and frequency comparison," *Frequency*, Mar. 1968.

METHODOLOGY FOR THE DETERMINATION OF THE COVER AND SECONDARY PHASE CORRECTION PLOTS FOR A LORAN-C CHAIN

K F Woodman

Admiralty Research Establishment, Slough, UK

R S Allen and D D Sandhu

CAP Scientific Ltd, New Malden, Surrey, UK

ABSTRACT

In order to fully characterise a Loran C chain it is necessary not only to predict the likely extent of the cover (both geometry limited and S/N limited) but also the so-called phase correction plots. The paper described in detail how these related aspects can be determined for any particular chain.

Both types of plot rely on similar computational techniques and data which have been derived from Millington's Method (to cope with the effects caused by boundary changes at land/land and land/sea interfaces) and the geometric-optics approach of Bremmer (which calculated the attenuation and phase variations with distance over a path with known electrical characteristics).

INTRODUCTION

It has been left to Europe to devise a post-1994 strategy after the USCG pull-out to maintain not just the existing Loran C cover in Europe but to extend coverage in areas where cover is either very poor or non - existent. Such areas presently exist around the British Isles, particularly in the North Sea, English Channel, Irish Sea and South-west Approaches.

The authors have chosen to avoid look-up tables whenever possible and have concentrated on a wholly computational approach. Given the hypothetical North European chain configuration, the station positions in latitude and longitude and the characteristics of the region over which the chain is intended to provide cover, the computer-based plotting routines allow the production of both S/N contour plots and the true hyperbolas which include secondary phase corrections (usually on a Mercator Projection).

While this computational approach is not intended to supplant full scale calibration of a chain (which will undoubtedly continue to be necessary for the foreseeable future), it is intended that the computational approaches described in the paper would form an integral part of any initial assessment of a proposed chain and before any turf is cut.

A number of possible options have been studied, ranging from a hypothetical chain covering most of the UK and French areas of interest (Lessay-Sylt-Soustons-Barra) to an approach which splits off the French stations to form a Southern European chain with an additional station in the Republic of Ireland at Mizen Head and leaves a station on the North-East coast of England to join with a newly proposed southern Norwegian Chain to improve cover in the North Sea and the northern half of the UK.

Loran C propagation is comparatively easy to characterise over ocean areas of cover. The propagational complexities become considerable when the system becomes effectively an island chain as would be the case for the area covering NW Europe, with its enclosed seas and islands. The paper seeks to characterise propagation over such a region using the above chains as examples.

LORAN-C

LORAN-C is a hyperbolic long-range navigational aid covering many strategically important areas around the world. It is operated by the US Coast Guard (USCG) but for stations located in countries other than the USA, arrangements are made with these governments where the system is set-up and calibrated by the USCG but often staffed and run on a local national basis. A glance at any number of handbooks listing navigation systems worldwide, such as the UK Admiralty list of Radio Signals (1) quickly shows that the area around the British Isles, particularly, the Southwestern Approaches, parts of the North Sea and Bay of Biscay either do not enjoy adequate LORAN-C cover or are completely lacking in any cover. Northern areas of Britain can make some use of the Norwegian Sea Chain (designation 7970) but the ground-wave cover at the 1/4nmi error contours effectively bisects the UK along 53°N with insufficient cover further south. The French Navy have for some time been operating a pair of stations at Lessay as the Master with Soustons as the only secondary in a rho-rho mode which has not been generally available to other civil users.

It might be supposed that any perceived future need to extend LORAN-C coverage may well be superseded by the arrival of the satellite-based, worldwide, all-weather, continuously available navigation system usually known as NAVSTAR-GPS (Global Positioning System). The unparalleled position-fixing capability available to those receiving

Reprinted from *Rec. IEEE Position Location and Navigation Symp.*, pp. 207–214, 1988.

Navstar-GPS will indeed render terrestrial aids such as DECCA, OMEGA and LORAN-C obsolete. But a good case has been made out for the retention of LORAN-C as a back-up radio navigation aid in the event that, for whatever reasons, Navstar-GPS might become unavailable. The most likely cause of loss of GPS would be from temporary outages and LORAN-C is seen as providing a very convenient back-up system (2).

From the West European standpoint, there is a more immediate problem. The USCG have widely advertised their intention to pull-out of the operation of the LORAN-C European Chains (Norwegian Sea, Mediterranean Sea) by the mid-1990's although they will continue to operate the LORAN-C chains covering the US Confluence Zone. The USCG on behalf of Congress have offered these European stations to Western Europe and in particular to those governments who currently have LORAN-C stations on their national territories. The LORAN-C Working Party, comprising members from Norway, Germany, Iceland, France, Belgium and Holland (not the UK, which only sent an observer), deliberated on the problem and have since published their conclusions and recommendations (3). In the main, it was recognised that LORAN-C working should continue in the West European area, subject to acceptable hand-over terms and costs from the USCG, but that for effective use, any Chain(s) in these areas must be operated on an international basis. This feature requires some formal agreement between the countries involved (UK, Iceland, Norway, West Germany, Denmark, Iceland, France and the Netherlands) which now seem to be forthcoming in the form of a draft Memorandum of Understanding.

PROPOSED LORAN-C CHAINS

No doubt due to a lack of LORAN-C cover in their areas of interest along their Atlantic Coast, the French Navy have operated for some time a pair of national "LORAN-C" transmitters, with the Master located near Lessay and the only Secondary situated near Soustons on the French Pyrenees; the system is being operated in the Rho-Rho Mode. Partly as a result of the LORAN-C Working Group deliberations, both the USCG and the French Navy were asked to co-ordinate a short test of the three-station chain consisting of Lessay (as Master), with Soustons and Sylt (dual-rated) as the two secondaries; the group repetition internal was 89400 µsecs, hence the designation 8940 for the (West European) Chain.

The experiment lasted, on and off, from late October 1985 to the end of June 1986 and despite some problems, was deemed successful (4). However, the published coverages diagrams showed that cover was still lacking in western ocean areas of interest to the UK. The addition of an hypothetical 3rd secondary somewhere in Northwestern Scotland, eg the Hebrides might fill in the coverage gaps and a short study (5) was undertaken by the authors to assess the feasibility of such a location. The location of another hypothetical secondary at Barra did indeed indicate an improvement in cover when part of the Lessay-Souston-Sylt chain but was still unable to provide adequate cover into the Bay of Biscay.

At the special 1987 IALA Conference (6), it was further proposed that the existing sets of LORAN-C chains be regrouped into seven new chains (Icelandic, Norwegian Sea, North Sea, Atlantic, Iberian, Mediterranean and Adriatic). Of direct interest to the UK is the performance and cover from the Atlantic and North Sea Chains. However, for the purposes of this phase of our study, and reflecting some more recent trends, the Atlantic Chain is now expected to include a secondary near Mizen Head rather than Barra and the North Sea Chain is expected to include an additional secondary somewhere in North East England, as listed in

Table 1: Proposed Atlantic and North Sea Loran-C Chains (all 250kw ERP except Ejde at 325kw)

Atlantic Chain		North Sea Chain	
Lessey	Master	Bergen	Master
Soustons	Sec	Bo	Sec
Sylt	Sec	Ejde	Sec
(Mizen Head)	Sec(New)	Sylt	Sec(upgrade)
		(NE England)	Sec(New)

Table 1, mainly to try and overcome the lack of cover off the southwest coast of Ireland. Figure 1 summarises our extent of cover for the Atlantic Chain (with a station in NE England near Scarborough and includes both noise-limited and geometry-limited considerations.

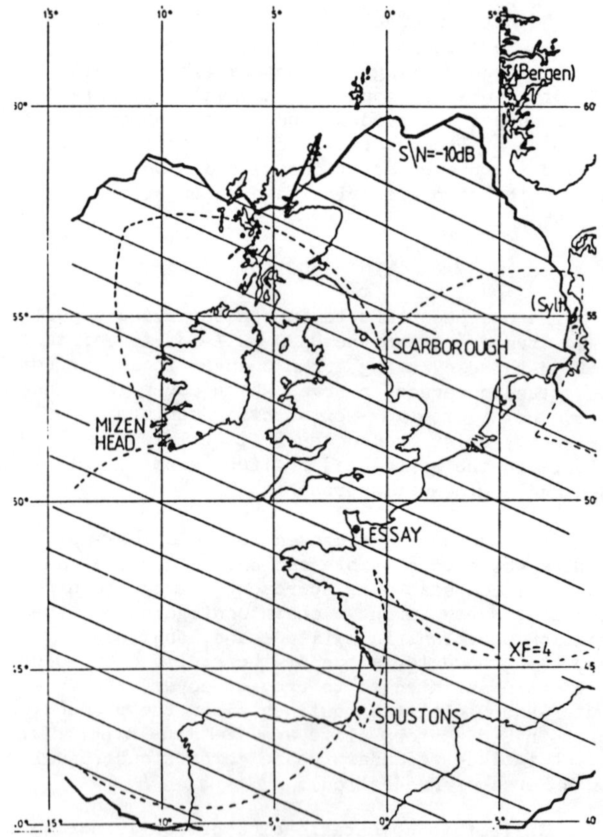

Figure 1 Full cover diagram for the proposed Atlantic Chain including a 3rd hypothetical secondary near Scarborough

Incorporating a station at Scarborough into the Atlantic chain considerably improves the area of cover.

CHARACTERISATION OF A CHAIN

In order to assess the usefulness of any particular chain configurations, it is necessary to determine a number of parameters over the expected area of cover from the chain. These parameters vary throughout the region of interest but can be visualised as a number of contours, such as:-

(a) expansion-factor loci,

(b) Constant signal/noise contours,

(c) reference hyperbolas for an all-sea path,

(d) additional secondary phase (ASF) corrections.

Expansion-factor plots are wholly station-geometry dependent and can be useful employed to highlight regions where the cover becomes geometry-limited (such as the area off the baseline-extended). Signal-to-noise ratio contours (usually at some specified level eg, -10dB) establish the distances from a transmitting at which the LORAN-C signal can be received. Regions where the S/N falls below the specified level are defined to be areas where Loran-C cannot be received due to excessive signal-path attenuation (field-strength limited). A full cover diagram for a chain combines both expansion-factor and signal/noise considerations. Time-difference hyperbolas (used in position-fixing) assume the propagation velocity to be that of a wave over an homogeneous sea surface. The ASF corrections are the differences between the times of propagation over a mixed terrain and those over an all-sea path of the same length.

Ground-wave Electrical Constants

The propagation conditions are very considerably influenced by the nature of the ground (and more so for sea) over which the signal is being propagated. The sea has been allocated the same set of electrical constants throughout but the ground can have quite a wide range of electrical constants dependent on the rock or soil type etc. Table 2 lists the conductivity and permittivity values used in the present analysis (as defined from USCG publications) with the main exception of urban areas. Sensitivity analysis has been carried out on the influence these changing variables have on the value of path-loss which demonstrates that conductivity is the most influential property on signal attenuation, with permittivity less so. Accounting for a wide variety of land types does not influence the attenuation particularly but does greatly increase the computing time. Consequently, it has been found sufficiently accurate to use only a few representative sets of parameters for land, as listed in Table 2. To render the calculation practical for a computer, the land masses (and the sea) in the areas of interest have been segmented into the nearest ½° x ½° in geographical co-ordinates, as shown in Figure 2, with a particular set of values for conductivity and permittivity allocated to each ½° x ½° square of land or sea (land has been designated one of eight surface types as listed in Table 2; the sea labels 'S' have been omitted from the figure for clarity). It can be seen from the figure that the ½° x ½° partition is a reasonable fit.

Table 2 Conductivity and Permittivity values for Land and Sea paths

Land Type	Designation	Conductivity /m	Permittivity (relative)
Mountains	M	0.0005	15
Rocky Soil	R	0.001	15
Fresh Water	F	0.005	75
Poor Soil	P	0.002	15
Cultivated Areas	C	0.005	15
Urban Areas	U	0.01	15
Sea Water	S	5.0	80

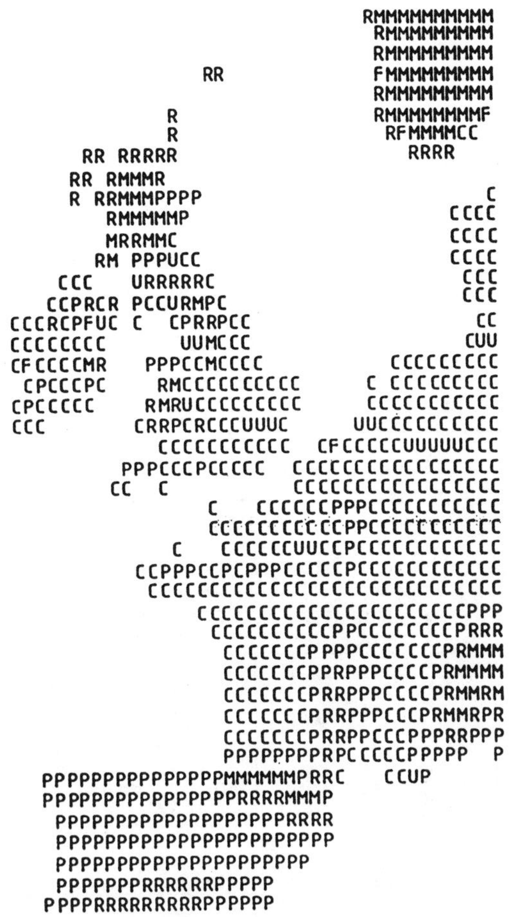

Figure 2 Subdivision of land is ½° x ½° geographical squares identified by land type (sea areas omitted for clarity)

Noise Levels

Data for European noise levels at or around 100kHz is rather sparse but the figures used by the USCG may be reasonably representative (as listed in Table 3). Some simple analysis reveals that the average noise level is 49.3 dBμV/m but that the 90%-ile figure of 56 dBμV/m may be a more representative figure (see Table 3) and is the noise level used in the present analysis, In practice, the noise level in urban industrial areas tends to be rather high; figures at ARE (Slough) have reached 75dBμV/m on occasions.

Table 3 Ambient Noise Levels for 100kHz at 55°N 00°W

TIME BLOCK	NOISE LEVEL dB ABOVE (1μV/m)	TIME BLOCK	NOISE LEVEL dB ABOVE (1μV/m)
00 - 04	48.9	00 - 04	53.6
04 - 08	49.1	04 - 08	45.5
08 - 12	37.1	08 - 12	50.4
12 - 16	27.1	12 - 16	60.5(MAX)
16 - 20	47.1	16 - 20	50.9
20 - 24	47.9	20 - 24	58.5
00 - 04	52.9	00 - 04	53.9
04 - 08	41.8	04 - 08	52.3
08 - 12	36.4	08 - 12	51.1
12 - 16	49.3(Av)	12 - 16	47.7
16 - 20	54.0	16 - 20	56.1
20 - 24	54.2	20 - 24	55.9

Table 4 (Calculations as prescribed in "Specifications of the transmitted LORAN-C signal" USDOT/USCG July 1981

Mean Noise Level	49.3 dB above 1μV/m
1σ Noise Level	7.6 dB above 1μV/m
90%-ile " level	56.0 dB above 1μV/m
Worst Case Level	60.5 dB above 1μV/m

Reference field at the unattenuated distance

The attenuation of field strength with distance from a transmitting antenna over imperfect ground has the general but simple form:-

$$E_d = \frac{E_o}{d} \cdot A \qquad (1)$$

where d is the distance from the transmitter, E_o is a reference field strength at a standard distance (1 km or 1 nmi) and takes into account the characteristics of the transmitting antenna (gain, directivity, effective radiated power etc). The attenuation factor (or path-loss) A is a complex function of wavelength, ground and sea permitivities as well as their corresponding conductivities. The value of E_o may not necessarily be a measured value for the electric field at 1km or 1nmi but is usually the so-called unattenuated field at the reference distance and used to convert the ERP of the transmitter into an electric field close to the transmitter but outside the near-field of the transmitting antenna. For the purpose of the study, E_o has been set at 1650mV/m (or 124.3dBμV/m) which the USCG consider to be the field strength at 1nmi appropriate to a transmitter whose stated ERP is 400kW. Using simple antenna analysis for a vertically polarised monopole at the Loran-C frequency of 100kHz it can be shown that the following parameters apply, viz:-

frequency,	f	=	100kHz
wavelength,	λ	=	3000m
antenna height,	h	=	225m
	h/λ	=	0.075
radiation resistance		=	2.22 Ω
Figure of Merit,	FoM	=	186 (for 'd' in nmi)
	(or FoM	=	300 if 'd' is in km)
Antenna Current	I_0	=	425 amps (rms)

On this basis, the value for E_o at 1nmi is 3720mV/m (or 6000 mV/m at 1km).

These values are the reference electric field for the condition d=1 nmi (or 1Km), providing these values are measured sufficiently close to the antenna such that the propagation losses are negligible and that the far-field condition is obeyed. This simple analysis assumes a theoretical vertical monopole over a high conductivity, infinite ground-plane. These conditions are never achieved in practice since the antenna is invariably erected over an imperfect and finite ground plane. To allow for practical circumstances, it is usual to accept lower values for the figure of merit, depending on the antenna type (the FoM is a convenient method of wrapping the antenna constants into a single figure).

If the theoretical FoM used in the above analysis is reduced to 50%, as suggested in the literature (7) for a practical monopole, the unattenuated field at 1nmi from the 400kW transmitting antenna become 1860mV/m, a figure more in keeping with the USCG estimate for this ERP.

Expansion factor

The error in a position-fix at the intersection of a pair of hyperbolic lines-of-position (LOP's) is usually described statistically in terms of an error ellipse and RMS error values but in the present instance, it is more convenient to express the errors in a rather different way using loci of constant expansion factors. The expansion factor XF is given by the simple expression:-

$$XF = \operatorname{Cosec} \tfrac{1}{2}\delta \qquad (2)$$

and while this definition may appear arbitrary, the angle δ subtended at the point of intersection by the baseline, is related to the distances to this point from each station and the fundamental constants of the hyperbolas passing through the point . The locus of constant expansion factor is an arc of a circle, erected on the baseline joining any pair of transmitting stations as chord.

Such loci appears circular in planar rectangular co-ordinate systems but would not necessarily appear so when accurately replotted on, say, the Mercator Projection.

If a typical receiver has a 3σ time uncertainty of 0.4 μs on the baseline, this translates to a positional error of about 120m. This timing error can be considered to be a lane in the hyperbolic lattice. As the width of a lane increases away from the baseline, the magnitude of the positional error for a given timing error increases in proportion. Thus, at a point where the lane expansion factor is, for example, 4, the 0.4 μs timing error now gives rise to a 480m positional error. It can therefore be seen that the lane expansion factor can be used as an index of likely positional error based purely on the geometry of the chain, divorced from a myriad of other factors which might also influence positional error. If curves are plotted of points having the same lane expansion factor, some idea can be gained of the areas in which a given chain would be unsuitable for navigation or provide no cover at all, as in the regions of the baselines-extended (where cover is geometry-limited).

In order to generalise this approach, only the constant expansion factor contours need be drawn. The implication of such a plot is that the arc of the circle (XF = 4 here) sets a bound on the errors inherent in a hyperbolic system due to the increasing separation of the hyperbolas away from the baseline. A position-fix requires two LOP's so that a pair of expansion factor loci are required in assessing such an error associated with the position-fix. If the error for either baseline is bounded to be less than or equal to some specified value, (eg XF 4) then one is interested in the area of overlap from a pair of expansion factor circles having the same set value (XF=4) for the expansion factors (it should be noted that each baseline has two expansion factor circles erected on it, one upper and one lower). Within this overlapping area, the time uncertainties are never worse than about x4 the error on the baseline (for positional uncertainties, the angle-of-cut must be considered).

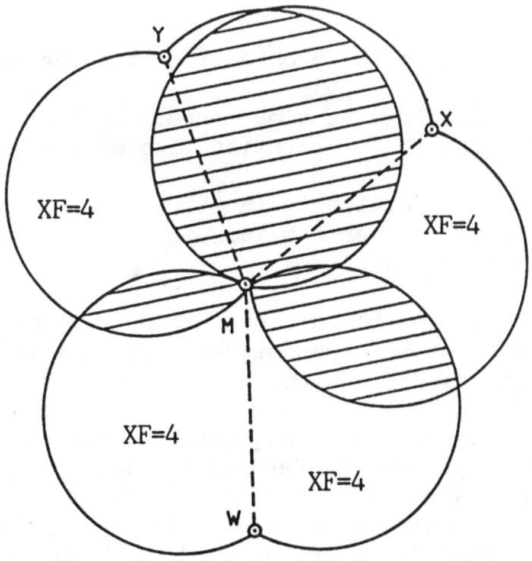

Figure 3 Expansion-factor plot for three baselines

Figure 3 illustrates a typical expansion-factor plot appropriate to a three baseline chain where the striped region represents the cover not limited by geometry considerations.

Attenuation of Field Strength

The attenuation of field strength (A) with distance is dependent in a complicated way on the wavelength, ground conductivity (σ) and permittivity (ε) and other parameters, such as the height-gain factor and Earth Curvature. In the particular analysis used during the course of this work the attenuation factor A is found as the modulus of the Hankel Function Approximation in the geometric-optic approach used by Bremmer (8) and is given by the following expression:-

$$E_d = \frac{E_o}{d} \cdot \sqrt{\chi} \underbrace{\left| \sum_{s=0}^{\infty} \frac{\exp(i\tau_s \chi)}{2\tau_s - 1/\delta^2} \right|}_{A \text{ (Modulus)}} \quad (3)$$

It is not intended to elaborate on equation (3) here as the reference can be consulted for a detailed description of the parameters, χ, τ_s and δ (not to be confused with the expansion factor angle, δ) and which are also functions of wavelength permittivity and conductivity. However, a very important feature of Bremmer's evaluation of A is that it can be approximated by series solutions which fall into two categories, known as the short range and long range solutions (the boundary between the solutions being set at 100 km). To evaluate the expression for the attenuation factor A in equation (3), it is ordinarily required to solve the differential equation where solutions are combinations of Hankel Functions and where corresponding arguments include the parameter τ_s. Fortunately, Bremmer provides a sufficiently good approximation which avoids the need for a direct and rigorous solution and which renders the computations more tractable when solved by a computer program.

At Loran C wavelengths it has not been found necessary to consider such effects as height gain factors since the antenna length above the ground is a small fraction of a wavelength (a 750' mast has an effective height of around λ/13). Groundwave propagation over a spherical Earth is assumed, with sky-wave effects absent or not considered. The appropriate values for A are inserted in equation (1) to yield field strengths at distances beyond the unattenuated field distance. Additionally, if the transmitter is not actually 400 kw, the normalisation factor (nf) must be applied ie.

$$nf = 10 \log_w \left(\frac{\text{reference ERP}}{\text{actual ERP}}\right) \quad (4)$$

Millingtons Method

This particular technique (9) leads to the calculation of attenuation over a mixed sea/land path or mixed land paths and allows for possible anomalies (or land-loss and sea-recovery as Millington termed them) which can arise at the boundaries of the different types of land or sea. Each segment of the path between transmitter and receiver is treated separately. An important part of the technique is the determination of the attenuation, firstly in the forward direction and then secondly, in the reverse direction with the positions of the receiver and transmitting antenna interchanged; the attenuation is the average of the two determinations. The result is a path comprised of several sections each of known length and electrical characteristics. As an example, consider the case where the transmitter and receiver are both on the coast, with the propagation path passing over an island, as shown diagrammatically in Figure 4.

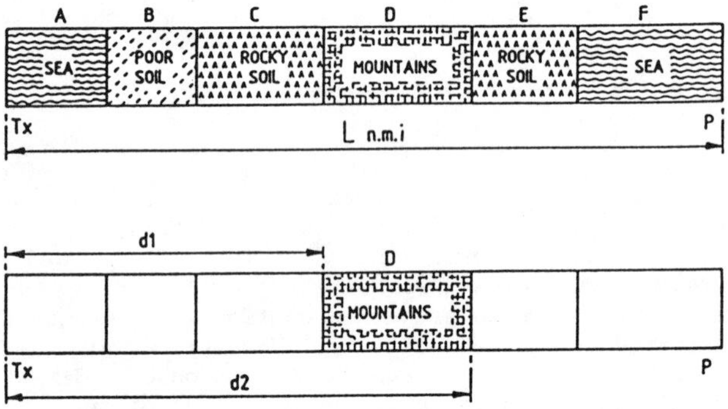

Figure 4 Attenuation along a mixed terrain path

If the attenuation over section D is A_D, then it follows that;

$$A_d = A_{d2} - A_{d1} \tag{5}$$

where A_{d1} is the attenuation due to a path length d1. as if it consisted entirely of mountains and A_{d2} is the attenuation due to a path length d2, as if it also consisted entirely of mountains. In a similar manner, the attenuation for each segment is determined so that the total attenuation along the the chosen pathlength is the sum of the individually determined contributions. By interchanging the transmitter and receiver positions, similar calculations for the reverse direction are carried out such that the total attenuation over the particular pathlength is the average of the values for both directions.

Millington's Method of coping with the anomalies observed at boundaries between land and sea, for example, is not without its critics. The main criticism is that the use of the technique tends to over-emphasise losses for complicated paths with many boundaries. The alternative view, shared by the authors, is that the method remains useful and valid and that there are other reasons why the overall attenuation sometimes appears greater than expected, such as a lack of knowledge of the true transmitter ERP.

Time Differences: Primary and Secondary Phase Corrections

Position-fixing by Loran C is based on the principle of measuring the time differences between the times-of-arrival of pulses from transmitters in a chain. Until quite recently, charts were produced which had been printed with overlaid hyperbolas (each of which is a locus of points having the same time difference). Nowadays, however, modern receivers provide a direct readout of position in latitude and longitude. But for the purposes of assessing the characteristics of a proposed new chain, it becomes necessary to derive the hyperbolic lattices over the region of interest in order to study the effects on time-differences of propagation over mixed terrain rather than just over a homogeneous all-seawater path.

Because the characteristics of the surface affect the propagation velocity of the Loran C pulse and therefore the propagation time and time differences, it is necessary to include corrections which account for these propagation velocity changes by calculating modified hyperbolas for the actual propagation path over the mixed terrain. Propagation in-vacuo represents the 'reference' so that the primary phase correction is applied for propagation through air while the secondary phase correction is made for propagation over a land or sea-surface (this latter term includes the ASF).

The time difference calculation is performed in the following stages:-

a. calculation of the phase difference between propagation over the given surface and propagation of a standard wave through air (the secondary phase correction),

b. conversion of the phase difference to a time correction,

c. finding the propagation time through air,

d. applying the time correction b to c to determine the propagation time over the path.

e. using Millington's method to allow for propagation over a mixed path comprising several surface types,

f. finding the difference in propagation time from each transmitter to the given point (known by definition).

Based on Bremmer's analysis the expression for the secondary phase difference is given by equation 6, where a is the earth's radius, and 'd' the distance from the transmitter. The remaining parameters in equation 6 are defined in the literature(10)

$$\Phi_c = \arg\left[\sum_{s=0}^{\infty} \frac{1}{(2\tau_s - 1/\delta^2)} \exp\{i[(K_1 a)^{\frac{1}{3}} \cdot \tau_s \cdot \alpha^{\frac{2}{3}} \cdot \frac{d}{a} + \frac{\alpha d}{2a} + \tfrac{1}{4}\pi]\}\right] \tag{6}$$

where,

$$\delta = \frac{i(K_2/K_1)^2 \cdot \alpha^{\frac{1}{3}}}{(K_1 a)^{\frac{1}{3}} \cdot ((K_2/K_1 - 1)^{\frac{1}{2}}}$$

and,

$$K_1 = 2\pi/\lambda; \qquad K_2 = \frac{\omega}{c_0}[\epsilon_2 + \frac{i(\sigma \mu_0 c_0^2)}{\omega}]$$

and will not be elaborated upon here. After some manipulation, it can be shown that the total propagation time over the pathlength is:

$$T = \frac{\eta d}{c_0} + \frac{\Phi_c}{2\pi f} \tag{7}$$

where η is the refractive index of the air and c_0 the velocity of light in-vacuo.

As the path from transmitter to the given point crosses several sections of different characteristics, Millington's Method must be applied to the calculation of the time propagation in a manner similar to the determination of the attenuation. The contribution from each section is the time taken for the wave to travel over that section. As in the example using the mountains (see figure 4) this is found as the difference between the time taken to travel from the transmitter to the beginning of the mountain section, and the time taken to travel to the end. The sum is calculated for each direction and the mean value becomes the propagation time for the wave over the given mixed path. The operation of Millington's Method in the time difference calculations is thus exactly analogous to the operation of it in the attenuation calculations. The time difference for the given point is found simply as the difference between the propagation times to the point from each of the transmitters.

Graphical Presentation

For the purposes of this analysis, the main characteristics associated with a Loran C chain need to be plotted by computer in a graphical format over a map. In order to accomplish this, a particular plotting method has been adopted for generating the expansion factor curves, the S/N contours, hyperbolic lattices, including those incorporating secondary phase corrections. The method is based on the fact that whichever property is being considered, the curve required is that connecting points of the same value. It is also true that if a radial line is drawn out from a transmitter then if it crosses one of the curves, it will only do so once. In order to draw the curve with a computer plotter it is desirable to find each point along the curve in turn. The method used is therefore to calculate a point on a radial line for each degree of a circle, giving a complete curve of 360 points. The correct point on the radial is found by finding the value for a very small distance, known to be inside the curve (d_0). This distance is successively doubled until a pair of points (d_1, d_2) is found where the value for d_1 is still inside the curve and the value for d_2 is outside the curve. The distance between these two is then successively halved (d_3) until a point is found having the correct value (d) but to within a small tolerance. This is shown diagramatically in Figure 5. However, at this stage the curve only exists as a series of points which the computer graphics connects together by a series of straight-line segments. Thus it is important to ensure that these are sufficient points (and hence radials) to accurately reproduce the curve in question. In the present analysis, radials at 1° intervals has been found to be adequate.

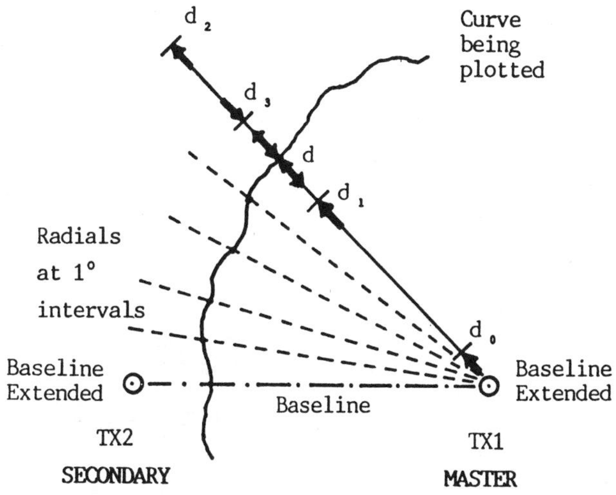

Figure 5 Iteration Method in Curve Plotting

The general description applies to the curves found for all three properties but there are idiosyncracies for each one. For the expansion factors, each curve is a line connecting one transmitter position to the other. The curve has two lobes, so the method of calculation is to treat each lobe separately by using radials iterated clockwise or anticlockwise. The radials are thus drawn from the southern transmitter and start 1° to the appropriate side of the baseline. Points are calculated for each radial until the last one is 1° from the southern baseline extension. The first and last points of the curve are, by definition, the transmitter positions themselves. The attenuation curves are calculated with the first radial being due north of the transmitter, iterating round 360°. The last point is set the same as the first point to produce a closed curve when these are plotted. The time difference hyperbolas are calculated in two halves. Radials are drawn around each transmitter in turn with one set of 358 radials producing one hyperbola. A given transmitetr is used to find the hyperbolas lying nearer to it than to the other transmitter. The first radial is drawn 1° off the baseline extension and iteration continues until that last one is drawn 1° before the baseline extension. There is, by definition, no solution along the baseline extension. To avoid excess calculation a maximum doubling distance of 10,000 miles is used. This is far outside the range of a transmitter or the chart and allows for faster calculation of hyperbolas which may be close to parallel to particular radials.

It may be noted that there has been no constraint on the speed and quantity of calculations involved. It is to be appreciated that the computation could be optimised considerably.

DISCUSSION AND RESULTS

The proposed Atlantic Chain is intended to consist of three stations, Lessay (master) and Soustons and another secondary somewhere on the southwest coast of Ireland; in our analysis we have chosen Mizen Head as a convenient location. However, study of the expansion factor and S/N contours immediately makes obvious the lack of cover over southern Britain, the Channel and the southern North Sea. It can also be demonstrated in the same way that the Norwegian Chain, even with the addition of a new secondary on the north-east coast of England (possibly near Scarborough) cannot fill-in the gap in cover left unprovided by the Atlantic Chain. But a very simple remedy exists, namely, the dual-rating of Scarborough to operate with both Chains. As can be seen in Figure 1, this strategy significantly improves the Atlantic Chain cover and effectively eliminates the previously severely geometry-limited region to the north-east. Expansion factor plots and S/N contours remain crucial elements in assessing areas of operability from any proposed chain.

In addition to the above mentioned characteristics, the effects of propagational variations on the time-difference contours (hyperbolas) have been addressed. As part of the analysis, it has been necessary to generate the true hyperbolas taking into account propagation over mixed terrain as well the reference hyperbolas over an all-sea path. On a scale sufficiently large to accommodate all the hyperbolas appropriate to a particular baseline (in our example, Lessay-Mizen Head as the inset to Figure 6) there is seldom any evidence of propagational anomalies, even when plotted at AO size. Only when an area of interest is greatly enlarged, as in Figure 6 (covering an area 1° x 1°) do differences between the all-sea path and the mixed terrain become apparant. In the example chosen variations in the order of 8 to 12 μsecs occur just to the west of the Cherbourg Peninsula. It is from considerations such as these which enable error contours to be derived, if required. Analysis of this type highlights other areas where time-difference corrections might need to be applied.

Finally, it is to be noted that for the purpose of making the analysis tractable, it has been necessary to define a given point so that the propagation path is known and a precise time-difference determined. In practice, this state of affairs hardly applies aboard ship where the position is not known exactly and therefore, the propagation paths cannot be accurately determined.

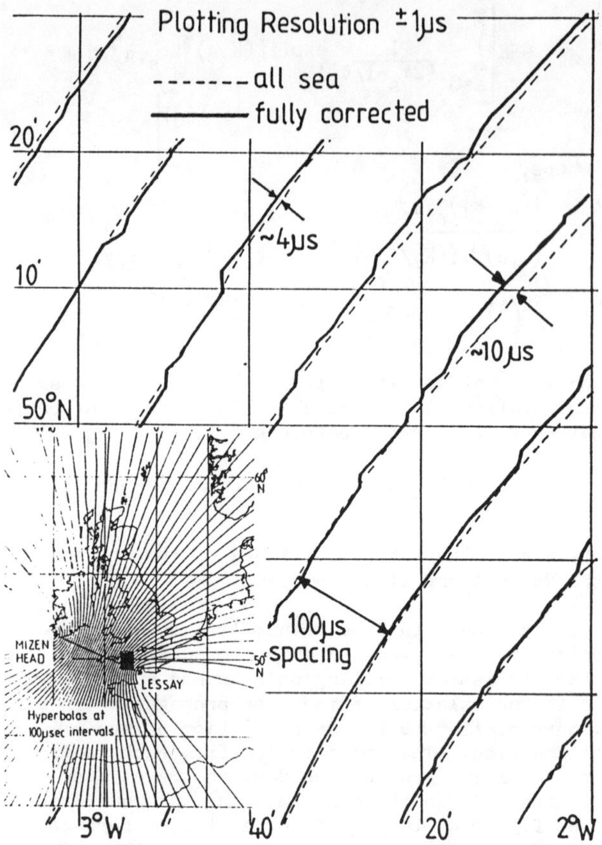

Figure 6 Hyperbolic fine resolution

It will be necessary to find some practical method of incorporating any corrections into the receiver software, perhaps by using the reckoned position as assumed value and some form of recursive technique. Otherwise reliance on tabulated corrections will continue as at present.

REFERENCES

1. Admiralty Handbook NP285a Chart B13, 1983

2. Braisted P, Eschenback, R and Tiwari A
J. Inst. Navigation 22(1) 20 (1986)

3. LORAN-C Working Group, Final Report (1985)

4. Wenzel R and Thrall W (USCG Europe)
Sylt Dual Rating Experiment (1986)

5. Woodman K, Juleff J and Allen, R
J. Navigation 51(2) 149 (1988)

6. IALA Conference, London March (1987)

7. Reference Data for Radio Engineers STC (1973)

8. Bremmer H. Terrestial Radio Waves (1949)

9. Millington G
Proc IEE Part 3, 53 (1949)

10. McCullough J, Irwin B and Bowles R
ION Proc. Nat. Tech. Meeting 127 Jan (1985)

Part 5
Introduction to Aircraft Navigation

5.1 Early Flights

THE history of aircraft navigation is brief compared to that of land and ship navigation. The earliest aircraft made demonstration flights in the 1900s without instruments. After World War I, surplus Jennies were bought by barnstormers who spread all over the United States and Europe. They flew by day along highways and railroads. They often buzzed towns to read identifying signs. In the 1920s, many midwestern and western American cities began patriotic campaigns to paint the town's name prominently on the roof of a large school or church. Such was air navigation.

Gradually, on-board instruments became available to permit the pilot to hold attitude and heading in all weather. The magnetic compass and airspeed indicator (a cup anemometer) were installed early. Then came a ''turn-rate indicator,'' a rate gyro with a vertical input axis (Fig. 5-1). It was co-mounted with a lateral accelerometer, built in the form of a bubble in a laterally mounted glass tube, which permitted the pilot to make a blind coordinated turn. Centering the bubble assured that the apparent gravity vector was in the plane and symmetry of the pilot's body, even when the aircraft was banked. At the end of the 1920s, the directional gyro (which showed heading angle relative to a hand-set ''zero'' heading) and the vertical gyro (which showed pitch and roll relative to the horizontal plane; Fig. 5-1) were invented. In 1929, James Doolittle made a blind takeoff and landing using these new gyro instruments.

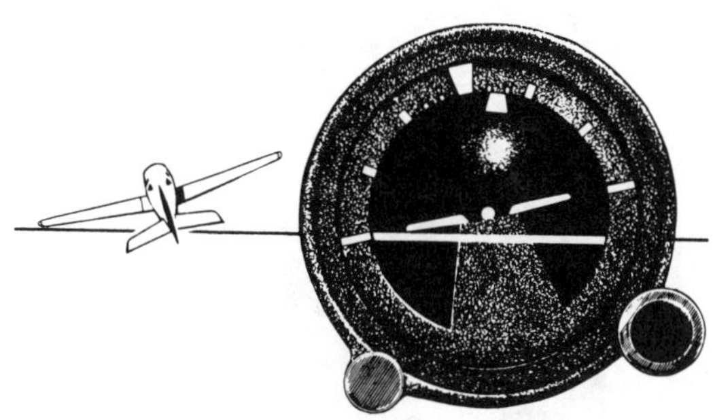

Fig. 5-1. Aircraft attitude displays, 1930–1980.

5.2 Airways and Navaids

In the 1920s, lighted beacons were established at key airports along widely used routes served by airmail contractors. In 1929, they were augmented by four-course radio ranges, direction-finding beacons, and vertical marker beacons. Figure 5-2, a 1961 chart, shows four-course ranges emanating from Boston north to Salem, west to Bedford, and southwest to Providence. The ground stations operated on 65 channels from 200 to 400 kHz, just below the broadcast band. Switched loop antennas on the ground emitted signals in space that created audible Morse tones aboard the aircraft. If the aircraft was on one of the four radial courses from the beacon (3° wide), a steady hum was heard. If the aircraft was right or left of the course, a Morse A (·-) or N (-·) was heard above the hum. The system was usable more than 1000 mi from the station over oceans (for example, on the route from San Francisco to Hawaii). Over land, transmitter power was reduced to prevent interference with other stations on the same frequency. Hence, overland range was deliberately reduced to about 100 mi. In the terminology of Part 1, the four-course range was a guidance system, not a navigation system. The last range, in Alaska, was decommissioned in 1974. No countries other than the U.S. installed 4-course ranges.

Many aircraft were equipped with HF direction finders (DF) that measured the bearing of broadcast stations or DF beacons relative to the axis of the aircraft. Omnidirectional DF beacons were located near major airports, allowing aircraft to fly an accurate heading to the beacon, then execute a ''nonprecision'' approach to the runway. By observing the gyro heading and relative bearing to the DF beacon, a skilled pilot could compensate for crosswinds during his approach and descend within a hundred horizontal feet of his runway. He used a precision barometric altimeter to hold the altitude sequence specified on the published ''approach plate'' during the descent.

Vertical beacons at 75 MHz, called Z-beacons, were installed along the four-course airways and along approaches to runways. They gave a positive indication of position along the airway by lighting a lamp and sounding a tone in the aircraft. They added navigation information to what was otherwise a guidance system. Like the four-course range and DF station, the Z-beacon transmitted a Morse-code identifier.

At the start of World War II, 312 four-course ranges and even more DF beacons were the mainstay of horizontal navigation in the United States. The barometric altimeter was the mainstay of vertical navigation. There were 12 traffic-

Fig. 5-2. Navigation chart of the Boston area showing A-N airways, 1961.

control centers and 24,000 miles of lighted airways in the United States.

During the 1930s, French mail routes were extended into Africa. French and U.S. mail routes were extended into South America, the French crossing from Dakar, Africa while the Americans flew via Florida and Texas. Chronologies of the period show flights entirely by visual aids, day and night, though with manned HF weather stations along the routes at major airports. These pilots were truly heroes, often dead ones; the writings of Antoine de Saint-Exupery recount their adventures [see (Saint-Exupery, 1931 and 1939)].

World War II produced a rapid advance in electronic navaids and made precision barometric altimeters (with 2 needles and a window to set sea-level pressure) widely available. Many bomb-navigation aids were invented of which Loran-A survived the war as a marine navaid (see Part 4.8, pg. 155). Few aircraft used it because of the weight of the receivers and the need to plot positions on the chart that showed the hyperbolas.

After the war, the International Civil Aviation Organization (ICAO) was formed as an agency of the United Nations. The ICAO was forced to select an international standard from the many electronic aids. It accepted the Very High Frequency Omnirange (VOR) that had been adopted by the United States and which was eventually installed throughout the world. Each VOR station creates signals in space that allow an aircraft to measure its bearing from the station, relative to magnetic north, with an accuracy of $\frac{1}{2}$ to 2°. Aircraft fly along certain radials that serve as numbered air routes. By taking cross-bearings from other VOR stations, position along a route can often be determined. Z-beacons were sometimes located on an airway to define key fixes for holding or transitioning to descent.

In 1955, the U.S. Navy sponsored an L-band substitute for VOR named Tactical Air Navigation (Tacan). It was to be nine times as accurate in bearing as the VOR and deployed on aircraft carriers as a beacon for returning aircraft. It included a ground-based ranging transponder (called Distance Measuring Equipment, DME) that replied to two-pulse interrogations from the aircraft. Bearing information was conveyed in the amplitude modulation applied by the ground station to the DME pulses. Today, Tacan is universally deployed in U.S. military and NATO aircraft, most of which do not carry VOR. It is smaller than the lower-frequency VOR sets but several times more expensive. Multipath reflections make its angular accuracy no better than that of VOR. Beginning in the 1960s, Tacans were colocated at many VOR sites in North America and Europe (the colocated station is called Vortac) making the DME feature available to civil users. Today, the most accurate means of civil radio navigation (0.2 to 0.4 nm) is the intersection of DME fixes from two or more Vortac stations. In 1989, more than 1500 Tacan stations were in operation on land and on aircraft carriers. Technical details about VOR and Tacan can be found in the reference by Kayton and Fried (1969). The reprint by Lang and Hundley (1980) pp. 303–312 describes the current generation Vortac transmitters, of which more than 800 are installed in the United States. They are remotely monitored and remotely controlled on digital telephone networks operated from the Air Route Traffic Control Centers. Their design has reduced the average maintenance frequency from two visits per week at each site to five visits per year.

By 1958, there was a network of civil air routes over the United States and the military and civil traffic control systems had been merged under common controllers located in more than 20 Air Route Traffic Control Centers throughout the United States (today, 23). A network of unmanned communication relays provided nearly 100% voice coverage throughout the continental United States. Routes were divided into 1000-ft-high altitude bands below 29,000 ft (2000-ft-high bands above 29,000 ft) and into horizontal blocks marked by VORs or Z-beacons. Aircraft reported entering a block, whereupon the controllers closed the block to all traffic until the aircraft reported leaving. This was a direct outgrowth of railroad practice (Part 3.2, pg. 66). In the absence of radar coverage, the system worked well though capacity was low since aircraft tended to fly 10 to 20 min apart. Even in 1989, on non-radar-monitored routes over oceans, aircraft fly 100-nm apart laterally at the same altitude, separated longitudinally by 20-min flying time (about 100 nm).

By the mid-1960s, ground-based surveillance radars (called Secondary Surveillance Radars, SSRs) covered much of the United States. They measure the horizontal positions of aircraft whose images are displayed to traffic controllers with superimposed identification codes and altitudes reported by airborne transponders (next paragraph). Stevens (1988) has described SSR in great detail. Its range error is about 100 ft; its azimuth error is about 0.04°, RMS. Radar coverage placed aircraft under "positive control" and allowed spacing as close as 3 nm on approach, within 40 nm of the antenna. Lower-power Airport Surveillance Radars monitor the airspace within about 50 nm of an airport. In 1989, they displayed skin-tracked aircraft, transponder replies, and weather returns. Efforts are being made to detect and display severe weather disturbances (e.g., microbursts). See Lopez and Wilson (1988).

In World War II, transponders were invented (called Identification Friend or Foe, IFF) that issued a 3-bit code in response to radar illumination. Ground gunners were supposed to shoot at any aircraft that did not reply but equipment failures discouraged such stringent rules of engagement. Price, in his memoirs of World War II electronic countermeasures (1978), tells that Allied crews habitually left their IFF transponders "on" when flying toward Germany, thus permitting smart German radar operators to detect them at longer ranges than skin reflections would have permitted. After the war, IFF technology was selected to enhance the skin-tracked image of civil aircraft. The ICAO standardized on a 3-bit interrogation code, emitted by the surveillance radar, and a 15-bit reply code emitted by an aircraft whenever it is interrogated. In the 1960s, two modes were added: Mode 3/A replies with a crew-set 4-octal-digit code when suitably interrogated, as a unique identifier; Mode 3/C replies with the digital pressure altitude when suitably interrogated. A new mode, Mode S, has been developed but not deployed. Mode S can send digital messages to aircraft at 40 bits/s on the civil IFF channel [see Leeper and Anderson (1979)]. It is intended for sending hard-copy clearances but may be preempted by future satellite communication links. Military aircraft have additional IFF modes. Military Mark-12 is similar to the civil transponder, but with an encrypted reply that makes it difficult for an enemy aircraft to masquerade as a friendly aircraft during wartime. The military Mark-15, now in development for 21st-century NATO forces, is similar to Mark-12 but has a hard-to-detect and hard-to-jam spread-spectrum mode at L-band and at 3 GHz [see (Hughes, 1988)]. In a spread-spectrum system, each bit of the reply message is composed of several hundred chips, produced by phase-shifting the carrier. It is the same modulation method used in JTIDS (Part 3.5) and GPS (Part 3.6). In an era of fast-moving attack vehicles, shoot-on-reply rules of engagement are again being proposed. The IFF must be good enough to prevent the destruction of friendly, neutral, or civilian vehicles. An example of poor IFF was the inadvertent shooting down of an Iran Air airliner by a U.S. surface ship in 1988, with the loss of 280 lives.

In 1975, the FAA established airways that connect phantom Vortac stations, to reduce the traffic on busy Vortac-to-Vortac transcontinental routes without adding real radio stations. This "area-navigation" (area-nav) technique was possible because the new wide-body jet aircraft carried airborne digital computers that were able to calculate the deviations from and distance along the area-nav airways from nearby Vortac stations. The reprint by Bobick and Bryson, pp. 250–257, shows detailed error models for VOR and DME sensors and shows mechanizations for mixing them with each other and with inertial sensors. Velocity errors of 4 kt and position errors of less than 1 nm were predicted and are being achieved.

Area navigation is being extended throughout the United States and will be in use until Vortac is replaced in the 21st century. Following the fuel-price increase in the mid-1970s, optimized vertical navigation became economical and several companies began selling three- and four-dimensional guidance computers and software that optimized climb and descent profiles. The four-dimensional computer targets predetermined points at predetermined times. Farmer (1981) described the on-board software and predicted fuel savings. Mohleji (1988) has described ground software that permits traffic controllers to issue clearances permitting on-board–optimized maneuvers. Lin and Hong (1989) describe the details of a traffic-separation algorithm for the airspace over Taiwan.

There are at least 60 air-traffic control agencies in the western world. The European national agencies are coordinated by Eurocontrol, a transnational agency that sets standards, develops software and procedures, and collects fees from users for payment to each country [see Sutton, Feb. (1989)]. In 1988, 12,000 flights, almost entirely made up of commercial and military aircraft (e.g., little general aviation), used Eurocontrol's airspace. Part 8.5 describes some of Europe's growing navigation requirements, driven by the need to make better use of already crowded airspace.

Aircraft flying over oceans are presently controlled from national centers on the seacoasts (e.g., New York and Gander). Five centers control the North Atlantic alone. Aircraft fly on organized tracks and report by HF radio every 10° of longitude, as measured by their onboard navigation systems. Future upgrades are described by Massoglia (1989) in which messages will be relayed by communication satellite at a probable charge to the aircraft of $10 per call. Aircraft using satellite communications will surely be alloted preferred routes since clearances would be easily issued and revised. The number of oceanic control centers could probably be reduced to one, worldwide, or to one per ocean.

Loran-C is described in Part 4.8, pg. 155; it is principally used by ships but its use by general-aviation aircraft in the United States is growing. The first Loran airport approach was approved in 1988. Loran is little used by military and air carrier aircraft.

Before 1975, the Soviet Union maintained civil navigation aids of its own design, based on HF beacons, en-route (military) radars, DME, and ground-controlled radar approaches. The Russians are well along toward achieving compatibility with ICAO standards in VOR, ILS, and MLS. Their on-board collision-avoidance system requires special equipment in all aircraft and hence is not compatible with U.S. TCAS (described below). Their design is possible because they do not have a general-aviation community that is reluctant to invest in expensive new equipment. Russian en-route traffic control is radar based, the en-route spacing being 30 km with radar, 10 min flying time without.

Omega is a 10- to 14-kHz navigation system that was invented for updating submarine inertial navigators because its very low frequencies penetrate tens of meters beneath the ocean surface. It is also used by many general-aviation and military aircraft because of its worldwide coverage, especially over oceans and underdeveloped areas. Some airlines (e.g., American Airlines) are equipping their fleets with Omega receivers. The reprint by Kasper and Hutchinson, pp. 245–249, describes the system. The reference by Wenzel (1988) describes its current status.

The U.S. Coast Guard maintains two Omega stations and other countries maintain six more. Each station is off the air 1 or 2 weeks per year, usually for scheduled maintenance. Unscheduled outages typically occur for 3 h per year per station.

Each station transmits eight consecutive sinusoidal tones at a 10-s repetition rate. Four are navigation tones common to all stations; the other four tones uniquely identify each station. Each station has a cesium clock that is calibrated within 1 μs by exchanging portable atomic clocks with other Omega stations and with the U.S. Naval Observatory. Approximately seven worldwide U.S. Navy VLF communication stations (at 16 to 24 kHz) are received by Omega sets. These stations are intended for sending messages to submerged submarines but serve as "unofficial" navigation stations because of their atomic-clock–stable carrier frequencies.

Omega receivers compute position in either of two ways:

(a) They measure the phase difference between pairs of stations, comparing the consecutive tones to the receiver's local oscillator. The receiver's position is at the intersection of hyperbolas formed by the station pairs.

(b) They measure the one-way ranges from any two stations. To do this requires an atomic clock in the receiver. The receiver's position is at the intersection of circles centered on the stations.

Either way, reception of additional stations removes ambiguity and improves accuracy. The phase-difference patterns repeat in "lanes" every 8 to 20 nm (when using one frequency) and every 144 nm (when using all five frequencies). Hence, position is ambiguous no matter how many stations are chosen. Thus, the usual method of navigation is to start at a known point and maintain continuous reception, counting lanes and measuring phase to interpolate within a lane. If coverage is lost while the receiver is moving, an independent navaid is needed to reinitialize. Therefore, Omega is a dead-reckoning radio Navaid.

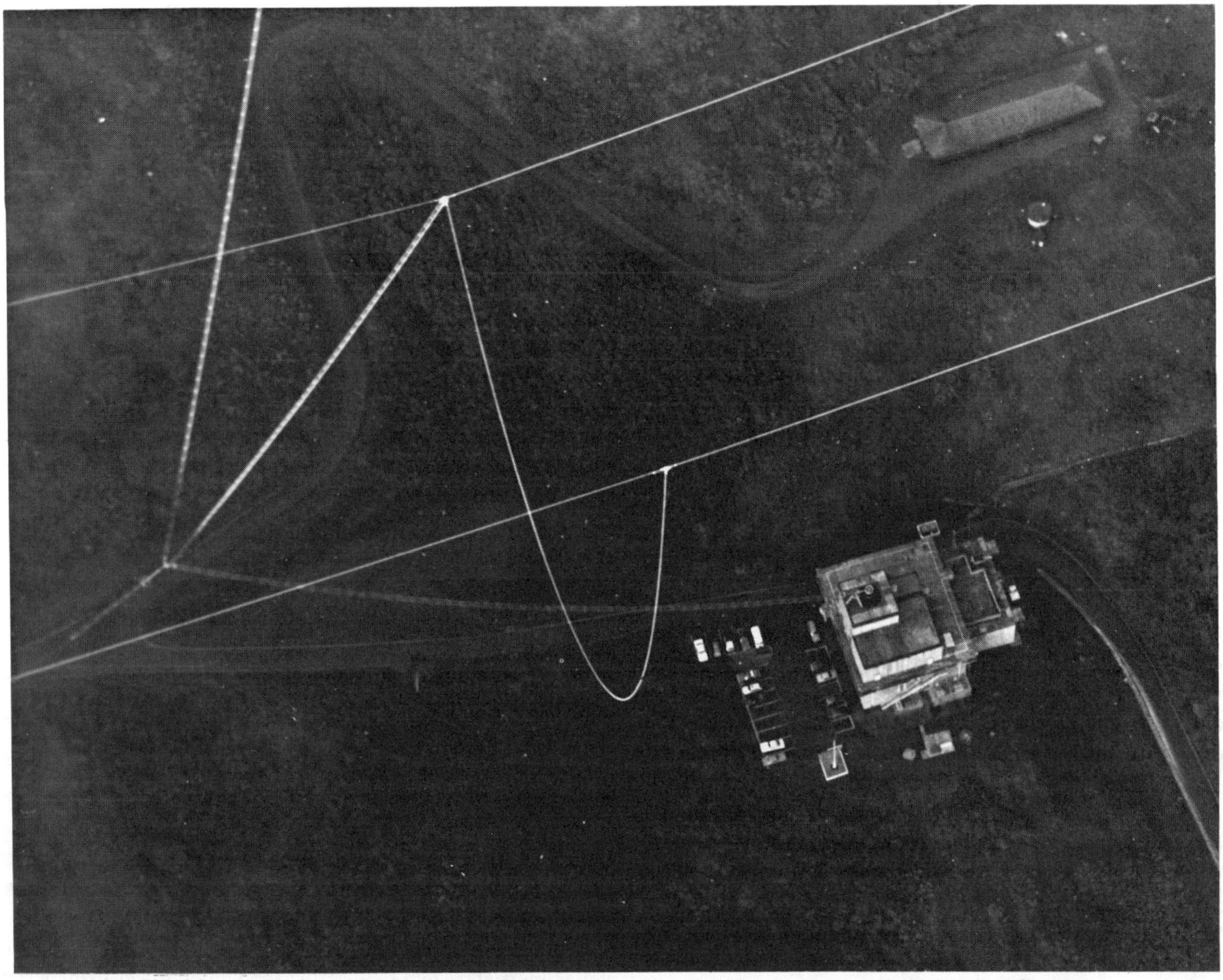

Fig. 5-3. Aerial view of the Omega transmitting antenna at Kaneohe, Hawaii. Courtesy of the U.S. Coast Guard.

Figure 5-3 shows an aerial view of the Omega transmitting antenna at Kaneohe, Hawaii. The antenna consists of six horizontal cables strung across a valley. The photograph shows the feed rising from the transmitter building to one cable. It also shows the suspended shunt that connects to the adjacent cable. The other four cables are not shown but their feed line is visible. A weight suspended from the feed line is faintly visible. The antenna at Norway is also horizontal, strung across a fjord. The other six Omega antennas are vertical masts. In most parts of the world, 4 or 5 stations can be received; in some places, 3 or 7 can be received. Propagation data are collected by 55 worldwide monitor stations. The magnitude of many propagation anomalies is broadcast on the U.S. Bureau of Standards' radio station, WWV, hourly. Predictable anomalies (e.g., those that occur when the signals cross the sunlight terminator) can be programmed into the receiver. They are as large as 8 nm and are complicated functions of position relative to each transmitter and of time-of-day. Unpredictable anomalies as large as 4 nm occur without warning. After correction, the absolute accuracy is 2 or 3 nm in most parts of the world. To improve accuracy, 17 ''differential-Omega'' sites have been installed, principally in France, Spain, and Portugal. They monitor the signals and transmit position offsets on 300-kHz marine beacons. Use of the ''differential'' signal allows positions to be corrected to 0.3 nm accuracy within 50 to 200 mi of the differential-Omega station. The poor accuracy of Omega and its lane ambiguities are offset by its global coverage and moderate receiver cost ($30,000) relative to Navstar or inertial navigation sets. The Coast Guard is still validating Omega coverage in various parts of the world. The polar regions (where conductivity of the ice is variable), for example, are not yet validated.

The first 5 operational Global Positioning System Satellites (Navstar-GPS) were launched in 1989, giving all vehicles on or near the Earth (even fast-moving aircraft) unprecedented navigation accuracy (see Part 3.6). At least one international airline has ordered combined GPS–Glonoss receivers (see Part 3.6). Airborne early warning aircraft carry JTIDS terminals (see Part 3.5) that permit cooperating fighters to be vectored to targets. Army close-air-support helicopters and aircraft may carry PLRS (see Part 3.4). Many experiments are now in progress to balance the cost versus accuracy of various combinations of inertial, Omega, Loran, GPS, and Transit equipment. Airborne receivers are being designed that combine navaids operating in a common radio band (e.g., GPS,

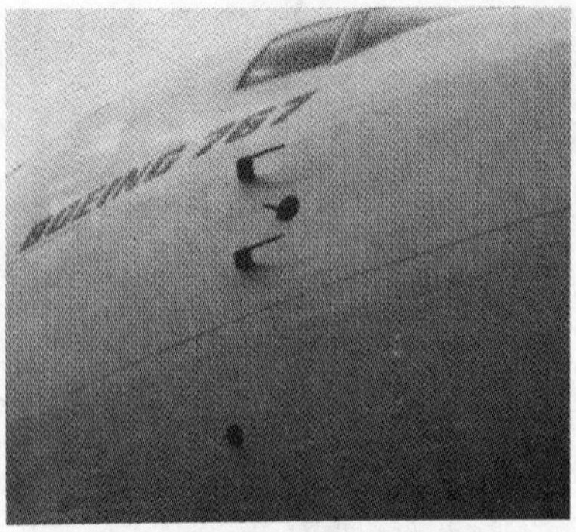

Fig. 5-4. Air-data sensors on a Boeing 767 airliner: two Pitot probes, one angle-of-attack vane between them, and one total-temperature sensor at the bottom. Courtesy of Rosemount Inc.

DME, JTIDS, and IFF). Cost savings are expected from common housings, power supplies, clocks, antennas, detector circuits, and output circuits.

INMARSAT's "Marecs" communication satellites (Part 4.10) serve ship traffic but do not have the capacity to handle air traffic or provide navigation functions. Several proposals have been made for high-capacity communication satellites (some of them with navigation capability) that would cover the oceans and entire continents, perhaps eventually displacing the ground-based air-traffic control network. One NASA proposal intends to serve 80,000 aircraft over the contiguous United States with 6000 uplink and 7000 downlink channels at L-band. The satellite would illuminate 30 regions of the country, each region having several radio channels. Radio frequencies would be reused in nonadjacent regions. Aircraft would need a 20-W transmitter. Some proposed nav–comm satellites are mentioned in Part 4.10. ARINC asked the U.S. Federal Communications Commission for permission to construct terminals in Annapolis, Maryland, and San Francisco for INMARSAT satellites. They will communicate their own messages directly to and from aircraft for approximately $10 per call. This represents the beginning of the oceanic traffic control improvement described on pg. 232.

5.3 On-Board Navigation

Dead-reckoning technology grew out of the gyroscopes and magnetic compasses of the early aircraft (see section 5.1 above). In early dead-reckoning systems, measured airspeed was resolved through gyro or magnetic heading to produce navigation information. The measurement of airspeed evolved from cup anemometers in the 1920s to venturi tubes in the mid-1930s, to Pitot tubes in World War II. Pitot probes drove panel instruments via pressurized tubes. By the late 1950s, aircraft speeds had surpassed Mach-1, where compressibility effects made it impossible to relate Pitot pressure to airspeed on a simple gauge. Therefore, Pitot and static pressures were converted into analog voltages and were input to an analog "air-data computer" that calculated true airspeed, free-stream pressure, and outside air temperature. In the late 1960s, air-data computers were digitized and solved a standard set of fluid-flow equations [see (Kayton and Fried (1969)]. Figure 5-4 shows two redundant Pitot probes on the side of the fuselage of a commercial airliner. In modern aircraft, the Pitot probe and static-pressure port are separated so each can be located where its measurements are least sensitive to the aircraft's attitude. The next step is to eliminate the drag-inducing Pitot probes and calculate airspeed, free-stream pressure, angle-of-attack, and angle-of-sideslip from pressure measurements at several flush-mounted static ports.

Early dead-reckoning computers resolved the calculated airspeed through azimuth, as measured by a directional gyro or magnetic compass, thus giving the north and east components of velocity. By integrating them in an analog computer, the north and east distances traveled were calculated. Such dead-reckoning systems had errors greater than 10% of the distance traveled, principally due to errors in the heading reference and forecast wind. Directional gyros and magnetic compasses are maneuver sensitive; the directional gyro drifted at 20°/h unless periodically reset. Over large distances, the curvature of the Earth must be included in the calculation of east distance.

The German World War II cruise missile, V-1, did not navigate; it was guided by directional and vertical gyroscopes and a magnetic compass. It held heading, attitude, and altitude until a nose-mounted propeller sensor made a predetermined number of rotations, whereupon it dove. Crude as it was by today's standards, it was enormously effective.

Improvements in gyro-based dead-reckoning led to inertial navigation, in which errors were compensated based on computed position and velocity. Inertial navigators resolve measured aircraft acceleration into east and north components and integrate them to produce ground speed. The air-data speed reference becomes a degraded-mode backup or a means of calculating instantaneous wind. In 1948, MIT's FEBE stellar-inertial system flew across the United States using a star tracker to monitor the drift of crude gyros. The system was 5-ft tall and contained seven servoed gimbals. In 1953, the six-gimbal SPIRE made a series of pure-inertial flights across the United States. Figure 5-5 shows the enormous gimbal set of SPIRE. It resembles a powered armillary sphere. By 1959, the A3J attack aircraft was equipped with an Autonetics inertial navigator for carrier operations. By 1960, the A2F was equipped with a Litton inertial navigator including the world's first central integrated digital computer. Aircraft inertial technology languished during the late 1950s and early 1960s as ship and ballistic missile technology advanced. Virtually all aircraft inertial navigators delivered before 1968 had analog computers. However, in the 1970s, inertial navigators with digital computers proliferated on civil and military aircraft, spurred by the wide-body jets and the demands of the Vietnam war. The newest inertial navigators use single-axis ring-laser gyroscopes that promise long life compared to spinning-wheel gyros. Most accelerometers retain the conventional pivoted-mass design. The reference by Kayton (1962) in the Introduction to Part 2 describes the fundamental physical limits on inertial measurements, approximately 0.1 nm/h.

Before the lunar module (pg. 325), all gyros and accelerometers were mounted on servoed, gimballed "stable platforms" that isolated the instruments from the angular velocity of the vehicle. In the past 20 years, increasing numbers of vehicles

Fig. 5-5. SPIRE gimbal set, 1954. Courtesy of Draper Laboratory.

have been employing "strapped-down" sensors mounted directly on the airframe. The resulting inertial measurement unit (IMU) is cheaper, lighter, smaller, and more reliable than a gimballed platform and therefore much desired by project managers. However, strapped-down IMUs expose the gyros and accelerometers to enormous angular velocities and angular accelerations, to the regret of the designers of mechanical instruments. The Boeing 757–767 and Aerospatiale A-310 are the first commercial aircraft to use strapped-down IMUs (with laser gyroscopes). They are less accurate than a gimballed inertial navigator built with the same instruments.

Strapped-down IMUs are difficult to align and calibrate, requiring a multiposition rotating fixture in place of a gimbal set, or a radio sensor and digital filters, or removal from the vehicle. The reprint by Wefald and McClary, pp. 258–266, shows how the strapped-down instruments are calibrated on the Boeing 757/767 by comparing measured landing coordinates with known landing position and by measuring the transverse velocity during takeoff on runways of known heading.

The fast-response but unstable vertical axis of an inertial navigator is often combined with the slow-response barometric altimeter to provide an accurate vertical velocity (2 ft/s), especially in military aircraft that use vertical velocity measurements for weapon delivery. The reprint by Widnall and Sinha, pp. 278–284, describes the optimization of the baro-inertial velocity channel of a military inertial navigator to give 0.65 m/s RMS vertical velocity error. Precise vertical speed measurements can be used to fly optimized climbs and descents, as shown in the reference by Farmer (1981). The "flight management system" computes optimum climb rate, cruise Mach number, and optimum time to start the descent and deploy flaps. Horizontal steering is by means of the usual radio and inertial navigation aids. If aircraft are to use their flight management systems, the air-traffic control computers must permit an on-board–selected altitude profile, as described by Mohleji (1988).

Perhaps the most important side benefit of inertial navigation was the invention of optimum filters that combine many sensors, usually an inertial navigator with radio aids. Some sensors are replicated (e.g., several gyros); others combine measurements from instruments of diverse design (e.g., air data, inertial, and radio aids). The filter's outputs are most probable position and velocity, and sensor errors. The reprint by Bobick and Bryson (pp. 250–257) and the references by Carlson (1988), DiFilippo *et al.* (1988), Harlow (1980), Hemesath (1971), Schneider and Maida (1988), and West-Vukovich *et al.* (1988) describe optimum filters.

Another outgrowth of inertial navigation and air-data technology was one form of wind-shear predictor that calculates the time-rate-of-change of the horizontal and vertical differences between inertial speed and airspeed. Most on-board wind-shear predictors observe the Doppler shift in the backscatter from a forward-directed laser or weather radar. On-board wind-shear predictors will be required in U.S. air transport aircraft beginning in 1991. The FAA is installing a network of airport-based Terminal Doppler Weather Radars, whose predictions will be sent to aircraft by voice and Mode-S communication link [see (Lopez and Wilson, 1988)].

Airborne mapping radar was invented for bombers in World War II, which found targets and dropped flares to mark them for the mass of non–radar-equipped bombers. The reference by Skolnik (1985) describes the history of airborne radar from World War II to the most current phased and conformal arrays, and synthetic apertures. Inexpensive weather radars were an outgrowth of the war, as were modern airborne mapping radars used for bombing and reconnaissance. Airliners and many multiengine general-aviation aircraft carry $100,000 weather radars that allow coarse navigation fixes, for example at distinctive land–water boundaries. By the 1970s, miniaturization permitted single-place military fighters to carry complex "multimode" radars that seek airborne targets and map the ground.

The highest-resolution airborne (and spaceborne) radars use a synthetic-aperture antenna. Radar returns are processed as the aircraft flies, giving an effective aperture several thousand feet long. At K-band, the angular resolution of a 1-m diameter dish is about 2°. However, the angular resolution of a 1000-m linear array is less than a minute of arc, if the aircraft motion is precisely known during the 10-s measurement interval. An inertial navigator usually measures attitude, velocity, and position changes while the radar data are being processed. The reference by DiFilippo *et al.* (1988) described the navigational measurements required to form a synthetic radar image and showed photographs of the raw radar image and the navigationally focused image. The reference by Kennedy (1988) described the algorithms.

From the 1930s to 1970 over-ocean aircraft, civil and military, had astrodomes in which human navigators took celestial fixes by hand with a bubble-sextant. Those sextants measured elevation angle relative to a bubble level (rather than to the horizon, as at sea), which was disturbed by aircraft motion. Long-distance radio aids and inertial navigation systems displaced these astrodomes and eliminated the human airplane navigator. A few high-altitude aircraft carry automatic star trackers, mounted on an inertial platform. The star tracker, a few inches in diameter, obtains continuous celestial fixes above the clouds, where stars are visible day and night to a narrow-field-of-view instrument. The measurement accuracy is better than 1 mi.

In 1958, a new form of navigation was installed on the MACE cruise missiles (MGM-13B). Aerial photographs and radar images had always been interpreted by skilled operators. MACE carried a navigation system called ATRAN, in which a sequence of expected radar images was stored on 35-mm film and compared to real-time images from the MACE-born radar. I believe this was the first map-matching system; coincidentally, I worked on its design as a very junior engineer in 1955. Later, about 1963, a system called Fingerprint was proposed that matches a measured terrain-altitude profile to a stored profile. The measured profile is obtained by subtracting radar altitude from barometric (or baro-inertial) altitude. That system, now called Tercom, is used on U.S. cruise missiles. It achieves an accuracy of less than 200 ft (50% of measurements) when the root-mean-square terrain height exceeds 30 ft (having the appearance of slightly hilly terrain). The reprint by Hinrichs, pp. 288–295, describes Tercom. The reference by Boozer and Fellerhoff (1988) describes the equipment and algorithms for a terrain-profile matcher that can use standard digital maps. In simulation, it achieved a 75-m mean radial error and a 17-m altitude error.

All pilots feel that they are at the mercy of faceless air-traffic controllers who try to assure their separation from other aircraft. Following a collision between two American commercial aircraft over the Grand Canyon in 1956, many attempts were made to design on-board collision-avoidance systems (CASs) that issue warnings and steering commands. The most desirable solution would allow each aircraft to avoid the other, even in a multi-aircraft encounter, without benefit of external aids. The first CAS was a time-slot ranging system called EROS built by Sierra-Wilcox and tested by McDonnell in 1966 to separate flight-test aircraft [see (Borrok, 1967)]. More recently, the FAA defined three levels of Traffic Alert and Collision-Avoidance Systems [TCAS, see (FAA, 1989)] that interrogate the civil IFF transponders in other aircraft (Mode 3/C, section 5.2). The lowest level, TCAS-1, alerts the crew of a nearby aircraft, based on transponder reply time. TCAS-2 issues climb-dive advisories based also on altitude, heading, and a coarse measure of the direction of arrival of the replies. TCAS-2 can display a map of transponding traffic on a CRT in the cockpit (e.g., the horizontal situation display). With advanced software, multiple aircraft in the vicinity can coordinate their TCAS-2 advisories on the Mode-S data link, to prevent their maneuvering into each other. Finally, the still-to-be-approved TCAS-3 will measure the direction of arrival of transponder replies accurately enough to issue altitude and steering advisories. TCAS interrogates aircraft that are closer than 8 nm (in dense traffic) to 14 nm (in light traffic). The accuracy of TCAS-2 is limited by the ability of its antenna to measure angle of arrival. A goal of the software design is to reduce false alarms in densely populated airspace. TCAS systems have been extensively flight-tested by Piedmont Airlines (which ordered 81 TCAS-2 sets in 1987). Tests by United Airlines in 1988 showed an alert occurring every 2.2 h and an altitude or steering advisory every 28 h, two-thirds of which occurred in descent and approach. The FAA Final Rule (1989) requires TCAS-1 and 2 to be installed by 1993 in U.S. air-carrier aircraft and, by 1994, in all jet aircraft operating in the United States. The reprint by Sobocinski, pp. 285–287, describes the results of TCAS-3 tests. Kolcum (1989) has described TCAS-2 equipment for sale to airlines. Leeper and Anderson (1979) have described the Mode-S data link.

A closely related problem is faced by cargo aircraft and helicopters that frequently fly in formations of 2 to 30 aircraft in all weather, and hence require relative navigation called "stationkeeping." The lead aircraft navigates relative to Earth, by any of the techniques described above. The others carry stationkeeping equipment (SKE) that measures position relative to the leader. The most common stationkeeper is a 3.4- to 3.5-GHz radio system that permits each follower to calibrate its range, bearing, and barometric altitude–difference with respect to the leader. Each aircraft transmits a ranging pulse and other data in one of 36 predetermined time slots on an omni-directional antenna. When it is not transmitting, each aircraft sweeps the sky with a rotating antenna. The rotating antenna identifies and locates the other aircraft in the formation by one-way ranging and in bearing. All aircraft in the formation synchronize their clocks from the transmissions. Aircraft fly 2000 to 4000 feet apart and maintain relative navigation errors of 200 ft in range and 2° in bearing (100 ft at 3000 ft range). Relative altitude is measured within 100–200 ft. Callahan (1990) describes the system in general terms. A ground-based transponder, called a Zone Marker, permits cargo and personnel to be dropped at night and in instrument weather with an accuracy of 100 m. The crew of the lead aircraft can signal an impending standard maneuver and all aircraft will lead into it. The system, made by LTV Sierra, is installed on many U.S. C-130, C-141, and C-17 aircraft and on foreign aircraft. It is shown on the cost-accuracy chart, Fig. 1-1, pg. 2.

Future stationkeeping systems may use GPS receivers, operating in a "differential" mode. The lead aircraft would broadcast its GPS-derived position and information about which satellites it is using. The other aircraft would each carry a receiver that measures its own position and would navigate relative to the position of the master. Accuracies of 2 to 5 m are predicted within the formation, if all aircraft use the same satellites. Inter-aircraft separations would be much larger to allow for the dispersion of each aircraft in holding the desired position and to allow for an escape maneuver in case of equipment failure.

Air Force bombers and Army helicopters have a special navigation problem concerned with low-altitude flight (often less than 200 ft above the mean terrain) to avoid detection by enemy radar. By 1970, the F-111 had a manual terrain-following system that displayed pitch cues on a HUD (section 5.4) based on airborne-radar scanning of peak terrain ahead of the aircraft. Terrain-following technology has advanced enormously to permit the airborne radar to measure terrain profiles at several azimuths and issue pitch and azimuth commands (called "terrain-following / terrain-avoidance," TFTA) that choose the best path and best vertical profile for maximum concealment, subject to a pilot-selected risk of ground contact. Mission-planning software is being developed that extends TFTA to replanning routes in flight, to avoid enemy guns and missile sites while still concealing the aircraft. Aircraft and helicopters flying within 50 m of the ground may soon use laser radars to detect power lines and masts [see (Busbridge and Puleston, 1989)] and even birds.

Some military aircraft drop bombs, fire guns, or launch missiles. The delivery accuracy depends heavily on the ability to measure the aircraft's velocity. Most aircraft measure ground speed within 1 to 3%. At 600 ft/s, the uncertainty is 12 ft/s which causes a 240-ft dispersion for a free fall of 20 s. Diesel (1987) showed that a GPS-inertial system can maintain 0.25 ft/sec error, even during maneuvers, thereby greatly improving the accuracy of delivering unguided weapons.

Unmanned aircraft are assuming a major role in the surveillance of battlefields and shorelines (looking for submarine-launched ballistic missiles, for example). They require precise, continuous navigation, usually in conjunction with a data link. GPS and JTIDS will probably be the preferred navigation systems, updating a low-quality inertial navigator. Leavitt and Peters (1988) have described a GPS-inertial system for a reconnaissance aircraft. The system is predicted to have 35-m accuracy, 300 nm from its base. Cruise missiles are unmanned aircraft that fly thousands of miles on preplanned routes. They navigate with Tercom-inertial systems.

No one likes to plan for accidents but if one occurs, everyone wants to analyze the preaccident trajectory. Thus, turbine-powered airliners are required to carry recorders (formerly all-analog, now often digital). In the past, they operated from the same pressure ports that serve the cockpit instruments. As aircraft have grown in size, separate air-data probes (static and dynamic pressure, temperature) have been installed in the tail, where the recorders have the best chance of survival.

A special form of navigation is required for aircraft being tested, whether civil or military. To test an aircraft navigation system requires an even better reference. Overflying a visual checkpoint is accurate to a few degrees of vertical cone angle across-track and to tens of degrees along-track. Formerly, ground-based radars tracked the aircraft and the trajectory was plotted manually; now digital tapes of radar-derived position are prepared. Especially on military ranges, a navigation pod is often attached to the test aircraft, as described by Hughes (1976) and by Solomon and Lewantowicz (1989). The pod contains a precision inertial navigator, DME-ranging equipment, recorders, and rate gyroscopes. The pad records all attitude and trajectory data. A typical navigation test reference achieves 13 ft and 0.1 ft/s error horizontally, and 40 ft and 0.4 ft/s error vertically (one sigma).

Figure 5-6 shows the antenna complement on an aircraft of the 21st century. Most of the antennas are flush-mounted; others, such as VHF, are blades. The Omega antenna is a stinger as are some HF antennas (mounted in the tail cone). Redundant antennas (e.g., for MLS, TCAS, VOR, and VHF communication) are provided when continuous signals are needed despite the shadowing of a banked aircraft. Note the presence of satellite-communication and GPS antennas on the top of the fuselage.

5.4 Landing Aids

The Instrument Landing System (ILS) was introduced in 1941 and selected by the ICAO in 1946 as the international all-weather landing aid (more precisely, as an aid to approaching within 200 ft above an airport, after which the landing is visual). The ILS establishes a VHF horizontal pattern and a UHF vertical pattern. The signals in space are detected by airborne receivers that measure up–down and left–right departures from the desired straight-line path (2.5 to 3° above the horizon) and display the errors on cockpit pointers. Modern ILS ground equipment is described in the references by Watts (1980) and McFarland (1981) where the complexities of design and installation are illustrated. They especially address the effect of reflection from the ground, buildings, and vehicles on the shape of the beam. Both authors show that modern ILS antennas can be adapted to virtually any site.

From World War II until 1967, the FAA offered Precision Approach Radar (PAR) monitoring of approaches. An operator in a radar hut would call out "high–low" and "left–right" instructions. PAR service was discontinued by the FAA in 1967 due to the prevalence and reliability of ILS. It is still offered by the U.S. military and by civil and military authorities in a few other countries. It is said to be the most widely used landing aid in the Soviet Union.

Nonprecision approaches use direction-finding beacons, VOR, or Loran. They usually require that the approach be

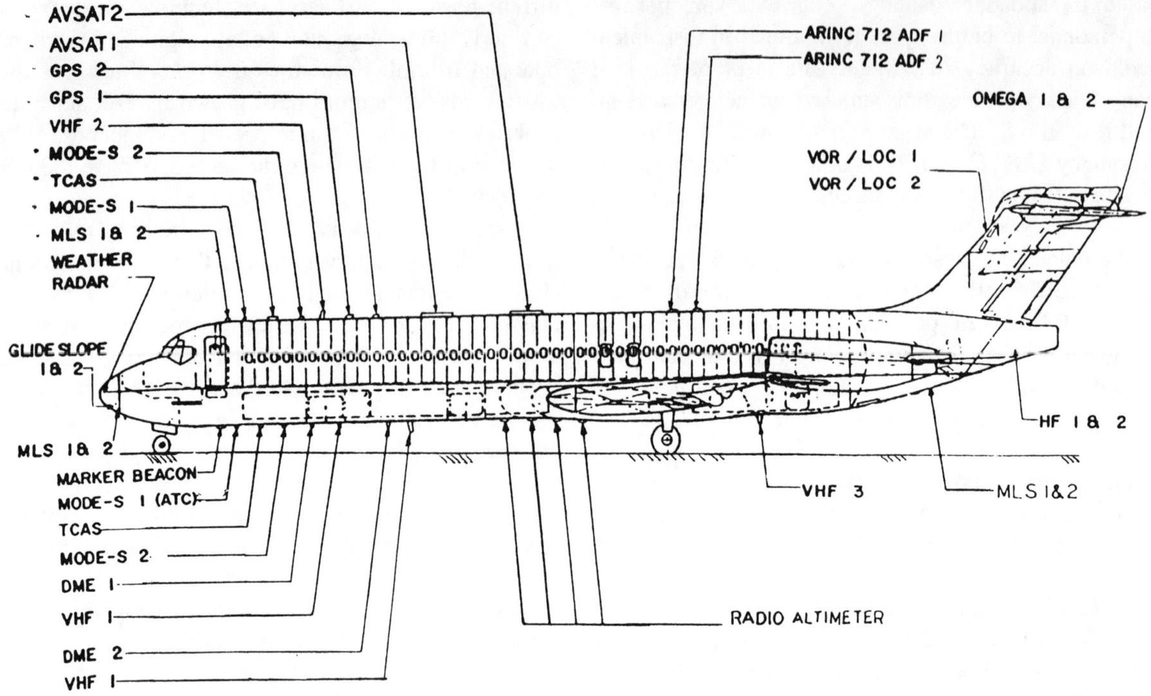

Fig. 5-6. Antenna complement of future commercial aircraft. Courtesy of McDonnell-Douglas Corp.

abandoned if the runway lights are not visible 700 ft above the ground. The crew need have no prior knowledge of horizontal visibility at the airport.

Precision approaches are made with ILS, MLS, or precision approach radar. They are prohibited unless horizontal visibility, called "runway visual range" (RVR, measured by transmissometers), exceeds a published threshold. The crew must execute a missed approach if the runway lights are not visible when the descent reaches a published "decision height" (DH). As visibility falls, the following problems must be solved:

(a) The transition from instrument flying to manual control with visual cues becomes more complex. Fully automatic landing to touchdown and manual landing with a head-up display have been solutions, as noted below.

(b) Rollout and taxiing become more difficult. The solution has been on-board rollout control using the localizer or MLS for lateral deviation from the centerline, and precision DME for distance to go. Traffic signals are being installed on the taxiways at some airports, to assist in taxiing in fog. Buried cables may someday be used to keep the aircraft on the taxiway. Airport Surface Detection Radar [described in (Schwab, 1985)] may be of assistance on taxiways not shadowed by buildings. It operates at 14–16 GHz and locates aircraft within 6 m.

(c) Failures in ground or on-board equipment must be accommodated. Solutions have been fail-passive systems (that fail-safe, without transients, but cannot continue the landing) and fail-operational systems (that continue to operate, without transients). To reduce the exposure time to multiple failures, that could be disastrous, an "alert height" (AH) has been introduced above which the landing must be aborted if a failure occurs. At present, the alert height is 100 ft above ground level.

(d) The aircraft risks touching the runway if the missed approach is begun too low, either on the pilot's initiative or at the request of the traffic controllers. The solution has been to provide enough fault tolerance that a pilot-initiated abort is unlikely for reasons of deteriorating visibility or failures below the DH. Traffic control procedures must not require an abort when a Category III aircraft is below the DH.

In 1964, the French government permitted ILS approaches to a "decision height" of 100 ft, at which runway lights must be seen or the approach aborted. In 1967, the U.S. FAA reduced the decision height from 200 to 100 ft for specially equipped "Category II" aircraft. The French government further reduced the decision height to 50 ft in 1966 and the FAA approved Category III (zero decision height) fully automatic landings in 1971. In 1967, a Pan Am Boeing 727 made a hands-off automatic touchdown in New York and a British Airways Trident did the same in London, both in good visibility. Templeman and Parker describe the Boeing 727 autoland system in the reprint on pp. 296–302. To permit landings at lower minimums, avionic designers had to permit a crew, flying on instruments, to transition suddenly to visual cues from runway lights. The British Blind Landing Experimental Unit (BLEU) solved that problem in 1964 with a head-up display (HUD) that projected instrument readings onto the windscreen so the pilot saw electronically generated cues superimposed on a natural view of the runway and lights. Today, HUDs are widely used in single-pilot military aircraft (and the Space Shuttle) and have been introduced into the McDonnell-Douglas MD-80 civil airliner. Low-visibility landings have always been of greatest interest to northwestern Europeans who have the worst flying weather in the developed world. Sutton (Apr. 1989) has stated that more than 100,000 automatic landings have been made worldwide, approximately 5000 of which have been in Category III weather.

In the United States, Category I approaches are permitted with one ILS receiver if RVR exceeds 2400 ft. The decision height is 200 ft. About 600 U.S. runways meet Category I standards. Category II approaches are permitted at specially equipped airports by aircraft with redundant equipment and specially trained crews. RVR must exceed 1200 ft and the DH is lowered to 100 ft. About 70 U.S. runways met Category II standards in 1988.

Category IIIA permits approaches with a RVR greater than 700 ft and allows a DH of 100 ft. The aircraft must have fail-passive flight control and rollout control systems. Category IIIB permits a RVR below 700 ft (eventually 150 ft) without a decision height. A fail-operational flight-control system and a rollout-control system whose fault tolerance increases as the DH decreases are required. Cotter and Cohen (1979) describe an automatic rollout system for the Boeing 747. In 1989, Category IIIB approaches had been approved at 33 European airports and 19 U.S. airports. When Category IIIC is approved, it will have no restrictions on RVR or DH, allowing the crew to land without ever having seen the airport. Category IIIC equipment requirements have not yet been defined. Most new civil airliners are available with avionics that meet Category IIIB. The Fokker 100, for example, is certified to 490 ft RVR.

The Microwave Landing System (MLS) was developed in the 1960s to permit greater flexibility in approach paths to airports than ILS, which requires a 5-mi-long straight-line descending approach along the runway centerline. MLS was also designed to reduce multipath errors caused by reflections from buildings and rough terrain and to be easily qualified to Categories II and III without the terrain sensitivity exhibited by ILS. The C-band (5 GHz) MLS transmits identification data and signals in space that cover a wedge, $\pm 60°$ wide by 30° elevation, at the approach end of the runway. Each aircraft decodes the signals and determines its own position relative to the runway within 0.05°. MLS allows each aircraft to select its own glide-path angle above the horizontal (versus a fixed 2.8° path for ILS). With a suitable on-board computer, MLS allows each aircraft to make a curved approach in the horizontal and vertical planes. A colocated precision DME measures range from touchdown. As does the ILS, MLS guides the aircraft throughout the approach and in azimuth on the runway. As does ILS, MLS guides the aircraft to a point 50 ft above the runway, after which an on-board altimeter controls the flare if full autoland is required. MLS receivers were on the market in 1989 and MLS/ILS combined receivers are being designed. The reprint by Evans, pp. 313–316, and the reference by Cox and Shirley (1980) describe MLS. Kelly (1982) has described the specifications for the associated precision DME.

The FAA ordered 178 MLS ground stations for civil airports but canceled the procurement in mid-1988. Airlines will be encouraged by the FAA to install receivers by preferential treatment in landing. The FAA will give preference to MLS-equipped aircraft in New York and some other cities where several airports have interlaced approach paths. The FAA will probably reduce Category I (1200-ft visibility) DH from 200 to 150 ft for MLS-equipped aircraft. Thus, the FAA hopes to replace ILS with MLS in air-carrier service over the next 10 years in order to:

—reduce the spacing of aircraft;

—increase the number of radis channels, especially in cities with many instrumented runways;

—allow parallel instrument approaches; and

—allow simultaneous helicopter and fixed-wing instrument landings.

The recent mergers of airlines, consumated by general-purpose businessmen concerned with short-term profits, have slowed the introduction of MLS. Installation will not be complete, even at major U.S. international airports, much before the turn of the century. The ICAO has accepted MLS as the standard landing aid for commercial aviation in the 21st Century. Nevertheless, ILS and MLS will co-exist for many decades.

Triple microwave sets at K-band (15 GHz) are the only landing aids carried by the Space Shuttle. The Shuttle's microwave landing system is the same as the Navy's SPN-41 carrier landing system (described below).

The U.S. Navy has been conducting aircraft operations from carrier decks since the 1930s. Carrier landing is a dangerous business because aircraft sometimes fail to find the ship, hit the superstructure, slide overboard, and improperly execute missed approaches (''waveoff''). Tacan solved the problem of finding the carrier. The other problems were addressed as follows: In World War II, a Landing Signal Officer (LSO) ''talked'' each plane down, based on his visual observations. A few years later, carriers were equipped with a Fresnel-lens system that displayed a yellow light bar (''meatball,'' still in use in 1989). The optic system is stabilized for the ship's roll and pitch but offers no azimuth cues. To an approaching aircraft that is on the desired glide path (usually 3.5° above the horizontal, relative to the moving carrier), the meatball is centered and lines up with a green light bar. A pilot above or below the Earth-stabilized glide path sees an offset meatball.

Development of the SPN-10 radar guidance system began in the 1950s. It was a carrier-borne radar that tracked landing aircraft on their last 10 mi of approach. It had ship-rotation sensors and an analog computer that generated steering signals which were radioed to the aircraft. In 1954, the first automatic landing was made on a carrier deck with it. SPN-10 was upgraded to the SPN-42, which is installed on all U.S. aircraft carriers in 1989. Its 33-GHz radar tracks a corner reflector or X-band beacon on the landing aircraft. The software in its digital computer allows for many subtleties, including corrections for the ship's angular and linear motion, the 500-ft-long burble (''rooster tail'') produced by the air flow over the deck, the wake of the island, the flow of stack gases, vertical translation (''heave'') of the deck, and the dynamics of each aircraft type. As the aircraft approaches the pitching, heaving deck, it receives commands by UHF radio that drive either the autopilot or the cross-pointers, at the pilot's option.

A wholly independent system, also installed on all carriers, allows the pilot to monitor the SPN-42 approach. SPN-41 is a carrier-borne, Earth-stabilized scanning beam instrument landing system operating at K_u-band (15 GHz). Its beam is stabilized for roll, pitch, and yaw of the deck, as measured by the Ship's Inertial Navigation System (Part 4.9, pg. 157). Its signals in space are detected by the aircraft and actuate cross-pointers in the cockpit that show the pilot how far his aircraft is from the desired glide path. A land version of SPN-41 is used as the landing aid for the Space Shuttle under the name "Microwave Scanning Beam Landing System" (MSBLS, Part 6.4, pg. 322).

In other than battle conditions, the combined SPN-41/42 permits approaches to a carrier in 200 ft ceiling, 0.5 nm visibility. A newer system, SPN-46 replacing SPN-42, is in development. It will have a combined X/K-band radar antenna and distributed digital computers.

5.5 Displays

In early aircraft and in most presently flying aircraft, each navigation sensor has its own display (e.g., a tuning head for VOR, a channel selector for ILS). In the newest aircraft, several navaids share common displays:

(a) attitude-director indicator (ADI), whose needles display VOR or ILS errors, as selected by the crew. Older ADIs are mechanical; the newer ones are CRT emulations of the mechanical ADI. Even aircraft as new as the Soviet MIG-29 still use mechanical ADIs ("8-ball");

(b) horizontal situation indicator (HSI), which shows bearings to targets. The bearings are obtained from a computer, from VOR, or from a direction finder. Older HSIs are mechanical; the newer ones are CRTs.

(c) map displays in which a film is projected onto a ground glass or a computer-generated map is displayed on a CRT. Aircraft position can be superimposed on a map display, whether position is measured by a sensor or calculated from multiple sensors. Map displays will eventually be integrated with HSIs;

(d) head-up displays (HUDs) in which a computer-generated image is projected onto the windscreen. With a HUD, a pilot can see key instrument indications without lowering his gaze into the cockpit. Map information and synthesized cues can be superimposed on the visual field. As noted in section 5.4, HUDs are used on single-place military fighters and the Space Shuttle and are being introduced into two-pilot commercial airliners. Their principal use has been to guide air-to-ground and air-to-air attacks by fighter aircraft (e.g., A-7, F-111, F-16, and F-18). Figure 6-14, pg. 330 shows a HUD in the Space Shuttle;

(e) synthesized voice system has been used to report failures but has not yet been used for navigation data.

The Air Force Digital Avionics Integration System (DAIS, started 1973) developed multipurpose displays and began the trend toward distributed computing. Keyboards have been used in flight to enter waypoints, initial position, and radio frequencies since the Apollo spacecraft. With more computers on-board, more information is prestored, requiring that the crew select among displayed alternatives. Nevertheless, the crew will sometimes enter new data, such as the coordinates of waypoints.

Figure 5-7 shows the incredibly clean cockpit of the MD-11 commercial airliner. The outer CRT for each crewman is normally a dedicated ADI that gives the pilot cues for pitch and roll steering. It shows attitude, airspeed, heading, steering errors, barometer setting, and other data; approximately 60 symbols. The next inboard CRT is normally the HSI that shows a plain view of heading and the directions toward key waypoints, airways, and landing beams. It also displays weather radar images and TCAS targets. The innermost CRTs show subsystem status, an engine format being visible in Fig. 5-7. ADI and HSI displays are transitioning from digital emulations of the older analog displays into displays that are specially designed to present information so it is absorbed as rapidly as possible. They are cleverly cross-strapped to the symbol generators so formats can be interchanged in case of failure. The center pedestal of the MD-11 contains the controls for the engine and autopilot and three flight-management CRTs that control the navigation equipment.

On single-place military aircraft such as the F-18, the cockpit is even more tightly integrated. Three CRTs, in the traditional T-configuration, have peripheral buttons that configure any one into a communication, navigation, subsystem, or weapon controller. Some systems, in development, display a digitally synthesized map on the HSI as an aid to human navigation. The F-18 also has an array of buttons on the center stick and throttle that control the subsystems and the CRTs, without requiring the pilot to take his hands off the controls. In case of total loss of its CRTs, the F-18 has backup conventional instruments on the front panel (mechanical attitude indicator, altimeter, and airspeed indicator) and backup control panels at the sides. Adam (1980) and Drummond (1981) have described the F-18 cockpit in detail.

The next step, especially in military fighters, may be the panoramic display [see (Haralson, 1989)], a reconfigurable flat panel covering the entire instrument area. It could show the computed position of friendly and hostile targets, and a map. Computer-generated pictorial inserts would display subsystem information and flying cues. A head-up display or helmet-mounted display would be the primary flight reference and would show light-amplified scenes at night. Simulations are required to determine the value of such a display. Technology improvements would be needed to build a 300-in^2 flat panel. Theorists are speculating about windowless cockpits that would protect the crews against nuclear radiation. The pilot's only outside view could be a panoramic display or stereo television, presented to each eye in a helmet.

Voice actuation of subsystems has been tested but not implemented, to my knowledge, on any vehicle. "Personalization" cassettes will probably be needed to adapt the

Fig. 5-7. MD-11 cockpit showing CRT attitude (outer), heading, and status (inner) displays. Courtesy of McDonnell-Douglas Corp.

computer logic to the speech patterns of each crew member. The crew will have to restrict its vocabulary when the voice system is armed.

Hence, from the crew's viewpoint, navigation systems are becoming ever more automatic as the crew need not decide which sensors to use nor manipulate dedicated navigation instruments.

5.6 Computation

The first vehicle-borne computers were naval analog gun-laying computers, invented by Hannibal Ford prior to World War I. They computed the desired azimuth and elevation of large-caliber turret guns from optical measurements of target range and bearing. In those computers, shaft-rotation angles were analogs of range, bearing, and elevation. The computer corrected the gun-laying direction for the ship's pitch and roll, as measured by inputs from the ship's gyrocompass. The analog computer was extended into the aircraft autopilot, first as a pneumatic analog computer, then as an electric analog computer. During World War II, simple analog computers were invented that predicted the lead angle for pointing an aircraft so its gunfire would hit an airborne target. The computer moved a ''disturbed reticle'' on the windscreen; the pilot flew the aircraft to place the ever-dancing reticle over the target. Draper (1981) has described World War II gunsights in detail. The Norden bombsight was the most famous World War II analog computer. It calculated release time from inputs of aircraft speed, altitude, and distance to the target (entered manually from an optical or radar image).

In aircraft of the 1950s, the only computing was for the autopilot and stability augmentation, both of which were executed in analog electronic circuits. The autopilot performed heading hold, altitude hold, speed hold, and control-wheel steering (the manual input of a turn-rate command to the autopilot). In an occasional high-performance fighter (F-101, F-106), a marginally stable airframe was stabilized by multiaxis electronic filters. Gain was sometimes varied as a function of instantaneous dynamic pressure.

The Concorde supersonic transport was the first passenger aircraft that was unstable and required electronic augmentation, which was implemented by a four-redundant-channel analog computer. Subsonic jet airliners usually have under-damped yaw axes that require a 1-axis analog computer for stability, though these aircraft are flyable without the yaw damper. In the late 1970s, unstable military aircraft became the rule because they could maneuver rapidly and save weight of empennage. These aircraft are often unflyable in all axes without the electronic stability-augmentation system that is now usually implemented in redundant digital computers.

Inertial navigators and air-data systems of the 1960s contained analog computers in which positions and speeds were represented by servo-driven shaft angles. The A2F, circa

1960, combined inertial navigation, air data, and autopilot calculations in a central digital computer. Fifteen years later, the Space Shuttle combined all computing functions, including those for safety of flight, into five redundant central digital computers. The DAIS test bed developed intercomputer interfaces that became the MIL-STD-1553 data bus. As a direct result, the newest aircraft have a multitude of on-board computers, each dedicated to a particular function (e.g., inertial navigation, air data, radar, Tacan, weapon delivery, communications) and interconnected to exchange data in a network that behaves like a large computer. Each computer is programmed by the experts in the relevant technology; only the intercomputer protocols are standardized. In 1989, fault tolerance in aircraft was provided by replicative redundancy with rare exceptions in which tasks were interchanged among computers after failures. The reprint by Wei and Huddle, pp. 267–277, describes fault detection to the level of the failed sensors. Fault tolerance has been a major characteristic of safety-related flight controls for decades and is now expanding to other aircraft subsystems. In the next generation, integration of the front-end sensors (especially radios) may force a return to centralized computers in which several subsystem functions execute in the same computer. Military test beds (Pave Pillar, Pave Pace, and Dames) are developing computer and sensor architectures and software for future aircraft.

References and Bibliography

Acharya, N. A. and D. P. Glasson, "Intelligent FDI management in the adaptive tactical navigator," in *Rec. IEEE Position Location and Navigation Symp.,* 1988, pp. 120–125. Failure detection and identification in an inertial, Doppler radar, GPS, altimeter system.

Adam, E. G., "F/A-18 Hornet crew station," in *Proc. IEEE NAECON,* 1980, pp. 1236–1240. Good description of multifunction displays including navigation and communications.

Anon., "Aircraft stationkeeping," *Space Aeronaut.,* pp. 83–91, Apr. 1967. APN-169 then in development for C-130 aircraft and CH-53 helicopters.

Anon., "TCAS fleet installation," *Aviat. Week and Space Technology,* Nov. 23, 1987. Piedmont to start installing 81 Bendix TCAS-2's in 1989. Two directional antennas. Upgradable to TCAS-3 with software change. Shows traffic information on CRT.

Anon., "T/CAS trials," *Aviat. Week and Space Technology,* Nov. 1988. Six-month evaluation of TCAS-2 in UAL B-737 and DC-8 for 2025 hours on 1224 flight segments.

ARINC Standards on various aeronautical navigation devices. Annapolis, MD: Aeronautical Radio, Inc. Used by all commercial airlines.

Bobick, J. C. and A. E. Bryson, Jr., "Updating inertial navigation systems with VOR/DME information," *AIAA J.,* vol. 11, no. 10, pp. 1377–1384, Oct. 1973. Reprinted in Part 5, pp. 250–257.

Boozer, D. D. and J. R. Fellerhoff, "Terrain-aided navigation test results in the AFTI/F-16 Aircraft," *Navigation*, vol. 35, no. 2, pp. 161–175, Summer 1988. Describes a TERCOM-like terrain-altitude-matching navigator, using the standard DMA digital data base. Concentrates on the algorithm. Calculates at 15 Hz. Achieves 50- to 75-m mean radial error and 17-m altitude error.

Borrok, M. J., "EROS in operation," *Navigation*, vol. 14, no. 3, pp. 315–320, Fall 1967. Early McDonnell tests on F-4 collision-avoidance system.

Busbridge, M. I. and D. J. Puleston, "A laser obstacle avoidance and display system," in *Proc. IEEE NAECON,* 1989, pp. 179–186. Description of laser radar and cockpit display designed to detect 10-mm-diameter cable at a range of 1–2 km, even in rain.

Callahan, J., "Rendezvous in the air," *Defense Systems International Air Systems 1990*, London: Sterling Pub., pp. 141–145. User-level description of SKE-169C electronics developed by LTV-Sierra for formation flights of cargo aircraft and for air drops. Pictures and top-level characteristics of the hardware.

Carlson, N.A., "Federated filter for fault-tolerant integrated navigation systems," in *Rec. IEEE Position Location and Navigation Symp.,* 1988, pp. 110–119. Describes dispersed Kalman filter in which each subfilter is associated with a sensor, with intercommunication.

Cotter, C. J. and G. C. Cohen, "Automatic rollout control of B-747 airplane," *AIAA J. Guidance and Control,* vol. 2, no. 1, pp. 25–30, Jan. 1979. Good detailed paper on rudder and nose-gear control for localizer-guided rollout-control system.

Cox, R. M. and J. M. Shirley, "MLS—A new generation landing guidance system is here," in *Rec. IEEE Position Location and Navigation Symp.,* 1980, pp. 208–215. History and pictures of flight sets and antennas (air and ground).

Diesel, J., "Integration of GPS/INS for maximum velocity accuracy," *Navigation*, vol. 34, no. 3, pp. 190–211, Fall 1987. Analyzes velocity error in GPS-inertial system, one-minute and ten-minutes after losing GPS data. Conventional 17-state Kalman filter. INS rate-aids GPS carrier tracking. Velocity error is always less than 0.25 ft/s.

DiFilippo, D. J., *et al.,* "Evaluation of a Kalman filter for SAR motion compensation," in *Rec. IEEE Position Location and Navigation Symp.,* 1988, pp. 259–268. Compensation of synthetic-aperture radar for aircraft motion. Description of a filter using strapped-down inertial, Doppler radar; air data; and barometric sensors.

Doubt, R., "OMEGA regional validation program," *Navigation,* pp. 115–164, Fall 1984. Cesium clock closures among stations. No numerical results.

Draper, C. S., "Fifty years of flight technology," *IEEE Trans. Aerosp. Electron. Syst.,* vol. 14, no. 4, pp. 708–712, July 1978. FEBE-SPIRE history and photographs.

Draper, C. S., "Origins of inertial navigation," *AIAA J. Guidance and Control,* vol. 4, no. 5, pp. 449–463, Sept. 1981. Heavy on fire control and gunsights (air and naval). FEBE; SPIRE; SPIRE, Jr.; MAST; SINS Mark IV.

Drummond, R. C., "F/A-18A weapon system: 1976 state of the art," in *Proc. IEEE Digital Avionics Conf.,* 11 pgs., 1981. Block diagrams of all subsystems.

Evans, T. E., "Microwave landing system," *IEEE AES Mag.,* pp. 6–9, May 1986. Reprinted in Part 5, pp. 313–316.

FAA, *Advisory Circulars.* For example: AC 120-20, 1966, Category II; AC 120-28, 1984, Category III; AC 90-45, 1975, Area Navigation.

FAA, *National Airspace System Plan,* May 1985. Chapters on demand, air-traffic control (en route, terminal, FSS), ground–air communication, interfacility communication, maintenance, and operations support.

FAA, *Plan for the Security Control of Air Traffic and Air Navigation Aids: SCATANA,* Aug. 1987. Control of air traffic in an air defense emergency.

FAA, *1986 Federal Radio Navigation Plan* (by DOT and DOD), Sept. 1987. Recommends mix of systems (GPS, MLS) and phaseout of TACAN, VOR, ILS, OMEGA, LORAN, and TRANSIT. Tabulates the estimated number of users.

FAA, "Traffic alert and collision avoidance systems," Final Rule, Jan. 1989. Requires installation of TCAS-1 in small jets by 1994 and TCAS-2 in large jets and U.S. air-carrier aircraft in 1992–93. Encourages all aircraft to carry Mode-C altitude-reporting transponders. Defines further tests of TCAS-3. The Aug. 1987 Notice of Proposed Rulemaking reviewed Piedmont tests of 1981–82 and 1986–87.

Farmer, R. H., "Flight management systems for modern jet aircraft," in *Proc. IEEE NAECON,* 1981, pp. 823–829. Vertical pitch and throttle controls on C-141 with projections for B-747 and DC-10.

Fischetti, F. A. and T. S. Perry, "Our burdened skies," Special issue of *IEEE Spectrum Mag.*, pp. 36–82, Nov. 1986. Articles on traffic volume and system requirements. Little on navigation.

Gorham, J. and H. K. Richter, "Where we are with all weather landing," AIAA paper 67-572, 1967. Status of BLEU tests and FAA experiments. Thirty-three CAT IIIA systems installed in 1974.

Hanson, P. O., "Correction for deflection of the vertical at the run-up site," in *Rec. IEEE Position Location and Navigation Symp.,* 1988, pp. 288–296. Alignment of SPN/GEANS inertial system, corrected for the deflection of the vertical at the takeoff site but not in flight.

Haralson, D. G., *et al.*, "Toward the panoramic cockpit and 3D displays," in *Proc. IEEE NAECON,* 1989, pp. 80–85. Reviews simulations intended to determine requirements and benefits of panoramic aircraft displays, more than 300 in^2 in area.

Harlow, R. A., "Flight measurements of area navigation performance using various combinations of ground aids and airborne sensors," in *Rec. IEEE Position Location and Navigation Symp.,* 1980, pp. 228–235. Test of VOR-DME, DME-DME with air data and INS velocity.

Hemesath, N. B., "Optimum complementation of VOR/DME with air data," *AIAA J. Aircraft,* pp. 456–460, June 1971. Mixing air data and VOR improves position error, estimates wind to 20 kt, and reduces time delay when maneuvering.

Hershey, W. R., *et al.,* "Ranging and processing satellite," in *Rec. IEEE Position Location and Navigation Symp.,* 1988, pp. 394–401. Traffic and bandwidth estimates for an air-traffic control nav–comm satellite based on Mode-S format. Two satellites cover the United States.

Hinrichs, P. R., "Advanced terrain correlation techniques," in *Rec. IEEE Position Location and Navigation Symp.,* 1976, pp. 86–96. TERCOM. Reprinted in Part 5, pp. 288–295.

Hirst, M., "Technology to reduce airspace congestion," *Interavia*, pp. 117–120, Feb. 1989. Describes European plans to increase aircraft density by reducing separation and clearing aircraft in "tubes," atop each other.

Hughes, D., "USAF selects Allied-Signal/Raytheon to develop NATO-compatible IFF," *Aviat. Week and Space Tech.*, p. 26, Dec. 5, 1988.

Hughes, P. F., "History of development of SCORE pod," in *Rec. IEEE Position Location and Navigation Symp.,* 1976, pp. 77–86. Range Test pod with DME to ground stations.

Hunt, V. O. and G. V. Kloster eds., "The FAA's advanced automation program," *IEEE Computer,* pp. 14–69, Feb. 1987. Automation of the ARTCCs, concentrating on software.

IEEE, Special Issue on VOR/DME, *IEEE Trans. Aerosp. Navig. Electron.,* Mar. 1965. Lots of detail; still usable.

IEEE, Special Issue on Airborne Collision Avoidance System Development, *IEEE Trans. Aerosp. Electron. Syst.*, pp. 234–314, March 1968. Good history from 1955–1968. Articles on time-frequency methods, synchronization, master clocks, message formats, biphase coding (then new), spectral energy distribution, slot length, warning algorithms, and system architecture.

Jackson, E., "Implementation and future of LORAN-C for general aviation," *IEEE AES Mag.,* pp. 7–8, Aug. 1987. LORAN was approved for nonprecision experimental approaches in 1987.

Jankovitz, J. and J. Perdzock, "Laboratory and flight evaluation of integrated inertial sensor assembly," in *Rec. IEEE Position Location and Navigation Symp.,* 1988, pp. 126–133. Lab test setup and some results.

Kasper, J. F., Jr. and C. E. Hutchinson, "OMEGA: Global navigating by VLF fix," *IEEE Spectrum,* pp. 59–63, May 1979. Reprinted in Part 5, pp. 245–249.

Kayton, M. and W. Fried, *Avionic Navigation Systems.* New York, NY: J. Wiley, 1969. Descriptions of all aircraft navigation technologies then in service or development. Most of it is still current.

Kelly, R. J., "DME/P, the new international standard," in *Rec. IEEE Position Location and Navigation Symp.,* 1982, pp. 42–55. Excellent treatment of precision DME for MLS.

Kennedy, T. A., "Strap-down inertial measurement units for motion compensation for synthetic aperture radars," *IEEE AES Mag.*, pp. 32–35, Oct. 1988. Gives SAR compensation equations.

Kolcum, E. H., "Bendix/King gears up to meet demand for T/CAS in 1990s," *Aviat. Week and Space Technology,* pp. 109–113, Feb. 20, 1989. Market is 5700 aircraft. Describes Bendix TCAS-2 emphasizing alerts and displays, reviews informal tests by airlines, and describes modifications to reduce "nuisance targets" for an aircraft leveling off from a climb. Brief mention of Mode-S.

Lang, A. and W. Hundley, "Solid-state Vortac with remote maintenance and monitoring," *IEEE PLANS 80,* pp. 198–207, 1980. Reprinted in Part 5 on pp. 303–312.

Leavitt, L. L. and B. L. Peters, "Low-cost high-accuracy integrated GPS-inertial navigator for reconaissance mission," in *Rec. IEEE Position Location and Navigation Symp.,* 1988, pp. 381–388. Strapped-down IMU packaged with Magnavox GPS. Error models and simulations of performance.

Leeper, J. L. and J. C. Anderson, "The discrete address beacon system data link," in *Proc. IEEE Digital Avionics Conf.,* 1979, pp. 117–121. Mode-S (DABS) operation.

Lincoln Laboratory Staff, "Special issue on air-traffic control development at Lincoln Laboratory," *The Lincoln Laboratory Journal*, vol. 2, no. 3, Fall 1989. Fourteen in-depth papers on the development of Mode-S, ASR-9 radar, TCAS, MLS, wind shear detection, and expert systems.

Lin, C. E. and M. C. Hong, "A knowledge-based en-route monitor for air traffic control," *IEEE Trans. Aerosp. Electron. Syst.*, vol. 25, no. 3, pp. 392–400, May 1989.

Lopez, R. L. and J. D. Wilson, "FAA moves out on solving wind shear problem," *Interavia,* pp. 260–263, Mar. 1988. Reviews available on-board sensors (passive infrared and laser radar) and ground-based anemometers, lapse-rate sensors, terminal weather radars, and FAA weather centers.

Maher, R. A., "Comparison of multichannel sequential and multiplex GPS receivers for air navigation," *Navigation,* vol. 31, pp. 96–111, Summer 1984. Link margins for one- and two-channel sequential and multiplexed receivers vs. five-channel receivers.

Massoglia, P. L., "Automatic dependent surveillance for oceanic air traffic control applications," *Navigation*, vol. 35, no. 4, pp. 397–406, Winter 1988–89. Reviews present oceanic air traffic control and describes proposed modernization that will use L-band satellite communication, probably via INMARSAT.

McFarland, R. H., "Application of endfire arrays at contemporary glide-slope problem sites," *IEEE Trans. Aerosp. Electron. Syst.*, pp. 261–267, March 1981. New antenna design to reduce multipath errors. Shows test data on a single installation.

Minzner, R. A., The U.S. Standard Atmosphere 1976. *NOAA Rept. 76-1562*, NASA SP-390. 227 pgs. Based on rocket measurements between 50–80 km; satellite and radar scatter above 140 km; little data between 80 and 140 km.

Mohleji, S. C., "ATC automation concepts to optimize flight management system utilization," in *Rec. IEEE Position Location and Navigation Symp.,* 1988, pp. 341–346. Air-traffic-control software that will allow aircraft equipped with flight management systems to use their preferred trajectories.

Morgan, T. A., "An overview of enhanced TCAS II," *IEEE AESS Newsletter,* p. 15, Oct. 1984. Difference between TCAS-1 and TCAS-2. Antenna requirements.

Negre, Y., "All weather automatic landing on civil aircraft," AIAA paper 75-1023, 1975, 6 pgs., Caravelle CAT II and CAT III tests, 1964 and 1966.

Pogust, F., "Microwave landing systems," *IEEE Spectrum,* vol. 15, no. 3, pp. 30–36, Mar. 1978. Describes MLS before ICAO selection; 200 channels.

Potocki, P., "A310/A300-600 electronic flight instrumentation system," in *Proc. IEEE Digital Avionics Conf.,* 1983, 7 pgs. Describes ARINC-429 interfaces and switching, color usage, and symbology.

Price, A., *Instruments of Darkness, the History of Electronic Warfare.* New York, NY: Scribners, 1978. Excellent account of early radar and radar countermeasures in World War II.

Saint-Exupery, A. de, Vol de nuit. Paris: Editions Gallimard, 1931, 188 pp. (Also *Terre des hommes,* 1939, 185 pp.) Excellent description of early mail routes from Europe to Africa and South America.

Sandretto, P. C., *Electronic Avigation Engineering.* New York: ITT, 1958, 772 pp. Circuit drawings of all radio navigation aids current at the time.

Schneider, A. M., and J. L. Maida, "Kalman filter for an integrated Doppler/GPS navigation system," in *Rec. IEEE Position Location and Navigation Symp.,* 1988, pp. 408–415. Eight-state filter over land, 10-state filter over water for Doppler-AHRS with GPS updates. Error models and simulation results.

Schwab, C. E., "Airport surface detection equipment," *Proc. IEEE,* vol. 73, no. 2, pp. 290–300, Feb. 1985. Description of Cardion ASDE-3; 14–16 GHz; 6-m position error. Radome rotates with antenna to avoid distortion.

Scott, W. B., "Wind shear systems provide recovery cues," *Aviat. Week and Space Technology,* pp. 347–351, June 15, 1987. Inertial-air-data calculation of horizontal and vertical wind shear. Reports Delta and U.S. Air to install system in early 1988.

Shifrin, C.A., "FAA completes transition to host computer system," *Aviat. Week and Space Technology,* p. 83, June 27, 1988. All ARTCC have switched to new computers. Old software rehosted with few changes.

Skolnik, M. I., "Fifty years of radar," *Proc. IEEE,* vol. 73, no. 2, pp. 182–197, Feb. 1985. Describes 1930s radar research by U.S., U.K., France, Italy, and the Soviet Union in VHF, UHF, and L-bands. Good discussion of World War II fire control and bombing radars. History of synthetic arrays and digital processing for airborne radars. Ground tracking radars, especially for spacecraft and ballistic missiles, including phased arrays. Many specific examples: Cobra Dane, Cobra Judy, Lantirn, Nexus, etc. By the master himself.

Sobocinski, R., "Flight test results with collision avoidance systems," in *Proc. IEEE/AIAA 5th Digital Avionics Systems Conf.,* 1983, pp. 20.2.1–20.2.3. Reprinted in Part 5, pp. 285–287.

Solomon, J. K. and Z. H. Lewantowicz, "Estimation of atmospheric and transponder survey errors with a navigation Kalman filter," in *Proc. IEEE NAECON Conf.,* 1989, pp. 140–147. Describes Kalman filter portion of a "Completely Integrated Reference Instrumentation System" used at Holloman Air Force Base to calibrate navigation devices under test.

Spitzer, C., *Digital Avionic Systems.* Englewood Cliffs, NJ: Prentice-Hall, 1987. Effect of Federal Air Regulations and military specifications on avionic design. ARINC and military packaging, ARINC 429/629 bus and 1553 bus. CRTs, electric power. BITE-ATLAS, life-cycle costs.

Stein, K. J., "Microwave landing system effort gaining momentum," *Aviat. Week and Space Technology,* pp. 120–126, Nov. 9, 1987. Eastern shuttles installing MLS in hope of saving 25 min/flight. Hazeltine producing 178 for FAA.

Stevens, M. C., *Secondary Surveillance Radar.* Norwood, MA: Artec House, 1988. An excellent detailed book on antennas, interrogators, reply decoders, and transponders. Covers system operation and Mode S, and has a few words about TCAS.

Sturza, M. A. and C. C. Richards, "Embedded GPS solves the installation dilemma," in *Rec. IEEE Position Location and Navigation Symp.,* 1988, pp. 374–380. Describes a Litton integrated GPS-inertial system using strapped-down or gimballed IMUs.

Sutton, O., "Eurocontrol—Back in favor," *Interavia,* pp. 113–115, Feb. 1989. European traffic statistics and plans to improve national air-traffic control systems in the 1990s.

Sutton, O., "Swissair is first with All-Cat-3 fleet," *Interavia,* April 1989. pg. 315–317. Statistics on automatic landings and Cat-3 landings. 33 Cat-3 airports in Europe; 19 in USA. New planes are Cat-3A, 3B depending on fault tolerance in autopilot. Swissair = 100% CAT3B.

Tang, W. and N. Johnson, "Precision trajectory reconstruction," in *Rec. IEEE Position Location and Navigation Symp.,* 1988, pp. 297–302. Movie camera photographs panels on the ground as a method of updating INS during a flight test.

Templeman, J. E. and R. H. Parker, "Boeing/Sperry automatic landing system 727 airplane," *AIAA J. Aircraft,* vol. 5, no. 3, pp. 285–291, May–June 1968. Reprinted in Part 5, pp. 296–302.

Watts, C. B., Jr., "A new end-fire ILS glide slope," in *Rec. IEEE Position Location and Navigation Symp.,* 1980, pp. 190–197. Describes vertical mast, flush, and two-antenna (end-fire) arrays for creating glide-slope signals in space. The end-fire array does not depend on reflections from the ground to create an image. Shows test data at a Wyoming mesa-top airport.

Wefald, K. M. and C. R. McClary, "Autocalibration of a laser gyro strapdown inertial reference/navigation system," in *Rec. IEEE Position Location and Navigation Symp.,* 1984, pp. 66–74. Reprinted in Part 5, pp. 258–266.

Wei, S. Y. and J. R. Huddle, "Sensor management for a fault tolerant integrated inertial flight control reference and navigation system," in *Rec. IEEE Position Location and Navigation Symp.,* 1986, pp. 445–455. Reprinted in Part 5, pp. 267–277.

Wenzel, R. J., "Omega: system status update—1988," *IEEE AES Mag.,* pp. 24–33, July 1989. Possibility of decommissioning is affecting capital expenditures. Efforts are underway on timely validation, new electronics and antennas, synchronization with GPS time.

West-Vukovich, G., *et al.,* "Honeywell/DND helicopter integrated navigation system," in *Rec. IEEE Position Location and Navigation Symp.,* 1988, pp. 416–425. Thorough description of Doppler-inertial-GPS navigation system for the Canadian Sea-King ASW helicopter.

Widnall, W. S. and P. K. Sinha, "Optimizing the gains of the baro-inertial vertical channel," *AIAA J. Guidance and Control,* vol. 3, no. 2, pp. 172–178, Mar.–Apr. 1980. Reprinted in Part 5, pp. 278–284.

Zywiel, J. Z., *et al.,* "Navpack—A package of simulation tools for integrated navigation applications," in *Rec. IEEE Position Location and Navigation Symp.,* 1988, pp. 470–475. Integrated tools for flight profile, sensor models, filters, and postflight data reduction.

Omega: global navigating by VLF fix

This internationally sponsored global VLF system is nearing full operational status

Omega is shaping up as the only system that meets the world's need in the rest of this century for simple, economical, moderately accurate navigation by sea and air. Now nearing full operational status, this VLF system will provide worldwide, all-weather position fixes with root-mean-square accuracy of 1-2 nautical miles. The minimum receiver cost is perhaps the lowest of any navigation system—just a few thousand dollars. Thanks to steady increases in microprocessor applications, a user just enters initial position, and the system can continuously read out positions anywhere on the globe.

Unlike other navigation systems, Omega is an internationally sponsored system. No security classification restricts use, development or ultimate accuracy, as it does in the Loran-C system, or in the U.S. Navy's Transit navigational satellite system. After initial development by the U.S. Navy, Omega responsibility shifted to the U.S. Coast Guard. International sponsorship came about because the United States realized that full benefits could best be achieved in partnership with friendly nations. This partnership has allowed the construction of transmission facilities around the world. Nations with strong maritime interests—Norway, Liberia, France, Argentina, Australia, and Japan—are sponsors.

Seven of the eight planned Omega transmitting stations are now operational and provide almost complete coverage (Fig. 1). With the completion of the final station next year, coverage becomes complete.

Early VLF experiments paved way

Omega evolved from World War II research on LF and VLF radiowave propagation conducted by J.A. Pierce of Harvard University and others. In this regard, Omega is a relative of Loran-A; more directly it is a refinement of Radux, a 40-kHz, phase-comparison navigation system that was service-tested in 1956. Pierce suggested the use of even lower-frequency signals to exploit the VLF potential of increased phase stability, greater range, and improved accuracy. A combined 10-kHz/40-kHz system known as Radux-Omega was tested and discarded in favor of the Omega system, which operates in the 10–14 kHz band.

In the early 1960s a master-slave network of VLF transmitting stations was established for experimental testing. Transmitter locations included Forestport, N.Y.; Haiku, Hawaii; and Balboa in the Canal Zone. The availability of precise time/frequency standards around 1965 prompted the design of a controlled system of eight individually synchronized stations, rather than a system based on a master station that prompted transmissions from slave stations at fixed times after receipt of the master signal.

Although antenna configurations vary, the electronic equipment used in Omega is almost identical at each site. Each station is designed for 10 kW of effective radiated power (Fig. 2).

Within the overall system, the control and timing subsystem most directly affects navigational accuracy. A major component is the signal format generator. Omega stations transmit semicontinuously in eight-segment patterns, with a basic repetition rate of 10 s. Three frequencies—10.2, 11.33, and 13.6 kHz—are used primarily for navigation. A fourth frequency, 11.05 kHz, as well as one discrete frequency for each station, have recently been added. These additional frequencies facilitiate conventional navigation and also support special applications, such as buoy-tracking, search-rescue operations and new techniques for global time transfer.

Signals are synchronized

To be useful for navigation, Omega signals must be synchronized—that is, all antenna currents must be in phase. Each transmitting station employs four cesium-beam standards in its control and timing subsystem. Interstation phase measurements are obtained twice daily. An optimal estimation and control algorithm generates corrections to reduce offset and divergence of individual transmitters from mean Omega system time.

Responsibility for system synchronization rests with the Japanese Maritime Safety Agency. Mean Omega system time is established with reference to U.S. Naval Observatory calibration sources. Supporting data are provided by satellite-TV links and visits from personnel with portable-clock standards. Individual station timing is held within 2 μs rms and is not a dominant error source in Omega navigational accuracy. Beyond system synchronization, individual receivers must be synchronized with the Omega signal format. Both the distinctive pattern within the 10-s repetition and the discrete frequency for each station facilitate automatic synchronization in most Omega receivers.

In the usual mode of operation, an Omega position fix is obtained at the intersection of two hyperbolic lines of position. Each line of position (LOP) is generated from the difference in phase of signals from two transmitting stations. Signals from at least three stations must be used to get two LOPs and a position fix. The hyperbolic mode is attractive because it eliminates use of a precise, expensive local oscillator. However, a circular mode is possible with

Joseph F. Kasper Jr.
The Analytic Science Corp.
Charles E. Hutchinson
University of Massachusetts

Reprinted from *IEEE Spectrum,* pp. 59-63, May 1979.

two stations, but a precision oscillator is needed. Position fixes are then defined as the intersection of circular rather than hyperbolic LOPs. An Omega receiver can operate in conjunction with other sources of positional information.

Let's examine Omega operation at 10.2 kHz. All distance measurements with the Omega system are in nautical miles. At 10.2 kHz the signal wavelength is approximately 16 miles and phase-difference values repeat (cycle through 360° of phase) approximately every eight miles along the baseline (the line that joins two transmitting stations). The region between the loci of 0° phase is referred to as a lane. Since the loci are hyperbolas (with foci at the two transmitting stations), the lane width increases farther away from the baseline. Baselines in the Omega system are on the order of 5000–6000 miles, and there are typically 600–700 lanes between two transmitting stations. Each lane is numbered for identification.

Omega users refer to each 360° of phase as 100 centicycles. At 10.2 kHz, one centicycle equals 0.08 miles.

Ambiguity in position-fixing occurs because an observed phase difference corresponds to an LOP in each of the 600–700 lanes for a given station pair. Thus one must initialize an Omega receiver to an accuracy of ± 0.5 lane width and continuously maintain count of lane changes. For 10.2-kHz navigation, the initialization process requires knowledge of the initial position to within approximately ± 4 miles. The lane counting is performed automatically in most Omega receivers. If signals are lost or the receiver malfunctions, the receiver must be reinitialized when normal operation resumes. If there is only a short signal outage and the craft is slow-moving, one might assume no change in lane count. Otherwise an external fix might be needed or a difference-frequency technique might be used.

The combined processing of 10.2-kHz and 13.6-kHz signals yields a 3.4-kHz phase difference value; the 3.4-kHz lane width is approximately 24 miles. As a consequence, the initial position need be known only to within ± 12 miles for initialization. If the 11.05-kHz signal is combined with an 11.33-kHz signal, the difference frequency is 0.28-kHz, the lane width approximately 288 miles and the initial accuracy requirement ± 144 miles.

The position-fixing process is quite straightforward. For hyperbolic implementation, the receiver is initialized with the proper lane counts for two different LOPs. The phase difference between two stations is measured. To this value are added Predicted Propagation Corrections (PPCs), or phase corrections, which account for daily, seasonal and geographic variations in the phase of a propagating VLF signal. The corrected phase difference measurements define the position fix. To obtain latitude and longitude, one may use an annotated plotting chart or convert LOP values by a geometric transformation. Most receivers handle almost all of the position-fixing steps automatically. Manual receivers are available at much lower cost.

Hyperbolic fixes with Omega (and also Loran-C) do not yield uniform error characteristics over the entire signal region. Position error at any site is a function of the LOPs that define the fix and also the lane width, since a phase uncertainty translates into a larger position error if the lane is wider. In short-baseline systems, such as Loran-C, the loss of precision can be significant at points distant from the baseline, yielding position errors double or triple those on or near the baseline. However, with Omega the baselines are approximately 6000 miles, and there is typically redundant signal coverage, so that in most locations two nearly orthogonal LOPs reasonably close to baselines can be found.

[1] Australian Omega transmitting station is almost complete. The station is scheduled for operation in 1980-81. An interim station operates in Trinidad.

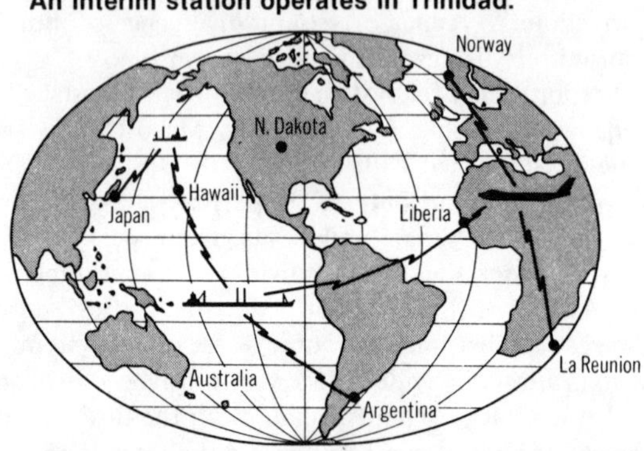

Station Desig.	Location	Latitude	Longitude	Antenna type	Cognizant agency
A	Aldra, Norway	66°25′N	13°08′E	Valley span	Norwegian Telecommunications Administration
B	Monrovia, Liberia	6°18′N	10°40′W	Grounded tower	Liberian Ministry of Commerce, Industry and Transportation
C	Haiku, Hawaii	21°24′N	157°50′W	Valley span	U.S. Coast Guard
D	LaMoure, N. Dakota	46°21′N	98°20′W	Insulated tower	U.S. Coast Guard
E	La Reunion	20°58′S	55°17′E	Grounded tower	French Navy
F	Golfo Nuevo, Argentina	43°03′S	65°11′W	Insulated tower	Argentine Navy
G	Woodside, Australia	38°29′S	146°56′E	Grounded tower	Australian Department of Shipping and Transport
H	Tsushima, Japan	34°37′N	129°27′E	Insulated tower	Japanese Maritime Safety Agency

VLF propagates via a global waveguide

In simplest terms, Omega signal propagation can be described by a waveguide model in which the earth's surface and the D Layer of the ionosphere, 50–90 km above the earth's surface, are the boundaries of the waveguide.

Two primary parameters describe VLF propagation with respect to the earth's surface: effective conductivity and average relative permittivity.

The effective conductivity at VLF is approximately equal to the average value within the first 50 to 100 m below the earth's surface. In regions such as oceans, these parameters are more or less constant. In other regions they vary over a wide range, even in small areas. This is especially true for land areas made up of several different topographical features. Also these parameters change abruptly in the vicinity of a coast line. The earth's surface in most regions is a reasonably good conductor at VLF, although for propagation over ice caps the ground cannot be considered a good conductor. The practical consequence of this is that VLF waves suffer heavy attenuation while propagating across such regions as Greenland and the Antarctic. At Omega frequencies the signal, for all practical purposes, is carried by low-order TM modes—that is, modes whose electrical fields are polarized perpendicular to the earth's surface. For these modes, the detailed characteristics of the topography are relatively unimportant.

Ionospheric conditions greatly affect VLF propagation. A key factor in Omega propagation is the variation of ionospheric height from day to night and from season to season. A typical day-to-night height variation is on the order of 20 km, which is a considerable change in the dimensions of the waveguide. The resulting daily variation in signal propagation is readily noticeable with Omega data (Fig. 3).

A number of different models are used to predict the behavior of VLF signals. The electromagnetic field characteristics of the signals propagating in the waveguide are derived from the solutions of Maxwell's equations for waveguides. Exact solutions are difficult because of complex waveguide properties or boundary conditions, such as the spherical shape of the waveguide, spatially varying ground conductivity, and the coupling of propagating wave signals caused by the anisotropy and inhomogeneity of the ionosphere.

Two factors determine the utility of Omega signals:

1. Signal-to-noise ratio below receiver threshold.
2. The degree of modal interference, in which several propagation modes occur.

Predictions of signal-to-noise ratio over the earth's surface are obtained from signal-strength estimates and noise values. In the VLF signal region, noise caused by lightning dominates other noise sources. But even allowing for poor reception and interference, predictions indicate that the signals from three or more Omega transmitting stations are available nearly continuously in nearly all areas of the world.

For increased navigational accuracy, certain Omega receivers can process more than the usual signals.

Accuracy continues to improve

It is evident that the position-fixing accuracy of Omega depends mainly on the ability to predict signal phase. Without propagation predictions, accuracy is on the order of 5 miles, rms. With phase corrections, the error reduces to 1-2 miles, rms. Error by the receiver and transmitter network is negligible—on the order of 0.1-0.2 miles, rms.

The best accuracy is being achieved in "mature" Omega operating areas—areas where signals have been available and measured for more than a year. Refinements in global prediction theory, which are under way, should substantially improve navigational accuracy.

There is an ongoing program for collecting Omega phase data to support system calibration. The information is being gathered from a network of monitor stations and processed by the U.S. Coast Guard.

Phase-correction values are composed of two elements (Fig. 4). One element is based on a global physics-propagation model; the second uses statistical estimation techniques based on monitor data to refine Omega accuracy. The final output of the prediction process is a set of tables that provide phase corrections for various regions and times.

Other propagational effects

Occasionally propagation conditions differ from the nominal and are not completely predictable. Solar phenomena have a major impact on Omega propagation, especially sudden ionospheric disturbances and polar-cap absorption events.

Sudden ionospheric disturbances are produced primarily by an increase in the energy and quanity of X-ray flux received from the sun as a result of eruptions, or "flares," near sunspots. Flares producing sufficiently

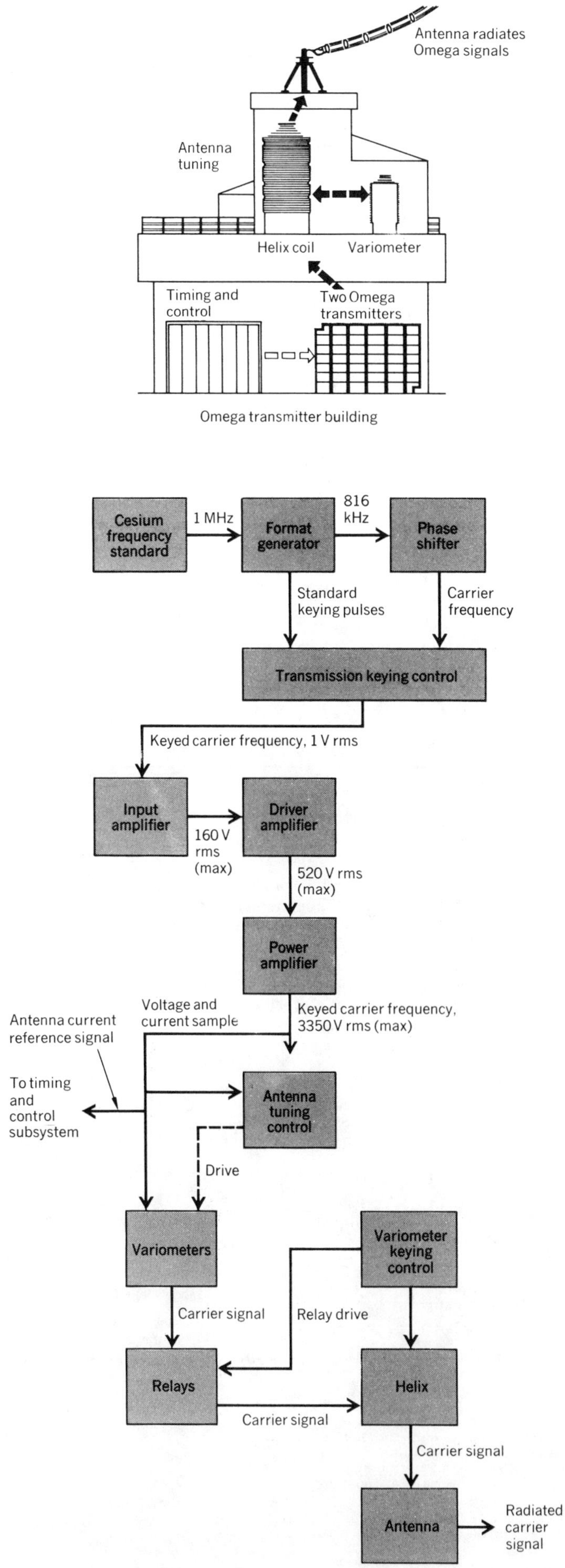

[2] Omega uses phase-controlled transmissions of eight-segment patterns. Four cesium-beam standards maintain internal synchronization accuracy. A variometer feedback loop maintains antenna-current phase.

energetic radiation in the X-ray portion of the spectrum cause radical changes in the ionization of the daytime D region of the ionosphere, resulting in a lowering of the effective height of the sunlit portion of the ionosphere. This produces a phase advance on Omega transmission paths that are exposed to the sun.

Polar-cap absorption events are caused by the precipitation of solar particles, predominantly protons, in the vicinity of the earth's magnetic poles shortly after large solar flares. They cause a reduction in the effective height of the ionosphere inside the polar regions above 60°N and below 65°S geomagnetic latitude. This, in turn, causes a phase advance along Omega transmission paths that cross the polar regions duiring these events.

Monitoring by earth satellites and various ground stations provides useful data for detecting sudden ionospheric disturbances and predicting polar-caps absorption events. Currently this type of information, in the form of user warnings, is broadcast on the National Bureau of Standards radio stations, WWV and WWVH, at 16 minutes past the hour. Navigational warning messages are also disseminated via conventional channels, such as *Notices to Mariners* and *Notices to Airmen*.

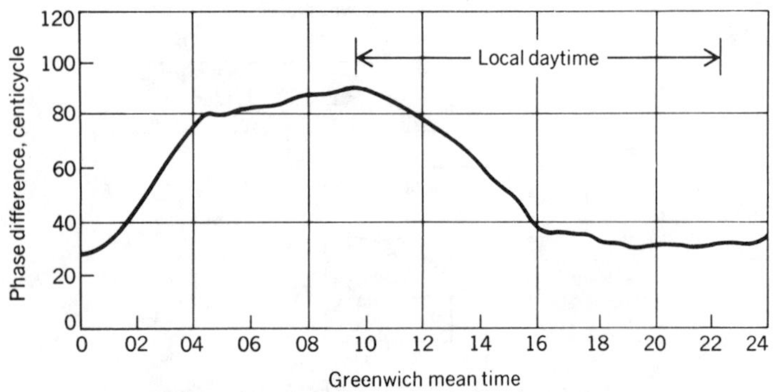

[3] Variations of ionospheric height, which occur daily and seasonally, produce readily noticeable deviations in Omega phase data.

[4] Phase-correction tables are prepared from a combination of statistical regression and model estimates.

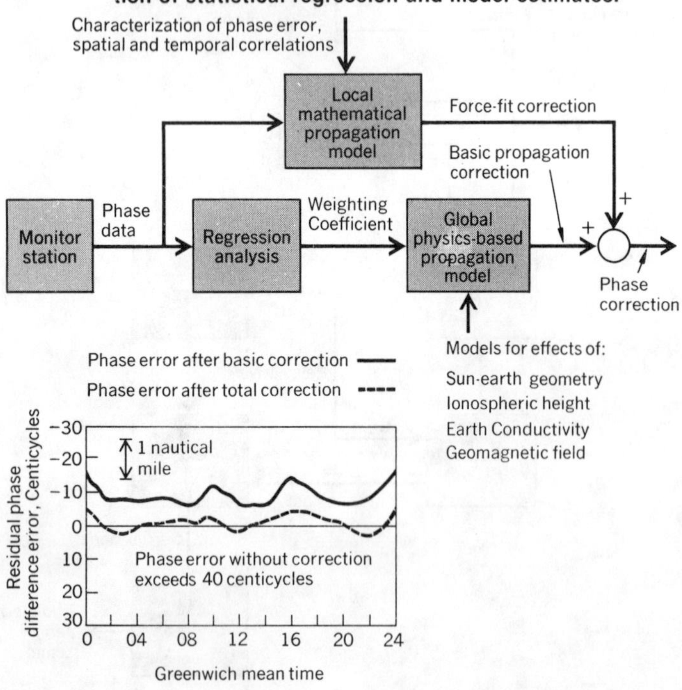

Not all disruptive propagational events are understood enough to be predictable. It has been noted that occasionally a signal traveling the long way around the earth will have greater amplitude than the signal that traveled the short way. Since the transformation from Omega coordinates to geographic coordinates is based on the geometry of the short-way path, long-way signals produce an erroneous position indication.

Also, anomalous propagation effects have been observed with Omega signals that traverse the geomagnetic equator in a westerly direction. Latitude-dependent terms in the propagation models may become distorted at the geomagnetic equator.

System equipment costs vary

The Omega system has a broad spectrum of users, and there is an equally broad spectrum of user equipment. Prices range from several thousand dollars for a manual, single-frequency receiver to $10 000 for an automatic marine receiver and to $25 000 for an automatic, three-frequency receiver capable of interfacing with complex avionics. Units that also perform computational functions, required by integrated multisensor navigation, cost in excess of $100 000.

The functions performed by an Omega receiver can, in most instances, be conveniently broken down into antenna/coupler, receiver/processor, and control/display. Depending upon the configurations, these functions may or may not correspond to separate hardware elements.

There are two major Omega user antenna configurations. The first is a whip, typically between 2.4 and 4.5 m long, which is used to sense the vertical component of the electric field of the Omega signal. Whip installactions predominate in ground-based and ship applications. The second configuration involves a pair of crossed orthogonal loops, typically mounted in 20 × 25 × 5-cm units. This antenna can be near-flush-mounted to the skin of an aircraft and is used to sense the horizontal component of the magnetic field of the Omega signal. In addition to geometric compatibility, this configuration has an advantage in aircraft applications because it is less susceptible to electric fields on and near the aircraft. Such fields are created by other aircraft equipment and by build-up of charge resulting from motion through the atmosphere.

As for Omega receiver/processors, they accept the available signal and extract phase information from it. In early receivers this was accomplished by analog and/or electromechanical means; in modern receivers digital techniques are used, typically in a phase-locked-loop configuration. In a simple manual receiver, raw phase-difference information is available from numeric displays or strip-chart recorder. The user must then apply phase corrections and plot the resulting LOPs on an Omega chart to determine latitude and longitude.

On a "full-up" automatic receiver, the processor performs the functions. A simplified algorithm is used to generate phase corrections in a manner similar to that used to generate published tables. The transformation from Omega LOPs to latitude and longitude is also performed automatically. The final output on the control/display device is the ship or aircraft's geographic position.

The computational capability in the automatic receiver can also be used to perform other portions of the navigational function—steering to a desired track, distance-to-go and time-to-go calculations, ground speed estimation, etc. Also, if other sources of navigation information are available (true airspeed, manual positions fixes, magnetic heading), the processor can combine all available inputs (including signals from more than the minimum requirement of three stations) and perform reasonable checks on the Omega data.

Other functions can be automated as well, including:

• Synchronization—matching the receiver timing pattern to that of the Omega system.

• Initialization—establishing proper lane count for the Omega LOPS.

• Station selection—choosing the best Omega stations, or those that are sufficiently strong and offer suitable geometry.

The first two functions are relatively straightforward; the third is considerably more complex. The most recent advances in receiver automation are based on microprocessor technology. Receivers are being programmed to avoid use of signals that might be affected by undesirable propagation characteristics. Differential Omega may increase accuracy of 1-2 miles to 0.25 mile. As proposed and tested, a local area monitor measures Omega phase relative to predicted values and communicates the resulting error values to users in the vicinity of the monitor. Users apply the error values as corrections to their received phase values. The errors measured at the monitor (mostly caused by unpredicted propagation phenomena) closely match those experienced by users within several hundred miles of the monitor. Analysis and actual experimentation indicate rms accuracies of approximately 0.25 mile within 200 miles of a monitor. Differential Omega offers promise of improved navigation informaton for such areas as congested shipping lanes. The Federal Aviation Administration in the United States is involved in a program to test the suitability of differential Omega for en route and terminal operations in Alaska. A limited number of monitor sites could offer equivalent performance at a cost savings compared to establishment of full coverage in Alaska by the conventional very-high frequency, omni-range radio and distance-measuring equipment. France is implementing differential Omega for service in its own coastal waters.

Another variation of Omega signal processing being examined is composite Omega. It takes advantage of the fact that abnormal, unpredicted Omega propagation conditions have similar effects on the signals of each of the Omega frequencies. By combining and analyzing the information obtained from the signals of two different frequencies, composite Omega is able to minimize the effect of a sudden ionospheric disturbance on position accuracy. At present an extensive base of operational data is not available to validate the concept of composite Omega. However, based on the theoretical analysis and limited data, several receiver manufacturers have adapted their equipment to handle composite Omega.

For further reading

Omega, A World-Wide Navigation System: System Specification and Implementation, second edition, Office of Naval Research, May 1966, DDC No. AD-630-900.

"Omega" by E.R. Swanson, *Navigation*, vol. 18, no. 2, summer 1971.

"Omega" by J.A. Pierce, *IEEE Transactions on Aerospace and Electronic Systems*, vol. AES-1, no. 6, December 1965.

"Omega Navigation System: Present Status and Plans 1977-1980" by E.R. Vass, *Navigation*, vol. 25, no. 1, Spring 1978.

"The Omega Navigation System—An Overview" by J.F. Kasper Jr. and C.E. Hutchinson, *IEEE Communications Society Magazine*, May 1978.

Omega Bibliography, International Omega Association, Arlington, Va. 1977.

Some concern has also been expressed about the reliability of Omega signal coverage in certain areas of the world under other than nominal propagation conditions. The performance validation effort being supported by the Coast Guard will help in this regard. Also, several manufacturers have augmented their Omega receivers with circuitry to take advantage of signals from U.S. Navy VLF stations. These signals in the 16–24 kHz band are used in the same manner as Omega signals, but a pure communication station position fix is not of the same accuracy as an Omega-derived fix, in large measure because less effort has gone into prediction of propagation behavior.

Omega support is strong

It is anticipated that Omega will be maintained and operated beyond 1990, with use of the system growing. User support will continue to be provided in the United States by the Coast Guard and the Defense Mapping Agency, in the form of plotting charts, phase-correction tables, user warnings, etc. Although nominally characterized as a system with 1-2 mile accuracy, Omega should offer rms accuracies of better than 1 nautical mile in mature operating areas as data from the worldwide monitoring program is refined.

The only feasible competitor to Omega as a worldwide, bounded-error navigation aid is the Global Positioning System (GPS/Navstar)—an advanced satellite navigation system scheduled for full operational capability in the mid-1980s. Although the accuracy potential of GPS/Navstar cannot be matched by Omega, no one can confidently say that a GPS/Navstar receiver will be available at the cost of the lowest-priced Omega receiver. ◆

Updating Inertial Navigation Systems with VOR/DME Information

J. C. Bobick*
Center for Naval Analyses, Arlington, Va.

AND

A. E. Bryson Jr.†
Stanford University, Stanford, Calif.

Updating an inertial navigation system (INS) with VOR/DME information (from one or two stations) by means of a maximum-likelihood filter is shown to result in substantial improvements in navigational accuracy over that obtained by the use of a single VOR/DME (current practice). When continuously updating, the use of a high-quality INS (0.01°/hr gyro drift) instead of a low-quality INS (1.0°/hr gyro drift) does not substantially improve position accuracy. In-flight alignment (or realignment) of an INS to an accuracy comparable to that of ground alignment can be accomplished by using two DME's. Several reduced-order suboptimal filters were found to perform nearly optimally.

Introduction

THE primary navigation aid for civil aircraft flying in the airspace of most of the developed countries of the world is the VOR/DME system. The VOR (Very high-frequency Omni-Range) and the DME (Distance Measuring Equipment) enable onboard determination of an aircraft's bearing relative to north at the fixed ground station and slant range from the station, respectively.

Current use of the VOR/DME system involves primarily radial navigation, i.e., aircraft fly directly to or from the ground stations. However, some beginnings have been made in using the VOR/DME system for area navigation, i.e., use of the system without being restricted to fly directly to or from the ground stations.

The number of commercial airliners equipped with inertial navigation systems (INS's) is steadily increasing. The systems now onboard aircraft utilize a gyro-stabilized platform on which the accelerometers are mounted. The platform is aligned before takeoff to the desired orientation. Due to alignment errors and in-flight random disturbances such as gyro drift, scale-factor errors, and accelerometer bias errors, errors in the desired orientation of the platform increase with time. This results in increasing position and velocity errors. Thus, improving navigational accuracy by in-flight realignment of an INS is an interesting possibility.

Position errors are generally greater for area than for radial navigation. This comes about because the position error resulting from a VOR angular error increases with distance from the station; and an aircraft is farther, on the average, from the VOR stations for area than for radial navigation. Hence, improved navigational accuracy is required to obtain an accuracy for area navigation comparable to that of present-day radial navigation. The availability of an onboard computer to do the triangulation computations required for area navigation suggests the possibility of obtaining improved navigational accuracy by using the computer to implement a filter to combine VOR/DME information with the information from a dead-reckoning system. Since air data (airspeed and heading) are already available onboard nearly all aircraft and an increasing number of aircraft are equipped with inertial navigations systems, air and inertial data are the foremost choices of dead-reckoning information. The use of air data has been previously considered[1]; the use of inertial data is of concern here.

Problem Statement

The use of external position information to update an INS has been previously considered. However, attention has been focused on the use of long-range radio systems such as LORAN and OMEGA. Because the accuracy of the position information derived from the VOR/DME system depends upon the relative location of the aircraft and the VOR/DME station as well as the number of VOR and DME stations used, the possibility of using VOR/DME information requires specific consideration. Although the use of VOR/DME information to update an INS has been mentioned in the literature,[2–6] no comprehensive study of the possibility of combining VOR/DME information and inertial data has been found.

Our objective here is to study the possibility of improving the accuracy of air navigation by combining VOR/DME information with inertial data by means of a maximum-likelihood filter. This is equivalent to in-flight alignment of the INS.‡ Since more than one VOR/DME station is nearly always in sight at jet altitudes, the simultaneous use of two VOR/DME stations and inertial data is considered. DeGroot and Larsen[7] have considered the use of two VOR's and two DME's (without inertial data). The possibilities of using 0, 1, or 2 VOR's and 0, 1, or 2 DME's with and without inertial data are considered in this work.

In-flight realignment, e.g., before or after a transoceanic leg of a flight, or periodically (every one or two hours) during a domestic flight, is also considered. Using VOR/DME information to simply reset the INS position display (no filtering involved) is compared with using the VOR/DME information to realign the

Presented as Paper 72-846 at the AIAA Guidance and Control Conference, Stanford, Calif., August 14–16, 1972; submitted February 23, 1973; revision received June 14, 1973. This research was supported in part by NASA Grants NgR-05-020-431 and NgL-05-020-007, U.S. Air Force Contract F33615-72-C-1297, and an NSF Traineeship to the first author.

Index category: Aircraft Navigation, Communication and Traffic Control.

* Research staff member.

† Professor of Applied Mechanics, Aeronautics and Astronautics. Fellow AIAA.

‡ Although alignment usually refers to physically rotating the INS platform to a desired orientation, in this work alignment simply means the estimation of the platform attitude errors. Compensation for these errors is then made in estimating velocity and position.

Reprinted with permission from *AIAA J.*, vol. 11, no. 10, pp. 1377–1384, Oct. 1973.

system. The possible improvement of navigational accuracy over land areas by using an INS continuously updated with VOR/DME information is investigated. Both radial and area flights are considered; the use of a high-quality INS is compared with the use of a low-quality system. The performances of several suboptimal filters, resulting when various states are neglected, are studied.

Error Model for the Inertial Navigation System

Three sets of coordinate axes enter into the INS error analysis: 1) platform coordinate axes, the axes lying along the nominal accelerometer sensitive axes, 2) locally-level coordinate axes, the axes representing the ideal alignment of the platform axes at the actual location of the platform, and 3) computer coordinate axes, the axes corresponding to the locally-level coordinate axes at the location indicated by the INS computer. For a perfectly operating system, the platform, locally-level, and computer sets of axes coincide. However, in actual operation these coordinate axes are generally rotated relative to each other through small angles.§

It can be shown[8,9] that the INS error equations may be put in the following form:¶

$$\delta\dot{\mathbf{R}} = \delta\mathbf{V} - \boldsymbol{\rho} \times \delta\mathbf{R} \tag{1}$$

$$\delta\dot{\mathbf{V}} = \boldsymbol{\alpha} - \omega_s^2\delta\mathbf{R} - \boldsymbol{\psi} \times \mathbf{a} - (2\boldsymbol{\Omega}+\boldsymbol{\rho}) \times \delta\mathbf{V} + 3(\omega_s^2/R)\,\delta R\mathbf{R} \tag{2}$$

$$\dot{\boldsymbol{\psi}} = \boldsymbol{\varepsilon} - \boldsymbol{\omega} \times \boldsymbol{\psi} \tag{3}$$

where the time derivatives are taken in locally-level coordinates, and $\mathbf{R}$ = radius vector from the center of the Earth to the vehicle, R = magnitude of $\mathbf{R}$, $\delta\mathbf{R}$ = error in computing $\mathbf{R}$, i.e., position error, δR = magnitude of $\delta\mathbf{R}$, $\mathbf{V}$ = velocity of the vehicle relative to the Earth, $\delta\mathbf{V}$ = error in computing $\mathbf{V}$, $\boldsymbol{\rho}$ = angular velocity of the locally-level coordinates relative to the Earth, $\boldsymbol{\Omega}$ = angular velocity of the Earth relative to inertial space, $\omega_s^2 = g/R$, the square of the Schuler angular frequency, $\mathbf{g}$ = gravity vector (including centripetal acceleration), g = magnitude of $\mathbf{g}$, $\boldsymbol{\alpha}$ = accelerometer bias error, $\boldsymbol{\psi}$ = vector angle relating the platform and computer coordinates, $\boldsymbol{\varepsilon}$ = platform drift rate, $\boldsymbol{\omega} = \boldsymbol{\rho}+\boldsymbol{\Omega}$, and $\mathbf{a}$ = accelerometer output.

For the simulations performed here, $\mathbf{a}$ is taken to be the accelerometer output along a nominal flight path, i.e.,

$$\mathbf{a} = \dot{\mathbf{V}} + (2\boldsymbol{\Omega}+\boldsymbol{\rho}) \times \mathbf{V} - \mathbf{g} \tag{4}$$

where the differentiation is taken in locally-level coordinates and nominal values of $\mathbf{V}$, $\boldsymbol{\Omega}$, $\boldsymbol{\rho}$, and $\mathbf{g}$ are used.

For this study, a locally-level, north-pointing mechanization is used, i.e., the platform is maintained as closely as possible to locally-level, with respect to the Earth, with accelerometer sensitive axes pointing east (x), north (y), and up (z). Since this mechanization breaks down near the poles of the Earth, it is assumed that the vehicle does not operate at high latitudes.

Since the vertical channel of the INS is unstable (Ref. 8, Sec. 4.6), it is assumed for this study that altitude information is available from another source such as a barometric altimeter. Thus, only the two horizontal channels of the INS are considered. A further assumption is that the Earth is spherical and that gravitational equipotential surfaces associated with $\mathbf{g}$ are spherical. Although the nonspherical character of the Earth and its gravitational field must be taken into account during actual operation, for the simulations performed in this work, this assumption is considered reasonable.

Under the preceding assumptions, it follows that

$$R_x = R_y = 0, \quad R_z = R, \quad \delta R_z \cong \delta R \cong 0 \tag{5}$$

$$g_x = g_y = 0, \quad g_z = -g, \quad g = g_0(R_0^2/R^2) \tag{6}$$

$$\Omega_x = 0, \quad \Omega_y = \Omega\cos\lambda, \quad \Omega_z = \Omega\sin\lambda \tag{7}$$

$$\rho_x = -V_y/R, \quad \rho_y = V_x/R, \quad \rho_z = V_x/R\tan\lambda \tag{8}$$

§ Hence, the theory of infinitesimal rotations is assumed to be applicable.

¶ A dot denotes differentiation with respect to time; a bold face quantity is a vector.

where g_0 is the magnitude of $\mathbf{g}$ at the surface of the Earth ($\cong$32.2 ft/sec^2), R_0 is the mean radius of the Earth ($\cong$3440 naut miles), Ω is the Earth rotation rate ($\cong$15.04°/hr) and λ is the latitude. The subscript x, y, or z denotes the component along the x, y, or z axis of the associated vector quantity.

The platform attitude error $\boldsymbol{\phi}$, i.e., the vector angle relating the platform and locally-level, north-pointing (L) axes, is given by

$$\boldsymbol{\phi} = \boldsymbol{\psi} + \delta\boldsymbol{\theta} \tag{9}$$

where $\delta\boldsymbol{\theta}$ is the vector angle relating the computer and the L axes. For the mechanization used, the components of $\boldsymbol{\phi}$ are given by

$$\phi_x = \psi_x - \delta R_y/R, \quad \phi_y = \psi_y + \delta R_x/R, \quad \phi_z = \psi_z + (\delta R_x/R)\tan\lambda \tag{10}$$

where ϕ_x and ϕ_y are the platform tilts about the east and north axes, respectively; and ϕ_z is the platform azimuth error.

The east and north accelerometer errors (α_x and α_y) will now be modeled. Assuming a well calibrated high-quality INS operating in cruise conditions, the main error in the accelerometers is due to null shifts (or bias). This error can be modeled as an exponentially-correlated process.[10,11] A reasonable value for the standard deviation of this process is $10^{-4}g$ while the mean value is assumed to be zero. The correlation time is very long, a reasonable value being 10 hr. Hence, shaping filters which generate α_x and α_y are given by (see, e.g., Ref. 12, Sec. 11.4)

$$\dot{\alpha}_i = -(1/T_{\alpha_i})\alpha_i + (1/T_{\alpha_i})n_{\alpha_i}, \quad i = x, y \tag{11}$$

where n_{α_x} and n_{α_y} are independent white noise processes with**

$$E[n_{\alpha_i}] = 0, \quad E[n_{\alpha_i}(t+\tau)n_{\alpha_i}(t)] = 2T_{\alpha_i}\sigma_{\alpha_i}^2\,\delta(\tau) \tag{12}$$

The quantities σ_{α_x} and T_{α_x}, and σ_{α_y} and T_{α_y} are the standard deviation and correlation time of α_x and α_y, respectively.

The platform drift rate $\boldsymbol{\varepsilon}$ is due mainly to uncompensated gyro drift. Hence, gyro drift is the only source of gyro error that will be modeled. For our purposes, the gyro drift can be modeled reasonably well as an exponentially-correlated process.[10,11] For high-quality gyros, a typical value of the correlation time is 5 hr. Hence, the models for east gyro drift (ε_x), north gyro drift (ε_y), and vertical gyro drift (ε_z) are given by

$$\dot{\varepsilon}_i = -(1/T_{\varepsilon_i})\varepsilon_i + (1/T_{\varepsilon_i})n_{\varepsilon_i}, \quad i = x, y, z \tag{13}$$

where n_{ε_x}, n_{ε_y}, n_{ε_z} are independent white noise processes with

$$E[n_{\varepsilon_i}] = 0, \quad E[n_{\varepsilon_i}(t+\tau)n_{\varepsilon_i}(t)] = 2T_{\varepsilon_i}\sigma_{\varepsilon_i}^2\,\delta(\tau) \tag{14}$$

The quantities σ_{ε_x} and T_{ε_x}, σ_{ε_y} and T_{ε_y}, and σ_{ε_z} and T_{ε_z} are the standard deviation and correlation time of ε_x, ε_y, and ε_z, respectively.

Error Model for the VOR/DME System

Error models for the VOR and DME systems have been previously derived.[1] Both the VOR and DME errors can be separated into two components, a component with a long correlation time which can be modeled as a random bias and a component with a short correlation time which can be modeled as white noise. The mean values of all of these error components are essentially zero.

Letting b_V and e_V denote the VOR error components with a long and a short correlation time, respectively, we have

$$\dot{b}_V = 0 \tag{15}$$

$$E[e_V] = 0, \quad E[e_V(t+\tau)e_V(t)] = 2\sigma_{e_V}^2 T_{e_V}\,\delta(\tau) \tag{16}$$

where σ_{e_V} and T_{e_V} are the standard deviation and correlation time of e_V, respectively. A reasonable value for σ_{e_V} and σ_{b_V}, the standard deviation of e_V and b_V, respectively, is 1.0°. The correlation distance of e_V appears to be about one naut mile, which corresponds to a correlation time of 7.2 sec for an aircraft traveling at a speed of 500 knots.

Similarly, letting b_D and e_D devote the DME error components with a long and a short correlation time, respectively, we have

** $E[\ ]$ is the expected value function and $\delta(\tau)$ the Dirac delta function.

$$\dot{b}_D = 0 \qquad (17)$$

$$E[e_D] = 0, \quad E[e_D(t+\tau)e_D(t)] = 2\sigma_{e_D}{}^2 T_{e_D}\delta(\tau) \qquad (18)$$

where σ_{e_D} and T_{e_D} are the standard deviation and correlation time of e_D, respectively. Realistic values of σ_{e_D} and T_{e_D} are 0.1 naut mile and 3.6 sec, respectively. A reasonable value for σ_{b_D}, the standard deviation of b_D, is 0.14 naut mile.

The INS, VOR, and DME error models are summarized in Table 1. The correlation times shown are for an aircraft flying at a speed of 500 knots. The mean values and rms (root-mean-square) values are reasonable values for a "typical" VOR/DME station, assuming the use of quality receivers, i.e., the type used on jet airliners. The values of the INS error parameters shown in Table 1 are for a high-quality system. For a low-quality INS, the rms value of the gyro drifts is taken to be 1.0°/hr.

Table 1 Summary of INS, VOR, and DME error models

System	Error	Mean value	rms value	Correlation time	Model
INS	$\varepsilon_x, \varepsilon_y, \varepsilon_z$	0	0.01°/hr	5 hr	Exponentially correlated process
	α_x, α_y	0	$10^{-4} g$	10 hr	Exponentially correlated process
VOR	b_V	0	1.0°	hours	Random bias
	e_V	0	1.0°	7.2 sec	White noise
DME	b_D	0	0.14 naut mile	hours	Random bias
	e_D	0	0.1 naut mile	3.6 sec	White noise

System Model

A model for a system using an inertial navigation system and the information from two VOR/DME stations is given below:

$$\left.\begin{aligned}
\delta\dot{R}_x &= \rho_z\delta R_y + \delta V_x \\
\delta\dot{R}_y &= -\rho_z\delta R_x + \delta V_y \\
\delta\dot{V}_x &= -\omega_s{}^2\delta R_x + (2\Omega_z+\rho_z)\delta V_y - a_z\psi_y + a_y\psi_z + \alpha_x \\
\delta\dot{V}_y &= -\omega_s{}^2\delta R_y - (2\Omega_z+\rho_z)\delta V_x + a_z\psi_x - a_x\psi_z + \alpha_y \\
\dot{\psi}_x &= \omega_z\psi_y - \omega_y\psi_z + \varepsilon_x \\
\dot{\psi}_y &= -\omega_z\psi_x + \omega_x\psi_z + \varepsilon_y \\
\dot{\psi}_z &= \omega_y\psi_x - \omega_x\psi_y + \varepsilon_z \\
\dot{\varepsilon}_x &= -(1/T_{\varepsilon_x})(\varepsilon_x - n_{\varepsilon_x}) \\
\dot{\varepsilon}_y &= -(1/T_{\varepsilon_y})(\varepsilon_y - n_{\varepsilon_y}) \\
\dot{\varepsilon}_z &= -(1/T_{\varepsilon_z})(\varepsilon_z - n_{\varepsilon_z}) \\
\dot{\alpha}_x &= -(1/T_{\alpha_x})(\alpha_x - n_{\alpha_x}) \\
\dot{\alpha}_y &= -(1/T_{\alpha_y})(\alpha_y - n_{\alpha_y}) \\
\dot{b}_{V1} &= 0 \\
\dot{b}_{D1} &= 0 \\
\dot{b}_{V2} &= 0 \\
\dot{b}_{D2} &= 0
\end{aligned}\right\} \qquad (19)$$

where the subscripts 1 and 2 indicate the VOR/DME station with which a quantity is associated. Equations (19) are in the state variable form

$$\dot{\mathbf{x}} = F\mathbf{x} + \mathbf{n} \qquad (19a)$$

where F is a 16×16 matrix with many zeroes, and the vector $\mathbf{n}$ has many zeroes.

The VOR/DME station configuration is shown in Fig. 1. The vector $\mathbf{R}_{12}$ is defined by

$$\mathbf{R}_{12} \triangleq \mathbf{R}_1 - \mathbf{R}_2 \qquad (20)$$

The easterly, northerly, and vertical components of $\mathbf{R}_1$, $\mathbf{R}_2$, and $\mathbf{R}_{12}$ are denoted by (x_1, y_1, h_1), (x_2, y_2, h_2), and (x_{12}, y_{12}, h_{12}), respectively. Denoting the VOR and DME measurements from station 1 and station 2 by V_1 and D_1, and V_2 and D_2, respectively, we have

$$\begin{bmatrix} V_1 \\ D_1 \\ V_2 \\ D_2 \end{bmatrix} = \begin{bmatrix} \arctan(x_1/y_1) + b_{V1} \\ [x_1{}^2+y_1{}^2+h_1{}^2]^{1/2} + b_{D1} \\ \arctan[(x_1-x_{12})/(y_1-y_{12})] + b_{V2} \\ [(x_1-x_{12})^2+(y_1-y_{12})^2+(h_1-h_{12})^2]^{1/2} + b_{D2} \end{bmatrix} + \begin{bmatrix} e_{V1} \\ e_{D1} \\ e_{V2} \\ e_{D2} \end{bmatrix} \qquad (21)$$

The state varibles of the system are errors in the inertial and VOR/DME systems. If Eq. (21) is expanded in a Taylor series about a nominal flight path (the nominal values of x_1 and y_1 being denoted by $\bar{x}_1$ and $\bar{y}_1$, respectively) and only first-order terms are retained, the following measurement equations result:

$$\mathbf{z} = H\mathbf{x} + \mathbf{v}$$

i.e. (22)

$$\begin{bmatrix} \delta V_1 \\ \delta D_1 \\ \delta V_2 \\ \delta D_2 \end{bmatrix} = \left[\begin{array}{cc:c:c} \bar{y}_1/A, & -\bar{x}_1/A & & \\ \bar{x}_1/[A+h_1{}^2]^{1/2}, & \bar{y}_1/[A+h_1{}^2]^{1/2} & 0 & I \\ (\bar{y}_1-y_{12})/B, & (x_{12}-\bar{x}_1)/B & (4\times 10) & (4\times 4) \\ \dfrac{\bar{x}_1-x_{12}}{[B+(h_1-h_{12})^2]^{1/2}}, & \dfrac{\bar{y}_1-y_{12}}{[B+(h_1-h_{12})^2]^{1/2}} & & \end{array}\right] \begin{bmatrix} \delta R_x \\ \delta R_y \\ \delta V_x \\ \delta V_y \\ \psi_x \\ \psi_y \\ \psi_z \\ \varepsilon_x \\ \varepsilon_y \\ \varepsilon_z \\ \alpha_x \\ \alpha_y \\ b_{V1} \\ b_{D1} \\ b_{V2} \\ b_{D2} \end{bmatrix} + \begin{bmatrix} e_{V1} \\ e_{D1} \\ e_{V2} \\ e_{D2} \end{bmatrix}$$

with

$$A = \bar{x}_1{}^2 + \bar{y}_1{}^2, \quad B = (\bar{x}_1 - x_{12})^2 + (\bar{y}_1 - y_{12})^2 \qquad (23)$$

where δV_1 and δD_1, and δV_2 and δD_2 denote the differences between the actual and nominal VOR and DME measurements from stations 1 and 2, respectively.

In the derivation of the measurement equations, it was assumed that the L axes, i.e., the set of axes with origin at the location of the aircraft and axes pointing east, north, and up, and the

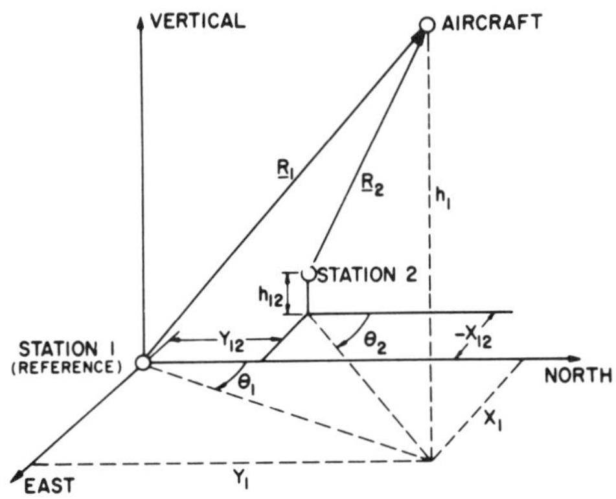

Fig. 1 VOR/DME station configuration.

reference station coordinate axes (see Fig. 1) are parallel when in fact they are not, due to the curvature of the Earth's surface. However, since an aircraft can use a VOR/DME station which is at most 200 naut miles away, the true and coordinate axes are rotated relative to each other through small angles when not at high latitudes. Hence, the differences between the components of $\mathbf{R}_1$, $\mathbf{R}_2$, and $\mathbf{R}_{12}$ in true and station coordinates are small and are therefore neglected. In other words, in the simulations, flight paths consisting of a series of straight line segments rather than true great circle paths are flown.

Letting t_0 denote the time that the system is initialized, it follows from the models derived previously that

$$E[\mathbf{n}(t)] = E[\mathbf{v}(t)] = E[\mathbf{n}(t)\mathbf{v}^T(\tau)] = 0 \tag{24}$$

$$E[\mathbf{x}(t_0)] = E[\mathbf{x}(t_0)\mathbf{n}^T(t)] = E[\mathbf{x}(t_0)\mathbf{v}^T(t)] = 0 \tag{25}$$

$$\begin{aligned} E[\mathbf{n}(t+\tau)\mathbf{n}^T(t)] &= Q\delta(\tau) \\ &= \operatorname{diag}\left\{0,0,0,0,0,0,0,\frac{2\sigma_{\varepsilon_x}^2}{T_{\varepsilon_x}},\frac{2\sigma_{\varepsilon_y}^2}{T_{\varepsilon_y}},\frac{2\sigma_{\varepsilon_z}^2}{T_{\varepsilon_z}},\right. \\ &\qquad \left.\frac{2\sigma_{\alpha_x}^2}{T_{\alpha_x}},\frac{2\sigma_{\alpha_y}^2}{T_{\alpha_y}},0,0,0,0\right\}\delta(\tau) \end{aligned} \tag{26}$$

$$\begin{aligned} E[\mathbf{v}(t+\tau)\mathbf{v}^T(t)] &= R\delta(\tau) \\ &= \operatorname{diag}\{2T_{ev1}\sigma_{ev1}^2, 2T_{eD1}\sigma_{eD1}^2, \\ &\qquad 2T_{ev2}\sigma_{ev2}^2, 2T_{eD2}\sigma_{eD2}^2\}\delta(\tau) \end{aligned} \tag{27}$$

The error covariance matrix P, at time t_0 is given by

$$P(t_0) = E[\mathbf{x}(t_0)\mathbf{x}^T(t_0)] \tag{28}$$

where all the elements are zero except those on the main diagonal, and possibly $P_{12}(t_0)$ (the element in the first row and second column) and $P_{21}(t_0)$. In particular

$$\begin{aligned} \operatorname{diag} P(t_0) = \{&\sigma_{R_x}^2(t_0), \sigma_{R_y}^2(t_0), \sigma_{V_x}^2(t_0), \sigma_{V_y}^2(t_0), \\ &\sigma_{\psi_x}^2(t_0), \sigma_{\psi_y}^2(t_0), \sigma_{\psi_z}^2(t_0), \sigma_{\varepsilon_x}^2, \sigma_{\varepsilon_y}^2, \sigma_{\varepsilon_z}^2, \\ &\sigma_{\alpha_x}^2, \sigma_{\alpha_y}^2, \sigma_{b_{V1}}^2, \sigma_{b_{D1}}^2, \sigma_{b_{V2}}^2, \sigma_{b_{D2}}^2\} \end{aligned} \tag{29}$$

$$\sigma_{R_xR_y}^2 = P_{12}(t_0) = P_{21}(t_0) = E[\delta R_x(t_0)\delta R_y(t_0)] \tag{30}$$

In Eqs. (29) and (30), the initial position error statistics may be those associated with a VOR/DME reading[1] or those determined from another estimate of position. The variances of the velocity and platform attitude errors will depend on the state of the INS when the filter is initialized, and the values of the remainder of the variances appearing in Eq. (29) are readily determined from the error models derived previously.

Discretization of the System Model

For the purposes of simulation, the continuous, time-varying, linear system under consideration will be approximated by a multistage system. Thus, a multistage process described by

$$\mathbf{x}_{i+1} = \Phi_i \mathbf{x}_i + \Gamma_i \mathbf{n}_i, \quad i = 0,1,2,\ldots \tag{31}$$

is sought such that

$$\left.\begin{aligned} &\mathbf{x}_0 \cong \mathbf{x}(t_0) \\ &\mathbf{x}_i \cong \mathbf{x}(t_i), \quad t_i = t_{i-1} + \Delta t, \quad i = 1,2,\ldots \end{aligned}\right\} \tag{32}$$

where ΔT is a time increment. The continuous process described by Eq. (19) can be approximated to first order in ΔT by the process described by Eq. (31) if (Ref. 12, Sec. 11.5)

$$\Phi_i = [I + F(t_i)\Delta T] \tag{33}$$

$$\Gamma_i = \Delta T \tag{34}$$

$$\mathbf{n}_i = \mathbf{n}(t_i), \quad t_i \leqq t < t_{i+1} \tag{35}$$

with

$$E[\mathbf{n}_i \mathbf{n}_j^T] = Q'\delta_{ij} = Q/\Delta T\, \delta_{ij} \tag{36}$$

where δ_{ij}, the Kronecker delta function, is equal to zero unless $i = j$, in which case it equals one. Furthermore, the discrete approximation to the continuous measurement system described by Eqs. (22) is given by

$$\mathbf{z}_i = H_i \mathbf{x}_i + \mathbf{v}_i, \quad i = 0,1,2,\ldots, \tag{37}$$

with

$$\mathbf{v}_i = \mathbf{v}(t_i), \quad t_i \leqq t < t_{i+1} \tag{38}$$

where

$$E[\mathbf{v}_i \mathbf{v}_j^T] = R'\delta_{ij} = R/\Delta T\, \delta_{ij} \tag{39}$$

The approximation holds only if ΔT is small compared to the characteristic times of the system.

Filter for Combining VOR/DME and Inertial Data

The Kalman filter equations[13] for the linear multistage approximation described by Eqs. (31–39) are:

Time update:

$$\hat{\mathbf{x}}_{i+1} = \Phi_i \hat{\mathbf{x}}_{i+} \tag{40}$$

$$P_{i+1} = \Phi_i P_{i+} \Phi_i^T + \Gamma_i Q' \Gamma_i^T \tag{41}$$

Measurement update:

$$\hat{\mathbf{x}}_{i+} = \hat{\mathbf{x}}_i + K_i(\mathbf{z}_i - H_i \hat{\mathbf{x}}_i) \tag{42}$$

$$P_{i+} = (I - K_i H_i) P_i (I - K_i H_i)^T + K_i R' K_i^T \tag{43}$$

$$K_i = P_i H_i^T (H_i P_i H_i^T + R')^{-1} \tag{44}$$

where $\hat{\mathbf{x}}_i$ and P_i, and $\hat{\mathbf{x}}_{i+}$ and P_{i+} denote the state estimate and covariance matrix at stage i before and after processing the measurement at stage i, respectively, and I is the 16×16 identity matrix.

Note that in the absence of measurements, the propagation of the state estimate and covariance matrix are described by Eqs (40) and (41). This would correspond to the propagation of errors in unaided-inertial operation.

Combining VOR/DME information from two stations without inertial data reduces to the problem of finding the maximum-likelihood estimates of x_1 and y_1 (see Fig. 1). An estimator that performs this task is designed in Ref.1.

The multistage filter described by Eqs. (40–44) estimates the easterly and northerly components of the position and velocity errors, the angles of rotation about the east, north, and vertical axes which relate the platform and computer axes, the gyro drift in each of the three gyros, the bias errors in the east and north accelerometers, and the bias in each VOR and each DME. In deriving the measurement equations, the rotation between the true and reference station coordinate axes, due to the curvature of the Earth, is neglected. Upon tuning in a new VOR DME station, the associated VOR and DME biases must be reinitialized.

Simulation Studies

A computer program was written to calculate the error covariance matrix for any nominal flight path consisting of a series of straight line segments. The inputs to the program are a) the model parameters, b) the latitude, longitude, and altitude

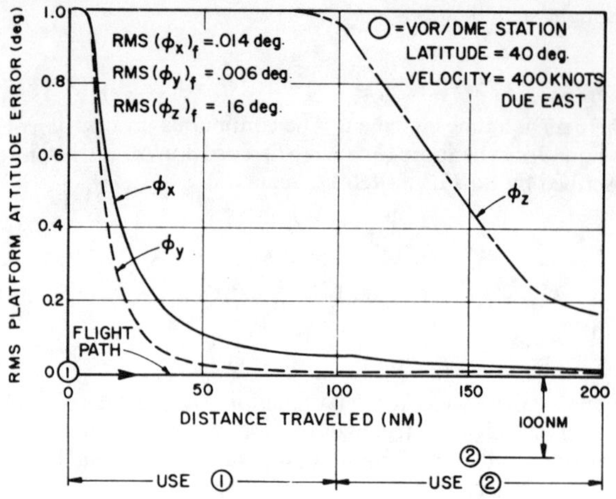

Fig. 2 Platform attitude errors for in-flight alignment using the information from one VOR/DME.

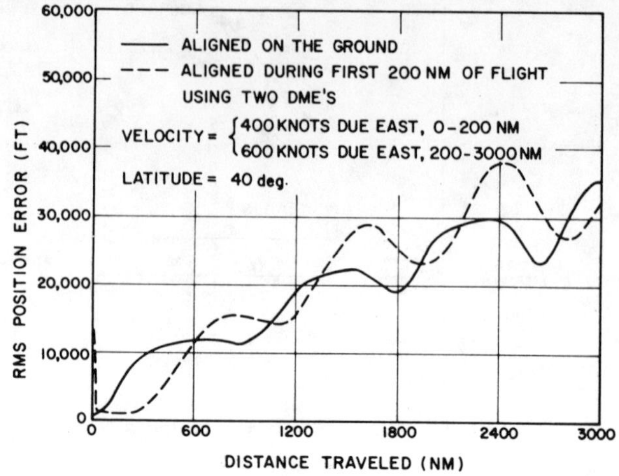

Fig. 4 The rms position errors resulting during an unaided-inertial flight preceded by ground and in-flight alignment.

of each of the VOR/DME stations to be used, and c) the latitude and longitude of each of the switching points, i.e., points along the flight path where the aircraft tunes in a new VOR/DME station or changes its speed or directional heading. The outputs of the program are the rms errors in the estimates of the states.

In-Flight Alignment

The possibility of performing the fine alignment of the INS platform while in the air by using VOR/DME information was investigated. Figure 2 shows the rms errors in the estimates of the platform attitude (ϕ_x = tilt about east axis, ϕ_y = tilt about north axis, and ϕ_z = aximuth error) for a 30-min flight (200 naut miles at 400 knots) using one VOR/DME to update a high-quality INS (0.01°/hr gyro drift). In Fig. 3 these same errors are shown for a flight using two DME's. The time between measurement updates was 60 sec and the altitude was 33,000 ft. The present method of ground alignment takes about 15 min with the final rms error in the platform tilts, rms $(\phi_x)_f$ and rms $(\phi_y)_f$, equal to about 0.005° and the final rms azimuth error, rms $(\phi_z)_f$, equal to about 0.05°. Thus, the accuracy of in-flight alignment using two DME's is about the same as for ground alignment, whereas in-flight alignment using one VOR/DME is less accurate by a factor of 2 or 3. In Fig. 4 the rms position errors during an unaided-inertial flight preceded by ground alignment are compared with errors during a flight with in-flight alignment (using two DME's, the station configuration being that of Fig. 3).

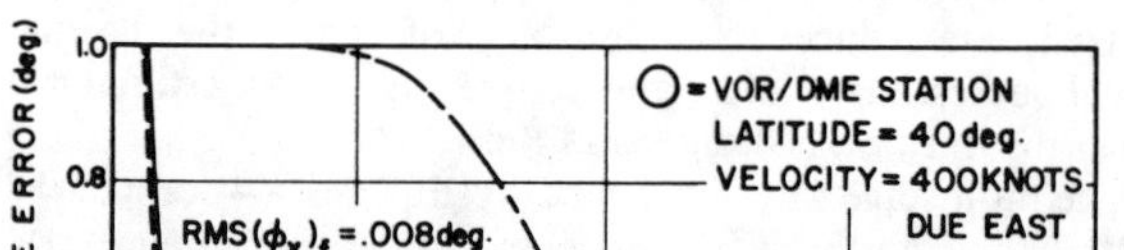

Fig. 3 Platform attitude errors for in-flight alignment using the information from two DME's.

In-flight alignment using combinations of VOR/DME information other than the two previously discussed were also considered. The use of only the DME measurements during the flight of Fig. 2 yields much less accurate alignment than when both the VOR and DME measurements are used. The addition of VOR measurements during the flight of Fig. 3 does not significantly improve the accuracy of alignment.

Station configurations other than the two simple configurations of Figs. 2 and 3 were also studied. It was found that more complex configurations with more frequent switching between stations do not result in a significant improvement in alignment accuracy. Note that the results of Figs. 2 and 3 are not restricted to the specific flight paths considered. For example, if during a 30-min flight, the information from one VOR/DME is used where the line-of-sight to the station used during half of the flight is, in general, orthogonal to the line-of-sight to the station used during the other half of the flight, the resulting alignment will essentially be that of Fig. 2. Similarly, if the information from two DME's is used where the crossing angle between the lines-of-sight from the aircraft to the stations lies between 60° and 120°, the resulting alignment will be about the same as that of Fig. 3.

Periodic Updating

The use of VOR/DME information to update a high-quality INS during a 5-hr transcontinental flight was investigated. The system is assumed to be initially aligned on the ground with a 0.005° rms error in the platform tilts and a 0.05° rms azimuth error. Furthermore, the initial position and velocity errors are taken to be 0.1 naut mile and 0.1 knot, respectively, in both the easterly and northerly directions. The altitude is 33,000 ft and the time between measurement updates is 2 min. The rms position errors are shown in Fig. 5 for the unaided-inertial case as well as for the cases where the system is realigned twice during the flight (from 800 to 1000 naut miles and from 1800 to 2000 naut miles). Realignment is done using one VOR/DME with the station configuration of Fig. 2 and using two DME's with the station configuration of Fig. 3. Obviously, periodic realignment of the INS results in more accurate position information than that resulting from unaided-inertial operation. There is not much difference between the errors resulting from the use of one VOR/DME and the use of two DME's.

Also shown in Fig. 5 are the position errors resulting if two position resets are performed during the flight (at 950 and 1950 naut miles). Here position reset means that the position display is reset to correspond to a position fix from two DME's (whose lines-of-sight are perpendicular). Thereafter, the increments of

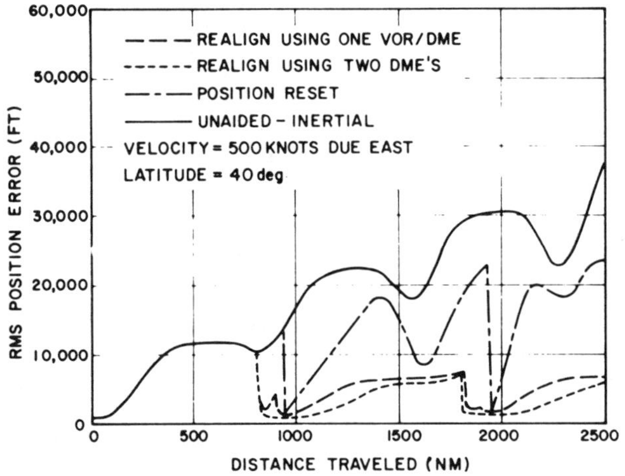

Fig. 5 The rms position errors for realignment, position reset, and unaided-inertial modes of operation during a transcontinental flight.

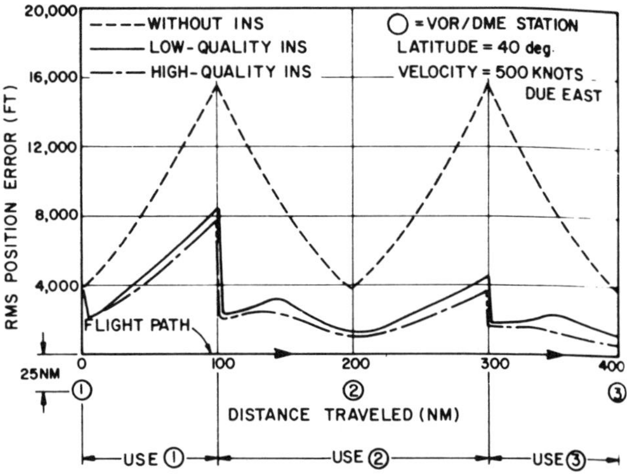

Fig. 7 The rms position errors for an area flight using one VOR/DME with a high-quality INS, a low-quality INS, and without an INS.

position change calculated by the INS are simply added to the display. The position used in the routine which integrates the outputs of the accelerometers is not changed because such a change causes large amplitude oscillations in the rms position and velocity error histories. Position resets must be performed frequently (every few minutes) to obtain position accuracy comparable to the case where the system is realigned. Obviously, position resets do not result in decreased velocity or platform drift errors.

Continuous Updating

Since the VOR/DME system provides nationwide coverage in the U.S., there exists the possibility of continuously updating an INS with VOR/DME information during domestic flights. For such flights it may be possible to use a low-quality INS instead of the high-quality system required for unaided-inertial operation.

In Fig. 6, the rms position errors for a radial flight using the information from one VOR/DME with a high-quality INS (0.01°/hr gyro drift), a low-quality INS (1.0°/hr gyro drift), and without an INS are shown. These same error histories are shown in Fig. 7 for an area flight. The time between measurement updates is 90 sec and the altitude is 33,000 ft. The rms value of the initial position error is that of a VOR/DME reading taken at the beginning of the flight while the rms velocity error is 10 knots in the northerly and easterly directions. The rms values of the initial platform tilts and azimuth error are 0.05 and 0.1°, and 0.5 and 5.0° for the high-quality and low-quality systems, respectively. As can be seen in Figs. 6 and 7, there is considerable improvement in position accuracy when inertial data are added to the information from one VOR/DME. However, the improvement in position accuracy when using a high-quality INS is not much greater than when using a low-quality INS.

The sharp decreases in rms position errors in Figs. 6 and 7 at the points where the aircraft switches from one VOR/DME station to another are of interest. The decreases which occur in the radial flight (as well as a portion of the decreases for the area flight) are due to a transient effect which is introduced when, upon tuning in a new station, the variances of the biases are reinitialized and the offdiagonal terms involving the biases are set to zero in the covariance matrix. The larger decreases occurring in the area flight are explained by the fact that the DME position information is more accurate than the VOR position information (except when very near the station), and the fact that the lines-of-sight to the new and old stations at the switching points for the area flight (Fig. 7) are not parallel.

In Fig. 8, the rms position errors are shown for various combinations of the information from two VOR/DME's and data from a low-quality INS. The use of a high-quality INS yields only a small improvement over the position accuracy

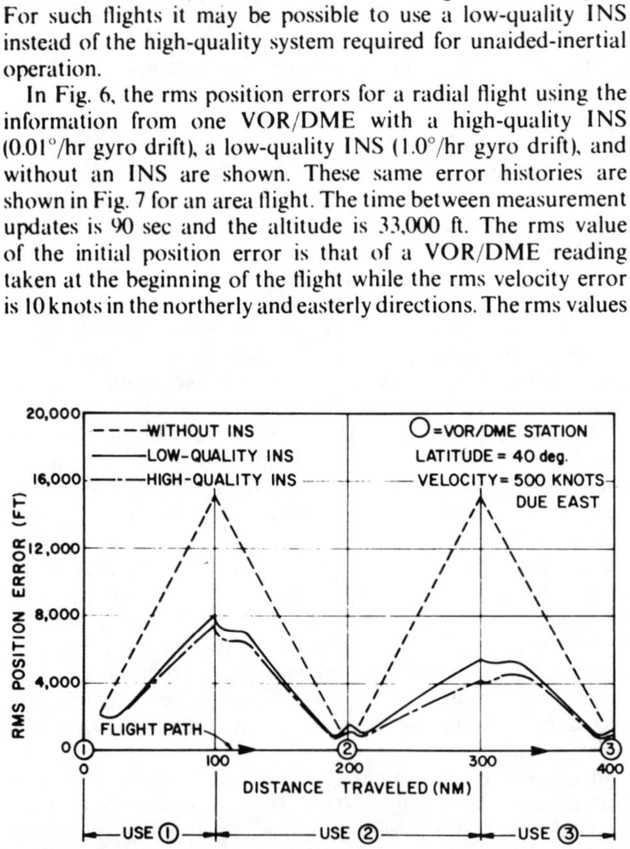

Fig. 6 The rms position errors for a radial flight using one VOR/DME with a high-quality INS, a low-quality INS, and without an INS.

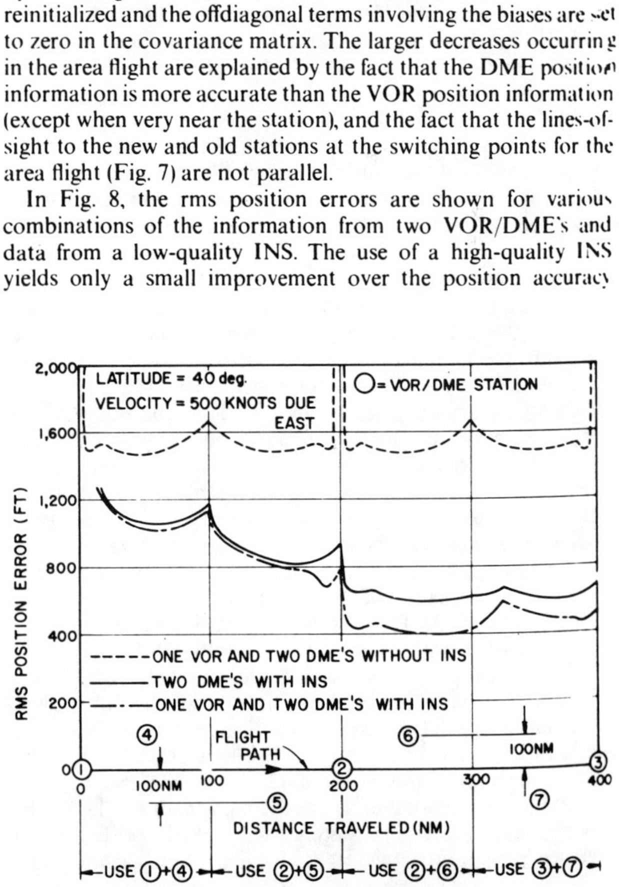

Fig. 8 The rms position errors for various combinations of the information from two VOR/DME's and data from a low-quality INS.

Table 2 **Approximate factors of improvement in rms position accuracy over the use of a single VOR/DME for various combinations of VOR/DME information and data from a low-quality INS**

Combination of information	Factor of improvement
1 VOR, 1 DME, INS (radial flight)	2.8
1 VOR, 1 DME, INS (area flight)	3.5
1 (or 2) VOR's, 2 DME's	9
2 DME's, INS	24
1 (or 2) VOR's, 2 DME's, INS	37

obtained when using a low-quality INS. When a second VOR is added to the case of one VOR, two DME's and inertial data, the decrease in the rms position error is negligible.

Factors of improvement in rms position error over the use of a single VOR/DME are shown in Table 2 for various combinations of information from two VOR/DME's and data from a low-quality INS. These factors were calculated for a point halfway between the second and third stations because at this point the error for the case of using a single VOR/DME is maximum; and hence, the factor of improvement at this point is of prime importance. The case where two DME's are used without inertial data is not included in Table 2 because there are generally two position fixes possible in this case. However, the addition of a VOR measurement to two DME measurements resolves the ambiguity, although it does not substantially improve the accuracy of the position fixes. From these results, it is seen that the navigational accuracy resulting from the use of a given combination of VOR/DME information is improved by roughly a factor of 3 or 4 by the addition of inertial data.

When using the information from one VOR/DME to update a low-quality INS, rms velocity errors of 7 or 8 knots can be expected, whereas, when using a high-quality INS, velocity errors of 3 or 4 knots occur. The use of the information from two VOR DME stations to update an INS yields rms velocity errors of roughly one and two knots for a high-quality and a low-quality system, respectively.

If the flight paths of Figs. 6–8 were extended and similar configurations of stations were encountered, the error histories between the second and third stations would repeat. Insight into the use of air vs inertial data can be obtained by comparing these results with those of Ref. 1.

Suboptimal Filters

For the alignment or realignment of a high-quality INS (which takes from 20 to 30 min), the suboptimal filter resulting when the gyro drifts, accelerometer errors, and DME biases are neglected performs nearly optimally.[9] Hence, when using one VOR/DME, the alignment accuracy shown in Fig. 2 can be obtained by using an eighth-order filter, whereas, when using two DME's, the alignment accuracy shown in Fig. 3 is obtainable with a seventh-order filter. This seventh or eighth order suboptimal filter also behaves nearly optimally when using VOR/DME information to periodically update a high-quality INS.

When using VOR/DME information to continuously update an INS, the performances of the suboptimal filters resulting when certain VOR and DME biases are neglected were investigated. For the case where information from one VOR/DME and inertial data from a low-quality INS are used (for the flight path shown in Fig. 6), the performance degradation†† of the suboptimal filter resulting when the DME bias is neglected is less than 1%, whereas if both the VOR and DME biases are neglected, the performance degradation is about 125%. For the case of using two DME's with inertial data (for the flight path of Fig. 8), the performance degradation resulting when both DME biases are neglected is approximately 100%. In general, neglecting VOR biases results in great performance degradation, while neglecting DME biases does not. Note that although the 100% degradation stated for the case of two DME's and inertial data is a significant percentage, the degradation in terms of feet of rms position error is small (see Fig. 8). The above performance degradations were calculated at the midpoint between the second and third stations.

†† In this context, performance degradation means the increase in rms position error over that resulting from the optimal filter.

Conclusions

Combining VOR/DME information (from one or two stations) with inertial data by means of a maximum-likelihood filter results in substantial improvements in navigational accuracy over that obtained by the use of a single VOR/DME, as is the current practice.

The use of a low-quality INS with one VOR/DME reduces the rms position error by a factor of roughly 3, and yields rms velocity errors of about 8 knots. When using VOR/DME information continuously throughout the flight, the use of a high-quality INS (0.01°/hr gyro drift) instead of a low-quality INS (1.0°/hr gyro drift) does not substantially improve position accuracy, but does reduce the rms velocity error to about 4 knots. Some of the improvement due to the use of inertial data results from the ability of the filter to estimate the bias errors associated with the VOR/DME measurements. Estimates of VOR bias error are better for area than for radial flights while the opposite is true for the DME bias error.

The use of information from two VOR/DME stations with inertial data yields large factors of improvement (roughly 25 to 35) in rms position accuracy over the use of a single VOR/DME station. In general, the rms position error obtained when using a given combination of the VOR/DME information from two stations is decreased by a factor of about 3 or 4 by the addition of inertial data. When using information from two VOR/DME stations with data from a high-quality and a low-quality INS, the velocity errors were about 1 and 2 knots, respectively. As far as position and velocity accuracy are concerned, at most one VOR station need be used.

Accurate in-flight alignment of an INS platform by using VOR/DME information is possible. The accuracy of in-flight alignment using two DME's is about the same as for ground alignment, whereas, when using one VOR/DME, alignment is less accurate by a factor of 2 or 3. In-flight alignment takes about 30 min. Although initial in-flight alignment of inertial systems onboard commercial aircraft is probably not needed, the in-flight realignment of the system before a transoceanic flight (e.g., from Los Angeles to London or Chicago to Hawaii) results in significant improvement in navigational accuracy. This might permit a reduction of separation requirements over the North Atlantic routes. Also, realignment of the system after a transoceanic flight could prove useful since the result would be accurate position, velocity, and attitude information in the terminal area and for approach and landing.

Periodic realignment (every one or two hours) of a high-quality INS during a transcontinental flight results in significant improvements in position and velocity errors over unaided-inertial operation or the use of periodic position display resets. The use of a periodically-realigned, high-quality INS would make it possible to fly precise great circle routes between cities since the locations of VOR/DME stations would not substantially restrict the flight paths (as they do when using a continuously-updated INS). If the final realignment is performed just prior to entering the terminal area, accurate position, velocity, and attitude information would be available for approach and landing without reliance upon VOR/DME information.

In general, when combining VOR/DME information with inertial data, the suboptimal filter resulting when the DME biases are neglected performs nearly optimally. However, neglecting VOR biases results in great performance degradation. During the period of time that it takes for in-flight alignment (or realignment) of a high-quality INS (20 to 30 min), the suboptimal filter resulting when the gyro drifts, accelerometer errors, and DME biases are neglected performs nearly optimally. Hence, alignment can be accomplished by using an eighth-order filter when using one VOR/DME, or a seventh-order filter when using two DME's.

The quantitative results presented here are, of course, dependent upon the validity of the error models assumed. In particular, the VOR/DME error model proposed here has not been as thoroughly checked experimentally as the inertial navigation system model.

References

[1] Bryson, A. E., Jr. and Bobick, J. C., "Improved Navigation by Combining VOR/DME Information and Air Data," *Journal of Aircraft*, Vol. 9, No. 6, June 1972, pp. 420–426.

[2] Hemesath, N. B., "Optimal and Suboptimal Velocity-Aiding for VOR/DME Systems," *AIAA Journal*, Vol. 10, No. 1, Jan. 1972, pp. 24–30.

[3] Hemesath, N. B., Meyer, D. H., and Schweighofer, H. M., "Complementing VOR/DME with INS An Improved Navigation System," Paper 690338, National Business Aircraft Meeting, Society of Automotive Engineers, Wichita, Kansas, March 26–28, 1969.

[4] Del Balzo, J. M., "Inertial Systems and Area Navigation in the U.S. Domestic Airspace," *Navigation*, Vol. 17, No. 4, Winter 1970–71, pp. 426–431.

[5] Holm, R. J., "Flight Evaluation of Inertial/DME/DME System," Rept. RD-70-24, May 1970, Federal Aviation Administration, Washington, D.C.

[6] Silver, M. and Greenberg, M., "Area Navigation—A Quantitative Evaluation of the Effectiveness of Inertial System Aiding," *Navigation* Vol. 18, No. 4, Winter 1971–72, pp. 360–368.

[7] DeGroot, L. E. and Larsen, J., "Extended Capability from Existing Navaids," *Navigation*, Vol. 15, No. 4, Winter 1968–69, pp. 391–405.

[8] Pinson, J. C., "Inertial Guidance for Cruise Vehicles," *Guidance and Control of Aerospace Vehicles*, edited by C. T. Leondes, McGraw-Hill, New York, 1963, Chap. 4.

[9] Bobick, J. C. and Bryson, A. E., Jr., "Improved Navigation by Combining VOR/DME Information with Air or Inertial Data," SUDAAR 442, May 1972, Dept. of Aeronautics and Astronautics, Stanford University, Stanford, Calif.

[10] Erzberger, H., "Application of Kalman Filtering to Error Correction of Inertial Navigators," TN D-3874, Feb. 1967, NASA.

[11] Faurre, P., *Navigation Inertielle Optimale et Filtrage Statistique*, Dunod, Paris, 1971.

[12] Bryson, A. E., Jr. and Ho, Y. C., *Applied Optimal Control*, Xerox-Blaisdell, Waltham, Mass., 1969.

[13] Kalman, R. E., "A New Approach to Linear Filtering and Prediction Problems," *Transactions of the ASME, Ser. D: Journal of Basic Engineering*, Vol. 82, March 1960, pp. 35–45.

AUTOCALIBRATION OF A LASER GYRO STRAPDOWN INERTIAL REFERENCE/NAVIGATION SYSTEM

By

Knut M. Wefald And Charles R. McClary

Laser Navigation Systems
Commercial Aviation Operations
1625 Zarthan Avenue South
St. Louis Park, MN 55416
(612) 542-4927

ABSTRACT

The Honeywell Laser Gyro Strapdown Inertial Reference/Navigation System used on the Boeing 757/767/737-300 airplanes and in a slightly modified form on the Aerospatiale A.310 aircraft is described. The system uses three ring laser gyros, three current rebalance accelerometers, and three microprocessors to perform all attitude, world wide, strapdown inertial navigation. This implementation using state-of-the-art unheated components provides fast alignment, accurate navigation, and reliable operation at low cost to the airlines. The system integrates all of the inertial sensors on the airplane, eliminating duplication on previous aircraft, and provides body-referenced angular rates and linear accelerations, attitude, heading, velocity, position, and various calculated track-related parameters for the primary flight instruments and the flight management system.

The inertial sensor initial calibration and in-service automatic calibration techniques used to provide long-term accuracy and sensor replacement without calibration are described. Automatic calibration of the three gyro axis and the two level accelerometer axis is accomplished using Kalman filter techniques. Practical filter implementation considerations as they relate to strapdown inertial auto calibration are reviewed and filter mechanization is discussed. Results from laboratory and flight test of the auto calibration feature are presented.

BACKGROUND

The Honeywell Inertial Reference System (IRS) has become the industry standard for reliability and performance(1). This system, with minor variations, is used in a variety of commercial and general aviation applications, all of which have the basic signal flow diagram shown in Figure 1. The inertial data is provided by three laser gyros and three inertial grade accelerometers. The gyro and accelerometer errors(2) are initially calibrated by a combination of measurements made at the sensor level and the system level. The sensor biases, scale factors, and misalignments are determined as a function of temperature and the data reduced to a set of calibration coefficients that are applied by the software to the sensor outputs. See Figure 1. A second, system level set of measurements determines the misalignments of the sensor mounting surfaces and the misalignment of the sensor triad to the mounting provisions. These measurements are reduced to a second set of coefficients that are used to correct the misalignment of the sensor mounting structure and the attitude/heading outputs(3). The autocalibration techniques described later in this paper are applied in addition to these initial calibrations.

The IRS uses the excellent stability of the Honeywell laser gyro and state-of-the-art electronics to provide reliability unequaled in the inertial reference/navigation field(4). The IRS, composed of three Inertial Reference Units (IRU) and one Inertial Reference Mode Panel (IRMP) has, as of June 1984, achieved a mean time between failure of 7,500 flight hours for the IRU and 144,000 flight hours for the IRMP. The units are flying in 126 aircraft operated by 19 airlines. Even so, in 1982, Honeywell recognized that the serviceability of the system would be enhanced by auto calibration and began development of auto calibration techniques for all three axes of the gyros and the level axes of the accelerometers.

INTRODUCTION

Automatic compensation of sensor bias errors is provided in order to permit some long term sensor bias growth without a corresponding degradation of navigation performance. By eliminating the removals of otherwise good sensors because of long term bias shifts, it is expected that automatic bias calibration (Autocal) will eliminate 10 IRU removals on every aircraft in airline service over a period of twenty years, thus achieving a significant savings for both the airlines and Honeywell.

The effect of gyro and accelerometer biases on navigation performance for typical bias errors is shown in Figure 2. Note that the long term position error is primarily due to gyro bias and that the velocity error is predominantly due to accelerometer bias.

Our first approach to automatic bias calibration (Autocal) was a five state Kalman filter using a simple position measurement at the end of flight to estimate all three gyro biases and the two level axis accelerometer biases. However, because the effects of gyro and accelerometer biases on navigation errors are quite different, it was found to be relatively easy to estimate gyro biases using a simple position measurement. It was also found, however, that estimating accelerometer biases required the addition of a velocity match at a considerable increase in complexity. Moreover, because velocity is affected by so many parameters, the integrity of the velocity match was judged to be

Reprinted from *Rec. IEEE Position Location and Navigation Symp.*, pp. 66–74, 1984.

less reliable than the position match. Also, accuracy of the accelerometer bias estimate was found to be somewhat degraded for longer duration flights.

Strapdown Associates Inc. of Minneapolis, MN, proposed a unique accelerometer automatic bias calibration (accel Autocal) configuration which is based on the fact that the aircraft lateral velocity is identically equal to zero on the ground. Hence the indicated body axis lateral velocity during taxi can be used as the measurement for the Kalman filter. By restricting the accel Autocal computation to a very short time interval (five minutes) during taxi, the effect of gyro bias is effectively eliminated from the measurements and the Schuler effect need not be considered. Note in Figure 2 that the accelerometer (accel) bias induced velocity is nearly linear for the first five minutes.

The Autocal configuration selected employs a three state filter using final position error as the measurement to estimate gyro biases, and a two state filter using body axis lateral velocity during taxi to estimate level axis accelerometer biases.

KALMAN FILTER OVERVIEW

Navigation system position and velocity errors due to sensor biases are trajectory dependent. If the sensor biases are known and if the trajectory flown is known, the nav system errors can be computed. In essence, the Autocal Kalman filters accomplish the inverse of this, i.e., the aircraft flight trajectory is determined during flight, the appropriate nav system errors (position or velocity) generated for that trajectory are measured, and the appropriate sensor biases are estimated based on the measured errors and known trajectory using Kalman filter techniques.

Kalman filters combine the system dynamic error characteristics (e.g., how bias errors are propagated into navigation errors), measurement error characteristics (i.e., the inaccuracies inherent in the source of the measurement and those resulting from mismodeling), and the process noise characteristics. These errors are characterized by their statistical descriptions (covariance) and are used to compute Kalman filter gains which establish a quantitative relationship between the measured navigation errors and the estimates of the bias errors. These estimates are continually refined through an iterative process.

Two different Kalman filters are employed for gyro bias error estimation and accelerometer bias estimation. The gyro bias estimation filter uses the operator supplied position input for the next flight as the measurement. Hence, a bias estimate reset is possible only once per flight. Thus, the gyro Kalman filter may be thought of as having a large iteration interval which spans the entire flight. The state transition matrix contains the trajectory history for the entire flight and is computed as the product of several matrices, each computed for the integration interval of 50 seconds. The computed correction is "reset" into the gyro bias coefficient in software once at the start of each "cold start" alignment, which is also the start of each filter iteration. The iterative process takes several (perhaps 20 or 30) individual flights to refine the bias estimate for an initial bias error of 2 or 3 sigma.

The accelerometer filter, on the other hand, uses body axis lateral velocity during taxi as the measurement. In this case, the measurement is constantly available during the taxi phase and the bias estimate can be continuously refined to produce a highly accurate bias estimate for each flight. This correction is also reset into the accel bias coef in software at the start of each next flight.

A top level conceptual diagram of the gyro automatic bias calibration (gyro Autocal) is shown in Figure 3. A model of the position error is continuously computed during flight. This model measures the aircraft trajectory and computes how the gyro biases propagate to nav errors for that trajectory. The Kalman filter computations are run once at the start of the next flight where the measurement is computed as the difference between the "last computed position" and the starting point for this flight. The Kalman gain computation takes into account the probable measurement error for the last flight in arriving at the bias estimate, i.e., large measurement uncertainties reduce the gains, small uncertainties increase the gains. The error state estimate is given as:

$$\delta G = CK * Y \qquad \text{EQ 1}$$

where: δG is the change in the gyro bias estimate vector
CK is the Kalman gain vector
Y is the measurement vector

A top level conceptual diagram of the accel Autocal is shown in Figure 4. A model of the body axis lateral velocity error is computed continuously during the first five minutes of taxi. This model computes how the accelerometer biases propagate to body axis lateral velocity error for that trajectory. Since actual body axis lateral velocity is continuously available during taxi, it is continuously compared with the estimate and the residual (the difference between the actual and predicted measurements) is applied to the Kalman gain matrix to compute the error state (bias) estimate. The error state estimate is given as:

$$\delta A = \delta A + K * (ZN - Z^*) \qquad \text{EQ 2}$$

where: δA is the bias estimate vector
K is the Kalman gain vector
ZN is the measurement based on body lateral velocity
Z^* is the predicted value of the measurement

Each successive iteration of the filter reduces the difference between ZN and Z^*. When Z^* equals ZN, the estimated bias vector is equal to the actual bias vector. The filter is iterated every four seconds for a total of 75 iterations during the five minute computation period.

GYRO AUTOCAL FUNCTIONAL DESCRIPTION

The gyro Autocal conceptual block diagram of Figure 5 shows the basic functions that go into the Kalman gain computations. A more detailed block diagram which includes the basic Autocal equations is shown in Figure 6.

Running estimates of the sensor bias variances are contained in the covariance matrix (P). This matrix is updated each flight for the long term sensor bias growth which is assumed to be 10 sigma in 10 years (modeled as a square root of time function which reaches 3.16 sigma in one year and 10 sigma in 10 years). The covariance (P) appears in the gain equation and it permits large corrections when sensor variances are estimated to be large, but only small corrections when variances are estimated to be small. The covariance matrix is initialized to .5 sigma squared to provide large initial gains.

An estimate of the position error, in feet, due to a one-sigma bias error for the trajectory of the last flight is contained in the scaled position sensitivity matrix (HP). This term has two functions in the gain computation:

1. It represents an uncertainty in the measurement due to a turn on (TOT) bias uncertainty of one sigma.

2. It can be thought of as a weighting factor vector which defines how the individual gyro errors would have propagated into position errors for the trajectory of the last flight.

An estimate of the uncertainty in the measurement is contained in the measurement covariance term (R). Any position errors which are not due to long-term sensor error growth must be ignored by Autocal in order to avoid miscalibration of the sensors. Hence an attempt is made to estimate the measurement noise i.e., position errors due to short-term affects such as TOT bias and gyro random walk. A measurement noise covariance term (RDM) is also included to account for measurement errors (as opposed to actual position errors) due to simplification incorporated into Autocal.

Errors included in the error models are:

1. TOT bias. This is the day to day bias uncertainty and is conservatively assumed to be 1 sigma.

2. Bias error growth. This is the long term bias growth that Autocal is intended to compensate for. It is conservatively assumed to be ten sigma in ten years.

3. Gyro random walk is conservatively estimated to be 1 sigma.

Non volatile memory (NVM) is provided in order to hold the propagated sensor covariance and to hold the data computed inflight for the next beginning of flight Kalman filter computations. This data includes the position sensitivity matrix, final position, and flight duration time of the last flight. The assumption is that the aircraft is powered up for the next flight in the same location as it was powered down. For several reasons the data in the NVM may not be consistent with the aircraft position; a different IRU may have been installed, an IRU may have been powered down for a flight leg, or the flight data may have been lost due to a procedural error. In any of these cases, the NVM would contain a totally random position. To avoid false corrections due to problems such as these (and others) an Autocal test is provided which rejects any data set with a position error larger than 60 nautical miles.

Re-initialization of Nav Outputs

Capability has been provided for the crew to re-initialize the nav outputs if the accumulated errors become significant. This re-initialization would affect the gyro Autocal accuracy since provisions for resets are not included in the position error model. The effect of resets on Autocal performance is minimized by treating the error in the position error model as increased measurement noise. The velocities are zeroed and the tilts re-leveled by engaging the align mode for a minimum of thirty seconds between flight legs. Provisions also exist to enter a new position and execute a full alignment between flight legs. Three possible results may occur as a result of the align/reset mode:

1. Velocity and tilts are zeroed
2. Velocity and tilts are zeroed and position is updated.
3. Full alignment.

The position error model does not expect these resets/updates. Consequently the measurement errors introduced by the resets/updates must be compensated. Several possible solutions were considered. The chosen solution has two parts:

1. Any time resets in any combination are applied (velocity, tilt, position), the measurement noise covariance is increased by 1.5 nm squared (8.1E7 ft2). Any number of resets will be tolerated provided that position is not also reset.

2. Whenever position is reset, the Autocal results are frozen at the point of position reset and the measurement noise covariance is increased.

Removal of the Schuler Frequency Terms

When the navigation error equations are written, it can be seen that sensor errors cause the system to oscillate at a characteristic (Schuler) frequency which has a period of 84 minutes. These Schuler oscillation effects can be seen in Figure 2, which shows the propagation of navigation errors with time due to gyro and accelerometer biases. The H matrix, which describes the relationship between gyro bias and final position errors, is propagated using two other matrices which describe the propagation of sensor errors to navigation ($\phi 12$) and the long term effect of navigation errors upon themselves ($\phi 11$). (See Figure 6, gyro Autocal functional diagram.)

In order to achieve a very significant savings in the software and computer throughput, the gyro Autocal error equations have been simplified to the extent that $\phi 12$ and $\phi 11$ are both 3×3 matrices. As a consequence, the short term (Schuler) frequency effects have been eliminated from the position error model. The effects of this simplification on the position error model are negligible for trajectories with little or no turning during the cruise portion of flight (which is characteristic of airline flights).

One of the verification test cases, however, revealed that trajectories with a large number of heading changes in a single flight (especially if synchronized with the Schuler frequency period) could cause significant errors in the Autocal position error model. (Up to 20 nm after 3.5 hours was noted.) While trajectories of this type are not expected, we felt it would be prudent to account for this effect.

The approach taken was to treat the Schuler oscillation as added measurement noise. The measurement noise covariance (R) was increased by (2.5 nm) squared (2.25E8 ft2). Analytical studies indicate that the steady state bias estimate errors are about 10 percent higher for short flights and negligible for longer flights.

Bias Update

The sensor bias compensation coefficients are adjusted by Autocal. The new coefficients are set into software at the beginning of align and no changes are permitted after that. Hence, there is actually a lag of two flights on the bias updates, i.e., the bias change is computed at the start of the "next" flight and is inserted into the software at power-up on the flight after that.

Sensor Replacement

Sensor calibration data is stored on sensor PROMS which are replaced when sensors are replaced. The fact that sensors have been replaced is determined by looking at the PROM sum checks before the Kalman filter computations are executed. When sensor replacement is detected, the Kalman filter computations are aborted, the covariance matrix (P) is re-initialized, and the replacement sensor bias coefficient is set at the new value.

ACCELEROMETER AUTO CAL FUNCTIONAL DESCRIPTION

The accel Autocal functional block diagram is shown in Figure 7. When the IRS changes mode from align to nav, the horizontal accelerometer biases are balanced by platform tilts. These biases will have no effect on nav performance until a heading change occurs to decouple them. Small residual tilt errors will exist, however, and accel Autocal engages a tilt and velocity measurement loop prior to engagement of the bias estimation process to measure the average velocity and the tilt error associated with any rate of change of average velocity.

The bias estimation process is engaged when "taxi" starts. The start of taxi is defined as a longitudinal velocity in excess of 0.5 ft/sec or a change in heading from the nav entry heading by .002 pirads (.36 deg). When the bias estimation process is initiated, the tilt/velocity measurement loop is disabled and a velocity estimate is computed as the previously measured average velocity plus the velocity drift associated with the measured tilt. The difference between this velocity estimate and the actual velocity is ascribed to the accelerometer bias errors. The velocity errors are computed in the navigation reference frame and resolved into the body frame where lateral velocity becomes the basis for the Autocal measurement.

A question often asked is "How can both of the horizontal accelerometer biases be estimated from a measurement based on body axis lateral velocity alone?" If the aircraft heading never changes after nav entry, the measurement will remain zero because the accelerometer bias errors are balanced by platform tilt errors in the align mode: Note that both body axis lateral and longitudinal velocity errors can be generated by align tilt errors, but Autocal can distinguish these tilt errors from real bias errors by measuring the tilt errors as described above. When the aircraft changes heading, both horizontal accelerometer biases are decoupled from the compensating tilt errors, and an error is produced in the total velocity vector. Autocal resolves this total velocity vector into the body axis coordinates. It would be possible to estimate the biases using either lateral velocity, longitudinal velocity, or both. The reason for using lateral velocity is, of course, the fact that the correct value is guaranteed to be zero. All that is needed to identify the contribution of each accelerometer to the total error is the heading trajectory which is embodied in the velocity sensitivity matrix (H).

The H matrix is derived from the aircraft heading trajectory, and it describes the manner in which accelerometer bias errors contribute to the body axis lateral velocity error. The observable measurement (ZN) is derived from body axis lateral velocity as shown in equation 3 below:

$$ZN = \int_{T\ell=1}^{T\ell=4} \delta VY^B dt \qquad \text{EQ 3}$$

where: VY^B = body axis lateral velocity

But we can write an expression in terms of the velocity sensitivity matrix and the actual accelerometer biases which is equivalent.

$$ZA\ell = HX * \delta ax + HY * \delta ay \qquad \text{EQ 4}$$

Where: $ZA\ell$ = accel bias contribution to lateral velocity average measurement over the ℓth 4 second averaging interval

$\delta ax, \delta ay$ = Actual accelerometer bias error states

HX,HY = Velocity sensitivity Matrix

The Kalman filter estimates the error states using the recursive equation EQ2 which is rewritten below for the individual axes.

$$\hat{\delta ax} = \hat{\delta ax} + KX * (ZN - Z^*) \qquad \text{EQ 5}$$

$$\hat{\delta ay} = \hat{\delta ay} + KY * (ZN - Z^*) \qquad \text{EQ 6}$$

where: $\hat{\delta ax}, \hat{\delta ay}$ = error state estimates

$$ZN = ZA\ell = HX * \delta ax + HY * \delta ay \qquad \text{EQ 7}$$

Combining equations 5 and 7

$$\hat{\delta ax} = \hat{\delta ax} + KX * (HX * \Delta\hat{\delta ax} + HY + \Delta\hat{\delta ay}) \qquad \text{EQ 8}$$

$$\text{Where: } \Delta\hat{\delta ax} = \delta ax - \hat{\delta ax} \qquad \text{EQ 9}$$

$\Delta\hat{\delta ax}$ = Error in the estimate

The error state estimate is iterated at 0.25 Hz for five minutes. In general, two separate moderate amplitude turns at any turn rate or one large amplitude turn at a relatively slow rate are required to produce a "good" estimate. Under these conditions the estimate converges rapidly to the correct value. When the turn trajectory is insufficient to produce an acceptable error state estimate, the covariance (P) will remain relatively large. This is used as a test which prevents use of the estimate if the sum of PX + PY exceeds 1E-6. (It appears that this value could be increased to 1E-5 without sacrificing much accuracy.)

The Kalman gain and covariance update equations are identical in form to the gyro Autocal equations. Bias corrections and sensor replacement handling are also similar to the gyro Autocal.

TEST RESULTS

Gyro Autocal and accel Autocal were developed more or less simultaneously, but gyro Autocal reached flight test first. In addition to exhaustive hardware/software testing in the lab, gyro Autocal flight testing was conducted in both the Honeywell Grumman GII executive aircraft and in Boeing test aircraft. Testing in the Honeywell GII aircraft consisted of offsetting the gyro bias coefficients and flying a random sequence of executive flights to assure that the offsets were removed. Later, a sequence of flights was flown with correct biases to assure that gyro Autocal did not upset the calibration.

Because of aircraft availability problems, only the first 18 flights were actually done in the GII aircraft. The testing was continued in the lab for a total of 40 tests. Figure 7 shows that the biases converged smoothly toward zero from the offset condition. A total of 40 flights was required to remove the horizontal gyro offsets of 0.05 degree/hour. Because the system is less sensitive to Z gyro biases, the Z gyro bias was offset 0.1 degree/hour. About 40 percent of that was removed during the 40 flights. Also, unlike the horizontal gyro bias, Z gyro bias is removed considerably faster for longer flights. Hence, the last few lab tests were increased from the range of 2 to 4 hours to 6 hours. (The GII aircraft tests were of random lengths, ranging from one to more than four hours in length.) Tests with zero bias offsets showed no tendency to create offsets in any axis.

In addition to extensive lab testing, accel Autocal testing included taxi tests in the GII aircraft and simulated taxi tests in a modified van. Because of non-volatile memory limitations, each sequence of aircraft and van testing consisted of only 10 or 12 runs before resetting the biases. This isn't enough to show that the bias remains at zero after converging to zero from the offset condition. A test software patch was written to get around this limitation and subsequent lab test sequences include at least 24 runs. Figure 9 shows results of a typical lab test sequence, the first part of which is also typical of the results obtained for the first 10 or 12 runs in the taxi and van tests. The heading trajectories used for these tests are also identified on the figure. The results show smooth, well-behaved convergence to zero bias followed by minimal bias changes during subsequent trajectories.

STATUS IN REVENUE SERVICE

Production inertial reference units (IRUs) with Autocal activated began to enter revenue service in October of 1983. One year later, approximately 80 percent of the fleet has been equipped with Autocal. While we do not have solid numbers at this time, we estimate that the Autocal-equipped fleet has accumulated about 250,000 IRU flight hours in actual revenue service.

REFERENCES

1. P.J. Fenner, "Requirements, Applications and Results of Strapdown Inertial Technology to Commercial Airplanes," presented to NATO in 1984 community as part of AGARD Lecture Series No. 133: Advances in Strapdown Inertial Systems, published in AGARD-LS-133.

2. P.G. Savage, "Strapdown Sensors," presented to the NATO in 1978 community as part of AGARD Lecture Series No. 95: Strap-Down Inertial Systems, published in AGARD-LS-95.

3. P.G. Savage, "Calibration Procedures for Laser Gyro Strapdown Inertial Navigation Systems," Ninth Annual Electro-Optics/Laser Conference and Exhibition 24-27 October 1977, Anaheim, CA.

4. R.R. Shoquist, "The ARINC 704 Ring Laser Gyro Inertial Reference System," National Technical Meeting of the Institute of Navigation (ION), January 17-19, 1984, San Diego, CA.

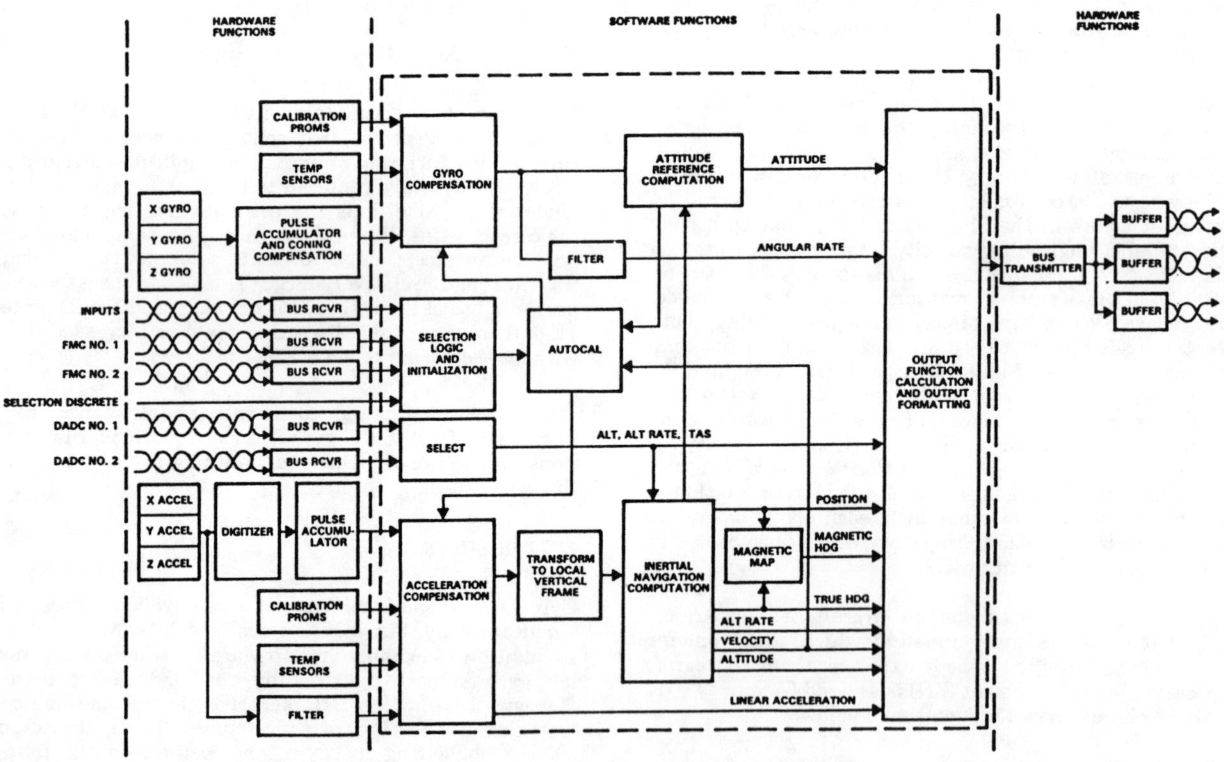

Figure 1. IRU Signal Flow Diagram

Figure 2. Nav Errors After 90 Degree Turn from Align Heading for Typical Sensor Biases

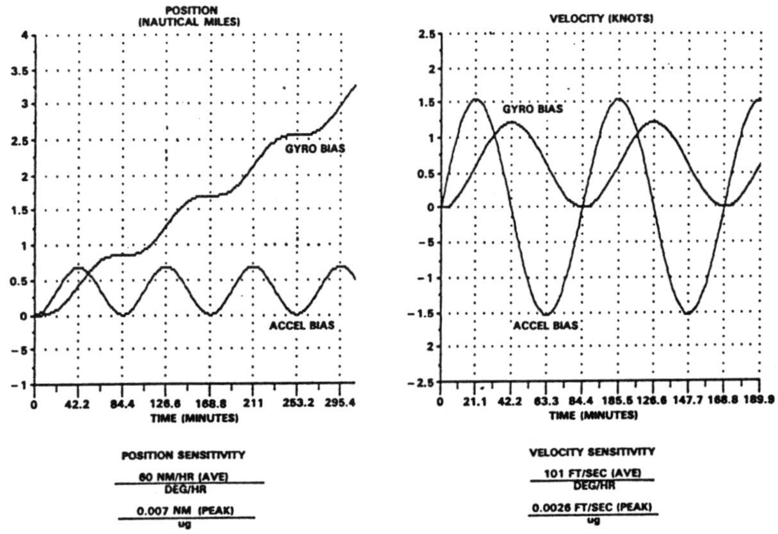

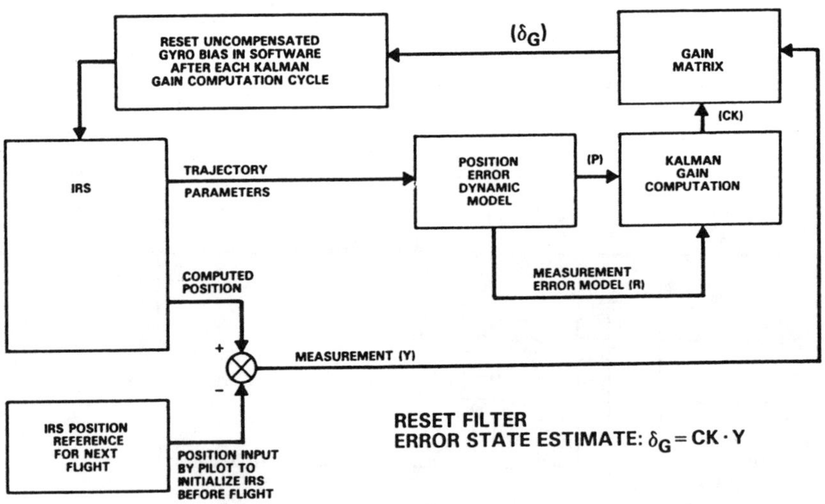

Figure 3. Gyro Autocal Top Level Conceptual Diagram

Figure 4. Accel Autocal Top Level Conceptual Block Diagram

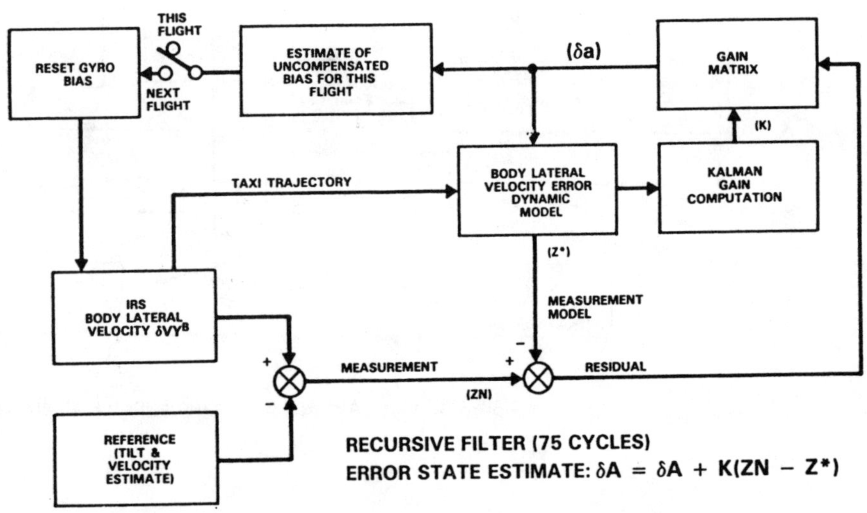

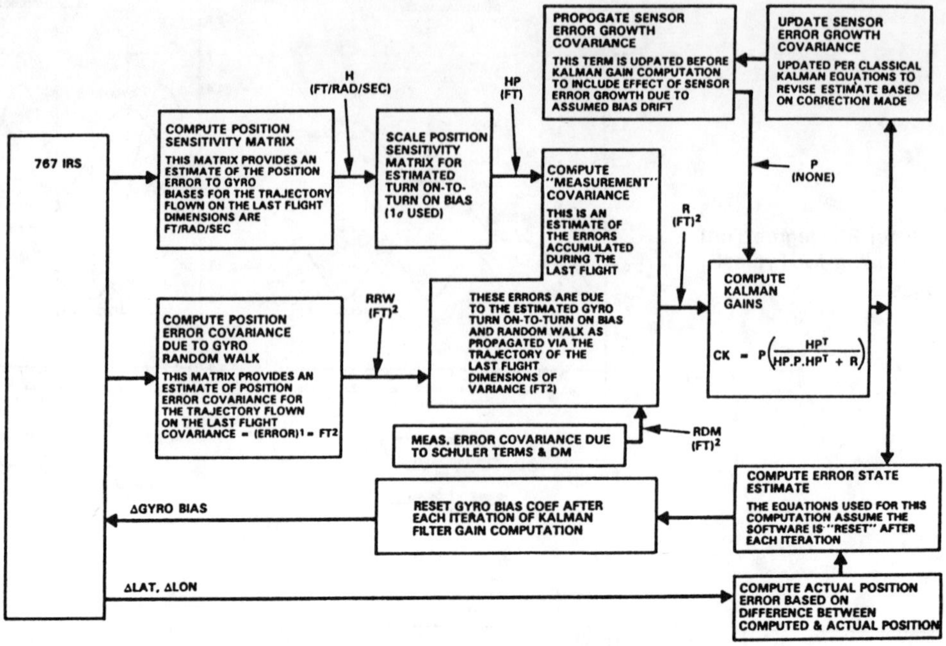

Figure 5. Gyro Autocal Conceptual Block Diagram

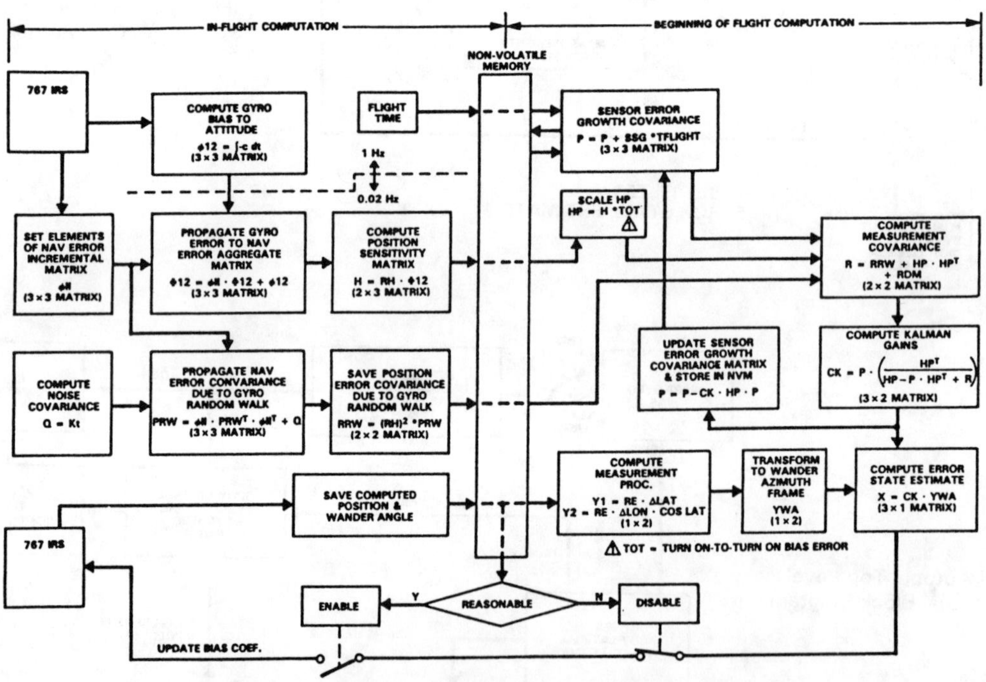

Figure 6. Gyro Autocal Simplified Functional Diagram

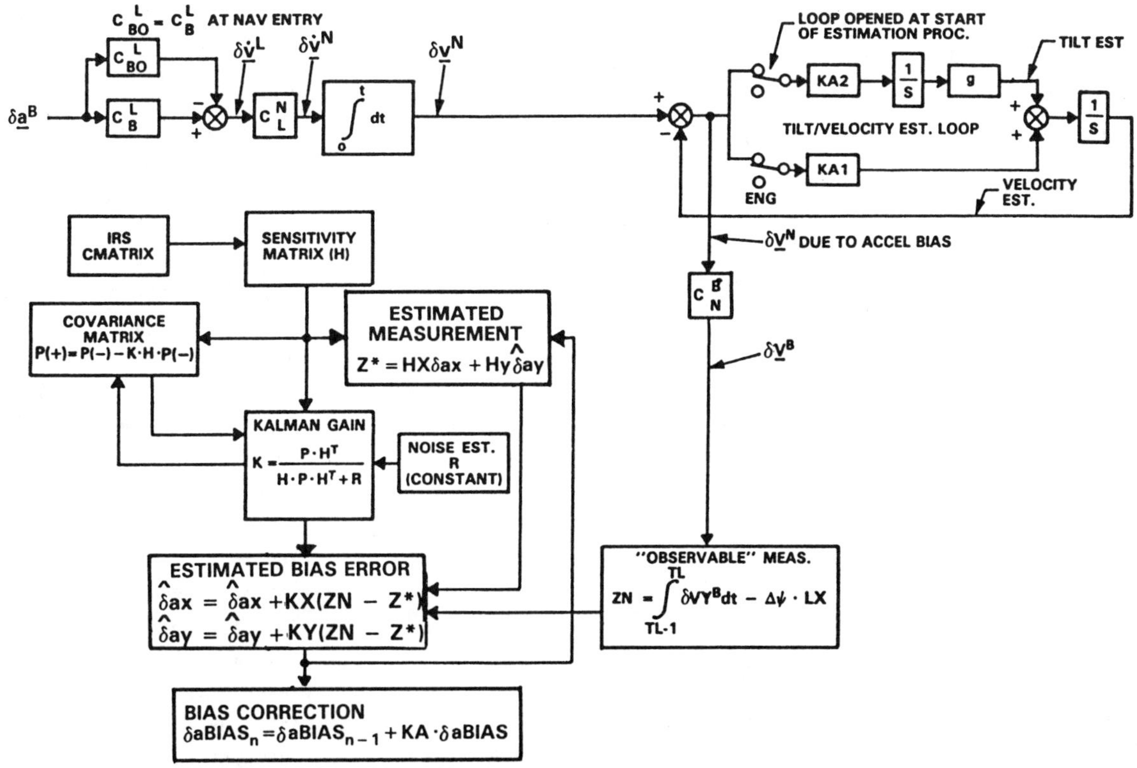

Figure 7. Accel Autocal Functional Block Diagram

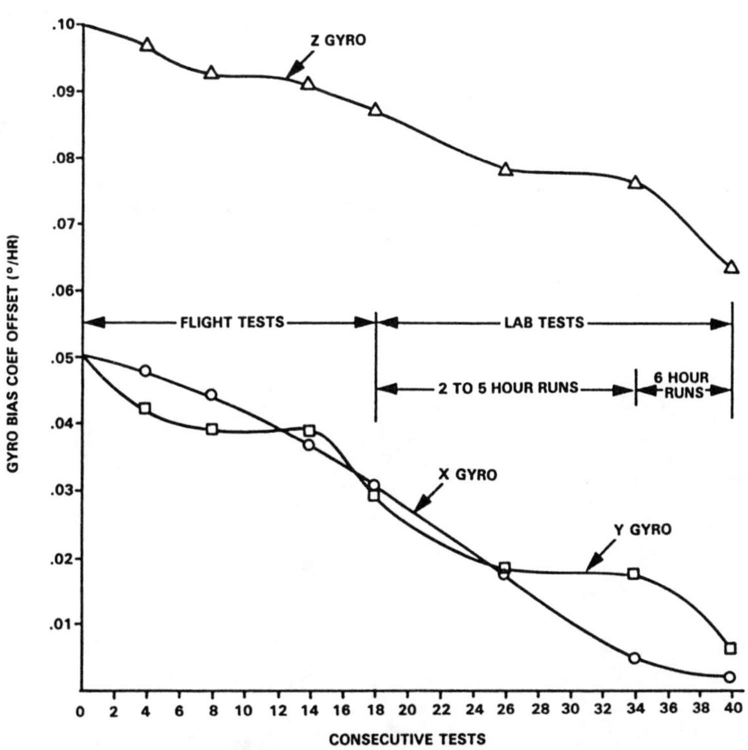

Figure 8. Gyro Autocal Test Results for IRU 354 with X, Y, Z Gyro Initial Offsets of .05°/hr, .05°/hr, and 0.1°/hr, Respectively

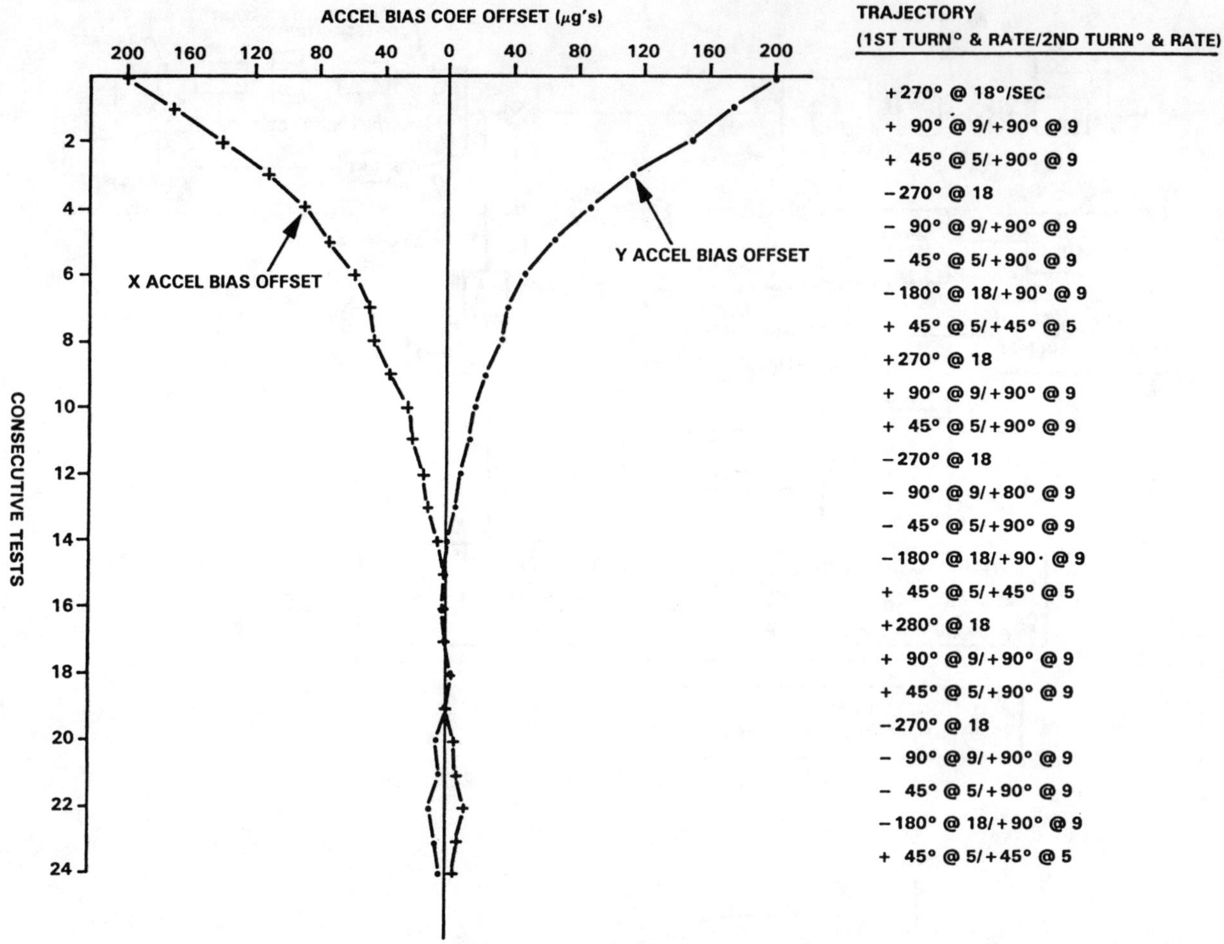

Figure 9. Accel Autocal Lab Test Results for IRU #91 With Initial Offsets 200μg on X & Y. Accelerometer biases with trajectory identified for each test.

SENSOR MANAGEMENT FOR A FAULT TOLERANT INTEGRATED INERTIAL FLIGHT CONTROL REFERENCE AND NAVIGATION SYSTEM

By

DR. S.Y. WEI and DR. J.R. HUDDLE

LITTON'S GUIDANCE & CONTROL SYSTEMS DIVISION
5500 CANOGA AVENUE
WOODLAND HILLS, CA 91365

ABSTRACT

One way to achieve higher reliability for an integrated flight control reference and navigation system is to employ redundant inertial measurement (IMU) units on the aircraft. Further, redundant instruments can be employed in each IMU. Provision of the flight control reference and navigation information for such a system in the presence of faulty sensor operation, or sensors in the case of multiple simultaneous failures, requires two general types of algorithms. The first type of algorithm is one which detects and subsequently isolates the failed or degraded sensors. The second type of algorithm is one which employs the remaining properly operating sensors to optimally construct three orthogonal components of acceleration and attitude rate which are required for the flight control system reference inputs and for the navigation of the aircraft.

The system design described in this paper obtains for the worst case failure pattern a fail-operate, fail-operate and fail-safe (FO-FO-FS) capability both as a flight control reference and navigation system. The paper first discusses highly-efficient algorithms for rapid detection and isolation of failed or degraded instruments for a system of two identical inertial measurement units each with redundant gyros and accelerometers. A specific relative orientation between these two inertial measurement units is assumed in the derivation.

The paper then discusses the second type of algorithm which utilizes the non-degraded sensor outputs to construct the optimal permissible solutions for both the flight control system reference inputs and the navigation function. The specific configuration that is selected by these latter design equations is identified for both of these functions for each possible failure detection and isolation condition up to that for FO-FO-FS.

1. INTRODUCTION

In recent system applications, attention has been focused on the incorporation of strapdown IMUs in redundant configurations as references for both flight control and navigation. The objective is a significant improvement in system reliability and hence flight safety and mission success. To achieve this objective effective redundant sensor management algorithms are required to detect and isolate failed or degraded sensor operation and to provide the required flight control reference and navigation information using the remaining non-degraded sensors.

One of the most common redundant configurations consists of two IMUs which are installed at two separate locations on an aircraft to improve their survivability during any mission hostilities. To achieve FO-FO-FS for flight control, a minimum number of six sensors (six gyros and six accelerometers) is necessary. However with six sensors, the relative sensor orientation is critically important to obtain the redundancy requirement. This in turn imposes constraints on the relative orientation of the two installed IMUs.

Additionally due to separation and structural flexure between the IMUs, accurate navigation cannot be achieved employing sensors from one IMU with those from the other. Navigation of reasonable accuracy therefore requires the employment of redundant gyros and accelerometers in each IMU to obtain any capability beyond fail-safe.

When one additional gyro and accelerometer is added to each of the two IMUs, reasonably accurate FO-FO-FS navigation is achieved. Simultaneously this approach eliminates any constraint on the relative orientation of the two IMUs installed in the aircraft to obtain FO-FO-FS in using the system as a flight control reference. This paper discusses the general characteristics of such a system comprised of two IMUs, each with four gyros and accelerometers.

In any redundant system configuration two tasks which must be addressed are:

1. Detection and subsequent isolation of failed or degraded sensors.

2. Construction from the remaining properly operating sensors of the flight control reference inputs and the navigation solutions.

The first task is accomplished by the failure detection and isolation algorithm (FDI). The second task is achieved by a set of design equations which implements the reconfiguration. In this paper, a highly efficient algorithm for rapid detection and isolation of failed or degraded instrument operation is first discussed. Subsequently, the design

Reprinted from *Rec. IEEE Position Location and Navigation Symp.*, pp. 445–455, 1986.

equations for optimally constructing the flight control reference inputs and the navigation solution is presented along with possible schemes for reconfiguration once the faulty sensor operation has been identified.

2. FAILURE DETECTION AND ISOLATION PARITY EQUATIONS

The FDI algorithm consists of two processes. The first process is the detection of any sensor failure when it occurs. The second process is a logic mechanization which isolates the failed sensor. To detect a failed sensor, a set of equations called the parity equations is employed. A parity equation is defined as the comparison of the measurements between several sensors such that the result is zero if the sensors involved are perfect. Mathematically the available measurement vector m, obtained from ℓ sensors is defined as:

$$m = H X + \varepsilon \tag{2-1}$$

where

X is the 3-dimensional system dynamic state vector
H is the measurement coefficient matrix (ℓx3)
ε is the sensor noise vector

A set of parity equations is defined as:

$$P = Vm, \tag{2-2}$$

where the parity matrix V satisfies the condition

$$VH = 0. \tag{2-3}$$

It can be seen that the parity vector P is simply the linear combination of the sensor noise components.

$$P = V\varepsilon \tag{2-4}$$

Some of the important characteristics of the parity equations are:

1. Any linear combination of the rows of V, V_i, is a parity equation

   $\bar{V} = \Sigma_i \lambda_i V_i$, as

   $\bar{V}H = (\Sigma_i \lambda_i V_i) H = \Sigma_i \lambda_i (V_i H) = 0.$

2. The maximum number of independent parity equations is given by ℓ-3, where ℓ>3 is the number of measurements (sensors) and 3 is the number of independent measurement components. The rank of H is 3 for any non-trivial case.

It is clear that if all the sensors are operating properly, the parity vector should remain within the sensor noise level. However if a failure occurs, some parity vector components will exceed the sensor noise level. Detection of a failure can occur with a test

$$|P| \begin{cases} \leq T, \text{ if no failure occurs} \\ > T, \text{ if failure(s) occurs} \end{cases} \tag{2-5}$$

where the threshold T is set suitably above the noise level. The setting of the threshold requires a balance between the probability of valid failure detection and the false alarm rate which is an invalid indication of failure.

Upon detection of a failure, isolation is required to identify the failed sensor. This can be obtained by "voting" which isolates the failure based on the knowledge of the failed parities. The numerous papers that have been published on FDI algorithms fall in two main categories:

1. Direct comparison of the measurements

2. The general likelihood ratio test.

The general likelihood ratio test is based on the assumption that the parity vector P of equation (2-2) is Gaussian. The log-likelihood ratio of the two hypotheses of Failure or No Failure is obtained and maximized to obtain the detection decision. The isolaton decision is obtained by projecting the parity vector along the column vectors of the parity matrix V.

The direct comparison method simply compares the projection of several measurements along a predetermined direction which achieves a balance between the selected sensors such that a "well conditioned" parity is obtained. The isolation is accomplished by a prior knowledge of the sensors involved in all the parity equations. The general likelihood ratio test requires more numerical computation especially in constructing the parity matrix V. It is physically straight forward to construct a parity matrix V with predetermined sensors in each parity equation in the direct comparison scheme. The direct comparison scheme which has a simple physical interpretation, and avoids the required Gaussian assumption for the parity vector and the additional computation required for the general likelihood ratio test, is employed in this paper.

3. SENSOR AXES AND IMU ORIENTATION

A system comprised of two IMUs each containing four gyros and four accelerometers satisfies FO-FO-FS for accurate navigation and flight control reference information with no constraint imposed on the relative orientation of the two IMUs. Within each IMU it is desirable to have a physically orthogonal triad of each instrument type such that with proper operation of these sensors, no geometric dilution of accuracy along any direction in inertial space occurs. Given such an orthogonal triad of instruments, the fourth, skew instrument should logically be placed such that it projects equally onto each axis of the orthogonal set so that no sensor is favored in a parity equation involving the four sensors within an IMU. The orientation of the cluster of four accelerometers relative to the four gyros is completely arbitrary with this reasoning. However, for simplicity it is assumed these clusters are coincident below. Additionally with no loss of generality it is assumed that the IMUs have a coincident orientation with the sensor axes numbering scheme depicted in Figure 1.

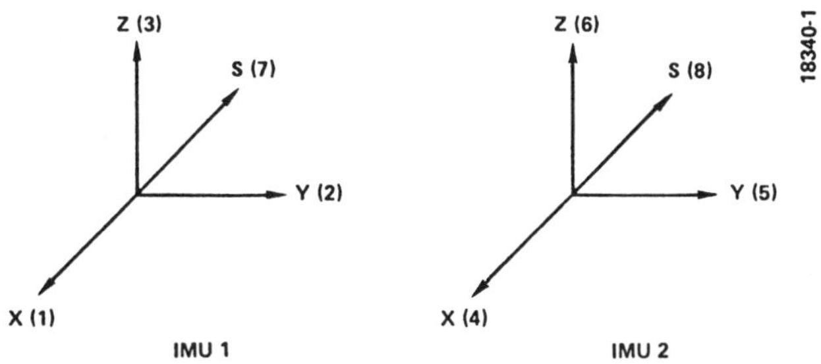

Figure 1. Two Parallel Inertial Measurement Units

4. INSTRUMENT PARITY EQUATONS FOR THE FLIGHT CONTROL FDI ALGORITHM

The FDI algorithm for the flight control reference input requires parity equations which directly compare the gyro measurements and in a similar fashion, the accelerometer measurements. The derivation of the parity equations for flight control is discussed next.

4.1 Instrument Parity Equations

For a redundant system, which consists of four gyros and four accelerometers per IMU as shown in Figure 1, there exists two types of instrument parities. One type compares the four sensors within each IMU. The other type compares sensors from both IMUs. The first type is called the Intra-IMU parities and the second type is called the Inter-IMU parities. The Intra-IMU parities simply compare four sensors within an IMU. There are two such types of parities per IMU, one for the gyros and one for the accelerometers. From the geometry of the sensor orientation, the Intra-IMU parity equation is given as

$$P = \frac{1}{\sqrt{3}} (m_X + m_Y + m_Z) - m_S \tag{4-1}$$

where the measurements are along the X, Y, Z and S (skew) axes.

The coefficient $1/\sqrt{3}$ results from the orientation of the sensor axes. Due to the fact the Intra-IMU parities will not be affected by the separation of the IMUs, and the associated flexure bending and lever arm effects, and the threshold for these type of parities is lower than that for the Inter-IMU parities.

The Inter-IMU parities are constructed by taking a number of sensor measurements from both IMUs for comparison. There are an excessive number of Inter-IMU parities. The most obvious parities are

$$P_{14} = m_1 - m_4 \qquad P_{36} = m_3 - m_6$$
$$P_{25} = m_2 - m_5 \qquad P_{78} = m_7 - m_8 \tag{4-2}$$

where the m_i is the measurement along the numbered axis indicated in Figure 1.

Two additional Inter-IMU parities can be obtained by comparing the skew sensor with three orthogonal sensors from the other IMU.

$$P_{1238} = \frac{1}{\sqrt{3}} (m_1 + m_2 + m_3) - m_8$$
$$P_{4567} = \frac{1}{\sqrt{3}} (m_4 + m_5 + m_6) - m7 \tag{4-3}$$

The maximum number of independent parities for the eight (gyro or accelerometer) sensor configuration is five. One of the Intra-IMU parities from equation (4-1) and four Inter-IMU parities from equation (4-2) forms a set of five independent parities for the system shown in Figure 1. The remainder of the parities can be obtained as linear combinations of these five independent parities. There are clearly many parities which can be derived. To satisfy FO-FO-FS conditions along with the capability of isolating simultaneous failure of sensors, a limited specific set of parities is required. For the system shown in Figure 1, there are 20 parities required to meet the requirements stated above. These parities are listed in Table I.

For the case of no prior failure, the first six parities P_1 to P_6, of Table I are required to detect and isolate a single failure. These parities also detect and isolate two simultaneous failures occuring in the same IMU (e.g., X(1), Z(3)) or two simultaneous failures occuring along the same axis but in each IMU (e.g., Y(2), Y(5)). For the case of two simultaneous failures one in each IMU but along different axes, the remaining 14 parities in the set are subsequently used to isolate the failures. Detection of three simultaneous failures can be achieved by testing all 20 parities.

The ICODE in Table I is used to isolate failures. This is simply an entry code. An entry is

TABLE I. 20 PARITIES REQUIRED FOR MULTIPLE FAILURE ISOLATION

Parity	ICODE	Parity	ICODE
P_1 = 14	(10010000)	P_{11} = 7135	(10101010)
P_2 = 25	(01001000)	P_{12} = 8135	(10101001)
P_3 = 36	(00100100)	P_{13} = 7246	(01010110)
P_4 = 78	(00000011)	P_{14} = 8246	(01010101)
P_5 = 7123	(11100010)	P_{15} = 7612	(11000110)
P_6 = 8456	(00011101)	P_{16} = 8612	(11000101)
P_7 = 7456	(00011110)	P_{17} = 7561	(10001110)
P_8 = 8123	(11100001)	P_{18} = 8561	(10001101)
P_9 = 7234	(01110010)	P_{19} = 7453	(00111010)
P_{10} = 8234	(01110001)	P_{20} = 8453	(00111001)

equal to one if the corresponding sensor is used in the given parity, otherwise the entry is zero.

To illustrate the use of the ICODE for isolation, the following examples apply:

Example 1: Sensor #1 and Sensor #3 fail

The unfailed parities and ICODEs from the first six parities are listed below:

Unfailed Parities	ICODE
P_2 = 25	(01001000)
P_4 = 78	(00000011)
P_6 = 8456	(00011101)
Sum =	(01012112)

It is seen from the sum of unfailed ICODE, that sensor #1 and sensor #3 have failed. This is clear since the unfailed parities have zero entry for the failed sensor and hence when the unfailed parities are summed the entry for the failed sensors cannot be other than zero.

Example 2: Sensor #1 and Sensor #5 fail.

The unfailed parities from the first six parities and the corresponding ICODEs are listed as follows:

Unfailed Parities	ICODE
P_3 = 36	(00100100)
P_4 = 78	(00000011)
Sum =	(00100111)

At this point, all we can conclude sensors #3, #6, #7, and #8 are working properly. By testing parities P_7, P_8, and P_9 the following unfailed parities are obtained:

Unfailed Parities	ICODE	
P_3 = 36	(00100100)	From Previous Test
P_4 = 78	(00000011)	
P_9 = 7234	(01110010)	
Sum =	(01210121)	

The sum of the unfailed parity ICODEs indicate Sensors #1 and #5 have failed.

After a failed sensor has been isolated, there are eight sets of five parities required to isolate the next failure(s) depending on the particular sensor involved. All these possibilities are listed in Table II. To detect a third failure after two failures have been isolated, three parities are required. Since there exists 28 possible combinations of fwo failures, 28 sets of three parities are required to detect the 3rd failure depending on which 2 failed sensors have been previously isolated. All these possibilities are listed in Table III.

TABLE II. EIGHT SETS OF PARITIES FOR ONE ISOLATED FAILURE

#1 Failed	#2 Failed	#3 Failed	#4 Failed
25	14	14	25
36	36	25	36
78	78	78	78
8456	8456	8456	7123
7234	7135	7126	8561
#5 Failed	**#6 Failed**	**#7 Failed**	**#8 Failed**
14	14	14	14
36	25	25	25
78	78	36	36
7123	7123	8456	7123
8246	8453	8123	7456

TABLE III. 28 SETS OF PARITIES FOR TWO ISOLATED FAILURES

2 isolated failed sensors	12	13	14	15	16	17	18
required parities for detecting the 3rd failure	36 78 8456	25 78 8456	25 36 78	36 78 7234	25 78 7234	25 36 8456	25 36 7456
2 isolated failed sensors	23	24	25	26	27	28	34
required parities for detecting the 3rd failure	14 78 8456	36 78 7135	14 36 78	14 78 7135	14 36 8456	14 36 7456	25 78 7612
2 isolated failed sensors	35	36	37	38	45	46	47
required parities for detecting the 3rd failure	14 78 7246	14 25 78	14 25 8456	14 25 7456	36 78 7123	25 78 7123	25 36 8123
2 isolated failed sensors	48	56	57	58	67	68	78
required parities for detecting the 3rd failure	25 36 7123	14 78 7123	14 36 7123	14 36 7123	14 25 8123	14 25 7123	14 25 36

The parity selection method shown in Figure 2 can be easily implemented in a real-time computer.

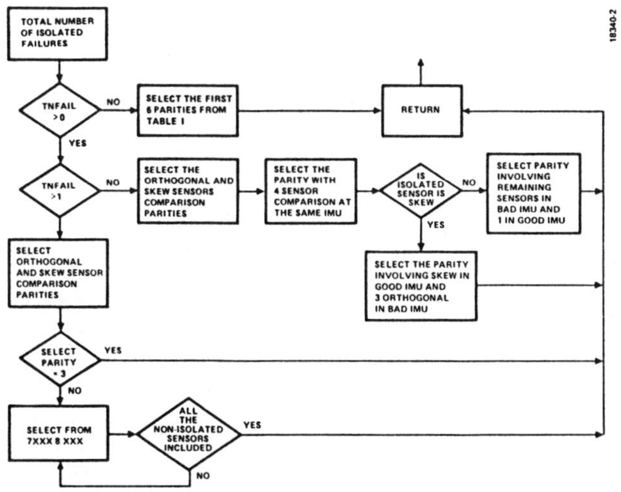

Figure 2. Flow Chart for the Parity Selection

The parity selection logic can be discussed depending on the failure/isolation status of the system:

1. A failure has not been isolated.

   In this case, there are six parities required.

(1)	Compare the coincident orthogonal sensors	3
(2)	Compare the 2 skew sensors	1
(3)	Compare the 4 sensors in the same IMU	2
	Total	6

2. One failure has been isolated.

   It can be seen from Table II that five parities are required to isolate the second failure or detect two simultaneous subsequent failures. The selection logic can be described as follows:

	Isolated Failure is an Orthogonal Axis	Isolated Failure is a Skew Axis
(1) Compare the coincident orthogonal sensors	2	3
(2) Compare the skew sensors	1	0
(3) Compare the 4 sensors in the good IMU	1	1
(4) Compare the remaining sensors from the bad IMU with one sensor from the good IMU	1	0
(5) Compare the Skew sensor in the good IMU with the 3 orthogonal sensors in the bad IMU	0	1
Total	5	5

3. Two failures have been isolated.

   In this case, 3 parities are required for detection of the third failure. The selection of parities can be accomplished by the logic described below until the 3 required parities are determined:

   (1) Compare the coincident orthogonal sensors.

   (2) Compare the skew sensors.

   (3) Select parities from P_5 to P_{20} as listed in Table I, such that the parities include all non-isolated sensors which are not shown in (1) and (2) stated above.

4.2 The Flight Control FDI Algorithm

The FDI algorithm utilizes the parity equations discussed above to detect and subsequently isolate instrument failures. The algorithm consists of 9 steps which are discussed below and illustrated in Figure 3.

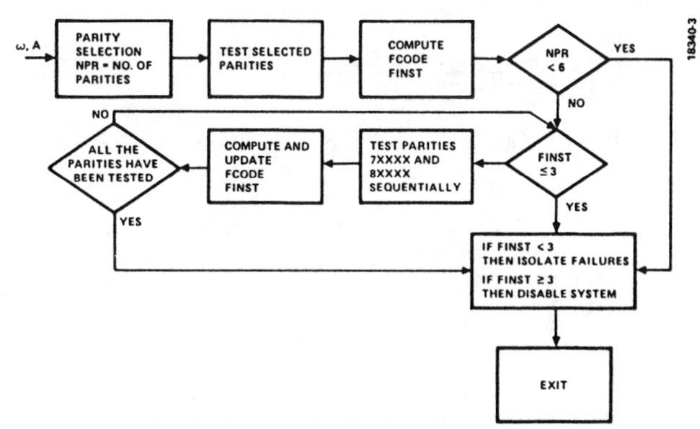

Figure 3. The FDI Algorithm

Step 1: Select parities according to the isolated failures as described.

Step 2: Test the selected parities.
With the sensor outputs, the selected parities can be computed and compared against a pre-determined threshold.

Step 3: Compute the sum called FCODE of unfailed parity ICODEs.

Step 4: Compute the total number of possible failures by summing up the number of zero entries of the FCODE.
FINST is the total number of possible failures.

Step 5: If one or more failures have been isolated previously then go to step 9.

Step 6: If FINST ≤ 3 then go to step 9.
Note in steps 5 and 6 when there is no failure, there are certain failed patterns of sensors which will cause more than 3 zero entries in the FCODE. When this happens, further testing of parities is required which is accomplished in steps 7 and 8.

Step 7: Sequentially test parities of the form of 7xxx or 8xxx.

Step 8: Compute and update FCODE and FINST.

Step 9: If FINST ≥ 3, then three or more failures have occurred and the system is disabled. If FINST ≤ 2, then the failures have been isolated.

The FDI algorithm described above has been successfully tested for all the possible combinations of failures with a digital computer simulation program.

It should be pointed out that the parities P_1 thru P_4 of Table I only involve two sensors in each parity. The misalignment between IMUs can cause problems in failure detection. However, this can be overcome by replacing P_1 thru P_4 with another set of four parities which has at least four sensors per parity and is compensable for the misalignment when it is known.

4.3 The Design Equations for the Flight Control Reference Inputs

The design equations can be defined as:

$$V_R = D\, V_{inst}$$

where

$V_R^T = [V_x, V_y, V_z]$ is the constructed measurement vector with three orthogonal components

$$V_{inst}^T = [\underbrace{V_1, V_2, V_3}_{3}, \underbrace{V_4, V_5, V_6}_{3}, V_7, V_8]$$

V_1, V_2, V_3: 3 orthogonal sensors of IMU 1; V_4, V_5, V_6: 3 orthogonal sensors of IMU 2; V_7: skew sensor of IMU 1; V_8: skew sensor of IMU 2

D is the design matrix, 3x8.

The design matrix D can be constructed based on the orientation of the sensor axes.

For the flight control reference inputs, several possible schemes for constructing the design equations have been examined. They are summarized as follows:

a. Use three properly operating sensors (orthogonal or not) from one IMU.

b. Use a least squares estimate from all the properly operating sensors in one IMU.

c. Use a least squares estimate from all the properly operating sensors in both IMUs.

Due to the FO-FO-FS requirement and the fact there are 4 sensors per IMU, there always exists a set of three sensors in one IMU for reconfiguration purposes. A diagram illustrating the possible schemes for the design equations is presented in Table V.

The selection criterion for the optimal reconfiguration scheme is the minimization of errors due to sensor errors, aircraft flexure, aircraft vibration, etc. Using least squares estimation the results of Table VI can be obtained.

TABLE V. POTENTIAL STRATEGIES IN FLIGHT CONTROL DESIGN EQUATION SELECTION

System Status	Options
No Failure	Least Square Estimation - 4 Sensors, One IMU 3 orthogonal sensors - One IMU Least Square Estimation - 8 Sensors, Both IMUs
One Failure	Least Square Estimation - 4 Sensors, Unfailed IMU 3 orthogonal Sensors - Unfailed IMU Least Square Estimation - Remaining 7 Sensors, Both IMUs
Two Failures At The Same IMU	Least Square Estimation - 4 Sensors, Unfailed IMU 3 orthogonal Sensors - Unfailed IMU Least Square Estimation - Remaining 6 Sensors, Both IMUs
Two Failures At Different IMUs	Orthogonal Sensors - If one of the Failed Sensors is Skewed 3 Non-Orthogonal Sensors - One IMU Least Square Estimation - Remaining 6 Sensors, Both IMUs

TABLE VI. LEAST-SQUARES ANGULAR RATE ERROR VARIANCE (WORST CASE AXIS)

	Sensors from One IMU Used			Sensors From Both IMUs Used
	4 Sensors	3 Orthogonal Sensors	3 Non-Orthogonal Sensors	
No Failure	$\frac{5}{6}\sigma^2$	σ^2	N/A	$\frac{5}{12}\sigma^2$+Flexure
One Failure	$\frac{5}{6}\sigma^2$	σ^2	N/A	$\frac{5}{7}\sigma^2$+Flexure
Two Failures At the Same IMU	$\frac{5}{6}\sigma^2$	σ^2	N/A	$\frac{4}{5}\sigma^2$+Flexure
One Failure In Each IMU	N/A	σ^2 ————OR———→	$5\sigma^2$	$\frac{5}{2}\sigma^2$+Flexure

5. COMPARISON OF NAVIGATION SOLUTIONS FOR FAILURE DETECTION AND ISOLATION

Due to aircraft flexure, misalignment between the IMUs, and lever arm effects, reasonably accurate navigation solutions can only be obtained by employing the instruments within an IMU. The navigation FDI algorithm compares the different navigation solutions resulting from all such combinations of gyros and accelerometers to detect and isolate degraded instruments. These comparisons require that any 2 solutions employ at least one different instrument between them. Since these comparisons incorporate no cognizance of the relative orientations of the sensors involved, except to accommodate natural accuracy variation due for example to geometric dilution of accuracy, the navigation redundancy capability is independent of the relative orientation between the 2 IMUs.

For an IMU with four gyros and four accelerometers, there exists 16 different navigation solutions per IMU which results in 120 possible Intra-IMU comparisons per IMU and 256 Inter-IMU comparisons as shown in figure 4. The implied cost of this theoretically complete system comprised of computing 32 navigation solutions and making 496 navigation comparisons in a real-time computer is significant. In this section the theoretical capability of this complete structure to detect and isolate degraded instrument operation is first defined. A minimized computational structure which obtains the minimal FO-FO-FS theoretical capability of the complete system under worst case failure patterns is then derived. Finally, a strategy which obtains the maximum FO-FO-FO-FO-FS theoretical capability of the complete system under certain failure patterns is determined using only the minimum computational structure.

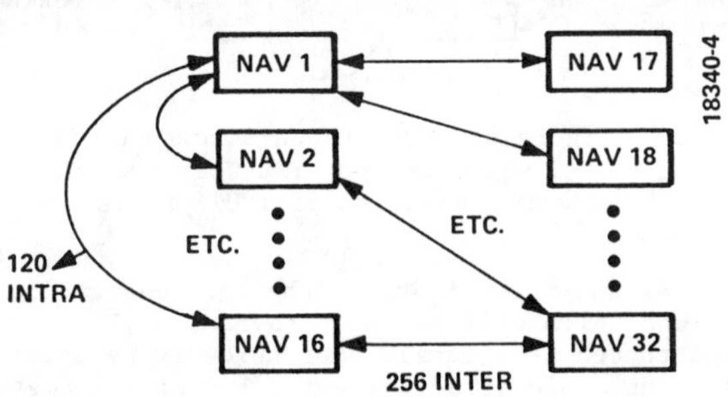

Figure 4. Overall Intra-IMU and Inter-IMU Navigation Solution Comparison for the Complete System

5.1 Theoretical Capability of the Complete System

For the complete system, a pattern of 2 instrument failures of the same type (e.g., 2 gyros) in one IMU disables that IMU. Consequently with a subsequent failure of one of either type of instrument in the second IMU, there will exist 4 valid navigation solutions each dependent on the 3 non-failed instruments of the second IMU. This is an unsafe condition as the failure of one of these instruments will affect the 4 navigation solutions in a similar fashion and hence go undetected. Therefore the minimum capability of the complete system is FO-FO-FS.

The maximum capability of the complete system is FO-FO-FO-FO-FS. This is achieved when each IMU has two failures but the instrument types are different (1 accelerometer and 1 gyro). This failure pattern results in one valid navigation solution from each IMU which can be monitored by a comparison between the 2 IMUs thus realizing the last stage fail-safe condition.

The 16 navigation solutions from each IMU are obtained by pairing each of the 4 possible triads of accelerometer measurements with each of the 4 possible triads of gyro measurements. Using the 16 solutions from each IMU the Intra-IMU comparison of the 16 navigation solutions serves the general purpose of determining if the IMU has any degraded instruments and the type of instrument(s) that has degraded. Isolation of the specific degraded instrument(s) is not possible with the Intra-IMU comparisons. The subsequent Inter-IMU comparison of the 2 sets of 16 navigation solutions isolates up to 1 specific degraded instrument of each type. The Intra-IMU comparisons do serve a purpose not achievable by the Inter-IMU comparisons. This is clear when 2 instruments of the same type degrade in one IMU resulting in disagreement of all Inter-IMU comparisons. Disagreement of the Intra-IMU comparisons in the affected IMU identifies the non-degraded IMU. The general behavior of the complete system is clarified by considering a few failure patterns below.

For the case when one instrument within an IMU degrades (e.g., gyro) the 4 possible triads of rate measurements will disagree. In this case the 4 navigation solutions for each gyro triad will approximately agree as the 4 different triads of accelerometer measurements used in obtaining them are not degraded. The set-wise disagreement of Intra-IMU navigation comparisons determines that there is degradation within the IMU and the type of instrument involved (gyro in this example) but not the specific instrument which has degraded. For a single instrument degradation within an IMU, the Inter-IMU navigation comparisons detect which one of the 4 triads of instruments (gyros in this example) contain the non-degraded instruments. This occurs as only the 4 navigation solutions associated with the non-degraded triad will approximately agree with those from the other IMU.

For the case of 2 or more degraded instruments of the same type within one IMU, the Intra-IMU navigation solution comparisons yield the same results as above with the conclusion that the IMU is degraded along with the type of instrument that is degraded. The Inter-IMU comparisons determine that none of the navigation solutions is valid with the overall conclusion that this IMU can no longer support navigation.

For the case of degradation of one accelerometer and one gyro in one IMU, the Intra-IMU navigation solution comparisons all disagree indicating degradation within the IMU and that the degradation involves both accelerometers and gyros. The Inter-IMU comparisons determine which triad of gyros and which triad of accelerometers is not degraded as these 2 triad sets provide the only accurate navigation solution.

For the case of one instrument degradation in each IMU, 4 navigation solutions from each IMU each involving the non-degraded triad of instruments will approximately agree in the Inter-IMU comparisons and isolate the 2 degraded instruments. For an extension of this situation with a gyro and accelerometer degradation in one IMU and one instrument degradation in the second IMU, the Inter-IMU comparisons determine the one valid navigation solution from the first IMU and the 4 valid navigation solutions from the second IMU. Consequently, the 3 degraded instruments are isolated. The further extension of this case in which both IMUs have one degraded accelerometer and gyro, results in the agreement of only 1 Inter-IMU comparison.

5.2 Minimum Computational Structure

Reduction of the computational burden of 32 navigation solutions and 496 comparisons associated with the complete system above is achieved by minimizing the number of navigation solutions per IMU while retaining its minimum FO-FO-FS capability. Once this minimum computational structure is determined, it is shown that as a function of the failure pattern that the structure can be dynamically reconfigured such that the maximum theoretical capability of the complete system which is FO-FO-FO-FO-FS under certain failure patterns is obtained without additional computation cost.

The minimum computational structure which achieves FO-FO-FS is realized by 4 navigation solutions per IMU. This is demonstrated by showing 3 navigation solutions per IMU is inadequate for FO-FO-FS while 4 solutions is adequate. The proof that 3 navigation solutions is insufficient is given by the example shown in Figures 5 and 6. The selection of 3 solutions will necessarily have at least one sensor axis involved in all 3 solutions, as is sensor number 3 of IMU 1 as shown in Figure 5. If sensor 3 fails or degrades then all the IMU 1 Intra-IMU comparisons and all the Inter-IMU comparisons fail. Since as shown in Figure 6, only Intra-IMU comparisons remain for the detection of a second failure, this system is only FO-FS.

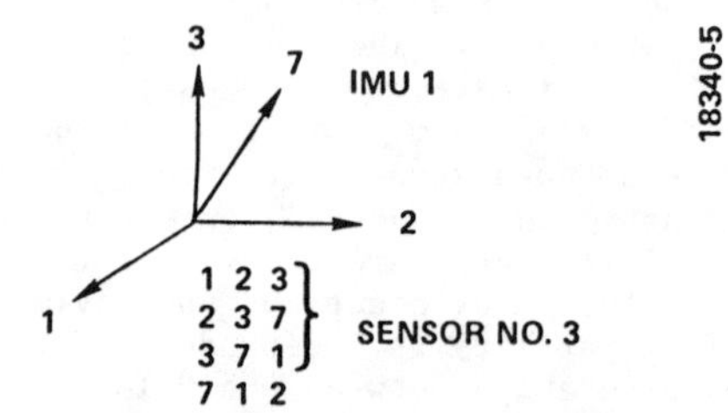

Figure 5. Selected 3 Navigation Solutions (Insufficient)

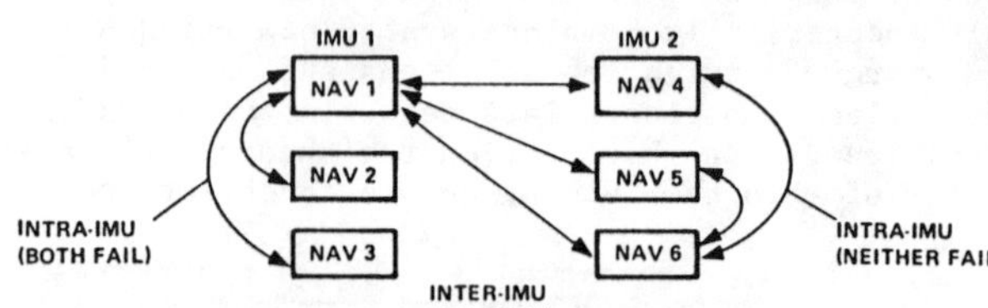

Figure 6. Comparison Structure with 3 Solutions per IMU When the Common Axis (#3 of IMU1) Fails

As shown in Figure 7, the four navigation solutions which will be shown to realize FO-FO-FS are constructed arbitrarily, but uniquely by pairing the 4 possible triads of gyros and accelerometers such that all combinations are represented in at least one navigation solution. This ensures that each instrument appears in at least 3 navigation solutions. Six Intra-IMU comparisons per IMU and

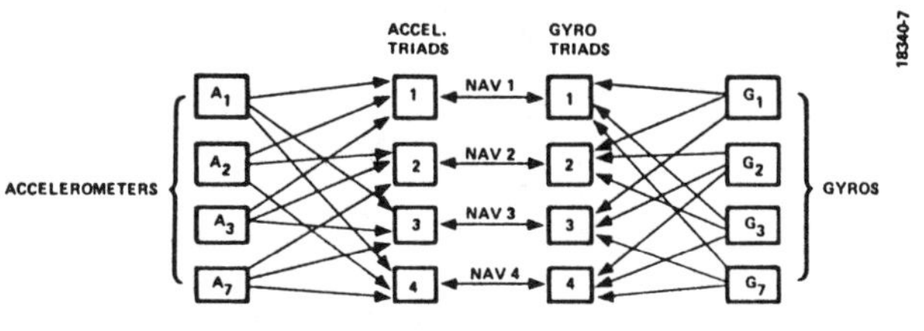

Figure 7. Construction of 4 Navigation Solutions per IMU

16 Inter-IMU comparisons can be obtained with 4 navigation solutions per IMU.

The role of the Intra-IMU comparisons with this minimum computational structure is to detect failure or degradation within an IMU without the capability of isolating the particular instrument or even identifying whether an accelerometer or a gyro has degraded. This is clear by considering the case of failure of any one instrument in the IMU. Since this results in disagreement of all navigation solutions within the IMU, all 6 Intra-IMU comparisons fail and the only knowledge gained is that the IMU has at least one failure among its instruments. This condition is detectable by using only 2 of the Intra-IMU comparisons as long as all 4 navigation solutions are considered. This is for example realized by comparing solutions 1 with 2 and comparing solutions 3 with 4. Failure of any 1 instrument again results in disagreement of these 2 Intra-IMU comparisons and provides the maximum knowledge that at least one failure has occurred among the instruments of the IMU.

The role of the Inter-IMU comparisons is to isolate the triad of gyros and the triad of accelerometers that provide a valid navigation solution when failure occurs. These Inter-IMU comparisons can at best only isolate failure to the level of a gyro-accelerometer pair. They cannot determine the type of instrument that has failed.

The proof that the minimum 4 navigation solution structure is sufficient to achieve a minimum capablity of FO-FO-FS follows by showing for any 2 failures, a navigation solution comparison remains valid, obtaining the FS stage. Two failures occur in only two possible ways:

1. Two in one IMU - In this case there exists Intra-IMU comparisons in the other IMU to detect the third failure.
2. One in each IMU - In this case there exists 1 valid navigation solution for each IMU which results in 1 valid Inter-IMU navigation comparison to detect a third failure.

In summary what has been shown here is that with 4 suitably configured navigation solutions per IMU, the minimum capability of the system is FO-FO-FS. Further this capability is achieved with 4 Intra-IMU comparisons (2 per IMU) and 16 Inter-IMU comparisons. Further with these comparisons, the type of instrument that has failed cannot be determined, so clearly the specific instrument cannot be identified.

5.3 Extended Capability with Re-configuration of The Navigation Solutions

One way to achieve the full capability of the complete system discussed previously with the minimum computational structure, is to re-configure the navigation solutions after specific instrument failures have been isolated. This is possible when the type of instrument that has failed can be identified. From the previous discussion, it is seen that the Intra-IMU comparisons only serve the purpose of detecting failures. These Intra-IMU comparisons can be replaced with 2 Intra-IMU instrument parities per IMU - 1 for the gyros and 1 for the accelerometers. These Intra-IMU instrument parities can detect IMU degradation and also identify the type of instrument failure. Since these Intra-IMU instrument parities do not suffer from the error effects due to aircraft flexure and lever arm as the Inter-IMU instrument parities do, their thresholds can be much lower allowing earlier identification of probable failure conditions.

As noted above the Inter-IMU comparisons can determine the failure of a gyro-accerlerometer pair in the 4 navigation solution structure. Consequently with the Intra-IMU instrument parities, identification of the specific instrument that has failed is possible.

The IDCODE concept introduced in section 4.1 is applicable in the navigation FDI algorithm. In this case, each navigation comparison will have two IDCODEs associated with it - one for the gyros and one for the accelerometers. To illustrate the isolation algorithm, an example is given. Using the navigation solution shown in Figure 7, if gyro #2 of IMU 1 and accelerometer #8 of IMU 2 fail, there is only one navigation solution remaining per IMU which generates the one valid navigation comparison.

$$\text{NAV}\begin{bmatrix} G_1 & G_3 & G_7 \\ A_1 & A_2 & A_3 \end{bmatrix} -\text{NAV}\begin{bmatrix} G_4 & G_6 & G_8 \\ A_4 & A_5 & A_6 \end{bmatrix}$$

The associated IDCODE for the unfailed comparison is given as follows:

For gyros	For accelerometers
(10110111)	(11111100)

The unfailed IDCODE indicates the possible failures G_2, A_7 of IMU 1 and G_5, A_8 of IMU 2. The intra instrument parities will indicate that gyro(s) of IMU 1 and accelerometer(s) of IMU 2 have failed, which leads to the isolation of G_2 of IMU 1 and A_8 of IMU 2.

Once a specific instrument in an IMU has been identified as degraded, it is possible to dynamically reassign the remaining valid instrument outputs such that 4 valid navigation solutions are provided by the IMU. For example, if a gyro degrades, the triad of valid gyros can be used in conjunction with the 4 possible triads of valid accelerometers to yield 4 valid navigation solutions as shown in Figure 8. This re-configuration can be extended such that for the pattern where a gyro and an accelerometer degrade in each IMU, one valid navigation solution will still be available for each IMU (See Figure 9). It is seen here that the capability of the minimum computational structure for this failure pattern is FO-FO-FO-FO-FS.

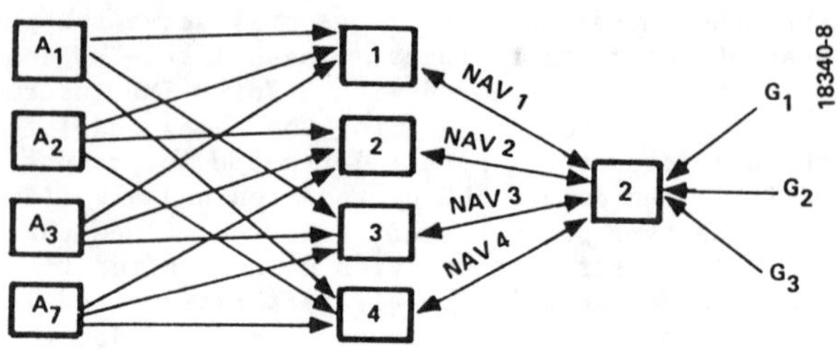

Figure 8. Re-Configuration When a Gyro (G_7) Degrades

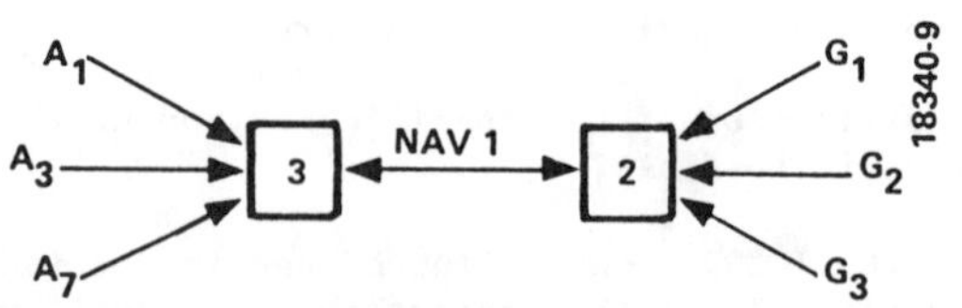

Figure 9. Re-Configuration When a Gyro (G_7) and an Accelerometer (A_2) Degrade

This approach of dynamic re-assignment of instrument outputs to navigation solutions as a function of the failure pattern as its evolution is detected, realizes the maximum capability of the complete system previously discussed. This is achieved with the minimum computational structure using 4 navigation solutions.

5.4 The Design Equations for the Navigation Function

The design equations for the navigation function select an optimal navigation solution from the 8 available solutions based on the knowledge resulting from the FDI algorithm. As shown in Figure 10, the navigation solution selection is affected by the following 3 inputs:

1. Sensor status from the flight control FDI.
2. The IMU status from the Intra-IMU parity checking.
3. The status of the 4 navigation solutions from each IMU.

The sensor status provides information on failed sensors. When Intra-IMU instrument parities have a high threshold, it is likely that IMU degradation will be detected by Intra-IMU navigation comparison. However, it is also possible that due to the slow degradation of the navigation solutions, degradation of the sensors will not be immediately detected by the navigation solution comparisons. In this case, the sensor status from the flight control FDI might provide early knowledge of potential sensor failure if a tight threshold is used for one set of the Intra-IMU instrument parities. The IMU status resulting from such an Intra-IMU parity is useful in operating the navigation FDI processor. The navigation FDI algorithm provides the status of the 4 navigation solutions. The navigation solution selection shown in overview in Figure 10 is based on the following criteria:

1. First use the navigation solution obtained from three orthogonal sensors in the non-degraded IMU.
2. Otherwise use a navigation solution obtained from three sensors in one IMU.

All possible combinations for the parity and comparison status are presented in Figures 11 through 15.

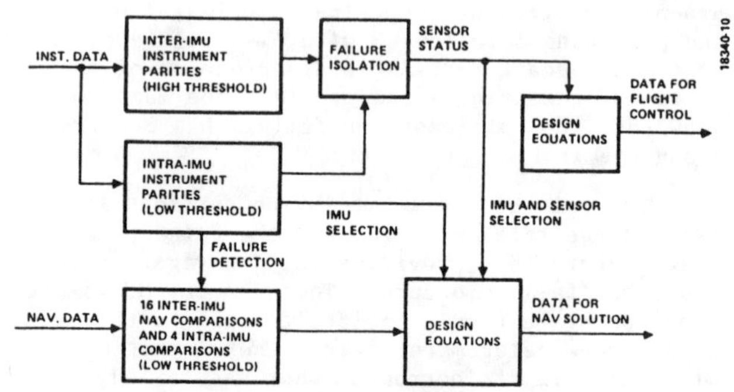

Figure 10. Block Diagram of Sensor Management Algorithm

6. SUMMARY

This paper has described a mechanization to realize a FO-FO-FS integrated flight control reference and navigation system. For the flight control reference a particular type of parity equation called the direct comparison parity, has been discussed. The direct comparison parities have clear physical interpretation and can be constructed easily as they depend directly on the orientation of the sensor axes. It is shown with the example of two parallel quad channel IMUs that FO-FO-FS condition is achieved. The potential strategies for constructing the flight control system reference inputs as a function of failure status has been presented with the associated least square error variances. An ID code concept which simplifies the failed instrument isolation problem has been discussed.

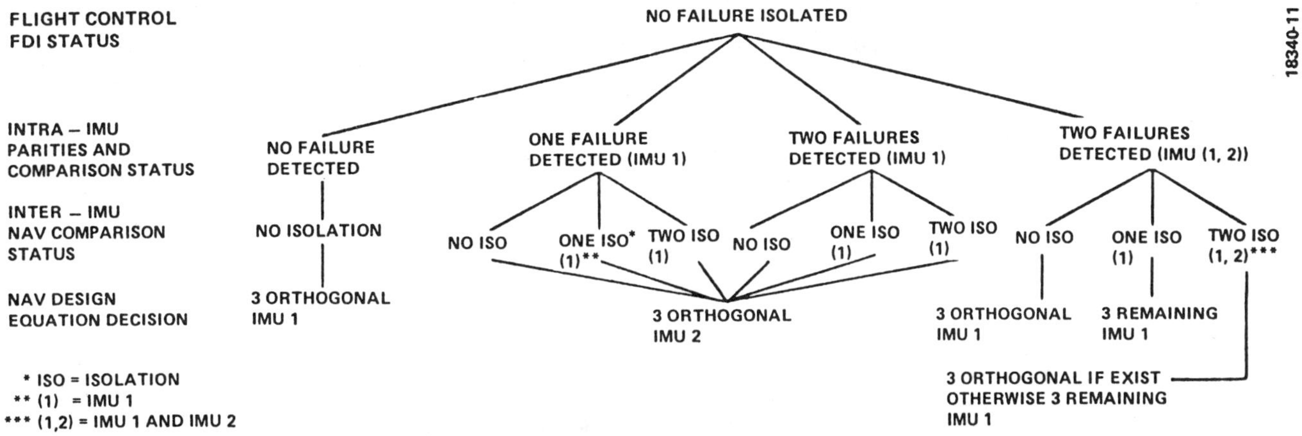

Figure 11. Case of No Failure Isolation from Flight Control FDI

For the navigation function the theoretical capability of what is called the complete system has been discussed. Also a minimum computational structure employing 4 navigation solutions which achieves FO-FO-FS has been derived. Finally using Intra-IMU instrument parities it was shown that the minimum computational structure obtains the maximum capability of the complete system which is FO-FO-FO-FO-FS for some failure patterns.

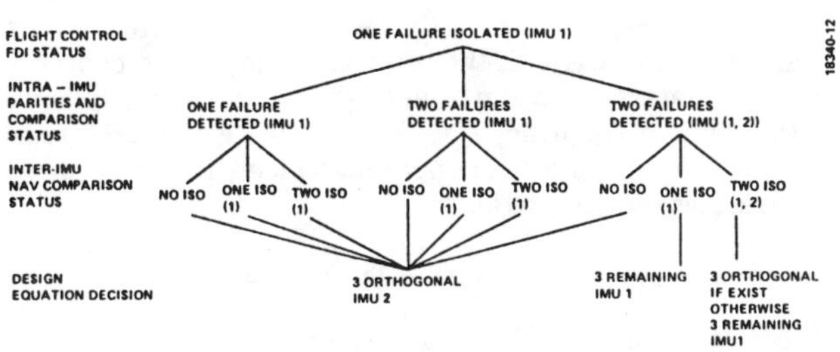

Figure 12. Case of One Failure Isolation from Flight Control FDI

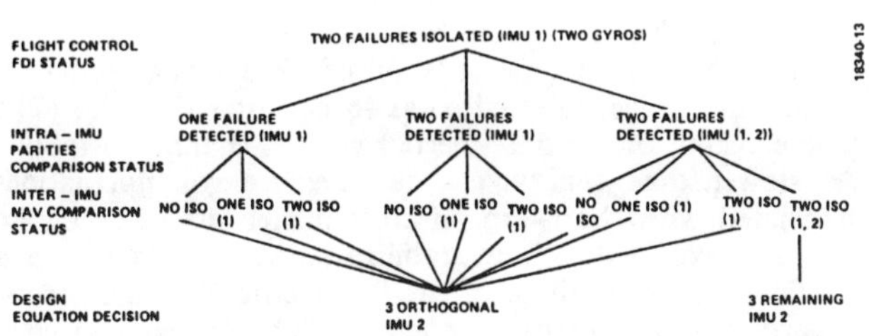

Figure 13. Case of Two Failure Isolations from Flight Control FDI (one IMU)

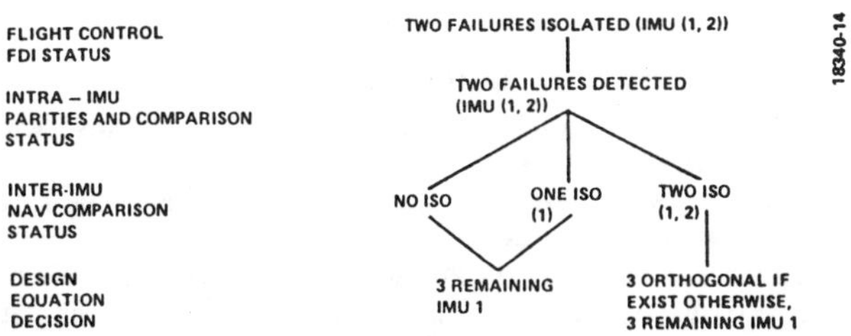

Figure 14. Case of Two Failure Isolations from Flight Control FDI (Both IMUs)

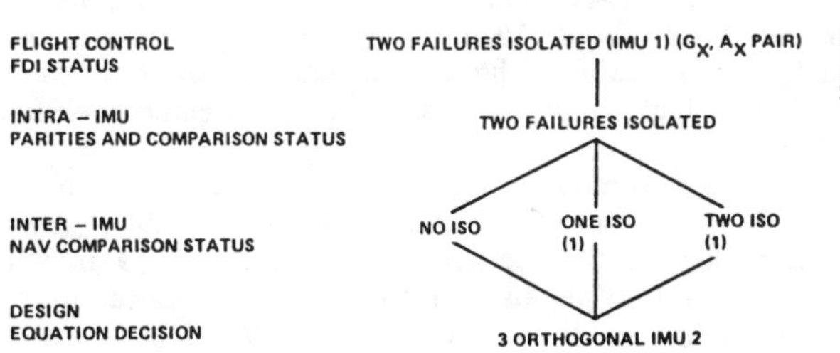

Figure 15. Case of Two Failure Isolations (Gyro - Accelerometer pair) from Flight Control FDI

Optimizing the Gains of the Baro-Inertial Vertical Channel

William S. Widnall*
Massachusetts Institute of Technology, Cambridge, Mass.
and
Prasun K. Sinha†
Intermetrics, Inc., Cambridge, Mass.

The selection of the three gains in the baro-inertial vertical channel has been formulated as a stochastic optimal control problem, where the objective is to minimize the mean-square error of the indicated vertical velocity. The optimal set of gains is surprisingly different from a conventional set of gains, and it provides a significant performance improvement. Sensitivity of the results to the statistical assumptions is explored. Approximate analytical formulas are presented giving the optimal gains and pole locations as a function of the assumed statistics of the sources of error. A time domain simulation also exhibits the performance improvement.

Introduction

THE first aircraft inertial navigation systems were primarily two-channel systems that provided horizontal navigation data.[1-3] Inertial navigators instrumenting three channels were introduced for missile navigation and guidance. In addition, the value of inertially derived vertical velocity was recognized in aircraft applications involving flight path angle determination and precision weapon delivery. It is well known that the altitude channel of a pure inertial navigation mechanization, in which gravity magnitude is computed as a function of the indicated altitude, is unstable.[4-6] Near the surface of the Earth, the time constant of this exponential instability is about 10 min. Hence, for typical cruise navigation durations, the vertical channel of a terrestrial inertial navigator must be stabilized by some external altitude reference.

The most commonly used external altitude reference in aircraft is a barometric altimeter. The optimal time-varying combination of the inertial and barometric data may be obtained using a Kalman filter.[7] However, in applications not demanding the minimum navigation errors or in which the computer capacity is severely limited, a simple mechanization is commonly used in which the difference between the indicated and barometric altitude is fed back through constant gains or simple transfer functions to stabilize the altitude navigation variables.[8] A typical set of constant-gain baro-inertial mechanization equations, which is analyzed in detail in this paper, is

$$\dot{h} = v_z - k_1 (h - h_b) \tag{1a}$$

$$\dot{v}_z = f_z - g(h,L) + \text{Coriolis terms} - k_2 (h - h_b) - \hat{\delta a} \tag{1b}$$

$$\dot{\hat{\delta a}} = k_3 (h - h_b) \tag{1c}$$

where h is the indicated altitude, v_z is the indicated vertical velocity, $\hat{\delta a}$ is the computed vertical acceleration error, f_z is the measured vertical specific force, g is the magnitude of gravity computed as a function of indicated altitude and latitude, and k_1, k_2, k_3 are the loop gains. This third-order vertical channel mechanization is superior to a second-order mechanization, which omits the $\hat{\delta a}$ equation, because it has zero vertical velocity error due to any bias vertical acceleration error such as accelerometer bias or gravity computation error.

The Litton CAINS (Carrier Aircraft Inertial Navigation System) implements such a third-order baro-inertial vertical channel. In the CAINS, the three gains have been chosen so that the characteristic equation of the errors has a triple pole at the complex frequency $s = -1/\tau$, where τ is the desired time constant. For such a triple pole placement, it can be shown one chooses the gains to be

$$k_1 = 3/\tau \qquad k_2 = 3/\tau^2 + 2g/R \qquad k_3 = 1/\tau^3 \tag{2}$$

where R is the geocentric radius. In the CAINS, the time constant has been chosen to be $\tau = 100$ s. We have no literature explaining the designer's choice of the triple pole and its location. Perhaps the triple pole configuration was arbitrarily selected to reduce the gain-setting problem from three parameters (k_1, k_2, k_3) to one parameter (τ). Speculating further, perhaps the time constant of 100 s was an order-of-magnitude choice, selected so as to be faster than the 571 s time constant of the pure-inertial vertical-channel instability yet slower than the typical barometric error fluctuations associated with short-term aircraft maneuvers. This choice would be expected to both stabilize the vertical channel and provide some smoothing of the barometric altimeter errors. Regardless of the reasoning, the CAINS has peformed well in its intended applications. We shall refer to the CAINS set of gains, given by Eq. (2) with $\tau = 100$ s, as the baseline set.

Some applications have more demanding vertical velocity requirements than the CAINS requirements. In such applications it may be necessary to optimize the vertical channel gains to reduce the vertical velocity errors. One such application was the use of the Magnavox X-set GPS navigator in the demonstration of pinpoint bombing on a target whose absolute coordinates were known. The X-set GPS navigator includes a barometric-inertial navigation subsystem and a GPS X-set receiver whose outputs are combined by a Kalman filter. An error analysis by Ausman[9] predicted that the two largest contributions to bomb miss distance would be due to

Received June 26, 1978; presented as Paper 78-1307 at the AIAA Guidance and Control Conference, Palo Alto, Calif., Aug. 7-9, 1978; revision received July 3, 1979.

Index category: Guidance and Control.

*Associate Professor, Dept. of Aeronautics and Astronautics. Associate Fellow AIAA.

†Senior Engineer, Navigation and Analysis Dept.

Reprinted with permission from *AIAA J. Guidance and Control,* vol. 3, no. 2, pp. 172–178, Mar.–Apr. 1980.

the altitude navigation error and the vertical velocity navigation error at bomb release. Flying at 450 knots, at 5000-ft altitude above ground level, with a level release, and using low-drag bombs, Ausman indicated that the anticipated altitude and vertical velocity navigation errors of 19 m and 0.75 m/s would contribute 24 m and 17 m, respectively, to downrange bomb miss distance. The absolute altitude error of the integrated navigation system is caused by the bias-like errors in the GPS measurements. The choice of baro-inertial gains has little or no effect on the absolute altitude accuracy. However, the absolute vertical velocity errors are a noticeable function of both the GPS measurement errors and the baro-inertial errors. The integrated system vertical velocity errors can be reduced if first the baro-inertial vertical velocity errors are minimized.

There is no reason to assume that the triple pole gain set provides the best performance. Other pole placements might provide superior performance. To obtain new insights into the effect of the gains, we have formulated the vertical channel gain setting problem as a parameter optimization problem for the control of a stationary stochastic process. This paper presents the optimization problem formulation, computational results, analysis of the results, and a time-domain simulation of the performance with the recommended gains.

Before proceeding, we comment that any constant gain set (including our optimized set) will be less optimal than appropriate time-varying gains that take into account the nonstationary nature of the inertial and barometric altimeter errors. The optimal time-varying combination of the inertial and barometric data may be obtained using a Kalman filter. Of course it is also possible to select effective time-varying gains with non-Kalman approaches. One example of the use of time-varying gains is provided by Whalley.[10] Whalley points out the large error in barometric altitude when a supersonic aircraft passes through Mach 1 due to shock waves moving past the static pressure port. Whalley suggests eliminating this source of error by switching out the air data automatically from say Mach 0.95 to Mach 1.1. This could be mechanized by programming k_1, k_2, k_3 to be zero in this Mach interval.

Another example of the use of time-varying gains is provided by Ausman and associates.[11,12] They note that in subsonic flight the largest source of barometric altimeter error is often the scale factor error due to the atmosphere not having the standard-day temperature-vs-altitude profile. In climbs and dives the scale factor error induces significant vertical velocity error into a constant gain baro-inertial vertical channel. Ausman and associates designed and implemented vertical channel mechanizations that reduce the gains during climbs and dives, while observing the baro-bias shift due to scale factor error. The estimated baro-bias shift is automatically subtracted from the baro-inertial error feedback so that loop transients due to the scale factor error are minimized.

Also before proceeding, we comment that additional external data may be useful in reducing the effect of nonstandard-day temperature. Blanchard[13] proposes using inflight measured temperature data, in place of the standard-day lapse-rate assumption, to relate more accurately pressure changes to altitude changes.

Formulation of the Gain-Optimization Problem

The error model for the vertical channel is shown in Fig. 1. The positive feedback with gain $2g/R$ is the destabilizing effect of normal gravity being calculated at the closed-loop altitude, which is in error by δh. The error state δa is a random walk modeling any bias or slowly varying error in the vertical acceleration due to accelerometer bias, gravity anomaly, or error in the Coriolis terms. The white noise w_{a2} into the integration provides the random walk. The white noise w_{a1} models short correlation time acceleration error, such as the effect during a short maneuver of vertical accelerometer scale

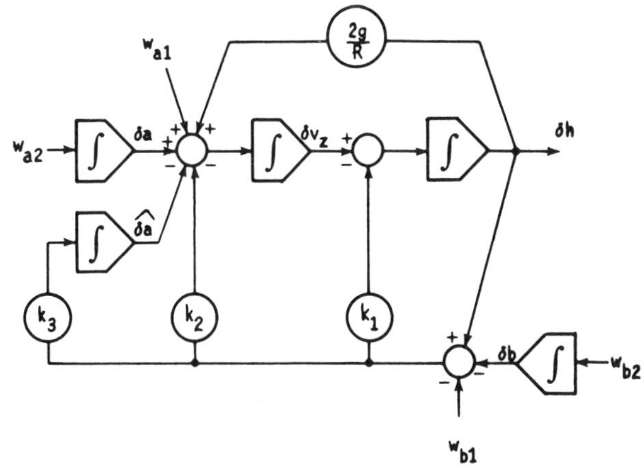

Fig. 1 Baro-inertial vertical channel error model.

factor error and input axis misalignment. The error state δb is a random walk modeling any bias or slowly varying error in the altitude indicated by the barometric altimeter. Physical error sources include zero setting error, static pressure measurement error, variation in the height of a constant pressure surface, and scale factor error due to nonstandard day temperature. The white noise w_{b2} into the integrator provides the random walk. The white noise w_{b1} models short correlation time altimeter error, such as due to changes in the angle of attack or sideslip angle during a maneuver, or due to altimeter quantization or other noise.

It is difficult to suggest appropriate values for the spectral densities of the four independent white noises in this stochastic model. Nevertheless, it must be done for the analysis to proceed. Table 1 shows the nominal values of the noise spectral densities that have been selected. These somewhat arbitrary numerical values have been arrived at by the following considerations.

For the short correlation time acceleration error, a typical amplitude could be 200 μg. This error could be caused by a vertical accelerometer input axis misalignment of 200 μ rad (41 arc-sec) together with a horizontal maneuver acceleration of one g. The typical duration of a horizontal maneuver is assumed to be of the order of 60 s. For a repeated series of random aircraft maneuvers, the autocorrelation function of the acceleration error would have area approximately

$$Q_{a1} = (200 \times 10^{-6} \times 10\,\mathrm{ms}^{-2})^2 \times (60\mathrm{s}) = 2.4 \times 10^{-4}\,\mathrm{m}^2\mathrm{s}^{-3} \quad (3)$$

The area of the autocorrelation function equals the low-frequency value of the spectral density. For the white noise, whose autocorrelation function is a Dirac delta function with area Q_{a1}, the spectral density Q_{a1} applies at all frequencies. Of interest is the response of the vertical channel at frequencies lower than the higher frequencies of the short-correlation acceleration error. So the low-frequency density of Eq. (3) is used for the spectral density of the white noise.

Table 1 Nominal values of noise spectral densities

White noise for	Noise density symbol	Noise density value
Short correlation time acceleration error	Q_{a1}	$2.4 \times 10^{-4}\,\mathrm{m}^2\,\mathrm{s}^{-3}$
Acceleration error random walk	Q_{a2}	$1.0 \times 10^{-9}\,\mathrm{m}^2\,\mathrm{s}^{-5}$
Short correlation time altimeter error	Q_{b1}	$100\,\mathrm{m}^2\,\mathrm{s}$
Altimeter error random walk	Q_{b2}	$100\,\mathrm{m}^2\,\mathrm{s}^{-1}$

The acceleration error random walk models the slowly varying error due to accelerometer bias shifts, changes in the gravity anomaly, and changes in the error in the Coriolis terms. If over a period of 1000 s, the accelerometer bias were expected to shift 100 μg, the appropriate noise spectral density for the random walk is

$$Q_{a2} = (100 \times 10^{-6} \times 10\,\mathrm{ms}^{-2})^2/(1000\,\mathrm{s}) = 1.0 \times 10^{-9}\,\mathrm{m}^2\mathrm{s}^{-5} \tag{4}$$

For the short correlation time altimeter error, it is assumed that repeated random fluctuations of the order of 10 m may be present in the baro-indicated altitude, and that these errors persist for correlation times of the order of 1 s. To match the low-frequency spectral density of this error, the white noise error model should have density

$$Q_{b1} = (10\,\mathrm{m})^2 \times (1\,\mathrm{s}) = 100\,\mathrm{m}^2\mathrm{s} \tag{5}$$

The altimeter error random walk models the slowly varying error due to: changes in the static pressure measurement error (due to speed changes), variations in the height of a constant pressure surface (the weather pattern of highs and lows), and scale factor error (related to nonstandard-day temperature and nonzero aircraft climb or descent rate). For an atmospheric scale factor error of 3% and an aircraft climb or descent rate of 33 m/s (6500 ft/min), the error rate of the baro-indicated altitude is 1.0 m/s. If the climb or descent continues for 100 s, the change in altimeter error will be 100 m. Assume that the aircraft trajectory is characterized by a random sequence of such climbs and descents. The appropriate noise spectral density for the random walk is then

$$Q_{b2} = (100\,\mathrm{m})^2/(100\,\mathrm{s}) = 100\,\mathrm{m}^2\mathrm{s}^{-1} \tag{6}$$

The mean-squared error of the indicated vertical velocity has been selected as an appropriate performance index. Note that with the random walk error models for acceleration error and for altimeter error, only the vertical velocity error has a stationary and finite mean-square value. All other error states have mean-square values that grow unbounded with time. Referring to Fig 1, $\hat{\delta a}$ tracks $\delta a + (2g/R)\delta h$ as this sum wanders off, and δh tracks δb as it wanders off.

The mean-squared vertical velocity error may be computed as an explicit function of the input noise spectral densities and of the loop gains. One first calculates four transfer functions $H_i(s)$ relating the vertical velocity error response to each independent white noise input. The power spectrum of the vertical velocity error is then

$$\Phi_{\delta v}(\omega) = \sum_{i=1}^{4} H_i(j\omega) H_i(-j\omega) Q_i \tag{7}$$

where Q_i is the spectral density of the ith white noise. The mean-square value of the vertical velocity error is the integral of the power spectrum

$$\overline{(\delta v)^2} = \sum_{i=1}^{4} \frac{Q_i}{2\pi j} \int_{-j\infty}^{j\infty} H_i(s) H_i(-s)\,\mathrm{d}s \tag{8}$$

The four integrals are evaluated using an appropriate table of integrals.[14] The result is

$$\overline{(\delta v)^2} = \frac{(k_1^2 + k_2 - c) Q_{a1}}{2[k_1(k_2 - c) - k_3]} + \frac{(k_3 + k_1^3) Q_{a2}}{2k_3[k_1(k_2 - c) - k_3]} + \frac{[k_1^2(k_2 - c) + (k_3 + ck_1)^2] Q_{b1}}{2[k_1(k_2 - c) - k_3]} + \frac{[k_1^2 k_3 + (k_3 + ck_1)^2 k_1] Q_{b2}}{2k_3[k_1(k_2 - c) - k_3]} \tag{9}$$

where near the surface of the Earth $c = 2g/R = 3.07 \times 10^{-6}\,\mathrm{s}^{-2}$. For the steady-state solution to exist and be equal to Eq. (9), the set of loop gains must yield a stable system. Therefore, the permissible values of the gains are in the regions defined by

$$k_1 > 0, \quad k_2 - c > 0, \quad k_3 > 0, \quad k_1(k_2 - c) - k_3 > 0 \tag{10}$$

The explicit computation of the mean-squared vertical velocity error is used in a computer program that seeks a set of gain values that minimizes the mean-squared error. The pattern search algorithm of Hooke and Jeeves[15] has been utilized. The algorithm does not require explicit gradient information.

The natural frequencies (poles) of the closed-loop portion of the baro-inertial vertical channel are the three roots of the characteristic equation

$$s^3 + k_1 s^2 + (k_2 - c)s + k_3 = 0 \tag{11}$$

For a candidate set of gains, it is often of interest to inspect the locations of the three poles p_1, p_2, p_3. In such a case, the roots of the cubic Eq. (11) are computed according to the known formulas for those roots. When the time constant of a pole is mentioned, it is defined to be the inverse of the real part of the complex frequency of the pole.

Optimization Results

To provide a baseline design and performance against which to compare the optimized performance, the mean-squared velocity error is evaluated for the set of gains, Eq. (2), which place a triple pole at $\tau = 100$ s

$$k_1 = 3.0 \times 10^{-2}\,\mathrm{s}^{-1} \qquad k_2 = 3.0307 \times 10^{-4}\,\mathrm{s}^{-2}$$
$$k_3 = 1.0 \times 10^{-6}\,\mathrm{s}^{-3} \tag{12}$$

The mean-squared velocity error, with the nominal (Table 1) values of noise spectral densities, is found to be

$$\overline{(\delta v)^2} = 0.818\,\mathrm{m}^2\mathrm{s}^{-2} = (0.904\,\mathrm{m/s})^2 \tag{13}$$

Using the Hooke and Jeeves pattern search procedure, the gains that minimize the mean-square velocity error, with the nominal noise densities, are found to be

$$k_1 = 1.003\,\mathrm{s}^{-1} \qquad k_2 = 4.17 \times 10^{-3}\,\mathrm{s}^{-2}$$
$$k_3 = 4.39 \times 10^{-6}\,\mathrm{s}^{-3} \tag{14}$$

The corresponding value of the mean-square velocity error is

$$\overline{(\delta v)^2} = 0.418\,\mathrm{m}^2\,\mathrm{s}^{-2} = (0.647\,\mathrm{m/s})^2 \tag{15}$$

This is a significant performance improvement relative to the baseline case. The rms velocity error is 30% lower.

The three poles associated with the gain set, Eq. (14), are located at

$$p_1 = -0.998\,\mathrm{s}^{-1} \quad p_2, p_3 = -2.082 \times 10^{-3} \pm j\,2.34 \times 10^{-4}\,\mathrm{s}^{-1} \tag{16}$$

They have time constants of

$$\tau_1 = 1.002\,\mathrm{s} \qquad \tau_2, \tau_3 = 480.3\,\mathrm{s} \tag{17}$$

The optimized gains and resulting pole placements (and time constants) are radically different from the baseline triple pole set. One time constant is a factor of 100 faster; the other two time constants are a factor of 5 slower.

Table 2 Contributions to mean-square velocity error of nominal noise densities

Noise density	Mean-square velocity error, $(m/s)^2$ Triple pole, $\tau = 100$ s	Optimized set
Q_{al}	0.018	0.0291
Q_{a2}	0.00175	0.0275
Q_{bl}	0.00018	0.00087
Q_{b2}	0.798	0.361
Total	$0.818 = (0.904)^2$	$0.418 = (0.647)^2$

The individual contributions of the various white noises to the mean-square velocity error are shown in Table 2, for the nominal values of the white noise spectral densities. The data of this table show that the mean-square velocity error is dominated by the altimeter error random walk (Q_{b2}), while the contribution of short correlation time altimeter error (Q_{bl}) is least.

To obtain further insight into the nature of the optimal solution and to exhibit the sensitivity of the optimal solution to the noise density assumptions, the optimal solution has been computed for various values of the noise spectral densities. Table 3 shows the results for four cases in which one of the noise densities is increased while holding the other three densities at their nominal (Table 1) values.

With the optimized gains being so different from the baseline gains, it is interesting to ask: for what set of input noise densities are the "triple pole" gains optimal? From Table 2, one notes that in the "triple pole" case, the contributions of the altimeter error noises to the mean-square velocity error would be more nearly equal if Q_{bl} were 10 times larger and Q_{b2} were 100 times smaller. The optimization program has been rerun with these altered values for noise density. The results are presented as the last case in Table 3. These results demonstrate that the baseline triple pole set is close to being an optimal set if the random walk component of the altimeter error is significantly smaller and if the short correlation time altimeter error is somewhat larger than the nominal assumed values.

Analysis of Results

In all cases presented in the previous section (except the greatly reduced Q_{b2} case) the dynamics of the optimal third-order vertical channel are that of a fast first-order loop nested inside a slower second-order system. From Table 3 it is clear that the fast pole frequency is simply related to the first gain

$$p_1 \approx -k_1 \tag{18}$$

With such a fast real pole, the characteristic equation of the third-order vertical channel can be factored as

$$(s+k_1)\left(s^2+\frac{k_2'}{k_1}s+\frac{k_3}{k_1}\right) \approx 0 \tag{19}$$

where

$$k_2' = k_2 - c \tag{20}$$

This can be shown by multiplying the two factors in Eq. (19) to obtain

$$s^3+\left(k_1+\frac{k_2'}{k_1}\right)s^2+\left(k_2'+\frac{k_3}{k_1}\right)s+k_3 \approx 0 \tag{21}$$

The correct characteristic equation of the third-order vertical channel, Eq. (11), in terms of k_2' is

$$s^3+k_1s^2+k_2's+k_3=0 \tag{22}$$

Comparing Eqs. (21) and (22), it is clear that sufficient conditions for the factorization of Eq. (19) to be true are

$$k_2' \ll k_1^2 \tag{23}$$

$$k_3 \ll k_1 k_2' \tag{24}$$

Inspecting the second-order factor of Eq. (19), one sees that the two slower poles are a function only of the two gain ratios k_2'/k_1 and k_3/k_1. The pole locations are

$$p_2, p_3 \approx \frac{-k_2'}{2k_1} \pm \sqrt{\left(\frac{k_2'}{2k_1}\right)^2 - \frac{k_3}{k_1}} \tag{25}$$

Equation (25) gives good agreement with the values of poles and gains in the computed results (Table 3).

The optimized gain k_1 in the nominal noise density case has a very curious numerical value of unity. This provides a clue that there exists a very simple relationship between this gain

Table 3 Changes in optimal solution with altered noise densities

	Optimal gains: k_1, s^{-1}, k_2, s^{-2}, k_3, s^{-3}	Pole locations, s^{-1}	Time constants, s
Nominal case (Table 1 values)	1.003 4.17×10^{-3} 4.39×10^{-6}	-0.998 -2.082×10^{-3} $\pm j2.34 \times 10^{-4}$	1.002 480.3 480.3
Increased Q_{bl} (5 nominal)	0.449 1.86×10^{-3} 1.96×10^{-6}	-0.445 -2.084×10^{-3} $\pm j2.6 \times 10^{-4}$	2.2 479.9 479.9
Increased Q_{b2} (5 nominal)	2.21 8.07×10^{-3} 7.46×10^{-6}	-2.20 -1.83×10^{-3} $\pm j2.6 \times 10^{-4}$	0.45 546 546
Increased Q_{al} (33 nominal)	1.018 9.84×10^{-3} 4.44×10^{-6}	-1.008 -9.26×10^{-3} -4.91×10^{-4}	0.99 108 2037
Increased Q_{a2} (33 nominal)	1.010 6.74×10^{-3} 1.82×10^{-5}	-1.003 -3.33×10^{-3} $\pm j2.63 \times 10^{-3}$	0.997 300 300
Reduced Q_{b2} (0.01 nominal) Increased Q_{bl} (10 nominal)	4.66×10^{-2} 5.86×10^{-4} 1.004×10^{-6}	-2.90×10^{-2} -1.05×10^{-2} -6.78×10^{-3}	34 95 147

and the assumed noise densities, two of which have equal values. A dimensionally correct expression that also gives the right numerical value is discovered to be

$$k_1 \approx \sqrt{Q_{b2}/Q_{b1}} \tag{26}$$

This formula is in excellent agreement with the numerical results in Table 3 (except for the greatly reduced Q_{b2} case). The formula appears valid under the same conditions that give rise to the nested fast loop. A remarkable conclusion is that the first gain is optimized based only on the relative strengths of the two noise densities in the assumed altimeter error model.

One may derive additional useful formulas for the other gains as a function of the noise densities as follows. The explicit formula for the mean-square velocity error J as a function of the noise densities was given in Eq. (9). Assume that the nested fast loop conditions Eqs. (23) and (24) apply. Also assume that $k_2 \gg c$. Note that the numerical results indicated that the contribution of the term proportional to Q_{b1} was negligible. Assume that the gradient of this term with respect to k_2 and k_3 is also negligible. Delete this term from the analysis. An approximate formula for the cost (mean square velocity error) is then

$$2J \approx \frac{k_1}{k_2} Q_{a1} + \frac{k_1^2}{k_2 k_3} Q_{a2} + \left[\frac{k_2}{k_1} + \frac{k_3}{k_2} + \frac{2ck_1}{k_2} + \frac{c^2 k_1^2}{k_2 k_3}\right] Q_{b2} \tag{27}$$

Necessary conditions for a set of gains to be optimal include

$$\partial 2J/\partial k_2 = 0 \tag{28}$$

$$\partial 2J/\partial k_3 = 0 \tag{29}$$

Using Eq. (27) in Eq. (29) one obtains

$$k_3/k_1 \approx \sqrt{c^2 + Q_{a2}/Q_{b2}} \tag{30}$$

Similarly, using Eq. (27) in Eq. (28) and applying Eq. (30) one obtains

$$k_2/k_1 \approx [2c + Q_{a1}/Q_{b2} + 2\sqrt{c^2 + Q_{a2}/Q_{b2}}]^{1/2} \tag{31}$$

The preceding formulas for the gain ratios are in excellent agreement with the numerical results in Table 3, for all cases that have the nested fast loop.

An approximate formula for the location of the slower poles as a function of the noise densities may be derived by assuming $k_2' \approx k_2$ and by using Eqs. (30) and (31) in Eq. (25).

$$p_2, p_3 \approx -\tfrac{1}{2}[2c + Q_{a1}/Q_{b2} + 2\sqrt{c^2 + Q_{a2}/Q_{b2}}]^{1/2}$$
$$\pm \tfrac{1}{2}[2c + Q_{a1}/Q_{b2} - 2\sqrt{c^2 + Q_{a2}/Q_{b2}}]^{1/2} \tag{32}$$

This formula also is in excellent agreement with the numerical results.

An interesting observation supported by the computer results is that when the nested fast loop is optimal, the optimal second and third gain ratios (k_2/k_1 and k_3/k_1) as well as the optimal second and third pole locations are not a function of the assumed density of the short-correlation-time altimeter error. The computer results showed a factor of 5 increase in Q_{b1} producing a shift in the optimal time constant of less than 0.1%.

When the strength of the altimeter error random walk Q_{b2} is sufficiently large relative to the strength of the other sources of error, that is when

$$Q_{a1}/Q_{b2} \ll 2c \tag{33}$$

$$Q_{a2}/Q_{b2} \ll c^2 \tag{34}$$

then the optimal gain ratios and their associated pole locations and time constants are simply

$$k_2/k_1 \approx 2\sqrt{c} = 3.5 \times 10^{-3} s^{-1} \tag{35}$$

Table 4 Error source initial values and statistics

Random walks	$\dot{x} = w$	
Error source	Initial value	Noise spectral density N
X, Y (level) gyro drift rates	0.003 deg/h	$(0.003 \text{ deg/h})^2$/h
Z (azimuth) gyro drift rate	−0.295 deg/h	$(0.005 \text{ deg/h})^2$/h
X, Y (horizontal) accelerometer biases	50 μg	$(10\ \mu g)^2$/h
Z (altitude) accelerometer bias	100 μg	$(10\ \mu g)^2$/h
First-order Markov processes	$\dot{x} = -\beta x + w$	$N_w = 2\beta\sigma^2$
Error source	Initial and 1σ value	Inverse correlation time, β
Barometric altimeter time-varying error	500 ft	v/(250 n. mi.)
East deflection of gravity	26 μg	v/(10 n. mi.)
North deflection of gravity	17 μg	v/(10 n.mi.)
Gravity anomaly	35 μg	v/(60 n. mi.)
Random constants	$\dot{x} = 0$	
Error source		Initial value
G-sensitive gyro drift coefficients		0.3 deg/h/g
G^2-sensitive gyro drift coefficients		0.04 deg/h/g^2
X, Y gyro scale factor errors		300 ppm
Z gyro scale factor error		1000 ppm
Gyro input axis misalignments		±40 arc-sec
Accelerometer scale factor errors		150 ppm
X, Y accelerometer input axis misalignment about Z		±180 arc-sec
Other accelerometer input axis misalignments		±30 arc-sec
Barometric altimeter scale factor error		0.03
Coefficient of static pressure measurement error		1.54×10^{-4} ft/(ft/sec)^2
Barometric time delay		0.25 s

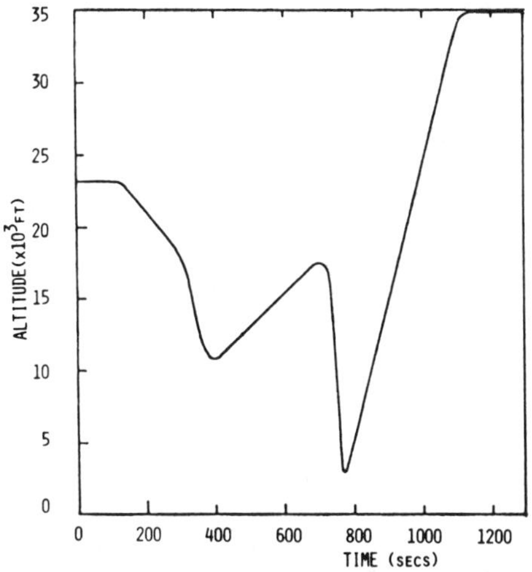

Fig. 2 Trajectory altitude history.

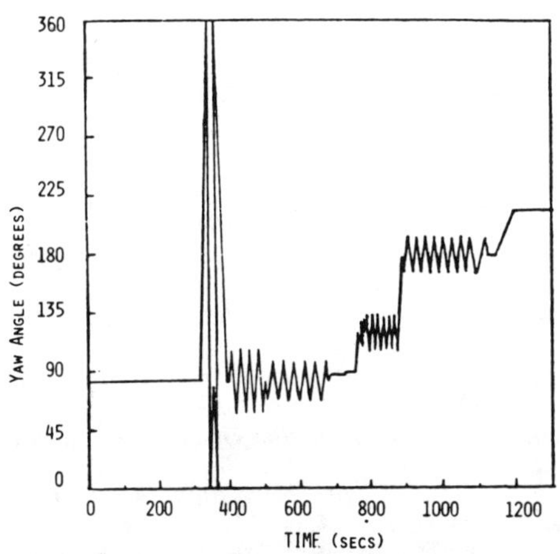

Fig. 3 Trajectory heading history.

$$k_3/k_1 \approx c = 3.07\times 10^{-6}\text{s}^{-2} \tag{36}$$

$$p_2, p_3 \approx -\sqrt{c} = -1.75\times 10^{-3}\text{s}^{-1} \tag{37}$$

$$\tau_2, \tau_3 \approx 1/\sqrt{c} = 571\ \text{s} \tag{38}$$

These limiting results are a function only of the destabilizing gravity gradient c. Note that the nominal case, the increased Q_{b1} case, and especially the increased Q_{b2} case have computed results (Table 3) approaching this limiting case. An important conclusion is that even if the measured specific force is perfect (zero Q_{a1} and Q_{a2}), the feedback gain ratios k_2/k_1 and k_3/k_1 must be maintained at certain nonzero values to stabilize the vertical channel and to minimize the effect of the gravity computation error. These required values correspond to an upper limit on the optimal double pole time constant of 571 s.

When the strength of the acceleration short correlation error Q_{a1} is important in the sense that

$$2c + Q_{a1}/Q_{b2} \approx Q_{a1}/Q_{b2} \gg 2\sqrt{c^2 + Q_{a2}/Q_{b2}} \tag{39}$$

then the optimal gain ratio k_2/k_1 simplifies to

$$k_2/k_1 \approx \sqrt{Q_{a1}/Q_{b2}} \tag{40}$$

and Eq. (32) yields two real poles at

$$p_2 \approx -\sqrt{Q_{a1}/Q_{b2}} \tag{41}$$

$$p_3 \approx -\sqrt{(Q_{b2}/Q_{a1})c^2 + Q_{a2}/Q_{a1}} \tag{42}$$

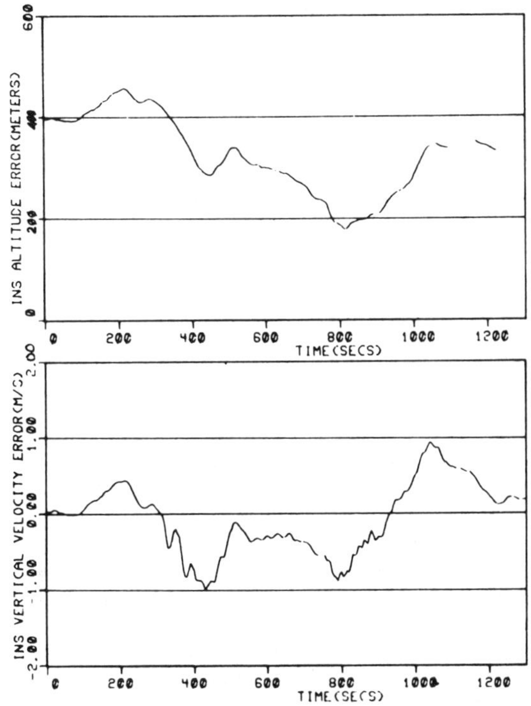

Fig. 4 Performance with baseline gains.

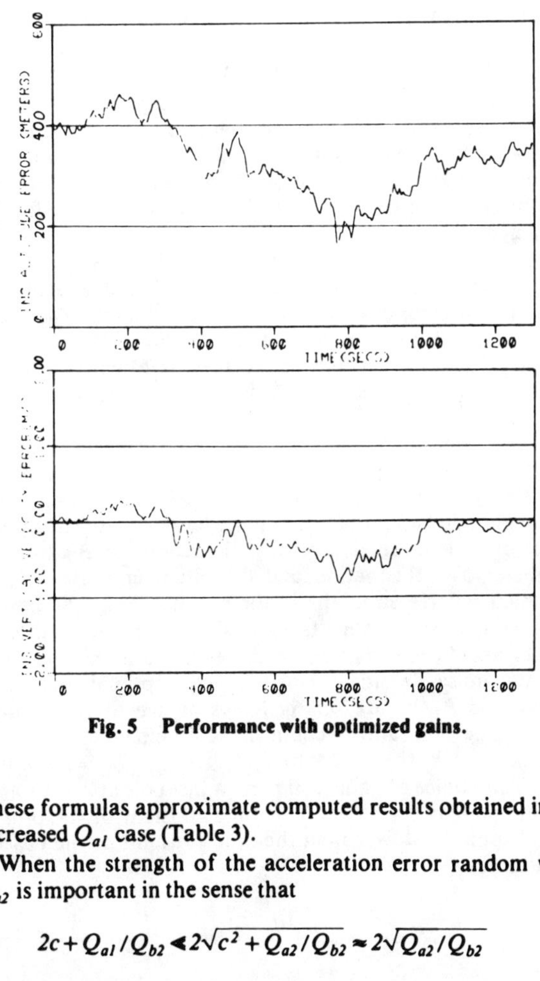

Fig. 5 Performance with optimized gains.

These formulas approximate computed results obtained in the increased Q_{a1} case (Table 3).

When the strength of the acceleration error random walk Q_{a2} is important in the sense that

$$2c + Q_{a1}/Q_{b2} \ll 2\sqrt{c^2 + Q_{a2}/Q_{b2}} \approx 2\sqrt{Q_{a2}/Q_{b2}} \tag{43}$$

then the optimal gain ratios are

$$k_2/k_1 \approx \sqrt{2}(Q_{a2}/Q_{b2})^{1/4} \tag{44}$$

$$k_3/k_1 \approx \sqrt{Q_{a2}/Q_{b2}} \tag{45}$$

and the associated pole locations are

$$p_2,p_3 \approx -(1/\sqrt{2})(Q_{a2}/Q_{b2})^{1/4}[1 \pm j] \qquad (46)$$

These formulas approximate the computed residuals obtained in the increased Q_{a2} case (Table 3).

Simulated Performance

The vertical channel gains obtained by the optimization procedure have been evaluated using a time-domain three-channel simulation that includes detailed models for the sources of acceleration and altitude errors and that exhibits the dependence of these sources of error on the aircraft trajectory. Error sources included in the simulation are listed in Table 4. The aircraft trajectory simulated represents a F4 tactical mission profile. Figure 2 shows the altitude vs time. Figure 3 shows the heading vs time. This is a high dynamic trajectory. During the first descent the aircraft executes a "figure-eight" maneuver. During the remainder of the flight the aircraft is performing rapid zigzag evasive maneuvers while climbing and diving. In some of these maneuvers, maximum bank angles of 70 deg are used, with associated load factors of $3g$. One pull-up has a load factor of $5g$.

The vertical channel performance results with the baseline gains, Eq. (12), and with the optimized gains, Eq. (14), are exhibited in Figs. 4 and 5. The optimized gains do provide a lower level of vertical velocity error. A noticeable consequence of the fast ($\tau = 1$ s) loop around the indicated altitude is the increased noise content in the indicated altitude due to the baro-error fluctuations. This may be a disadvantage of the optimized gain set in some applications.

Conclusions

The fundamental assumptions underlying the results of this analysis are that the most important sources of error in the third-order vertical channel may be adequately modeled by random walks and white noises as presented earlier. If these assumptions are correct, then the following conclusions are obtained.

The most significant source of error in the vertical channel is the fluctuation in the altimeter bias (such as due to altimeter scale factor error and nonzero vertical velocity). The second most significant source of error is the short-correlation time acceleration error (such as due to specific force measurement error during a maneuver). Altimeter noise at the assumed level has negligible effect.

The optimal choice of the gain set is radically different from the baseline gain set, which provides a triple pole at $\tau = 100$ s. The optimal gains include a k_1 that is two orders-of-magnitude larger than that in the baseline set. The optimal value of k_1 is a function only of the relative strengths of the altimeter bias fluctuation and the altimeter noise. For the assumed relative strengths in the nominal case, the optimal gain value recommended is $k_1 = 1.0\ s^{-1}$. The optimal values for k_2 and k_3 are sensitive to the assumed numerical values for the noise densities. However, the optimal gain *ratios* k_2/k_1 and k_3/k_1 are relatively *in*sensitive to the assumed noises. The gain ratios recommended by the nominal optimal solution are $k_2/k_1 = 4.2 \times 10^{-3}\ s^{-1}$ and $k_3/k_1 = 4.4 \times 10^{-6}\ s^{-2}$. This choice of gains will place a fast pole at $\tau = 1$ s and a double pole at $\tau = 480$ s. This time constant of the optimal double poles is slower than the time constant of the baseline triple poles.

The optimal gains produce a significant performance improvement compared with the baseline case. The rms vertical velocity error is reduced 30%.

The recommended value for k_1 perhaps should be accepted with some degree of skepticism. However, the recommended values for the gain ratios k_2/k_1 and k_3/k_1 *can* be adopted with some confidence, because of their low sensitivity to the assumed noise values. The low sensitivity is a result of the gain ratios approaching a fundamental limit imposed by the destabilizing feedback of the gravity computation error.

If the short-correlation time acceleration error is more important than assumed in the nominal case, the optimal gain ratio k_2/k_1 increases and the optimal double pole splits into two real poles. On the other hand, if the acceleration error bias (such as due to accelerometer bias and gravity anomaly) is shifting more than assumed in the nominal case, both optimal gain ratios are increased and the optimal double pole splits into a complex conjugate pole pair.

A detailed baro-inertial error simulation has exhibited the reduced vertical velocity errors that can be obtained with the optimized gains. It provides confidence that the fundamental assumptions of the stochastic analysis are sound.

Because of the very long settling time associated with the recommended optimized gains, one should also implement a faster set of gains for use in the ground alignment mode.

References

[1] Draper, C.S., Wrigley, W., and Hovorka, J., *Inertial Guidance*, Pergamon Press, New York, 1960.

[2] McClure, C.L., *Theory of Inertial Guidance*, Prentice-Hall, Englewood Cliffs, N.J., 1960.

[3] Markey, W.R., *The Mechanics of Inertial Position and Heading Indication*, John Wiley & Sons, New York, 1961.

[4] Pitman, G.R., Jr. (ed), *Inertial Guidance*, John Wiley & Sons, New York, 1962, p. 41.

[5] O'Donnel, C.F., *Inertial Navigation, Analysis and Design*, McGraw-Hill, New York, 1964, pp. 35-36.

[6] Broxmeyer, C., *Inertial Navigation Systems*, McGraw-Hill, New York, 1964, p. 148.

[7] Farrell, J.L., *Integrated Aircraft Navigation*, Academic Press, New York, 1976, pp. 21-22.

[8] Kayton, M. and Fried, W.R., *Avionics Navigation Systems*, John Wiley & Sons, New York, 1969, pp. 317-319.

[9] Ausman, J.S., "GPS Bombing Demonstration Error Analysis," Litton Guidance & Control Systems Division presentation to General Dynamics Electronics Division, May 1976.

[10] Whalley, R., "Inertial Techniques in Height Measurement," *Journal of the Institute of Navigation*, Vol. 19, No. 1, 1966, pp. 61-67.

[11] Ausman, J.S. and Kouba, J., "Airborne Range Instrumentation System (ARIS)," Vol. II (Mechanization Equations), Litton Guidance and Control Systems Document No. 402029, Woodland Hills, Calif., Nov. 1972, pp. 25-28.

[12] Ausman, J.S., et al., "Close Air Support System (CLASS) F-4D Flight Test," Vol. I (Summary), Air Force Avionics Laboratory, Wright-Patterson Air Force Base, Ohio, TR-73-363, Sept. 1973, pp. 96-102.

[13] Blanchard, R.L., "A New Algorithm for Computing Inertial Altitude and Vertical Velocity," *IEEE Transactions on Aerospace and Electronic Systems*, Vol. AES-7, Nov. 1971, pp. 1143-1146.

[14] Newton, G.C., Gould, L.A., and Kaiser, J.N., *Analytic Design of Feedback Controls*, John Wiley & Sons, New York, 1961.

[15] Hooke, R. and Jeeves, T.A., "Direct Search Solution of Numerical and Statistical Problems," *Journal of the Association for Computing Machinery*, Vol. 8, April 1961, pp. 212-229.

FLIGHT TEST RESULTS WITH COLLISION AVOIDANCE SYSTEMS

Richard Sobocinski

Vice President Research & Development
Dalmo Victor Operations, Bell Aerospace Textron
Belmont, California

Abstract

Dalmo Victor has developed and extensively flight tested prototypes of the ARINC 735 TCAS-II. This includes over 900 hours in Piedmont in-service operations. Results show that MOPS requirements will be met for the future projected high traffic density conditions with a simple, four element directional transmit antenna. In-flight testing shows that external, dynamic errors caused by aircraft influences limit angular rate accuracy. However, absolute accuracy is adequate for traffic advisory display.

Flight Testing

Dalmo Victor has developed a prototype ARINC 735 TCAS-II Collision Avoidance System under multiple FAA contracts over the past three years. As part of the development effort, extensive flight tests have been conducted at the following facilities:

1. FAA Technical Center at Atlantic City, flight tests of a system in a Boeing 727 aircraft.

2. MIT Lincoln Laboratory, test flights in a twin-engine Cessna aircraft.

3. Piedmont Airlines, 900 hours of flight test of scheduled in-service flights in a Boeing 727 aircraft.

4. FAA High Density Operation Verification, Los Angeles Basin utilizing a Boeing 727 aircraft.

The first flight test on Piedmont Airlines had the display situated in an observer position such that they were not viewable by the flight crew. The TCAS-II System has been updated to provide display viewable by the pilots. The software has been upgraded to the complete TCAS-II level as defined by the MOPS. This is now in flight test on a Piedmont Aircraft Boeing 727. Dalmo Victor, with the FAA, will deliver production configuration ARINC 735 TCAS-II Systems for airline limited production deployment evaluation in 1984.

TCAS-II CONCEPT

As shown in the figure below, TCAS-II operates by interrogating aircraft in a 90° sector to allow operation in the high density conditions postulated for the 1990 traffic environment. The system will work with Mode S TCAS-II equipped aircraft along with a mix of equipped general aviation aircraft. Current ATCRBS transponders (with or without encoding altimeters) will be tracked and displayed. Also, TCAS-I equipped aircraft will be displayed and communications provided over the Mode S data link.

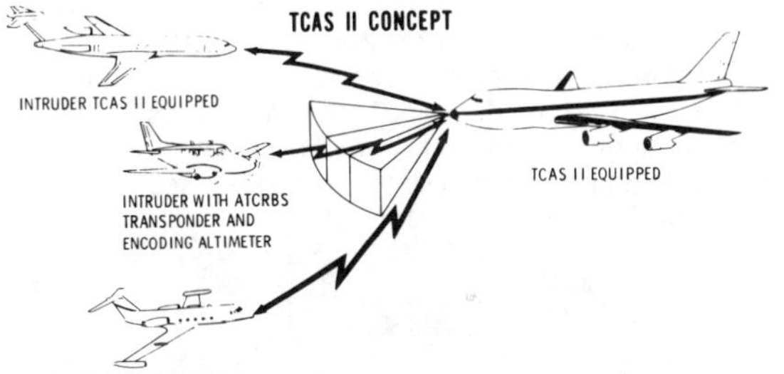

Production System

Shown below is the ARINC 735 TCAS-II System to be flight tested during the airline limited production deployment program. The system displays the resolution advisories on the Instantaneous Vertical Speed Indicator which provide the vertical advisories.

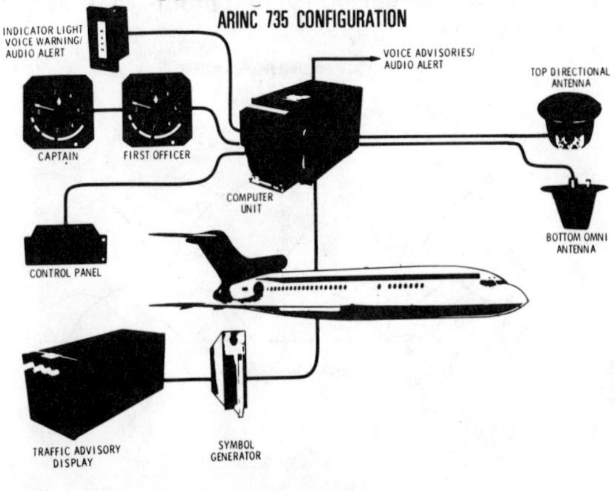

Reprinted from *Proc. IEEE/AIAA 5th Digital Avionics Systems Conf.*, pp. 20.2.1–20.2.3, 1983.

Also shown is a typical traffic advisory display provided on the weather radar. This presents the bearing and relative altitude of Mode C equipped aircraft. It will provide display of aircraft without altitude descriptors for aircraft equipped only with Mode A.

The photograph below shows the TCAS-II Directional Transmit Antenna configuration that has been flight tested extensively by the FAA. This is the same antenna which will be used in a production system. It should be noted that it is flush mounted and has an exterior diameter of only 8". This antenna transmits in four 90° sectors to interrogate intruder aircraft.

The received signals are processed by quadrant monopulse systems. The measured antenna patterns for an antenna currently in flight test are shown in the figure below.

TCAS-II DIRECTION FINDING PATTERNS

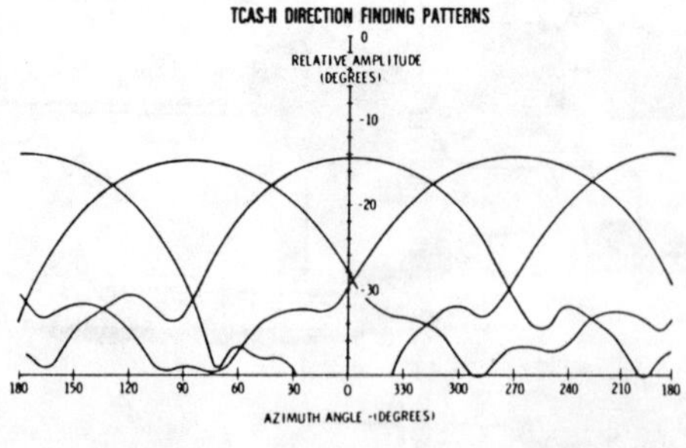

The direction finding errors of this antenna indicate that the forward sector of the aircraft (±60°) has a measured error of approximately ±3°. The error is symmetrical to the crossover points between the two lobes for errors up to 10° at the crossovers. The errors past 60° are predictable and repeatable in the antennas. Therefore, these are compensated for in the computer. It should be noted that the antenna is simply installed onto the aircraft. No alignment or cable adjustments are required to adapt the antenna to the aircraft. Aircraft calibration factors do not need to be entered into the computer to correct the aircraft influences.

Probability of Track

The system was tested in the high density of Los Angeles in December of 1982. The flights were flown East and West at the same VFR altitude to increase the probability of encounters. During the flight test, 19 close encounters were recorded. Aircraft density varied from .06 to .23 aircraft per nautical mile squared with an average of 0.15. It should be noted that the MOPS requires operation up to 0.3. At no time did the system generate a false track which satisfied the conditions for producing a resolution advisory or a traffic advisory. Therefore, no false tracks were observed.[1]

The TCAS-II system provided timely traffic advisories on all 19 aircraft. No degradation of the probability of detection versus traffic density was observed. The probability of track is shown in the figure below. As can be seen, the actual results tracked very closely to the probability of track predicted by the directional antenna in the MOPS.

DIRECTIONAL INTERROGATION

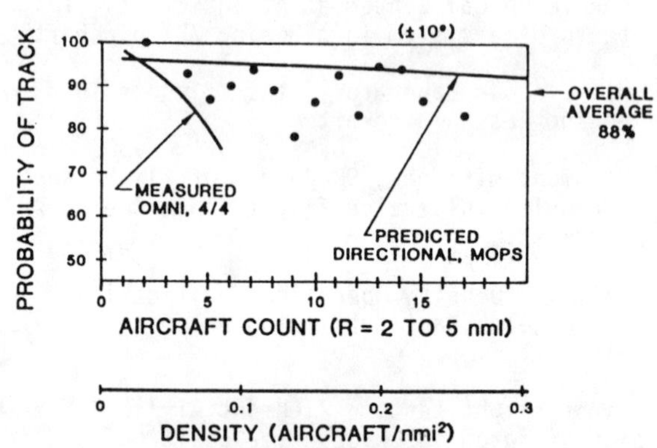

TCAS-II PERFORMANCE MEASURED IN LA.

This data demonstrates that the current TCAS-II range/altitude processing algorithms with multiple level Whisper/Shout interrogation will operate in the highest postulated densities without angular rate filtering.

In-Flight Bearing Accuracy

In-flight bearing accuracy is degraded due to multiple factors. This can be shown in the data below gathered during January 1983 in the Teterboro traffic area.

TETERBORO FLT 16 RAW ANGLES

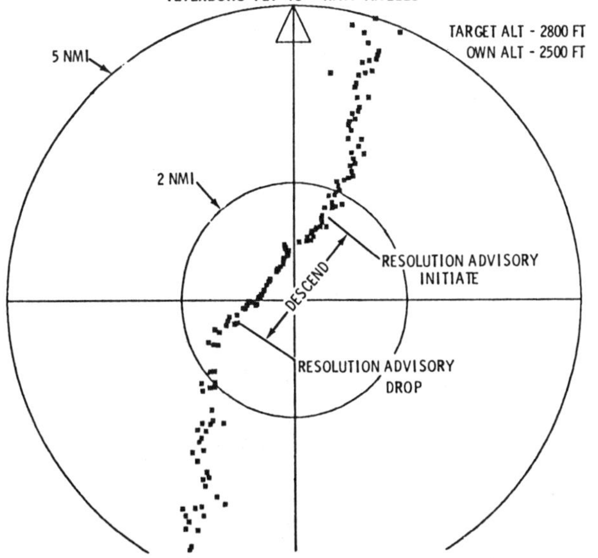

Distortion was observed around the nose due to aircraft shadowing of the antennas. The raw data typically shows a dispersion of 5° to 10° from each one second interrogation group to the next. The probable causes of these short term fluctuations are ground multipaths, reflections from the own aircraft surfaces and phasing variations due to short term aircraft pitch and roll affecting the beam patterns at different elevation angles. The raw angles from about 5 miles into about 1-1/2 to 2 miles have a significant scatter in them.

The next figure shows the effect of the scatter on the CAS tracker even after significant filtering.

TETERBORO FLT 16 - TCAS TRACKER ANGLES

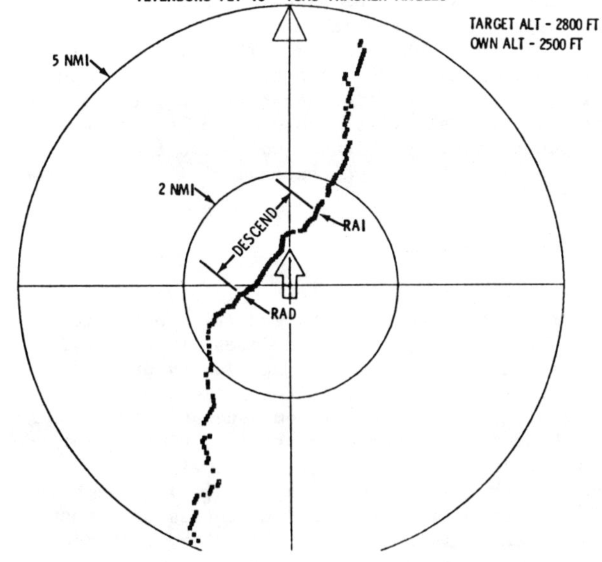

The pattern of the scatter tends to cause the bearing track to build at angles which are up to 5° variant from the actual track for short periods of time. It can be seen at approximately 2 miles the track settled down to relatively good angle if you were simply trying to display bearing. However, it should be noted there are a number of deviations in apparent short term track due to the scatter. Thus, the only use for this presentation would be on advisory display because of the short term bearing variations. Even though the actual angle error is small, the deviation from the actual straight line flight of the aircraft would cause false alarms if these rates were utilized for horizontal maneuver advisories. In this case, the resolution advisory was generated at approximately 1.3 nmi in a dead-ahead position on range and altitude data only.

It should be noted that the MOPS requires that TA and RA advisories be generated on range and altitude data inputs, not bearing. The DV TCAS-II consistently tracked range and altitude data well.

Summary

The Dalmo Victor TCAS-II Prototype System has performed to the desires of the MOPS. It demonstrated consistent range/altitude track viability in densities approaching the MOPS requirement of .3 aircraft per square nmi. Also, it provided traffic advisory displays better than the MOPS required value ±9° RMS bearing accuracy for traffic advisories.

[1] Massachusetts Institute of Technology Lincoln Laboratory, "Traffic Alert Collision Avoidance System (TCAS) Quarterly Technical Letter," 42QTL-TCAS-83-02, 26 May 1983.

Advanced Terrain Correlation Techniques

Paul R. Hinrichs

E-Systems Inc., Garland Division
P. O. Box 6118, Dallas, Texas 75222

Abstract

The requirements for an ideal positioning system are enumerated and compared with terrain correlation systems. Models for terrain and performance characteristics are introduced. Properly implemented, a terrain correlation system can be very accurate and tolerate large navigational uncertainties prior to the update. Error estimates are made for state-of-the-art components and reference source data, and predictions using the performance models and these error characteristics indicate that accuracies rivaling optical systems can be achieved.

INTRODUCTION

The concept of using terrain elevation signature to position update a navigation system was introduced over fifteen years ago. Considerable success was achieved using off-the-shelf radar altimeters and Class A contour maps as the source of elevation data. When the terrain elevation standard deviation was 30 feet or more, these systems had accuracies of 200 feet (CEP) or less and could tolerate navigational uncertainties of several square miles. Like any system, the performance is limited by noise. This paper presents an investigation of these limitations and develops the accuracies which can be achieved using current technology.

The characteristics of the ideal updating system are first contrasted with a terrain correlation system. In order to investigate the performance of a terrain correlation system, the statistical characteristics of terrain are discussed in terms of theoretical models and experimental data. This is followed by the introduction of performance models for accuracy and the probability of successfully finding the correct matchpoint. Finally, error characteristics are hypothesized for state-of-the-art sensors and reference source data, and performance based on these error magnitudes is estimated.

IDEAL SYSTEM

The criteria for an ideal position locating system for operation in a wartime environment would be as follows: adequate accuracy; autonomous all-weather, day/night operation; and counter-measure resistant. Note that systems which rely on ground and satellite stations are eliminated due to their vulnerability. If it is to be a "correlation" positioning system, then additional features are desired. First, the signal should have certain characteristics. It should be stable with time and weather. It should be a physical quantity that is directly measurable. Note that systems using radiometric and radar reflectivity signals are now eliminated. Second, it should be economically feasible and practical to generate the reference data and the signal must have adequate character to provide a sufficient signal in the area of interest. The signal should be measurable with practical sensors with the necessary precision to provide the accuracy. The system should not be overly sensitive to the rest of the navigation system. That is, it should be able to operate with large navigational uncertainties prior to the update and still fix correctly. Finally, the total system costs to produce, operate and support the system must be competitive.

Positioning systems which utilize variations in terrain elevation as the signal closely approximate the ideal system. The positioning accuracy is limited only by the accuracy of the source data and sensors. The on-board system operates autonomously. The signal (terrain undulation plus ground cover, foliage, and cultural features) are relatively stable. The signal is a physical quantity and directly measurable using the parallax from stereo photo pairs. The coverage is limited to land areas which have some elevation character. For position accuracy, the degree of character necessary is a function of source data and sensor error magnitudes which can be made arbitrarily small at the expense of increasing the system cost.

TERRAIN CHARACTERISTICS

Before any system design is considered, it is necessary to understand the general stochastic character of the terrain. It goes without saying that terrain is not a space-stationary process; however, the terrain can often be separated into regions which are locally homogeneous. It is only for these locally homogeneous areas that terrain characteristics such as standard deviation, σ_T, and autocorrelation function, $C_T(x, y)$, have been investigated. Terrain characteristics have been studied extensively in the past using digitized terrain contour maps. Contours on these maps strictly represent terrain, i.e., elevation of ground cover and cultural objects are excluded.

Reprinted from *Rec. IEEE Position Location and Navigation Symp.*, pp. 89–96, 1976.

Furthermore, since the contours are irregular, a smoothing routine is necessary in order to have data at regular intervals such that autocorrelation functions can be generated. This results in additional smoothing, especially if the contours are widely spaced and, consequently, do not adequately describe terrain high frequency spatial characteristics. In the past, this smoothing has not created a problem because the radar sensor smoothed the acquired data considerably unless the beamwidth was very narrow.

In order to gain insight into terrain characteristics, it is of interest to consider the theoretical basis for two-dimensional space-like stochastic processes. To this end, let R' be a region of the earth's surface and consider T(x, y) to be the value of the elevation (above sea level) of that surface at the location (x, y) in R'. In general, R' can be sectioned into disjoint subregions $R_1, \ldots, R_k$ such that the process T(x, y) has certain homogeneous characteristics within each subregion. Consider a particular one of these subregions, say $R = R_j$. T(x, y) can be considered to be a homogeneous random field in the following sense: if (x, y) and (x', y') are in R then

(a) $ET(x, y) = m_R$, m_R a constant depending only on R.

(b) $C_{R,T}(|x-x'|, |y-y'|) = E[T(x',y') - m_R][T(x, y) - m_R]$.

This homogeneous property (b) is a natural generalization of the covariance stationary concept useful in one-dimensional applications. In two dimensions, it is often useful to suppose that the process is also isotropic, i.e.,

(c) $C_{R,T}(x, y) = C_{R,T}(r)$, $r = \sqrt{x^2 + y^2}$.

For the model with properties (a), (b), and (c), we say that the process is both homogeneous and isotropic (i.e., statistically invariant (over R) under both translations and rotations). Property (c) is seldom observed - even approximately - in practice. However, it greatly simplifies any detailed analysis and many results obtained under this assumption can be readily modified to remove this constraint. In the remainder of this section, all three properties are assumed for terrain. In addition, whenever needed to obtain explicit results, the requirement

(d) T(x, y) a normal random field

will be imposed.

The covariance function $C_{R,T}(r)$ and the mean m_R provide a complete statistical description for T(x, y) when the process is an isotropic normal random field. Knowledge of the functional form of $C_{R,T}(r)$ then allows calculations to be made of the average properties of the surface. Examples of such calculations are illustrated later in this section. The first task, however, is to find a reasonable physical model for terrain so as to obtain the required functional form for $C_{R,T}(r)$.

The most successful models for continuous stochastic processes have proved to be Markovian type equations, the most important example of these being the one-dimensional process U(t) defined by

$$\frac{dU(t)}{dt} + \mu U(t) = N(t)$$

where N(t) is white normal noise such that

$$EN(t) = 0,\ EN(t)N(t') = 2D\delta(t-t').$$

Then U(t) is a normal process such that (assuming the process started at $t = -\infty$)

$$EU(t) = 0,\ EU(t)U(t') = \frac{D}{\mu} e^{-\mu|t-t'|}.$$

An important point to be noted here is that U(t) only depends on the values of N(t') for $t' \le t$, i.e., the past values of N(t'). The parameter t is said to be "timelike" since a distinction must be made between the past, present, and future values of this quantity. Terrain, on the other hand, depends on parameters x and y that are distinctly "space-like", there being no natural distinction between, say, east and west of a location (x, y) as there is between before and after an epoch t. It was this consideration that led P. Whittle[1] to question the use of exponential autocovariance functions as models for two-dimensional space-like processes. Whittle proposed that the simplest acceptable model would be described by the stochastic partial differential equation

$$\left[\alpha^2\left(\frac{\partial^2}{\partial x^2} + \frac{\partial^2}{\partial y^2}\right) - 1\right] T(x, y) = \eta(x, y)$$

where $\eta(x, y)$ is a two-dimensional continuous and normal random field described by

$$E\eta(x, y) = 0,\ E\eta(x, y)\eta(x', y') = 4\pi A^2\delta(x-x')\delta(y-y')$$

The covariance function associated with T(x, y) is found to be

$$C_T(r) = \sigma_T^2 \left(\frac{r}{\alpha}\right) K_1 \left(\frac{r}{\alpha}\right),\ \text{for } \sigma_T^2 \equiv \frac{A^2}{\alpha^2}$$

where $K_1(\cdot)$ is the modified Bessel function of the second kind of order one. We refer to a normal random field described by this covariance function as Whittle's process.

Heine[2] following a suggestion of Whittle's examined the class of linear second order partial differential equations of the form

$$\left(a_0\frac{\partial^2}{\partial x^2} + a_1\frac{\partial^2}{\partial y^2} + a_2\frac{\partial^2}{\partial x dy} + a_3\frac{\partial}{\partial x} + a_4\frac{\partial}{\partial y} - a_5\right) T(x, y) = \eta(x, y).$$

He was able to show that only the form

$$\left[\left(\frac{\partial}{\partial x} - \mu\right)^2 + \frac{\partial^2}{\partial y^2} - \gamma^2\right] T(x, y) = \eta(x, y)$$

gives rise to a process depending on the input $\eta(x, y)$ for all x and y and hence is space-like. This corresponds to Whittle's process for $\mu = 0$ and $\eta(x, y)$ white normal noise.

These results indicate that Whittle's process can be regarded as the simplest two-dimensional model for a continuous random field just as the one-dimensional Markovian process is the simplest in one-dimension.

It would not be surprising if terrain corresponds to such a model and, in fact, study of estimated covariance functions for a wide range of terrain types do show good agreement with the $\frac{r}{\alpha} K_1 (\frac{r}{\alpha})$ covariance function. Figure 1 shows three examples of this agreement. The values of the correlation parameter α for these data span the complete range of this quantity encountered in terrain. The standard deviations of these data sets are also typical of the range encountered for terrain. The sampling interval in each case was $\Delta x = 400$ feet. The estimates used in this example were obtained by averaging estimates from several rows of data from the same area (for the number of rows averaged see the legend accompanying the figure). The values of α were chosen so that $\left(4 \frac{\Delta x}{\alpha}\right) K_1\left(4 \frac{\Delta x}{\alpha}\right)$ equaled the empirical value at $4 \cdot \Delta x$. The agreement is obviously excellent. Many other terrain areas have been examined in this manner and in almost all cases - even when only a single line of data was available - they showed this same agreement with Whittle's process. This procedure does not prove that Whittle's process is the correct model for terrain, but it does indicate that it at least provides a useful approximation. In contrast, Figure 1 shows the same curve fitting procedure applied to the data assuming in (a) an exponential covariance and in (b) a Gaussian form. It is obvious that these two functional forms are not in agreement with the empirical data. This further suggests the worth of Whittle's process as a model for terrain.

Although the Whittle process appears to most closely approximate the terrain stochastic process, the associated covariance function has the disadvantage that it is not analytic at the origin which limits its usefulness when it is desired to make expansions near the origin.

Terrain standard deviation, σ_T, will range from zero to several hundred feet. Regions with $\sigma_T < 30$ feet generally appear to be quite smooth; regions with $\sigma_T \approx 100$ feet appear hilly; and regions such that $\sigma_T > 300$ are generally classified mountainous. If a terrain correlation length, X_T, is arbitrarily defined as the distance in which $C_T(r)/C_T(0) = e^{-1}$, then the range of X_T is generally from one thousand to several thousand feet. Furthermore, X_T generally (not always) increases with σ_T. A typical value for X_T might be 2500 feet using contour map equivalent data; however, X_T values of a few hundred feet have been measured from data with higher resolution.

TERRAIN CORRELATION SYSTEM CHARACTERISTICS

Heuristic arguments and simulation are used in this section to develop the general characteristics of a terrain correlation system. Parameters of operation will be further quantified in later sections after methods of dealing with system's errors and false fix probability has been introduced.

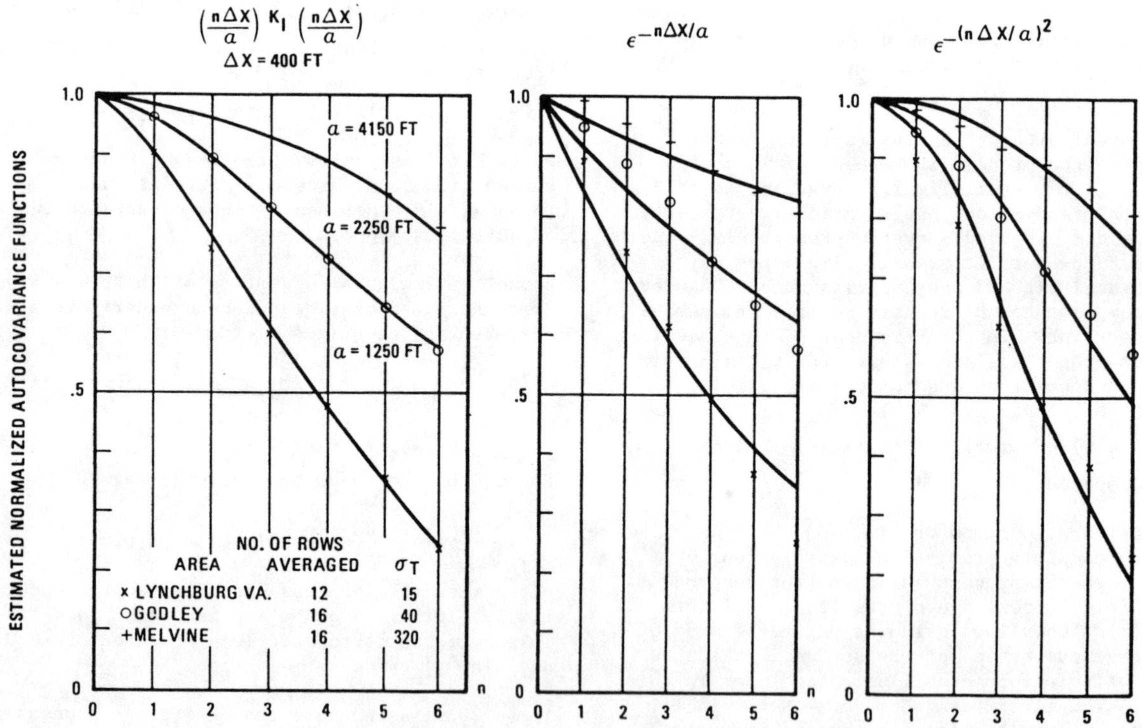

FIGURE 1. COMPARISON OF WHITTLE, EXPONENTIAL, AND GAUSSIAN AUTOCOVARIANCE FUNCTIONS WITH EMPIRICAL ESTIMATES

It is desirable to design a system to work over the maximum possible land area. A terrain correlation system will operate over all areas where the ratio of the standard deviations of terrain to vertical noise (σ_T/σ_N) is adequate. Thus, there is always a boundary that generally separates terrain into regions which will work and regions in which the system will be noise limited. If there were no practical limits on the length or area of data to be used for matching, then the ratio σ_T/σ_N could approach an arbitrarily small number. However, practical limits are generally encountered, such as:

(a) The amount of data or time required to perform the match will become excessive.

(b) Terrain will not maintain a homogeneous characteristic indefinitely.

(c) Dead reckoning errors will eventually distort the gathered data and degrade the match accuracy.

Experimental data and analytical results[3] indicate that the limits for matching a line of data correctly (without a false fix) are:

$$\sigma_T/\sigma_N \geq 2.5$$

when

$$L/X_T = \beta \geq 4$$

where L is length of the acquired profile used for correlation. For accuracy[4] it is desirable for these limits to be

$$\sigma_T/\sigma_N \geq 3$$

and

$$\beta \sim 10$$

This means that an acquired linear profile needs to be 20 to 30K feet. This approach of gathering a line (not necessarily a straight line) of data is obviously the most simple since the antenna beam can be fixed and the vehicle motion provides the scan. The obvious penalty is the relatively long acquired profile required. For a fixed σ_T/σ_N ratio, the length of the acquired profile can be shortened by scanning in crosstrack. Because of the relatively long terrain correlation lengths, collecting two parallel paths of data a few hundred feet apart will not improve the fix accuracy very much. However, if the two paths are separated by a terrain correlation length, then they are relatively independent and an equivalent β can be achieved in half the distance. Continuing this argument, the required profile length as a function of the scanning width, W, can be approximated as

$$L(W) = \frac{L(0)}{\frac{W}{X_T} + 1}$$

The important thing to note is that the physical width of the ground area scanned must be large (several thousand feet) in order to appreciably reduce the acquired profile length. Such scanning becomes impractical at low altitude because of terrain masking, beam smearing, sidelobe returns, etc. In addition, system costs will increase considerably due to the expense of a scanning antenna and the complications of the computer algorithms. Because of these penalties and limitations, it will be assumed that the system gathers a single line of data.

Three candidate algorithms have been studied for terrain contour matching (TERCOM). They are cross correlation, mean absolute difference (MAD), and mean square difference (MSD). For finite data line intervals, they are defined as

$$\text{cross correlation} = \frac{1}{L}\int_{-L/2}^{L/2} T_{ACQ}(x,y) \cdot T_{ST}(x+\tau_x, y+\tau_y)\,dx$$

$$\text{MAD} = \frac{1}{L}\int_{-L/2}^{L/2} \left|T_{ACQ}(x,y) - T_{ST}(x+\tau_x, y+\tau_y)\right| dx$$

$$\text{MSD} = \frac{1}{L}\int_{-L/2}^{L/2} \left[T_{ACQ}(x,y) - T_{ST}(x+\tau_x, y+\tau_y)\right]^2 dx$$

where T_{ACQ} is the acquired or sensed terrain data profile and T_{ST} is the stored or referenced terrain source data.

There are advantages of each function. Cross correlation yields the best results if unknown scale-factors exist in the data. In fact, the maximum (which is the match criterion) is independent of scale-factor errors. However, in the absence of scale-factor errors, which is the case for terrain correlation, the other algorithms are more accurate. Because of the finite lengths of data used, cross correlation will not generally have the maximum value when the data have zero shift even in the complete absence of noise. MAD is more accurate than cross correlation and is often the easiest to implement. MSD is the more accurate of these algorithms, but not much better than MAD. The match criterion of MAD and MSD is for the shift which results in the minimum residue, and, unlike cross correlation, will result in a perfect match in the absence of noise with finite data lengths.

A simple experiment was carried out to determine the relative accuracies of the three algorithms. One-dimensional exponentially correlated gaussian zero mean random noise was used as independent reference and noise signals with equal correlation lengths. The three algorithms operated on the reference signal and the reference signal plus noise. The results are shown in Figure 2 where the fix error standard deviation, σ_ε, normalized to a terrain correlation length, X_T, is plotted as a function of the normalized length of data, β, used for matching. Note that cross correlation in the absence of noise is still worse than MSD and MAD with noise. Since MSD is slightly superior to

MAD, it will be assumed to be the algorithm used in later analysis.

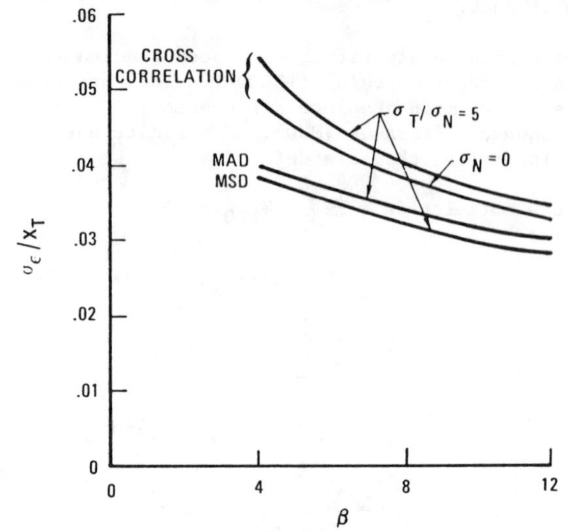

FIGURE 2. FIX ALGORITHM ACCURACIES

Similar experiments with the three algorithms were run to determine false fix performance with the results that MSD and MAD performance was about the same, but cross correlation was much worse. This result is attributed directly to the absence of scale-factor errors.

It is not intended from this discussion to infer that these three algorithms exhaust the possibilities. On the contrary, numerous possibilities and trades exist when whole systems are taken into consideration. For example, when Kalman filtering is utilized to update the system states it may be advantageous to have fixes that are less accurate, but at more frequent intervals. Carrying this concept to the limit, every radar terrain clearance measurement could be used to update the system state variables. That is, given the radar terrain clearance measurement, the vehicle's estimated position, and the stored terrain elevation, there corresponds a most likely vehicle position error estimate. If the process were linear, then implementation of such a system should result in an optimal position estimate since it makes the maximum use of each measurement.

As noted from these discussions, there are two performance factors to be considered, viz, fix accuracy performance and the probability of fixing at a location near the match point as opposed to a fix on noise. Performance models for estimating these two factors are introduced in the following sections.

ACCURACY PERFORMANCE

As previously noted, the Whittle autocorrelation is an excellent description of the salient stochastic properties of terrain as described by contour maps. However, early fix performance models utilizing non-analytic autocorrelation functions was beset with difficulties. Generally, the fix error was significantly underestimated, and there was a non-physical dependence on cell size. These non-analytic difficulties also precluded the treatment of velocity error sources on an equal basis with other noise sources.

It has become clear from consideration of these difficulties that they can be obviated in a relatively straightforward way if terrain, considered as a random function of position, has a Taylor series representation. This will be the case if this function has an analytic correlation function. An analysis based on the assumption that the terrain correlation function is gaussian and, therefore analytic, viz, $C_T(r)=\sigma_T^2 e^{-\left(\frac{r}{X_T}\right)^2}$, has been made. The relative flatness of the gaussian correlation function at the origin, as compared to a Whittle function, should result in conservative estimates of performance.

As stated, comparison of the acquired and stored profile data is effected by use of the MSD function where

$$\text{MSD}(\tau_x,\tau_y) = \frac{1}{L}\int_{-L/2}^{L/2} [T_{ACQ}(x,y) - T_{ST}(x+\tau_x, y+\tau_y)]^2 dx.$$

The purpose is to find the value of τ_x and τ_y which minimize $\text{MSD}(\tau_x,\tau_y)$. If τ_{xo} and τ_{yo} denote these particular values, then the location of the minimum is determined by

$$\frac{\partial\ \text{MSD}(\tau_{xo},\tau_{yo})}{\partial\ \tau_x} = 0 \text{ and } \frac{\partial\ \text{MSD}(\tau_{xo},\tau_{yo})}{\partial\ \tau_y} = 0$$

Subject to small perturbations and other reasonable limitations (such as, terrain and noise correlation lengths are equal and $\beta \geq 4$) the downtrack variance σ_x^2 and the crosstrack variance σ_y^2 can be estimated in terms of the terrain variance (σ_T^2), the terrain correlation length (X_T), the noise variance (σ_N^2), the length of data used for matching (L), and the normalized velocity error components $\mu = \delta\dot{x}/\dot{x}$ and $\phi = \delta\dot{y}/\dot{x}$. The development of these variances is quite lengthly[4] and will, therefore, not be shown. The results are as follows

$$\sigma_x^2 = \frac{X_T^3}{4L\sigma_T^2}\left(\frac{\pi}{2}\right)^{\frac{1}{2}}\left\{\sigma_N^2 + \frac{2\sigma_T^2}{X_T^2}\left[\frac{1}{4}\left(1 - \frac{4}{\sqrt{2\pi}\ \beta}\right)\left(L\sigma_\mu\right)^2\right]\right\}$$

$$\sigma_y^2 = \frac{X_T^3}{2L\sigma_T^2}\left(\frac{\pi}{2}\right)^{\frac{1}{2}}\left\{\left(1 - \frac{1}{\sqrt{2\pi}\ \beta}\right)\sigma_N^2 + \frac{2\sigma_T}{X_T^2}\left[\frac{1}{6}(L\sigma_\varphi)^2\left(1 - \frac{3}{\beta\sqrt{2\pi}}\right)\right]\right\}$$

One thing to be noted from these equations is that the fix error magnitude is not symmetrical in the downtrack (x-direction) and crosstrack (y-direction). This phenomenon was apparent from flight test data prior to the generation of these accuracy performance estimates and is a result of utilizing a line of data for the fix.

In the absence of velocity errors, it is immaterial as to what point of the flight path the fix error is associated with. However, when velocity errors are introduced, the variance is with respect to the midpoint of the data collected that results in the best match. Thus, there is an additional error at the time of update due to the velocity errors and the elapsed time from overflight of the midpoint of the match to the time of update.

The equations presented above estimate the horizontal plane fix error as a result of errors in the terrain measurement process, e.g., the radar terrain sensor, the reference altitude measurement, the reference data, and velocity errors. Other error sources must also be included to complete the analysis. These include errors in antenna pointing, reference source data location, etc.

FALSE FIX PERFORMANCE

It is a matter of concern for terrain correlation that the surface representing the MSD match function exhibits many minima, any one of which may have a magnitude less than the magnitude of the minimum at the matchpoint. This situation is presented in Figure 3 where $M(\tau)$ denotes the match function at a distance τ from the matchpoint $\tau = 0$.

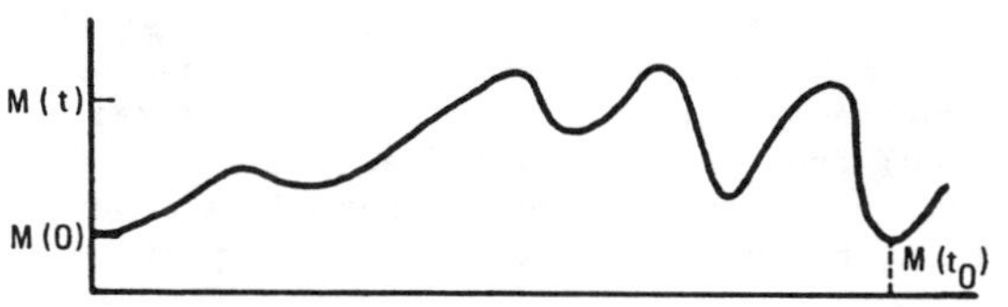

FIGURE 3. MATCH FUNCTION VS DISTANCE FROM THE CORRECT MATCH POINT

As τ increases, the magnitude of the oscillation of $M(\tau)$ increases and it becomes increasingly likely that, e.g., $M(\tau_o) \leq M(0)$. Such an occurrence is termed a false fix and the probability P_F of this event is called the false fix probability. P_F increases with τ so that P_F for $\tau = \infty$ is a pessimistic estimate of P_F for any finite value of τ.

A distinction is made in the nature of the fix error. If the error, i.e., the location τ_o of the absolute minimum of $M(\tau)$ is less than the terrain correlation length X_T, it is termed the fix error. If the error is greater than X_T, it is called the false fix error.

The problem of estimating the false fix probability for an entire search area is approached in a manner analogous to estimating radar false alarm rates. First, the search area is divided into intervals in which the false fix probabilities are approximately independent. The problem is thus divided into two parts, estimating P_F for each interval and division of the search area into independent intervals.

An exact expression for the probability that $M(\infty) \leq M(0)$ has been developed[3] using the following assumptions.

(1) terrain and noise are stationary normal processes with zero means,

(2) terrain and noise are independent,

(3) terrain and noise are each exponentially correlated, and

(4) the correlation lengths of terrain and noise are equal.

However, the exact result is unwieldy and will not be shown here. A solution that gives almost identical results for $\beta \geq 4$ is as follows

$$\text{Prob}[M(\infty) \leq (M(0)] = \frac{\beta e^{\beta}}{\pi} \int_0^a \frac{K_1[\frac{\beta}{2}(\xi + \frac{1}{\xi})]\, d\xi}{\xi}$$

where

$$a = \left[1 + 2\left(\frac{\sigma_T}{\sigma_N}\right)^2\right]^{-\frac{1}{2}}$$ and $K_1[\,]$ is a Bessel function of the second kind of order one. A parametric plot of this equation is shown in Figure 4, and a comparison of theoretical and simulation results is shown in Figure 5. Note that the "single trial" false fix probability is of the order of 10^{-4} when $\beta = 10$ and $\sigma_T/\sigma_N \to 2$.

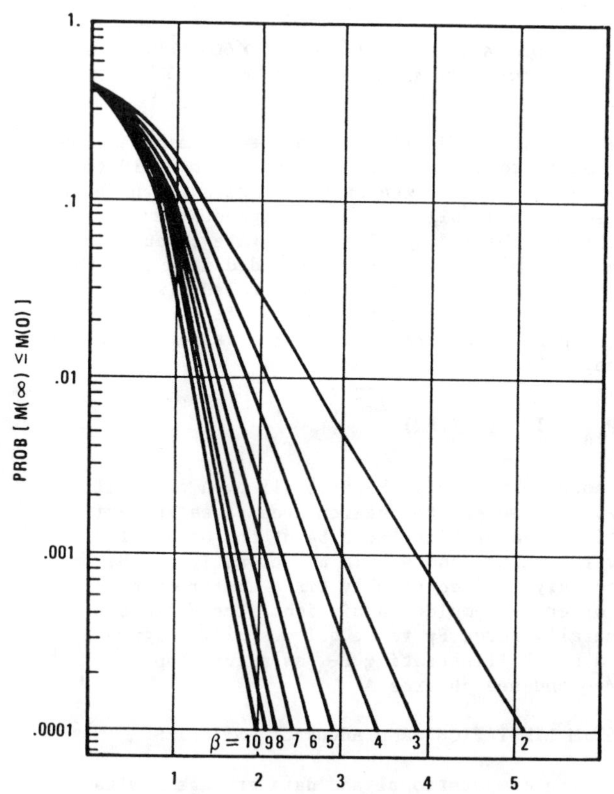

FIGURE 4. PROBABILITY OF A FALSE FIX

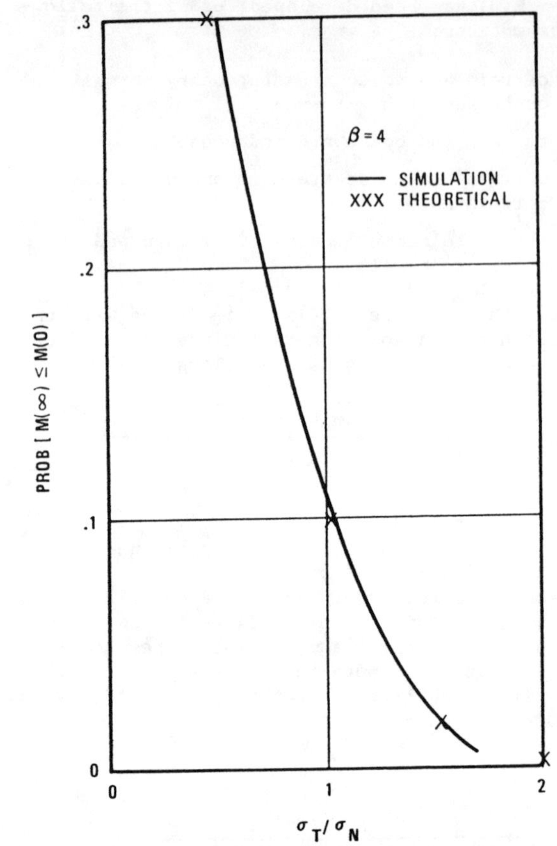

FIGURE 5. FALSE FIX PROBABILITY COMPARISON OF THEORETICAL AND EXPERIMENTED RESULTS

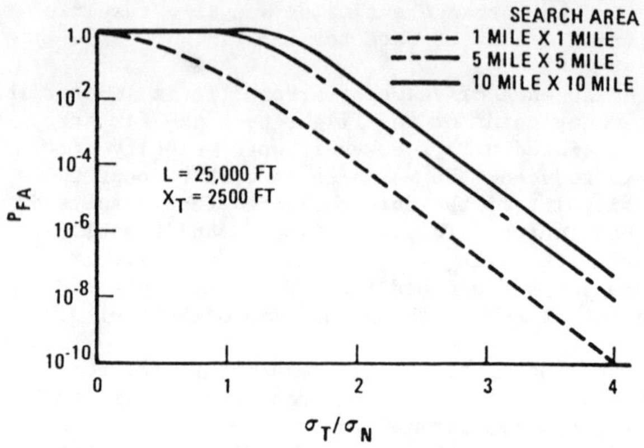

FIGURE 6. PROBABILITY OF A FALSE FIX AS A FUNCTION OF σ_T/σ_N AND NAVIGATIONAL UNCERTAINTY (SEARCH AREA)

The Prob[M(∞) $\leq$ M(0)] will be used as a pessimistic estimate of P_F. The next step, then, is to divide the search area into sub areas in which the false fix probability is an independent event. Following the radar false alarm rate calculation analogy, the search area is divided into regions $X_T * X_T$.. Since by definition a false fix is a fix at a distance greater than X_T from the correct match point, the probability of a false fix for an area, A, is

$$P_{FA} = 1 - (1 - P_F)^{\frac{A}{X_T^2} - \pi}$$

As noted previously, P_F is small even for small σ_T/σ_N. However, the search area (area of navigational uncertainty) can be of the order of several miles square. In addition, it may be necessary to successfully fix a number of times in order to complete a mission. These factors generally force P_F to be quite small. Figure 6 is a graph illustrating P_{FA} as a function of σ_T/σ_N and search area A.

SYSTEM LIMITATIONS AND ACCURACY ESTIMATES

Ignoring computational and data storage limitations, which are generally nominal, the system's performance is generally limited by the accuracy of the navigational sensors, the reference altitude sensor, the radar terrain sensor and the source data. The navigational sensor dictates the search area size which, in turn, determines the false fix probability. Velocity errors also degrade fix accuracy and because the acquired profile data tend to fix best at the midpoint of the data interval when velocity errors are present, an effective lag is introduced. The reference altitude system can be a sensitive barometric altimeter, the vertical channel of the navigational sensor, or a combination of the two. For example, an undamped vertical INS channel which generally has a near linear ramp output can be used provided the ramp information is removed (filtered) from both the acquired and stored data. This will reduce the effective terrain standard deviation, but may enhance the overall performance.

The radar terrain sensor (RTS) is generally the most important sensor from a system limitation point of view. Ideally, the RTS is designed to sense the objects which comprise the reference source data which include culture and foliage in addition to the terrain undulations. Thus, the r.f. frequency of operation should be such that these objects will be adequately reflected. The RTS beamwidth, bandwidth, and video data processing techniques should be such that the acquired data profile is not materially distorted or smoothed. The beamwidth, for example, should illuminate an area of the terrain with a diameter somewhat less than a terrain correlation length X_T for maximum accuracy. Radar state-of-the-art is such that almost any accuracy requirement can be met; however, size, weight, operational conditions, and cost limit accuracies to about 5 feet (1σ).

The reference source data accuracy is another system limitation. If the source data is from standard Class A contour maps, a generally recognized formula for estimating the error in feet is

$$\sigma_{SD} = .304 \text{ Contour Interval} + 1.013 \times 10^{-3} \times \text{Map Scale} \times \text{Tan (Avg Slope)}$$

Unfortunately, contour intervals are often quite large and map scale factors are nominally 50 000 or greater. In addition, cultural objects and foliage are generally omitted. Low altitude stereo photo data can easily produce elevation data with an accuracy of one foot, provided adequate ground control points are available. Without these control points, relative elevation data can still be measured with one foot accuracy, but a ramp error may be introduced. Furthermore, highly accurate reference source data can reasonably be expected to have a higher frequency content (shorter correlation lengths), i.e., smaller objects would not be smoothed out.

Taking all of the various errors into account, a reasonable estimate of system noise, σ_N, that can be achieved is about 5 to 10 feet, and terrain correlation lengths can be expected to be about 1000 feet when obtained from high resolution photography. Based on this estimate, the system CEP range is shown in Figure 7 as a function of σ_T

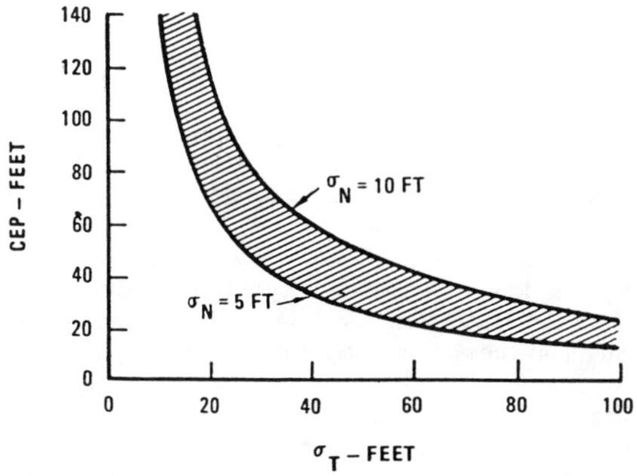

FIGURE 7. CEP RANGE FOR STATE-OF-THE-ART SYSTEM

for a data collection interval, L, of 10 000 feet. These accuracies rival optical systems, and, furthermore, the terrain contour matching system can achieve these accuracies when the initial navigational uncertainty is several miles.

CONCLUSIONS

It has been pointed out that a terrain correlation updating system can be built with state-of-the-art components which results in accuracies comparable to optical homing sensors. In addition, because scale-factor errors are not present, algorithms can be used which allow large areas to be searched with a high probability of successfully determining the correct matchpoint. These factors plus the advantage of being all-weather, autonomous, and having a stable reference source data base result in a near ideal updating system.

REFERENCES

[1] P. Whittle, "On Stationary Processes in the Plane," Biometrika, 41, (1954), pp. 434-439.

[2] V. Heine, "Models for Two-dimensional Stationary Stochastic Processes," Biometrika, 42, (1955), pp. 170-178.

[3] E-Systems, Inc., Report No. 416-05512/4, False Fix Probability, dated 1 July 1970.

[4] E-Systems, Inc., Technical Note GS-70-10, TERCOM Performance, dated 18 September 1970.

Boeing/Sperry Automatic Landing System 727 Airplane

J. E. TEMPLEMAN*
The Boeing Company, Renton, Wash.

AND

R. H. PARKER†
Sperry Rand Corporation, Phoenix, Ariz.

A milestone in the history of American commercial aviation was reached on the night of February 27, 1967. A Pan American Boeing 727 jetliner made a fully automatic landing at JFK International Airport in New York. This was a scheduled flight with 98 fare-paying passengers onboard. With visibility restricted by snow flurries, and despite crosswinds gusting to 12 knots, the touchdown was smoothly executed within 4 ft of runway centerline. This relatively unheralded feat was the culmination of years of effort by many related groups in the field of aviation. This paper considers the changing role of the automatic flight control system in commercial aviation relative to function and safety. It describes the evolution of the 727 system with respect to its use in obtaining progressively lower operational weather minimums for the airlines. The over-all landing system is then described, explaining details of the dual-pitch channels of the automatic pilot, and the unique methods used to implement a system that is both safe and dependable. Following a discussion of the 727 automatic landing certification program, touchdown dispersions are shown to meet tentative FAA criteria.

Introduction

THE role of the automatic flight control system (AFCS) in commercial aviation had not, until recently, changed very dramatically since the first automatic pilot was introduced into fleet operation nearly 35 years ago. The early models of AFCS did nothing more than maintain a fixed air plane attitude for the pilot. The typical AFCS found in commercial transports today provides most of the functions listed in Table 1. However, the role of the AFCS did not change; it was not a dispatch item, but served primarily to reduce pilot workload and add to the comfort of the passengers. The safety aspects of the AFCS were passive in nature. It was necessary to demonstrate only that the AFCS could not in any way compromise the safety of the airplane. The interface between the automatic and manual control systems was designed so that the automatic system could be readily and positively disconnected from the manual controls in the event of a malfunction of the automatic controls.

Table 1 AFCS modes/functions

Lateral/directional control	Longitudinal control
Yaw damping	Pitch attitude hold
Roll attitude hold	Barometric altitude hold
Heading hold	Pitch knob maneuvering
Turn knob maneuvering	Airspeed hold
Preset heading control	ILS glide slope tracking
VHF omnidirectional range (VOR) radial tracking	Autothrottle control
Instrumental landing system (ILS) localizer tracking	

With the development of the commercial jet transport, the AFCS was called upon to perform some new functions, and for the first time its role began to change (Table 2). The swept-back wing of the jet transport accentuated the lateral directional instability known as "Dutch roll." The safety of an airplane is directly related to its handling qualities, and less than satisfactory handling qualities resulted from the Dutch roll instability. It was highly desirable to augment the characteristics of the basic airplane with a means for damping the Dutch roll. In some commercial jet transports, this damper is mandatory for safe operation in some portions of the altitude/airspeed flight envelope. In effect then, a portion of the AFCS has become a dispatch item.

Another phenomenon demonstrated by the high-speed jet transport is that known as Mach tuck-under. This effect manifests itself as a divergent mode whereby an incremental increase in speed (Mach number) causes a nose-down pitching moment on the airplane, which in turn causes a further increase in speed. Although the amount of Mach tuck-under, and the particular range of Mach number over which it occurs, will vary from one model of airplane to another, it typically occurs in the region of Mach 0.8–0.9. This phenomenon again represents a compromise of handling qualities and is of such a magnitude in some models as to be considered unsafe. The AFCS was again called upon to come to the rescue and to provide automatically a compensating pitching moment that would eliminate the divergent mode. This is typically accomplished by programming elevator deflection as a precise function of Mach number over the critical Mach range. Again, a subsystem of the AFCS was required to provide the margin of safety deemed necessary for high-speed operation of commercial jet transports.

Thus, we find that the role of the AFCS relative to safety is changing from passive to active, and two of its subsystems

Table 2 Evolution of AFCS role

Provide pilot relief and passenger comfort.
Enhance safety by improving handling qualities.
Allow operation at lower weather minimum.

Presented as Paper 67-573 at the AIAA Guidance, Control, and Flight Dynamics Conference, Huntsville, Ala., August 14–16, 1967; submitted August 3, 1967; revision received January 2, 1968.

* Research Engineer, Automatic Flight Controls, Commercial Airplane Division.

† Department Head, Transport Flight Controls, Sperry Flight Systems Division.

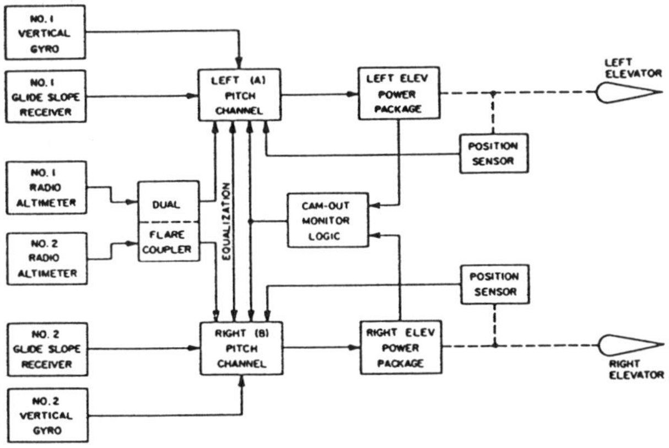

Fig. 1 Dual-pitch axis control systems.

have already become dispatch items. It is not only the handling qualities of the jet transport that have acted as stimuli in changing the role of the AFCS. Rapidly increasing density of air traffic in major terminal areas has provided the incentive to find the means by which it will be possible to operate into and out of these terminal areas in spite of adverse weather conditions. Economic penalties and personal inconveniences resulting from delayed arrivals and diversions to alternate airports are becoming intolerable.

The task of defining acceptable means by which operational weather minimums may be safely reduced has proven to be monumental. Notwithstanding the massive effort already put forth, there is still much to be done before "zero-zero" or all-weather operation can be realized. One thing has become apparent, i.e., the lowering of operational weather minimums must be accomplished in realistic steps. The steps have been defined by the FAA, with the help of the aviation industry, in terms of weather minimums categories. The equipment requirements and the associated performance and safety criteria for each of these categories have been only partly defined. Briefly, the weather minimums categories are as shown in Table 3.

One of the major deterrents to defining the equipment and facility requirements for lower operational weather minimums is the lack of operational experience in those environments. It was decided that in taking each of the steps toward lower minimums we would impose design objectives that, in our best judgment, would provide capability for operating to the next lower weather category. We felt that only by doing this would the pilots be able to gain the operational experience so vital to defining the associated performance and safety criteria for that next category. Let us trace this progression to date, with reference to Table 4.

The AFCS originally certified in the 727 was designed to meet the safety and performance requirements for Category I, but with sufficient margin to provide stable and accurate path control on the instrument landing systems (ILS) localizer and glide slope beams down to altitudes below 100 ft. Furthermore, its elevator authority was limited so that a malfunction would not result in significant altitude loss which would otherwise limit its use to some altitude above 100 ft. As a result, in October 1963, the 727 AFCS was certified for clear-weather operation down to an altitude of 78 ft. By demonstrating performance and pilot proficiency, the airlines soon obtained FAA authorization to operate under Category I weather minimum conditions and, having established visual contact at the 200-ft minimum decision altitude (MDA), to continue under automatic control down to 78 ft. Operational experience gained through this and other programs led to the formulation of criteria for Category II operation. These new accuracy criteria required that certain improvements be made to the 727 AFCS, primarily in the glide slope control circuits of the automatic pilot. This improvement program led to a recertification of the 727 AFCS in October 1965 for Category II operation, and these improvements were delivered to the airlines. By early 1966, through demonstrations of system performance and pilot proficiency, the airlines began obtaining FAA approval to operate in Category II weather minimum conditions.

The step to Category IIIA operation is not as simple. First of all, performance and accuracy criteria are not defined, to say nothing of safety requirements. However, one hurdle that was defined and had to be overcome was the so-called 50-ft rule. That is, the autopilot could not be used below an altitude corresponding to 50 ft plus the amount of altitude lost during manual recovery from a nose-down, hard-over malfunction of the autopilot. Since the manual recovery must allow for a 1-sec fault recognition time by the pilot, it was obvious that the autopilot had to be made "fail-passive." This, and other practical considerations, led to self-imposed design criteria that can be rather simply stated as follows: the pilot shall be immediately warned of a malfunction, and no malfunction shall disturb the flight path of the airplane to the extent that the pilot cannot take over manually and complete the landing safely or execute a go-around. Furthermore, consistent with the philosophy of providing operational margin below the Category IIIA objective, it was deemed necessary to provide automatic flight path control throughout the flare maneuver to touchdown. By certification of such a system, we would again be in a position to obtain airline operational experience that would aid in defining the criteria for a further reduction in operational weather minimums.

727 Automatic Landing System Description

Equipment added to the 727 to enable certification of the airplane for automatic approach in Category II weather minimums consists of improved ILS receivers, a radio altimeter and indicator for minimum decision altitude (MDA) determination, an improved rain-repellant system, and autopilot modifications to provide tight ILS beam tracking. Additional equipment necessary for the 727 automatic landing system includes 1) a second radio altimeter, 2) a second autopilot pitch channel, 3) an autopilot flare coupler (dual), 4) an autopilot roll monitor channel, and 5) autopilot approach progress display modifications.

In pitch axis control, two pitch control channels are used to independently drive separate elevator hydraulic actuators.

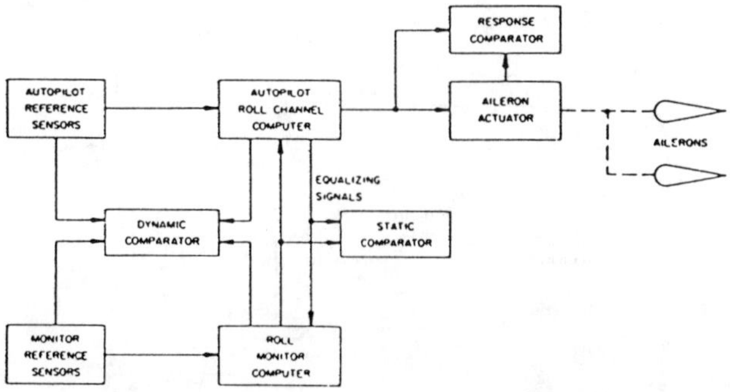

Fig. 2 Roll control and monitor system.

Table 3 FAA weather minimums categories

Category	Minimum decision altitude, ft	Runway visual range, ft
I	200	2600
II	100	1200
IIIA	0	700
IIIB	0	150
IIIC	0	0

Figure 1 is a simplified block diagram of the dual pitch system. Complete duality is maintained in the pitch axis system from the input sensors to the elevator control servos, inclusive. The only inter-tie between the two pitch control channels is an equalization signal path. The equalization signal has limited authority, and enables the two channels to track within system accuracy requirements in the presence of normal component tolerances. A dual flare coupler provides flight path computation between the glide slope control phase and touchdown. The flare coupler commands an exponential flare profile as a function of altitude rate and displacement, and is programmed to control the touchdown rate of descent to $2\frac{1}{2}$ fps. The radio altimeter signal is used in the glide slope mode to program the glide slope gain as a function of altitude above the terrain, with the unique feature that depressions in the terrain which appear as an increase in altitude are ignored. In the flare mode, the radio altimeter signal is the primary source for the flare coupler command computation.

Tracking of the two pitch channels is continuously monitored by fail-safe amplifiers. If a malfunction causes the tracking error to exceed a predetermined amount, the pitch channels disconnect automatically.

Roll axis control is based upon a different dual concept. In this case, the second channel is a roll monitor that duplicates the computations of the roll channel, but it does not drive a separate aileron control servo (Fig. 2). Fundamentally, the roll monitor is a comparison monitor. It receives signal information from a separate set of sensors, operates on these signals to compute aileron command, and compares these commands with the corresponding command signals in the roll channel to verify their integrity. Having validated the command signals, it also computes the required response of the aileron servo, and compares this computed response with that of the actual servo. The various signal comparisons are monitored by a fail-safe amplifier; if static or dynamic errors exceed prescribed thresholds, automatic disengagement occurs, with pilot warning. This system also utilizes an equalization technique to minimize the effect of tolerances.

The autothrottle system (Fig. 3) used on the 727 airplane was developed specifically to enhance the performance of the automatic approach and landing system. It reduces the pilot's workload and allows more time for his role as the master monitor of the automatic approach progress. However, it is not a required subsystem of the 727 automatic landing system. Upon engagement, the autothrottle system operates the thrust levers to control the airplane indicated airspeed to coincide with the airspeed indicator speed command "bug" which is manually set by the pilot. The airspeed error signal is combined with a longitudinal accelerometer signal through a low-pass filter to minimize throttle activity in gusty air conditions. A pitch attitude signal is used to augment the airspeed error computation to reduce transient airspeed errors during normal airplane maneuvers. The autothrottle system complements the automatic flare maneuver by retarding the throttle levers at a fixed rate during the flare mode of autopilot operation.

The approach progress display is an array of annunciators used to inform the pilot of the autopilot mode of operation. Figure 4 illustrates the sequence of events during an automatic approach to touchdown and the associated indication displayed to the pilot. Six cues are given by amber and green indications from three light annunciators. An automatic approach to touchdown begins with the autopilot engaged and the airplane course set to intercept the localizer. When the automatic coupling mode is selected, the following occurs:

1) The localizer annunciator will be amber, indicating that the roll axis control system is in an armed state for automatic acquisition of the localizer beam.

2) The glide slope annunciator will be amber, indicating that the pitch axis control system is armed for automatic acquisition of the glide slope beam.

3) Upon intercepting the localizer beam, the localizer annunciator turns green, and the roll control system initiates capture of the localizer beam.

4) Upon intercepting the glide slope beam, the glide slope annunciator turns green, and the pitch control system initiates capture of the glide slope beam. Up to this point in the approach, only one pitch channel has been engaged. Upon intercepting the glide slope beam, the second pitch channel is engaged automatically, and fail-safe monitoring begins.

5) At approximately 1500 ft of altitude, the autopilot initiates localizer and glide slope gain programming in response to radio altimeter signals. Simultaneously, the roll monitor is activated, which in turn arms the autopilot flare mode. This is indicated to the pilot by the flare annunciator turing amber. The approach is now, and will remain, fully monitored to touchdown.

6) At approximately 50 ft of altitude, the pitch axis control system switches from the glide slope mode to the flare mode, indicated by the flare annunciator turning green. Normal time between flare initiation and touchdown is approximately 8 sec.

Table 4 Certification history of 727 AFCS

Category	Certification date	Minimum altitude, ft: Visual contact required	Minimum altitude, ft: Manual takeover required
I	Oct. 1963	200	78
II	Oct. 1965	100	78
II with automatic landing	June 1966	100	0

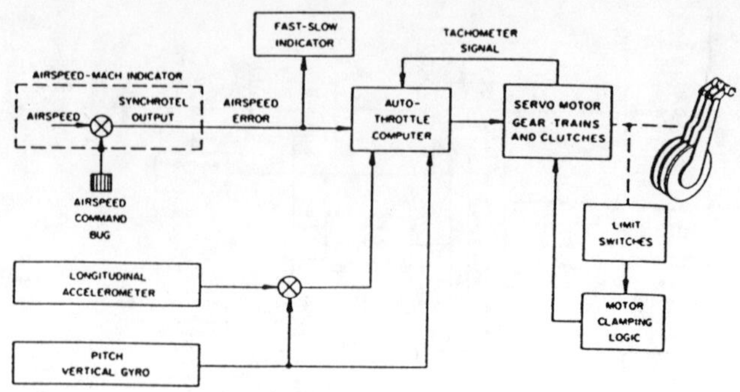

Fig. 3 727 autothrottle system.

Dual-Pitch Channel Description

In order to implement a fail-passive system, it is necessary to employ a means for reliably detecting both active and pas-

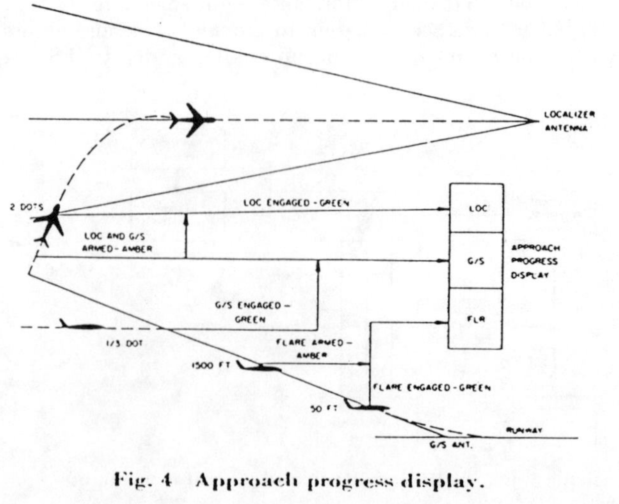

Fig. 4 Approach progress display.

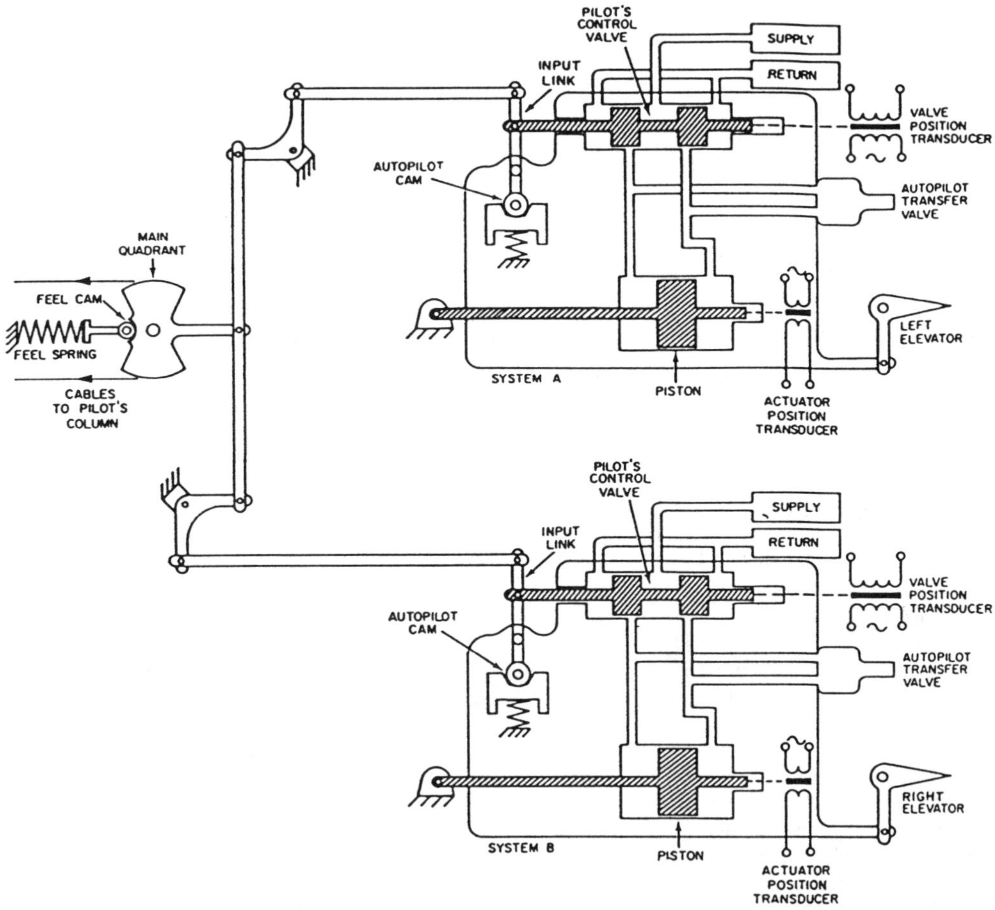

Fig. 5 Elevator control system.

sive failures, and a certain amount of duality is required, particularly at the sensor level.‡ Once the sensor signals are validated, the response of the remaining components to the signals can be validated by comparison monitoring techniques. Alternatively, the system can be completely dualized so that a failure of one channel is counteracted by the opposing channel. As indicated by the previous description of the system, the 727 AFCS autopilot uses both techniques. The simultaneously operating dual pitch channels are considered unique and are explored in more detail.

The 727 basic elevator control system contains dual electrohydraulic actuators. To appreciate fully the fail-safe protection of the 727 dual pitch channels, it is necessary to investigate the elevator actuators, control system linkages, and associated artificial feel system. This is most clearly done with reference to the mechanical schematic diagram of Fig. 5. For simplicity, the actuators are shown as single-stage devices driven by a pilot's control valve (manual operation). The feel system is shown as a simple mechanical spring. The 727 elevator actuators are actually two-stage devices, and the feel system varies the effective feel spring gradient as a function of flight condition.

It can be seen that the right and left elevators are not bussed together, but they are connected together through the actuator input linkages of the control system. Each elevator is powered from a separate actuator. In manual operation, when the pilot moves the control column, the control cables rotate the main quadrant. The feel spring provides an opposing force consisting of an initial preload plus an additional force proportional to quadrant rotation as it compresses the feel spring. The quadrant rotation is transmitted through the bell cranks and linkages, and displaces the pilot's control valves with respect to the actuator housings. As a result, hydraulic pressure is ported to the main pistons, causing motion of the actuator housings and corresponding deflection of the elevators. Motion of the actuator housings recenters the pilot's control valves.

In single-channel autopilot operation, hydraulic supply pressure is connected to one autopilot transfer valve. Also, the input link that drives the pilot's control valve of the same actuator is caged by a spring-loaded cam (autopilot cam); this prevents movement of the pilot's control valve with respect to the actuator housing. When the autopilot amplifier supplies d.c. current to the transfer valve, the valve ports hydraulic pressure to the main piston, causing motion of the actuator housing and a corresponding elevator deflection. As the actuator moves, it forces the control system linkages to follow it, thus rotating the main quadrant and compressing the feel spring. Note that the pilot's column follows the movement of the actuator, and the linkages of the opposite actuator cause the pilot's control valve of that actuator to move; that is, the opposite actuator is "slaved" to the autopilot-driven actuator, and both elevators deflect an equal amount. The actuator position transducer provides a feedback signal to the autopilot amplifier that cancels the command signal at the amplifier input and rebalances the transfer valve when the elevator has reached the commanded position. If a large command signal is present at the amplifier input, the actuators will move and deflect the elevators up to the point where the reaction force of the feel spring is greater than the spring preload force at the autopilot cam. Then a small increase in actuator displacement will cause the input link to rotate (since the autopilot cam spring compresses), and the pilot's control valve will move. The mechanical advantage and relative valve authorities are such that a small motion of the pilot's control valve will completely cancel the hydraulic input from the autopilot transfer valve, and further actuator

‡ In some cases, equivalent reliability may be obtained by in-line monitoring of a single sensor, but some sensors have failure modes that cannot be practically covered by this technique.

displacement is prohibited. Thus, it is seen that autopilot elevator authority is determined by the feel spring gradient relative to the autopilot cam spring preload.

During dual-pitch channel operation, both actuator input links are caged by their respective autopilot cams, and each autopilot transfer valve is driven from a separate autopilot channel. As long as the two autopilot commands agree, both actuators (and elevators) will respond together. Also, since the reaction force of the feel spring is distributed equally between the autopilot cams, the autopilot authority is doubled.

Assume now that the two autopilot commands do not agree. In a typical case both systems are initially balanced, with the elevators faired. If a command is transmitted to system A transfer valve only, then actuator A will start to move, and in so doing, it will attempt to rotate the main quadrant as well as the input link to system B actuator. However, to do so it would have to overcome the preload force of the feel spring plus the preload force at B autopilot cam. Since the sum of these preload forces is greater than the preload force on A autopilot cam, that cam spring will yield, allowing the input link of actuator A to rotate and move its pilot control valve. This control valve input cancels the transfer valve input for a very small elevator deflection. The airplane will start to respond to the small elevator deflection, but as it does so, the sensors of the autopilot B supply an opposing command to transfer valve B. As actuator B tries to move and rotate the main quadrant, the preload forces of the main quadrant and A autopilot cam must be overcome. Again, the sum of these forces is greater than the preload force on B autopilot cam, and its cam spring yields. This allows the input link to the actuator B to rotate and move its pilot control valve. Now the input from the transfer valve B is cancelled; the two elevators are deflected slightly in opposing directions, and the airplane continues on its path undisturbed. Therefore, the dual system is inherently fail-passive.

In the following discussion the phrase "cam-out" will be used with reference to an actuator, when its autopilot cam spring preload has been overridden and its transfer valve flow is cancelled by flow from the pilot's control valve. Also, it will be seen in Fig. 5 that linear transducers are connected to the pilot's control valves. The polarity and magnitude of their signal outputs are indicative of pilot valve travels, and supply intelligence when cam-out occurs, as well as the direction of commanded surface which caused the cam-out. These transducers will be referred to as "cam-out sensors."

Equalization

In operation of dual autopilots, it is to be expected that tolerances will be such as to result in unequal commands to the two transfer valves. This does not affect performance, as it can be shown that this system will respond to the smaller of the two commands, with one actuator controlling. The other actuator will be cammed-out and slaved to the controlling actuator. However, the cam-out transducers provide intelligence to a failure-warning system, and nuisance failure warnings cannot be tolerated. Therefore, means were devised for equalizing the commands from the two autopilot channels (Fig. 6). There is a certain amount of compliance in the control system linkages, so that relative elevator positions can disagree by approximately 1° before the force unbalance of the system is enough to initiate cam-out of an actuator. An equalization reference signal is obtained by subtracting the two actuator position transducer signals. This error signal is fed back to the two autopilot channels in a sense to increase the command to the elevator of least deflection, and decrease the command to the elevator of greatest deflection. The equalization signal is fed back to the output amplifier of each channel for short-term correction, and to an integrator in each computer signal path for long-term correction. The "authority" of the equalization signal is very important

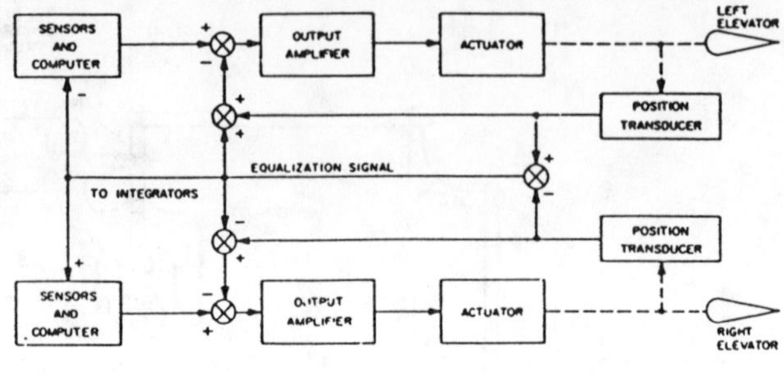

Fig. 6 Dual-pitch channel equalization.

since it tends to propagate a failure from one autopilot channel to the other channel. With the implementation chosen for the 727, the equalization signal authority is inherently limited at the sensor level by the mechanical compliance of the system. (The elevators can never disagree in excess of the compliance range without causing actuator cam-out.) Gain authority is controlled by using redundant passive circuit elements whose failure modes cannot result in a gain increase. Hot-wire fault protection is inherent in that a fault at any point in the signal chain will cause dual opposing actuator cam-outs, shown previously to be a passive failure. The authority of the equalizer signal is restricted then to only that required to compensate for tolerance build-up in the two pitch channels, and thus keeps the elevators tracking within the compliance range below the cam-out threshold. For example, in the 727 a hardover failure at one autopilot output amplifier will result in dual opposing cam-outs of the actuators for a pitch attitude transient of less than 1 deg, and automatic disconnect will occur in less than 1 sec, warning the pilot and leaving the flight path essentially undisturbed.

A most desirable goal has been achieved through the previously described implementation: the components of the system can be maintained within practical tolerances and will provide dependable performance that is both safe and free from nuisance disengagements. It is not enough for a system to be reliable in terms of the probability of detail part failures, and safe in terms of fault detection. If it is necessary to maintain the components to unreasonably close tolerances in order to preclude nuisance disengagements, then such a system will not be dependable and will not contribute to lowering the operational weather minimums.

Monitoring of the Dual-Pitch Channels

An exhaustive failure analysis has indicated that all active and passive modes of failure are manifested by the relative states of only four signals in the dual channel system. The signals of significance are the two cam-out transducer signals and the two signal outputs from the autopilot output amplifiers. Certain relative states of these signals are indicative of malfunctions, whereas other relative states are indicative of normal performance. Fail-safe monitoring, with automatic disengagement and pilot warning, required the transposition of the states of these four signals into corresponding states of redundant relays, in a fail-safe manner. The mechanism for achieving this transposition has been referred to as a fail-safe amplifier. Under normal operation, the output of this amplifier energizes two pairs of relay coils. In order that these relay coils remain energized, it is necessary that the amplifier properly transmit an 800-cycle signal, applied at its input. In this way, the fail-safe amplifier is self-checking. A signal that is to be monitored is connected to the input of the fail-safe amplifier. If the level of the monitored signal exceeds a prescribed threshold, it will cancel alternate pulses of the 800-cycle reference signal and result in de-activating the associated pair of output relays. The pair of output relays which is de-activated corresponds to the phase of the monitored signal. The integrity of the monitored signal path is

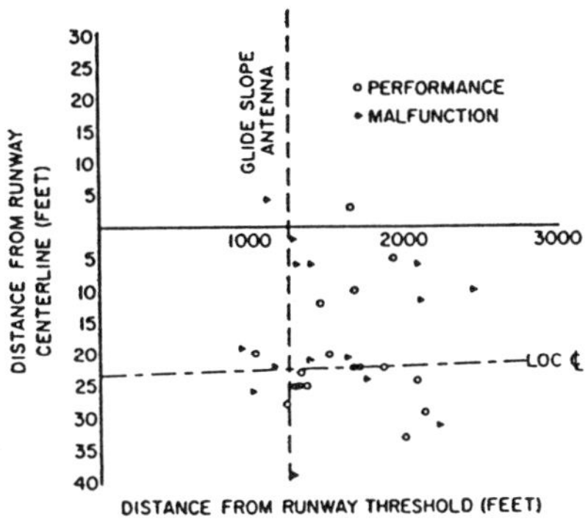

Fig. 7 Touchdown dispersions, Runway 29—Oakland.

established by tracer signal techniques. This is not intended to be a failure analysis of the monitoring system, but rather to indicate the basic philosophy used in monitoring the dual pitch channels. By transposing signal states into corresponding states of redundant relays, and then by interconnecting the contacts of these redundant relays in the proper fashion, the monitor provides automatic disengagement in the event of a malfunction in the system. At the same time, it ignores those conditions resulting from normal operation.

The redundant relay contacts of the fail-safe monitor are interconnected to perform the following functions:

1) Disengage the system if a single cam-out is indicated unless the cam-out agrees with the command from the autopilot output amplifier. This protects against a failure in the actuator which would cause it to revert to the slaved mode, or a failure of the transfer valve (active or passive).

2) Disengage the system if dual opposing cam-outs are indicated. This protects against any active or passive failure in the sensors and electronics or an active failure of the actuator.

It is significant to note that the monitor ignores a single cam-out when it agrees with the sense of the autopilot command, as well as a dual cam-out if the cam-outs are in the same direction. Neither of these conditions reflects a fault. Thus, the monitor has met the objective of detecting true failures and eliminating nuisance disengagements, thereby providing both the safety and dependability required for lower weather minimums.

Certification Program—Laboratory Phase

Prior to flight testing the fail-passive automatic landing system, laboratory tests were carried out with the autopilot hardware in a simulated airplane environment. The test setup consisted of coupling the preproduction autopilot hardware to the 727 Flight Controls Test Rig and an analog computer simulation of the 727 airframe equations of motion. The objectives of the laboratory tests were to evaluate the effect of system tolerances on both performance (including safety) and dependability (susceptibility to nuisance failure alarms) and to verify the monitor responsiveness to simulated malfunctions.

System failure modes investigated in previous autopilot programs involved only passive or instantaneously saturated malfunctions of signal paths. Since the automatic landing system is operating in one of the most critical phases of flight, the failure detection monitor was required to detect time dependent failures (ramp-type malfunctions), e.g., a slowly changing vertical gyro reference caused by a failure in its erection circuit. The failure mode investigation included taking into account those tolerance distributions that would be most significant in reducing the monitor sensitivity for detecting each failure condition.

The laboratory test results were combined with a detailed description of the monitoring techniques to form a "failure analysis" document of the automatic landing system. This document was submitted to the FAA along with a proposed flight test demonstration specification document in request for an FAA Type Inspection Authorization (TIA). The flight test specification called for demonstrating the failure modes which the laboratory study showed to be most critical, as well as demonstrating the system automatic touchdown performance.

A TIA for the Model 727 Automatic Approach and Landing System was issued by the FAA, dated December 28, 1965. Basically, the TIA approved testing of an automatic approach and landing system configuration to meet performance requirements of Category II and automatically land the airplane under visual flying rules (VFR) conditions below the Category II weather minimums. The automatic landing performance was to be evaluated using best known criteria at that time without benefit of established approved criteria by the FAA.

FAA Flight Test Evaluation

In order to demonstrate the automatic landing system to the FAA, the autopilot had to be in a production configuration, and the test airplane had to reflect production standards for all equipment that interfaced with the autopilot. The TIA required that specific ground checks be made and witnessed by FAA personnel, to substantiate that the automatic landing system installed on the test airplane conformed to the drawings and diagrams presented to the FAA in request for a TIA. Bench and airplane ground tests also substantiated that the monitor threshold levels were at their least sensitive values with respect to malfunction detection, and that the control system power packages reflected the worst case tolerance situation established by the system failure analysis for the most critical system failures.

In-flight demonstration of the system to the FAA began on March 15, 1966. All simulated malfunction testing and the majority of performance testing was conducted at the Oakland International Airport in California, a Category II ILS facility. Performance testing was also conducted at Portland International Airport in Oregon, Seattle-Tacoma International Airport in Washington, and Boeing Field in Seattle to provide cross-comparison data of the system operation at different ILS ground facilities. FAA demonstration flight tests, involving 72 approaches, were 90% complete after 3 days of flying. The FAA required that satisfactory system performance be demonstrated under moderate turbulent, windshear, and crosswind air conditions. These air conditions were finally found at Stapleton International Airport, Denver, Colo., where the flight tests were completed on April 29, 1966.

Thirty-four approaches were conducted at Oakland with simulated malfunctions. All malfunctons were inserted at altitudes below 100 ft (Category II MDA). At no time was the FAA pilot made aware whether a given approach was to demonstrate performance or to demonstrate the system responsiveness to a failure. The types of malfunctions demonstrated were as follows: 1) autothrottle advance and retard hardovers during flare, 2) elevator nose-down and nose-up hardovers at approximately 80, 50, and 30 ft of altitude, 3) elevator nose-down and nose-up ramp fault commands at approximately 80, 50, and 30 ft of altitude, 4) failure of the system to provide a flare command, 5) aileron hardover at flare, 6) roll vertical gyro ramp fault at approximately 80 and 50 ft of altitude, and 7) both yaw dampers inoperative and one operative with a hardover at flare.

The ramp-type malfunctions were inserted both above and below the monitor threshold. At no time did the FAA pilot respond to an inserted malfunction by visual cues before the autopilot monitor disconnected the system. Most approaches were accomplished with the pilot remaining under the hood well below the Category II MDA, and in some cases, below 50 ft. Recognition of a system malfunction by a disconnect of the autopilot was followed by a manually con-

Fig. 8

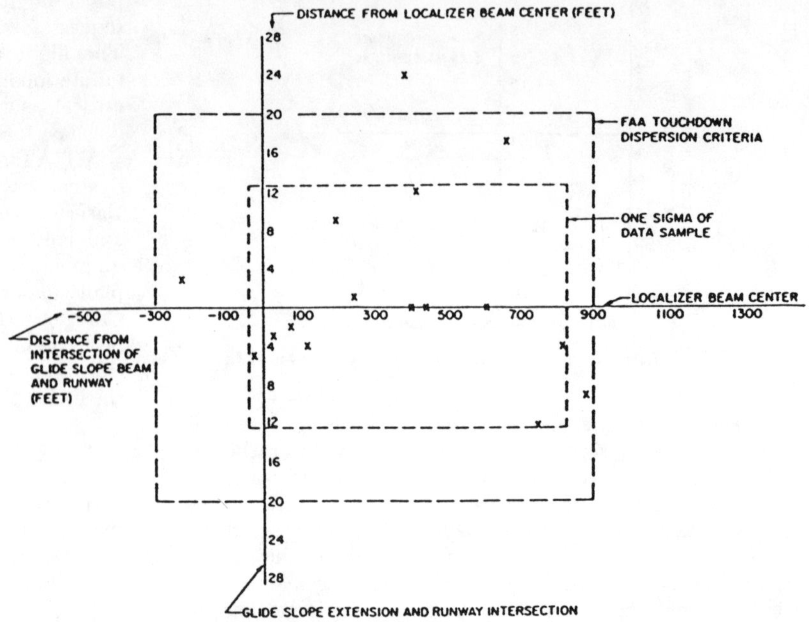

trolled go-around or landing at the discretion of the pilot. Failure of the flare mode was demonstrated at the request of the FAA to assure that the airplane remained in a nonhazardous attitude following fade-out of the glide slope control mode.

Sixteen approaches were conducted at Oakland to demonstrate the automatic approach and landing system performance to touchdown. All met the Category II performance criteria to 100 ft, and were continued to touchdown. Figure 7 illustrates the main gear longitudinal and lateral touchdown dispersion relative to a line on the runway designating the glide slope antenna location and the runway centerline. Included in Fig. 7 are the touchdown points for approaches where simulated malfunctions were undetected or were detected by the autopilot monitor just prior to touchdown. No specific FAA criteria existed with which to compare the performance results of the automatic landing system. However, the flight test results were compared with the results of laboratory studies, and all results agreed within 15% of predicted error values.

The approaches conducted at fields other than Oakland, to demonstrate system performance at several ILS facilities, exhibited touchdown dispersions comparable to those shown in Fig. 7. The Denver approaches satisfied the wind conditions of moderate turbulence, as classified by the FAA pilot, and were conducted with crosswind components between 13 and 26 knots as calculated from the airport tower reported winds. All approaches, whether for performance or malfunction detection, were considered by the FAA pilot to provide the controllability desired of an automatic approach and landing system, with an adequate margin of safety when the autopilot automatically disconnected upon recognition of a malfunction. Type certificaton for the 727 automatic approach and landing system was received from the Federal Aviation Agency in June 1966.

Conclusions

As far as Boeing and Sperry are concerned, the 727 automatic approach and landing system met the self-imposed design goals established at the outset of the program. Industry and the FAA have yet to detail an acceptable performance criteria for Category IIIA automatic landing systems.

However, an Agency Advisory Circular draft was issued last spring specifying tentative performance criteria for automatic landing under Category II weather minimums. In all probability this circular will, with some modifications, be the basis for performance criteria for Category IIIA automatic landing systems.

Let us now compare the results of the 727 performance testing at Oakland with the FAA first draft criteria for automatic landing under Category II weather minimums. The FAA circular states that the aircraft centerline and main gear touchdown dispersion should be shown on at least a one-sigma§ probability to lie 1) laterally, within ±20 ft of the ILS localizer beam center, and 2) longitudinally, within −300 and +900 ft of the intersection of the glide slope beam center and the runway.

Figure 8 shows the performance touchdown data, presented in Fig. 7, replotted with respect to the Oakland ILS localizer beam center. For the sample shown, the one-sigma distribution falls well inside the FAA boundary. In fact, the FAA boundary can well represent a two-sigma distribution. However, note that this data reflects only a small sample of touchdowns, and an accumulation of in-service data must be gathered under normal operational environment in order to substantiate any conclusions drawn here.

The 727 automatic landing system represents the AFCS state-of-the-art as it exists today. The system has the capability both in performance and safety to extend the weather minimums to Category IIIA. As new techniques are developed in the area of "fail-operational" approach systems, whether the philosophy be complete automatic control or specific roles for both pilot and autopilot, the 727 system has the potential for growth to an all-weather blind-landing system.

§ These are likely to become two-sigma criteria.

SOLID-STATE VORTAC WITH REMOTE MAINTENANCE AND MONITORING

A. Lang

ITT Avionics Division
390 Washington Avenue
Nutley, New Jersey 07110

W. Hundley

Wilcox Electric Company
14th and Chestnut Streets
Kansas City, Missouri 64127

ABSTRACT

The existing VORTAC and VOR/DME equipment in the National Airspace System is being replaced by the Federal Aviation Administration (FAA) with new equipment utilizing modern technology to provide remote maintenance and monitoring capabilities and improved system availability. The introduction of microcomputers, together with the latest solid-state technology provides a highly reliable, 100% solid state VORTAC system. Details of such a VORTAC beacon system, with the VOR and Tacan equipment designed for low cost, high reliability and optimum maintainability, are described. Incorporated with the VORTAC is a remote maintenance and control system which provides the capability for the system certification parameters to be remoted via existing communication lines from each installation to a central maintenance facility. Thus, the numerous trips now being made to the VORTAC installations for routine checking or for certification are eliminated. Use of microcomputer techniques permits the remote maintenance and control system to be used for all station control functions, as well as for remote diagnostic and/or trend monitoring to further improve maintenance. The paper describes how the use of this system offsets the initial procurement costs by savings in maintenance costs.

INTRODUCTION

VORTAC and VOR/DME have been widely implemented as prime navigation aids for the present National Airspace System. Also, numerous foreign countries are installing or expanding their use of these navigational aids. Because of this broad use and the high capital investments in the system, indications are that both the USA and ICAO will extend these services beyond the 1995 date that this service is presently protected to. The extension of the operational time of these systems, however, is incurring an increasingly high cost for maintenance and is posing problems with obtaining replacement parts for the beacons. To resolve this situation, the FAA has let a contract to a team consisting of the ITT Avionics Division and the Wilcox Electric Company to replace all of the old tube-equipped VORs, VOR-DMEs, and VORTACs with a total system architecture that embodies modern technology. This paper describes the system being produced for the FAA.

The primary objective for this new equipment is to reduce the overall life-cycle costs for the VORTAC - VOR/DME equipments. This is accomplished by a reduction in maintenance costs resulting from:

- Improved reliability of an all solid-state design.
- Improved maintainability through modular construction with automatic fault isolation.
- Reduced maintenance by use of solid state and digital circuits.
- Reduced routine trips by maintenance technicians to each facility, from one or two per week to once per 3 months, by remote certification and maintenance testing.

The resulting savings in maintenance costs due to these reductions will more than offset the acquisition costs and, therefore, result in a cost-effective program with an estimated 5 to 6 year payback.

The new equipment continues to provide present VORTAC navigation service and performance. Also, improvements in technical performance are made where new technology permits such improvement at low initial costs. In addition to the new maintenance features mentioned above, other added capabilities include:

- Remote control of VOR and DME/Tacan operating parameters.
- Remote adjustment of VOR and DME/Tacan monitor alarm limits.

Reprinted from *Rec. IEEE Position Location and Navigation Symp.*, pp. 198-207, 1980.

- Solid state VOR voice identification.
- Facility battery backup to provide uninterrupted operation after primary power failure.
- Backup modes of operation such as fall-back from Tacan to DME operation.
- Y-channel operation.
- Increased DME/Tacan traffic handling capability.
- Correction of problems limiting present system performance through features such as remotely adjustable DME/Tacan Echo Suppression and Increased Traffic capabilities.
- Operation at shelter temperatures of $-40^{\circ}C$.

In addition to reducing the need for trips to the facility and providing improved performance and energy savings, these added capabilities will provide increased on-air availability of the navigation service.

VORTAC SYSTEM DESCRIPTION

Figure 1 shows a typical VORTAC facility as it exists today. This is one of approximately 725 VORTACs located throughout the National Airspace System that are operated and maintained by the FAA. These VORTACs, together with approximately 145 VOR/DME and 80 VOR-only facilities, provide navigation service to civil and military aircraft over the continental U.S., Alaska, Hawaii, and other U.S. territories. The FAA contract to ITT Avionics Division/Wilcox Electric Company provides for the replacement of the equipment in all of these facilities by 1984.

Figure 1. VORTAC Facility

When the new equipment is installed, there will be a change in the appearance of the VORTAC facilities as shown by figure 1. The existing VOR and Tacan antennas will remain. However, an array of 16 VOR monitor antennas will be added around the circumference of the counterpoise to enable a completely automatic VOR ground check. This will eliminate the need for a lengthy test now made once per month. Also, a new DME antenna will be added to the VORTAC facilities. This will enable backup DME coverage in case of a Tacan azimuth function failure or a partial loss of output RF power from the Tacan. Thus, DME navigation service, utilized by civil aircraft, is maintained.

Inside the VORTAC shelter, the 16 cabinets presently comprising a typical VORTAC installation will be removed and replaced with the 3 cabinets shown by figure 2. This reduction in equipment size permits the partitioning-off of a large volume of the shelter, thereby reducing the heating and cooling requirements. This energy savings is another aspect in the reduction of system operational costs.

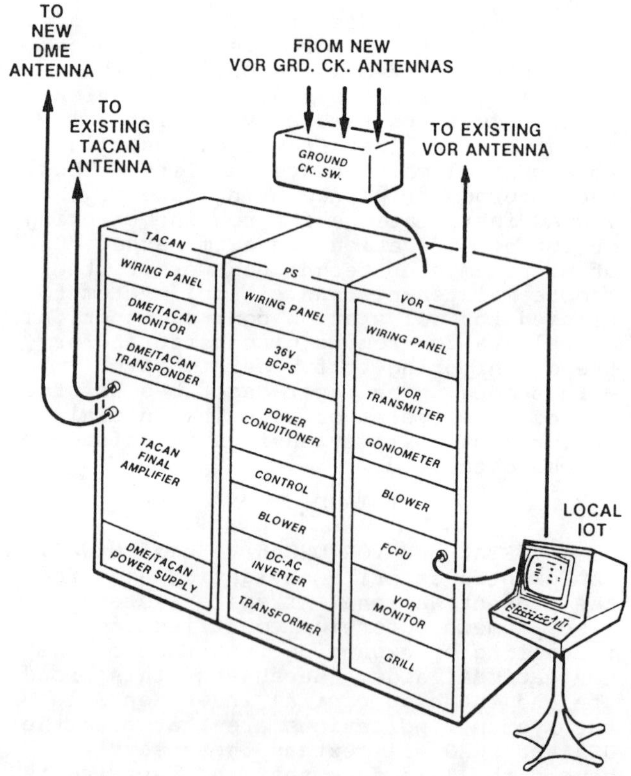

Figure 2. New VORTAC Equipment

The Tacan transponder/power amplifier and monitor equipments are enclosed in the single left-hand cabinet. The VOR transmitter, goniometer, and monitor are contained in the right cabinet. These equipments are completely solid-state, including the 5 KW Tacan final amplifier. Computer technology has been used extensively in the design to provide automated monitoring, control and reporting of system functions. Both the Tacan and VOR include dual monitors to assure reliable navigation service with an extended mean-time-between-outages. Each monitor contains a microcomputer that, in addition to controlling monitoring of the key system navigation functions, also provides complete system control, adjustment of equipment operating parameters, all maintenance measurements, and the storage and reporting of all data.

The control and data from the monitors are remoted by the Facility Central Processing Unit (FCPU) contained in the VOR cabinet. This unit, also microcomputer-controlled, includes an interface with a video input-output terminal (IOT) and a teleprinter. All local control and maintenance operation of the system is provided by these equipments. The FCPU also provides other facility control and status monitoring functions and communications with equipment at remote locations.

The center cabinet of figure 2 is the Power Supply (PS) which supplies the main d.c. voltages to the VOR, Tacan, and FCPU. This unit, in conjunction with the associated float-type battery bank, also enables uninterrupted service during the switchover to the standby generator operation after failure of the a.c. primary power. Finally, two hours of battery operation is available after complete loss of a.c. power.

The equipments are designed for maximum commonality of the VORTAC, VOR/DME, and VOR facilities. At a VOR/DME, the only significant differences from the VORTAC are that the 5 KW Tacan Power Amplifier is replaced by a 1 KW DME Power Amplifier and that the primary a.c. power is single phase. A VOR-only facility is the same as a VOR/DME except that the DME cabinet is eliminated. Thus, operation and maintenance of all three types of facilities is almost identical.

The operational configuration of this next generation VORTAC system is illustrated by figure 3. All of the units shown, apart from the basic Tacan and VOR Units themselves, are added to fulfill the basic objectives of the program; that is, to reduce maintenance costs by providing a remote maintenance and monitoring capability and to provide improved system availability.

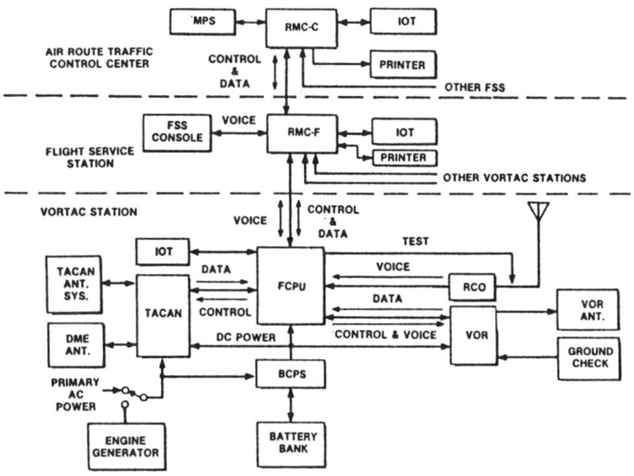

Figure 3. VORTAC System

The equipment at the VORTAC facility communicates, via the FCPU, to a Remote Monitor and Control Unit (RMC-F) located at the Flight Service Station (FSS). Up to 8 VORTAC facilities can communicate with an RMC-F. Each RMC-F, in turn, is connected to an RMC-C, located at an Air Route Traffic Control Center (ARTCC), via a communications network which permits up to 128 VORTACs to report to a single RMC-C. Twenty-three such communication networks (one corresponding to each of the 23 ARTCCs), will ultimately be implemented. All of the system control and data reporting, as discussed later, is available at any terminal within this network. The system allows for all status reporting and facility maintenance to be implemented at the FSS (as configured with the present VORTAC system), as well as a centralized status and control at each ARTCC. This complete Remote Maintenance and Monitoring (RMM) capability is projected to reduce the need for routine on-site trips to once per 3 months and potentially to once per 6 months, significantly reducing maintenance costs.

The functional navigation subsystems (i.e. Tacan/DME and VOR) and the power supply equipment located at each VORTAC facility are described in the following paragraphs.

Tacan/DME Subsystem. The Tacan subsystem, which operates with the existing Tacan antenna system and a new DME antenna, consists of a dual monitor, a transponder, and a 5 KW power amplifier unit. A DME subsystem, used in the VOR/DME facilities, contains essentially identical dual-monitor and transponder units; however, the power amplifier is a 1 KW unit. Figure 4 is a block diagram of the Tacan subsystem. As shown, portions of this system are dualized. Additionally, the Tacan power amplifier has a parallel redundancy of 10 PA modules. This selective redundancy of system elements provides an MTBO of 9600 hours for the Tacan subsystem, which is greater than that provided by the presently installed fully dualized system.

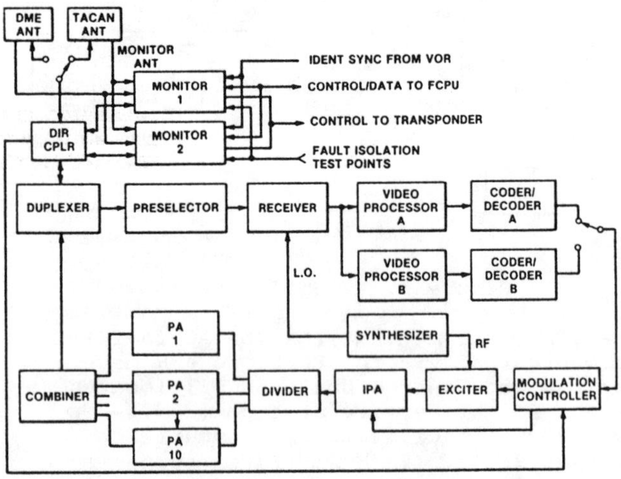

Figure 4. Tacan System

The RF signal received from the antenna is applied through the antenna switch, directional coupler, and duplexer to the receiver input circuits. A low-noise RF preamplifier at the input of the receiver provides a sensitivity of better than -94 dBm. This sensitivity can be remotely reduced to levels between -60 and -94 dBm to tailor a system to a particular site.

The video processor and coder/decoder circuits following the receiver are redundant. This is part of the selectively redundant design to improve system reliability. Programmable digital techniques are used in these circuits to provide flexibility in operation. Upon commands from the operator and/or monitor, X/Y timing and the Tacan/DME mode are selected. Also, the echo suppression threshold and gate width, together with the dead-time gate width, are adjustable to "trim" a system to the echo conditions existing at a partgicular site. The reply delay is operator-adjusted and then automatically controlled by use of the system reply delay measurements made in the monitor, thus providing precise system delay performance. In a time-shared arrangement, the programmable counters are used to generate the azimuth reference bursts, identity pulses, and random squitter pulses. The replies and squitter are added with the proper priority to the azimuth reference bursts and the identity code signals to form the modulation pulse train.

The pulse train from the coder/decoder is received by the modulation controller, which generates gating pulses for the exciter and video waveforms to modulate the Intermediate Power Amplifier (IPA). The major function of the modulation controller is to develop modulation pulses that result in an output RF signal that meets all of the requirements for a Tacan/DME system (i.e. MIL-STD-291B and ICAO Annex 10). Unique methods are employed to achieve an RF output of 5 KW, a spectrum such that the power contained in a 500 KHz band centered on frequencies ±800 KHz removed from the on-channel frequency is more than 50 dB below that contained in a 500 KHz band centered on the on-channel frequency (and more than 70 dB at ±2 MHz), and an output pulse train whose amplitude remains constant from pulse-to-pulse and throughtout an azimuth reference group. The problems introduced by solid-state circuits require special methods of correction to achieve these characteristics. For instance, thermal heating of transistors can cause the operating characteristics to change during a pulse. A feedback loop is used in the exciter to correct this problem. Also, the transfer characteristics exhibited by a solid state power amplifier require unique modulation waveform generation and RF drive. This video modulation waveform is generated in the modulation controller and applied to a modulator located in the IPA. The precise waveform needed to ensure compliance with the required spectrum emission levels is controlled by a feedback loop to the modulation controller from the final output.

The RF carrier is generated by a frequency synthesizer. The frequency of this unit is remotely selected under computer control. The CW output of the synthesizer is amplified and gated by the exciter to generate rectangular-shaped RF pulses. The exciter provides the RF drive to the IPA, which in-turn, drives the final amplifier with the modulated RF pulse.

The 5 KW power output from the final amplifier is derived by combining the amplified output of ten PA amplifier modules using low-loss ten-way, divider and combiner networks. The final amplifier design provides full power over the complete Tacan/DME frequency bandwidth with no tuning. Each of the PA modules in the final amplifier contains one high-power transistor driving four parallel identical transistors in the final stage. This configuration provides sufficient gain and power capability to attain the 5 KW over the full band. Each PA module includes an efficient array of 2 way RF power dividers and combiners which provide equal power distribution across the frequency band with low insertion loss, and an isolation in excess of 20 dB between transistors, allowing for fail-soft operation (i.e., a failure of one transistor does not affect the operation of other transistors). The outputs of the PA modules are then combined (in a 10-way combiner which is identical to the divider) to produce greater than 5000 watts. The combiner exhibits extremely low loss and an interport isolation exceeding 20 dB. This characteristic is important to prevent interaction between PA modules in case of degradation of any PA module.

The final amplifier for a DME system differs from that of the Tacan in that only 2 PA modules are used to derive the 1 KW required power output. These 2 PA modules are idential to the 10 PA modules used in the Tacan. This, together with the expansion capability of the equipment rack that houses these modules, allows conversion to a Tacan system with minimum modification.

Each PA module, and the IPA module, include a sensor for each output transistor. The sensor signal is applied to the monitor where the status of each transistor is determined. The degradation of any module can thus be readily detected and pin-pointed. Failure information is then sent to the remote site via the FCPU. Because of the redundancy provided by the power amplifier configuration, approximately 25 percent of the transistors can fail in the final PA stages before the power drops 3 dB to the alarm limit. Therefore, there is no immediate need to repair a failed module. The remotely indicated failure information provides the technician with an up-to-date status of the equipment and permits him to determine whether he can wait until the next normally scheduled maintenance trip for repair.

A backup mode of operation is provided by the RF antenna switch, which connects the system to either the Tacan antenna or DME antenna. This switch is normally activated to the Tacan position. It selects the DME antenna when a failure occurs in the Tacan antenna or when the RF power output drops below the monitor alarm limit (nominally 3 KW). Simultaneously, the Tacan azimuth function is removed and the output power is dropped to 1 KW. Thus, normal DME operation can continue until maintenance action is completed.

The Tacan/DME monitor unit consists of two functional monitors and provides executive monitoring, maintenance testing, and control of the system. Both monitors are continuously and independently monitoring the system operation on a once-per-second basis. One of the monitors is designated as "controller" and controls the operation of the system. Each monitor also continually exercises a fail-safe test routine. In case a failure is detected in the "controller" monitor, the redundant monitor will assume control of the system. Thus, complete redundancy of the system control, as well as the monitor functions, is provided.

General monitor operation is initiated by RF signals which are applied to the receiver input for monitoring and test. The RF signal originates in a frequency synthesizer which is identical to the synthesizer in the transponder. On-channel, as well as ± 200 KHz, and ± 900 KHz test frequencies can be selected by the microcomputer for receiver selectivity measurements. The RF output is pulse modulated and fed to the transponder receiver via the directional coupler. This equipment also serves as a versatile built in RF signal generator. The RF output can be either CW or pulsed, and can be adjusted from -110 dBm to -30 dBm on a pulse-pair to pulse-pair basis upon command from the microcomputer. The pulse width, pulse pair spacing, and the pulse repetition rate can also be varied. For test purposes, pulse rates of from 10 to 10,000 ppps having an exponential or regular distribution are selectable.

During normal executive monitoring, interrogations at a repetition rate of 30 ppps are applied at two different RF levels to the transponder to monitor receiver sensitivity reply and system delay. The replies to the monitor interrogations, as well as other transmissions, are detected and multiplexed to a counter to measure the system delay time. Other pulse parameters, such as peak amplitude, are also measured. A quantitative value is obtained for each of the monitored parameters. These values are compared against limits which are stored in the microcomputer, and if out of tolerance, a fault condition is established and the system is transferred to a backup mode of operation or shutdown. The quantitative value of each measurement is retained in memory for future analysis. The radiated Tacan signal, basically used for azimuth monitoring, is received from the Tacan (or DME) monitor antenna. It is detected in the monitor, and multiplexed to circuits used for other monitor tests which are time-shared with the azimuth measurements. A computer program utilizes a digital tracking loop to make the azimuth error calculation.

System test and certification functions are conducted using techniques similar to those for normal executive monitoring. However, in these functions, interrogation pulse spacing, amplitude, and repetition rates are varied depending upon the test. Quantitative measurements are made on all parameters and the data is forwarded to the FCPU. The monitor includes built in test equipment to generate azimuth calibration, clock signals, and other signals necessary to certify correct monitor operation and alarm limits. Normal monitoring operation continues unaffected by these tests.

VOR Subsystem. The VOR subsystem, which operates with the existing VOR antenna system, consists of the Transmitter, Goniometer, Monitor, and Automatic Ground Check Antenna Array.

Figure 5 is a block diagram of the VOR subsystem which illustrates the manner in which these major units function together.

The Tranmsitter, starting with a precise frequency, low level output from a temperature compensated crystal oscillator (TCXO), modulates and amplifies that signal to provide a properly modulated high level VOR carrier to the antenna bridge. The carrier power is 150 watts minimum into an unbalanced 50 ohm resistive load at the end of 50 feet of RG-214U coaxial cable with a VSWR up to 1.5. The power level can be adjusted remotely or locally via keyboard over a range from 150 watts to 25 watts with 0.5 db stability. The carrier is amplitude modulated in a low level PIN diode modulator with the 9960 Hz subcarrier to form the VOR reference signal. In addition the carrier is also amplitude modulated by the 1020 Hz keyed identification signal and voice (including voice identification). Maintaining a low level on the sideband harmonics of the 9960 Hz subcarrier is an important operational consideration particularly if the new 50 kHz VOR channel spacing should be widely implemented. Sideband harmonics are specified to be 2 db better than the ICAO Annex 10 requirement.

The Goniometer, which also starts with the TCXO output, provides the antenna bridge with two balanced double-sideband suppressed-carrier signals offset ± 30 Hz

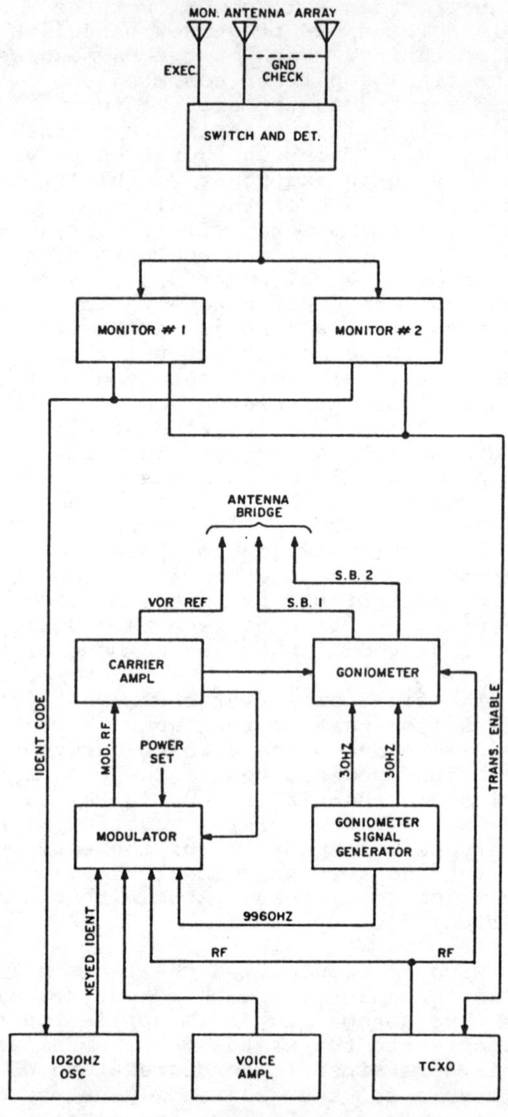

Figure 5. VOR Subsystem

from the carrier. These sidebands automatically track the carrier power level through feedback control in order to provide constant modulation percentage. The audio components of the two double sideband pairs are in phase quadrature and are switched 180° in RF phase on alternate half cycles of the 30 Hz navigation signal. These RF double sidebands are combined in the antenna system to produce a 30 Hz amplitude modulation on the carrier which is the variable signal for conventional VOR. When combined in space with the reference signal the result is the basic VOR navigational signal. In order that this be radiated with the required accuracy it is essential that the parameters which define the sideband pairs developed by the Goniometer be tightly controlled. For the new VOR system this includes specifications on the following sideband parameters:

- Phase quadrature of audio components - $\pm 0.2^{\circ}$
- RF phase tracking with carrier - $\pm 3^{\circ}$
- Sideband amplitude balance - 1%
- 30 Hz harmonics - Close-in harmonics better than 30 db.

The Goniometer unit also provides the Transmitter with the 9960 Hz subcarrier, frequency modulated with the 30 Hz reference at a modulation index of 16.

An additional attribute of the new VOR is that the relative phase of the 30 Hz variable and the 30 Hz reference signals can be adjusted from the keyboard; thereby providing remote control of station alignment to North. Granularity of the adjustment is better than 0.01° over a range of $\pm 10^{\circ}$.

The VOR Monitor utilizes a microcomputer design approach, and in conjunction with certain elements of the automatic ground check system, (specifically the monitor radial antenna and detectors) performs the VOR executive monitoring function. In the interest of reducing the mean time between outages the monitoring function is implemented with dual monitors. The detector output from the executive monitor antenna is provided to both monitors which examine the executive monitor parameters of the signal on a once per second basis for possible out of tolerance conditions. Where two monitors are functional it is required that both monitors agree that an out of tolerance condition has persisted for the prescribed time before the VOR is disabled. If there is a monitor disagreement, the FCPU runs an integrity test on the "happy" monitor, utilizing the VOR test generator. If the "happy" monitor should fail, the system will be shut down. In addition to the integrity test each monitor periodically undergoes a microprocessor controlled self-test routine.

VOR signal monitoring does not occupy the VOR microcomputer anywhere near full time. This enables the VOR microcomputer, usually in conjunction with the FCPU microcomputer, to perform a number of other tasks. Included among these tasks are automatic ground check; automatic certification tests; station parameter control; identification keying and data collection for purposes of status reporting, fault history files, fault isolation, trend analysis and special tests.

The automatic ground check is particularly interesting. The VOR ground check equipment consists of 16 monitor antennas with radomes and a single pole, sixteen throw solid state switch. The output of the switch operates into the monitor detectors and processing circuits. VOR ground check may be programmed to be performed periodically or may be performed upon operator request. The ground check antennas are switched to the VOR monitor in a synchronized manner, under control of the Monitor microcomputer, and an azimuth phase measurement is performed for each antenna radial. During ground check, only azimuth phase measurements are made on signals received from the ground check antennas. The process of VOR ground check is performed on an interlaced basis where the executive monitoring antenna is sequenced for measurement every one second period. Therefore, the facility is not without executive monitoring for longer than one second during the ground check routine.

<u>Power Supply Subsystem</u>. There are two types of Power Supply Subsystems which are associated with the solid state VORTAC. The Type I system operates from 120/240 VAC, single phase power and the Type II operates from 120/208 VAC, three phase power. Type I Power Supplies will be used at VOR and VOR/DME installations while Type II will be used with the full VORTAC installation. Functionally the two systems are similar with the principal hardware differences occurring in the input power transformer and the details of the regulator circuits.

The major units of the Power Supply subsystem are illustrated by Figure 6.

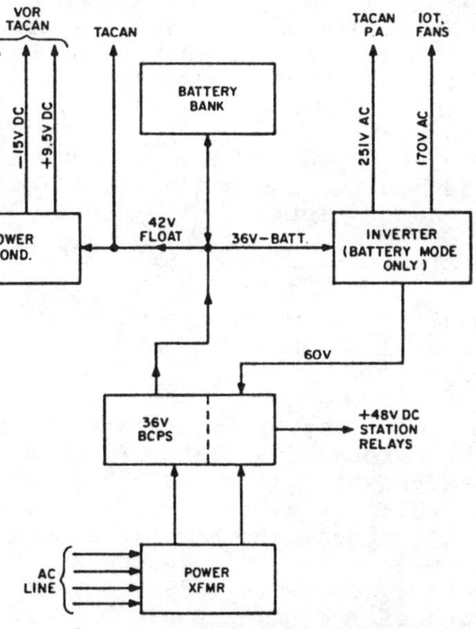

Figure 6. Power Supply Subsystem

Operation occurs in either one of two modes: normal mode, or battery mode. The net result is that the Power Supply is a source of uninterruptible power. During normal operation the 36 volt BCPS (Battery Charger Power Supply) is driven through the Power Transformer from the AC input to the station. The 36 volt BCPS develops the DC voltage (42 volts nominal) with sufficient power capability (3 kw) to drive the Power Conditioner, 30 Volt Tacan/DME Regulator, and float charge a 590 ampere hour battery bank. The 36 Volt BCPS also provides a regulated +48 volt DC to power the station relays.

The Power Conditioner consists of four high efficiency switching regulators which provide the VORTAC system with regulated DC voltages of +28, +15, -15 and +9 volts. During normal operation the 50 VDC power required by the TACAN/DME Power Amplifier is derived from a power supply in the TACAN/DME equipment which is powered directly from the AC line. The IOT and blowers are also driven directly from the line.

When the commercial AC power fails the floated 36 volt Battery Bank automatically assumes the load. In addition to driving the Power Conditioner, the batteries also provide power to the DC-AC Inverter which now provides the input power to the TACAN/DME Power Amplifier power supply, the +48 volt relay supply, the IOT and the blowers. Under this condition the 50 volt DC requirement to the TACAN/DME Power Amplifier is reduced. The floated battery concept provides the VORTAC system with uninterrupted power - no glitches or transients - during the interval between commercial AC power failure and the time the standby engine generators assume the load. Where the engine generators are not present or not functioning, the Battery Bank will provide a minimum of 2 hours additional service. The fact that the facility is operating on batteries is automatically conveyed to the remote monitor and control station as a maintenance alert.

Communications, Control, Monitoring and Reporting

Communication System. A block diagram of the overall communications system is shown in Figure 7.

The primary elements are the FCPU, which is located at the VORTAC facility, the RMC-F which is at the FSS, and the RMC-C which is at the ARTCC. Associated with the system and also located at the ARTCC will eventually be the Maintenance Processor Subsystem (MPS). Human interface to each of the primary elements is accomplished through keyboard controlled video Input/Output Terminals and Teleprinters.

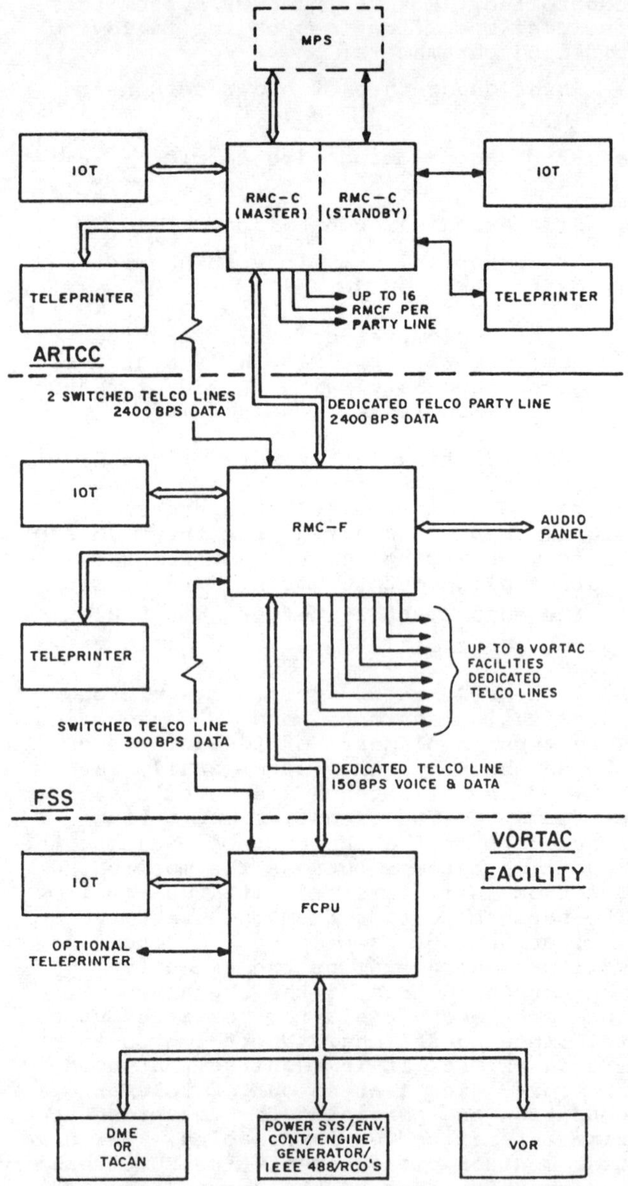

Figure 7. Communications System

The communication link between the FCPU and the VORTAC is through the VOR and TACAN/DME Monitors. These are identical, binary digital, serial by bit and byte, asynchronous, full duplex links operating at 1200 bits per second. Signalling across these interfaces is by voltage deviation on balanced lines in accordance with the requirements of EIA-RS422.

The primary communication link from the FCPU to a remote location is the dedicated telephone line between the FCPU and the RMC-F. This line carries data via AFSK tones at 150 bits per second and is asynchronous. In addition, the AFSK signal frequencies share the dedicated line with voice and a push to talk (PTT) tone. The voice (together with the PTT tone) can originate at the audio panel of the FSS for transmission with the VOR navigational signal or it can originate at the VORTAC facility either from the VOR Monitor (detected VOR voice and ident transmissions) or from the Remote Communications Outlet Receiver (RCO) which receives air to ground transmissions. There is also a switched telephone line, providing a 300 bit per second link for data only, which can serve either as a backup to the dedicated line from the RMC-F, or as a port for any remote compatible terminal to communicate with the FCPU.

At the RMC-F there is capability for providing up to eight of the voice-plus-data dedicated-line links described above, permitting the RMC-F to be in continuous communication via a polling mode with up to eight VORTAC facilities. The 300 bit per second, switched line, data only link is also available for backup communication. Communication between the RMC-F at the FSS and the RMC-C at the ARTCC is also handled over dedicated telephone lines. In this case, the dedicated line from the RMC-C is connected as a party line and can service up to sixteen RMC-Fs. Back-up to the dedicated line is again via the use of switched telephone lines. Both these links are for data only, utilize synchronous differential phase-shift-keying (DPSK) transmission and operate at a 2400 bit per second rate.

At the ARTCC there are two RMC-Cs operating in a redundant master-slave relationship. Each RMC-C has four dedicated party-line links and two switched-line back up links. The RMC-Cs also serve to link the VORTAC network with the Maintenance Processor Subsystem. This is a hard wired RS499 interface in which data is transferred in a binary bit stream at a synchronous 2400 bit per second rate. Should this link fail, there is provision for automatically establishing a link to an alternate MPS at a remote location through the switched line back-up port of the RMC-C.

Facility Central Processing Unit (FCPU). The Facility Central Processing Unit (FCPU) is the central core of a distributed, computerized control, monitoring and reporting system with both local and remote capabilities. The microcomputer in the FCPU, in conjunction with the microcomputers in the VOR and TACAN Monitors, contain the VORTAC data base in nonvolatile RAM. This data base consists of the following files: programmable system operating parameters; programmable executive monitor alarm limits; programmable certification limits; programmable test generator settings; current operating status; certification data; executive monitor data; fault isolation test results; special test results; maintenance data; and environmental data. The data files are tagged with the time and date at which the data was taken. All data files are available for display at the local video terminal or at the remote terminals by means of the communications system previously described.

The central core of the FCPU is its microprocessor and its associated memory, A/D converters, VOR Test Generator and communications interface controls. This central core oversees the following functions:

- Automatic fault isolation to the module level
- Initiation of automatic ground check and analysis of results for duantal, quadrantal and octantal errors
- Programmable parameter control
- Certification testing
- Monitor integrity testing
- System security management and control
- VORTAC shutdown and reset control
- Communications control to local operator and remote stations
- Facility environmental measurements
- Facility relay function controls
- Maintenance data collection
- Collection and dispatching of real time status information.

In addition to the above, the FCPU provides the capability for testing the remote communication outlet receivers and also supplies the voice identification to the VOR.

Of particular significance are the certification testing and automatic fault isolation functions. Certification testing of the VORTAC can now be accomplished by remote command through the FCPU and the results reported to the remote station in a matter of minutes, thereby eliminating the many costly and time consuming visits to the site which are presently required. Similarly when a fault does occur as indicated by an alarm, test routines which are designed to isolate the problem to the replaceable module level are automatically enabled or if the operator suspects a fault he can initiate the fault isolation routines. Both these features alone are expected to result in substantial maintenance savings.

Remote Monitor and Control (RMC). Remote monitoring and control of the VORTAC resides in the RMC-F units at the FSS and the RMC-C units at the ARTCC. It is possible by means of these units, their human interfaces in the person of the video terminals (IOT's) and teleprinters, and the communications system to accomplish all FCPU functions remotely from these stations that can be accomplished locally by the FCPU. The human interface design for all the terminals in the VORTAC system employs an interaction technique commonly known as "menu selection". This technique provides the operator with a display of numbered options in English text at each step of a control sequence. The operator selects the desired option by entering its number and in some cases such information as "V" for VOR or "T" for TACAN or "1" for Monitor 1 and "2" for Monitor 2 and the display program automatically executes the command and displays the next appropriate menu in the control sequence. The operator is prompted at each point in the sequence.

By means of a split screen technique the upper half of the RMC display provides the operator with a real time indication of the navigational status of the TACAN, DME, and VOR signals for each station monitored by that RMC plus an indication of any maintenance alerts at these stations. The input data for the status displays is provided by the FCPU at each of the VORTAC facilities.

In addition to their monitoring and control functions both the RMC-F and the RMC-C serve as data concentrators and message switches. Each RMC-F can service up to 8 VORTAC facilities and each RMC-C through its RMC-F network can service up to 128 VORTAC facilities. The RMC-C also interfaces the VORTAC systems it is monitoring with the Maintenance Processor Subsystem (MPS) which is planned for installation at each of the Centers. The MPS will have the capability for performing a variety of maintenance functions such as trend analysis, record keeping, message transmission, alarm histories, etc.

MICROWAVE LANDING SYSTEM

Lt. Thomas E. Evans
Aeronautical Systems Division, Wright-Patterson AFB, OH 45433

ABSTRACT

The National Microwave Landing System (MLS) program is a joint DOT/DOD/NASA effort to implement a common civil/military precision landing system to replace the current Instrument Landing System (ILS).

The MLS will be capable of providing precision landing guidance down to Category III minimum while allowing for complex approach paths in both the horizontal and vertical planes. The system is based on the Time Reference Scanning Beam (TRSB) technique which was selected by the International Civil Aviation Organization (ICAO) in April 1978 as the new international landing system standard. MLS is less susceptible to interference from the surrounding area and provides a greater signal coverage area than ILS.

INTRODUCTION

ILS has been in use since the 1940's, but has limitations that make it unusable in some airport environments. The localizer beam is a wide beam (6–10 degrees) that is subject to interference from nearby structures. The glideslope beam requires reflection from the airport surface to form the guidance signal. This requires that at least a 1500 ft. area in front of the antenna is flat and free from obstacles. Due to these problems, ILS cannot be installed in sites with irregular terrain, greatly limiting the usefulness of some airports. (1:312) ILS is also limited to site location due to the size of the antennas. Another problem that ILS is experiencing is the availability of frequencies. ILS has 40 channels allotted, but this is not enough channels for highly populated areas with many airports. The last problem is the ability of ILS to handle the large air traffic loads of the future.

These concerns prompted the Radio Technical Commission for Aeronautics (RTCA) to meet in 1967 and form a Special Committee, SC-117, to develop an operational requirement for a new landing system. The results of the RTCA SC-117 led to the preparation of a five year development plan and a proposal of two formats for the new MLS standard; the "scanning beam" MLS and the "doppler" MLS. The FAA was the lead Government Agency and with the support of DOD and NASA, led to a feasibility study of the two systems. In December 1974, the results of this study led to the Time Referenced Scanning Beam (TRSB) being chosen as the United States recommendation to the International Civil Aviation Organization (ICA0). (1:318)

Based on a paper presented at NAECON 85.

ICAO had decided at the Seventh Air Navigation Conference in 1972 to proceed with MLS. After evaluating five proposals from various countries, ICAO selected the TRSB in April 1978 as the new international standard for landing systems. (1:318) ICAO has prepared a specification called Standards and Recommended Practices (SARPS) which will assure a standard guidance signal among the 172 member nations. According to the MLS SARPS, ICAO Annex 10, Vol. 1, ILS has a projection date of 1995, at which time MLS will become the only ICAO recognized landing system.

ICAO compatible MLS calls for a single accuracy standard to be implemented worldwide. This standard guarantees, in most cases, that the quality of the MLS signal will be the quality necessary for CAT III ILS and autoland capability. This insures the pilot that he can expect the same guidance performance at each runway.

MLS CONFIGURATION

The Basic MLS configuration consists of an approach azimuth signal, approach elevation signal, a set of basic data words, and a Precision DME (DME/P) signal capable of being transmitted over a 200 channel range. This basic configuration can be expanded to include a back azimuth signal and a set of auxiliary data words.

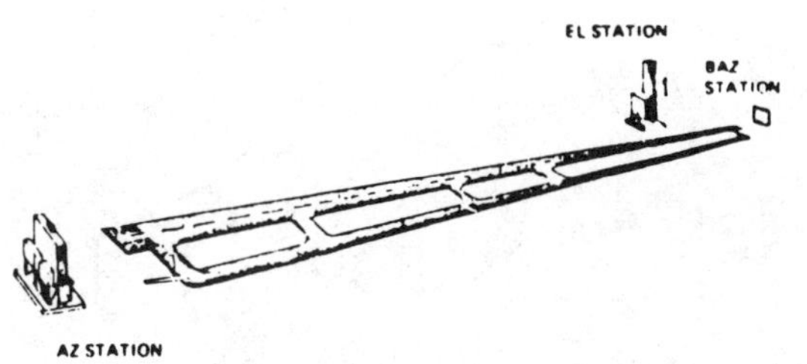

Figure 1. MLS Site Configuration

The MLS site configuration is as shown in Figure 1. The approach azimuth (AZ) antenna is located at the stop end of the runway, as is the ILS localizer, and scans a beam up to ± 60 degrees in azimuth to either side of the runway centerline (Figure 2). Most sites in the U.S. will scan ± 40 degrees. The elevation (EL) antenna is located approximately 1,000 ft. from the runway threshold at the Glide Path Intercept Point (GPIP), as is the ILS glideslope antenna. The EL antenna has a scan coverage of 0.9 degrees to 30 degrees (Figure 3), but most sites will scan to 20 degrees in elevation. The DME/P transponder has a 360 degree coverage up to 22

Reprinted from *IEEE AES Mag.*, pp. 6–9, May 1986.

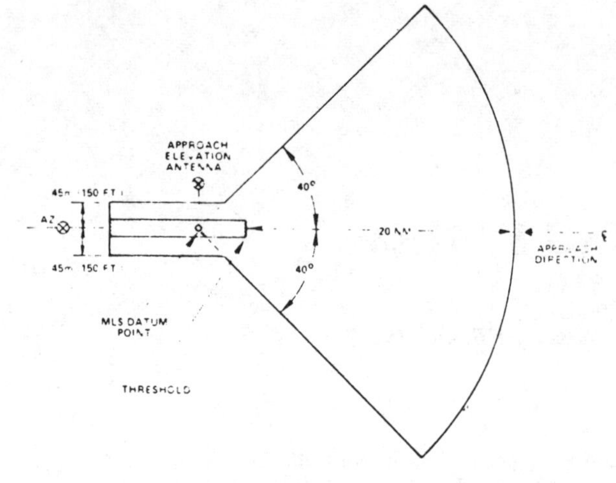

Figure 2. Azimuth Antenna Coverage

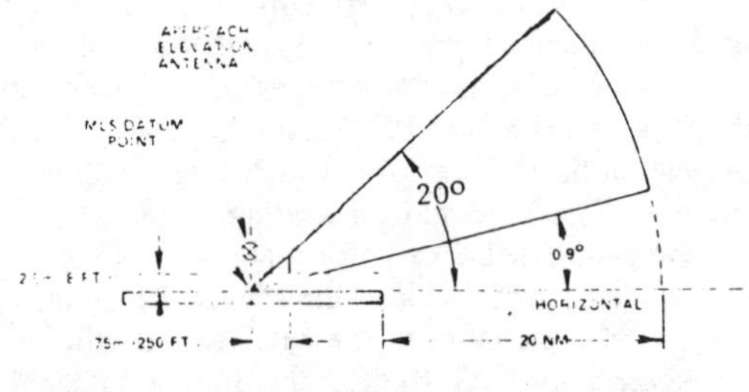

Figure 3. Elevation Antenna Coverage

nm and is usually collocated with the AZ antenna. By using these three signals, AZ, EL, and DME/P, the three dimensional position of the aircraft, in relationship to the runway, can be determined. Adding a back azimuth (BAZ) antenna to the approach end of the runway, (Figure 4), provides guidance to the aircraft for take-off and missed approaches. (1:321–325)

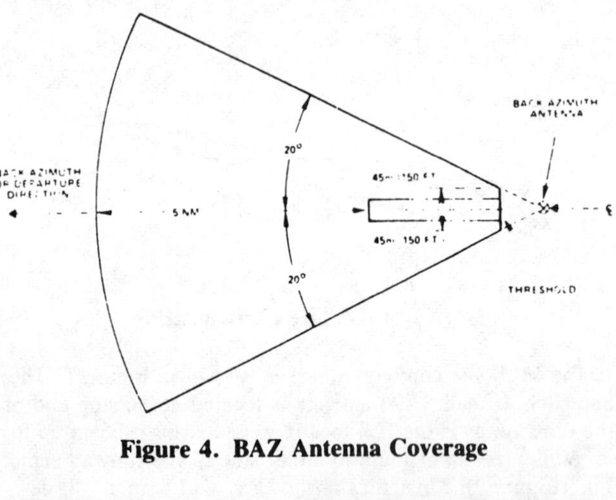

Figure 4. BAZ Antenna Coverage

The Basic Data function provides information which is directly used by the MLS receiver to process the angle functions and provide landing guidance. This consists of six data words transmitted from the AZ antenna. According to the latest revision of the MLS SARPS, ICAO Annex 10, Volume 1, these six data words provide information on coverage area, signal status, beamwidths, runway heading, and station I.D. (2:2A-16–2A-22) The SARPS also provides for three sets of auxiliary (AUX) data words. Presently, four words identify the site geometry of the AZ, EL, BAZ and DME antennas. The rest of the data words are not yet defined, but will most likely provide information necessary for MLS Area Navigation (RNAV) calculations.

All of the MLS signals (AZ, EL, BAZ, Basic and AUX data) are transmitted over the same frequency in a multiplexed fashion as shown in Figure 5. Each signal has a digital preamble preceding it to identify the signal. DME/P is transmitted and received over one of the 200 channels that is paired with the MLS channels.

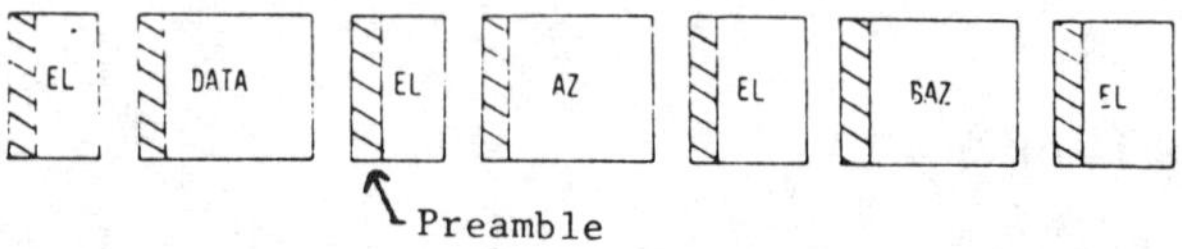

Figure 5. Time Multiplexing of Signals

POSITION DETERMINATION

MLS position is determined by timing the passes of the scanning beam; thus Time Referenced Scanning Beam. Using the AZ signal as an example, it begins its scan at the +40 degree radial of the AZ coverage area. This beam scans at a constant rate through the runway centerline (0 degrees) to the −40 degree radial (Figure 6). This scan is called the "TO"

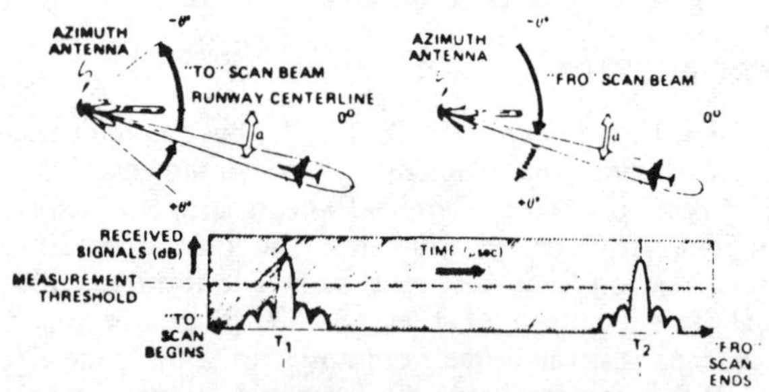

Figure 6. Angle Measurement

scan. After a predetermined pause, the beam scans back from the −40 degree radial to the +40 degree radial. This is called the "FRO" scan. The aircraft's MLS receiver starts a timer when it detects the TO scan and stops the timer when it detects the FRO scan (Figure 6). By using the linear relationship shown in Figure 7, the receiver can calculate the azimuth

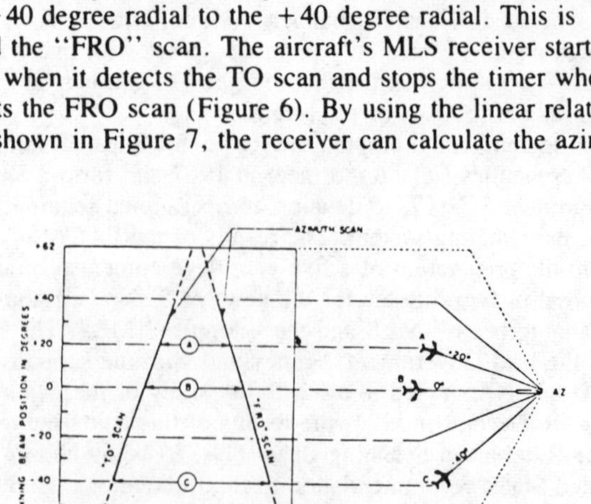

Figure 7. Angle Measurement

angle on which the aircraft is located (1:329). The same process occurs with the EL scan, starting at 0.9 degrees elevation, scanning to 20 degrees elevation, and returning back to 0.9 degrees. These two signals can determine the two dimensional location of the aircraft.

The third signal required to determine the three dimensional position of the aircraft is Precision DME (DME/P). ICAO desired a signal format that was compatible with the standard DME (DME/N) associated with TACAN and VOR, but could still distinguish itself as a precision interrogation. (The accuracy of DME/N is only 0.17 nm which does not meet the accuracy requirements for precision approaches and landings.) The proposed format was two-pulse/two-mode. (3:2) This format uses the same pulse-pair format that is used in DME/N, but also provides a second mode in which DME/P interrogations can occur. The first mode of the DME/P signal is the initial approach (IA). IA interrogations occur frm 22 nm to 8 nm (figure 8). These interrogations are fully interoperable with DME/N and provide accuracies similar to DME/N. At 8 nm, the DME/P interrogator begins to transition to the final approach (FA) mode. This transition is complete by 7 nm and

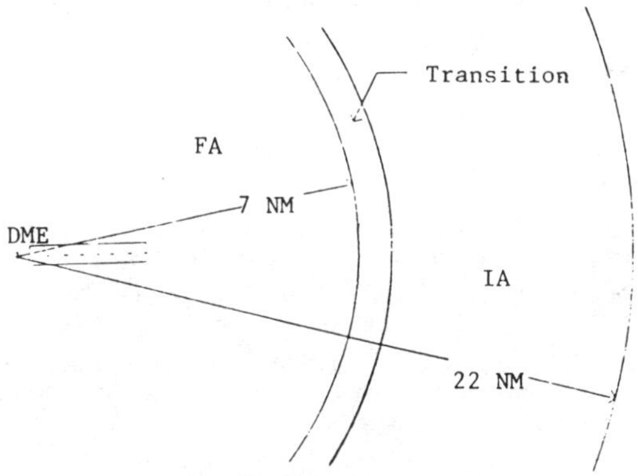

Figure 8. DME/P Coverage

the FA mode is used through landing. The transition for the FA mode is signaled by changing the spacing between the DME pulse-pair interrogation. The DME/P ground transponder recognizes this new spacing and begins to process the signal as a FA interrogation. The increased accuracy in the FA mode is derived by pulse shape modification, lowering the pulse detection threshold, and increasing the rate of interrogation. (For more information on this subject, refer to Reference Three.) Two accuracy standards are provided in DME/P. Standard 1 provides 100 ft. accuracy which is adequate for conventional take-off and landing (CTOL) operations. Standard 2 accuracy is 40 ft. which is needed for STOL operations and CAT III and autoland capability. With this third signal, the position of the aircraft can be accurately determined.

MLS OPERATIONS

The operational advantages of MLS are numerous. It provides a distinct advantage over ILS by providing command guidance in a much greater coverage volume. Because of this increased guidance coverage, the pilot has much more flexibility in his approaches to landing, missed approaches, and take-offs. Presently, many approaches can only be flown to non-precision decision heights due to the complexity of the approach and the lack of positive guidance. Complex approaches are required when conducting noise abatement and obstacle clearance maneuvers and flying in restricted airspace. Several examples of this problem are the River Approach into

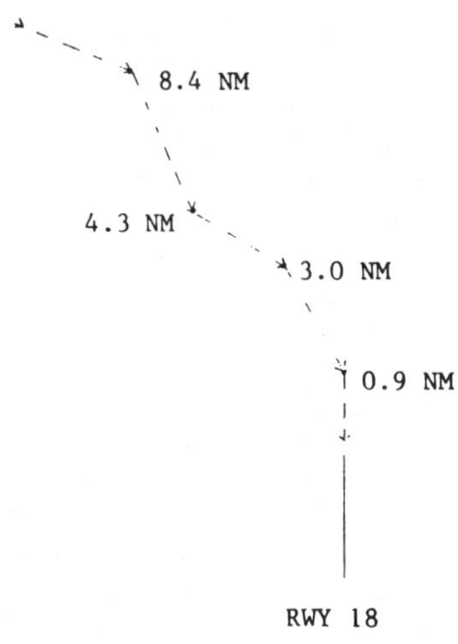

Figure 9. River Approach

Runway 18 at Washington National Airport (Figure 9), the Canarsie Approach into Runway 13 at JFK Airport in New York City (Figure 10) and the Quiet Bridge Approach into

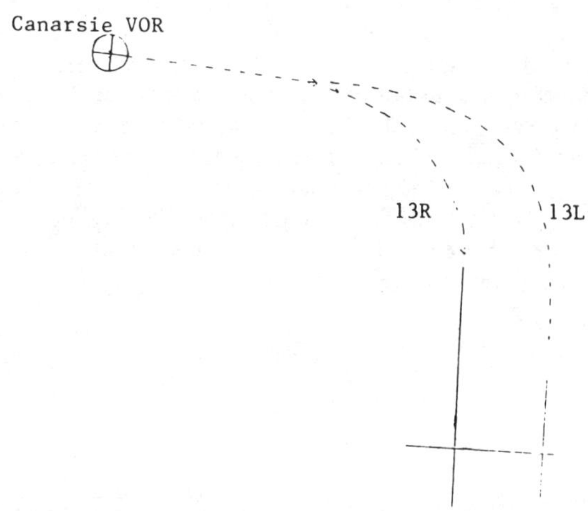

Figure 10. Canarsie Approach

Runway 28 at San Francisco International Airport (Figure 11). These three approaches cannot be flown to precision decision heights. Therefore when weather minimums become low at these airports, these approaches are shut down, either closing the airport or causing a large back-up in air traffic. Neither of these situations are desirable to the airlines or the airports. Since MLS can provide guidance throughout these approaches, they can be flown as precision approaches to at least CAT 1 minimums.

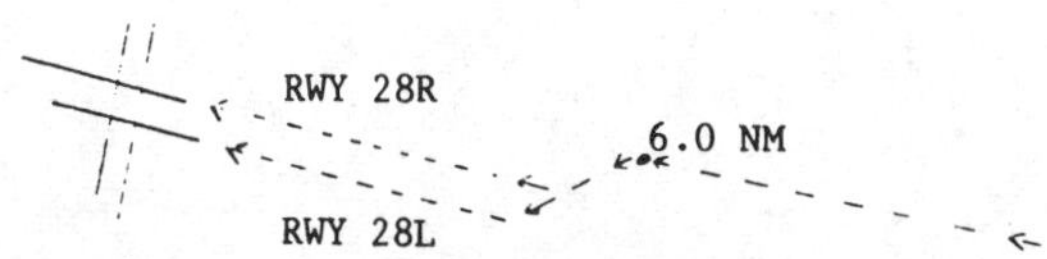

Figure 11. Quiet Bridge Approach

MLS is also advantageous to the Air Traffic Controllers. Once an aircraft is in the terminal and MLS coverage area, the pilot will fly a prescribed approach to landing. The controller will no longer need to provide continuous radar vectors to the pilot for interception of the ILS localizer beam or aircraft separation. This will greatly reduce his workload.

It is anticipated, that with the greater capability of MLS, aircraft separation can be reduced, simultaneous instrument approaches can be flown into parallel runways, and complex approaches can be flown to precision heights. These added capabilities will increase the rate at which aircraft can be recovered and deployed and decrease time in the traffic pattern.

MLS also provides advantages for special operations such as STOL and decelerated approaches. For STOL and tactical applications, a high glideslope angle is desired, but present systems do not have the vertical range necessary to provide instrument guidance for these approaches. Helicopters doing steep approaches need range-rate information for decelerated approaches. Range-rate can be derived from the DME/P range at Standard 2 accuracies.

AVIONICS

Present MLS avionics configurations cannot provide the capability required to fully implement MLS operations. Some kind of navigation computer (MLS RNAV) is needed to fly curved/segmented approaches. This computer must provide the accuracy necessary to fly in the terminal area and conduct approaches to landing. Present RNAV computers used in the enroute structure do not provide the accuracy necessary for terminal procedures. The Radio Technical Commission for Aeronautics (RTCA) has formed a special Committee, SC-151, to define the Minimum Operational Performance Standard (MOPS) for MLS RNAV equipment. This equipment will be required to accept MLS waypoints and provide positive guidance in the terminal area.

Another area that will require further research is in the display of MLS information to the pilot. Previous studies have shown that present electromechanical displays are not adequate to provide information to the pilot for flying curved/segmented approaches. The greatest concerns in this area are with situational awareness (i.e., where the runway is relative to the aircraft) and turn anticipation. Minor modifications have been done to present instruments to supply this information. But as the dynamics of the approaches increase, further cues and transition anticipations are required for the pilot to accurately follow a prescribed course. This problem can hopefully be eliminated by taking advantage of the flexibility provided by an Electronic Flight Instrumentation System (EFIS). Programming the CRTs to display MLS information in new formats should provide the pilot with the guidance commands necessary for high dynamic approaches.The Air Force will be conducting research with different display formats to meet this need.

There has been very little operational implementation of MLS to this date. Therefore the avionics requirements for flying MLS complex approaches have not been totally defined. The FAA and USAF will be involved in operational tests of MLS over the next couple of years to more fully define MLS capabilities and avionics.

REFERENCES

1. Redlien, H.W. and Kelly, R.J., "Microwave Landing System: The New International Standard." New York: Academic Press, Inc., 1981.
2. "Report of the Tenth Meeting of the All Weather Operations Panel (AWOP) 1984," ICAO, Montreal, September 4–20, 1984.
3. Kelly, R.J. and Jensin, G., "DME/P—The New International Standard." IEEE 1982 Position Location and Navigation Symposium (PLANS 82), Atlantic City, NJ, December 1982.

Part 6
Introduction to Space Navigation

6.1 Objectives

THE reasons for navigating a spacecraft are:

—to measure position and velocity prior to changing from one orbit to another (for example, to stationkeep in synchronous orbit, to rendezvous with another spacecraft, or to enter the atmosphere) and

—to measure the position where data are collected (for example, for an Earth-observation satellite).

The art of navigating spacecraft evolved from the art of navigating aircraft. Virtually all unmanned spacecraft are ground-tracked and have no on-board navigation sensors or computation. By contrast, all manned spacecraft are ground-tracked and all but the earliest also have on-board navigation systems.

Spacecraft navigation is simplified by the fact that once an orbit has been determined, whether by ground-based or on-board sensors, the trajectory is predictable within hundreds of meters for several revolutions in low orbit. Hence, spacecraft navigation consists of periodic fixes in coasting flight and continuous measurements during thrusting flight.

Periodic fixes have traditionally been made by ground-tracking stations. In the mid-1980s, experimental on-board fixes were made with the Tracking and Data Relay Satellite (TDRS; see section 6.3.4 below) and GPS satellites (Part 3.6, pg. 68). Measurements during thrusting flight are made with accelerometers, usually to guide an orbit-change maneuver. The on-board computer calculates the instantaneous orbit while the spacecraft is thrusting, using the integrated accelerometer outputs. The computer shuts off the rocket motor when the correct orbit is achieved, after which a radio fix can verify its accuracy.

6.2 Booster Navigation

A booster is a rocket-propelled vehicle that lifts spacecraft into orbit or throws ballistic warheads onto their targets. The origins of ground-tracked navigation systems lie in ballistic-missile guidance systems. In 1944, the German V-2 ballistic missile flew a preset pitch–azimuth profile while its engine was firing. The V-2's engine ceased thrusting when the output of a longitudinal integrating accelerometer reached a preset value. The missile then coasted ballistically to its target.

The earliest space boosters were navigated by ''radio-inertial'' guidance systems that uplinked attitude-steering commands from ground-based X-band radars (at the launch site) to an on-board autopilot. Once per second, the trajectory was measured on the ground and a new steering command uplinked. Status telemetry was returned by VHF radio but was not used for real-time navigation. Beginning with Thor and Minuteman, most boosters have been all-inertially guided and have transmitted S-band telemetry. A few ballistic-missile boosters contain a star-tracker that measures accumulated gyro drift during powered flight so the booster can correct its own trajectory. A few others contain a GPS receiver that measures position and allows the correction of accumulated accelerometer errors before thrust terminates. Some boosters (e.g., Titan-3) still radio-guide the first and second stages (after a 10-s vertical rise and programmed pitchover). Even on radio-guided boosters, the upper stages are inertially guided because they burn downrange, beyond the radar's horizon. Haeusermann (1981) has described the guidance of boosters through the Saturns. Wright and Sponnick (1988) have described, in detail, a ring-laser-gyro inertial guidance system being installed on the Centaur upper stage, replacing a 20-year-old inertial system. The Space Shuttle is guided by the same on-board, all-inertial navigation system during boost as in space and during entry. Range safety for all launches has always been assured by an independent C-band radar, based at the launch site. Therefore, all boosters carry a C-band range-safety transponder.

6.3 Ground-Based Navigation

6.3.1 U.S. Military Tracking

When the Discoverer reconnaissance satellite was developed in the late 1950s, the Air Force used existing booster-tracking facilities and added a UHF system that tracked the satellites on-orbit, transmitted downlinked telemetry, and transmitted uplinked commands. This worldwide network became operational in 1957 and was called the Satellite Control Facility (SCF); now called the Satellite Control Network (SCN). Starting in 1966, the Air Force replaced the UHF radios with the S-band (2 GHz) Space Ground Link System (SGLS). Today, the SCN is used for the Air Force's unmanned spacecraft; it is completely independent of NASA's net. SCN consists of at least 10 manned, self-contained remote tracking stations (RTSs), connected by diverse communication paths to the Consolidated Satellite Test Center (CSTC) in Sunnyvale, California, and, since 1986, to the Consolidated Space Operations Center (CSOC) in Colorado Springs, Colorado. Each RTS has at least two steerable antennas and electronics that track the satellites' SGLS signal. Figure 6-1 is an aerial view of the New Boston, New Hampshire RTS showing a 60-ft-diameter antenna and a newer, more sensitive 46-ft antenna. The range and angle measurements are sent to the control centers via common carrier and the Defense Satellite Communication System. Because of the large number of military satellites in orbit, each RTS antenna normally

Fig. 6-1. Antenna complex of the remote tracking station in New Boston, New Hampshire. Courtesy of the U.S. Air Force.

services a different spacecraft, unlike NASA's policy of assigning redundant antennas to the same spacecraft. The CSTC and CSOC have many mission rooms, each of which services a spacecraft. These two centers back each other up in case of damage, loss of communication, or overload. The SCN is becoming a distributed network. All stations are scheduled to be automated by 1994, when RTSs will employ 3 people per shift, instead of today's 15 per shift.

6.3.2 Early NASA Tracking

In 1958, the predecessor agency to NASA established the Goldstone tracking station in California for tracking, receiving telemetry, and transmitting commands to interplanetary spacecraft. By 1963, Goldstone was part of a Deep-Space Instrumentation Facility (DSIF) that included large antennas in Spain, South Africa, and Australia. Goldstone, Spain, and Australia now have 85- and 210-ft steerable antennas (Fig. 6-2). An independent network was created for low-orbit unmanned civil spacecraft, called the Space Tracking and Data Network (STADAN), initially based on interferometer tracking. Still another, called the Manned Space Flight Network (MSFN), was created in the early 1960s, based on radar tracking. STADAN and MSFN were eventually combined into the unified S-band (USB) system, renamed the Spaceflight Tracking and Data Network (STDN) (see section 6.3.4). DSIF and STDN employ phase-locked loops in their spaceborne and ground receivers. The phase-locked loop, with its narrow bandwidth (5 Hz), tunes itself to the received carrier frequency, thus avoiding the noise that would have resulted from a receiver whose bandwidth was wide enough to accommodate the full range of Doppler shifts.

When the Mercury manned spacecraft flew in 1961, it was navigated by a mixed bag of four NASA tracking sites, two Australian-owned sites, and four Air Force missile ranges (Atlantic, Pacific, White Sands, and Eglin). Two NASA ships were positioned to fill gaps (that would have been longer than 10 min) between tracking stations. Eight sites were also equipped with C-band instrumentation radars (RCA's FPS-16) that tracked a transponder on Mercury. Seven sites had S-band radars (Reeves' VERLORT) that tracked another transponder on Mercury. Table 6:1 shows the tracking accuracy of these radars during the four Mercury orbital missions: $\frac{1}{4}$-mrad, in angle and 10-m in range at orbital distance. The C-band elevation error dropped sharply on the last flight when an atmospheric refraction correction was added to the processing software. Today's state of the art is more than twice as good.

The Mercury spacecraft communicated on HF and UHF radio with a capsule communicator (Capcom) at each station. The Capcom was an astronaut who observed telemetry, talked to the crew, made local decisions, and issued instructions to the crew based on predetermined mission rules. The Mercury Capcoms communicated to the Cape Canaveral center by teletype; telemetry summaries were also sent by teletype.

Fig. 6-2. Deep-space instrumentation antenna, 70 m in diameter. Courtesy of NASA.

TABLE 6:1
PERFORMANCE OF MERCURY TRACKING RADARS*

	Standard Deviations—Mission Averages					
	FPS-16			VERLORT		
Mission	Range (yd)	Azimuth (mrad)	Elevation (mrad)	Range (yd)	Azimuth (mrad)	Elevation (mrad)
MA-6	8.5	0.23	0.44	29.0	1.63	1.35
MA-7	9.8	0.22	0.40	33.7	1.62	1.72
MA-8	8.6	0.25	0.36	39.6	1.22	1.34
MA-9	11.2	0.27	0.26	20.2	1.36	1.42

* From (NASA SP-45, 1963).

Network synchronization was with WWV receivers at each site. Time at each site was maintained within about 1 ms; this was needed to synchronize computed trajectories with radar observations and to annotate telemetry. This was the Manned Space Flight Network.

For the Gemini-4 flight (1965), the NASA spacecraft control center (''Mission Control'') was moved from Florida to Houston, leaving the booster control center at Cape Canaveral. Real-time telemetry was transmitted on the NASA Communication System (NASCOM) at 40.8 kbits/s, whereas radar data were sent at 2.4 kbits/s. At first, orbit calculations used data from one radar at a time. By the end of Gemini, the orbit-prediction software could make best-estimates based on observations by several radars, using a model of the Earth's gravity field that was symmetric in longitude.

For the Apollo program, NASA's worldwide MSFN tracking and communication network was augmented to consist of 14 ground stations supported by five ships and eight Apollo Range Instrumentation Aircraft (ARIA) that were positioned for injection and entry. The major MSFN sites had a 30-ft

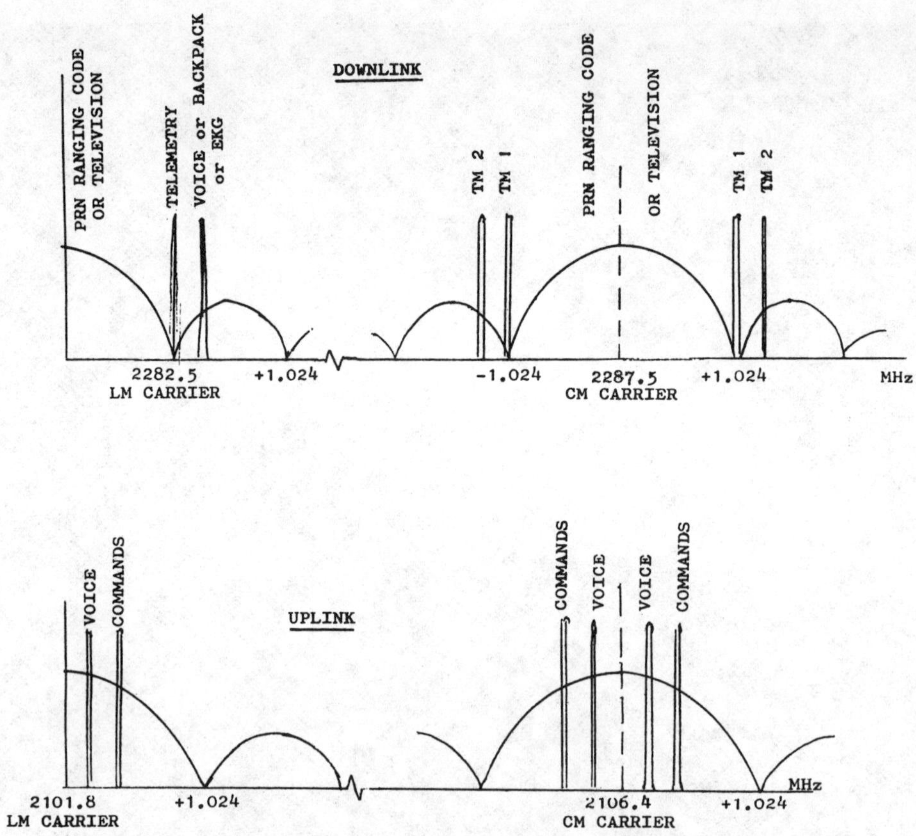

Fig. 6-3. Unified S-band frequencies for Apollo.

(now called 9-m) dish for Earth-orbit tracking and an 85-ft (now called 26-m) dish for lunar-orbit tracking. Major sites could handle two spacecraft within the antenna's beam. The uplinked commands, downlinked telemetry, voice, television, and precision distance measurements (''ranging'' or ''tracking'') were combined into a unified S-band (USB, at 2 GHz) system that avoided duplicate radios on the ground and in the spacecraft. The USB ground equipment began to be installed in 1967 and was first used in 1969, just before the lunar landing. Figure 6-3 shows the USB spectrum as it was used during Apollo. The same frequencies were still in service in 1989.

USB range was measured by uplinking a pseudo-noise (P-N, sometimes PRN, for ''pseudo-random-noise'') pulse train and retransmitting it from the spacecraft, on the USB downlink, after a calibrated delay. The downlink's carrier frequency was 240/221 × the uplink frequency, thus permitting the ground to compare the transmitted frequency with the received frequency independent of the drift of the spacecraft's oscillator. In other words, the returned frequency and transmitted frequency were ''coherent.'' The ground computer calculated range (from the time delay between transmission and reception of the P-N code) and range rate (from the ''two-way'' Doppler shift between the uplinked and downlinked carriers). Eight code lengths were used, depending on spacecraft range. At lunar distance, the P-N pulse rate was 1 MHz. As the calculated estimates of position and velocity improved, the code length was (and is) shortened for faster convergence. The accuracy was 5 ft in range and less than 1 ft/s in range rate. These same downlink and uplink frequencies and turnaround ratio were used for Skylab and the Shuttle. The Air Force uses the same downlink frequencies, but, because it chose a turnaround ratio of 256/205 (versus 240/221 chosen by NASA), its uplink frequencies are lower than NASA's. Several studies in the early 1960s recommending compatibility between the two networks were not heeded.

NASA's remote sites served only as data-collection points: the communication computers were at Goddard; the trajectory analysis computers, flight controllers, and Capcoms were in Houston.

6.3.3 Interplanetary Navigation

Planetary-exploration spacecraft are navigated and controlled by the Deep Space Network, one of whose antennas is shown in Fig. 6-2. These 70-m diameter antennas are located at Goldstone, California; Madrid, Spain; and Canberra, Australia. They are sometimes augmented by other U.S. and foreign tracking stations. Navigation and communications with the spacecraft are at S-band (2.3 GHz) and X-band (8.4 GHz), transmitted and received by the 70-m dish. Interplanetary spacecraft have no on-board navigation but they hold their own attitude, relative to the Sun and stars, with star trackers and rate gyros. Star-tracker data are sent to the ground where human controllers identify the constellations and calculate the spacecraft's orientation. These controllers, at NASA's Jet Propulsion Laboratory, then transmit commands for attitude-hold, rotation to a new attitude, or pointing of appendages.

Once the spacecraft is properly oriented, thrust commands may be sent to it.

At interplanetary distances, range rate is measured to 1 cm/s by observing the two-way Doppler shift of the carrier. Range is usually known approximately; it is fine-tuned within 30 m by observing the time delay in the reception of a pseudo-noise pulse train. The length of the pseudo-noise pulse train corresponds to the *a priori* uncertainty in range. Range and range-rate measurements are typically averaged for 1 min. The measurements are filtered into a mathematical model of the orbit to solve for all seven orbit elements. The components of velocity normal to the line of sight are not measurable directly. However, the components of position normal to the line of sight are measurable; a 10-ns error in rotation of the Earth at 10^9 nm would give a 50-ft transverse error. In practice, the astronomical declination of a deep-space vehicle is measured by observing the amplitude of the daily oscillation in Doppler shift. The right ascension is measured by observing the phase of the daily oscillation in Doppler shift, averaged over several days. Once position and velocity have been measured, the ground stations uplink the durations and times of desired translational burns for orbit corrections. A correction is made every few months. Melbourne (1976) has described the system and Taylor (1984) has described some of the algorithms used in navigating interplanetary spacecraft.

Prior to flight, the position error relative to Earth is hundreds (even thousands) of kilometers, due to uncertainties in the astronomic unit and in the planet's orbit. For example, Neptune has not yet made one revolution around the Sun since it was discovered; hence, its orbit is still uncertain. Once the spacecraft sights the planet, on-board observations of its disk relative to the stars are transmitted to Earth. There, flight controllers measure the angles and calibrate the spacecraft's position relative to the planet. As the spacecraft approaches the planet, the relative geometry becomes more precise. On the 1989 Voyager encounter with Neptune, observation of the planet began 9 months before encounter. On the Saturn encounter, the prelaunch predicted position of the spacecraft relative to the planet was in error by 5000 km and arrival time by 300 s (2-sigma). After the first optic observations, the error fell to 2000 km. A month before the encounter, after further optical observations of the planet at closer range, the relative position error had fallen to 300 km. At encounter, errors were less than 10 km. Voyager, for example, made 20 midcourse corrections between Earth and Neptune in 12 years, the last of which occurred 4 days before closest approach.

6.3.4 Present-Day Tracking Networks

When Apollo ended, NASA's Earth-orbit tracking facilities were combined into the Spaceflight Tracking and Data Network (STDN), under the management of Goddard Space Flight Center. By 1971, the network used satellite communication links and had rubidium clocks at all sites. Five years later, the frequency stability of the S-band transceivers had improved to the point (10^{10}) that velocity could be derived from the absolute Doppler shift of the received carrier (''one-way Doppler''). The ''two-way Doppler'' shift is measured by comparing the turnaround carrier with the transmitted carrier, which are related by a known multiple (240/221 or 256/205). The two-way comparison is ''coherent'' and does not depend on the absolute stability of the oscillators.

In 1988, all NASA spacecraft (including interplanetary space explorers and the Shuttle) were served by nine STDN ground stations, three DSIF stations, and the Tracking and Data Relay Satellites (TDRSs). The first TDRS was operational in 1984, the second in late 1988, and the third (replacing TDRS-1, which became a spare) in 1989. TDRS provides nearly continuous unified S-band coverage to as many as 24 satellites at once, below 2000 nm altitude. Its K-band link relays 50 Mbits/s downlink telemetry from one spacecraft to the White Sands ground terminal. When TDRS is fully operational, NASA will decommission nearly all of the STDN sites, thus saving money.

To navigate a spacecraft via TDRS, the ground station at White Sands, New Mexico, uplinks a series of tones and a P-N pulse train (at 3.12 MHz) to the spacecraft, via the TDRS S-band link. The spacecraft transponds the tones and code to White Sands, also via TDRS, whereupon White Sands calculates the satellite's orbit from the range and range-rate histories. The NASA references show a ranging accuracy of 1 to 10 m. The carrier and P-N code are acquired in 4s. TDRS is also the communication link by which White Sands can uplink the ground-measured state vector or the size and direction of the burn (''external-delta-*V* burn'') that a satellite will execute. The NASCOM, which interconnects all sites, includes diverse communication paths from STDN sites to the centers, including links by commercial satellites. NASCOM data rates have been increased to 1.54 MHz and will be increased further.

The current gravity model in NASA's ground computers is 7th order (vs. 180th order for geodetic measurements, pg. 11). The on-board models are discussed in section 6.4.4, pg. 328. In 1989, NASA's orbit prediction software included gravity harmonics beyond 28th order that allow for asymmetry in latitude and longitude. NASA also included luni-solar perturbations and nominal models for atmospheric drag. The U.S. military apparently uses a 41st-order model in its ground computers, whose coefficients above 18th-order are not publicly disclosed. The distances among tracking stations are readily calibrated within 1-m error from extended tracking data. The migration of the Earth's poles must be accounted for if site errors less than 10 m are significant. A 1-m error in the location of a tracking station causes a 1-m error in a low-altitude satellite and a 6-m error in a synchronous satellite.

Since the 1960s NASA has had a net of 12 Baker–Nunn cameras that have photographed Earth-orbiting space objects against the star background. They have been supplemented by the Air Force's Ground-Based Electro-Optical Deep-Space Surveillance System (GEODSS), which points its telescopes along an inertial line of sight (so the stars are fixed in the field of view), records images digitally, and automatically detects moving objects (satellites) in its field. A network of half a dozen stations is planned. They track objects as small as 1-m in diameter in synchronous orbit. Johnson and McKnight (1987) state that 80% of orbiting objects are below 2000-km altitude. At these low altitudes, objects larger than a few centimeters in diameter are tracked by the military radars of the United States and the Soviet Union. Collision-avoidance with space debris is discussed in Part 8.6, pg. 439.

Privately owned communication satellites are spaced in geosynchronous orbit, 2° apart in longitude. Position determination is done by the user with ranging tones. Ha (1986) has stated that ±0.05° tolerance (±37 km at synchronous altitude) is typical, with a north–south correction made every month and an east–west correction every 3 weeks, due to orbit perturbations caused by the Sun and Moon.

The Soviet Union maintains at least seven tracking stations on its own soil and many ships at sea during manned launches. The unmanned flight test of Buran ("snowstorm," the Russian shuttle) in 1988 was supported by 4 ships, 2 off South America, 1 off Senegal, and 1 off Algeria. There are improbable reports that their early manned flights were supported by 30 ships. From time to time, the Soviet Union has had tracking stations in Cuba, Chad, Egypt, Somalia, and elsewhere but its national policy is apparently to minimize the number of foreign treaties. Their Cicada (equivalent to U.S. Transit) and Glonass (equivalent to U.S. Navstar-GPS) also require ground sites to track and control the satellites.

The European, Japanese, and Indian space agencies have well-instrumented launch sites and tracking stations in their homelands. The European Space Agency maintains a master control center at Darmstadt, West Germany, and satellite-specific centers elsewhere. It operates tracking stations in Spain, Sweden, Italy, Kenya, Australia, and Japan. Range and range-rate measurements to their satellites are at VHF. Japan has foreign stations in Australia, Canada, and Luxembourg.

6.4 On-Board Navigation

The navigation of unmanned spacecraft was discussed in section 6.3. Few of them have on-board navigation sensors. Even fewer, if any, calculate their orbits on-board. Commands to maneuver and commands to reconfigure the hardware and software after failure are uplinked.

6.4.1 Early Manned Spacecraft

On-board navigation systems in manned spacecraft have their origin in manned aircraft, especially in the X-15 which flew from 1959 to 1968. The X-15 (Fig. 6-4, shown sitting in its cradle) was a precursor of the Shuttle in glide landing, fly-by-oil flight control, split-rudder, speed-brake, and blended-reaction/aerodynamic controls. It was dropped from under the wing of a specially modified B-52 bomber at high altitude, whereupon the X-15's rocket motors accelerated it to Mach 6.7 and 350,000 ft altitude. The X-15 could not have evolved into the Space Shuttle until fault-tolerant navigation and control technologies were developed in the late 1960s, because precision navigation is essential for flight safety.

Fig. 6-5 shows a block diagram of the X-15's all-analog navigation system. The upper part of the figure shows the equipment in the B-52, while the lower part shows the equipment in the X-15 itself. The X-15's four-gimbal inertial measurement unit (IMU) measured attitude and acceleration. It contained three floated single-degree-of-freedom gyroscopes and three analog-rebalanced accelerometers. The IMU was coarse-aligned from the B-52's magnetic compass within 2°. Then it was fine-aligned, using the B-52's Doppler radar to gyrocompass the X-15's inertial platform. The platform settled at a 0.5° azimuth error after approximately 20 min [see Kayton and Fried (1969) in the Part 5 Introduction for details of gyrocompassing]. Once the platform was aligned, the distance from the landing site and the barometric altitude were manually entered and the X-15 was dropped. The gyrocompass equations were solved in the X-15's analog computer, which also calculated position, velocity, inertial altitude, the gyro-torquing signals (at the fixed latitude of Edwards Air Force Base), and the Coriolis forces (also at fixed latitude). There was no barometric damping of the divergent inertial altitude for these 10-min flights; barometric and inertial altitudes were displayed separately.

Fig. 6-4. X-15 research aircraft in its cradle. Courtesy of NASA.

The gimbal angles were measured by analog synchros (rotary three-phase transformers) that drove the flight director ("8-ball") display, on which three-dimensional attitude was pictured. Dynamic pressure, angle of attack, and angle of sideslip were measured by a servoed sphere ("Q-ball"), mounted on the nose of the X-15. The outputs of the sphere fed an analog air-data computer that calculated true airspeed, altitude, and altitude rate. The X-15 had no radio navaids; it was radar-tracked from Edwards Air Force Base and sent its telemetry to Edwards.

The Mercury spacecraft (Fig. 6-6) was designed slightly later than the X-15. It was a manned ballistic missile "nose cone" that flew from 1961 to 1963, was launched by a Redstone or Atlas booster, and landed in the ocean. The crewman could only rotate the spacecraft in attitude (three axes) and fire the retrorocket (fixed-delta-V). There was no on-board computer, no on-board state-vector calculation, and no uplink of position. Horizontal navigation was calculated by the Mercury ground-tracking network (section 6.3.2) based on radar observations. Vertical navigation during ascent and descent was by barometric altimeter.

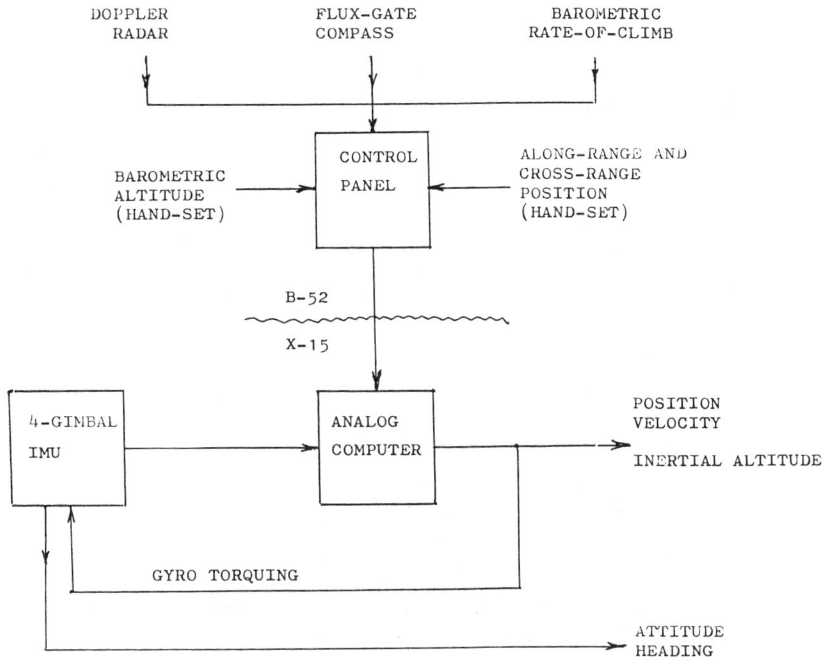

Fig. 6-5. Block diagram of the X-15 navigation system.

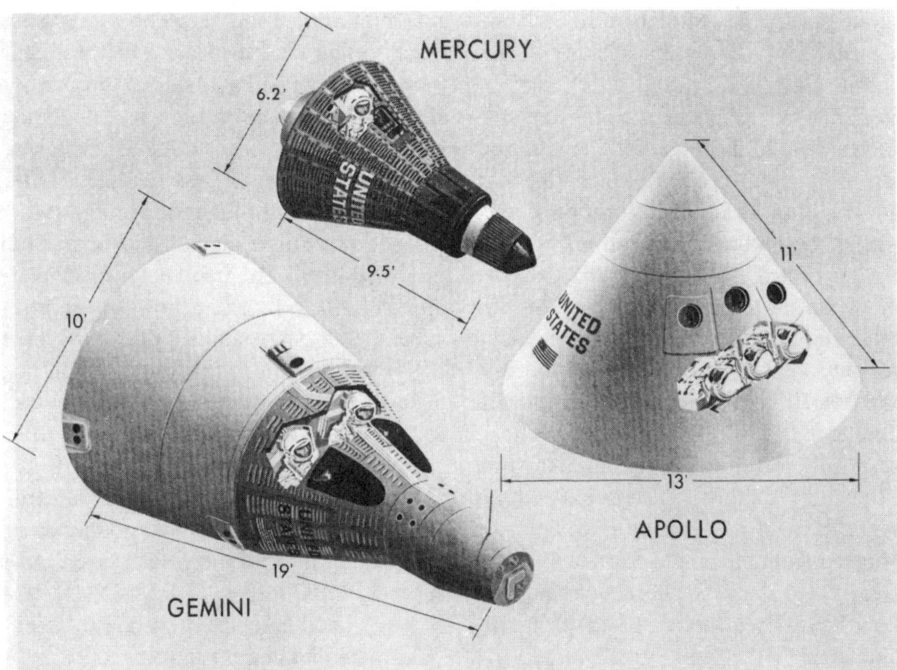

Fig. 6-6. Mercury, Gemini, and Apollo spacecraft. Courtesy of NASA.

Mercury had an optic horizon scanner that leveled a vertical gyro which, in turn, gyrocompassed a directional gyro into the orbit plane within 5°. Looking through a periscope, the crew measured the drift angle relative to the ground track. A clock displayed Greenwich Mean Time and crew-set event times. An Earth globe, rotating at a preset rate, was intended to show the position of the spacecraft along the ground track but, as it served little purpose, it was removed.

Deorbit was initiated by the pilot at a ground-uplinked time. The landing accuracy depended on the timing of retrofire, the attitude error at retrofire, and atmospheric dispersions. The retroburn was backward and nominally 37° below the horizon, the angle at which the landing error was least sensitive to attitude errors. The retroattitude was held manually, relative to the gyros, and confirmed by the astronaut's observing marks on the window and periscope. The symmetrical spacecraft was rolled manually at 10°/s during entry to average-out residual lift. The parachutes were deployed at 10,000-ft barometric altitude. Mercury landed on the world's biggest airport, the ocean, so entry error was important only insofar as it caused a delay while the Navy found the spacecraft after it "splashed-down." Landing dispersions

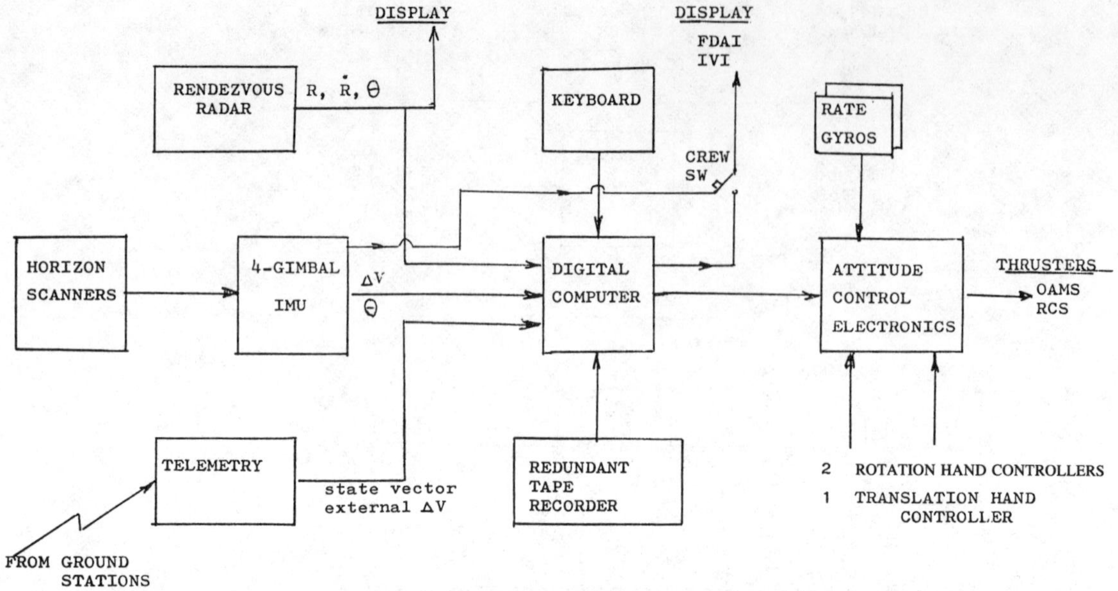

Fig. 6-7. Block diagram of Gemini guidance and control.

were 4.5 to 200 nm, suggesting that the X-20 Dynasoar would have been premature in the late 1960s without a precise, redundant navigation system. NASA's Mercury Project Summary (Kleinknecht and Bland, 1963) contains many details about the spacecraft and its support services.

The first man in space was Yuri Gagarin, who made a one-orbit flight on Vostok (''east''), 23 days before Alan Shepard flew a suborbital lob on Mercury. The early manned Russian spacecraft, Vostok (first flight in 1961) and Voskhod (''sunrise,'' first flight 1964), had no on-board navigation, paralleling Mercury practice. I suspect that they were tracked only over the Soviet Union, when the orbit was observed on the ground. I also suspect that ships and foreign stations described in section 6.3.4 were only for communication. The crew communicated from orbit via HF teletype and VHF voice until L-band voice was added later.

They landed on Russian soil, the crew bailing out of the first four spacecraft before they hit the ground. The two Voskhods had a radar altimeter and a solid-rocket motor to slow the spacecraft just before impact from parachute-descent speed to a soft landing. As of early 1989, every subsequent Russian manned spacecraft was a Soyuz (see section 6.4.2) that soft-landed on the Soviet homeland.

6.4.2 Middle-Period Manned Spacecraft

Gemini was a two-place manned ballistic missile with a large attached service module (Fig. 6-6) that flew from 1965 to 1966. Its navigation-related goals were to develop a space-rendezvous technique and a precision-entry technique to be used during Apollo missions. Gemini's navigation was an enormous advance over Mercury's. It carried a four-gimbal IMU with dual gimbal synchros and three pulse-rebalanced accelerometers (Fig. 6-7). Dual horizon scanners leveled the IMUs. Gemini had a 4K digital computer, by IBM, the largest airborne computer of its day, that calculated the state vector, the rendezvous burns, and deorbit time and drove displays. It had a 39-bit word length and a 250-kHz clock. Its computer program was so large that six in-flight overlays were needed in different mission phases. The overlays were stored in a three-track voting tape recorder and were called by the crew. The program contained algorithms for in-space gyrocompassing to find the orbit plane and for the calculation of each rendezvous burn from radar measurements, for example.

Gemini developed the Coelliptic Flight Plan for rendezvous that was used throughout Apollo (when it was called the ''concentric flight plan'') and is still used on the Shuttle in modified form. The interceptor maneuvered itself into an orbit approximately 15 nm below the target and coplanar with its orbit, based on ground-supplied $\overline{\Delta V}$. There it coasted (''phasing'') while catching up. When the on-board–measured line of sight to the target was approximately 26° above the interceptor's local level, the final maneuver to intercept was begun.

For each rendezvous burn, the crew had three solutions:

(a) on-board calculated burns, from rendezvous radar data via the on-board digital computer;

(b) uplinked burns, based on ground-tracking of the spacecraft and target; and

(c) hand-calculated on-board burns, based on displayed radar data and charts.

The crews chose among the three, favoring the ground solutions.

The rendezvous radar was a four-antenna phased array on the front face of the interceptor. The radar transmitted at 2 W average power, 1150 W peak at C-band. With its 1-ft baseline, it measured angles from the spacecraft's axis of symmetry to the target. Range was measured by timing the pulse returned from the target's transponder.

About an hour before entry, the crew aligned the inertial platform. At an uplinked time, the crew fired the solid-rocket motors in the service module holding the 37° nozzle-up attitude relative to the 8-ball display. The mass center of the first three Geminis was on the axis of symmetry. Those spacecraft executed rolling (10°/s) unguided entries whose 75-

Fig. 6-8. Soyuz-19 spacecraft, photographed from Apollo. Courtesy of NASA.

mi average landing error was caused by the initial dispersions in the deorbit burn, an unpredictable atmosphere, and an uncertain drag coefficient.

The next six Geminis were built with offset mass centers so they could develop lift (about one-fourth of the drag force, maximum). Their entries into the atmosphere were guided. The digital navigation computer predicted the instantaneous landing location (from inertial navigation measurements) and rolled the spacecraft so its own lift adjusted the location of the landing point. When the predicted landing point was at the desired geographic position, the guidance system began the 10°/s roll that resumed the ballistic trajectory. Landing dispersions on the guided flights averaged 4 nm with no guided Gemini landing more than 7 nm from its intended target. Further information about the Gemini vehicle can be found in (Malik and Souris, 1968).

Soyuz (''union,'' as in Soviet Union; first flight 1967) included a Gemini-like horizon scanner, a rendezvous radar, and an x-axis accelerometer that measured translation burns. The world's first unmanned automatic rendezvous and docking were done with the unmanned version of Soyuz, called Progress, in 1967. The docking over the Soviet Union was probably guided by remote control from the ground. Apparently not until 1980 did the first of the Soyuz-T series carry an IMU. I believe that improvement was a result of the Apollo–Soyuz joint rendezvous exercise. Fig. 6-8 shows Soyuz-19, photographed from the Apollo command module in 1975.

While the United States was exploring the Moon and, later, abandoning its manned space program, the Russians were operating a series of seven Earth-orbiting space stations called Salyut. Recent Salyuts and the new Mir (''peace'') space stations carry an autonomous navigation system that includes an inertial navigator, a star tracker and a horizon scanner. It may be needed to fill in tracking gaps.

The Apollo command and lunar modules (the first manned LM flight was in 1969) were extraordinarily complex for their time, using navigation and control technology that led the way for an entire generation of aircraft and for the Space Shuttle. Both the command module (Fig. 6-6) and the LM (Fig. 6-9) had identical 38K central digital computers that executed all navigation, flight control, and telemetry functions without overlays (Fig. 6-10). The computer, designed by MIT and built by Raytheon, had an add-time of 23 μs. A three-gimbal IMU measured attitude and three-axis acceleration. It was built by GM-Delco. Aircraft inertial platforms had always had four gimbals in order to prevent gimbal-lock (total loss of attitude reference), no matter how the crew rotated the vehicle. Nevertheless, to save weight, NASA chose three gimbals for the command and lunar modules and trained the crews not to rotate the middle gimbal more than 70°.

The command module's unique navigation sensor was a manually operated space sextant that could measure the angle from the Earth's limb or the Moon's limb to a star. The onboard computer stored the right ascension and declination of 37 stars. Battin (1987) reported that the Earth-horizon–to-star measurements made on Apollo-8, 30,000 nm from Earth, predicted the perilune to be −32 nm to +93 nm. The variation was due to errors in predicting the exact altitude of the Earth's visible horizon. Measurements made later, 50,000 and 35,000 nm from the Moon, recomputed perilune within 2%. The crew used the sextant to align the IMU prior to each translation burn and from time to time in coasting flight.

The lunar module had a more complex avionic system (Fig. 6-11) because it landed on and took off from the Moon. During its descent to the Moon, a landing radar (four-beam, X-band, radiating 200 mW) measured altitude and altitude rate while the guidance law, executing in the central digital computer, steered and throttled the descent engine. There were three distinct mission phases, each with a different guidance law, the last of which was a slow vertical descent from 200 ft above the Moon [see (Kayton, 1967)]. The two-man crew could take control manually at any time, though, in reality, they could not land manually within the propellant budget if they took over more than 1000 or 2000 ft above the surface.

The ascent to lunar orbit was inertially guided without radio updates. Once in orbit, the LM's burns were based on measurements by its rendevous radar and by Earth tracking. The three-tone, X-band, monopulse rendezvous radar measured range, range rate, and angle to the command module (waiting in lunar orbit) during the rendezvous with the command module. An optical sight allowed the crew to monitor the rendezvous quantitatively. The rendezvous followed the Coelliptic Flight Plan that was developed on Gemini.

Because the ascent and rendezvous were necessary to return the crew and because an abort during landing was time-critical, the LM carried an independent inertial navigator, the ''abort guidance system.'' Its computer contained 4K, 18-bit, ferrite-core words of independently written software for the ascent and rendezvous. The Abort Guidance System contained the world's first operational strapped-down IMU.

By today's standards, the Apollo control panels resemble those of a process plant. Two rotational hand controllers drove the primary digital computer and, for backup modes, the

Fig. 6-9. Lunar module on the Moon's surface. S-band steerable and omni antennas are visible at the upper left; rendezvous radar antenna is at the top right, pointing downward. Attitude-control motors are at the extreme left. Stairs to the surface are at the right. The Lunar Roving Vehicle is at the far right. Courtesy of NASA.

Fig. 6-10. Apollo primary guidance computer and keyboard. Courtesy of Draper Laboratory.

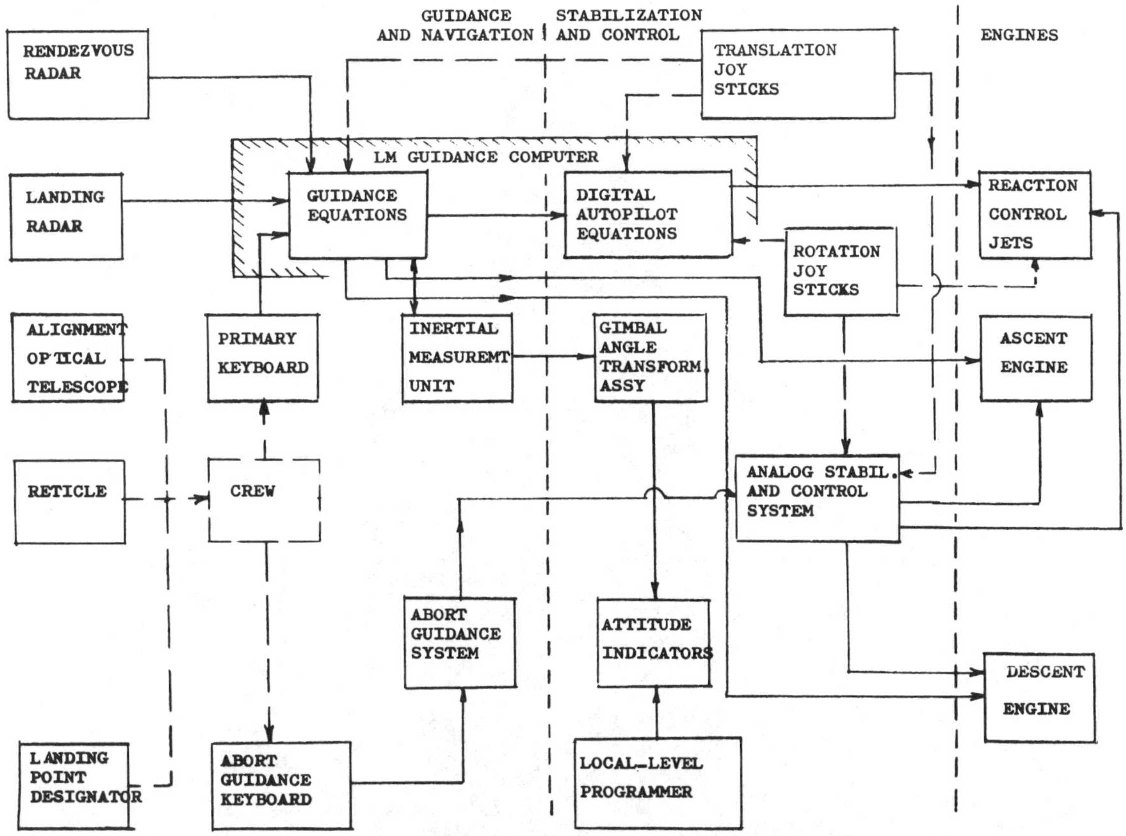

Fig. 6-11. Block diagram of the lunar module avionics.

analog flight control circuits. In accordance with the practice of its time (still often followed), the flight controls had their own rate gyro package, despite the presence of two IMUs. In order to achieve diverse redundancy in the lunar module, the keyboard-addressing systems of the primary and the abort guidance systems were totally different, imposing added training tasks on the crews. The hardware, software, and mission sequence are described in detail in (Kayton, 1967).

The Skylab space station, which flew in 1973, was put together from leftover Apollo parts. Its state vector was measured on the ground and uplinked. Skylab's on-board computer had algorithms to predict the orbit for a few revolutions with a second-order symmetric gravity model that did not account for air drag. It used one new navigation technique: the active rendezvous sensor was placed on the command module. It emitted a P-N pulse train on VHF radio that measured range and range rate to Skylab (versus a LM-mounted rendezvous radar on Apollo). The command module's software was upgraded to include the Kalman filter and rendezvous targetting that were in the LM.

6.4.3 On-Board Interplanetary Navigation

Planetary spacecraft are navigated entirely by ground tracking, as explained in section 6.3.3. Attitude commands that point the translation motor (and sensors) are uplinked by ground controllers, based on their interpretation of telemetry from the on-board attitude sensors. Wong and Lai (1982) have described the on-board recognition of star patterns that will allow the Sun and stars to be used for autonomous orientation, in a future spacecraft, prior to each burn.

The long travel time of radio waves from the planets has encouraged autonomy in landing probes where quick real-time decisions must be made. The reprint by Euler, pp. 366–373, describes the landing sequence and navigation sensors for an unmanned landing of the Viking on Mars in 1976. The lander was initialized by ground update and landed autonomously. The upcoming Mars Sample Return Vehicle, still in preliminary design, will probably be the prototype of a nearly autonomous surface-roving robot (see Part 3.3, pg. 67, and Part 8.6, pg. 438).

6.4.4 Modern Spacecraft

The Shuttle (Fig. 6-12) finally achieved regular X-15–like glide landings from space at an airport, thanks to a fault-tolerant design (e.g., five redundant computers and triple four-gimbal IMUs). Dual star trackers assure realignment of the IMU's attitude in space. Four of the central computers execute redundant calculations, calling eight overlays in different phases of the mission (preflight, ascent, abort, orbit, entry, etc.). The fifth computer is diverse-programmed for an abort to a landing site. All five computers contain 106K 32-bit words of RAM storage. Sensors, electronics, actuators, and displays are interconnected by a 1-MHz data bus that was a precursor of the MIL-STD-1553 bus. The Shuttle's avionic system (Fig. 6-13) was the most complex in existence at the

Fig. 6-12. Shuttle on the launch pad with a superimposed view of the Spacelab in the payload bay. Courtesy of NASA.

time it was designed, having the best features of manned aircraft and spacecraft avionics [see (Kayton, 1989)].

The state vector is measured and calculated by the tracking network (ground STDN and TDRS, section 6.3.4). It is calculated on-board, using the IMU to measure thrusting translations in order to extrapolate an uplinked, ground-derived state vector. NASA guarantees the on-orbit navigation accuracy to be better than 4 nm in position and 6 ft/s in velocity (3-sigma) between radio navigation fixes. The orbit is extrapolated (''propagated'') with an on-board algorithm that includes a fourth-order gravity model and luni-solar perturbations (the program stores the orbits of the Moon and Sun). With its zonal and tesseral harmonics, the on-board model accounts for large-scale asymmetry of the gravity field in latitude and longitude.

The full on-board state is updated every 2 s and a simpler extrapolation is made more rapidly in order to annotate experiments. The accelerometers are not read except when deliberate translation burns are executed. Hence, small position and velocity errors accumulate because of the inability to measure venting, asymmetric attitude-motor firings, aerodynamic drag, and solar radiation pressure. Though these errors can accumulate to 300 m per revolution, the Shuttle is usually updated from STDN several times per revolution so that the instantaneous navigation accuracy is closer to 100 m and 2 ft/s. When GPS receivers are added to the Shuttle, its on-board–

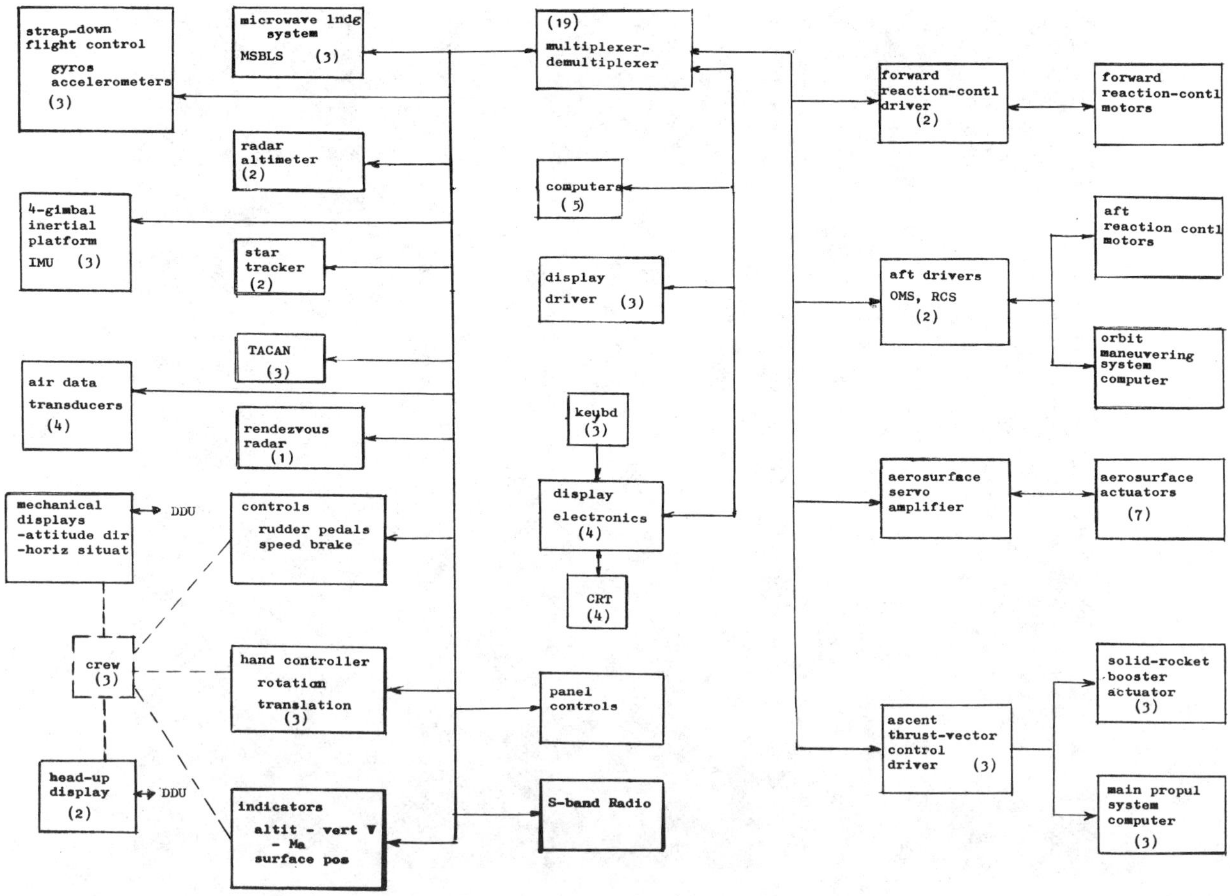

Fig. 6-13. Block diagram of Shuttle avionics. Courtesy of Rockwell Corporation.

computed navigation accuracy will improve by a factor of 3 or more.

For rendezvous, the velocity change required for each translational maneuver is calculated on-board and is also uplinked from ground-derived calculations. The on-board calculations use the Shuttle-borne K-band rendezvous radar. It measures range, range rate, and angles to find a cooperative or disabled spacecraft. The radar has been described by Griffin *et al.* (1988). Its peak power of 50 W permits it to skin-track to about 12 nm range and track a transponder to about 300 nm. The Coelliptic Flight Plan of Gemini and Apollo evolved into a double coelliptic rendezvous in which the interceptor (Shuttle) is guided into a phasing orbit, 20 nm below that of its target, entirely with ground-uplinked maneuver instructions. Then, measurements by the optics and rendezvous radar are made on-board to calculate the transfer burn to an orbit 10 nm below the target. Finally, after more measurements, an on-board–controlled burn is made onto an intercept trajectory. Docking maneuvers, from a range of a few thousand feet, have always been done manually on U.S. spacecraft. The reprint by Collins, pp. 388–394, describes the navigation, targetting, and guidance during the Shuttle's rendezvous. The reference by Gustafson and Kriegsman (1973) shows the algorithms required for automatic stationkeeping.

The mechanization of the Shuttle's boost-phase navigation is described by Olson and Sunkel in the reprint on pp. 374–387. They include block diagrams and review the accuracy achieved on the STS-1 and STS-2 flights (STS-*n* is the *n*th Shuttle flight to orbit).

During entry, navigation is inertial, the state vector having been radio-updated on orbit. Above 100,000 ft altitude, angle of attack is controlled to match the inertially measured drag to a nominal profile. The angle of attack starts at 40° and gradually diminishes to 10° at 80,000 ft. The guidance law nulls the in-plane forces normal to the velocity vector by rolling around it, hence flying S-turns at ± 80 to $\pm 90°$ bank angle. The S-turns are slightly asymmetric in order to steer the ground-speed vector toward the estimated position of the airport. The bank angle is modulated in order to steer downrange. These maneuvers reduce the dispersions due to faulty execution of the deorbit burn, nonstandard atmosphere, and uncertain drag coefficient. Preflight simulations assure that key structural temperatures are not too high for $\pm 3°$ variations in the angle of attack, while the real-time guidance law prevents wider excursions. Altitude is computed from the measured drag until Mach 2.5, when barometric altitude is substituted. Bennett (1983) describes the entry dynamics in detail.

The first external information about the position of the airport is from Tacan, beginning at 130,000 ft altitude (Mach 7). Thereafter, Tacan updates the inertial position and the spacecraft is guided to a vertical target cylinder upwind of the runway.

At Mach 5, the air-data sensors are deployed but their

Fig. 6-14. Right pilot station of the Shuttle. Hand controller, rudder pedals, head-up display, mechanical instruments, and two of the three status CRTs are shown. Courtesy of NASA.

measurements are used only for flight control and postflight calculation of wind until 85,000 ft altitude when barometric altitude substitutes for density-inferred altitude. Energy is dissipated by activating the split-rudder speed-brake. Tacan updates continue until the microwave landing system is acquired, about 10 mi from touchdown and 14,000 ft altitude. The Shuttle uses a Microwave Scanning Beam Landing System (MSBLS), a derivative of the Navy's SPN-41 (15.6 GHz, Part 5.4, pg. 237), not the civil MLS. Triple radar altimeters and microwave landing receivers permit automatic approach and landing during foul weather. The reprint by Ewell, pp. 366–373, describes the mechanization equations and the accuracy achieved for entry and landing of STS-1 and STS-2. Though automatic landing is designed into the Shuttle, all approaches have been automatic and all landings have been manual through 1988. STS-8 (1983) was the first night landing.

Shuttle displays use a conventional mechanical attitude indicator (ADI) and a horizontal situation indicator (HSI) at each pilot station plus three cathode-ray tube (CRT) status monitors located between the pilots (Fig. 6-14). The hand controller, rudder pedals, mechanical instruments, and reflector of the head-up display are visible in Fig. 6-14, as are two of the three CRTs. Steering information, from the navigation calculations or from the direct instrument outputs, is presented on the ADI and HSI needles and on the HUDs (see Part 5.5.d, pg. 240). The crew can elect to ''couple'' steering information to the flight controls, though steering and flight-control calculations are executing in the same central computer.

The Shuttle's hardware interfaces and software are verified by the developer, the vehicle integrator, NASA's integration laboratory, and in the crew trainers. By coordinating tests in these facilities, all likely dispersions and astronaut inputs are tested before flight. Most software constants (e.g., gravity model, calibrations, display formats, and telemetry formats) are embedded in the program so a change requires recompilation of the code, hence verification. A few constants (e.g., mass properties and targetting) are loaded on magnetic tape at flight time, into reserved memory locations. This is the same method by which Apollo constants were loaded. The first ''live'' test of the complex hardware interconnections and software algorithms was on the successful first flight to orbit.

The reference by Klass (1988) gives some information about Buran (''snowstorm''), the Soviet shuttle. Its entry navigation is clearly modeled after the Shuttle's. The guidance is purely inertial until 130,000 ft altitude, when at least three of six DMEs (located within 50 km of the runway) are acquired and update the inertial system by trilateration. At an estimated 10 km from touchdown, Buran's civil MLS (5 GHz, not 15 GHz as on the Shuttle) is acquired for final approach and lateral guidance during rollout. A precision DME assists on-board energy management during approach and rollout, as on the Shuttle and the civil MLS.

The European-funded Spacelab was initially conceived of as a free-flyer with autonomous navigation. After many budget cuts in 1973–75, it was designed to remain fixed in the Shuttle's cargo bay and depend on the Shuttle for navigation and communication. It remained so in 1989. Lord (1987) and NASA (undated reference) have described Spacelab in detail.

Autonomous navigation has been much studied and little implemented. Pure autonomy (called "passive self-contained" in Part 1, pg. 1) implies no external radio, radar, or star measurements and implies on-board failure detection and reconfiguration. Purely autonomous navigation is practiced by boosters and by short-time-of-flight upper stages that are extensions of the boosters. Examples are the Titan Transtage and the Inertial Upper Stage (IUS, first flight 1982). Both vehicles boost spacecraft, launched by the Shuttle or Titan, into synchronous orbit. IUS has on-board instrumentation and software that detect and correct failure in the dually redundant computers (one is required to work) and five redundant gyroscopes (three are required to work). It deliberately underburns its solid-propellant motors and achieves the proper orbit with precisely measured accelerations produced by small liquid-fueled engines.

The less stringent definitions of autonomy that I propose in Part 1 would allow ground communications (e.g., via TDRS) while not requiring a large staff of subsystem specialists and mission specialists on full-time duty on the ground. This form of autonomy will probably be practiced by the NASA Space Station and newer unmanned spacecraft. The reprint by Chory *et al.*, pp. 403–414, discusses autonomous navigation for spacecraft, as does section 8.6, pg. 439.

The GPS satellites (Part 3.5) are useful for navigating spacecraft in orbits below approximately 6,000 nm. Special receivers are being designed to take advantage of the large number of operational satellites that are visible to a spaceborne GPS receiver. Experiments on the Trident booster, Landsat, and the Shuttle have shown real-time accuracy of 5 to 10 m in orbit using the GPS' C/A-code. The reprint by Heuberger, pp. 347–352, and the reference by Ananda and Chernick (1982) show the Landsat results.

The future of space navigation is described in Part 8.6. The trend will continue for avionic costs to rise as a fraction of total spacecraft cost because the functions demanded of the avionics are growing so quickly.

References and Bibliography

Ananda, M. P. and M. R. Chernick, "High accuracy orbit determination of near-Earth satellites using GPS," in *Rec. IEEE Position Location and Navigation Symp.*, 1982, pp. 92–98. Results of GPS on Landsat-D show 5–10 m error after post-flight data reduction. Expect 1 m.

Baker, D., *The History of Manned Space Flight.* New York, NY: Crown, 1981. Excellent popular book. Appendices with lots of numbers.

Battin, R. H., *An Introduction to the Mathematics and Methods of Astrodynamics.* Washington, DC: AIAA Press, 1987. The most complete description of interplanetary guidance ever written; a storehouse of algorithms and past experience by a giant in the field. Emphasis is on on-board navigation.

Bennett, D. E., "Space Shuttle entry flight control overview," in *Advances in Astronautical Sciences, Progress in Guidance and Control*, pp. 393–402, 1983. Quantitative treatment of Shuttle entry with block diagrams in each phase.

Bourne, C. A. and P. W. Kirsten, "Approach and landing characteristics of Space Shuttle Orbiter," Shuttle Performance Lessons Learned, NASA Conf. Publ. 2283, pp. 637–651, 1983.

Burghduff, R. D. and J. L. Lewis, "Man-machine interface and control of Shuttle digital flight control system," *Proceedings of Space Shuttle Tech. Conf.*, NASA Tech. Conf. Pub. 2342, 1985, 7 pp. Shuttle crew station, including HUDs.

Carson, L., *et al.*, "Design and predicted performance of the GPS demonstration receiver for the NASA TOPEX satellite," in *Rec. IEEE Position Location and Navigation Symp.*, 1988, pp. 442–454. Reprinted in Part 6, pp. 353–365.

Chaffee, N. (Ed.), NASA Space Shuttle Tech. Conf. Publ. 2342, Parts 1 and 2, 1985. Papers on all aspects of Shuttle design: integrated avionics, guidance, navigation, control, aerodynamics, heating, structure, propulsion, communication and tracking, and ground operations.

Chory, M. A., *et al.*, "Satellite autonomous navigation—Status and history," in *Rec. IEEE Position Location and Navigation Symp.*, 1986, pp. 110–121. Reprinted in Part 6, pp. 403–414.

Collins, D. J., "Flight operations for Shuttle rendezvous navigation and targeting," in *Rec. IEEE Position Location and Navigation Symp.*, 1984, pp. 106–112. Reprinted in Part 6, pp. 388–394.

Euler, E. A., "Navigating the Viking landers," *Navigation,* vol. 25, no. 1, pp. 11–18, Spring 1978. Reprinted in Part 6, pp. 366–373.

Ewell, J. J., Jr., "Space Shuttle orbiter entry through landing navigation," in *Rec. IEEE Position Location and Navigation Symp.*, 1982, pp. 110–117. Reprinted in Part 6, pp. 395–402.

Green, G. B. and P. Axelrad, "Space applications of GPS," *Navigation*, vol. 36, no. 3, pp. 239–251, Fall 1989. Reviews current spaceborne GPS receivers. Discusses possible use of GPS interferometer for measuring spacecraft attitude. Expects 80 m SEP for standard positioning service. Repeats JPL estimate of 10-cm accuracy with 2-h arcs.

Griffin, J. W., *et al.*, "Shuttle rendezvous radar performance evaluation and simulation," in *Rec. IEEE Position Location and Navigation Symp.*, 1988, pp. 236–245. Block diagram, displays and controls, external interfaces, modes, performance calculations, test results, and mission operations on STS-7.

Gustafson, D. E. and B. A. Kriegsman, "A guidance and navigation system for automatic stationkeeping in Earth orbit," *AIAA J. Spacecraft*, vol. 10, no. 6, pp. 369–376, June 1973. Development of guidance equations and error equations using Wilshire–Clohessy approximation.

Ha, T. T., *Digital Satellite Communications.* New York, NY: Macmillan, 1986. Good survey of orbit slots and frequency assignments. Stationkeeping 0.05°; usually east–west every 21 days, north–south every 28 days. Most of text is mathematics of communication.

Haeussermann, W., "Developments in the field of automatic guidance and control of rockets," *AIAA J. Guidance and Control,* vol. 4, no. 3, pp. 225–239, May 1981. V-1, V-2. Army and NASA boosters: Saturn, Pershing, Jupiter.

Heffron, W. G. and F. LaPiana, "The navigation system of the lunar roving vehicle," NASA CR-116268, Dec. 1970. Describes the mission, navigation system, and error analysis.

Heuberger, H., "Performance of the GPS package on Landsat," in *Rec. IEEE Position Location and Navigation Symp.*, 1984, pp. 147–152. Compares Landsat-4 and Landsat-5 results; 50-m error, 6 cm/s with satellites visible; nine-state Kalman filter.

Hoag, D. G., "The history of Apollo on-board guidance, navigation, and control," *AIAA J. Guidance and Control,* vol. 6, no. 1, pp. 4–13, Jan. 1983. Emphasizes command module.

Johnson, N. L., *Handbook of Soviet Lunar and Planetary Exploration.* San Diego, CA: American Astronautical Society, 1979. Drawings of various spacecraft.

Johnson, N. L. and D. S. McKnight, *Artificial Space Debris.* Melbourne, FL: Orbit Books, 1987. Describes NORAD tracking of 60,000 objects per day by radar and Geodss.

Kayton, M., "Apollo lunar ascent and descent," ION Conf. on Space Navigation, Mar. 1967. Detailed description of landing navigation and guidance, lift-off and rendezvous. Flight software and mission sequences.

Kayton, M., "Avionics for Manned Spacecraft," *IEEE Trans. Aerosp. Electron. Syst.*, pp. 786–827, Nov. 1989. Describes navigation, flight control, crew station, subsystem failure detection, and reconfiguration of each of the world's manned space programs.

Kessler, D. T. and S. Y. Su, "Orbital debris," NASA Conf. Pub. 2360, 441 pp., 1982. Good review of space debris that are tracked (larger than 1-m diameter) and not tracked. Estimates 0.1 chance of low-orbit collision between 100-m space station and tracked objects in 10 years. Estimates 0.3 chance, if unobserved objects are included.

Klass, P. J., "Details of Soviet shuttle auto-land system disclosed," *Aviation Week,* pp. 33–34, Dec. 12, 1988. Three DMEs 20 km from each end of runway. MLS for landing.

Kleinknecht, K. S., and W. M. Bland, "Mercury project summary," NASA SP-45, Oct. 1963. Official report. Description of boosters, spacecraft, and tracking net.

Kullas, M. C., Handbook on Astronaut Crew Motion Disturbances for Control System Design. NASA Res. Publ. 1025, May 1979. Categorizes torques and accelerations on spacecraft caused by crews as two-impulse "soaring" maneuvers and as statistical models of long-term noise. Validates math models with Skylab data. Proposes design accommodations.

Lord, D., "Spacelab, an international success story," NASA SP-487, 1987. Development of requirements; emphasis on NASA reviews.

Malik, P. W. and G. A. Souris, "Project Gemini, a technical summary," NASA CR-1106, June 1968. Lots of detail with appendices.

Mancina, A. J., "Space Shuttle primary on-board software: STS-1 to operational use," *AIAA J. Guidance,* vol. 6, pp. 135–142, June 1983. Operating system and application software. Testing and test facilities. Nothing on module sizes.

McSwain, G. C., "Shuttle ascent GN&C postflight results," Shuttle Performance Lessons Learned, NASA Conf. Publ. 2283, pp. 581–591, 1983. Compares predictions to STS-1 through 4.

McWhorter, L. B. and O. R. DeVall, "Shuttle entry GNC performance during STS-1 and STS-2," in *Advances in Astronautical Sciences.* San Diego, CA: American Astronautical Society, 1982, pp. 443–456. Block diagrams of navigation system and test results.

Melbourne, W. G., "Navigation between the planets," *Sci. Amer.,* pp. 58–74, June 1976. Reprinted in Part 6, pp. 333–346.

NASA, "Ground STDN users' guide," Goddard Space Flight Center, June 1980. Describes sites and antennas.

NASA, "Shuttle payload accommodations handbook," JSC Document 0770, May 1983. Includes navigation accuracies.

NASA, "TDRS users' guide," Goddard Space Flight Center, Sept. 1984. S- and K-band communication and navigation.

NASA, "Spacelab news reference," Marshall Space Flight Center Rep. 14M983, undated. Semiquantitative description of experiment provisions.

Nessmith, J. T., "Range instrumentation radars," *IEEE Trans. Aerosp. Electron. Syst.,* vol. 12, no. 6, pp. 756–766, Nov. 1976. Specifications, error model, and measured performance of FPS-16, FPQ-6, and Tradex radars.

Oberg, J., *Red Star in Orbit.* New York, NY: Random House, 1981. Vostok, Voskhod, Salyut, and Soyuz.

Olson, L. and J. W. Sunkel, "Evaluation of the Shuttle powered ascent flight phase," in *Guidance and Control 1982,* vol. 48 of *Advances in the Astronautical Sciences.* San Diego, CA: American Astronautical Society, 1982, pp. 429–442. Reprinted in Part 6, pp. 374–387.

Romero, P. O. and A. M. Whitnah, "Space Shuttle entry longitudinal aerodynamic comparisons of flights 1–4 with preflight predictions," in *Shuttle Performance Lessons Learned.* NASA Conf. Publ. 2283, pp. 283–307, 1983. Body flap for pitch trim, speedbrake for energy management. Roll reversals to control entry dispersion. Lots of numbers.

Taylor, T. H., "Performance of differenced range data types in Voyager navigation," *AIAA J. Guidance and Control,* pp. 301–306, May 1984. Measure range and range rate to 30 m and 1 mm/s after 1 min averaging. Method of measuring transverse components.

Wong, E. E. and S. Y. Lai, "Celestial referenced attitude determination of Galileo spacecraft," *AIAA J. Guidance and Control,* vol. 5, no. 3, pp. 307–312, May 1982. Identifies catalogued stars on-board by observing pulse outputs from two angled slits as spacecraft spins. Calculates rotation rate. Identifies attitude and angular momentum.

Wright, R. J. and J. V. Sponnick, "RLG-based navigation for space launch vehicle guidance," in *Rec. IEEE Position Location and Navigation Symp.,* 1988, pp. 271–279. Description of Honeywell ring-laser strapped-down INS replacement for ancient Centaur guidance system. Block diagrams, photographs, software details, mission profile, alignment, packaging, parts selection, performance.

Navigation between the Planets

Advances in tracking and computing methods are such that when the first of the two Viking spacecraft reaches Mars late in June, its position should be known to within 50 kilometers

by William G. Melbourne

On June 19 *Viking 1*, the first one launched of two identical spacecraft now heading toward Mars with capsules designed to land on the surface and search for evidence of life, will reach the planet after a 10-month journey of 700 million kilometers. *Viking 1* is aimed at a target point 1,500 kilometers above the planet's surface. As the spacecraft encounters the target zone it will execute commands stored in its on-board computer that were sent earlier from the earth. Those commands will cause the spacecraft's retropropulsion system to fire for 43 minutes, subtracting enough velocity to place the spacecraft in an elliptical orbit around Mars that will vary in altitude from 33,000 kilometers at apoapsis down to 1,500 kilometers at periapsis. The orbital period will be 24.6 hours, so that *Viking 1* will circle the planet exactly once per Martian day.

The accuracy of the retropropulsion maneuver and ultimately the accuracy of the landing will depend on how well the position of the spacecraft is known at the time the commands are generated on the earth during the several days before the spacecraft arrives in the vicinity of Mars. The National Aeronautics and Space Administration flight team at the Jet Propulsion Laboratory that has been tracking *Viking 1* since its launching last August is confident that if the spacecraft continues to operate satisfactorily, the probability is greater than 50 percent that the error in determining its position before the encounter with Mars will be less than 50 kilometers. Since the first successful flight to Mars by *Mariner 4* in 1964 the accuracy of interplanetary navigation has improved by a factor of 100. It is realistic today to design planetary missions with a delivery accuracy of a few tens of kilometers for the terrestrial planets (Mercury, Venus and Mars) and a few hundred kilometers for Jupiter and Saturn.

The improvement in navigational accuracy, with the concomitant increase in scientific returns that was brilliantly demonstrated in the most recent photographic missions to Mars and Mercury, is the result of cumulative advances in key technologies. Radar measurements have greatly refined our knowledge of planetary distances and orbits. Over the past 15 years the stability of clocks has been improved by nearly four orders of magnitude. In the same period the area of the parabolic antennas for tracking space vehicles has increased sixfold, the sensitivity of radio receivers has been enhanced by a factor of 100 and antennas separated by nearly the diameter of the earth are now being used in pairs for sharpening tracking accuracy. There are also much improved techniques for calibrating the factors affecting the propagation of radio signals through the earth's atmosphere and over interplanetary distances. Finally, there have been steady refinements in the computer programs that process the tracking data and model the diverse and subtle forces acting on vehicles during their long free fall through space.

Navigation still means, as it did in the days of the Phoenicians, guiding a ship. It calls for a map, a travel plan, a means of periodically determining the ship's location and a method of selecting a new route when the ship has drifted off course. Interplanetary navigation satisfies this description by continually refining the map of the solar system, by devising a mission plan, by designing suitable spacecraft trajectories and by determining the trajectory en route and making corrections that enable the spacecraft to reach its interplanetary goal.

The Map

For interplanetary flight the map corresponds to a description of the time-varying positions, velocities and rotations of the earth and the other planets, together with such physical quantities as the planets' masses, sizes and shapes. A description of the orbital motion of the planets in the solar system is an ephemeris. Until about 1960 the accuracy of the planetary ephemerides was limited by the astronomical techniques for determining the positions of the planets (and in some cases the positions of asteroids) with respect to the background stars. From these measurements and gravitational theory the orbital positions of the planets were calculated and were commonly expressed in terms of the astronomical unit (essentially the mean distance between the sun and the earth). The resulting ephemerides were accurate to within a few parts per million, the equivalent of a few hundred kilometers for the terrestrial planets. Surprisingly, as late as 1960 the value assigned to the astronomical unit was in error by 66,000 kilometers, or nearly one part in 2,000. Since the astronomical unit is the scale factor needed to convert ephemeris positions into kilometers, if the value had been left uncorrected for the flight of *Mariner 2* to Venus in 1962, the spacecraft would have been aimed at a point on the wrong side of the planet.

The magnitude of the error was uncovered in 1961 when, with the development of large radio telescopes, high-power transmitters, ultrasensitive receivers and sophisticated signal-detection techniques, it became possible to detect radar echoes from Venus. The basic radar observation measures the two-dimensional distribution of the power of the returning echo signal with respect to its frequency and time delay. The rotation of the planet spreads the echo signal in frequency through the Doppler effect, and reflections from regions of the planet located at different distances from the earth cause the echo to be stretched out in time. From an analysis of this distribution of echo power the range to the spot on the planet closest to the earth (the subearth point) is obtained.

Radar systems have greatly improved over the past 15 years, so that today under favorable conditions the range measurement is accurate to about 50 meters for the terrestrial planets. Optical observations provide no direct means of determining range with any accuracy. Historically telescopes have been used to measure a planet's location along two axes perpendicular to the direction between the earth and the planet. For the terrestrial planets measurements of this kind have an accuracy of 500 kilometers. The ephemerides were formerly computed from a large number of such measurements over many decades.

The remarkable accuracy with which radar echoes yield range is somewhat misleading. As the planet rotates, the distance of the subearth point from the center of the planet changes because of the planetary topography. These variations affect the range measurement and must be mapped and accounted for to fully utilize the radar obser-

Reprinted with permission from *Sci. Amer.*, pp. 58–74, June 1976.

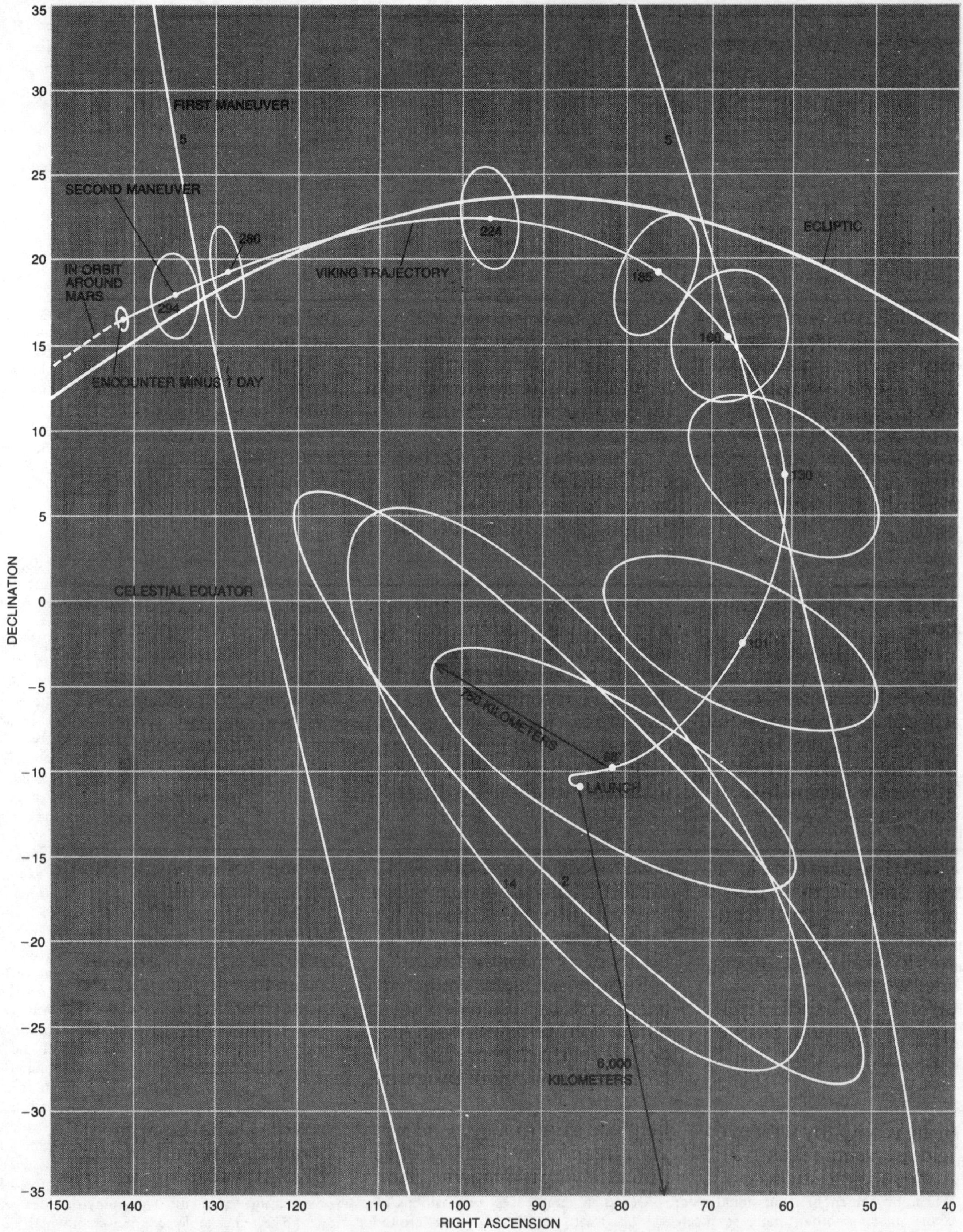

NAVIGATION UNCERTAINTIES in the position and the velocity of *Viking 1* at a given point on its journey to Mars (traced here on the celestial sphere) can be translated into position uncertainties at the target planet, visualized as ellipses drawn in a plane perpendicular to the incoming trajectory. The numbers inside the uncertainty ellipses, all drawn to the same scale, indicate the number of days after launching. There is a 50 percent probability that the actual trajectory error at the target will fall within the computed ellipse. Velocity errors are the chief source of target error early in the flight but become less significant later on. The first mid-course correction, on the fifth day (designed to shift the aiming point at the target), introduces such uncertainties in velocity that the long axis of the uncertainty ellipse extends far beyond the edges of the illustration. The spacecraft appears stationary on the celestial sphere for several weeks after launching because it is traveling nearly radially away from the earth. Target-error uncertainties become smaller as more radio-tracking measurements are made. The target-error uncertainties continue to diminish until the effects of small random forces on the spacecraft begin to cancel information gained from the additional tracking. Total length of the spacecraft's journey from the earth to Mars is 700 million kilometers.

vation for improving the ephemeris. The ephemeris describes the motion of the center of the planet, whereas radar measures range to the surface. An equivalent range measurement to the center would be accurate to about one kilometer. The range measurement is blind to certain combinations of displacement in the positions of the planets. For example, a uniform shift in the orientation of the entire solar system with respect to the background stars would not be detected by the radar measurement. As a result radar observations must be supplemented with other types of observations that give the orientation of the solar system.

Also of concern is the fact that the terrestrial planets follow orbits around the sun that are tilted anywhere from 1.9 to seven degrees with respect to the plane of the ecliptic (the plane of the earth's orbit). Radar measurements are relatively insensitive to a planet's displacement perpendicular to the plane of the ecliptic because such displacement alters the range only slightly. Nevertheless, on the basis of several thousand radar observations of the terrestrial planets made over the past decade the accuracy of the planets' positions with respect to the earth has been improved to one or two kilometers in the components lying in the plane of the ecliptic and to 10 or 20 kilometers in the direction perpendicular to the plane of the ecliptic. The astronomical unit is now known to be 149,597,871 kilometers, with an uncertainty of about one kilometer.

Keeping Time

The problem of accurately determining the ephemeris of a planet is analogous to the problem faced by early explorers trying to determine the position of land masses on the earth's surface, which is apparent in the distortions in 16th-century maps of the New World. A close examination of such maps reveals two interesting characteristics. The latitudes assigned to places in the New World are usually accurate to better than one degree, or about 100 kilometers. The longitude values, on the other hand, are often badly in error, sometimes by as much as 30 degrees. Latitude, or distance from the Equator, is readily determined by observing the position of the celestial pole in the night sky: the point in the sky near the star Polaris (in the Northern Hemisphere) around which the stars appear to revolve because of the earth's rotation. The elevation of the celestial pole above the horizon is equivalent to the latitude of the observer.

The determination of longitude is more difficult because of the earth's rotation. Imagine that the earth is at the center of a hollow spherical shell of infinitely large radius and that the stars are fixed to the surface of the sphere [*see illustration below*]. The angular position of a star on the sphere is given by its right ascension and declination; right ascension is analogous to longi-

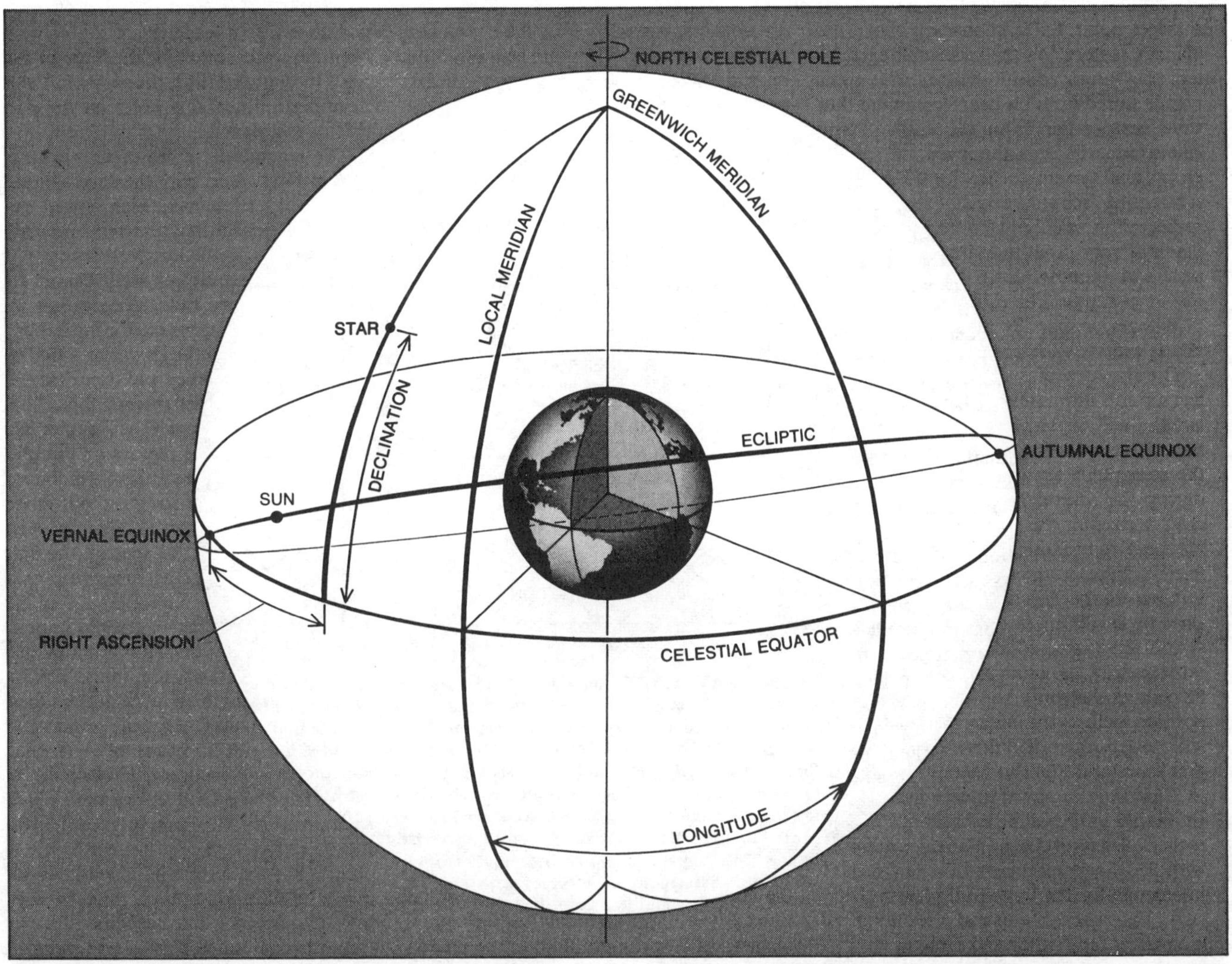

CELESTIAL SPHERE, the conceptual basis for describing the traditional astronomical coordinate system, is a westerly-rotating shell of infinite radius on which the stars are assumed to be fixed. The celestial equator is a projection of the plane of the earth's Equator. A plane through the earth's axis of rotation and Greenwich Observatory in England defines the Greenwich meridian. The angular position of a star (or of a spacecraft) is given by its right ascension and declination. Zero right ascension and declination, also referred to as the vernal equinox, is the point where the sun on its apparent northward journey across the sky crosses the celestial equator in the spring. The vernal and autumnal equinoxes therefore mark the points where the plane of the earth's orbit, the plane of the ecliptic, intersects the celestial equator. Right ascension is measured in an easterly direction along the celestial equator from the vernal equinox to the meridian passing through the star. Declination is measured along the meridian from the celestial equator to the star. The orientation of the celestial sphere at a given instant is provided by the right ascension of stars crossing the Greenwich meridian, which is equivalent to sidereal time.

tude and declination to latitude. The celestial sphere rotates around the celestial pole once a day, reflecting the earth's rotational motion. The orientation of the celestial sphere around its axis of rotation at any one instant is customarily given by Greenwich sidereal time, the equivalent of the right ascension of the stars just crossing the Greenwich meridian.

To obtain the longitude at night one observes the direction of one star or more and the time of the observation. With these observations and a catalogue giving the stellar positions, the right ascension of the stars crossing the local meridian at a given instant can be calculated. If the observer at that instant also knows the time in Greenwich, he knows the right ascension of the stars crossing the Greenwich meridian. The difference in right ascension between the stars on the Greenwich meridian and those on the local meridian is the longitude of the observer. Thus a critical element in navigating by the heavens is knowing what time it is in Greenwich when one is not there. Until the middle of the 18th century, when the first chronometers were developed and an alternate technique was introduced that made use of the moon's position on the celestial sphere, mariners had no way to keep time with any accuracy during voyages lasting several months; hence the large longitude discrepancies seen in the early maps.

The accuracy of timekeeping has improved by nine orders of magnitude over the past two centuries, with more than half of the improvement coming in the past 25 years. A good 18th-century chronometer could run for a week or two without gaining or losing more than a second. By 1945 quartz-crystal clocks had been developed that would maintain one-second accuracy for several years. Today's hydrogen masers are stable to better than two parts in 10^{15} over 24 hours. If the error in the rate of such a maser remained constant, a gain or loss of one second would take more than 10 million years.

Interplanetary navigation makes use of time in two different ways. Time gives the epoch associated with an event, such as a specific observation. For that purpose it is sufficient to know the correct time to the nearest tenth of a millisecond, since planets and spacecraft move only a few meters during such an interval and the earth rotates through only a few centimeters. The more critical role of time is in the making of radio observations of the range and radial velocity of a spacecraft. That calls for keeping track of the motion of a radio signal traveling at the speed of light with an accuracy of a few decimeters (10^{-1} meter). For this purpose one must be able to measure time intervals to an accuracy of nanoseconds (10^{-9} second) for periods of several hours. The task requires clocks that are stable to about one part in 10^{13}.

Space Communications

The measurements used to navigate U.S. spacecraft are derived from the radio link

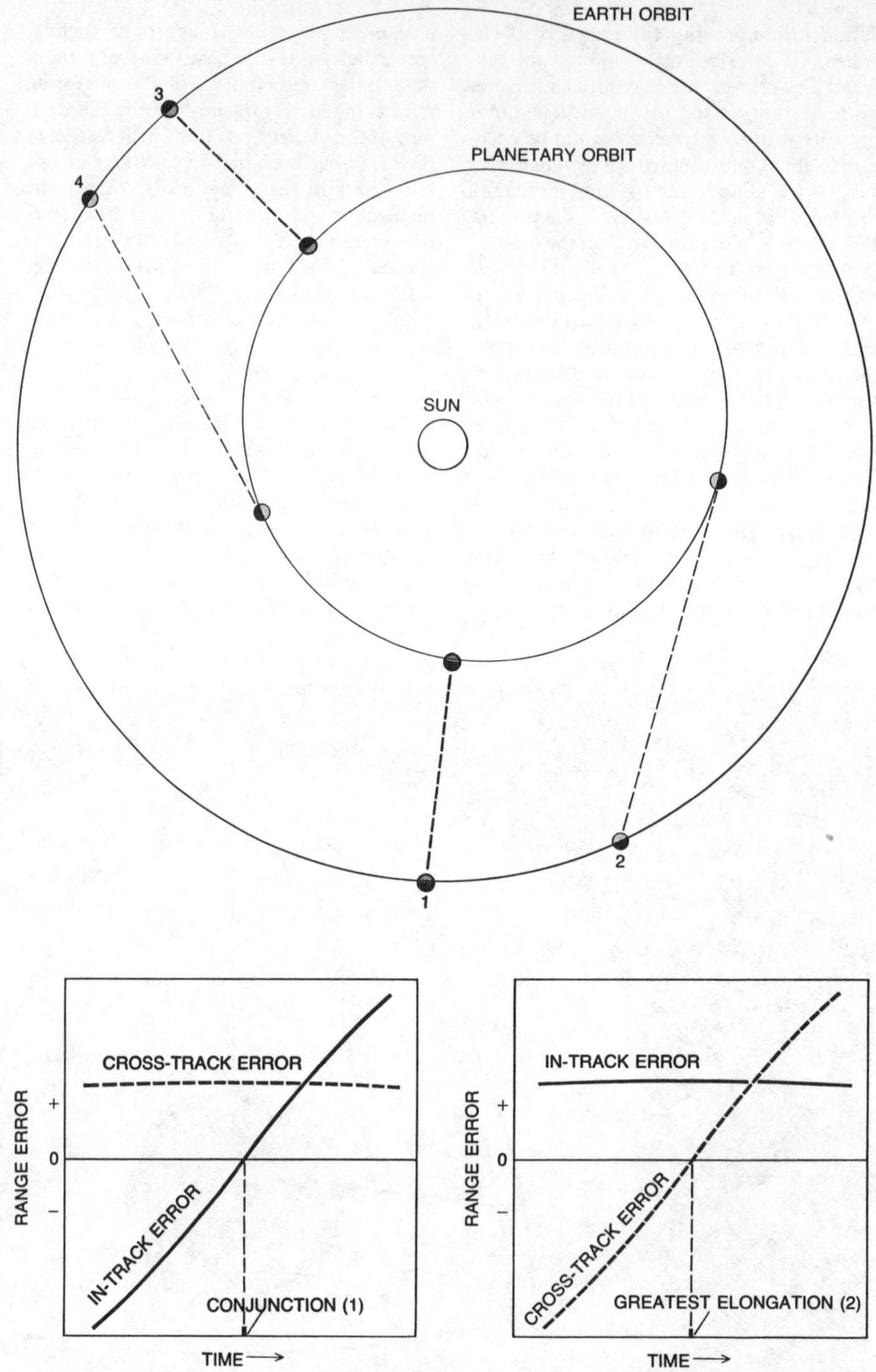

PLANETARY ORBITS have been determined with great accuracy over the past 15 years by measuring the round-trip travel time of radar echoes from the planets at many points in their orbit. The range to the point on a terrestrial planet's surface that directly faces the earth is now measured to an accuracy of 50 meters. In practice a weighted least-squares procedure is used to find the values of well over 100 parameters that describe the orbital motion and physical properties of the planets. The computation leads to the best fit between the theoretically predicted observations and the actual ones. An error in a given parameter gives rise to a corresponding error in the predictions that develops a characteristic "signature" over a period of time. The magnitude of the signature and its degree of distinctiveness compared with the signatures of other parameters determine the accuracy with which the parameter can be evaluated from the observations. In this simplified case the orbital parameters and the radius of an inner planet in the same plane are to be determined. When the planet and the earth are at conjunction (*1*), an in-track error (an error along the planet's path) is perpendicular to the earth-planet line of sight and would not give rise to a range error. As the planets move away from conjunction the in-track error begins to show up as a range error that develops a unique *S*-shaped signature over a few weeks of tracking (*bottom left*). The slope of this signature is very sensitive to an in-track error and provides an accurate determination. A cross-track error (one perpendicular to the path) gives rise to a nearly flat signature that resembles a radius error. At the position of "greatest elongation" (*2*), when the angle between the planet and the sun is at a maximum, the signatures of the in-track and cross-track errors are reversed, allowing the latter to be isolated. Later, when the planets have lined up again at *3* and *4*, signatures are slightly modified by eccentricity and orientation of orbit and provide additional information about these parameters.

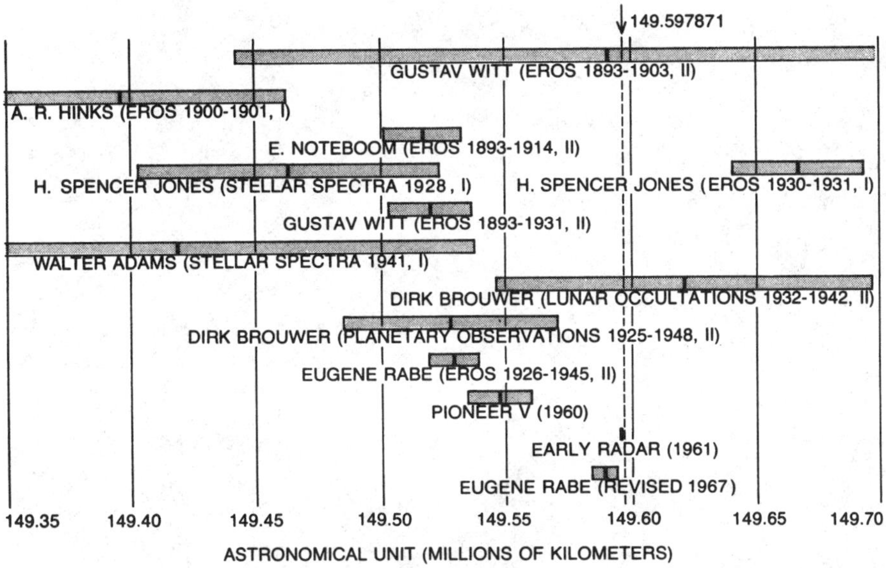

ASTRONOMICAL UNIT (A.U.), which is essentially the mean distance between the sun and the earth, was difficult to measure accurately before the advent of planetary radar. Preradar measurements were based either on direct (*I*) or dynamical (*II*) methods. One direct method involved triangulating the distance to the asteroid Eros on occasions when it passed within about 20 million kilometers of the earth. Another direct method exploited the Doppler shift in stellar spectra to determine the velocity of the earth. Dynamical methods use the observed perturbations caused by the earth in the orbits of other bodies to obtain the ratio of the sun's mass to the earth's mass. This ratio plus the known value of the universal gravitational constant times the earth's mass yields a value for the astronomical unit through a defining relation. The most successful application of the perturbation method involved observations of the orbit of the asteroid Eros. Successive determinations culminated in the value obtained in 1950 by Eugene K. Rabe. Subsequently Rabe discovered a computational error, which led to the revised 1967 value. The value based on the flight of *Pioneer 5*, the first spacecraft determination of the astronomical unit, was probably corrupted by effects of solar radiation pressure. Planetary radar, which determines to an accuracy of one part in 10^8 the time required for radio waves to travel one astronomical unit, yields the value of 149,597,871 kilometers, with an uncertainty of one kilometer.

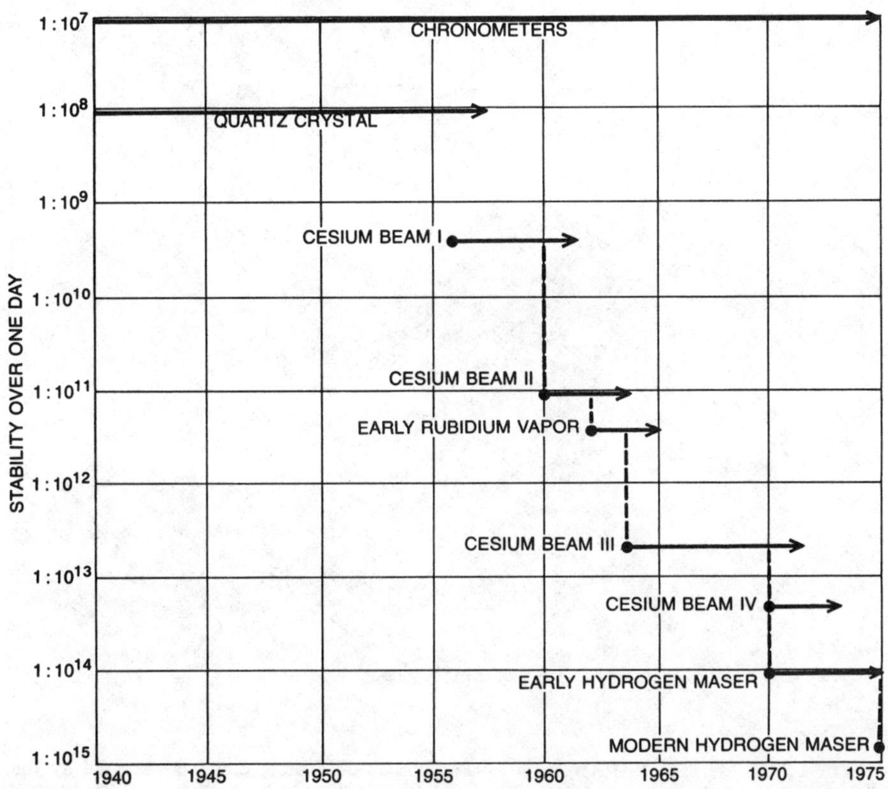

STABILITY OF CLOCKS has increased by about six orders of magnitude, or a factor of one million, over the past 30 years. Stability is expressed as the relative change in frequency averaged over 24 hours. Recent advances in timekeepers have exploited the extremely sharp radiation frequencies of certain molecules to regulate vibrational frequency of crystal oscillators.

between the spacecraft and the earth. The current Viking missions to Mars provide a typical example. In flight the Viking spacecraft are stabilized in orientation so that the panels that convert sunlight into electric power face the sun. The directional antenna of the spacecraft's radio system points toward the earth. The antenna transmits a phase-modulated radio signal that carries scientific and engineering data from the spacecraft to the radio tracking stations on the earth. The antenna also receives commands and information carried by radio signals sent from the tracking stations.

The earth-based part of the radio-communication system is a network of large radio telescopes, or tracking stations, located geographically so that the spacecraft is continuously in view. There are three principal tracking-station sites: one in California in the Goldstone Lake region of the Mohave Desert, one in Spain near Madrid and one in Australia near Canberra. Each site is equipped with antennas that are 26 meters and 64 meters in diameter. The tracking stations maintain a radio-communication link with the Space Flight Operations Facility at the Jet Propulsion Laboratory in Pasadena, Calif. The facility originates commands to the spacecraft and receives data from the spacecraft by way of the tracking stations. Thus the planetary mission is actually remotely controlled from Pasadena.

The need for the large antennas is twofold: to radiate a signal strong enough to be picked up across interplanetary distances by the spacecraft's small antenna and to collect the weak signal returned by the craft's necessarily low-power transmitter. For the purposes of transmission the larger the antenna, the narrower the beam of electromagnetic energy radiated toward the spacecraft. For example, when the Goldstone 64-meter antenna is operated at a frequency of 2.2 gigahertz (2.2 billion cycles per second) in the *S* band of the radio spectrum, half of the power in the transmitted beam falls within a circle with a diameter of about 1,000 kilometers at the distance of the moon. At 300 million kilometers, which will be the approximate distance between the earth and Mars when the Viking spacecraft have reached their goal, the diameter of the beam is about 700,000 kilometers. Although the total power contained in the beam is still half of the amount originally transmitted, which can reach 400 kilowatts, the spacecraft's 1.5-meter antenna can intercept only a minute fraction of it, about 10^{-12} watt.

The radio signal emitted by a Viking spacecraft is only 16 watts, and since the dish on the spacecraft is small, the return beam has spread out to a diameter of many millions of kilometers by the time it has arrived at the earth. Nevertheless, a Viking spacecraft could be successfully tracked even if it were 50 astronomical units away, or beyond the orbit of Neptune. A 64-meter antenna with a cryogenically cooled ruby maser in the first amplifier of its receiver is capable of tracking the phase of carrier signals as weak as 2×10^{-21} watt, which is roughly equivalent to the radiant power the antenna would collect from a lighted match on Mars.

The tracking stations are designed to transmit and receive simultaneously. The downlink frequency from the spacecraft to the earth is about 10 percent higher than the uplink frequency of 2.2 gigahertz in order to eliminate interference between the outgoing and the incoming signals.

Navigation Measurements

The navigation measurements that are obtained by the radio system are range and the rate of change in range, or radial velocity. The radial velocity is obtained from the Doppler shift in the frequency of the radio signal. The simplest Doppler measurement is the downlink, or one-way, mode, in which the frequency of the signal received by the tracking station is compared with the best estimate of the frequency of the signal sent by the spacecraft. The oscillator a spacecraft can carry is not as stable as it might be and its frequency is not accurately known; hence it cannot be relied on for measuring the craft's radial velocity. In the two-way coherent mode, which is the one that is normally used, the tracking station continuously transmits and receives. The signal transmitted from the earth is received by the spacecraft's radio system, which faithfully replicates the received phase, increases its frequency by 10 percent and transmits it back to the earth. In this mode the downlink signal is said to be coherent with the received uplink signal.

Because the spacecraft is moving with respect to the earth the craft's radio system "sees," or receives, an electromagnetic wave whose frequency differs from the one transmitted from the earth. Thus if the distance between the spacecraft and the tracking station is increasing, the craft's radio system retransmits a signal whose frequency is shifted to a slightly lower value by the Doppler effect. On the downlink trip from the spacecraft to the earth the frequency of the signal received by the tracking station is again shifted downward because of the increasing downlink distance. As a result a double Doppler shift is registered in the two-way mode. The difference between the frequency received by the tracking station (compensated for by the 10 percent downlink offset in frequency) and the frequency transmitted one round-trip time earlier is the actual Doppler frequency. In practice the measured Doppler frequency is obtained by comparing the frequency of the received signal with the frequency of the signal currently being transmitted instead of the signal transmitted one round-trip time earlier. The error this introduces is negligible because the stability of the transmitting frequency, being better than one part in 10^{13}, prevents any significant buildup of frequency error during one round-trip time interval.

The Doppler tone that results from comparing the received and the transmitting frequencies is typically on the order of 100

TWO VIKING SPACECRAFT now en route to Mars were launched on August 20 and September 9, 1975, with Titan III-E Centaur vehicles. *Viking 1* is scheduled to reach Mars and go into orbit on June 19; *Viking 2* will follow in orbit on August 7. Each craft weighs 3,500 kilograms, measures 9.7 meters between the tips of its solar panels and carries a 1,060-kilogram lander capsule. The two vehicles will conduct an intensive reconnaissance of the planet from orbit before releasing their landers, which will examine the Martian soil for signs of life and conduct other experiments. The Viking project is managed by the Langley Research Center of the National Aeronautics and Space Administration. Orbiters were built by the Jet Propulsion Laboratory of the California Institute of Technology, landers by the Martin Marietta Corporation.

SIXTY-FOUR-METER PARABOLIC ANTENNA at Goldstone, Calif., is operated by the Jet Propulsion Laboratory for radio communication with interplanetary spacecraft. Similar tracking antennas located near Madrid and Canberra enable the Jet Propulsion Laboratory's Space Flight Operations Facility to maintain an unbroken communications link with space vehicles. Antennas are designed to transmit at *S*-band frequencies (2.2 gigahertz) while simultaneously receiving signals from spacecraft at *S*-band and *X*-band frequencies (8.5 gigahertz).

kilohertz. A Doppler-cycle counter keeps track of the number of successive wave crests of the Doppler tone that have accumulated with time, and it resolves the phase with a precision of a hundredth of a cycle. During the time interval between two successive Doppler wave crests the phase of the received radio wave has slipped one full cycle with respect to the phase of the transmitting radio wave or, what is equivalent, the combined uplink and downlink distance has changed by exactly one wavelength of the transmitting signal. Since the wavelength in the *S* band is about 13 centimeters, the spacecraft has moved radially about 6.5 centimeters during the full cycle. By reading the Doppler-cycle counter at successive times one obtains the change in two-way range over that time interval. Doppler accuracies are often given in terms of radial velocity, although the fundamental measurement in this Doppler system is range change. The accuracy of the average radial-velocity measurement over a one-minute counting interval is about .1 millimeter per second under good signal conditions.

The other radio-navigation measurement, range, is obtained by determining the round-trip time of the radio signal. The current ranging accuracy is about 20 nanoseconds in round-trip time, which is equivalent to about three meters in one-way range. Although this accuracy is remarkable, it does not supplant the Doppler measurement. The Doppler system continuously tracks the relative change of phase of the received radio wave over a period of time, whereas the ranging system at discrete times measures the round-trip time that is determined by the velocity of the radio signal. For measuring "high frequency" range-change behavior over a short time interval, say several hours, the continuous phase-tracking provided by the Doppler system tends to be more accurate, to at most a few decimeters of error in range change over 12 hours. On the other hand, successive range observations made once a day over a period of many days or weeks provide a more accurate measure of "low frequency" range-change behavior.

Orbit Determination

The spacecraft and the planets follow orbits through space that almost exactly obey the gravitational laws of motion. If a spacecraft's position and velocity in space (specified by a set of six numbers) are known for a particular instant of time, a unique trajectory can be computed. For every combination of position and velocity values the range and the Doppler shift vary with time in a unique way that is characteristic of the resulting trajectory. From the range and Doppler observations of a particular spacecraft one obtains curves of range and radial velocity plotted against time from which one can compute the actual trajectory of the spacecraft. The process of determining that trajectory, together with the values of any relevant parameters affecting either the trajectory or the observations, is termed orbit determination. The accuracy of orbit determination depends on the accuracy of the measurements, the geometric characteristics of the trajectory, the time span over which the observations are made and the accuracy of the earlier information about the values of the parameters appearing in the mathematical structure of the dynamical and observational models used to analyze the observations.

The dynamical model must include, in addition to the flight path of the spacecraft, the orbital positions and physical characteristics of the planets that affect the trajectory, the acceleration caused by the pressure of solar radiation on the spacecraft, the velocity changes produced by the propulsion system of the spacecraft and so on. The observational model must include the positions and velocities of the tracking stations

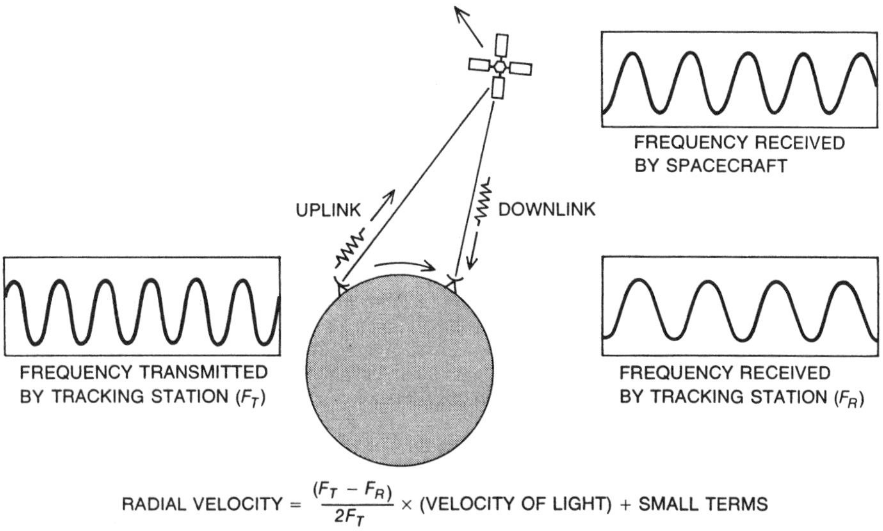

DOPPLER DISPLACEMENT of the uplink and downlink signals is used to determine the spacecraft's radial velocity, that is, its velocity along a line drawn between it and the earth. If the distance between an outbound spacecraft and the tracking station is increasing, the uplink signal is Doppler-shifted to a slightly lower value when it is received by the spacecraft. The radio system on board the craft amplifies the received signal and coherently retransmits it back to the earth, where it is received with a second Doppler displacement. The spacecraft's radial velocity is given by the equation. The phrase "small terms" refers to the fact that the radial velocity also depends on small but important effects that have been omitted in this illustration.

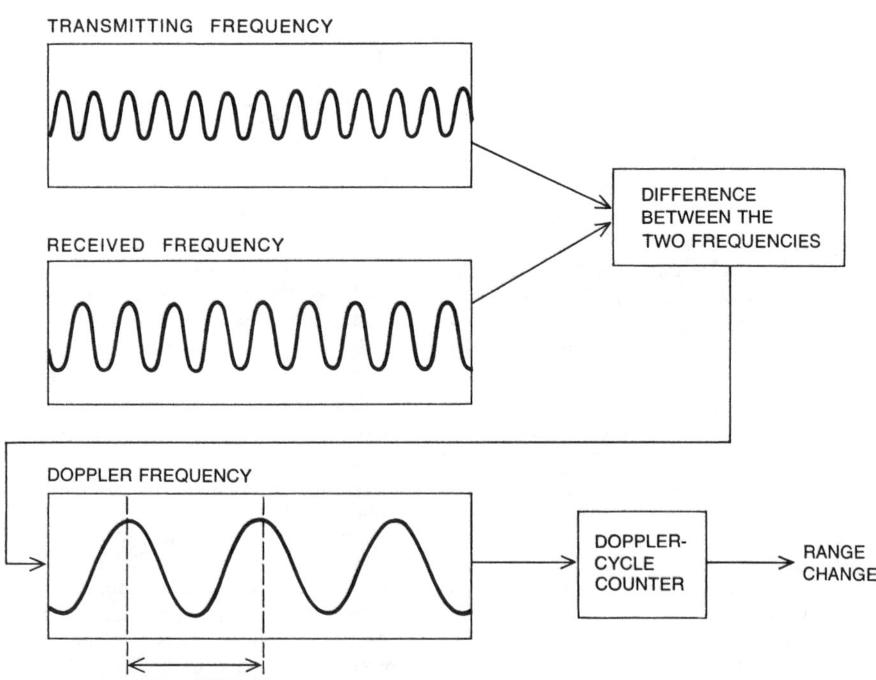

DOPPLER-CYCLE COUNTER keeps track of the number of cycles, and precise fraction of cycles, by which the phase of the received signal falls behind the phase of the transmitting signal in the course of time. The wavelength of an *S*-band signal is about 13 centimeters. Accordingly if the received wave falls exactly one cycle behind the transmitting wave during some measured time interval, the two-way change in range between the tracking station and spacecraft is one wavelength, or 13 centimeters. The one-way change is half that, or 6.5 centimeters.

and the spacecraft, together with small factors such as relativistic effects and the time delays incurred by the radio signal as it travels through the radio system of the spacecraft and through the ionosphere and troposphere of the earth. The parameters of the dynamical and observational models are adjusted to generate theoretical range curves and Doppler curves that match the set of observations as closely as possible. The difference between an actual observation and a theoretical value is termed a residual. If the residuals are small and have a fairly random pattern, the agreement is considered satisfactory and the models are used to predict the values of the observations that are to be made in the future.

This is the "moment of truth." If the agreement with later observations is poor, one must adjust the model and try the fitting process again. Usually only the trajectory parameters of the spacecraft need to be adjusted, but occasionally the residuals can be satisfactorily reduced only after other parameters of the model are reexamined. For example, the trajectory of a spacecraft in orbit around a planet is gravitationally affected by the nonspherical shape of the planet and by the slightly irregular distribution of its mass. Conversely, the residuals that result from those effects can be analyzed to gather information about such planetary characteristics.

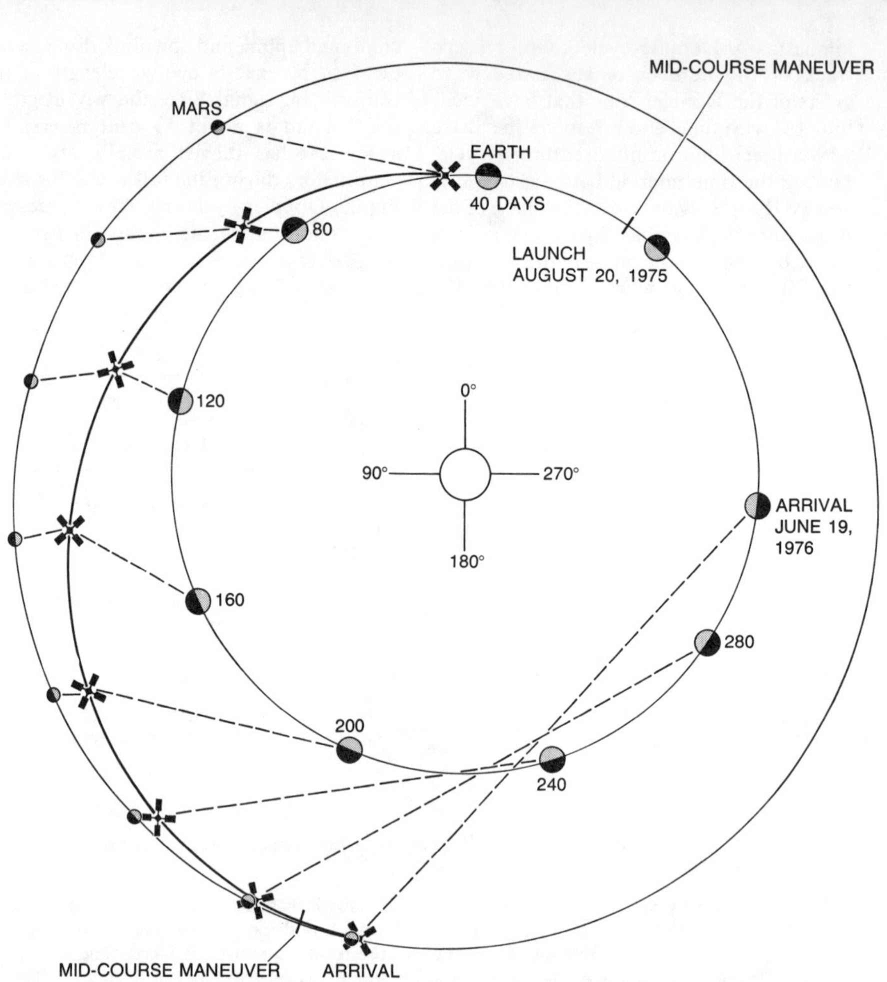

INTERPLANETARY TRAJECTORY OF VIKING 1 shows the positions of the earth, Mars and the spacecraft at 40-day intervals following the launching last August 20. *Viking 1* will reach Mars on June 19, after traveling 700 million kilometers in 305 days. The distance between the earth and Mars on the arrival date will be 313 million kilometers. One mid-course maneuver was made a few days after launching. A small velocity correction will be made June 9.

Trajectory-Correction Maneuvers

The next step in the navigation process is optimizing the flight path. If the residuals continue to behave satisfactorily, one acquires confidence that the model is generating a trajectory that closely coincides with the actual trajectory. One must then decide whether or not to change the course. If the spacecraft is close to the desired flight path, nothing needs to be done. If, however, there is substantial uncertainty about the actual path, additional tracking may be required before further action is warranted. Even if the spacecraft is close to its preplanned course, the mission team may decide that a modified flight path may yield a higher return of scientific information or may better satisfy other operational considerations. Such analyses continue throughout the flight, and the activity becomes intense as the spacecraft approaches its target.

Once a new trajectory is selected the correction in spacecraft velocity needed to achieve the new trajectory is calculated. For example, a few days after the launching of *Viking 1* the first trajectory-correction maneuver had as its objective the removal not only of the normal velocity errors resulting from launching but also of errors intentionally introduced in order to eliminate the risk that the Centaur stage of the launching vehicle, which trails the spacecraft, might collide with Mars and biologically contaminate the planet. If the velocity errors introduced at launching were left uncorrected, they would cause the spacecraft to miss Mars by several hundred thousand kilometers. The velocity adjustment needed to correct a miss of this magnitude is about 10 meters per second. The maneuver, which takes less than a minute, will normally achieve better than 99 percent of the required correction.

The residual velocity error, together with remaining uncertainties involved in orbit determination, is likely to cause the spacecraft to miss its target by several hundred kilometers. Although prior knowledge of the spacecraft's position is not affected at the time of the maneuver, knowledge of the spacecraft's velocity is corrupted because the maneuver is imperfectly executed, and it cannot be exactly reconstructed from engineering data sent from the spacecraft. Immediately before the first trajectory correction the spacecraft's velocity is usually known to an accuracy of about one kilometer per day. Immediately after that maneuver the velocity accuracy may be degraded by a factor of 10 or more [*see illustration on page* 334].

The results of a maneuver need to be evaluated as soon as possible both to verify that it was executed within normal limits and to obtain a measure of the resulting trajectory errors that must be corrected by maneuvers later in the flight. The new velocity components must be remeasured along three axes. It is convenient to choose mutually perpendicular axes along the line between the earth and the spacecraft (the radial direction), along the direction in which the spacecraft appears to be moving across the celestial sphere (the cross-velocity direction) and along the direction perpendicular to the other two.

The abrupt change in radial velocity during the maneuver can be directly measured with Doppler data to an accuracy of about .01 kilometer per day. For maneuvers executed early in a mission this uncertainty translates into an error at the target planet on the order of one kilometer. The component of the maneuver in the cross-velocity direction can be observed because it changes the centripetal acceleration of the spacecraft in relation to the tracking station through the change in cross velocity. The change in centripetal acceleration causes the radial velocity to drift linearly with time away from its behavior before the maneuver. After several hours of Doppler tracking, the linear drift in the residuals is usually sufficient to establish the slope of the drift, and hence the cross-velocity component of the maneuver, to an accuracy of about one kilometer per day. After several days of continued tracking the accuracy improves to about .1 kilometer per day. The change in the third component of the ma-

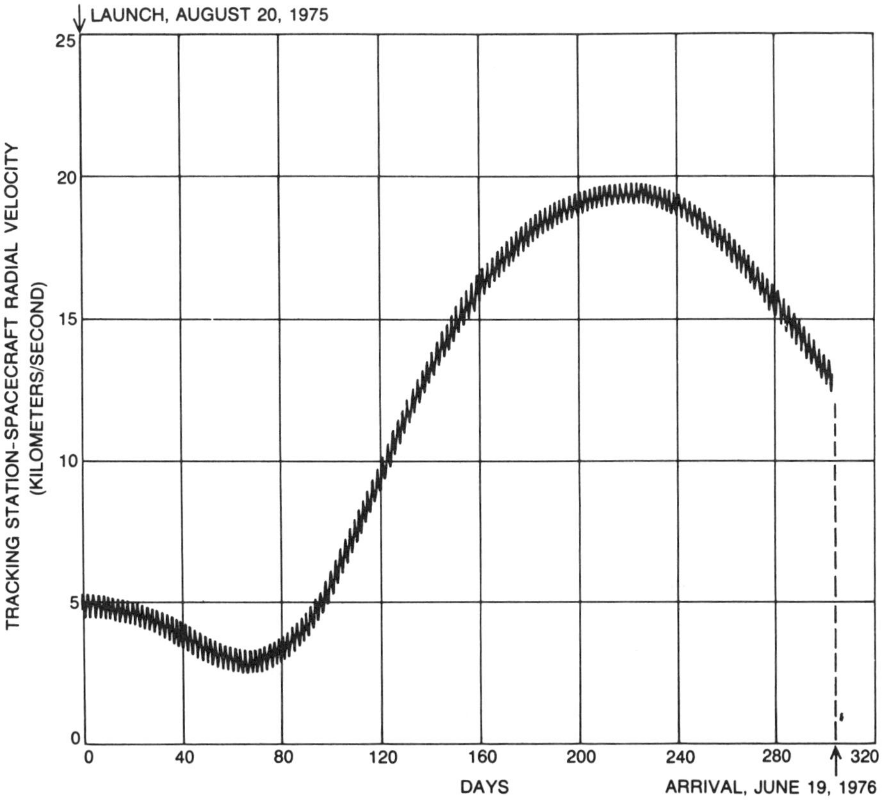

VIKING 1'S RADIAL VELOCITY has varied over a considerable range during its 10-month voyage. The rapid increase between Day 80 and Day 200 primarily resulted because the earth in its orbit is traveling faster than *Viking 1* in its larger orbit, and in effect is leaving the spacecraft behind. Around Day 220 the radial velocity of the spacecraft began to decrease because an increasing fraction of the earth's higher velocity appeared as a component perpendicular to the earth-spacecraft direction, as is shown in the preceding illustration. A small sinusoidal variation is imposed on radial-velocity curve by the motion of tracking station around the earth's axis. On May 26 (Day 280) *Viking 1*'s heliocentric velocity was 19.7 kilometers per second.

neuver, which is nearly perpendicular to the plane of the ecliptic, is the most difficult to establish. Since this velocity component is perpendicular to both the radial direction and the cross-velocity direction, it does not initially change the radial velocity or the radial acceleration. Many days of tracking are needed before its effect becomes observable.

The determinations of the three components of spacecraft position vary widely in accuracy. The spacecraft's radial position, that is, its location on a direct line drawn from the earth, can be determined from one range measurement to an accuracy of about three meters, or three orders of magnitude more accurately than our knowledge of the positions of the planets. The determination of the two position components perpendicular to the radial position calls for considerably more effort because the immense distances involved make the range measurement insensitive to perpendicular displacements. For example, when a spacecraft is at a distance of 100 million kilometers, a displacement of 1,000 kilometers perpendicular to the earth-spacecraft direction results in an increase of only five meters in range. Fortunately the gravitational field of the sun and the diurnal motion of the tracking station furnish important information about these two perpendicular components.

Determination of Position

The gravitational attraction of the sun is responsible for an acceleration of the spacecraft's radial position that varies slowly over the several months of the flight. This acceleration depends on the spacecraft's cross velocity and on the distance of the craft from the sun and from the earth. The three bodies lie at the vertexes of a triangle, two sides of which (the line from the earth to the sun and the line from the earth to the spacecraft) are known almost perfectly from the ephemeris of the earth and from one spacecraft range measurement. It follows that only the angle between the two known sides is required to calculate the length of the third side: the distance of the spacecraft from the sun. Thus the acceleration and its rate of change with time obtained from successive range measurements over many days furnish information about the sun-earth-spacecraft angle and about the cross velocity.

On the other hand, the behavior of the range with time is insensitive to a change of the spacecraft's position in the direction perpendicular to the plane of the triangle. That direction is usually almost perpendicular to the plane of the ecliptic because trajectories between the earth and the other planets cannot deviate very much from that plane. As a result where distant spacecraft are concerned the components of position and velocity perpendicular to the plane of the ecliptic cannot be accurately determined from solar gravitational effects. If it were not for the information provided by the rotation of the earth, as I shall now explain, these trajectory parameters could not be satisfactorily determined without many months of observation.

Since the tracking stations are on the earth's surface, the direct measurements they provide of the spacecraft's radial parameters (range, velocity and acceleration) are topocentric rather than geocentric, that is, the values are related to a point on the surface of the earth rather than to a point at the center of the earth. To obtain the geocentric equivalent of these values, which are normally more fundamental for trajectory calculations, one must allow for the tracking station's location on the surface of the rotating earth. The topocentric values can nonetheless serve for securing important information that would be difficult to get any other way. The topocentric radial velocity of the spacecraft includes a small daily sinusoidal variation because the tracking station is being carried toward the spacecraft by the earth's rotation when the spacecraft is ascending in the eastern sky (like a planet or a star) and away from it when it is descending in the western sky.

From Doppler-tracking measurements of the spacecraft from its rising to its setting it is possible to extract accurate estimates of the spacecraft's right ascension and declination. To do this one in effect plots against time the observed sinusoidal topocentric radial-velocity curve on top of the essentially smooth curve for the geocentric radial velocity [*see top illustration on page 342*]. The point where the two curves intersect gives the time the spacecraft crossed the tracking station's meridian. Knowing the time of crossing and the longitude of the tracking station is equivalent to knowing the right ascension of the stars also crossing the tracking station's meridian at that instant and therefore is equivalent to knowing the right ascension of the spacecraft itself. The amplitude of the topocentric curve (its maximum deviation from the geocentric curve) depends on the spacecraft's declination. Actually the amplitude is the product of three quantities: the rotational rate of the earth, the distance of the tracking station from the axis of rotation of the earth and the cosine of the spacecraft's declination.

Doppler tracking therefore complements range tracking by providing information about the spacecraft's position in the direction perpendicular to the plane of the ecliptic. It also provides information about position in the plane of the ecliptic, and in particular about the sun-earth-spacecraft angle, much earlier than such information could be obtained from range tracking alone. Just two days of radio tracking usually yields estimates of a spacecraft's angular position to an accuracy of a few tenths of a microradian, or a positional accuracy of

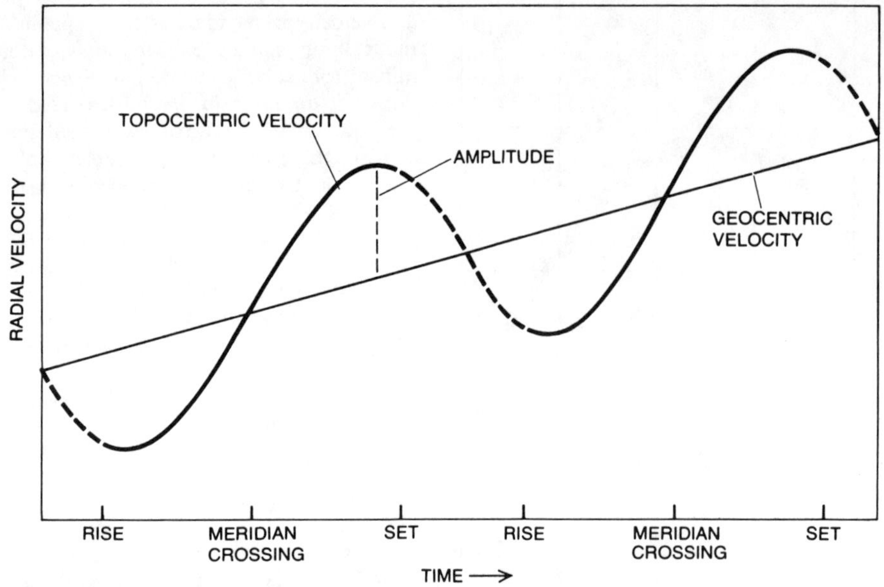

DAILY SINUSOIDAL VARIATION in the earth-spacecraft radial-velocity curve results because the tracking station is being carried toward the spacecraft as it rises in the east and away from the spacecraft after it crosses the meridian and sets in the west. The sinusoidal curve is the topocentric radial velocity, from which the geocentric radial velocity is computed. For determining the spacecraft's orbit the observed phase and amplitude of the topocentric curve provide information about the spacecraft's right ascension and declination. The slope of the geocentric curve is related to the spacecraft's cross velocity (the component of velocity perpendicular to radial velocity) and to position of spacecraft in relation to the sun and the earth.

MEASUREMENTS OF SPACECRAFT'S RANGE must be corrected for variations in the speed of radio signals as they pass through the earth's troposphere (*broken curve*), the earth's ionosphere (*middle*) and the plasma of interplanetary space (*bottom*). The variation in the resulting range errors with time degrades the Doppler measurements from which the spacecraft's right ascension and declination are computed. The curves show typical effects on *S*-band signals.

better than 100 kilometers for flights to the terrestrial planets.

This positional information also makes it possible to derive the cross velocity from the observed acceleration of the spacecraft. Of the six trajectory parameters (three of position and three of velocity) usually only the component of the spacecraft's velocity perpendicular to the plane of the ecliptic remains insufficiently determined after two days. It is derived primarily from successive determinations of declination and usually requires a few weeks of tracking. It is ironic that the principal function of Doppler tracking is to furnish angular information and that range measurements over successive days provide the best information about radial velocity.

Tracking-Station Locations

The accuracy of values for a spacecraft's right ascension and declination obtained by Doppler tracking is highly dependent on accurate knowledge of the tracking station's longitude and its distance from the axis of rotation of the earth. Any error in the assumed value for this distance introduces a corresponding error in the declination value. The locations of the tracking stations have been determined by tracking spacecraft that are "anchored" to a planet, either by passing close to it or by being permanently in orbit around it. Under such conditions one can make a good determination of the spacecraft's position in relation to the planet. Since the ephemerides already provide an accurate planetary direction, the spacecraft's right ascension and declination can usually be computed to a few tenths of a microradian. The amplitude of the topocentric Doppler curve and the intersection time are then implicitly utilized in reverse to obtain the locations of the tracking stations with an accuracy that is typically of one meter.

At that level of accuracy it is necessary to take into account variations in the rate of rotation of the earth and in the geographic location of the poles that result from changes in the earth's moments of inertia. For example, if variations in the location of the poles are neglected, the error in the assumed location of a tracking station can be as much as 10 meters. Classical astrometric techniques make it possible to determine the orientation of the earth to within a meter. New techniques now being introduced should improve on this level of accuracy by an order of magnitude over the next few years. For several years large radio antennas throughout the world, including those of the tracking network, have been used in pairs as very-long-baseline interferometers to study the structure of extragalactic radio sources. A by-product of this work is the determination of the baseline vectors between pairs of antennas, which is equivalent to a determination of the orientation of the earth. By the same token the relative locations of the antennas have also been accurately determined. Improvements in key elements of the interferometer system are now being made and should yield new determinations with an accuracy measured in centimeters.

Another promising technique is measuring the round-trip travel time of ultrashort laser pulses beamed to earth satellites and particularly to the special mirrors, known as corner reflectors, placed on the moon by American astronauts and by unmanned Russian spacecraft. Laser ranging to these reflectors from the McDonald Observatory of the University of Texas has been done for the past six years and appears now to have an accuracy of a few centimeters. A corresponding accuracy for the determination of the earth's orientation should be realized as soon as one or two additional widely separated laser sites become operational.

When the spacecraft is very close to the celestial equator, the declination is difficult to determine. Because the amplitude of the topocentric Doppler curve depends on the cosine of the declination it is insensitive to variations of the spacecraft's declination when the declination has a value near zero. Moreover, in such cases small errors in tracking-station locations that would change the amplitude of the topocentric radial velocity require a large change in the estimated declination to compensate for them. In other words, when the spacecraft is near the celestial equator, the declination determination is extremely sensitive to errors that have not been taken into account. Fortunately the direction of the spacecraft usually changes rapidly on flights to the terrestrial planets, so that it rarely remains near zero declination for more than a few days. For missions to the outer planets, however, this sensitivity is a more serious problem because the spacecraft slows down considerably as it recedes from the sun and can stay near zero declination for months. Indeed, that is what will happen in 1980, when the Mariner flight to Saturn will be in the critical two months just before the spacecraft encounters the planet.

For the zero-declination case a promising approach is simultaneous or nearly simultaneous ranging from the Northern and Southern hemispheres, making use of the Goldstone and Canberra tracking stations. The difference between the round-trip times of radio signals received at these two stations gives a direct measurement of the angular position perpendicular to the line of sight in the plane containing the spacecraft and the two tracking stations. The large north-south separation of Goldstone and Canberra makes the differential time-delay measurement particularly strong in the declination direction. If the difference in range errors at the two stations can be kept to within three meters, it should be possible to measure the declination of a spacecraft in the vicinity of Saturn with an accuracy of about 400 kilometers, or about .3 microradian in angular accuracy. Angle measurements from a single antenna are limited by diffraction effects to an accuracy of about one milliradian.

Propagation Effects

Another source of error in radio tracking is the variation in the speed of the radio signal as it propagates through space and through the earth's ionosphere and troposphere. The variations in speed cause small distortions in the Doppler curve with corresponding errors of up to a few parts per million in the determination of the right ascension and the cosine of the declination. The tropospheric effect is the largest, but it can usually be corrected with considerable accuracy by taking temperature, pressure and humidity measurements at the tracking station. The variations in the propagation velocity caused by free electrons in the ionosphere and in outer space are more difficult to correct for because of the partly random variation in the number of such electrons along the path of the radio signal.

The error in range caused by the electrons is inversely proportional to the square of the carrier frequency of the radio signal. Accordingly most spacecraft now transmit to the earth a second radio signal at an *X*-band frequency (8.5 gigahertz) that is coherent in phase with the primary *S*-band signal of 2.2 gigahertz. Phase coherence means that the phase of the transmitted *X*-band wave is at all times a precise and constant multiple of the phase of the transmitted *S*-band wave. Any departure from phase coherence shows up as a difference in the time delay and the Doppler shift between the two signals received at the tracking station. Such discrepancies provide a measure of the number of electrons and its rate of change along the downlink path from the spacecraft to the earth. With this information one can make suitable corrections in the round-trip time and in the Doppler measurements.

Nongravitational Forces

In general, knowledge of a spacecraft's position and velocity slowly improves over many weeks of continuous tracking. This improvement tends to be offset by a variety of small nongravitational forces acting on the spacecraft. Such forces are difficult to include in the orbit-determination models. It is not difficult, however, to include the pressure of solar radiation on the spacecraft; indeed, the effect of radiation pressure was allowed for in the earliest planetary missions. On a flight to Mars, for example, the pressure of solar radiation will have added about 10,000 kilometers to the spacecraft's heliocentric distance by the time the target is reached. What causes trouble is that the precise value of the solar pressure is altered by the steady bombardment of protons and other particles in the solar "wind." During the flight these particles change the reflectivity of the various materials covering the spacecraft by as much as 1 percent, and the changes are difficult to predict.

In order to maintain a desired orientation in space the spacecraft itself expends gases that give rise to small accelerations. Such force fluctuations and variations in solar-pressure acceleration are partly random and cannot be completely modeled in a deterministic way. The fluctuations are usually small, causing a displacement from the gravitational path of only a few meters per day, or only a few kilometers for the entire flight. These random forces effectively lead to a loss of information by breaking down the deterministic link between the old and the current observations that is provided by the gravitational laws of motion. The concern here is not how much these effects drive the spacecraft off course but how they affect the determination of the spacecraft's orbit.

The estimate of a poorly or indirectly observed component of position or velocity is particularly vulnerable to unmodeled effects. Typical random accelerations can cause trajectory errors at the target planet ranging from 100 to 1,000 kilometers if they are ignored. Serious errors in the estimation of trajectory parameters can arise in a direction that is not easily accessible to observation because the dynamical model "thinks" it knows the path of the spacecraft in the well-observed direction on the basis of past observations and the deterministic properties of the gravitational laws. As a result the model strongly resists being tinkered with in the well-observed direction. In compensating for the small changes in the observed range and the Doppler shift induced by the unmodeled accelerations the orbit-determination process is likely to greatly magnify the error in the estimate of the spacecraft's position or velocity in a poorly observed direction, because a large adjustment in that direction is needed to significantly reduce the error in the residuals and because the model does not resist change in that direction.

Some relief is obtained by relaxing the model so that it is not as rigid about predicting the future on the basis of the past. The relaxation is accomplished by restructuring the model so that the possibility that the spacecraft is being subjected to stochastic, or random, forces is taken into account. One tries to include a priori information on the statistical characteristics of these stochastic forces, such as their expected magnitude and their correlation with time. The result is that instead of generating only one estimated trajectory the model develops a probability corridor in position-and-velocity space. The model assigns a probability measure that the spacecraft is at any one place in the corridor at any one time. On the basis of the observations it also provides estimates of the statistical parameters of the

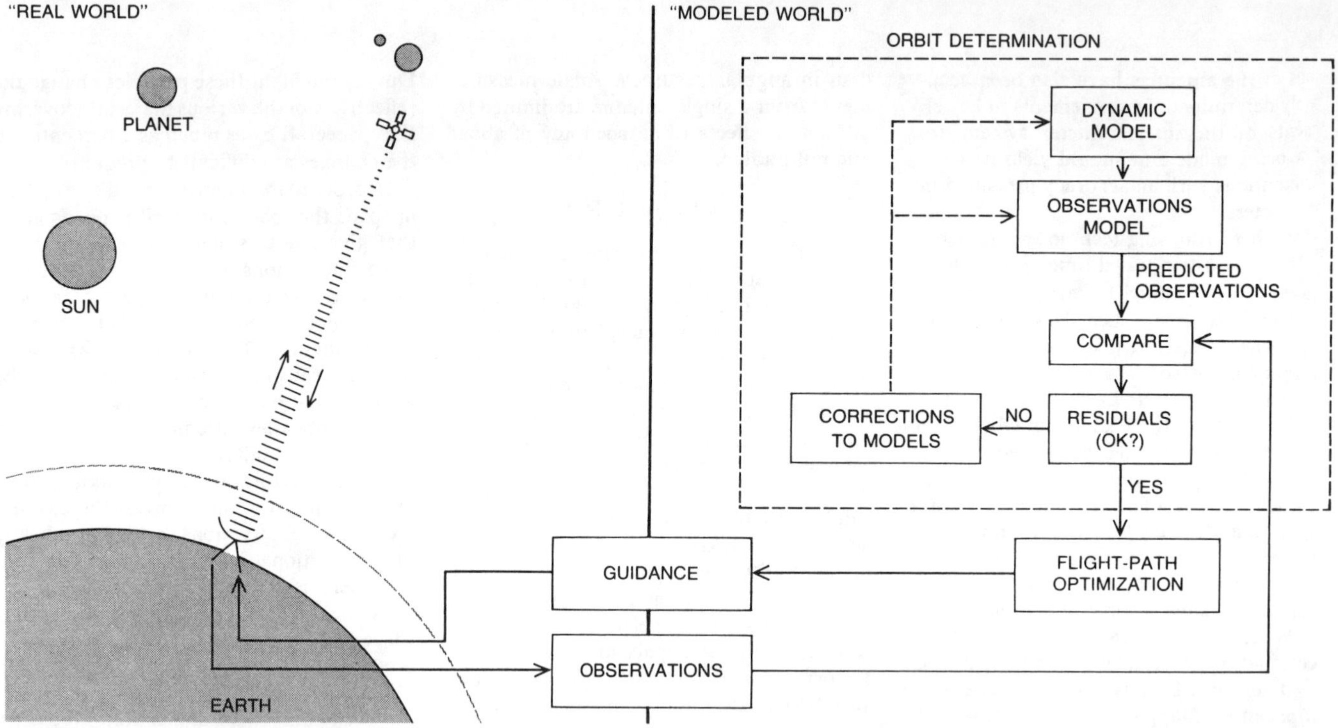

NAVIGATION PROCESS includes the flow of observations into computer data files, orbit determination, flight-path optimization and trajectory correction. Orbit determination is an iterative procedure for testing hypotheses aimed at obtaining the most likely values of all physical parameters affecting the observations. The procedure includes models of the physical processes that affect the motion of the spacecraft, the motion of the observer and the path of the radio signal. Flight-path optimization is a subset of the mission optimization effort, a large and complex iterative activity that is designed to maximize scientific information that can be obtained by spacecraft.

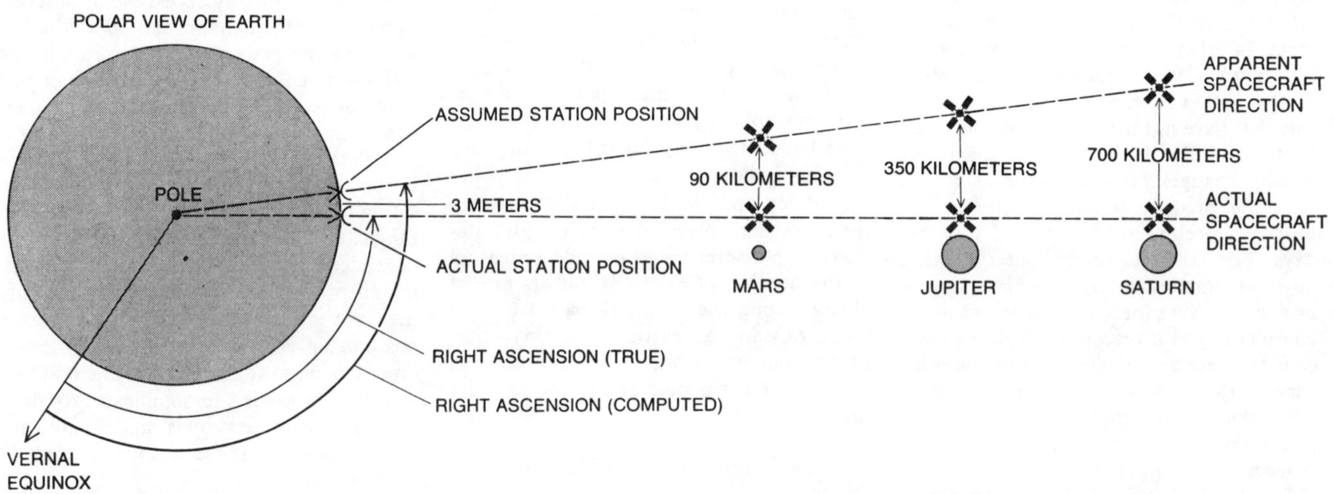

POLAR VIEW OF EARTH shows how a small error in the longitude assigned to the tracking station leads to a much magnified error in computing the spacecraft's right ascension. If the error in the location of the tracking station is only three meters, when the spacecraft is at the distance of Mars, its actual position will be 90 kilometers away from its inferred position, based on two days of tracking. Similarly, the error would be 350 kilometers at Jupiter and 700 kilometers at Saturn. Tracking over many weeks reduces such errors by means of the information furnished by the solar gravitational effects. Tracking-station locations are known to an accuracy of one meter.

actual stochastic forces acting during the observational period.

This approach, known as sequential estimation, is not a cure for the vulnerability of poorly observed parameters to unmodeled effects, but it helps greatly. The technique demonstrated its value in the flight of *Mariner 10* past Venus and then on to Mercury. That flight was the first where the gravitational field of one planet was harnessed to deflect a spacecraft toward another. The angle of deflection depended critically on the spacecraft's position as it reached Venus; any error would be magnified several thousandfold by the time the craft arrived at Mercury. In order to conserve trajectory-correction propellant so that there would be enough left to accomplish the three successive encounters with Mercury, the potentially very large trajectory correction just after the Venus encounter had to be minimized by navigating the spacecraft as accurately as possible before it arrived at Venus. Postflight studies have shown that *Mariner 10* was navigated by Venus within 17 kilometers of the preplanned path. Without sequential estimation the error could have been between 100 and 200 kilometers.

Planetary Encounter

So far I have been discussing the strictly interplanetary phase of the flight, where the gravitational effect of the target planet is small. As the spacecraft approaches the target, however, one can detect in the tracking data that the planet's gravitational field is beginning to accelerate the spacecraft along its trajectory. For the terrestrial planets the acceleration begins between one day and three days before encounter; for Jupiter it begins at about a month. Because at this point the spacecraft is traveling in a nearly straight line toward the planet, the change in acceleration due to the planet's gravity provides very accurate and early information on the distance remaining between the spacecraft and its target. As the spacecraft gets closer the planetary field begins to bend the trajectory, and the bending rate can be well determined.

With this information and knowledge of the spacecraft's velocity with respect to the planet (based on the planetary ephemerides and on orbit determination during the interplanetary phase) one can predict the distance of the spacecraft's closest approach. The third position coordinate, perpendicular to the plane of motion of the spacecraft in relation to the planet, is not well determined by the planet's gravity until the spacecraft is in the immediate vicinity of the planet. Information acquired during the approach phase is valuable in making final plans for operating the spacecraft's scientific instruments. The refined position information acquired by radio tracking during the planetary-approach phase usually comes too late to be useful in calculating the final velocity maneuver needed to place the spacecraft in the desired position at encounter with the planet. In the Viking missions, for reliability reasons, the retromaneuver at encounter that will put the spacecraft in orbit around Mars will be based on commands sent from the earth at least a day earlier; the information obtained during the last day before encounter will not be used for this critical maneuver. These are the reasons why accurate planetary ephemerides and the precise determination of a spacecraft's interplanetary orbit are important.

The orbit of the Viking spacecraft around Mars is determined on the basis of information in the tracking data that is different from the information used to determine the spacecraft's interplanetary orbit. The dominant factors are the known gravitational attraction of Mars and the relative motion of the earth and Mars along their respective orbits. The gravitational field of Mars is used to determine the size and shape of the spacecraft's orbit (to an accuracy of about one part per million) and the position of the spacecraft along the orbit. It also makes it possible to determine the orientation angle of the plane of the orbit around any axis that is perpendicular to the line of sight between the earth and Mars to within one microradian, or 10 meters at periapsis. The relative motions of the two planets cause the aspect of the spacecraft orbit as it is viewed from the earth to change slowly with time, making it possible to determine the orientation angle of the plane of the spacecraft's orbit with respect to the line of sight between the earth and Mars to within a milliradian, which corresponds to an accuracy of 10 kilometers in the position of the spacecraft at periapsis.

The irregular distribution of a planet's mass gives rise to perturbations in the spacecraft's orbit, particularly near periapsis, that degrade our ability to predict the position of the spacecraft along the orbit. The accurate prediction of position several orbits ahead is obviously important for landing the spacecraft at a specified location on the surface of the planet. One of the navigation activities during the two weeks when the Viking spacecraft will be in orbit around Mars before landing will be to determine the effects of mass perturbations in the regions over which the spacecraft will fly and thereby to improve our ability to make orbital predictions. Judging by our experience with the Mars orbiter *Mariner 9* in 1971, we expect to predict the Viking orbital period several orbits in advance to within one second per revolution. Insofar as errors in the determination of orbit are concerned that should introduce an error of no more than 10 kilometers in the landing point of the spacecraft.

Flights to the Outer Planets

The first flights to the outer planets began with the spacecraft *Pioneer 10* and *Pioneer 11*, which flew by Jupiter in 1973 and 1974 respectively. Next year two Mariner spacecraft will be launched toward Jupiter, and with gravitational assistance from the planet at encounter the spacecraft will be deflected on toward Saturn. The second spacecraft will be targeted to fly past Saturn and on toward Uranus. Some of the larger satellites of Jupiter and Saturn will be investigat-

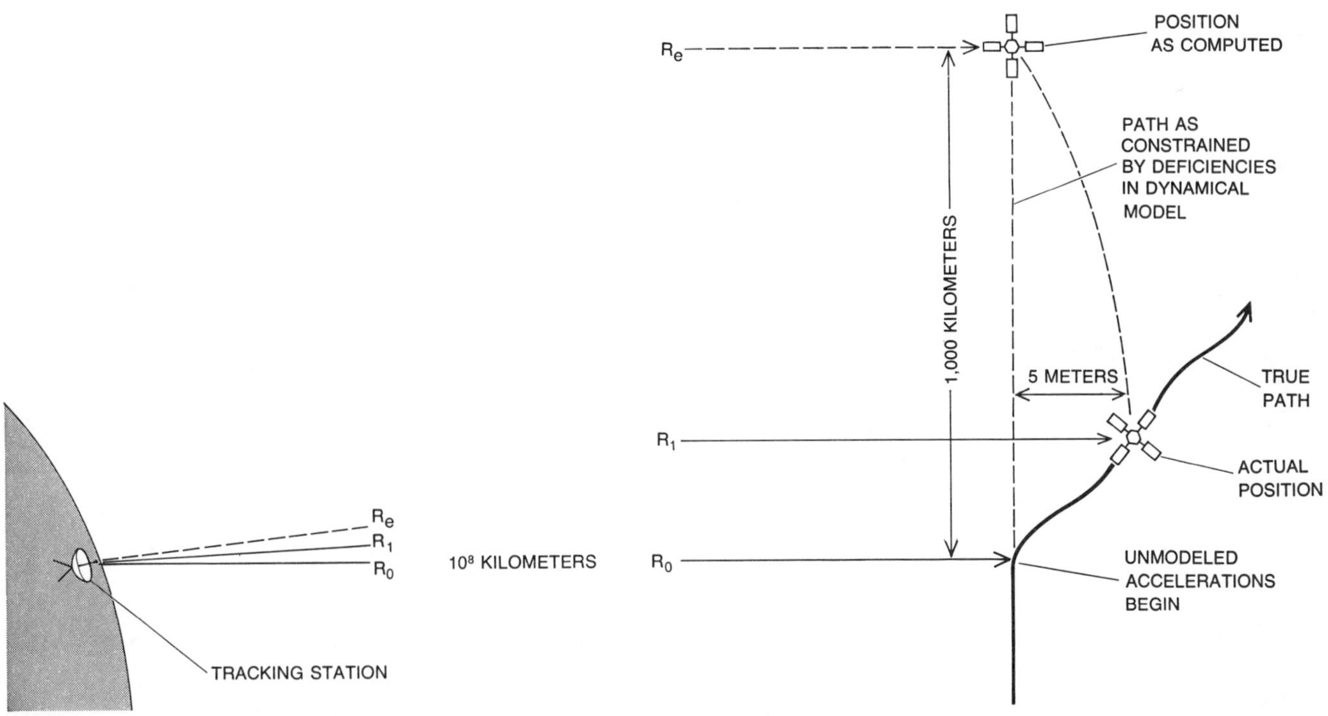

MAJOR ERROR IN COMPUTED POSITION can result if a spacecraft is unexpectedly subjected to small random forces that are not modeled in the orbit-determination program. The program is particularly sensitive to error in a poorly observed direction if the unmodeled forces shift the spacecraft along a well-observed direction. The hypothetical spacecraft depicted here has been traveling along a straight path whose perpendicular distance from the earth is 10^8 kilometers (R_0). Before the disturbing forces begin, the orbit-determination model from preceding observations correctly computes R_0 as 10^8 kilometers. After a while the small unmodeled forces cause the spacecraft to deviate from the previously straight path by five meters, and a new range observation is taken (R_1). The only way the deficient model can simultaneously satisfy the observation and the constraint of straight-line motion resulting from the preceding observations is to greatly magnify the error in new estimate (R_e) of spacecraft's position, misplacing it by 1,000 kilometers in perpendicular component.

ed on these missions. At present the ephemerides of the two satellite systems are not good enough for accurate targeting. The satellites interact with one another gravitationally in such a way that they give rise to resonance phenomena that make it hard to calculate their motions years in advance. Because the satellites are so far away from the earth (between five and 10 astronomical units) earth-based optical telescopes are unable to supply direction measurements accurate enough for the navigation of spacecraft. Although radar range measurements of some of the largest satellites of Jupiter and Saturn can now be made with the recently upgraded 305-meter radio telescope at Arecibo in Puerto Rico, the ephemerides of the satellites are likely to remain a problem for several years to come.

To compensate for these difficulties it is now planned to employ the television system on the Mariner spacecraft to augment earth-based radio tracking. The television system, whose primary function is to return detailed pictures of the planets and the satellites, will record the images of the satellites and the background stars when the spacecraft is still a long way from the target planet. Since the stellar positions are known, the observations will provide the right ascension and declination of the center of the recorded satellite image. By combining a series of such observations with the orbit-determination information obtained during the interplanetary phase of the spacecraft's flight, it should be possible to make corrective maneuvers early enough to navigate the spacecraft to within a few hundred kilometers of the desired position in relation to the satellites of the two giant planets. To lay the groundwork for making use of this navigational technique it was tested with *Mariner 9* and the two small Martian satellites, Deimos and Phobos. It will be used with the Viking spacecraft.

Prospects

Interplanetary navigation presents something of a paradox. On the one hand nature can be regarded as being somewhat perverse in requiring that many dynamical and observational processes be modeled with great precision if high navigational accuracy is to be attained. On the other hand a windfall of scientific information about the solar system has been extracted from the small effects that many of the same processes have on the spacecraft radio-tracking data. Knowledge of unprecedented accuracy has been obtained about the mass of the planets, the distribution of mass within the planets and the overall configuration of the planets. Valuable information has also been gained on the physical properties of the planets' atmospheres and ionospheres from their effect on radio signals.

Refinements in interplanetary navigation are planned over the coming five years that will improve its accuracy by an order of magnitude. Thus it should be possible to extend to Jupiter and Saturn the navigational capability already demonstrated with the terrestrial planets. Improved navigational accuracy should also make it possible to conduct definitive tests of the validity of some of the alternative theories of gravitation and in particular to refine by an order of magnitude the determination of effects predicted by the general theory of relativity.

PERFORMANCE OF THE GPS PACKAGE ON LANDSAT-5

Howard Heuberger

NASA/Goddard Space Flight Center, Greenbelt, Maryland 20771

ABSTRACT

The Landsat-5 spacecraft, launched in March 1984, carries an experimental Global Positioning System Package (GPSPAC) as did Landsat-4 in 1982. The primary objective of this experiment was to characterize and improve the performance of GPSPAC as an onboard navigation tool. In order to validate the accuracy of GPSPAC orbit solutions, definitive Landsat-5 ephemerides, derived from ground based tracking data, were generated and compared with GPSPAC estimates. This paper presents the results of such analysis as well as a summary of GPSPAC operation on Landsat-5. A description of the Landsat-5 spacecraft, GPSPAC and its navigation algorithms are also included.

INTRODUCTION

Landsat-5 is the second NASA spacecraft to carry an experimental GPSPAC. An identical GPSPAC (except for software) on the Landsat-4 spacecraft was the first spaceborne navigation system to use the NAVSTAR GPS. Since the launch of Landsat-4 in July 1982, two Navy satellites have also carried GPSPACs, but these Low Attitude Host Vehicles (LAHVs) are both classified.

Reference 1 details the performance of the GPSPAC on Landsat-4. Comparisons with definitive ephemeris indicated that errors in Landsat-4 position and velocity from GPSPAC were consistently less than 50 meters and 6 cm/sec respectively, during periods of good GPS Space Vehicle (SV) visibility. In the latter part of 1982, when the Landsat-4 data was accumulated, the GPS Space Segment consisted of four operating SVs clustered in one orbital plane. When GPS measurement gaps occurred the orbit errors grew exponentially, but the peak position errors during a Landsat–4 revolution were generally less than 1500 meters.

The GPSPAC on the Landsat-5 spacecraft was first activated on March 23, 1984, and it has been navigating intermittently since then. A number of things have interrupted GPSPAC operation on Landsat-5 (hardware failure, GPS-Space Segment anomalies, etc.), but we have received a good deal of navigation data. Enough data to say that the navigation performance of GPSPAC on Landsat-5 closely matches that on Landsat-4.

This paper describes the GPSPAC and its navigation algorithms, the actual operation of GPSPAC on Landsat-5, and the GPSPAC navigation performance. Because of operational problems with GPSPAC, there simply has not been enough time to carry out detailed analysis tasks (e.g., investigate modifications to the Kalman filter), so in a sense this paper represents only an interim report.

BACKGROUND

NAVSTAR GPS

The NAVSTAR GPS is an advanced satellite based navigation system, being deployed by the Department of Defense, that will provide extremely accurate position, velocity, and timing information continuously to a variety of users. In its operational configuration, the GPS satellite constellation will consist of 18 SVs in six nearly circular orbits of 12-hour periods (20,200 km altitude) each inclined 55 degrees to the equator. Since the launch of Landsat-5 on March 1, 1984, the GPS constellation has consisted of five operating SVs in two orbit planes with nodes at 120 and 240 degrees, respectively.

Position determination with the GPS is based on the measurement of transit time of RF signals from SVs whose position and velocity are known to the user. Each SV transmits a composite signal at two L-band frequencies (L_1 = 1575 MHz and L_2 = 1227 MHz) to allow users to correct for ionospheric delays in signal propagation time. The signals are modulated with two pseudo-random noise (PN) codes: precision (P) for precise measurement of time, and clear/acquisition (C/A) used to aid acquisition of the SV signals and to provide coarse navigation. A GPS navigation message is superimposed on both the L_1 and L_2 signals. This message contains parameters for computing the ephemeris and clock correction of the transmitting SVs.

A GPS user measures the apparent transit time by measuring the phase shift of identical P or C/A codes generated by the SV and the user receiver, each synchronized with its own clock. The receiver code is shifted until maximum correlation is achieved between the two codes; the time magnitude of the shift is the receiver's measure of transit time and hence, range to the SV.

The measurement of range (ρ) to the SV made by the user is called "pseudorange" since it is biased by an amount equal to the user clock error divided by the speed of light. For this reason, a user with an imprecise clock must solve for his own clock error in order to accurately estimate position. Since there are four unknowns (X, Y, Z, Time), a position fix requires four independent sets of measurements.

Reprinted from *Rec. IEEE Position Location and Navigation Symp.*, pp. 147–152, 1984.

Although the NAVSTAR system is designed primarily for precision position estimates, it also provides precision velocity measurements. Delta-pseudorange ($\Delta\rho$) to a SV can be obtained by integrating the Doppler shift in the carrier frequency of the navigation signal over a fixed time interval. Precision measurements are possible because of the precise knowledge of SV ephemeris and short wavelength of the carrier frequency (19 cm). Just as pseudorange data is used to solve for position and clock offset, delta-pseudoranges to four SVs determine the user velocity and clock frequency offset.

Milliken and Zoller provide an excellent discussion of GPS system usage (2).

Landsat Spacecraft

The Landsat-5 spacecraft, identical to Landsat-4, is an Earth resources monitoring system in a near circular sun-synchronous orbit with nominal altitude of 705 km and inclination of 98.25 degrees. This orbit was chosen to provide repeat coverage of the Earth surface every 16 days with the Thematic Mapper (TM) and Multi-spectral Scanner (MSS) image instruments. The system is designed to return data from both instruments either directly to a ground station or through a Tracking and Data Relay System (TDRS) satellite to its ground station and then by a commercial communications satellite link to GSFC.

The Landsat spacecrafts require accurate ephemeris information in real-time to define the nominal Earth-pointing attitude, and to point the high-gain antenna to the TDRS relay satellite. This same ephemeris data is downlinked and used as ancillary data to support TM and MSS data processing on the ground. Normally predicted ephemeris is uplinked to the spacecrafts based on conventional ground tracking. However, the Landsat onboard computers have the ability to utilize the GPSPAC supplied ephemeris for these purposes if the GPSPAC ephemeris is available.

GPSPAC DESCRIPTION

The Receiver/Processor Assembly (R/PA) is the principle GPSPAC subsystem and was developed by Magnavox Government and Industrial Electronics Company for NASA and the Defense Mapping Agency under subcontract to the Applied Physics Laboratory (APL) of Johns Hopkins University. The R/PA consists of two independent receiver channels, a single processor used for receiver control and for obtaining the navigation solution, a time code generator (TCG) and a power supply. An external oscillator provides an ultra-stable 5.115 MHz reference, and an L_1/L_2 antenna and preamp provide the RF input.

The GPSPAC R/PA selects the best geometrically situated SVs to view, acquires and tracks their signals, and makes the pseudorange (ρ) and delta-pseudorange ($\Delta\rho$) measurements. The positions, velocities and clock errors of SVs needed to compute ρ and $\Delta\rho$ are obtained by decoding the navigation message data blocks contained on the signal. The R/PA navigates on sequential measurements from up to four SVs in order to provide the Landsat ephemeris and time estimates in Earth-Centered, Earth-Fixed (ECEF) coordinates. When no SVs are in view, the R/PA performs dead-reckoning propagation of the last measured Landsat position and velocity according to a stored dynamics model.

If more than four SVs are in view, the R/PA determines which four to track based on a geometric dilution of precision (GDOP) parameter:

$$P = \text{trace}\,(H^TH)^{-1}$$

where H is the 4 x 4 position-time matrix based on line of sight. If the ranging measurement errors from each SV are equal and independent, then GDOP is the amplication factor of pseudorange error variance to the combined user position and time error variance. A low GDOP for a set of four SVs indicates that they exhibit good geometry as seen by the user.

The inclusion of two receiver channels, each capable of switching between L_1 and L_2, provides redundancy and allows for two modes of receiver operation. In single-channel (sequential) mode, one receiver channel dwells for 6 seconds on each SV, making sequential L_1 and L_2 measurements of ρ and $\Delta\rho$. In the dual-channel mode both receiver channels are powered and make simultaneous measurements to the same SV, one on L_1 and the other on L_2. In either case the R/PA uses the L_1 and L_2 differences to make ionospheric corrections to the L_1 pseudorange measurements. The dual-channel mode provides simultaneous ionospheric correction at the expense of additional power.

The root mean square (RMS) error in the ρ measurements passed to the processor is approximately 1.5 meters for P and 15 meters for C/A code. The $\Delta\rho$ measurements, in effect integrated Doppler measurements over a nominal 0.6 seconds period, have a precision of 0.02 meters RMS. These ρ and $\Delta\rho$ outputs still contain the offset between local user time and GPS "system time."

The R/PA accepts the uploading of time, Landsat orbital parameters, and one or more SV almanacs for initial acquisition of SVs. The processor is reprogrammable from the ground allowing for algorithmic and data base changes. Finally, the R/PA can output digital data files at selected intervals to the Landsat spacecraft for downlinked transmission. This data includes the navigation estimate of position, velocity and time, filter variance parameters, SV ephemeris and almanac data, raw and processed measurement data, as well as health and status information. Hoffman and Birmingham give a more detailed description of the GPSPAC (3).

NAVIGATION FILTER

The R/PA navigation module processes the ρ and $\Delta\rho$ measurements with a nine-state extended Kalman filter to determine Landsat position and velocity. Uyeminami describes the R/PA filter in detail (4). A block diagram of the navigation filter is shown in Figure 1. The filter state consists of three-dimensional ECEF position and velocity, GPSPAC time offset, frequency offset (first derivative of time offset), and a drag parameter. The drag factor is included in the filter state to aid in accurately propagating the position and velocity when no GPS measurements are available.

A modified Euler integration algorithm with nominal stepsize of 3 seconds propagates the state vector in the ECEF coordinate frame. Gravitational acceleration is modeled with a 4 x 4 geopotential field and an exponential atmospheric density model is used for computing drag.

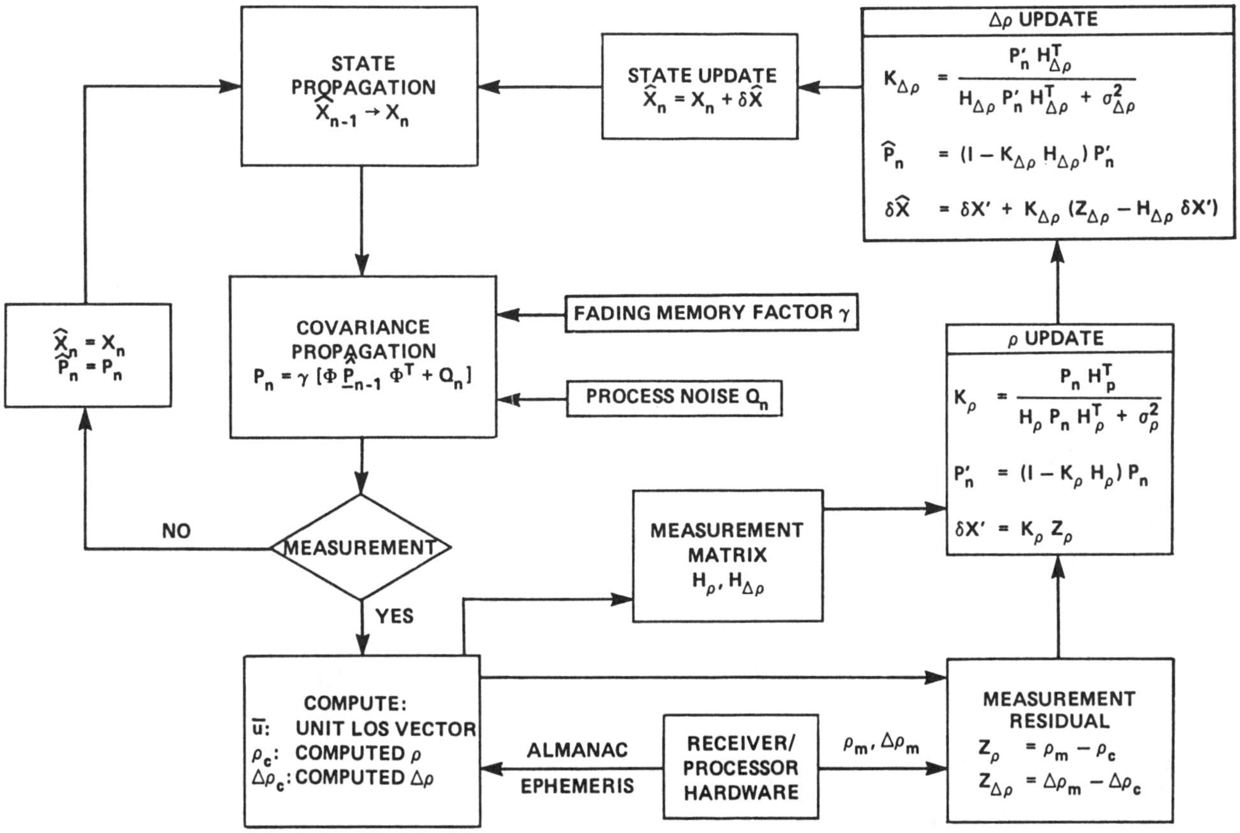

Figure 1. R/PA Navigation Filter

The filter is implemented with an upper triangular-diagonal (U-D) factorization of the error covariance matrix (P) to ensure that the covariance matrix remains positive definite in the presence of numerical errors. The state transition matrix (Φ) is a truncated Taylor series expansion of the analytic state transition matrix. The filter sequentially processes first the ρ residual and then the $\Delta\rho$ residual, using the gain matrix (K) computed from the propagated covariance, measurement partials matrix, and measurement noise.

There is always the possibility of filter divergence when the bounds of the linear assumption made in the R/PA filter state error model are exceeded. So methods must be employed to control divergence and keep the Kalman filter tracking new measurement data. To prevent the error covariance matrix (P) from becoming saturated, a state process noise covariance matrix (Q), which compensates for unmodeled forces, is computed and added to the covariance matrix when it is propagated. The R/PA filter deweights past measurements as a function of time and smoothed measurement residuals (fading memory). This deweighting is only applied when four SVs are in view and is achieved by multiplying the propagated covariance matrix by a factor that varies between one and two. Measurement noise adaptivity is also accommodated in the filter by making the measurement noise variances a function of measurement quality. The quality depends on the receiver state at the time of the measurement, whether the measurement was derived with SV almanac or ephemeris data, and if the ionospheric correction was made.

GPSPAC OPERATION

On Landsat-4 the biggest operational problem was the R/PA memory's susceptibility to cosmic-ray radiation. These charged particles would pass through the electronics and cause single bits to change their values from zero to one or vice-versa. Thirty-eight such "soft" errors occurred in the first 180 days after Landsat-4 launch.

Whenever such an error occurred, the GPSPAC located the portion of memory that contained the error, halted navigation, and awaited instructions from the ground. Ground personnel then had to dump an error block from GPSPAC, evaluate the error, uplink a software correction and reinitialize navigation. Given station-pass schedules and control-center schedules, recovery from these soft errors often took several days.

In October 1983, Magnavox provided a version of R/PA software that detects and corrects the soft single-bit errors autonomously in real-time. But in the meantime, Landsat-4 had developed severe bus power problems (two out of four solar panels became useless). In that situation, not only did operation of GPSPAC have to take a back seat to other spacecraft priorities, it was switched off a lot of the time to conserve power for operation of the TM, a big power consumer.

Thus the launch of Landsat-5 gave us another shot at developing a viable onboard navigation tool. Naval Surface Weapons

Center (NSWC) had confirmed that the soft-bit error correction and detection software was working on their LAHV, but had also warned that the R/PA software still contained some errors. A "tiger team" was established in March 1984, at Magnavox to shake down the R/PA software, and they quickly identified a number of bugs.

We started operating the GPSPAC on Landsat-5 on March 23, 1984, but the GPSPAC could only navigate 1 or 2 days before crashing. (By a "crash," I mean that the GPSPAC halts navigation and reverts to the Standby operating mode.) Interspersed with crashes due to software errors were failures due to GPS Space Segment anomalies. More than once SV clock irregularities caused large navigation and time errors, and occasionally SVs were set unhealthy leaving less than four SVs available for measurement acquisition. These situations usually resulted in filter divergence and hence a crash to Standby.

In the latter part of April 1984, we received a new version of R/PA software that corrected all of the known software discrepancies. Shortly after Landsat-5 flight engineers uploaded the new software, the R/PA experienced a "hard" memory failure (as opposed to a "soft" bit error caused by cosmic ray radiation). A single 1024-word RAM chip latched up, and became unusable. At that point Magnavox had to generate a version of software that bypassed this 1 K section of memory.

Again with the new build we initialized navigation a number of times in the single-channel receiver mode, but the R/PA still couldn't navigate longer than a day or two before crashing. Given the GPSPAC file data and non-program memory dumps from these crashes, the tiger team later identified a major timing/scheduling flaw in the system-monitor software that adversely affected operation in the single-channel mode. Apparently this particular bug had been responsible for most of our grief from the very beginning of operations on Landsat-5.

On June 11, 1984, we activated the GPSPAC in the dual-channel receiver mode, and it navigated for over 14 days until an anomaly with the clock data from SV 4 caused the Kalman filter to diverge on June 26, 1984. Fourteen days of uninterrupted operation represented a new record for spacecraft navigation with GPS. Until that time the longest running time was 9 days in the dual-channel mode with the GPSPAC on Landsat-4 in 1982.

Our jubilation was short lived however as the very next day, the R/PA experienced another hard memory failure. Again Magnavox had to rebuild the software to avoid a failed 1024-word chip.

We received this new version of R/PA software at the end of July, uploaded it, and initialized navigation in early August. As this paper goes to the printer, the GPSPAC R/PA on Landsat-5 has been navigating for over 2 weeks without any interruptions.

GPSPAC NAVIGATION ACCURACY

Definitive Ephemeris

To calibrate the navigation accuracy of the R/PA, we routinely make comparisons with definitive ephemeris, which are independently derived. The definitive orbit solutions used in these comparisons are generated by performing batch-least-squares orbit fits to Unified S-Band (USB) range and range-rate data from the NASA Ground Spaceflight Tracking and Data Network (GSTDN). Range and range-rate measurements are available once every 10 seconds for 10 minutes per station pass with at least one station pass per Landsat revolution.

These orbit fits are performed over 24 to 32 hour tracking arcs with some overlap between successive arcs. Comparing the two sets of GSTDN definitive ephemeris over the time span in common gives a measure of consistency between the two orbit solutions. If the RMS of the USB range (range-rate) residuals from the two least squares fits was small (i.e., if the data fits were good), then it was assumed that the maximum error in position (velocity) over the two definitive arcs was less than the maximum position (velocity) difference in the overlap.

Ephemeris Comparisons

Ephemeris comparisons are presented here from four 10-hour data arcs taken from the 14-day period of uninterrupted GPSPAC operation in June 1984. Table 1 shows the start times and maximum definitive ephemeris errors over each of these data spans. During all these arcs the R/PA was operating in the dual-channel receiver mode

Table 1
GSTDN Definitive Ephemeris Maximum Errors
Over Selected 10 Hour Data Arcs

Arc #	Start Time	Position (m)	Velocity (cm/sec)
1	June 13, 1984, 00^h	33.8	3.4
2	June 17, 1984, 00^h	30.3	2.2
3	June 22, 1984, 07^h	51.2	3.9
4	June 25, 1984, 00^h	46.4	4.2

Table 2 gives the maximum and RMS of the GPSPAC/GSTDN definitive ephemeris differences over these four arcs. The RMS of ephemeris differences during periods of optimum geometry (RMS-OG) is also shown for each data arc. We derived the RMS-OG statistics by selecting 10 minute segments from each Landsat-5 revolution and taking the RMS of the differences over all such segments within the 10-hour data arc. The 10 minute segments are selected around the time of the minimum ephemeris differences for each revolution, which always occurs when three or four SVs are visible. Therefore these RMS-OG values preview the expected accuracy with which the GPS, in its operational configuration, could provide onboard orbit estimates to a Landsat type user over its entire orbit. Actually one could expect to do even better than this since a spaceborne user of the operational GPS will view more than four SVs at all times, and can certainly expect better geometry than what is available to Landsat-5.

During all Landsat-5 revolutions there are periods of up to 40 minutes when no SVs are in view, and the R/PA must propagate its orbit solution. The R/PA has limited dynamic modeling capability and a simple numerical integration scheme for orbit propagation, so orbit errors grow quite rapidly. Furthermore these errors are likely to increase when the GPSPAC begins acquiring measurements from the first SV that comes into view. Generally,

Table 2
GPSPAC/GSTDN Definitive Ephemeris Differences

Arc #	Position (m)			Velocity (cm/sec)		
	MAX	RMS	RMS-OG*	MAX	RMS	RMS-OG*
1	1004	227	46.1	104	32	6.9
2	2099	284	38.2	293	46	5.1
3	1203	533	49.8	155	66	5.2
4	2599	534	41.7	323	85	3.8
Total		418	44.2		60.7	5.4

*RMS During Period of Optimum Geometry

not until two or three SVs are visible will there be enough observability for the Kalman filter to reconverge, and for the ephemeris errors to begin to decrease. During any Landsat-5 revolution the peak errors of the GPSPAC position estimates varied between 100 and 5000 meters depending on the accuracy of the orbit solution during the previous pass through the SV constellation, and on the amount of time this solution was propagated. On most of the Landsat-5 revolutions, the peak position errors were less than 1500 meters.

Figure 2 presents a plot of the GPSPAC/GSTDN definitive position differences over data arc #3. The visibility of each SV to the Landsat-5 spacecraft for this same period is shown in Figure 3. Also, Figure 4 plots the GDOP parameter over this data arc. In this plot a zero GDOP value implies that less than four SVs were in view of Landsat-5. One can see from these plots that Landsat-5 had to put up with some very poor SV geometry, but the GPSPAC was still able to produce a relatively decent orbit fix.

Accuracy of Clock Estimates

Direct measurement of the precision of the R/PA estimates of clock offset and drift (frequency offset) is impossible since no reference is available to compare them against. Nevertheless one can appraise the accuracy of these terms by inspecting the R/PA pseudorange and delta-pseudorange residuals (observed minus computed measurements). A clock offset error multiplied by the speed of light translates into an error in the computed pseudorange and hence will be factored along with other errors into the residual. Similarly a frequency offset error reveals itself in the delta-pseudorange residual.

The RMS of the pseudorange residuals over these four arcs is approximately 5 meters. So a very conservative measure of the one-sigma error in clock offset estimate would be about 17 nanoseconds. The actual mean error in the clock offset estimate is probably well below 10 nanoseconds since clock offset error is just one component of the pseudorange residual. The RMS of the delta-pseudorange residual over these arcs is approximately 4 cm. Given the 0.6 second interval over which the pseudoranges are differenced, this translates to a 7 cm/second range-rate residual. Thus a gross measure of the mean error of the clock drift estimate would be 0.2 nanoseconds/seconds. Examination of measurement residuals over other arcs produces almost identical assessment of the errors in the clock estimates.

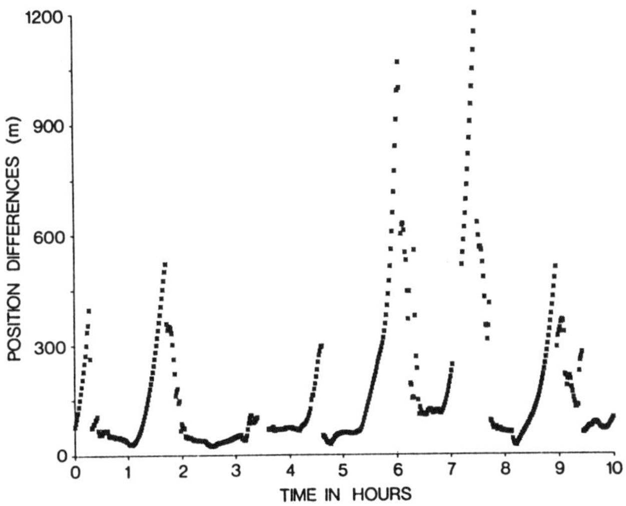

Figure 2. GPSPAC/GSTDN Definitive Position Differences – Arc #3

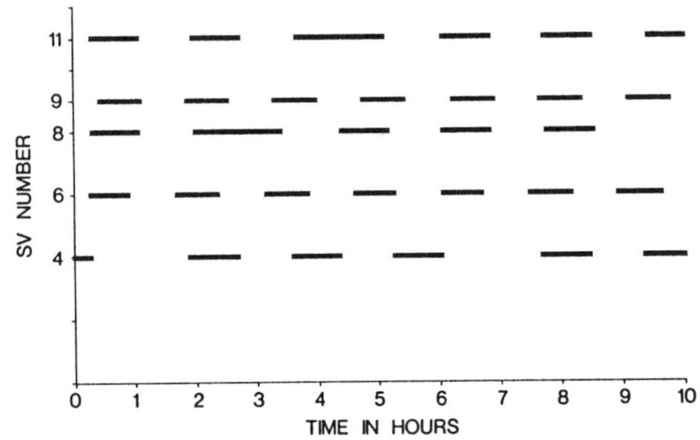

Figure 3. GPS-SV Visibility to Landsat-5 – Arc #3

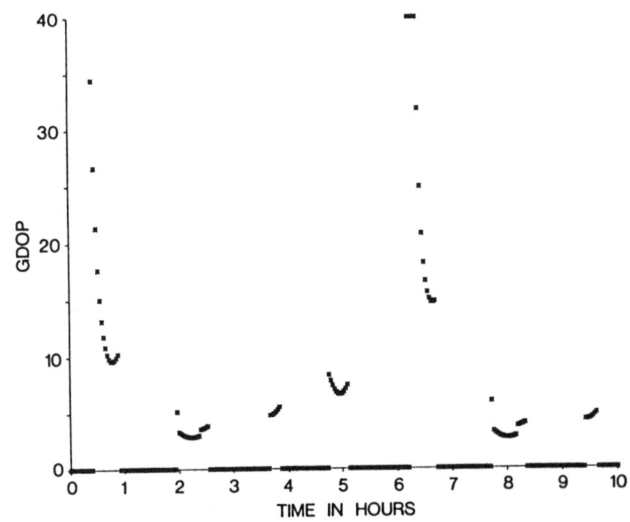

Figure 4. Geometric Dilution of Precision – Arc #3

PLANS

Even though the Landsat spacecrafts have been turned over to National Oceanic and Atmospheric Administration (NOAA), we at GSFC will continue to monitor the performance of the GPSPAC on Landsat-5. In fact we are considering turning the Landsat-4 GPSPAC back on. Since all systems on Landsat-5 are running smoothly, the TM on Landsat-4 will be operated infrequently if at all. That means there is power available to operate the GPSPAC as sort of a backup experiment.

We plan to use one or both of the GPSPACs as a test-bed to validate future software builds containing operational enhancements or error corrections. We would also like to test possible modifications to the navigation processing software. For instance, there has been some discussion about redesigning the covariance-extrapolation portion of the Kalman filter. After validating these algorithms with ground simulators, the next step would be to have Magnavox generate a new build and try it on Landsat-5.

One sub-experiment we have not yet executed is to see how the GPSPAC performs using C/A code only. The results from such a test should be of great interest to potential users who will not have access to the more accurate P-code.

Aside from such sub-experiments, the long-term goal is to calibrate the ultimate accuracy with which the GPS can provide real-time onboard estimates of orbit and time to a Landsat-type user. Hopefully the Landsat spacecrafts and the GPSPACs will hold out long enough so that we can monitor the GPSPAC performance as the operational GPS system is deployed beginning in 1986.

ACKNOWLEDGEMENT

I would like to thank Mssrs. Richard Stauffer and John Speer of GE Space Systems Division, and Ernest Wilson of Ford Aerospace Company for their considerable efforts to provide care and feeding of the GPSPACs on Landsat 4 and 5.

REFERENCES

1. Heuberger, H., and L. Church, "Landsat-4/Global Positioning System Navigation Results," *AAS/AIAA Astrodynamics Conference*, AAS 83-363, August 1983.

2. Milliken, R. J., and C. J. Zoller, "Principle of Operation of NAVSTAR and System Characteristics," *NAVIGATION: Journal of the Institute of Navigation*, Vol. 25, No. 2, Summer 1978, pp. 95-106.

3. Hoffman, E. J., and W. P. Birmingham, "GPSPAC: A Spaceborne GPS Navigation Set," *Proc. IEEE Position Location and Navigation Symposium*, November 1978, pp. 13-20.

4. Uyeminami, R. T., "Navigation Filter Mechanization for a Spaceborne GPS User," *Proc. IEEE Position Location and Navigation Symposium*, November 1978, pp. 328-334.

"DESIGN AND PREDICTED PERFORMANCE OF THE GPS DEMONSTRATION RECEIVER FOR THE NASA TOPEX SATELLITE"

Lance Carson and Lenard Hailey, Motorola Government Electronics Group, Phoenix, AZ

G. Jeffrey Geier and Robert Davis, TAU Corporation, Long Beach, CA

Gaylord Huth, Axiomatix Corporation, Los Angeles, CA

Timothy N. Munson, Jet Propulsion Laboratory, Pasadena, CA

ABSTRACT

NASA's Ocean Topography Experiment (TOPEX) satellite will perform topographic ocean mapping. TOPEX will carry an experimental tracking system based on the use of GPS. An onboard GPS Demonstration Receiver (GPSDR) will be used with a network of up to ten receivers at precisely known ground sites. The positioning technique will simultaneously adjust the TOPEX and GPS orbits to demonstrate sub-decimeter TOPEX altitude determination. The Jet Propulsion Laboratory of the California Institute of Technology has devised the GPS tracking concept, and awarded a contract to the Motorola Government Electronics Group for development of the GPSDR. TAU Corporation and Axiomatix Corporation were awarded subcontracts by Motorola. Following a brief overview of the TOPEX mission and its derived accuracy requirements, this paper will focus on the design of the GPSDR. The major hardware and software components are reviewed and discussed. Projected receiver tracking and navigation performance measures are presented and discussed.

INTRODUCTION

Satellite altimetry has a history of providing information useful in understanding the earth's surface, especially in measuring sea level and the ocean's topography. Several satellite altimeters of increasing accuracy have been flown on various earth orbiters such as Skylab, GEOS-3, SEASAT, and GEOSAT and have demonstrated the scientific usefulness of altimeter data. In 1984, the National Aeronautics and Space Administration (NASA) and the French Centre National d'Etudes Spatiales (CNES) proposed to launch a dedicated satellite altimetry mission for the study of the earth's oceans. The joint mission called the Ocean Topography Experiment (TOPEX)/POSEIDON will provide radar mapping of ocean topography with a precision of approximately two centimeters. The operational orbit determination will be done with laser tracking, which is projected to give 13 cm altitude accuracy, after the gravitational model has been tuned. The GPS-based geocentric altitude measurement is expected to be better than ten centimeters, a five-fold improvement over SEASAT. The nominal altitude of the TOPEX satellite is 1334 km. Scheduled launch for TOPEX is in late 1991 (1-4).

Aboard the TOPEX satellite, depicted in Figure 1, is an experimental package called the Global Positioning System (GPS) Demonstration Receiver (GPSDR) that tracks GPS satellites and provides precision radiometric data for transmission to the ground. These data are used to generate the highly accurate satellite position needed to extract ocean wave height from the altimetry data. The system requirements for the GPSDR were developed at the Jet Propulsion Laboratory (JPL) and a contract was awarded to the Motorola Government Electronics Group for design and implementation. Motorola brought together a diverse group to assist in various aspects of the development. TAU Corporation was contracted to design and implement the Control and Navigation Software. Axiomatix was contracted to perform systems analysis to help define and optimize the performance of the GPSDR.

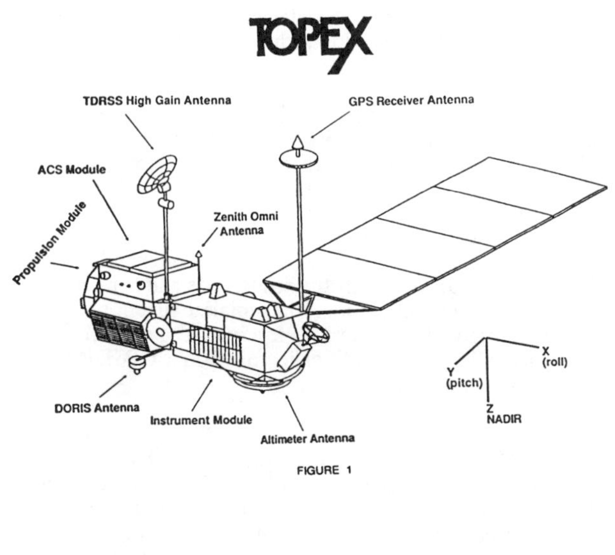

FIGURE 1

Once TOPEX is launched, JPL will collect, reduce, and archive the navigation data from the GPSDR. GPSDR data received from TOPEX and from a global network of ground stations, including data from the three Deep Space Network (DSN) stations, will be brought together and processed at JPL. By combining the pseudorange and carrier phase data available from GPS in an optimal way, a TOPEX position is obtainable to 10 cm or better (5-7).

Reprinted from *Rec. IEEE Position Location and Navigation Symp.*, pp. 442-454, 1988.

GPSDR Design Requirements

The GPSDR consists of a dual-band antenna mounted on a mast, a coaxial cable, and redundant receiver/processor units. According to (8), the receiver is required to:

1) automatically acquire and track the signals emanating from at least four GPS space vehicles (SVs).

2) demodulate and decode the navigation message from each GPS satellite being tracked and use the information to update the GPSDR's GPS ephemeris, monitor GPS satellite health, and update the GPS-SV clock correction terms.

3) make continuously accumulated carrier phase measurements from the dual-band signals (L1 and L2) whether the selective availability (SA) function is activated or not. When the GPS anti-spoofing (AS) function is activated on a given satellite, it is only required that accumulated carrier phase be produced from the L1 carrier. The random error on the carrier-phase measurements shall be less than 1 cm (1-sigma) for 1 second integration times. The systematic error between satellites shall be less than 0.5 cm for a two hour time interval.

4) make pseudorange measurements from the dual-band signals whether the SA (10) is activated or not. Pseudorange from only the L1 band coarse acquisition (CA) code is required when AS (10) is on. The random error (1-sigma) on the precision (P) code pseudorange measurements integrated for 1 second shall be less than 60 cm and the systematic error between satellites shall be less than 10 cm for a two hour time interval. The random error (1-sigma) on the CA code pseudorange measurements integrated for 1 second shall be less than 300 cm.

5) measure the difference between GPS time and the local clock. The offset shall be measured at least every 10 minutes and sent down in the telemetry stream. It shall be accurate to 0.5 microseconds with a precision of 0.01 microseconds.

6) be controlled from the ground via commands sent with the TOPEX uplink data. These commands allow the GPSDR operator to initiate functions that test, initialize, and control the receiver. The test functions include the ability to perform a memory checksum, dump all of memory contents to telemetry, and modify individual or groups of words. Also included in the test commands is the ability to start GPSDR self-diagnostic code and perform an inter-channel bias calibration. The initialize function provides a method to re-start the receiver. The control commands allow user modification of various GPSDR operating parameters.

The remainder of this paper will focus on the design of the GPSDR itself. In the next section, a high level description will be provided, including an identification of the major hardware and software functions. The following section addresses the receiver architecture and signal processing design. The next two sections address certain key hardware and software design features in more detail, respectively. The final sections summarize the GPSDR performance predictions and the major conclusions of the paper.

GPSDR DESIGN OVERVIEW

System Level Design Drivers

Several aspects of the TOPEX platform and the unique mission objectives of the experiment have played a significant role in the design tradeoffs and system configuration of the GPSDR. For example, the centimeter level measurement accuracies required dictate minimizing multipath effects due to reflections from the spacecraft structure. This requires locating the antenna well away from the structure and using a substantial ground plane beneath the antenna to reduce backlobe gain. The antenna gain pattern resulting from the minimum multipath constraints has relatively low gain at the lower elevation angles and results in a total dynamic range of 12 - 14 dB from elevations of 10 to 90 degrees.

The signal power variations due to the antenna and to expected variations in GPS satellite radiated power require a C/A code acquisition strategy which combats false locks due to the cross correlation properties of the C/A code. A two threshold sequential detection search strategy has been devised which avoids false locks and improves typical acquisition times.

Another primary driver in the system design is the carrier tracking loop. Since carrier phase provides the primary metric for the experiment, the system objective was to obtain as accurate a phase measurement as practical with low probability of a cycle slip. An unaided second order loop was found to be adequate for minimizing cycle slipping rate across the expected signal power ranges and Doppler rates. A carrier phase sampling rate of once per second was selected as a compromise between downlink telemetry data rate restrictions and random carrier phase measurement error.

Interchannel bias considerations in carrier phase and pseudorange measurements strongly influenced the signal processing approach toward an all digital implementation of the code and carrier loops with a common clock for processing samples of the composite GPS signal. The approach described herein achieves essentially zero bias between all channels in an L1 or L2 measurement set except for predictable carrier phase shifts as a function of Doppler.

The implementation of the code loop presented a unique challenge in that a completely digital common clock approach was necessary for bias considerations, but quantization error objectives dictated an approach which resulted in negligible error relative to thermal noise. The resulting approach achieves quantization errors of 0.001 chips at the P code rate while using a clock of only four times the chip rate.

The effects of single event upsets (SEU) on the integrity of the receiver measurements have also been an influencing factor on the selection of the signal processing and memory technologies. The design objective was set at 99% probability of no upset in a two hour tracking interval. Projected aggregate SEU rates for the sensitive parts in the receiver are in the range of 10^{-3} to 10^{-4} errors per day.

Signal Processing Architecture

A functional representation of the GPSDR is shown in Figure 2. Like most GPS receivers the GPSDR consists of an L-Band antenna and four distinct signal processing functions which include the RF signal processing, hardware digital signal processing, software signal processing, and navigation processing.

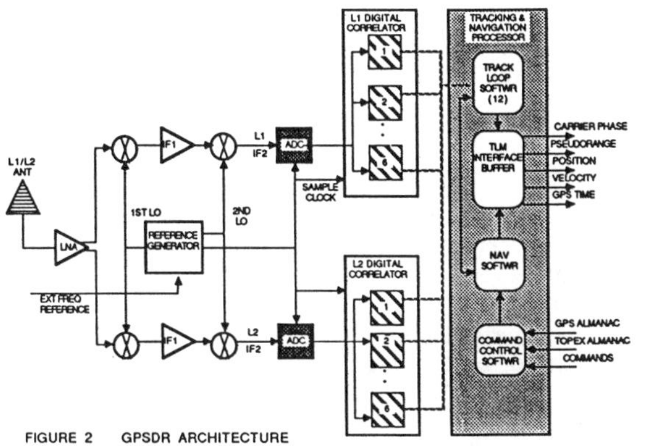

FIGURE 2 GPSDR ARCHITECTURE

The RF signal processing amplifies, frequency converts the two GPS carriers to intermediate frequencies, and provides the filtering for optimal recovery of the GPS signal. A double conversion receiver has been implemented with common LO injection signals for both L1 and L2 signal paths. The LO signals and the sample clock for the digital processing are all derived from an ultra-stable oscillator provided by the TOPEX spacecraft.

The digital hardware signal processing commences with the analog-to-digital conversion of the L1 and L2 IF signals using two high speed, four bit flash A/D converters. Simultaneous independent tracking of 6 GPS signals is accomplished first by separating the available signals by employing digital correlation and time domain decimation techniques. An Application Specific Integrated Circuit (ASIC), the Coder/Correlator, provides all the higher and medium speed signal processing functions required to recover pseudorange, carrier phase and navigation data from a single assigned GPS satellite.

Lower speed signal processing is accomplished in software resident code in a single MIL-STD-1750A Tracking & Navigation Processor. Program memory is contained in radiation hardened RAM which is uploaded during receiver initialization. Start-up code and certain diagnostics are implemented in ROM. Serial digital interfaces are provided to accept memory upload, almanac data, and mode control commands from the TOPEX spacecraft. Similarly, a serial telemetry output is provided which contains 10 seconds worth of accumulated L_1 and L_2 measurements from the 12 channels.

Lower speed tracking and acquisition functions, control and programming of the Coder/Correlator, and navigation data subframe synchronization are accomplished by the Tracking Loop Software. By restricting software processing to only low speed functions at or below 50 Hz, the Tracking Loop Software is able to process each of the 12 channels in sequence with adequate time to spare for the Navigation Software processing.

The Navigation Software processing consists of a number of highly interdependent tasks which include managing the receiver channels, collecting and storing almanac and ephemeris data, estimating GPS satellite positions, selecting satellites to track, computing an on-board solution of position, velocity and GPSDR clock offset and outputing measurement data to the TOPEX telemetry subsystem.

Channel Signal Processing

The four principal channel related signal processing functions are C/A Code acquisition, code tracking/pseudo-range measurement, carrier tracking/carrier phase measurement and navigation data synchronization. Figure 3 shows the channel related processing in simplified form.

After the receiver has been initialized the navigation software makes an estimate of expected Doppler based on TOPEX and GPS almanacs. The receiver channel sets this Doppler offset into its carrier and code clock frequency references and commences C/A code acquisition with a dual threshold sequential test. The first search interval uses a high threshold which is set to detect satellites above 25-30 degrees elevation. This higher threshold also prevents false acquisition of other satellites with higher power levels. If the first search fails, a second threshold, which is set to detect signals at the worst case power levels corresponding to low elevation angles, is activated. False detection of stronger signals is now provided by Doppler discrimination in that low elevation satellites will have relatively large values of Doppler while the stronger signal zenith satellites will have smaller values.

The code tracking loop implementation is an offset sampled early minus late configuration; sometimes known as a Modified Code Tracking Loop (MCTL)(9). The loop is second order with carrier aiding in the final measurement states. The second order term is required to prevent acceleration dependent biases due to different update rates between the code and carrier loops. For a given early minus late separation this approach provides optimum random error performance with the benefit of only one correlator error channel. The offset sampling technique, together with a digital single band translation achieves negligible quantization error without excessive digital clock rates. To obtain the pseudorange measurement the receiver measures the time-of-transmission of the GPS signal. The pseudorange is then the difference between the local receiver clock and the time-of-transmission multiplied by the speed of light. The three components of the time-of-transmission measurement include the fractional chip, whole chip, and Z-count. The fractional chip phase is obtained directly from the code numerically controlled oscillator (NCO) to a resolution exceeding .0001 chips.

The suppressed carrier tracking loop is implemented as a data estimation type of Costas loop. The loop is a second order, type II configuration implemented completely in digital form. Higher order loops were not given serious consideration due to additional hardware complexity, acquisition issues and the predictable TOPEX accelerations. Commandable loop bandwidths provide a mechanism for the experiment to vary the emphasis between random and systematic error contributions. Programmable predetection bandwidths for the arm filters are also possible but an optimum value of 1 khz has been selected for the final tracking states. Accumulated carrier phase measurements consist of a fractional cycle component and a whole cycle count. The fractional part is read directly from the carrier NCO accumulator to a resolution exceeding .01 degrees. The whole cycle count is kept partially in hardware to the 20 millisecond time epoch and then in software for the entire tracking pass of a given GPS satellite.

A number of channel related navigation data processing functions are required. First, since the 50 Hz navigation data clock will be asynchronous with the receiver local clock, a means of bit timing recovery is required. A conventional data-transition-tracking loop is employed for maintaining bit timing and for providing matched filter integration of the data. Secondly, subframe synchronization of the data is accomplished by data which is contained in subframe 1. Out of this subframe synchronization function also comes the proper timing for accomplishing the C/A to P code handover. Finally the message data must be validated by checking the various parity bits contained in the message. The parity check function is contained in the navigation software.

Navigation Processing

The Navigation Solution Function consists of two subfunctions: a propagation subfunction, performed each second, and a measurement update subfunction, to be performed every ten seconds. The propagation subfunction, once provided an initial three dimensional position and velocity in an Earth Centered Earth Fixed (ECEF) frame, propagates the position and velocity estimates, along with the user clock phase and frequency error estimates ahead in time. The position and velocity estimates are propagated using the estimated velocity and an estimate of the gravitational, Coriolis, and centrifugal accelerations.

The solution is refreshed at 1 Hz so that current position and velocity estimates are available for computing the predicted signal Doppler on all receiver channels in an acquisition state. The gravitational acceleration is computed using a fixed magnitude calculated at the nominal TOPEX altitude, and a direction computed from the estimated TOPEX position. Coriolis and centrifugal accelerations are computed using the estimated TOPEX position and velocity and the known magnitude (and direction) of the earth's angular velocity.

The measurement update subfunction processes the current available pseudo range measurements (if at least four are available) to compute corrections to the extrapolated TOPEX position and clock offset estimates. Similarly, the difference between the current available carrier phase measurement and the previous carrier phase measurement (i.e., one second ago) are used to compute corrections to the extrapolated TOPEX velocity estimates and the clock frequency error estimate (again, only if at least four carrier phase differences are available). Prior to making use of any pseudo range or carrier phase measurement, the corresponding status word is examined. If a particular channel is not in lock, or a valid ephemeris is not available for the corresponding satellite, its measurements are not used in the update process. Given that four or more measurement sets are available, either a matrix inverse or "pseudo inverse" is computed to permit calculation of the corrections. Both the inverse and the pseudo inverse are found from the measurement gradient vectors calculated for each satellite (as part of measurement processing). If less than four measurement sets are available, or if the absolute value of the determinant is less than a specified threshold no corrections shall be generated. This relatively straightforward measurement update method was selected over a Kalman update because of the comparatively coarse accuracy requirements for the onboard solution and to preserve memory and throughput for other functions.

Almanac Update Processing. To alleviate the imposed burden on the ground of periodically revising the TOPEX almanac stored onboard the spacecraft, the Navigation Software recomputes the almanac using its current best estimates of TOPEX position and velocity. This almanac update function is executed as a low priority, background task performed every 15 minutes. Once the earth referenced velocity vector is converted to an inertial velocity, it can be used in updating the almanac by making use of the conservation of energy and angular momentum equations. The semi major axis can be found directly from the energy equation, while the inclination and longitude of the ascending node can be found from the components of the angular momentum vector. The remaining elements (eccentricity, argument of perigee, and mean anomaly) are found through manipulations involving the position and velocity vectors and the angular momentum vector.

Satellite Selection Processing. The Satellite Selection Function chooses a group of satellites to track, based on the input of GPS satellite and ground station visibility data from the Satellite Position Function. The Satellite Selection Function must also choose several backup satellites to track, if one of the primary satellites chosen cannot be acquired.

The satellite selection algorithm, designed by JPL, can run at any one of three levels. At all levels, the set of satellites corresponding to the minimum weighted/modified Position Dilution of Precision (PDOP) (or pseudo PDOP) is selected: in the first two levels, a pseudo PDOP is computed from the remaining view periods for each of the satellites. In level three, a PDOP with diagonal weightings based on remaining view periods shall be used. Given that a PDOP or pseudo PDOP has been formulated, it is modified by adding two penalties. The first penalty is added (for all levels) whenever a change in the current active tracking set is mandated. The second penalty is added in levels two and three only when double difference data cannot be formed with a common ground station.

HARDWARE DESIGN OVERVIEW

RF Frequency Plan and Hardware Implementation

The significant aspects of the RF hardware implementation are shown in Figure 4. Rejection of out-of-band signals is accomplished by a bandpass diplexer tuned to the two carriers with bandwidths selected to minimize insertion loss while providing sufficient rejection of other

TOPEX interference sources. A single broad band, low noise amplifier provides L-Band gain and establishes the receiver effective noise temperature. The L1 and L2 signals are next split into two paths by image filters, identical to those used in the diplexer.

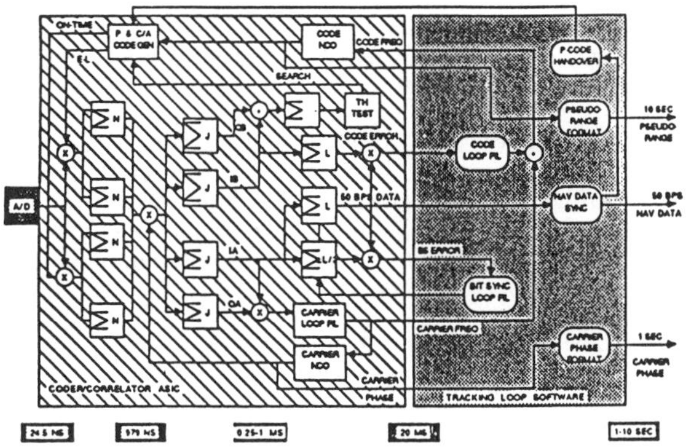

FIGURE 3 HARDWARE/SOFTWARE PARTITIONING OF CHANNEL PROCESSING FUNCTIONS

One of the objectives in the frequency plan selection was to preserve maximum frequency symmetry in the RF processing to minimize differential group delays between L1 and L2. A single LO is used for both the first and second conversion which provide low side conversion of L1 and high side conversion of L2. Note that the first and second IF frequencies are nearly symmetrical and allow common filter characteristics to be implemented in both channels. The bandpass filter in the first IF sets the precorrelation bandwidth to a value, 16 Mhz in this case, which optimizes correlation loss and aliasing noise.

The second LO is also common to both L1 and L2 conversions and results in second IF frequencies that satisfy several conflicting constraints. On the one hand, the second IF frequency must be high enough to preserve the 16 Mhz bandwidth and prevent baseband foldover. On the other hand, the frequency must be sufficiently low that it could be sampled at approximately four times per cycle to implement the complex bandpass decimation technique used in the Coder/Correlator ASIC. The resulting frequencies and sample rates are shown in Figure 4.

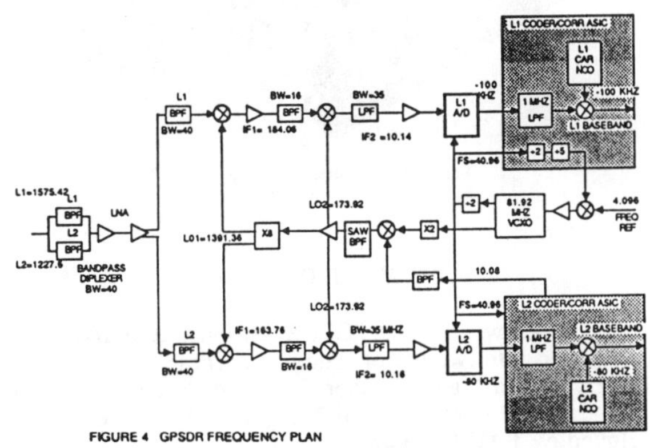

FIGURE 4 GPSDR FREQUENCY PLAN

Several other constraints were imposed on the frequency plan and LO chain generation. To prevent self jamming it is desirable that the IF frequencies not be harmonically related to any LO or reference signal produced in the receiver. Phase noise and spurious signals are of utmost importance because of their effect on the precision carrier measurements and the experiment's sensitivity to cycle slipping. Finally to minimize receiver complexity the LO signals and sample rates should be easily synthesized from the on-board frequency reference at 4.096 Mhz. This synthesis function is implemented by phase locking an 81.92 Mhz crystal oscillator to the reference to produce an effective frequency multiplication of 20. Another reference at 10.08 Mhz is digitally synthesized in the Coder/Correlator ASIC from the 40.96 Mhz sample frequency. This reference is mixed with the second harmonic of the crystal oscillator and filtered to produce the second LO. A simple times eight step recovery diode multiplier is then used to derive the first LO.

The last conversion in the frequency plan is to baseband and takes place in the Coder/Correlator ASIC. After correlation and decimation, the recovered signal is at a nominal frequency of 100 kHz for L1 and 80 kHz for L2. These offset frequencies are sufficiently high to handle the worst case Doppler frequency variation but low enough to pass through the 1 Mhz decimation filter without significant loss. Reconstruction of the complex baseband signal samples is accomplished by a single sideband transformation of the difference frequency between the carrier numerically controlled oscillator (NCO) and the complex signal samples.

Coder/Correlator ASIC Overview

As previously indicated, most of the signal processing functions dedicated to a receiver channel are contained in the Coder/Correlator ASIC. The ASIC is being developed in a radiation hardened 1.5 µm CMOS gate array technology. The functions implemented in the ASIC and their approximate gate count are shown in Figure 5. These functions total approximately 34,000 gates and are implemented in an array which is capable of 40,000 implemented gates.

The division of processing between the ASIC and the software was previously shown in Figure 3. Interfaces with the software are accomplished via the memory address and data bus of the Tracking & Navigation Processor at a basic interrupt rate of 20 milliseconds.

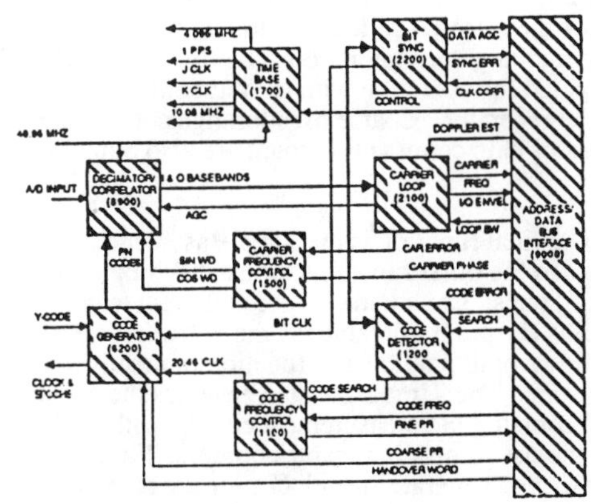

FIGURE 5 CODER/CORRELATOR ASIC FUNCTIONAL ORGANIZATION

The Decimator/Correlator provides the high speed correlation processing and the conversion to in-phase and quadrature baseband signals of the on-time and code error channels. This is accomplished by multiplication of the signal samples with the NCO reference and accumulation to the desired predetection bandwidth. These samples are also multiplied by an AGC word to provide gain normalization of signals with varying power levels.

The Carrier Loop block provides a data estimation type of suppressed carrier recovery, the carrier loop filter with selectable loop bandwidths, and integrated envelope estimates for AGC and lock detection. The Carrier Frequency Control Function uses the carrier loop filter output to drive the carrier phase accumulator or NCO. The phase is then mapped to sine and cosine functions of the carrier offset frequency for use in the single sideband transformation to baseband. The carrier phase is also output and consists of the accumulated number of whole cycles and the fractional cycle estimate obtained directly from the phase accumulator output.

The Bit Sync Function is engaged after coherent carrier recovery and provides synchronization of the local 50 Hz bit clock with the transitions of the navigation data using a data transition tracking loop. After synchronization is established an accumulate and dump matched filter makes the bit decisions at the 50 bps rate.

The Code Detector provides several types of code discriminators depending on the acquisition or tracking mode in use and the degree of carrier coherence. The code error is output to the tracking loop software. The code detector also provides the envelope estimator and sequential detector for establishing synchronization to the C/A code during initial search. Detection parameters and search step size are both programmable from the software.

The Code Frequency Control Function provides the phase accumulator (NCO) for the code clock loop and implements the quantized phase corrections to the clock to maintain code track. The loop filter for the code loop is in software where the carrier aiding term is also added prior to being received by the code NCO. The code phase accumulator output provides the fractional chip component of the pseudorange.

The Code Generator produces the GPS codes, both C/A and P, for the selected satellite. On-time and early-late replicas of the selected code are produced for the carrier and code tracking correlators respectively. An interface is also provided for operation with an outboard anti-spoof (AS) code generator. Coarse pseudorange components consisting of chip count and Z-count are also output to the software.

The final function is the Time Base which provides the clocks to the other functions for establishing applicable processing rates. Loop update rates and predetection filter bandwidths are programmable via the software interface which allows optimization as a function of acquisition or tracking state. The Time Base also provides the capability for synchronizing the measurement epochs with GPS time through the use of programmable counters. Finally the Time Base also generates the clocks for the receiver and a self test signal for injection into the receiver IF amplifier.

Processor and Spacecraft Interface Hardware

The processor module contains a single MIL-STD-1750A architecture processor. The processor was chosen for its total dose radiation hardness to meet the TOPEX mission and the built in floating point instructions. The hardware floating point instructions are necessary to meet the throughput requirements for the Navigation Software. The processor board contains 32K words of memory of which 2 to 8K words may be read only memory.

The spacecraft interface module contains 30K words of memory and the interface to the spacecraft. The spacecraft interface consists of a serial digital command input interface, a serial digital telemetry output interface, and a processor enabled one pulse per second output for other experiments to use.

Mechanical and Packaging Design Overview

The mechanical configuration, which is shown in Figure 6, was derived from other Motorola high reliability spaceborne equipment. It features eight individual module assemblies mounted together with common side, top, and bottom shear panels. Mounting feet are an integral part of the side panels. In its deliverable configuration the GPSDR is 5.5 inches in height by 5.5 inches wide by 8 inches in length and weighs 8.8 lbs. This rugged packaging design has been qualified to shock and vibration levels well beyond those produced by all current launch vehicles.

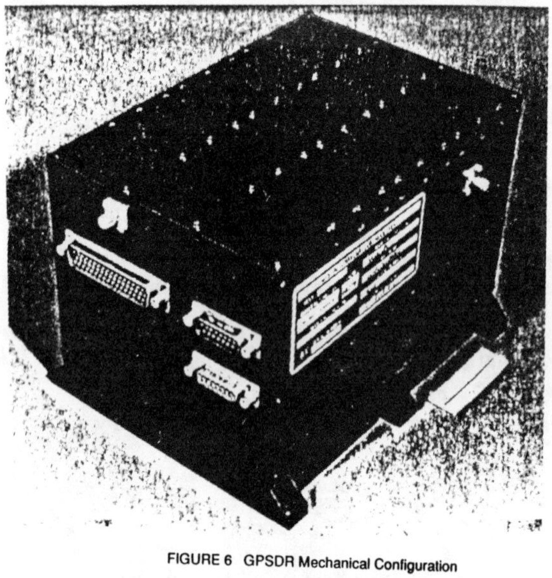

FIGURE 6 GPSDR Mechanical Configuration

SOFTWARE DESIGN

The software of the TOPEX demonstration receiver is partitioned into two major functions. These major functions are the Tracking Loop Software (TLS) being developed by Motorola and the Navigation Software being developed by TAU Corporation.

The Tracking Loop Software is responsible for controlling the receiver digital signal processing hardware (coder/correlators) to perform the acquisition and tracking of GPS satellite signals on all six channels. Tracking Loop Software accepts satellite numbers and Doppler estimates for each of the channels. Outputs from the Tracking Loop Software are pseudorange, carrier phase measurements, demodulated navigation data, and the current state for each of the channels which is sent to the Navigation Software. The Tracking Loop Software also accepts commands from the Navigation Software to initialize the software or put the software in a self test mode. The Tracking Loop Software is implemented as an interrupt service routine which is executed every 20 milliseconds.

The Navigation Software is responsible for accepting and decoding serial digital commands, buffering and outputting telemetry data, performing the onboard navigation solution, selecting satellites to track, and performing receiver channel management which controls the Tracking Loop Software.

TLS Software Overview

The Tracking Loop Software consists of the following functions:

- Sequencing Function
- Data Input Function
- Data Output Function
- Channel Processing Function
- Clock Correction Function

The Sequencing Function controls the transition between modes of the Tracking Loop Software. The mode transitions of the Tracking Loop Software are shown in Figure 7 and described below.

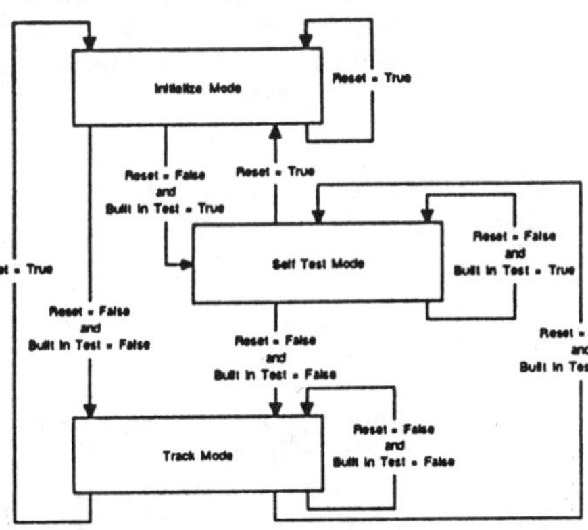

FIGURE 7 TRACKING LOOP PROCESSOR MODE TRANSITION DIAGRAM

1) Initialization Mode is commanded by the Navigation Software and initializes the receiver hardware and software. The software initializes the coder/correlator chip registers and selects a master chip to generate the timing. Software variables such as code loop accumulators and AGC values are also set in initialization mode.

2) Self Test Mode is commanded by the Navigation Software and sets the master coder/correlator chip to generate a test signal. The test signal is injected in the IF and all channels are commanded to track the signal.

3) Track Mode processing will sequence through the channels calling the channel processing function for each channel.

The Sequencing Function also calls the data input function to receive data from the Navigation Software every second. If the input data for a channel indicates a new satellite vehicle ID the software sets the channel state to zero to start the satellite acquisition. The output data function is also called every second to output pseudorange, carrier phase, state, and other information to the Navigation Software.

The Channel Processing Function sequences through all the channels and executes the proper state processing for each channel. The state transition diagram for the channel processing function is shown in Figure 8. The eight states are subfunctions of the channel processing function and are described below.

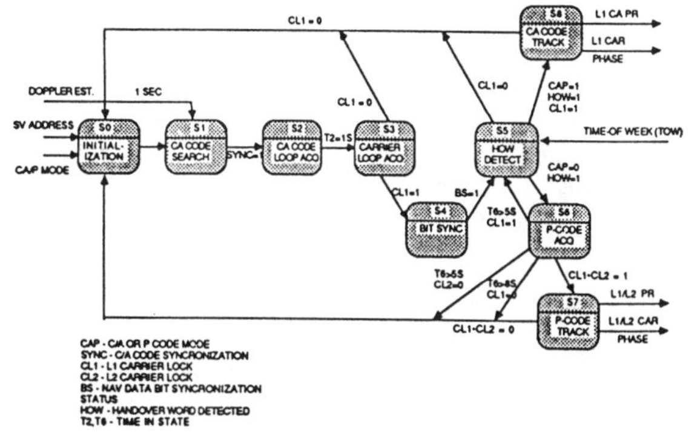

FIGURE 8 TRACKING LOOP SOFTWARE STATE TRANSITION DIAGRAM

1) State 0 - Channel Initialization performs the necessary initialization of the coder/correlator chip registers and software variables when changing satellites or loosing lock on a satellite. State 1 is entered after initialization and when the satellite ID is not zero.

2) State 1 - L1 C/A Code Search releases the coder/correlator to perform the code search. The carrier tracking loop is inhibited and the Doppler estimate from the Navigation Software is used for the carrier NCO word. The search is implemented as a two threshold search, as previously described. The software monitors the coder/correlator to determine a code detection and advances to state 2 when detected.

3) State 2 - L1 C/A Code Acquisition closes the code loop while maintaining the carrier NCO frequency at the value received from the Navigation Software. The AGC loop is also closed. This state is executed for a fixed time to allow the code loop to settle.

4) State 3 - L1 Carrier Acquisition releases the coder/correlator to perform the carrier tracking loop by setting appropriate gain factors for the loop. The AGC loop, code tracking loop, and carrier lock detection algorithm are active. The state is exited when carrier lock is detected.

5) State 4 - L1 Bit Synchronization performs the carrier tracking, code tracking, AGC calculation, carrier lock detection, and bit synchronization algorithm. The bit synchronization algorithm is performed for a fixed number of transitions and the state is exited.

6) State 5 - Handover Word (HOW) Detection performs carrier tracking, code tracking, AGC calculation, carrier lock detection, and bit synchronization. This function also accumulates navigation data and continually compares it with the preamble and the truncated handover word calculated from the GPS time received from the Navigation Software to detect the subframe timing. When subframe sync is detected the software synchronizes the L1 and L2 coder/correlator P codes. State 8 (C/A code tracking state) is entered if the HOW synchronization flag in the navigation message is set false or if the receiver is commanded to C/A code tracking only, otherwise state 6 (P code acquisition) is entered.

7) State 6 - P Code Acquisition performs P code acquisition on the L2 channel and then performs handover from C/A to P code on the L1 channel. If P code acquisition on the L2 channel is not accomplished in a fixed time period the channel is returned to state 5 to perform another handover word detection. When L1 and L2 carrier lock is detected, which indicates successful P code acquisition, the state is exited.

8) State 7 - P Code Tracking performs P code tracking, pseudorange measurement, carrier tracking, bit synchronization, carrier phase accumulation, and carrier lock detection on both L1 and L2. The function also performs data accumulation and bit synchronization on the L1 channel only.

9) State 8 - C/A Code Tracking performs code tracking, pseudorange measurement, carrier tracking, bit synchronization, carrier lock detection, carrier phase accumulation, and data accumulation on the L1 channel only.

The Clock Correction Function is executed on command from the Navigation Software. This function calculates and writes the appropriate coder/correlator register values to move the master coder/correlator one second epoch an amount equal to the Navigation Software clock offset to align the local one second epoch with the GPS system one second epoch. This feature is implemented to generate a hardware clock which is accurate to GPS time and also to synchronize the signal processing of the coder/correlators to GPS time.

Navigation Software Overview

The Navigation Software consists of eight functions:

- the Moding and Sequencing Function,
- the Navigation Solution Function,
- the Telemetry/Uplink Function,
- the Receiver Channel Manager Function,
- the Satellite Position Function,
- the Satellite Selection Function,
- the Satellite Data Manager Function,
- and the BIT/AFI Function.

The Moding and Sequencing Function controls the execution of the other functions within the Navigation Software. The modes are described below, and illustrated in Figure 9.

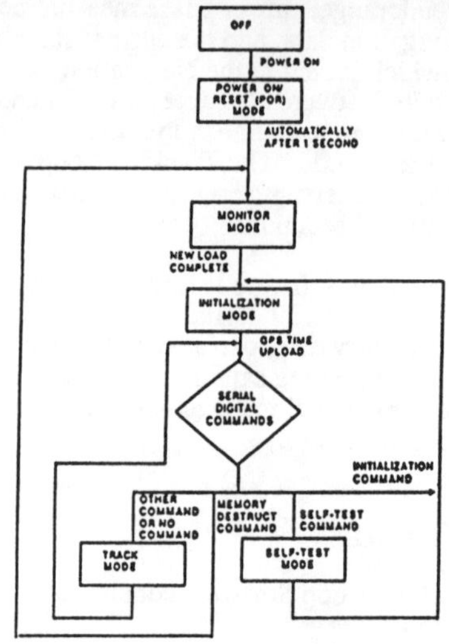

Figure 9 Mode Flow Diagram

1) Power On/Reset (POR) Mode - This mode will be entered only when the processor hardware is turned on or reset. There will be no software actions performed in POR Mode. After approximately one second, the processor will go automatically into Monitor Mode.

2) Monitor Mode - This mode will be entered from POR Mode or when a destructive memory test has been ordered. The navigation ROM software will perform destructive memory tests before performing any other actions in Monitor Mode. This mode will not be exited until the Navigation Software has been loaded.

3) Self-Test Mode - This mode will be entered when the operator commands a self-test to be done. One self-test will be performed, then the software will go to Initialization Mode.

4) Initialization Mode - This mode will be entered the first time tracking is to be done or after a self-test, memory modification, or initialization command, has taken place. If other commands are sent up by the ground, they will be processed. Then the software will run the Satellite Position Function and Satellite Selection Function to generate a list of satellites to track. Following this, an automatic transition will be made to Track Mode.

5) Track Mode - Track Mode will be the normal operating mode of the navigation processor. This mode will handle small scale user commands. Periodically, the serial digital command words will be checked to see if there are commands present which require a transition out of Track Mode.

The Navigation Solution Function consists of two subfunctions: a propagation subfunction, to be performed each second, and a measurement update subfunction, to be performed every ten seconds. The Telemetry/Uplink Function handles the command and telemetry interfaces with the spacecraft. This function communicates ground commands to the Navigation Software and outputs the telemetry data to the ground.

The Receiver Channel Manager Function is the interface between the Navigation Software and the Tracking Loop Software. The Receiver Channel Manager Function supplies Tracking Loop Software inputs from the Navigation Software, such as satellite assignments, clock corrections, and receiver commands. This function also processes Tracking Loop Software outputs before use by other Navigation Software functions, including carrier phase data, pseudo range data, and receiver status indicators.

The Satellite Position Function computes a satellite's (either TOPEX or GPS) position and velocity at the requested time from the currently available almanac data. It also outputs lists of the viewing periods over a given time frame for which the GPS satellite is in view of the TOPEX satellite and the ground stations. The Satellite Selection Function chooses a group of satellites to track, based on the input of GPS satellite and ground station visibility data from the Satellite Position Function. The satellites to be tracked have transition times associated with them so that the Receiver Channel Manager Function knows when to order the receiver to switch satellites.

The Satellite Data Manager Function maintains up-to-date storage of GPS and TOPEX almanac data used in satellite position calculations and satellite selection processing. The BIT/AFI (Built-In Test/Automatic Fault Indication) Function checks the operation of the processor.

The development environment consists of a MicroVAX* II running the VMS operating system. Ada and C were considered as languages for developing the Navigation and Tracking Loop Software. Ada was chosen because an Ada compiler exhibited throughput and memory usage advantages over the available C compiler for the 1750 host architecture.

There are three main tasks in the processor; the first task is the Tracking Loop Software, which runs every 20 milliseconds. The Navigation Software consists of two tasks, one of which runs every second and the other which runs every five minutes. The five minute task includes the GPS satellite visibility software and the Satellite Selection Function. The one second task has all of the other functions in it, including the Navigation Solution Function, and all of the interface functions.

Ground Command Processing

There are a number of ground commands which can be sent from the ground to cause changes in the Navigation software. The categories of commands are:

- Satellite selection related commands, which perform actions such as disabling a satellite from being chosen.

- Initialization related commands, such as updating GPS time.

- Navigation related commands, for instance, changing from P code to C/A code.

- Memory modification commands, where certain areas in RAM are uploaded with new values.

- Diagnostic commands that cause BIT operations to be performed.

- Navigation information commands, such as changing the loop bandwidth.

Built-In Tests (BIT)

There are three types of BIT tests which can be run: a destructive memory test, a checksum, and a memory dump. When any of these tests are run, the complete results of the test are transmitted to the ground through the telemetry link. The destructive memory test is automatically executed when the processor is powered up, or it can be executed through a ground command. This test causes preset memory patterns to be written to the RAM and then the data is read back to make sure that there have been no memory or addressing failures. After a destructive memory test, the entire RAM memory must be uploaded again with the operational program.

The checksum operation is initiated by a ground command, and runs a checksum over the area of memory selected by the ground. The results are then transmitted to the ground, where the same test can be run on an identical processor and software, and then a simple difference program on the ground can compare the results to search for incorrect code or memory failures.

The memory dump operation is also initiated by a ground command, and transmits to the ground the area of memory that the operator has selected in the command. The combination of these three commands gives the operator on the ground a powerful set of tools to investigate the health of the hardware and software on the satellite.

PREDICTED PERFORMANCE

Signal Dynamics

Two requirements that are unique to the space environment are the need to accommodate relatively large GPS signal power variations and relatively large Doppler due to motion between the GPS satellites and the TOPEX satellite.

The net signal power level available to the receiver is a function of (1) GPS satellite transmitter power, (2) system geometry, (3) receiver antenna gain in the direction of the satellite, and (4) receiver circuit losses. The satellite transmitter power may be higher in some cases due to satellites running "hot", i.e., radiating more power than needed to satisfy the specification. This condition is not detrimental to the receiver operation, but it must be defined for the purposes of the AGC design. The system geometry factor accounts for the difference between the signal strength for the satellite at the horizon and the satellite at zenith.

Figure 10 presents a comparison of the C/N_0 values available at the input of the GPSDR for the nominal, the

* MicroVAX, VAX and VMS are trademarks of the Digital Equipment Corporation (DEC).

best and the worst-case link budget scenarios. Because of the multipath dictated low antenna gain at low elevation angles there is a 10.3 dB spread between the nominal and the worst-case set of C/N_O values. The corresponding spread between the best and the worst cases is 23.3 dB. The maximum spread between the strongest received signal (L1-C/A, best-case) and the weakest received signal (L2-P, worst-case) is 29.3 dB. Because the worst-case C/N_O values are unrealistically low for optimizing receiver performance a minimum value of 37 dB-Hz (C/A) was established for acquisition and 34 dB-Hz for carrier tracking loop optimization.

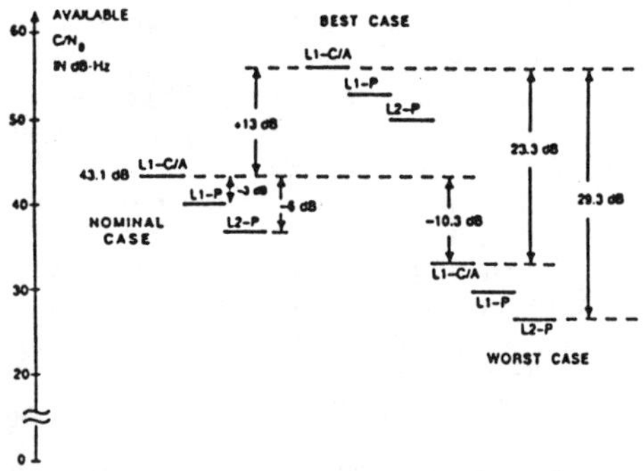

Figure 10 Nominal, Best-Case, and Worst-Case C/N_0's Comparison

A full simulation utilizing the actual TOPEX orbit and the full constellation of 18 GPS satellites predicted maximum relative velocity and accelerations of 7.71 km/sec and -10.17 m/sec^2, respectively.

These predicted maximum values have been confirmed by analysis of worst-case relative TOPEX/GPS geometries. The corresponding Doppler and Doppler rate seen by the receiver is:

Carrier	Doppler	Doppler Rate
L1	$\pm$40 KHz	55 Hz/sec
L2	$\pm$31 KHz	43 Hz/sec

Acquisition Performance

A sequential detector is used to acquire the C/A code. The C/A code length is 1023 chips and since it is a Gold code, there exist some 250 sidelobes spread throughout the 1023 chips. These sidelobes are 24 dB down from the main correlation point. The design of the sequential detector must be such that when searching for the main correlation point of a weak signal, it will not acquire the sidelobes of the best-case L1-C/A code with C/N_O of 56.1 dB-Hz or sidelobe C/N_O of 32.1 dB-Hz.

Monte Carlo simulation was used to establish the acquisition probabilities of the dual threshold sequential detector as a function of C/N_O. Figures 11 and 12 show these simulation results as cumulative acquisition probability for the appropriate design point and for Doppler frequency error extremes. False acquisition probability of sidelobes for threshold 1 was essentially zero.

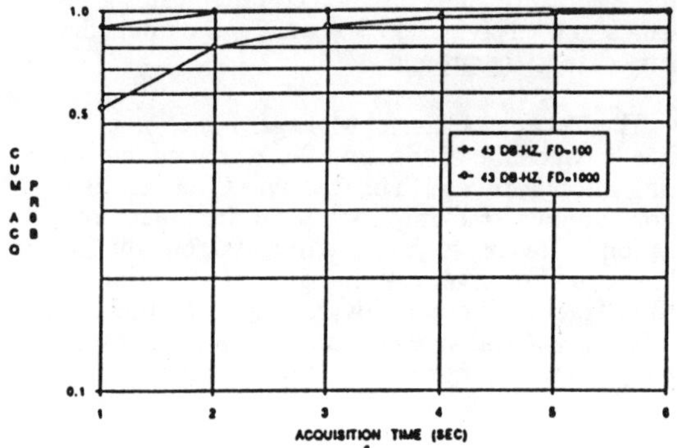

FIGURE 11 THRESHOLD 1 ACQUISITION TIME

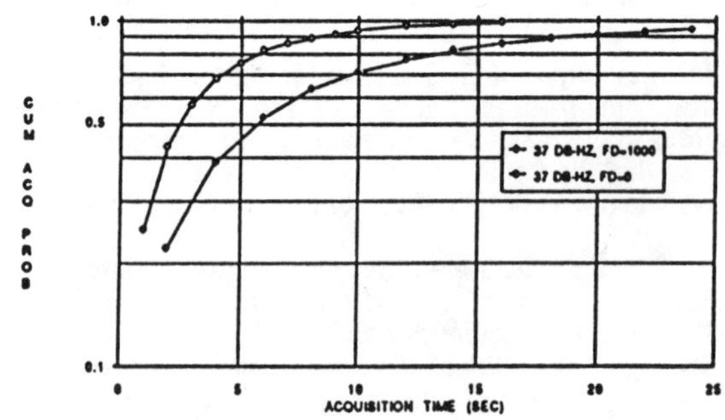

FIGURE 12 THRESHOLD 2 ACQUISITION TIME

Carrier Tracking Performance

The tracking performance discussed in this section concerns carrier and code tracking. A digital Costas loop is used for carrier tracking. There are two systematic errors that affect the carrier and need to be predicted and subtracted from the phase estimation. The first source is the dynamic errors in the carrier tracking loops due to different orbital dynamics of the GPS satellites. Since the Costas loop design uses a perfect second order loop filter (the digital loop filter has an ideal integrator), Doppler offsets will not create steady state phase errors in the loops but Doppler rates will cause a steady state phase error. The second source of systematic error is the carrier phase shift in the code correlation integrate-and-dump filters prior to NCO multiplication and Doppler removal. The RSS of the systematic errors and the effect of thermal noise on the L1 carrier tracking loop are illustrated Figure 13 for various C/N_O's as a function of the closed loop bandwidth B_L.

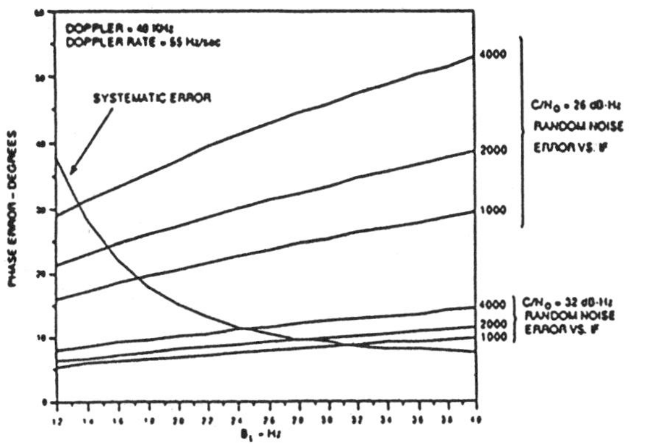

FIGURE 13 L1 CARRIER OPTIMIZATION FOR VARIOUS C/NO'S

Since the Doppler offset and Doppler rate are precisely known, both ϕ_{ss1} (steady state phase error due to Doppler rate) and ϕ_{ss2} (steady state phase error due to a phase shift in the code correlation output, caused by the Doppler offset) can be calculated and removed to get a corrected phase error accurate to less than 1 degree. The steady state phase error, however, does affect loop bandwidth selection and the performance of the loop, in particular, in terms of its mean slip rate.

Figure 14 shows the mean time to slip for a Costas Loop tracking L1, with maximum Doppler rates (55 Hz/sec) and the Doppler rates in the nominal acquisition range of zero to ten degrees elevation angle (12 Hz/sec). These curves are for a predetection filter bandwidth of 1000 Hz. The optimum loop bandwidth for Doppler rates near acquisition is close to 10 Hz. But this bandwidth must must be expanded to handle the highest Doppler rates experienced during a tracking pass of a GPS space vehicle. For maximum Doppler rates, a bandwidth close to 24 Hz (22 Hz for L2) minimizes the chance of cycle slips. At the optimum loop bandwidths for maximum Doppler rate, the probability of a cycle slip is less than 0.0001 per hour at a $C/N_O \approx 34$ db-Hz. (MTBS $\approx 3.6 \times 10^7$ seconds). With these loop bandwidths, the random error performance of the carrier phase measurement is 0.11 cm for C/A, 0.15 cm for L1-P, and 0.16 cm for L2-P at the nominal specified values of C/N_O.

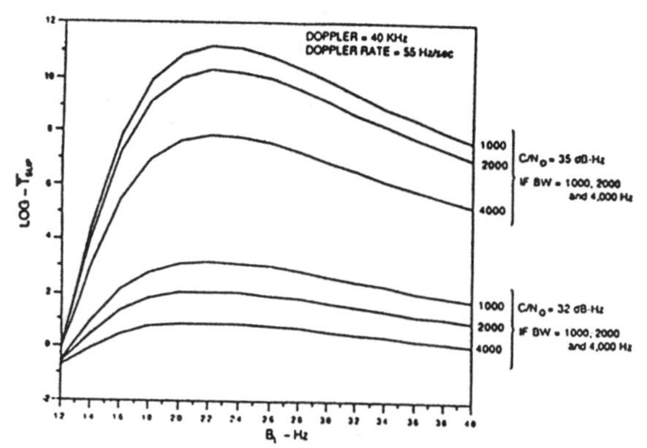

FIGURE 14 L1 CARRIER SLIP TIME OPTIMIZATION FOR C/NO'S OF 32 & 35 dB-Hz

Code Tracking Performance

The optimization of the code loop tracking bandwidth centers on restricting the number of pseudorange measurements that are required to be transmitted over the telemetry link consistent with no loss of measurement accuracy when the data is further smoothed in the POD solution. Because the telemetry system is operating at a 10 second dump rate the loop bandwidth was chosen to be smallest value consistent with independent pseudorange samples, i.e., $B_L = 0.1$ Hz.

Pseudorange random errors, including quantization effects, at the specified value of C/No are 50 cm, 7 cm, and 10 cm respectively for C/A, L1-P, and L2-P. Interchannel bias errors related to signal dynamics are effectively zero with the largest one, due to imperfect aiding in the carrier loop due to a jerk component, being on the order of 0.3 cm.

Doppler Prediction Accuracy

Once the GPSDR has been initialized and acquired a sufficient number of satellites to support a measurement update, the estimated TOPEX velocity and clock frequency error are used to estimate the signal Doppler for all subsequent acquisitions. As illustrated in the next section, the accuracies are well within the tolerable Doppler error of 500 Hz. In addition, GPS velocities can generally be estimated with almanac data to better than 2 m/sec ($\approx$ 10 Hz), so this error contributor is not significant. It is only during initial acquisition, then, that the Doppler error can become significant. The initial time tagging error of up to 2 seconds can produce 100 Hz Doppler error worst case, and errors in the initial TOPEX almanac based velocity can contribute an additional 50 Hz. These errors and other contributors to acquisition frequency are provided in Table 1.

Table 1 Acquisition Frequency Uncertainty Estimates

SOURCE	Δf (Hz)	COMMENTS
Oscillator Frequency	158	1×10^{-7} stability
Doppler Rate	53.4	
Doppler Estimate Error	160	
SA Clock Perturbation	50	
Total	421.4	
To allow for some design margin, use $\Delta f = \pm 500$ Hz		

Navigation Solution Accuracy

The accuracy of the onboard navigation solution was demonstrated using Monte Carlo simulation. The simulation included a detailed representation of the TOPEX satellite trajectory, with both non-spherical gravitational forces and drag modelled, an 18 satellite GPS constellation, and realistic error models. The modelled errors include multipath, as specified by JPL, and SA. Representative simulation results are provided for P code updates with SA off (Figs. 15-17) and SA on (Figs. 18-20).

The shifts in the plotted error levels are induced by changes in GDOP caused by changes in the tracked GPS constellation. Accuracies are generally well within the required levels for time tagging (0.5 microseconds, or approximately 150 meters) without SA, but come close to the allowable error levels when SA is activated.

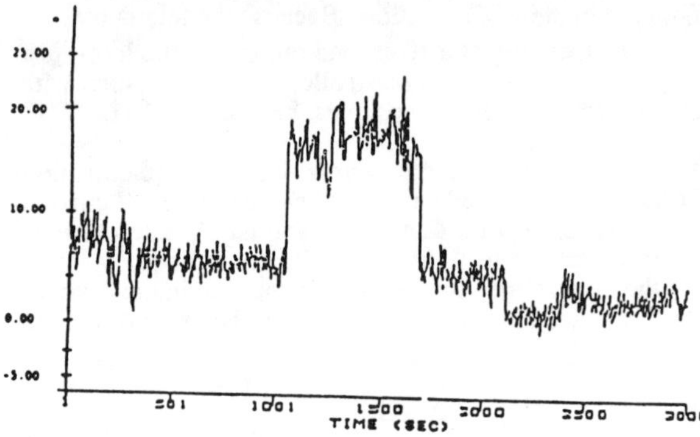

Fig. 15 RSS Position Error (meters) for P-Code, SA off

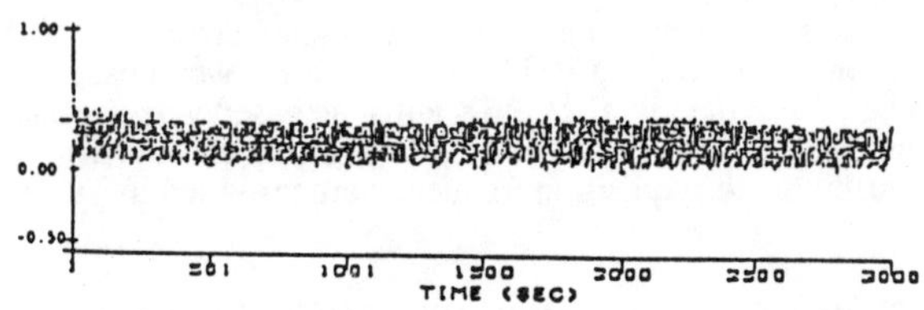

Fig. 16 RSS Velocity Error (m/sec) for P-Code, SA off

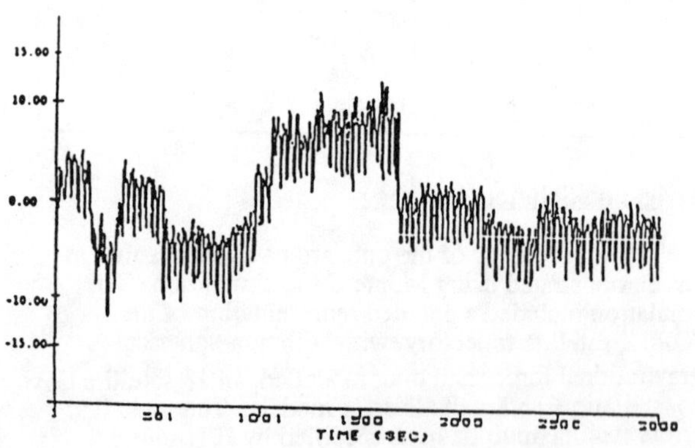

Fig. 17 Clock Offset Estimation Error (meters) for P-Code, SA off

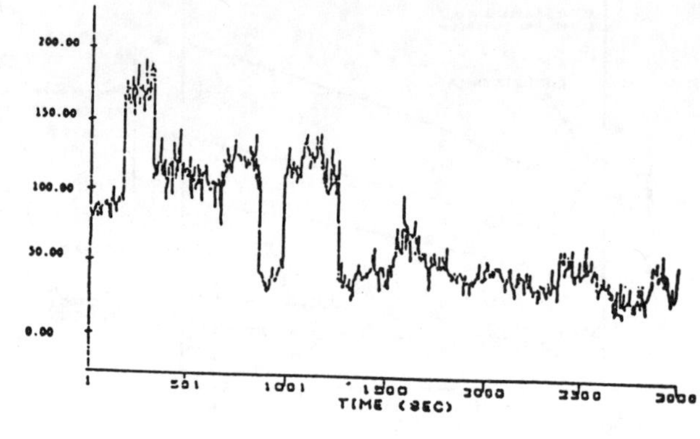

Fig. 18 RSS Position Error (meters) for P-Code, SA on

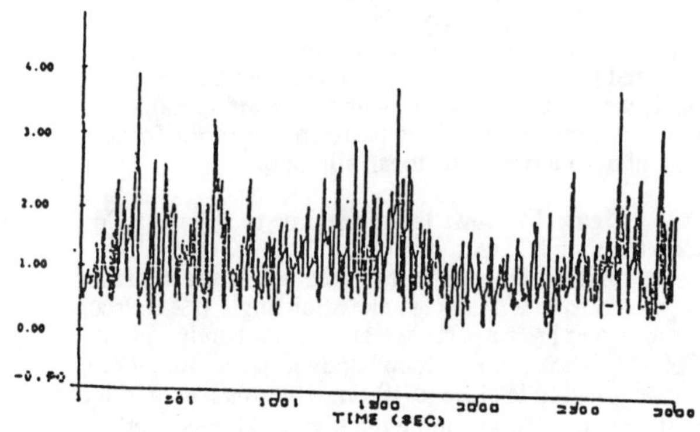

Fig. 19 RSS Velocity Error (m/sec) for P-Code, SA on

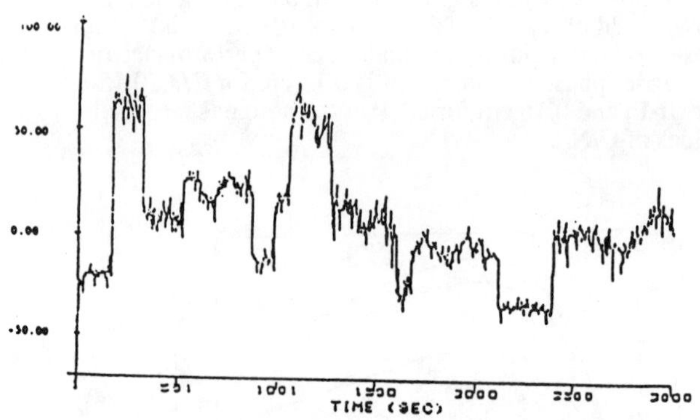

Fig. 20 Clock Offset Estimation Error (meters) for P-Code, SA on

SUMMARY AND CONCLUSIONS

The design of the GPSDR for NASA's TOPEX satellite has been described, and has been demonstrated by simulation and analysis to satisfy the performance requirements imposed by JPL. The GPSDR detailed design and testing efforts are progressing towards an expected 1991 launch date. This receiver should then enable unprecedented accuracies in orbit determination.

ACKNOWLEDGEMENT

The work described in this paper was performed for the Jet Propulsion Laboratory, California Institute of Technology, and sponsored by the National Aeronautics and Space Administration. The authors acknowledge the contributions of E.S. Davis, TOPEX GPS Experiment Manager for JPL, to this paper.

Reference herein to any specific commercial product, process, or service by trade name, trademark, manufacturer, or otherwise, does not constitute or imply its endorsement by the United States Government, or the Jet Propulsion Laboratory, California Institute of Technology.

REFERENCES

1. Born, G.H., Stewart, R.H., and Yamarone, C.A., "TOPEX - A Spaceborne Ocean Observing System," AIAA Vol. 97 of Progress in Astronautics and Aeronautics Series, 1985

2. Stewart, R., Fu, L.-L., and Lefebvre, M., "Science Opportunities From the TOPEX/POSEIDON Mission," Jet Propulsion Laboratory Publication (JPL internal document) 86-18, July 15, 1986

3. JPL/NASA, CNES, "TOPEX/POSEIDON Mission Description," Jet Propulsion Laboratory Publication,1635-5 Rev B, (JPL internal document), December 1985

4. Wunsch. C, "Satellite Altimetric Measurements of the Ocean - Report of the TOPEX Science Working Group," Jet Propulsion Laboratory Publication, (JPL internal document), March 1, 1981

5. Melbourne, W.G., and Davis, E.S., "GPS-based Precision Orbit Determination: A Topex Flight Experiment," AAS/AIAA Astrodynamics Specialist Conference, paper AAS 87-430, August 10-13, 1987

6. Yunck, T.P., and Wu, S., "Non-Dynamic Decimeter Tracking of Earth Satellites Using the Global Positioning System," AIAA 24th Aerospace Sciences Meeting, January 6-9, 1986

7. Bertiger, W.I., and Lichten, S.M., "A Demonstration of Sub-Meter GPS Orbit Determination and High Precision User Positioning," abstract accepted by the IEEE 1988 Position Location and Navigation Symposium, 1988

8. Galbraith, A.R., "GPS Demonstration Receiver Subsystem Requirements", Jet Propulsion Laboratory Publication 633-472, (JPL internal document), February 1986

9. Yost, Y.A., and Boyd, R.W., "Modified PN Code Tracking Loops: Its Performance Analysis and Comparative Evaluation," IEEE Transactions on Communications, Volume COM-30, May 1987.

10. Navstar GPS Selective Availability and Anti-Spoofing Requirements (SECRET), ICD-GPS-203, GPS Joint Program Office, USAF Space Division, El Segundo, CA.

Navigating the Viking Landers

E. A. EULER

ABSTRACT

ON JULY 20, 1976, Viking Lander I became the first unmanned spacecraft to land and operate successfully on the planet Mars. This was followed by a second successful landing on September 3, 1976. This paper gives a detailed description of the autonomous on-board navigation process to perform the guidance, control, and entry navigation functions. Also, the functions performed on the ground to generate and validate the guidance commands sent to the spacecraft prior to descent are described. In-flight measures of navigation system performance are compared against a-priori error estimates to show that the descents were completely nominal in all respects. The final landing errors were 25 km and 10 km for Missions 1 and 2, respectively.

INTRODUCTION

The successful landings and subsequent post-landed operation of the Viking Landers represent major milestones in this country's planetary exploration program. At the time the Viking Lander separated from the Viking Orbiter and descended to the surface, the range to Mars was over 350 million kilometers, resulting in a one-way light time of about 20 minutes. This long time delay, coupled with the relatively short descent time (~3 hours) dictated that the landing would have to be made autonomously in a planetary environment not well understood.

From the early stages of design up until the actual event, this factor had a strong influence on the design of the hardware, on-board software, and ground operations and software systems. The navigation system (defined here to be both the ground and on-board processes required to determine and control the vehicle position, velocity, and attitude from separation to touchdown) was designed to maximize accuracy, provide redundancy and reliability, and to cope with stacked worst case Mars environment and vehicle subsystem conditions during the landing sequence. More information on the vehicle subsystems can be found in Ref. 1, and a description of the in-flight subsystem performance is contained in Ref. 2. The following sections of this paper describe in detail the operation of the on-board and ground navigation system elements, the pre-flight accuracy predictions, and measures of the actual in-flight navigation system performance.

Mr. Euler is with Martin Marietta Corp., P. O. Box 179, Denver, CO 80201. He presented this paper at the National Aerospace Meeting in Denver, CO, on April 13, 1977.

LANDING SEQUENCE SUMMARY

The landings on Mars were made from a highly eccentric orbit with a radius of periapsis near 4891 km. and a Mars-synchronous period of 24.623 hours. The key events are shown pictorially in Figs. 1 and 2. After separation from the Viking Orbiter at an orbit true anomaly of about 271°, the Viking Lander (VL) maneuvers to the proper attitude and performs the deorbit burn. Following deorbit, the Lander stays in an inertial hold mode until shortly before the defined atmosphere entry point. For the Viking Mission, entry is defined to be 243.8 km. (800,000 ft.) above the mean surface, although significant aerodynamic forces on the VL do not actually occur until about 91.4 km. (300,000 ft.) above the mean surface. The time from separation to the entry point is about three hours. At entry, the vehicle is maneuvered to the proper attitude with the expected trim angle of attack.

Reprinted with permission from *Navigation,* vol. 25, no. 1, pp. 11-18, Spring 1978.

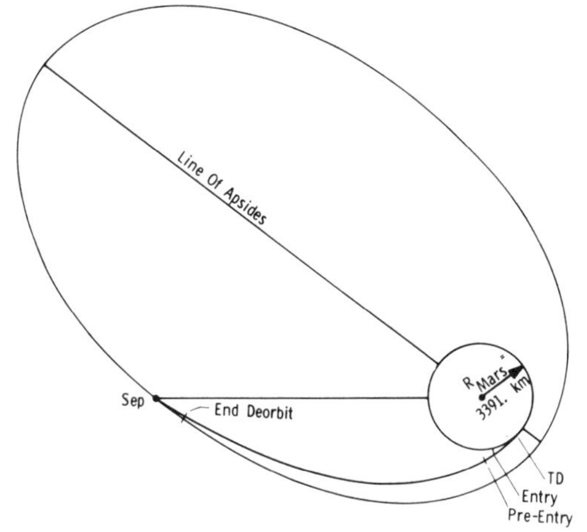

Fig. 1 – Nominal VL Descent Trajectory.

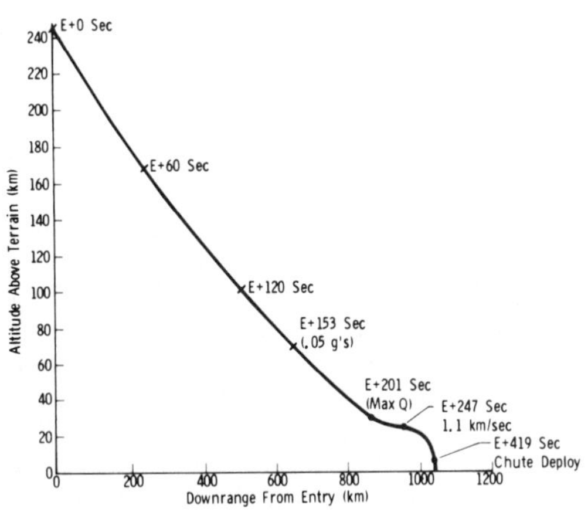

Fig. 2 – VL Entry Trajectory Profile.

When the on-board accelerometers sense .05 earth g's, the VL releases inertial hold in pitch and yaw and the aerodynamic characteristics of the vehicle determine the attitude. At 5.79 km. (19,000 ft.) above the local terrain, the parachute is deployed and seven seconds later the aeroshell is jettisoned. At 1.49 km. (4900 ft.) above the terrain the terminal descent propulsion system is activated and set to minimum throttle for two seconds at which time the parachute and basecover are jettisoned. The vehicle then follows a gravity turn to about 16.8 m. (55 ft.), and then descends vertically to the surface at 2.44 m/sec (8 fps).

VIKING LANDER ON-BOARD NAVIGATION

The guidance, control, and navigation functions on-board the VL actually begin during the pre-separation sequence that begins at Separation (SEP) – 30 hours and lasts for about 5. hours. During this time period, the inertial reference unit (3-axis accelerometers and gyros) is calibrated to determine biases, bias stability, and provide a ground check on scale factor. Also, all other descent subsystems, including radars and propulsion valve electronics, are fully checked out. At about 3 hours prior to separation, the IRU is turned on to warm up and verify spin motor operation. The Flight Software performs an on-board check of vehicle subsystems operation 45 seconds prior to separation.

The vehicles are separated by springs resulting in a total ΔV of about 0.17 m/sec (.56 fps). After separation the on-board software tracks the attitude by continuously computing the transformation matrix (direction cosines) between the current attitude and the attitude at SEP, which is the inertial system until entry. For the first four minutes after SEP, no attitude control is performed so that a sufficient distance can be obtained between VO and VL before engine firings. During this period the accelerometer biases are determined and the computed values are used throughout the remainder of the mission. At 241 seconds after SEP the vehicle maneuvers to the desired deorbit attitude. Three minutes are allowed for all attitude maneuvers at the nominal maneuver rate of 1°/second.

The deorbit maneuver is accomplished with the propulsion system attached to the aeroshell. Four clusters of 3 engines (pitch, yaw, and roll) are operated in a pulsed mode. The system is a monopropellant hydrazine blowdown system with an initial thrust of 6 lbs./engine and a thrust of 4 lbs./engine at the end of the maneuver. The on-board software controls the maneuver by sensing accumulated velocity in inertial coordinates and nulling lateral components while accumulating the desired ΔV in the axial direction. The software commands the accumulated velocity to follow a curve of ΔV vs. time, (implemented as a polynomial in time) which is designed for a 75% engine-on duty cycle and 3σ low propulsion system performance (Fig. 3). This implementation essentially removes any errors due to thrust misalignment

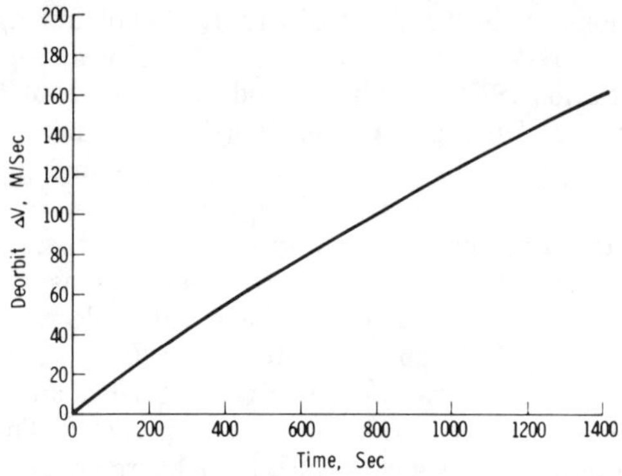

Fig. 3 – Deorbit ΔV vs Time.

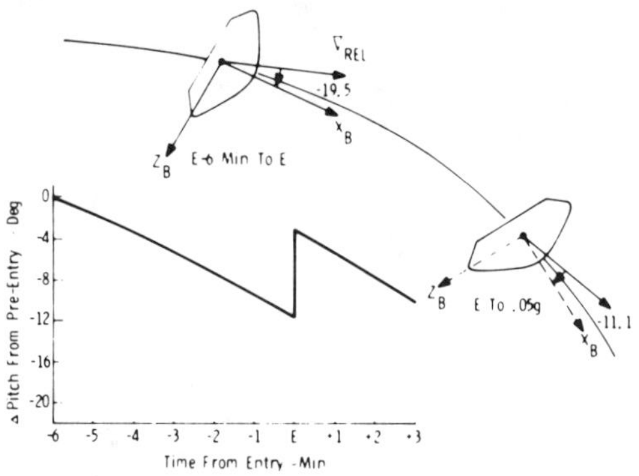

Fig. 4 – Pre-Entry to .05G Pitch Profile.

and cutoff time error and protects against single engine failure. This results in the ΔV accuracy being almost completely dependent on the accelerometer accuracy. During the burn, the VL is maintained in a .25° limit cycle and the burn shutdown is made on ΔV, with a timed backup. The nominal ΔV for both missions was 156 m/sec (511 fps), resulting in a burn time of 1336 seconds.

After the completion of the deorbit maneuver, another attitude maneuver is commanded to prepare the vehicle for the long coast phase prior to entry. The attitude limit cycle is 7° during this phase. Midway through the long coast, a 180° roll maneuver is performed to equalize thermal gradients within the VL due to solar radiation. Also, the software monitors and commands engine pulses such that 4 pulses are achieved per engine every 70 seconds to prevent freezing of the engines.

At entry −9 minutes (6 minutes, Mission 2) the vehicle is again commanded to another attitude to put the c.g. in the relative trajectory plane with an angle of attack of −20°. Also, a pitch rate is initiated to follow the flight path angle change and maintain the desired angle of attack. This is implemented by following a polynomial of pitch rate vs. time. The total pitch change is shown graphically in Fig. 4.

At entry, a number of different functions are initialized to prepare for the final phase of descent. A step change in pitch is made to put the vehicle at the desired angle of attack of −11.1° and the pitch rate is maintained to keep this value. The on-board inertial coordinate system is changed to a local vertical frame defined at the predicted touchdown point and the inertial navigation system is initialized to values provided in the uplink load prior to separation. From this point on the state is estimated by integrating the equations of motion including gravity and sensed accelerometer outputs. The attitude is still actively controlled (with the pitch rate defined above) until .05g is sensed. The radar altimeter is turned on, but the data is not used.

When .05 g is sensed by the accelerometers at an inertial velocity of about 4.8 km/sec (16,000 fps), active control is stopped and the vehicle is allowed to trim to the equilibrium angle of attack and sideslip. Pitch and yaw rates are controlled to be less than 1°/sec and roll is controlled by keeping the yaw axis perpendicular to the local vertical direction, thereby keeping the c.g. in the actual trajectory plane. At this time the radar data is used to update on-board altitude estimates, but it does not feed back into the computation of inertial position and velocity.

The next key event occurs when the inertial navigation computations sense a planet relative velocity of 1.1 km/sec (3609 fps). During preflight analyses, the potential presence of radar blackout between .05g's and 1.1 km/sec and the time to relock the radar altimeter after blackout showed that large altitude errors might exist in the on-board navigation process. To prevent the premature deployment of the parachute (deployed on navigator-sensed altitude) a time delay of +25 sec was instituted from this event to allow the navigator sufficient time to converge on the altitude estimate before the search for the parachute deployment altitude commenced.

At an altitude of 5.79 km. (19,000 ft.) above the local terrain, as sensed by the on-board navigation process, the parachute mortar is fired to deploy the parachute. The relative velocity at this point is about 0.24 km/sec (787 fps). Seven seconds later the aeroshell is jettisoned, including the deorbit propulsion system. At this point, the terminal descent propulsion system roll thrusters are used for roll control while pitch and yaw remain uncontrolled until the initiation of the terminal descent thrust phase. At twenty two seconds after parachute deployment, the lander legs are deployed and the navigation process changes from inertial to relative velocity and altitude only. The four-beam terminal descent and landing doppler radar data is used to obtain direct measurements of the vertical and lateral relative velocity. Also, the attitude navigation process is changed to keep track of body attitude with respect to local vertical only. An open-loop roll maneuver is performed to achieve the desired orientation at touchdown. During this phase, the VL approaches terminal velocity.

At a sensed altitude of 1.49 km. (4900 ft.) the terminal descent propulsion system is activated to guide the vehicle to touchdown. This system is also a blowdown monopropellant hydrazine system with three throttlable engines that provide 600 lbs. of thrust at ignition. The engines are warmed up for two seconds at minimum throttle, at which time the parachute and basecover are separated and the VL begins a pitch-up maneuver in order to align the thrust vector with the relative velocity vector to perform a gravity turn to touchdown. The total relative velocity at engine ignition is compared to pre-programmed altitude-velocity contours shown in Fig. 5. If the total velocity is to the left of the curves shown, minimum throttle is commanded until the interior contour is reached and then the appropriate thrust is commanded to follow this contour until near-touchdown (Trajectory A). If the initial state is between the contours, the on-board software interpolates a new contour and guides the VL along it (Trajectory B). If the initial velocity is to the right of curve 2, full throttle is commanded until curve 2 is reached, which is then followed to touchdown (Trajectory C). Curve 1 corresponds to the maximum thrust/maximum

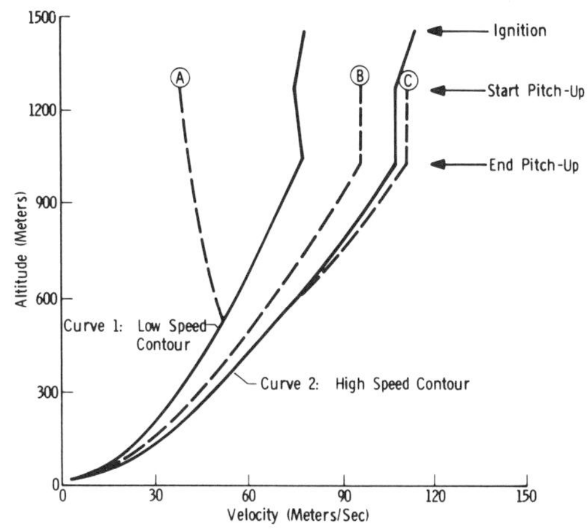

Fig. 5 – Terminal Descent Contours.

vertical velocity case and Curve 2 is sized for the maximum vertical velocity and maximum wind (lateral velocity). Differential engine throttling is used to control pitch and yaw attitude.

At 138 ft. the last radar altimeter measurement is used and at a sensed condition of 16.8 m. (55 ft.) or 2.44 m/sec (8 fps) relative velocity, the vehicle descends vertically at constant speed to touchdown. Neither vehicle encountered severe velocity conditions at the start of terminal descent, so that curves similar to Trajectory A were actually followed.

The engines cut off and the propulsion system is sealed when one touchdown sensor, located on each footpad, is activated. After touchdown, the IRU remains on for five minutes, and this data is used by the software to determine the spin vector and gravity vectors in body coordinates, and subsequently, the landed attitude and areographic latitude.

GROUND CONTROL FUNCTIONS

The ground activities leading to descent began about 3 days prior to separation when the final landing site coordinates were selected. VL-1 was targeted for an areographic latitude of 22.6° N and 47.5° W longitude and the corresponding values for Mission 2 were 47.89° N and 225.86° W. By SEP-64 hours, the trajectory design was complete and the guidance parameters for descent were computed in engineering units. The mission-dependent set of guidance parameters included the major event times relative to separation, the attitude matrices for all inertial attitudes prior to sensing the atmosphere, the deorbit ΔV, initial conditions at entry for the entry inertial navigation process, polynomial coefficients for the entry pitch rate, and some miscellaneous parameters such as back-up times for sensed events. The bulk of the guidance, control, and navigation parameters, such as limit cycle magnitudes, steering loop gains, and polynomial coefficients describing the deorbit and terminal descent contours were primarily dependent on subsystem operation and were loaded into the computer memory prior to launch. These values were not changed during the mission due to nominal subsystem performance.

The mission-dependent guidance parameters described above, along with post-landed data base parameter changes dictated by mission circumstances, were converted to the appropriate on-board computer data base elements for command transmission at approximately SEP-39 hours.

Prior to command transmission, a number of verifications were performed to assure the validity of the data base entries for descent (Fig. 6).

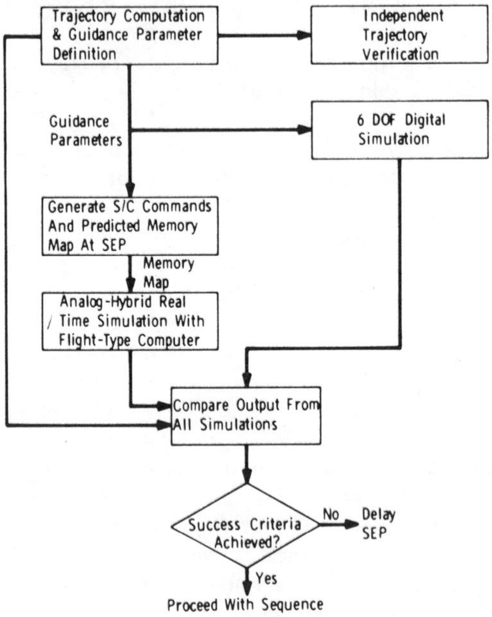

Fig. 6 – Descent Validation Process.

First the trajectory design and deorbit and attitude commands prior to entry were verified independently using software designed for precision trajectory generation for planetary missions (DPTRAJ) and in use for many years. Secondly, a full six-degree of freedom digital simulation was performed that functionally models subsystem and on-board software operation. Finally, and most importantly, the predicted memory map at separation was sent from Mission Operations at JPL back to Martin Marietta in Denver where a complete real-time analog-hybrid simulation of descent was performed with a development unit flight computer in the loop. The results were sent back to JPL where a comparison of selected data from all three simulations was made and judged against a set of pre-defined success criteria. The successful completion and comparison of these simulations was a prerequisite for commanding the vehicle at SEP-39 hours.

From SEP-30 hours to SEP-25 hours, the vehicle underwent the pre-separation checkout described earlier. The ground processing of this data continued to about SEP-16 hours at which time decisions were made on further updates to the computer memory prior to separation. For instance, during the checkout of VL-2, an anomaly was discovered in the terminal descent and landing radar beam 2 test. Since only three out of four beams must function properly during descent, the decision was made to send a command instructing the software to ignore this beam output during descent. Also, at this time the capability existed to make a small change to the separation time to account for the latest orbit solutions of the separation orbit. For both missions, the time of the desired separation state vector computed 3 days prior to separation was within one second of the later estimates and no update was required.

The complete descent verification process described above was repeated using the latest predicted separation state and any new memory changes. Due to the short time period, the command updates were sent at SEP-9 hours prior to completion of the verification process. This last verification was completed by about SEP-4 hours and successful comparisons between simulations were required prior to uplinking the final GO command at SEP-1 hour. From this point on, no further ground action was required, but checks were made on-board both the VO and VL shortly before separation and SEP could have been inhibited due to anomalous subsystem operation.

FLIGHT TRAJECTORIES

Due to the precise navigation of the Viking Orbiter, discussed in a companion paper (Reference 3), the descent trajectories chosen in flight were nearly identical to the pre-flight nominals. The deorbit ΔV for both vehicles was 156 m/sec (511 fps) and the inertial flight path angle at the entry point was targeted to $-16.9°$ ($-17.0°$ for Mission 2). The relative flight path angle when the atmosphere is encountered is about $-12.5°$. Due to the L/D characteristics of the entry vehicle, the flight path angle levels off to zero and then sharply decreases to near vertical by the time of parachute deployment. The nominal relative velocities were 4.6 km/sec (15,000 fps) at atmosphere encounter, 0.24 km/s (787 fps) at parachute deployment and 55.0 m/sec (180 fps) at terminal descent ignition. Both vehicles were targeted to land within .5° of the VO orbit plane, although the VL was capable of crossrange angles at touchdown of up to 6°. It was desired to minimize the crossranging because of increased sensitivity of landing site dispersions due to deorbit pointing errors.

ACCURACY PREDICTIONS

The contributions to the total landing dispersion for the flight trajectories are shown in Table 1 in terms of downrange and crossrange. The overall dispersion ellipse is composed of error contributions that can be characterized statistically, and other non-statistical effects that were conservatively handled on an additive basis.

Table 1—Contributions to Total Landing Dispersions

Statistical Contributions (3σ): Error Source	DR ~ km	XR ~ km
Orbit Determination	10.2	1.3
Deorbit ΔV Magnitude	39.2	~0
Deorbit Pointing	11.5	7.7
Map/Pole	45.0	45.0
Entry Aerodynamics	17.4	7.6
Winds	18.1	19.3
RSS	66.5 km	50.2 km
Non-statistical Contributions: Error Source	**DR ~ km**	**XR ~ km**
Atmosphere	±30.0	~0
Thermal bias shift	0 → 17.5	−2.1 → 0
Maximum Error	114.0 km	52.3 km

The deorbit ΔV error is primarily a proportional error attributable to accelerometer bias change during the maneuver. The calibration performed on-board between separation and the beginning of the deorbit maneuver helped minimize this error source. The pointing errors were due primarily to initialization errors at separation caused by Viking Orbiter limit cycles and sensor offsets. The accuracy of the VL gyros was not a major contributor to deorbit errors, but was important in the analysis of guidance and control performance during entry. The target landing site location uncertainties in inertial space were due to uncertainties in the direction of the Mars pole and the location of features with respect to map coordinates. The entry aerodynamics contributions arise from uncertainties in the lift to drag ratio, c.g. location, and lift and drag coefficients. Finally a statistical model of the winds near the surface of Mars was adopted by the Project and resulted in the error contributions shown. The non-statistical effects include the spread of potential atmosphere density profiles likely to be encountered and an unknown bias shift in the accelerometers during deorbit due to deorbit engine plume heating.

IN-FLIGHT NAVIGATION RESULTS

The performance of the VL systems during both descents was essentially flawless (Reference 2). All systems operated within expected limits, and due to the high quality data returned to earth during descent, it was possible to reconstruct the trajectory and evaluate systems performance to a fairly high degree of accuracy. Fig. 7 shows the altitude vs. time comparison which, at the scale of the plot, was nearly the same for both landings. The longer time between entry and touchdown was due primarily to the combined effect of an atmosphere density profile difference from nominal and a slightly higher L/D than expected. Although both of these effects were identified after analysis of the data from the first mission and prior to the second descent, the effects were

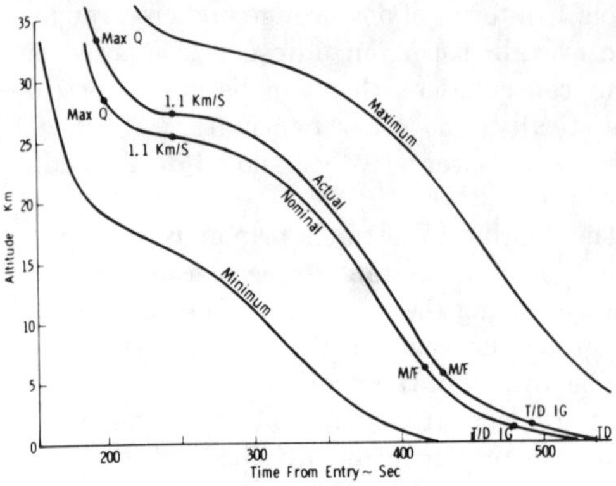

Fig. 7 – VL1 Altitude vs Time From Entry.

not well enough understood and were small enough such that no nominal model changes were made between missions.

Table 2 shows the best estimate of the individual error contributions to the total landing error in terms of downrange and crossrange errors. Not including the map and pole errors, the total landing errors were 25. km. for Mission 1 and 10. km. for Mission 2. The estimates of the total deorbit execution errors are .199% ΔV magnitude and 0.24° pointing for Mission 1, and .055% ΔV magnitude and 0.37° pointing for Mission 2.

The on-board navigation performance during entry is a function of both the initial condition errors and the guidance and control hardware performance. The initial condition errors at the defined entry time are shown in Table 3. Both of these error sets are significantly less than the maximum errors that the system was capable of handling. The altitude errors at parachute deployment and terminal descent ignition are indicated in Table 4. Again, these errors were well within requirements. The relative velocity errors at the time of first use of the terminal descent and landing radar data in the parachute phase were about 6 mps (20 fps), which were easily eliminated when the doppler data was used. Finally, the terminal descent guidance performance and propellant consumption were very near the predicted values, except that the constant velocity phase appears to have begun 2.7 m (9 ft.) higher than the expected value of 16.8 m (55 ft.) on both vehicles, suggesting a small systematic error in the on-board navigation system. One last measure of navigation system performance is obtained after touchdown in the on-board estimate of Mars gravity and landed latitude. Table 5 shows the on-board estimates of areographic latitude and landed radius (derived from the gravity measurement) compared to the esti-

Table 2 – Actual Landing Site Errors

Error Source	VL1 DR ~ km	VL1 XR ~ km
Orbit Determination	−2.6	~0
Deorbit Execution	−20.3	−2.0
Atmosphere	−1.2	~0
Winds	−8.9	−5.0
VL Aerodynamics	+7.6	~0
Unexplained	+1.1	+1.4
Total	−24.3	−5.6

Error Source	VL2 DR ~ km	VL2 XR ~ km
Orbit Determination	+0.9	~0
Deorbit Execution	−6.5	+1.8
Atmosphere	+4.7	~0
Winds	−1.9	−0.7
VL Aerodynamics	+10.0	+0.6
Unexplained	+1.9	−1.5
Total	+9.1	+0.2

Table 3 – Entry State and Attitude Errors

	VL1	VL2
RSS Position	14.5 km	6.2 km
RSS Velocity	8.8 m/s	3.1 m/s
Pitch	0.72 deg	0.25 deg
Sideslip	0.59 deg	0.14 deg
Roll	~0.deg	−0.26 deg

Table 4 – Altitude Errors at Parachute Deployment and Terminal Descent Ignition

	VL1	VL2
Parachute Depolyment (M/F)	273 feet	224 feet
T/D Ignition	3 feet	6 feet

Table 5 – Post-land Touchdown Estimates

	On-board Estimates	Estimates Based on Tracking Data
VL1:		
Areographic Latitude	22°.68	22°.46
Radius	3389.8 km	3389.4 km
VL2:		
Areographic Latitude	47°.71	47°.96
Radius	3381.5 km	3381.4 km

mates obtained from Viking Lander tracking data.

In summary, the performance of the on-board and ground navigation systems was completely nominal in all respects and the actual errors experienced on both flights were well within statistical expectations. Also, no major modelling discrepancies were evident in comparing the pre-flight and post-flight results. The success of this mission has provided another milestone in the continuing development of planetary navigation systems.

ACKNOWLEDGMENTS

The work reported herein was conducted under Contract NASl-9000 under the management of Langley Research Center of the National Aeronautics and Space Administration. The author wishes to acknowledge Mr. G. L. Adams of Martin Marietta Corporation, who helped in the preparation of this paper.

REFERENCES

1. Anon., *"Viking 'As-Built' Performance Capabilities,"* Contract NASl-9000, Martin Marietta Corporation, Denver, Colorado, 1976.
2. Goodlette, John D., *"From Separation to Touchdown: The Performance of Lander Capsules,"* AIAA Paper No. 77-269, January 1977.
3. O'Neil, William J., *"An Overview of Viking Navigation"* The Institute of Navigation National Aerospace Meeting, Denver, Colorado, April 1977.

EVALUATION OF THE SHUTTLE POWERED ASCENT FLIGHT PHASE

Lee Olson

McDonnell Douglas Technical Services Co.

and John W. Sunkel

NASA Johnson Space Center

The successful first flight test of the Space Shuttle provided additional confidence that the Shuttle's digital fly-by-wire GN&C system will be able to satisfy the demanding requirements across the spectrum of planned mission profiles. An overview of the ascent trajectory and GN&C system design is followed by a summary of flight test results for the ascent phase of STS-1. The most notable variance from nominal pre-flight predictions was the lofted trajectory observed in first stage due to an unanticipated shift in pitch aerodynamic characteristics from those predicted by wind tunnel tests. The GN&C systems performed as expected on STS-1 throughout powered flight.

Following a discussion of the software constants changes implemented for Flight 2 to provide adequate performance margin, a summary of test results from STS-2 is presented. Vehicle trajectory response and GN&C system behavior were very similar to STS-1. Ascent aerodynamic characteristics extracted from the first two test flights will be included in the data base used to design the first stage steering and pitch trim profiles for STS-3.

INTRODUCTION

Launch of the STS-1 from launch complex 39A on April 12, 1981 marked the beginning of a nearly flawless first Shuttle test flight. Data collected on vehicle subsystem performance and operating environments will be used to allow incremental expansion of the flight envelope in subsequent test flights leading ultimately to verification of the fully operational Shuttle capabilities.

First flight came after more than a decade of design, development and verification activities leading to certification of a rather complex vehicle avionics system. The quad-redundant, fly-by-wire digital guidance, navigation and control (GN&C) system provides automatic steering for ascent but also includes man-in-the-loop capability which may be required in contingency abort situations. Four general purpose computers (GPC) each perform identical computations in a synchronized manner, using inputs from

Reprinted with permission from *Guidance and Control 1982,* vol. 48 of *Advances in the Astronautical Sciences,* San Diego: American Astronautical Society, 1982, pp. 429–442.

redundant sensors. Thrust vector control (TVC) and aerosurface actuators thus receive four parallel commands which are force-summed at the power valve. Since the avionics system is designed to be two-fault tolerant (fail operational/fail safe) redundancy management is an integral part of the GN&C system. Redundant sensor inputs are sampled at the user rate (25 hz for rate gyros and accelerometers) and the middle value is selected for use by the flight control system. Fault detection and isolation algorithms are processed at lower rates and once a sensor is declared failed, the high rate selection process is moded accordingly and fault annunciation is displayed to the crew and telemetered to the ground. Since the GPCs do not have adequate memory capability for the entire mission profile, the software has been structured into several memory loads including ground checkout, ascent, on-orbit, on-orbit checkout and entry. These load blocks are stored on redundant mass memory units and GPC memory is reloaded at appropriate points along the mission profile. The ascent load provides GN&C software for use throughout the terminal count starting at t-20 minutes through orbit circularization. A portion of the GPC memory is preserved across the phase transitions which includes the operating system software, state vector, sensor and effector configuration, and failure status and initialization data.

ASCENT TRAJECTORY DESCRIPTION

Figure 1 provides an overview of the Shuttle mission profile. The ascent phase begins with ignition of the Solid Rocket Boosters (SRBs) and terminates at orbit insertion or at landing at KSC in the case of an

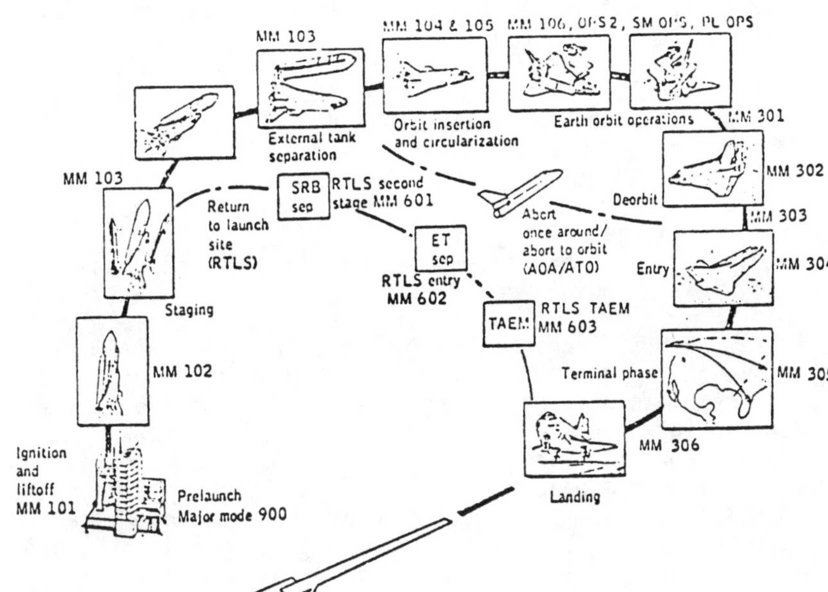

Fig. 1 Shuttle Mission Profile

Return to Landing Site (RTLS) abort which would be required with an Orbiter engine failure early in powered flight. Three subphases of a normal ascent include stage 1 from liftoff to SRB separation, stage 2 from separation to main engine cutoff (MECO), and a coasting phase which includes two Orbital Maneuvering System (OMS) engine burns for orbital circularization.

The Space Shuttle is launched from a vertical position with the tail south on the launch pad. Following a short vertical rise to clear the launch tower, a 3-axis maneuver is performed to orient the vehicle to the desired flight azimuth. A programmed pitch rate continues for the remainder of first stage flight until SRB staging when a constant attitude is maintained to minimize the probability of recontact with the spent solids. During ascent the Shuttle flies in a "tail-down" attitude so that the Orbiter thrust, inclined 12 degrees to the vehicle centerline, lifts the vehicle trajectory at zero angle of attack. Dynamic pressure ($\bar{Q}$) reaches its maximum value about 50 seconds after liftoff and follows a plateau for approximately 30 seconds before decreasing at higher altitude. The characteristic $\bar{Q}$ profile is achieved by shaping the stage 1 pitch command profile appropriately and throttling the main engines back to near their minimum power level (65%) through the high $\bar{Q}$ region. Nominal $\bar{Q}$ maximum was designed to be less than 600 lb/ft^2 for STS-1 to minimize aerodynamic loads and will be increased incrementally through the remainder of the flight test program. Orbiter aerodynamic surfaces are not used for control during ascent but the elevons follow an open-loop position profile in first stage to avoid excessive hinge moment and wing bending and torsion loads. A closed-loop elevon load relief feature is available to command the elevons away from the open-loop profile, if required, based on measured actuator differential pressure.

Approximately 6 seconds after SRB staging the stage 2 closed-loop guidance is initiated to steer the vehicle to the desired MECO target conditions on velocity, flight path angle, altitude and orbital plane. Late in stage 2, the guidance commands throttling of the main engines to limit vehicle acceleration to 3 g's. MECO occurs at an altitude of approximately 385,000 feet at an inertial velocity of 25,668 ft/sec. At this point, the vehicle is still on a sub-orbital trajectory to allow disposal of the external tank (ET) in the Indian Ocean.

Eighteen seconds after MECO, the Orbiter is separated from the ET and the coast phase transitional digital autopilot (TRANS DAP) commands the down-firing RCS thrusters on to provide a vertical velocity increment of 4 ft/sec, providing comfortable clearance from the separated tank. At MECO plus two minutes, a short OMS burn (approximately two minutes duration) is performed to raise both apogee and perigee. Thirty minutes later, a second OMS burn raises perigee to circularize at the desired orbital altitude. During the first OMS burn, residual propellants in the main engines and manifolds are dumped overboard and the main engines are commanded to a stowed position prior to deactivation of the Orbiter hydraulic systems.

ASCENT GN&C OVERVIEW

A simplified GN&C software overview is shown in Figure 2. The sensors used for ascent include rate gyros, normal and lateral accelerometers and Inertial Measurement Units (IMU's). GPC computations of flight control inner-loop errors are performed at 25 hz and transport delay is kept to less than 20 milliseconds to preserve adequate closed-loop autopilot phase margin. Less time critical navigation and guidance functions are processed at lower sample rates to keep the overall computing demand within the capabilities of the GPC.

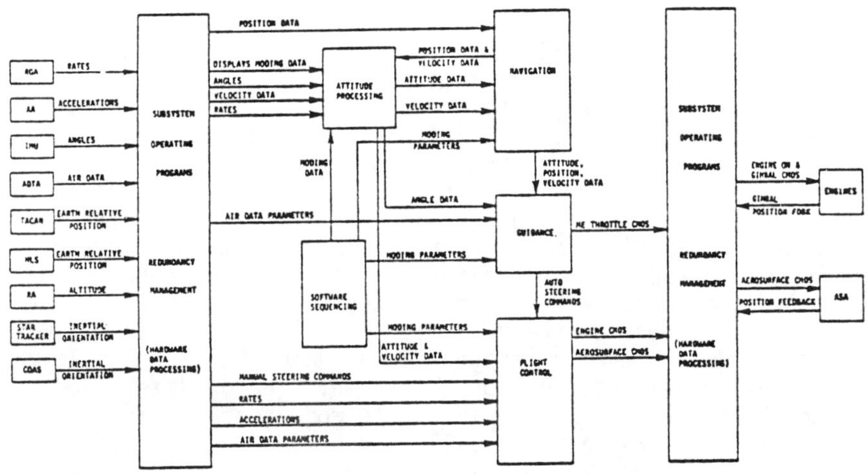

Fig. 2 Ascent GN&C Software Overview

The ascent navigation function uses selected IMU data and a model of the earth's gravitational acceleration to maintain a current estimate of the Shuttle state vector. State initialization occurs at t-8 seconds and a precise state computation is calculated approximately every 4 seconds. IMU accelerometers are sampled at a 1 hz rate and higher rate user demands are fulfilled by a propagation algorithm so that a velocity estimate is available for guidance at 6.25 hz. During the coast phase following MECO, the IMU sensed acceleration is not used if below a noise threshold and an atmospheric drag model is used instead. Also, a more accurate model of the earth's gravity is used by including higher order terms in addition to those used in powered flight.

First stage guidance computations are processed at 6.25 hz and consist of a table look-up of pitch, yaw and roll attitudes and main engine throttle commands as a function of navigated relative velocity (VREL). Velocity is used as the independent variable to provide a pitch profile which produces a desired angle of attack vs Mach number relationship dictated by aerodynamic loads requirements in the high $\bar{Q}$ region. Alternate pitch tables provide the necessary steering changes required for loss of a main engine in first stage.

Closed-loop stage 2 guidance consists of equations which cyclically compute the difference between the targeted MECO state and the current navigated state to produce a velocity and range to go to MECO. Stage 2 guidance is processed at 0.5 hz and consists of three major subtasks of predictor, corrector and steering parameter computation. A constant thrust is assumed early in second stage and constant 3g acceleration in the latter portion. Stage 2 guidance provides desired thrusting direction and thrust turning rate information to the flight control system and issues throttle commands to the main engines late in second stage to limit acceleration to no more than 3 g's. As MECO is approached, the position constraints are released and guidance computational rate is increased to provide an accurate velocity cutoff.

The ascent digital autopilot (ASC DAP) design is based on a conventional two-loop closure on attitude and attitude rate. A simplified software overview is shown in Figure 3. During the high $\bar{Q}$ region of first stage, an additional loop closure on normal and lateral acceleration is used to provide structural load alleviation. Attitude errors are calculated at 12.5 hz and rate and acceleration errors at 25 hz. Gains are scheduled against relative velocity for stage 1 and updated at 6.25 hz as are the reference thrust vector control (TVC) trim angle, normal acceleration, and elevon schedule profiles. Gain and trim profiles for stage 2 are scheduled as a function of calculated vehicle mass provided by stage 2 guidance. The rigid body closed-loop control frequency is slightly greater than 1.0 rad/sec to provide adequate steering response to guidance commands while maintaining sufficient propellant slosh phase margin. The oxidizer slosh mode near 3 rad/sec, is phase stabilized. Flexible mode compensation is provided by multiple digital filters (as high as 9TH order end-to-end in stage 1) and the filter coefficients are switched at transition to stage 2. The bending modes are gain-stabilized across their frequency spectrum beginning at 12 rad/sec early in first stage. A software module called RECON provides the necessary moding of TVC biases, mixing logic changes, limiters and faders based on mission timeline events and subsystem failure status. An example of the control law signal flow topology (stage 1 pitch channel) appears in Figure 4. Note that a forward path integration of attitude error is used as an autotrim function to wash out the effects of sensor bias or steady state mistrim due to thrust level imbalance or thrust vector misalignment.

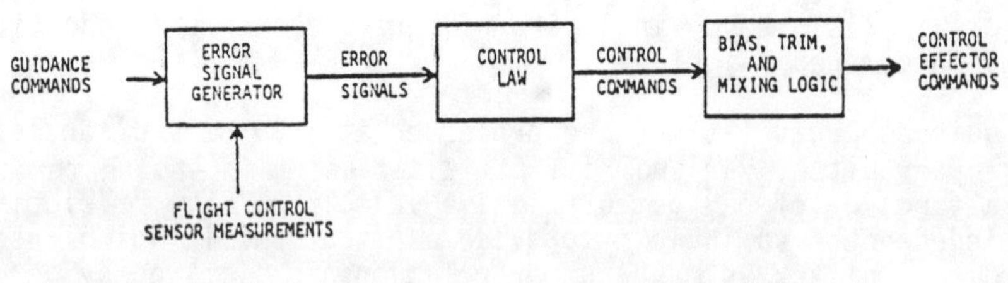

Fig. 3 Simplified Flight Control Software Overview

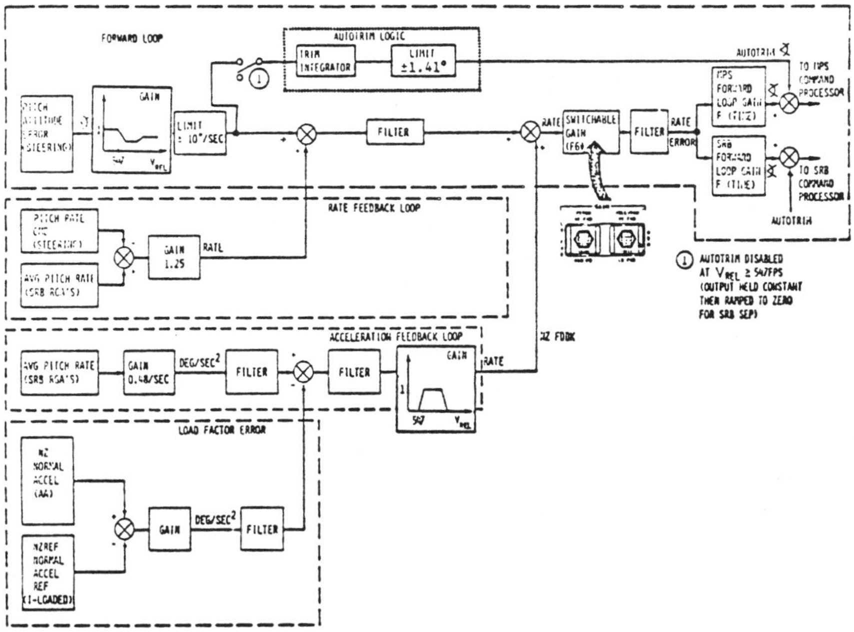

Fig. 4 First Stage Pitch Loop

RESULTS OF FIRST FLIGHT

Ascent guidance and control systems performed as designed on STS-1. No significant anaomalies were noted but several interesting aspects of the dynamic behavior of the mated vehicle were observed, most notably the lofted trajectory that resulted from an unanticipated pitch attitude error during the high dynamic pressure region of first stage flight.

Prelaunch countdown operations went smoothly and included Orbiter aero-surface and TVC actuator slew checks following hydraulic system activation at t-4.5 minutes. The SRB hydraulic systems were activated at t-18 seconds and an automatic SRB actuator slew check confirmed the health of the remainder of the actuation subsystems.

During the MPS engine thrust rise transient prior to SRB ignition, the stacked vehicle is subjected to a significant pitch bending moment. Structural response (primarily in the fundamental cantileverd pitch mode) is easily noted in attitude, rate gyro, accelerometer, and ground camera data. The initial "sway-over" was slightly larger than predicted and as a result the autopilot attitude error was more than anticipated at the time of control system activation immediately following SRB ignition. The structural dynamic transient response due to SRB ignition produced sensed pitch rates of ± 1.0 deg/sec which damped out rather quickly (Figure 5). Since the autopilot bending filters provide more than 25 db of attenuation across the bending spectrum, no significant TVC actuator command transients were developed as a result of the liftoff structural response. The crew

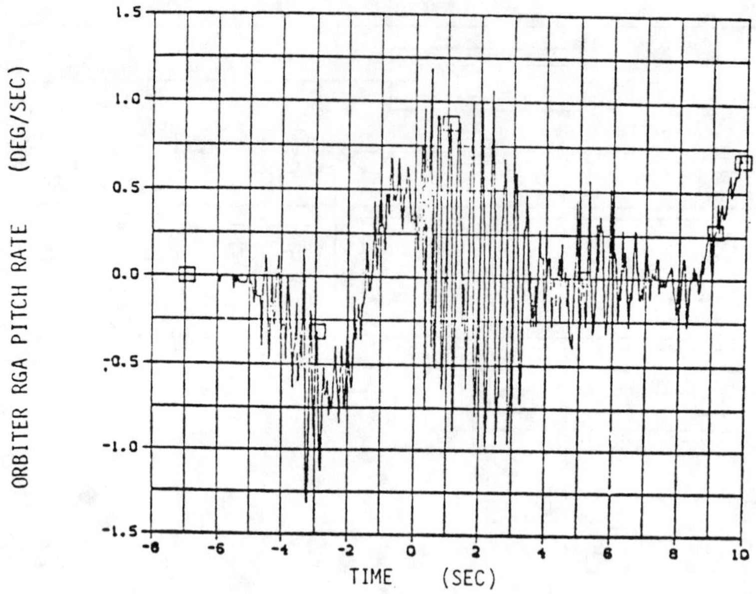

Fig. 5 STS-1 Sensed Pitch Rate at Liftoff

reported some blurring of the cockpit instrumentation initially which damped out by the time the tower was cleared and the roll maneuver began. The remainder of powered flight was unmarred by instrument vibration.

Post flight reconstruction of the vehicle lift-off dynamics and autopilot response indicated near-nominal facility clearance geometry. Northerly translation during the vertical rise was as expected and the absence of any discernable lateral drift confirmed the absence of any significant dispersions in SRB thrust mismatch, autopilot sensor bias, or ground winds.

Following the vertical rise, a simultaneous three axis maneuver is programmed in stage one guidance to re-orient the vehicle from its tail-south attitude on the pad to the heads down attitude in the desired flight azimuth. All sensor and effector data indicated nominal vehicle response and the crew reported no discernable lateral g-forces during the acceleration to the 10 deg/sec roll rate.

Winds aloft measured pre-launch were near the statistically predicted mean of the month for which the first stage open-loop steering profile was designed. As a result, roll and yaw attitude errors were minimal. A negative pitch attitude error began to develop after 40 seconds, peaked at -5.2 degrees and then gradually diminished toward the end of first stage (Figure 6). Post-launch wind measurements showed little change in the in-plane (pitch) wind compared to pre-launch data, indicating the attitude error was not wind-induced. Post-flight trajectory reconstruction efforts indicated a significant shift from wind tunnel predictions in the pitch aero coefficients. Specifically, the differences included a more positive normal force, more negative pitch moment, and less than predicted base drag.

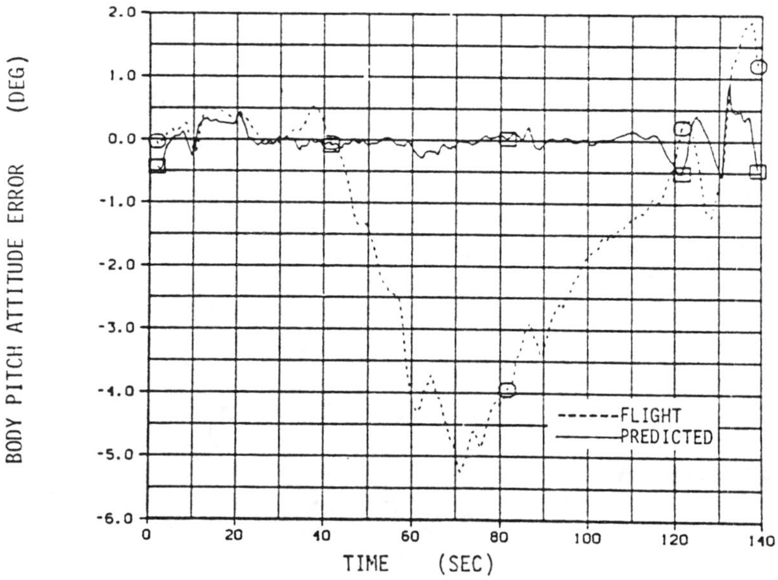

Fig. 6 STS-1 Pitch Attitude Error

The incremental pitch trim required of the TVC peaked at approximately 9 million ft-lbs. As a result of the steady state attitude error the first stage trajectory was lofted significantly. Table 1 is a comparison of flight with predicted state conditions at SRB staging.

Table 1

SRB Staging Conditions

	Predicted	STS-1	Difference
Time, sec	131.68	130.82	-0.86
Altitude,ft	164736	173957	+9221
Relative velocity, ft/sec	4178.56	4110	-68.56
Inertial velocity, ft/sec	5216.77	5127	-89.8
Inertial flight path angle, deg	26.13	28.73	+2.60

Although the pitch attitude error was rather large (the crews' ADI needle was saturated at 5 deg for more than 20 seconds) the rate of error build-up was rather gradual and the vehicle angle of attack did not deviate significantly from the predicted profile in the critical region near Mach 1 (Figure 7).

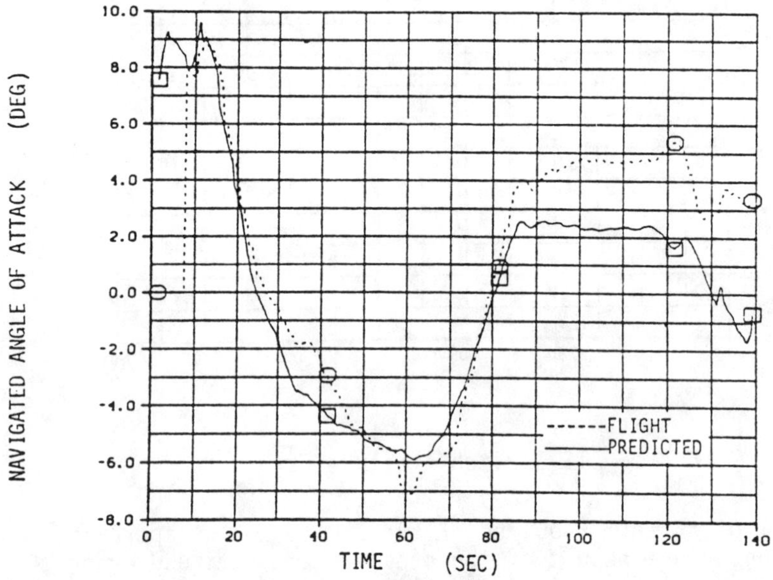

Fig. 7 STS-1 Navigated Angle of Attack

The trajectory design philosophy for early Shuttle flights included a constraint on maximum dynamic pressure which was 620 lb/ft^2 (PSF) for STS-1. Maximum navigated $\bar{Q}$ was 615 PSF (Figure 8) but the actual $\bar{Q}$, including wind effects, is now projected as 606 PSF. Note that as a result of the lofting the flight dynamic pressure during the latter part of stage 1 was less than pre-flight predictions.

A review of the elevon actuator position and differential pressure time histories showed elevon hinge moments to be reasonably close to pre-flight predictions. The elevons tracked the velocity dependent schedules and closed-loop load relief (based on sensed pressures) was not activated.

Reconstruction of SRB performance indicated both motors were near nominal predictions with relatively little thrust mismatch (propellant grain segments are poured in matched pairs to minimize thrust differential). Body rates were all less than 0.5 deg/sec at SRB staging and the transient response following separation was nominal as the main engines were re-trimmed in pitch to the shifted c.g. position.

There was evidence of Orbiter thrust vector misalignment in second stage based on examination of attitude error histories. The thrust was expected to deflect to some extent under load and flight data indicate somewhat larger misalignment than predicted. The autotrim function in the auto-

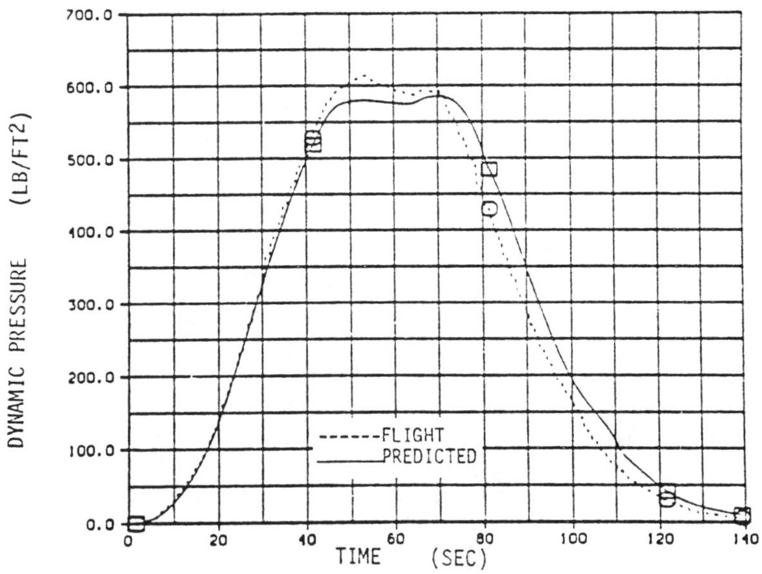

Fig. 8 STS-1 Dynamic Pressure

pilot gradually washed out the steady state attitude errors following the initial steering transient associated with stage two closed-loop guidance initiation. This misalignment effect was again observed near MECO when the main engines were throttled to their minimum thrust level (65%) just prior to MECO. The structural relaxation during the engine shutdown left the vehicle with a small residual rate in pitch which the RCS quickly nulled as control was transfered from the ascent TVC to the TRANS DAP RCS thrusters following MECO. There was no discernable bending acitvity observable in the stage 2 autopilot parameters and the propellant slosh activity was barely noticable in the rate gyro data. (A small amout of oscillation at the slosh frequency was anticipated due to rate gyro quantization and TVC nonlinearites).

Reconstruction of vehicle performance indicated that the MPS thrust levels were very close to predictions with some thrust increase trending noted on engines 1 and 3. This was found to be due to temperature effects on sensed engine chamber pressure used by the engine control computers to maintain a constant thrust level. Overall vehicle I_{sp} was approximately 2 sec less than predicted.

The GN&C system steered the vehicle to a MECO state which was well within the predicted accuracy at cutoff. Inertial velocity was only 1.4 ft/sec low which is less than the average error expected due to the 25 hz computation rate for MECO predictions.

Following MECO, the main engines are commanded their full-down position limit for the dumping of the main propulsion systems' residual propellants which is sequenced concurrently with the first OMS burn. Following the burn and dump completion, the MPS engines are commanded to a stow position for the remainder of the mission. Just prior to ET separation, an unexpected RCS thruster firing was noticed which was due to transient pitch rate exceedance of the TRANS DAP deadbands. The software processing rate for TVC actuator commands is reduced from 25 hz down to 1.0 hz following MECO and as a result the engines are slewed to their dump and stow positions in one degree increments. A pitch bending oscillation of about 0.5 deg/sec peak-to-peak at 4 hz was noted during these periods of engine slew. Fortunately, the TRANS DAP rate limits were large enough to avoid serious RCS interaction but the extra jet firing used an additional 10 lbs of RCS propellant. A software change to avoid this response is planned for later flights.

The right OMS pitch actuator froze at the null position about midway through the OMS-2 circularization burn. During the burn the engines were well trimmed and the OMS actuator malfunction caused no significant transient. The actuator performed sluggishly during slew checks following OMS-2 and the secondary drive motor in the actuator was used for the remainder of the mission. Post-flight trouble-shooting isolated the problem to insufficient rotor/stator clearances to accommodate dynamic deflections encountered. The actuator was replaced for STS-2. Adequate redundancy exists in each actuator should the failure repeat. Further redundancy exists in the OMS/RCS pods themselves as the RCS thrusters will correct for any significantly offset thrust vector. Design changes are in progress for OMS actuators to be installed on subsequent vehicles.

GN&C CHANGES FOR SECOND FLIGHT

As a result of the excellent performance of the GN&C system on STS-1, only minimal software changes were required for the second test flight; specifically a small compensation for the lofting phenomenon observed in first stage. The performance predictions for STS-2 when the design trajectory was evaluated using reconstructed ascent aero, propulsion predictions and seasonal environmental effects indicated poor payload performance margin. To partially compensate for the first stage lofting the open-loop first stage attitude profile was adjusted to compensate for half the flight path angle error st SRB staging. Orbital inclination and targeted orbital altitude were decreased somewhat to provide increased performance margin as STS-2 carried an additional 10,000 pounds of payload comapred to STS-1. In addition, the trajectory profile scribe lines provided on cockpit CRT displays to allow crew monitoring of trajectory performance were adjusted based on the aero-induced lofting observed on STS-1

RESULTS OF SECOND FLIGHT

Launch of STS-2, the second Shuttle test flight occurred on November 12, 1981. GN&C system performance and response compared well with that observed on STS-1. The launch facility clearance margins were again determined to be near nominal, this time with the assistance of data provided by improved engineering camera coverage. The aerodynamic characteristics of STS-1 first stage flight were reconfirmed as the pitch attitude error again grew to -5.2 deg (Figure 9). The vehicle encountered a significant component of crosswind in the altitude region of 35,000 ft which produced the roll and yaw attitude errors seen in Figure 10 and 11. Note also in Figure 11, the yaw attitude error in the region of 115 to 120 seconds which was due to a slight SRB mismatch as the SRB thrust is tailing off near the end of the first stage.

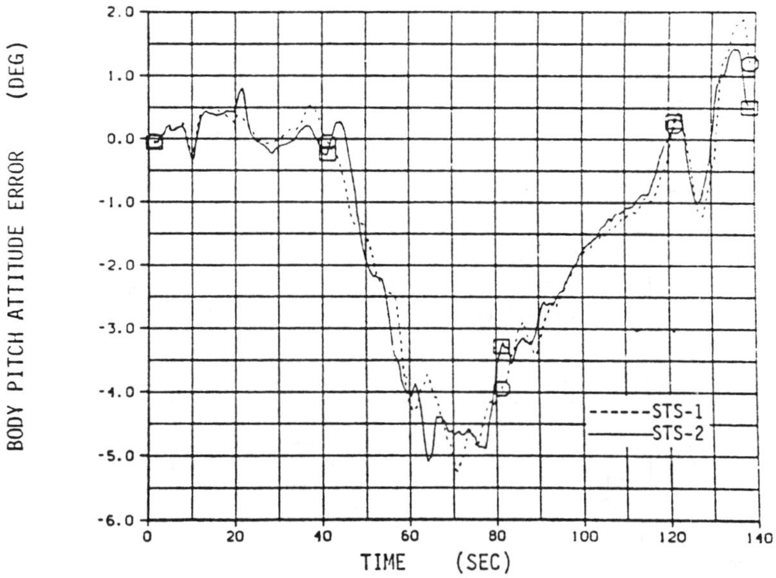

Fig. 9 STS-1 and STS-2 Pitch Attitude Error

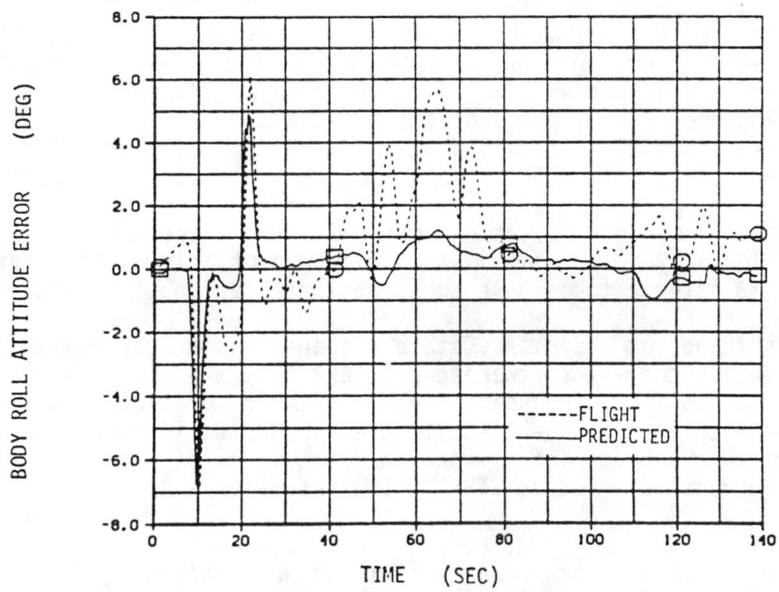

Fig. 10 STS-2 Roll Attitude Error

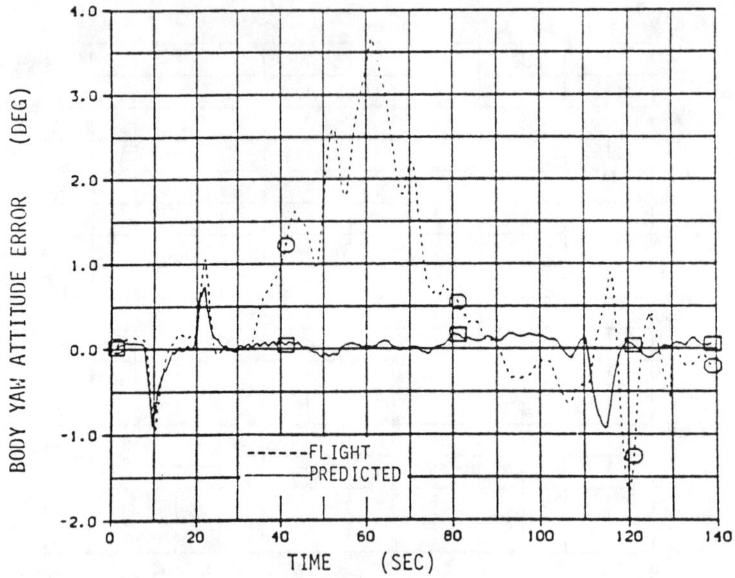

Fig. 11 STS-2 Yaw Attitude Error

Close examination of redundant rate gyro data for STS-2 first stage revealed the existence of a small amount of pitch angular misalignment of all three left hand SRB rate gyro assemblies. During the roll maneuver following tower clearance, the selected left SRB yaw rate signal was approximately 0.2 deg/sec less than either the selected signal from the right hand SRB or Orbiter rate gyros. During the remainder of the first stage, there was no disagreement, thus indicating a cross coupling of roll rate into the sensed yaw rate signal rather than a bias from null. Post-flight trajectory simulations incorporating a postulated 1.2 deg angular misalignment in pitch of the sensors in question produce simulated dynamic body rate histories which agree extremely well with flight response. Although the total error was small as seen by the autopilot (the selected left hand and right hand SRB rate signals are averaged), the suspected alignment error is slightly larger than allowed by specification. At the time of this writing, efforts are underway to attempt to measure the gyro installation alignment.

Stage 2 G & C transient and steady state response was comparable to first flight. Several autopilot software parameters from the GPC were included in the variable downlist, including the autotrim integration signals. Post-flight reconstruction of stage 2 attitude error, autotrim integration and TVC actuator command histories re-confirmed the level of Orbiter thrust vector misalignment observed on STS-1. The estimated alignments are shown in Table 2.

Reconstruction of vehicle performance again showed vehicle I_{sp} to be 2 sec below expected. For Flight 2, the engines' chamber pressure transducer harnesses received additional insulation and the thrust trending observed on STS-1 was eliminated. A minor change in the engine controller software caused the main engines to operate very close to the nominal 6.0:1.0 mixture

Table 2

Stage 2 MPS Thrust Misalignment

Pitch 1 = +0.1 deg	Yaw 1 = -0.4 deg
Pitch 2 = -0.4 deg	Yaw 2 = -0.25 deg
Pitch 3 = +0.8 deg	Yaw 3 = -1.45 deg

ratio compared to the 6.03:1.0 observed on STS-1.

MECO accuracy was again excellent on STS-2 and overall GN&C performance for ascent was as predicted.

Minor adjustments were made to the reconstruction of stage 1 aero coefficients as a result of the STS-2 flight data. This updated data base was used as a baseline to develop stage 1 steering, SRB trim, and reference N_z profiles for Flight 3.

CONCLUSIONS

In summary, the ascent GN&C system performed in an excellent fashion on the first two Shuttle test flights. Data gathered on subsystem response is being used to improve analytical math models to provide more accurate predictions of vehicle behavior for future missions. Minor modifications to hardware and software will continue as increased flight capabilities are added as a result of the expanding data base obtained in the Shuttle flight test program.

FLIGHT OPERATIONS FOR SHUTTLE RENDEZVOUS NAVIGATION AND TARGETING

by
Daniel J. Collins
Rockwell International

ABSTRACT

An important mission for the Space Shuttle is orbital rendezvous with free-flying satellites for repair and refurbishment. The recent successful rescue of the Solar Maximum Mission (SMM) satellite by the Space Shuttle Challenger emphasizes this fact. The task of completing a rendezvous is one that requires the capability to track the target satellite and autonomously determine the maneuvers necessary to complete the rendezvous. This paper describes the method by which Shuttle rendezvous is accomplished. An overview of Shuttle subsystems that are directly relevant to the rendezvous task is presented. Potential improvements to the Shuttle rendezvous capability are outlined.

INTRODUCTION

The Space Shuttle is the foundation of America's space transportation system. With the Space Shuttle, NASA is able to transport heavy payloads into space and return them to earth. Because it is a system that can return men and equipment to a predetermined point on earth, a potentially important mission for the Shuttle is that of rendezvousing with orbiting satellites. The ability to perform orbital rendezvous is necessary for the success of any Shuttle mission that involves orbital rescue, resupply, or repair. An example of the latter mission is the rendezvous and repair of the Solar Maximum Mission (SMM) satellite that was accomplished by the Space Shuttle Challenger in April 1984. The SMM repair mission was the first demonstration of the Space Shuttle's rendezvous capability. Future Shuttle missions will require rendezvous with the space station, space telescope, long duration exposure facility (LDEF), and other satellites. This means that the rendezvous task for Shuttle will become increasingly important as the program evolves.

Any orbital rendezvous mission begins at liftoff. There are three phases: ascent, ground-controlled orbital, and a later orbital phase in which the vehicle operates semiautonomously. Shuttle orbital rendezvous involves the active maneuvering of the vehicle from thousands of nautical miles of a target satellite to within a few hundred feet of that satellite. This must be accomplished in spite of potentially large errors in the initial estimates of the two vehicle's positions and velocities (i.e., state vector errors). To accomplish a semiautonomous orbital rendezvous in the presence of state errors, the Shuttle orbiter must perform three functions:

1. Progressively improve its estimate of position and velocity with respect to the target satellite (relative navigation)
2. Determine the maneuvers necessary to close in on the target satellite (targeting)
3. Perform the necessary maneuvers (i.e., guidance and propulsion)

To accomplish these tasks the Shuttle uses a combination of navigation sensors, propulsion, and computer processing.

The solution of the rendezvous problem for the practical cases of the Space Shuttle and its precursor vehicles (Apollo and Gemini) has been the responsibility of NASA's Johnson Space Center and a small number of its contractor organizations. A literature search finds only one publication that specifically discusses the operational aspects of orbital rendezvous (1). As members of the aerospace technical community, we have an interest in understanding the methodology of orbital rendezvous for Space Shuttle. This paper discusses Shuttle flight operations for orbital rendezvous. Emphasis will be placed on the later phase of orbital rendezvous when the navigation and targeting tasks of Shuttle are performed autonomously. This paper will discuss the relationship among the vehicle targeting, navigation, propulsion, and sensor systems as they are used in rendezvous flight operations. The role of the crew in the Shuttle flight system will also be discussed. This discussion includes the constraints and requirements on the Shuttle rendezvous problem as well as those Shuttle systems that are directly relevant to rendezvous, which will be presented first. Following that, the first rendezvous mission for Space Shuttle (April, 1984) will be discussed.

MISSION CONSTRAINTS AND REQUIREMENTS

Shuttle rendezvous flight operations and the rendezvous trajectory itself are the result of constraints imposed by current spaceflight technology and mission objectives. This is demonstrated by the fact that the Shuttle has only about 340 fps of ΔV available for orbital rendezvous maneuvers (2). Furthermore, errors in the estimated state vectors of both vehicles require that relative navigation be performed for Shuttle rendezvous missions. This requires sensor tracking of the target satellite by the Shuttle crew and navigation system. Unless the target is equipped with beacon lights, it must be illuminated by the sun at certain critical times in the mission so that the Shuttle orbiter's relative navigation sensors can detect it. Additionally, the rendezvous must be performed within a given amount of time so that other mission goals may also be met. Typically, a Shuttle rendezvous task is designed to take two days or less. In short, there are three constraints on Shuttle rendezvous operations:

1. ΔV available for maneuvers
2. Solar illumination of the target
3. Time available

Reprinted from *Rec. IEEE Position Location and Navigation Symp.*, pp. 106–112, 1984.

It is these constraints that determine the method by which orbital rendezvous is performed by the Space Shuttle. The major consequence of the first constraint is that at orbit insertion, the orbiter must be almost there in terms of additional propulsive maneuvers required to accomplish rendezvous. This means that the Shuttle must be launched into an orbit that is nearly coplanar with the orbit of the target satellite (e.g.; the two vehicles must have the same orbital inclination and approximately the same ascending node). For the SMM rescue mission, the Shuttle was launched into a lower (chaser) orbit with initial apogee and perigee of 251 nmi and 115 nmi, respectively (3). The apogee and perigee of the SMM were 268 nmi and 266 nmi, respectively (3).

OVERVIEW OF SHUTTLE FLIGHT SYSTEM

Propulsion System

The initial orbital trajectory of the Shuttle does not intercept with the target satellite. For this reason, the orbiter must perform a series of propulsive maneuvers to match its position and velocity with the target. There are two subsystems for performing the necessary maneuvers: the orbital maneuvering subsystem (OMS) and the reaction control subsystem (RCS). See Figure 1. The OMS consists of two gimbaled bipropellant engines located at the rear of the Shuttle. Each is rated at 6,000 pounds of thrust. These engines are used for any orbital maneuvers that require more than a few feet per second of velocity change. The RCS consists of 44 nongimbaled bipropellant engines located in both the nose and tail sections of the orbiter. These engines are oriented to provide thrust in all three axes. Of these engines, 38 make up the primary engines that deliver 870 pounds of thrust each and are used for both attitude control and small translation maneuvers (typically, less than 6 fps). The remaining six RCS engines are of the vernier type and are used only for fine attitude control.

Relative Navigation

For the nominal rendezvous case, the orbiter makes use of three relative navigation sensors. These are the star trackers (ST), rendezvous radar (RR), and crewman optical alignment sight (COAS). The ST is an image dissector electro-optical tracking device that operates at visual wavelengths. The primary function of the star trackers is to track stars for alignment of the inertial measurement units (IMU's); however, they are also used to track orbiting satellites from the orbiter during rendezvous operations. The RR is actually a radar function of the Ku-band communication system whose primary use is in data transmission through the tracking data relay satellite (TDRS). COAS is an optical sighting device with a reticule. Its primary use is as a backup instrument for IMU alignment. During rendezvous operations, the crew uses it to view the target. External views of these sensors are illustrated in Figure 1. During the rendezvous sequence the primary role of the star trackers, radar, and COAS is to provide relative navigation data to the crew and general-purpose computers (GPC's) of the Shuttle flight system to reduce the errors in estimates of the position and velocity of the Shuttle vis-a-vis the target satellite. These sensors are used during the middle and later phases of the rendezvous trajectory. The choice of which relative navigation sensor to use is based on such factors as distance to the target and lighting (if any) on the target. RR is the most useful relative navigation sensor since it provides range and range rate data as well as line-of-sight data, whereas ST and COAS provide only line-of-sight data; however, the useful range for the RR is only about 17 nmi for the SMM. For a target such as the SMM, the ST has a maximum useful range of about 220 nmi provided the target is reflecting sufficient sunlight in the direction of the orbiter (4). Because of the optical sensitivity of the ST it cannot be used at ranges below about 4 nmi or when the line-of-sight to the sun, moon, or earth limb is within a specified angle (5). The maximum useful range of the COAS has not been determined; however, the crew of the SMM rescue mission reported sighting the SMM with the COAS at a range of about 100 nmi (6). Figure 2 illustrates the crew member's use of this sensor at the aft work station. The COAS is an important tool for sighting and identifying the target.

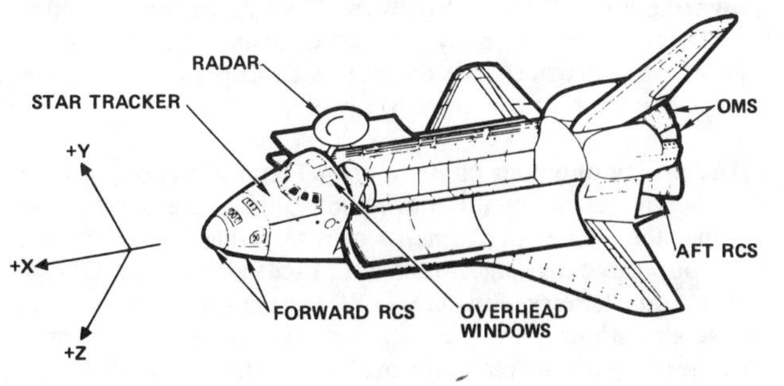

Figure 1. Space Shuttle Orbiter

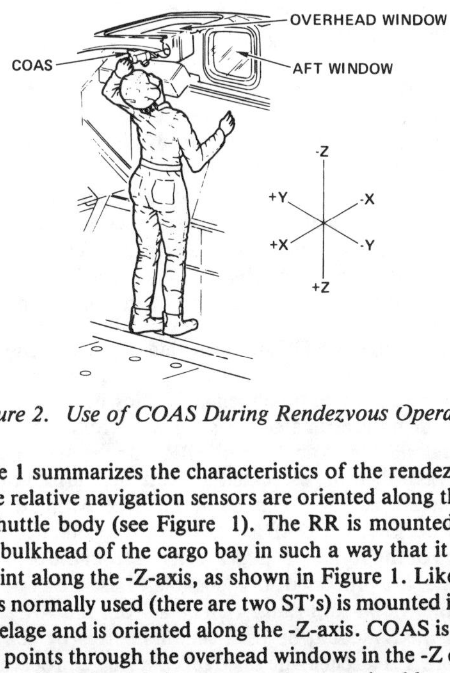

Figure 2. Use of COAS During Rendezvous Operations

Table 1 summarizes the characteristics of the rendezvous sensors. The relative navigation sensors are oriented along the -Z-axis of the Shuttle body (see Figure 1). The RR is mounted near the forward bulkhead of the cargo bay in such a way that it can most easily point along the -Z-axis, as shown in Figure 1. Likewise, the ST that is normally used (there are two ST's) is mounted in the forward fuselage and is oriented along the -Z-axis. COAS is mounted so that it points through the overhead windows in the -Z direction. With this orientation of the sensors, the crew is able to visually verify through the overhead windows the acquisition of the target by the RR and ST sensors by using the COAS or the unaided eye.

Table 1. Rendezvous Sensor Characteristics (SMM target)

	Data Types			Range Limits (nmi)	
Sensor	**Angle**	**Range**	**Range Rate**	**Maximum**	**Minimum**
ST	Yes	No	No	220	4
RR	Yes	Yes	Yes	17	0.01
COAS	Yes	No	No	100?	0.01

Rendezvous Navigation and Targeting Software

As stated in the Introduction, to complete the rendezvous the Shuttle system must determine its navigation state and calculate the velocity change maneuvers necessary to match position and velocity with the target satellite. Although the relative navigation sensors have provided information about the target (line-of-sight direction, range, and range rate), these data must be processed by the GPC navigation software to improve the state vector estimates. During rendezvous navigation processing, the software system cyclically propagates the inertial state vectors of both vehicles and the relative state vector. Although there are errors in the estimates for both states, the navigation system assumes that the state vector for one of the vehicles is perfectly known. Because the SMM had been tracked for several years, its state was known more accurately than the orbiter's. Thus for the SMM rescue mission, the rendezvous navigation software was configured to update the orbiter state vector with respect to the SMM state vector.

It is possible for the relative navigation instruments to accidentally track the wrong object such as space debris or (in the case of the ST) a star. Because of the imperative of maintaining the integrity of the estimate of the updated state vector, two copies of the state are maintained by the navigation system during rendezvous navigation: one state is numerically propagated and filtered (called the FLTR state) based on sensor data and one state is numerically propagated only (called the PROP state).

The Kalman filter uses the relative navigation sensor data to operate on one of the vehicles' state vectors to reduce the estimated error in that state. The Kalman filter can use data from whichever relative navigation sensor has been selected by the crew. The Kalman filter estimates up to 13 sources of navigation measurement error (defined to be the expected 1-sigma error). These sources include vehicle position error, vehicle velocity error, and unmodeled acceleration biases. The remaining sources of error are caused by the navigation sensor measurements.

There are range (RNG) and range rate ($\dot{R}$) measurement errors for the RR. The angle measurement errors processed by the Kalman filter depend on which sensor is being used. If the RR sensor is used, then the estimated measurement errors of the elevation (EL) and azimuth (AZ) angles of the RR antenna are processed. If the ST is used, then the estimated measurement errors of the vertical (V) and horizontal (H) angles of the target within its field-of-view are processed. If COAS is used, then the estimated measurement errors of the X- and Y-angles (derived from the orbiter's current attitude) are processed. It is possible to use a mix of sensors. For example, range and range rate data from the RR and angle data from the ST could be used simultaneously by the Kalman filter. Figure 3 is a schematic diagram of a single-processing cycle of the rendezvous navigation logic for the case in which the Shuttle state vector is the operant. As can be seen by referring to the figure, the inputs to the filter system are the relative navigation sensor data and the state vector estimates for both vehicles. If the relative navigation measurement is accepted, then the output is the filtered (and hopefully, improved) estimate of the state vector. This improved estimate of the Shuttle state is output to the rendezvous targeting and guidance logic along with the target state. Because of the possibility of corruption of the Shuttle state vector estimate by bad sensor data, the crew has the capability to overwrite the filtered (FLTR) state with the propagated (PROP) state. Likewise, the crew can overwrite the propagated state with the filtered state. The crew can also select which of the two states to output to the other software users.

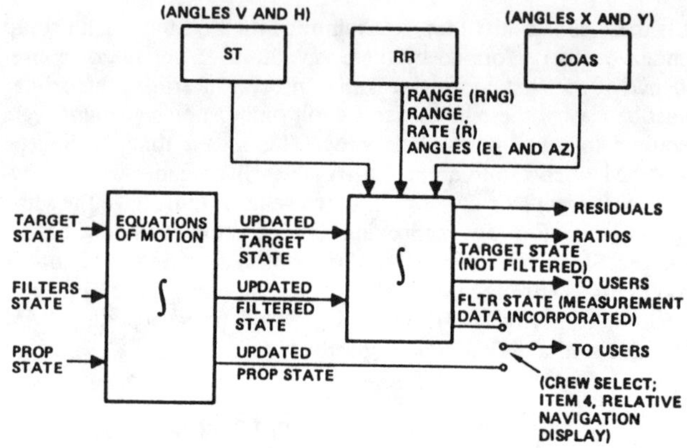

Figure 3. Rendezvous Navigation Schematic

In order to determine the quality of the relative navigation sensor data and the filtering process, the system also outputs the sensor measurement residuals and ratios. A residual in this sense is the difference between the actual parameter measurement (range and range rate or angles) and the measurement expected by the navigation system. For example, if the RR is the relative navigation sensor selected, then one measurement parameter would be the range to the target. The residual for the RR range would be the difference between the range measured by the radar and the range computed from the state vector at that time prior to incorporating that range measurement. The ratio is simply the absolute value of the measurement residual divided by the square root of a scaled value of the residual variance for that measurement type. It is thus a fraction of the maximum allowable residual. These measurement residuals and ratios are displayed to the crew and are used to prevent the filter from using bad sensor data and to judge the quality of the filtered state estimate. When authorized by the crew, the rendezvous navigation system will automatically incorporate any sensor measurement whose ratio is less than one.

Although the Shuttle flight system is able to estimate the positions and velocities of both vehicles during rendezvous navigation, it must also target the necessary velocity change maneuvers and guide the Shuttle through those maneuvers. As used here, to target a maneuver means to determine, via software algorithm, the required ΔV of the maneuver and the time at which it should occur. The task of targeting velocity change maneuvers to effect rendezvous is inherently dependent on knowledge of the current and predicted states of the two vehicles, i.e., rendezvous navigation data. Because of the limited range of the relative navigation sensors, the orbiter is not able to track the target vehicle during the early phase of the orbital rendezvous. This means that the best estimate of the vehicles inertial and relative states must be obtained through the ground-tracking network. Because of this limitation of the flight system, the early rendezvous maneuvers are targeted on ground-based mainframe computers and uplinked to the orbiter for execution of the maneuver.

The semiautonomous phase of orbital rendezvous begins after the orbiter moves into relative navigation range of the target. After this time, the orbiter performs its own targeting. Two different rendezvous targeting algorithms are used within the GPC's: Lambert and Clohessy-Wiltshire (C-W) targeting algorithms. Both of these algorithms solve the Gaussian time-of-flight problem in which the software calculates the orbiter velocity change that is required to transfer from one orbital position to another in a specified time. Each of these algorithms has advantages and disadvantages: the Lambert algorithm performs this calculation and cyclically updates its solution during the maneuver. This cyclic update increases the accuracy of the maneuver execution; however, the Lambert targeting algorithm is unable to generate a converged solution for maneuvers whose orbital transfer takes more than one orbital revolution. The C-W algorithm does allow for orbital transfers of more than one revolution; however, it does not update its solution during the maneuver. The Lambert algorithm has one other advantage that is directly relevant to flight operations. Because it continuously updates the maneuver solution, the Lambert algorithm allows a maneuver to be performed late. This is useful in a situation where the crew workload does not permit a maneuver to be performed on time. Figure 4 is a schematic diagram of the rendezvous navigation, targeting, and guidance logic. The targeting software can accommodate up to 40 separate rendezvous maneuvers. Each on-board targeted rendezvous maneuver is specified prior to the mission as a series of parameters in the targeting software. These parameters are:

1. Time of engine ignition (TIG)

2. Transfer time to the next maneuver (ΔT)

3. Elevation angle of the target (EL)

4. Position with respect to the target at the end of the transfer period (ΔX, ΔY, and ΔZ)

5. Lambert mode (yes or no)

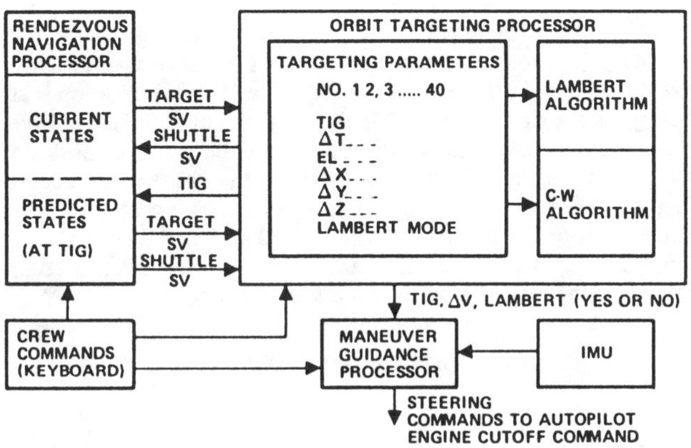

Figure 4. Rendezvous Navigation, Targeting, and Guidance Schematic

The time at which the maneuver is to begin can be set either by the TIG or EL parameters. If the EL parameter is used, then the system will set the maneuver start time to the predicted time at which the line-of-sight elevation of the target is at this value. Regardless of which algorithm is used, the output (ΔV and time of engine ignition) is sent directly to the maneuver guidance processor as shown in the figure. The output of the maneuver guidance processor is steering commands to the autopilot and the time of engine cutoff.

Displays and Controls

The presence of man in the Shuttle flight system greatly enhances its operational versatility. In order to exploit the mission capabilities provided by human abilities, the crew must be able to interact with the Shuttle system. This interaction occurs through the cockpit displays and controls. The crew has both a forward cockpit and an aft crew work station for conducting flight operations during rendezvous. Since the target is beyond sensor range during the early phase of the orbital rendezvous, flight operations are performed at the forward station during this phase. The panel controls for the Ku-band RR are located at the aft station. Because of the need to operate the COAS and the RR, most crew tasks are conducted at the aft station during the middle and late phases of orbital rendezvous. The aft work station is shown in Figure 5. The operation of the rendezvous displays and controls is designed to enable the crew to do four things:

1. Control the relative navigation sensors

2. Control navigation and targeting functions

3. Perform vehicle maneuvers

4. Monitor the approach trajectory

These crew functions are accomplished through a variety at cockpit switches, hand controls, and cathode ray tubes (CRT) displays, as shown in Figure 6. Of the hand controls, there are two types: the rotation hand controller (RHC) and the translation hand controller (THC). The RHC is used by the commander to manually control the attitude of the Shuttle. The THC enables manual control of small velocity changes using the RCS; the Shuttle can be thrusted in any axis using the THC. The keyboard unit is used by the crew to enter commands into the CRT display for controlling GPC functions. Keyboard commands are either selections of numbered display items on the CRT or data entries. The CRT's are important instruments of control. Through them, the crew is able to automatically perform each of the four tasks enumerated previously. To illustrate this, the CRT displays used by the crew for rendezvous navigation and targeting are shown in Figure 6 (display parameters not directly related to rendezvous operations have been "X'ed" out for clarity). These displays are the relative navigation display (Figure 7) and the orbit targeting display (Figure 8). The following is a brief description of some of the features of these displays.

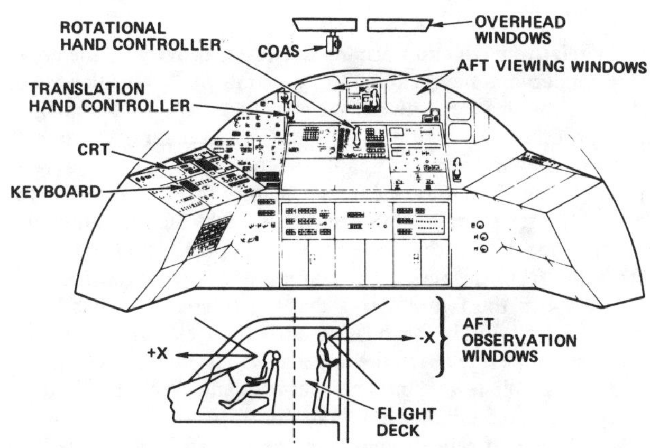

Figure 5. Aft Crew Work Station

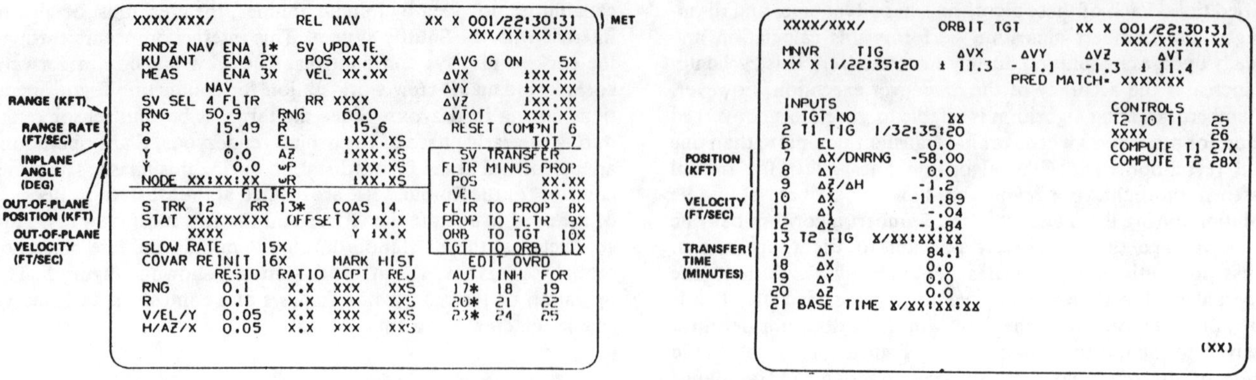

Figure 6. Crew Displays for Rendezvous Navigation and Targeting

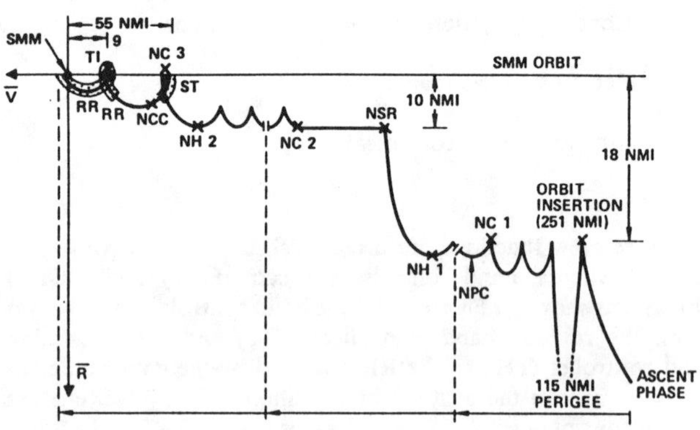

Figure 7. Far-Field Rendezvous Profile

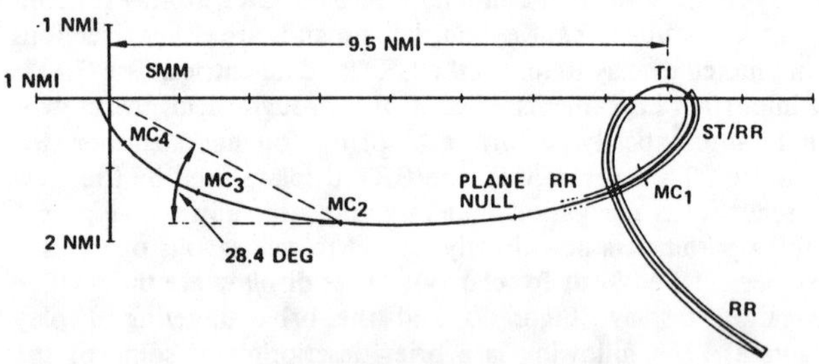

Figure 8. Near-Field Rendezvous Profile: Radar Acquisition to $+\bar{V}$ Stationkeeping

The relative navigation display is used extensively by the crew during rendezvous navigation. The function of this display is to enable communication between the crew member and the navigation algorithms that reside in the GPC's. Execution of Item 1 allows the crew member to enable two body rendezvous navigation processing. Execution of Item 4 enables the crew member to select the filtered state that has been modified by incorporating the target tracking data. Reexecution selects the propagated state that is protected from the results of incorporating the target tracking data. Item 4 in the figure shows that the filtered state has been selected. Below this is the relative navigation estimate of range (RNG) and range rate ($\dot{R}$) to the target as well its line-of-sight angle in the orbital plane (θ) and its out-of-plane position (Y) and velocity ($\dot{Y}$). In the middle column of this display is the readout of the RR estimate of range, range rate, azimuth (AZ) and elevation (EL) of the target satellite. By selecting Items 12, 13, or 14 of this display, the crew member is able to select the desired relative navigation sensor (ST, RR, or COAS) for angle tracking of the target. (The COAS can be used as a backup to the ST but its primary use is as a visual aid to the crew.) The crew member can then monitor the residuals (RESID) and ratios (RATIO) for each measurement (RNG, $\dot{R}$, or angles) and thereby judge whether to allow incorporation of this relative navigation data into the Shuttle state vector estimate. The crew can enable, disable, or force navigation measurements into the Kalman filter for range, range rate, or any angle measurement. Execution of any of Items 18, 21, or 24 will inhibit data incorporation. Execution of any of Items 17, 20, or 23 will enable automatic data incorporation. Finally, execution of any of Items 19, 22, or 25 will force sensor data into the filter under relaxed tests for reasonableness.

The orbit targeting display calculates the change in Shuttle velocity (ΔV) and maneuver start time (TIG) necessary to complete maneuvers in the rendezvous sequence. This information is at the top of the display. Item 2 is the time at which the propulsive maneuver is to start. Items 7 through 12 are the predicted position and velocity of the orbiter with respect to the target at the beginning of the maneuver. Item 13 is the time at which the next maneuver is to begin. Items 17 through 20 are the desired transfer time (ΔT) to the next maneuver and position (ΔX, ΔY, and ΔZ) relative to the target at the end the transfer time.

RENDEZVOUS FLIGHT OPERATIONS

To understand the process that culminates in a successful Shuttle rendezvous one needs to examine the flight operations for a particular rendezvous mission such as the SMM rescue of Shuttle flight 41C. An additional objective of flight 41C was deploying a satellite payload known as LDEF. This task was a factor in the rendezvous operations only in that it required the Shuttle to spend additional time in phasing with the target. This discussion will ignore the LDEF-related flight operations and concentrate on rendezvous flight operations. Figures 7 and 8 show the trajectory of the Shuttle during the orbital rendezvous with the SMM. This trajectory plot is in the target centered coordinate system (5). This is a rotating coordinate system with the target centered at the origin. It is a two-dimensional coordinate system in the orbital plane of the target satellite. The horizontal axis is called the $\bar{V}$ axis. It points along the velocity vector of the target. The vertical axis is the $\bar{R}$ axis. It points toward the center of the earth.

The trajectory shown in the figure is a stable orbit rendezvous profile. The principal characteristic of this trajectory profile is that it allows a comparatively slow approach to the target; thereby, giving more time for relative navigation. Table 2 describes the maneuvers performed by the Shuttle during the orbital rendezvous. As can be seen by referring to the table, 13 maneuvers were performed; every maneuver through NC-3 was targeted on the ground. All subsequent maneuvers were targeted on-board.

Early Rendezvous Phase

At orbital insertion, the Shuttle orbit was nearly coplaner with the SMM orbit. As stated earlier, the initial Shuttle orbit was fairly elliptic with a 251 nmi apogee and a 115 nmi perigee. Following orbit insertion, the Shuttle phased with the SMM for a period of three orbits as can be seen by referring to the figure. This phasing period allowed time for ground tracking of the Shuttle. The purpose of the first four maneuvers (called NC-1, NPC, NH1, and NSR) was to place the Shuttle into an orbit coelliptic with the SMM and 10 nmi below it with the proper phasing. During this phase of the rendezvous profile, the maneuvers performed were based on ground-tracking data. The NC-2 maneuver occurred about 8 hours after completion of the coelliptic phase. It was a pure posigrade maneuver using the RCS. Its purpose was to adjust the phase rate in order to set up lighting conditions for the first relative navigation period with the star tracker.

Table 2. Rendezvous Maneuvers

	MANEUVER	TIG (MET)	ΔV X	ΔV Y	ΔV Z	Engine Select	HA	HP
Ground Targeted	Insertion						251	115
	NC-1	0/5:17:58	11.2	0	0	R OMS	252	121
	NPC	0/6:57:35	0	9.3	0	L OMS	252	121
	NH-1	0/22:50:09	11.7	0	0.1	R OMS	258	122
	NSR	0/23:30:53	239.5	0	15.4	B OMS	258	255
	NC-2	1/7:40:02	1.5	0	0	RCS	259	256
	NH-2	1/20:13:38	14.2	0	0.1	R OMS	267	258
	NC-3	1/21:00:51	0.3	0	0	RCS	267	258
Shuttle Targeted	NCC	1/21:36:13	-0.5	-0.2	1.8	RCS	268	258
	TI	1/22:35:20	11.3	-1.1	-1.3	L OMS	268	264
	MC-1	1/23:01:26	-0.21	-0.07	0.35	RCS	268	264
	MC-2	1/23:30:35	-0.7	0	1.4	RCS	268	264
	MC-3	1/23:40:55	-1	0	-0.2	RCS	268	264
	MC-4	1/23:50:55	0.1	0	-0.4	RCS	268	264

HA = Apogee
HP = Perigee

Middle Rendezvous Phase

An SMM state vector based on ground tracking data was uplinked to the Shuttle at 1 day, 19 hours mission-elapsed time (MET). After verifying the good uplink the commander executed Item 1 on the relative navigation display. This began the cyclic processing of both vehicle states for rendezvous navigation and displayed to the crew the software estimate of the Shuttle's position and velocity with respect to the SMM. The Shuttle's -Z-body axis was pointed in the direction of the SMM to allow proper orientation of the sensors. The NH-2 maneuver occurred about 1 hour later. It was a posigrade OMS burn targeted to raise the Shuttle apogee to the altitude of the SMM and to intersect the SMM orbit at 8 nmi uprange after two orbits. This maneuver also slowed the closing rate of the Shuttle so that more time would be available for relative navigation tracking. The NC-3 burn used the RCS and was the last ground targeted maneuver; it was a phasing adjustment designed to set up the correct position offset with respect to the SMM preparatory to tracking the target with the ST. The required ΔV for this maneuver necessitated the use of the RCS; however, if the +X RCS had been used then an attitude maneuver to align the Shuttle thrust vector would have been necessary. This would have required an interruption of the tracking of the SMM and would have inhibited use by the crew of the COAS to sight the SMM. For this reason, an axis-by-axis translation technique using the THC was used to adjust the Shuttle velocity with respect to the SMM without losing sight of the SMM. This is how the NC-3 and all subsequent RCS maneuvers were performed. Following this maneuver, 20 minutes of ST relative navigation tracking was performed beginning at a range of 50 nmi from the SMM. After the ST had detected the SMM, the Kalman filter began processing the angle data from the ST. Residuals for both angles were about 0.1 degree and the ratios were very small. Though the ST residuals and ratios were within acceptable limits, an oscillation in their values was detected (3,6). After verifying that the residuals and ratios were within acceptable limits, the crew authorized automatic incorporation of the ST data into the filtered navigation state by executing Item 23 on the relative navigation display. They then executed Item 4 on the relative navigation display. This selected the filtered-Shuttle state for use by the guidance targeting functions as shown in Figures 5 and 6. This improved the relative state accuracy by several thousand feet in position and several feet per second in velocity (3). Following this ST-tracking pass, the first on-board targeted maneuver was performed. This was the NCC maneuver and it was performed using the RCS. Had the rendezvous trajectory been perfectly nominal, this burn would have been unnecessary; however, the improved state estimate resulting from the ST data showed a slight trajectory deviation. Therefore, a velocity correction burn of about 2 fps was performed. After the NCC maneuver, the RR was powered up to acquire the SMM from the aft flight deck. The RR acquired the SMM at a range of 16.5 nmi. Because of the previously observed oscillations in the ST residuals and ratios, it was decided to use the RR angle data (AZ and EL) for this relative navigation pass. This was done by selecting Item 13 on the relative navigation display. The residuals and ratios for all four RR measurement parameters (RNG, $\dot{R}$, AZ, and EL) were nominal. The RR improved the estimated state by 5,000 feet and 5 fps (3).

Terminal Rendezvous Phase

The initiation of the terminal rendezvous phase began with the terminal initiation (TI) maneuver. Figure 6 shows the relative navigation display as it may have looked during the TI phase of the

SMM rendezvous trajectory (7). Figure 6 shows the orbit targeting display as it may have looked during the TI phase of the SMM rendezvous (7); the time of the TI maneuver is 1/22:35:20 MET. The required ΔV is 11.4, -1.1, and -1.3 fps in the X-, Y-, and Z-axis, respectively. The TI maneuver targeted the Shuttle trajectory below the SMM in an arc that would cross the $\bar{R}$ axis and loop back toward the SMM. This would enable almost continuous tracking of the SMM by the RR following the TI maneuver without any danger of Ku backscatter from the earth. Four midcourse correction (MC) maneuvers were scheduled during this time. These MC maneuvers were performed with the RCS. Their purpose was to correct for any targeting or OMS maneuver errors that might have occurred in the TI maneuver in order to assure intercept of the SMM. During all of the these MC maneuvers, the Shuttle tracked the SMM through the -Z body axis so that the crew could maintain visual as well as radar contact with the SMM. Following the MC-1 maneuver, data from the relative navigation display were used to null the remaining out-of-plane velocity of the Shuttle with respect to the SMM. When the Y-parameter measured zero, this told the commander that the Shuttle was passing through the plane of the SMM orbit. The commander used the aft station THC to manually null the remaining out-of-plane velocity as measured by the $\dot{Y}$ parameter using the RCS. The targeting of the MC-2 maneuver was keyed on an EL to the target of 28.4 degrees. This maneuver was designed to place the Shuttle on a final trajectory such that the line-of-sight velocity toward the SMM would be zero following the MC-4 maneuver. This would enable manual operations to complete the transition to the +V station keeping position. Following the MC-4 maneuver, the commander took manual control of the rendezvous from the aft flight deck. The Shuttle crossed the $\bar{R}$ at a range of 3,100 ft (3). Keeping the SMM centered in the overhead COAS with the RHC, the commander used the THC to reduce the Shuttle closing rate on the SMM using range rate data from the RR. Braking in this way also caused the Shuttle to move forward from the SMM as it moved toward the $\bar{V}$ from the $\bar{R}$. The Shuttle crossed the $\bar{V}$ at a range of 800 feet from the SMM as intended. Because the Shuttle altitude on the $\bar{V}$ is the same as the SMM altitude, it is a stable position with respect to the SMM (neglecting drag effects). This maneuver completed the rendezvous phase of the mission. From this point, the commander used the aft THC to translate the Shuttle back toward the SMM to begin repair operations.

CONCLUSIONS AND FUTURE IMPROVEMENTS

The successful rendezvous with the SMM proved the viability of the Shuttle rendezvous flight system. The demonstration of operational capability to ground target to within relative navigation range and to complete the rendezvous using autonomous guidance and navigation capabilities affirms the Shuttle rendezvous concept. The oscillation in the values of the ST residuals were determined to be caused by the IMU selection criteria of the redundancy management system within the GPC's. This problem will be avoided in future rendezvous missions by deselecting one of the three IMU's during ST operations.

Adding a GPS receiver to the Shuttle fleet is a pending decision. If approved, this would significantly improve on-board estimates of the Shuttle state vector. This would obviate the need for ground-based tracking of the Shuttle. This new Shuttle capability will undoubtedly lead to greater autonomy for the Shuttle in performing rendèzvous missions. This may allow other rendezvous trajectories to be flown that are more optimum in terms of time of flight or propellant consumed. The efficiency of Shuttle rendezvous can be further advanced in any mission in which the target possesses a transponder that is compatible with the Shuttle Ku-band radar.

REFERENCES

1. Gemini Summary Conference. NASA SP-138, Houston, Texas (February 1-2, 1967).

2. STS Operational Flight Rules Annex 41-C(STS-13) Flight. (final revision) JSC-18308.

3. "STS 13 Solar Maximum Mission Rendezvous Reconstruction." Rockwell Internal Letter FSD&P/AFS84-218.

4. Communication with D. Danemiller; McDonnell Douglas Technical Services Co., Houston, Texas.

5. "Rendezvous/Proximity Operations." JSC RNDZ 2102, crew training manual (March 1, 1983).

6. "STS 41-C National Space Transportation Systems Program Mission Report." NASA JSC-19642.

7. STS-13 (41-C) Flight Data Downlist. Data Display and Review (DD&R) room, Rockwell International, Downey, California.

SPACE SHUTTLE ORBITER ENTRY THROUGH LANDING NAVIGATION

James J. Ewell Jr.

MCDONNELL DOUGLAS TECHNICAL SERVICES CO., HOUSTON, TEXAS 77058

ABSTRACT

The Space Shuttle Orbiter navigation system must be capable of determining its position and velocity throughout a variety of operational regimes. The design and operation of the entry through landing navigation system is described as it operates during a nominal end of mission from the orbital coasting phase throughout atmospheric flight and landing.

Design and operation of the Kalman filter is described. Stabilization of the altitude channel prior to acquisition of external measurement data is described. Utilization of the Tactical Air Navigation (TACAN), barometric altimeter, and Microwave Scan Beam Landing System external measurement data is described. A comparison is made between predicted performance and the navigation accuracy observed during flight.

INTRODUCTION

The Space Shuttle Orbiter entry through landing navigation system must maintain accurate knowledge of the Orbiter's position and velocity throughout several flight regimes. This mission phase begins during orbital flight about two hours prior to the deorbit burn, continues through the maneuver into the subsequent atmospheric flight, and down to and including landing and roll-out on the runway.

During the majority of this mission phase a three-state configuration is employed to provide protection against Inertial Measurement Unit (IMU) failures. In this configuration a position and velocity vector is maintained for each of the three IMUs. Integration of the equations of motion and measurement data processing is performed at .26 Hz (1/3.84 sec). During the final portion of this mission phase (approach and landing trajectory), a single state configuration is employed to permit processing of the Microwave Scan Beam Landing System (MSBLS) data at .52 Hz (1/1.92 sec). The criteria for moding from three-state to single-state is availability of valid MSBLS range and azimuth data. The mode to single-state navigation is permanent; no provision is made to mode from single-state navigation to three-state navigation.

During both configurations a Kalman filter is employed to blend IMU and external measurement data to determine the orbiter position and velocity. The sensors available to the filter are the IMU's, Tactical Air Navigation (TACAN), barometric altimeter, and MSBLS. In addition, the capability to update the onboard position and velocity with ground-computed data is provided.

FUNCTIONAL DESIGN

The complete, detailed navigation filter design is documented in References 1 and 2.

Integration of Equations of Motion

A "super G" numerical integrator is employed to integrate the position and velocity (state) vector. This integrator is third order in position and second order in velocity. Coasting vs. accelerated flight is determined by testing the IMU accelerometer data. If coasting flight is determined (IMU-measured acceleration below threshold test value), then sophisticated gravity and attitude-dependent atmospheric drag models are employed in the integration (3). If accelerated flight is indicated then the IMU-measured acceleration and simplified (central force plus first order oblateness) gravity models are employed in the integrator.

Kalman Filter Description

The onboard navigation employs a six element, extended, sequential Kalman filter to blend external measurement data into the estimate of the state vector(s) (1 and 2). The state vector(s) consist of three components of position and three components of velocity in the Aries, Mean of 1950 (M50) inertial coordinate system.

Additive state noise is employed in the propagation of the covariance matrix to account for unmodeled state errors and uncertainty in the propagation of the state. The magnitude of noise is selected depending upon the usage of IMU data for integration of the state.

Measurement data underweighting is employed to account for nonlinear effects and prevent the state error covariance matrix from shrinking more rapidly than the actual state errors. The underweighting is accomplished by testing the sum of the filter position variances against a threshold value. If the estimated position errors are larger than the threshold, the value of the a priori measurement

Reprinted from *Rec. IEEE Position Location and Navigation Symp.*, pp. 110–117, 1982.

variance is increased by 14 percent of the estimated state error before usage in the Kalman filter update equations.

Since a very limited time is available for processing measurement data and accuracy requirements are stringent, the filter has been designed to automatically edit measurement data if the measurement residual exceeds a multiplicative number of sigma values. This feature is included to prevent incorporation of erratic measurement data and the subsequent filter transient response. Edit multipliers are chosen for individual measurement data type.

Except for MSBLS, the orbiter crew can control processing of each measurement source. For each source there is an auto-inhibit-force (AIF) control. In the inhibit mode of operation the filter is precluded from using that data source to update the state vector, although certain display parameters may be computed. In the force mode, that measurement data is employed to update the state vector without consideration of data edit criteria. However, the force mode does not override the test of data good indicators. In the (preferred) auto mode of operation, measurement data are selected for processing per pre-mission determined criteria and employed to update the state vector if that data agrees with the filter's estimate of the measurement to within the limits of the data edit criteria.

No explicit crew controls are provided for MSBLS data processing; the automatic mode of data processing is always employed. During this flight regime little time is available for crew assessment of anomalies, and guidance and flight control are very sensitive to navigation performance.

Measurement Data Processing

While in the three-state configuration (prior to MSBLS availability) the sources of measurement data are the IMUs, TACAN, and barometric altimeter.

In addition to its usage in the integration of equations of motion, IMU acceleration data, an atmosphere model, and vehicle aerodynamic characteristics are employed to compute an estimate of vehicle altitude. This procedure is referred to as drag measurement processing. It is used to bound state error growth in the vertical direction. This data processing begins at a measured vehicle drag acceleration of 11 ft/sec^2 (approximately 235 kft altitude) and is terminated at approximately 85 kft altitude. The initiation criteria is chosen to avoid incorporation of IMU accelerometer instrument error sources and the lower limit is chosen to avoid extremely complex atmospheric variation modeling in the lower atmosphere. The weight applied to this measurement is 10 kft, 1 sigma. The data edit criteria is 8 sigma.

TACAN data consists of range and bearing measurements from the vehicle to a ground station. Reliable reception of these data are expected to occur about 140 kft vehicle altitude, with filter data processing initiated at approximately 130 kft altitude. Processing is terminated at acquisition of MSBLS data or 1500 ft altitude, whichever occurs first. The TACAN range data weight is 600 ft, 1 sigma. The TACAN bearing data weight is .1 radian, 1 sigma. TACAN range data edit criteria is 8 sigma, and TACAN bearing data edit is 1 sigma.

Air Data Transducer Assembly (ADTA) probes are deployed at a vehicle speed of approximately 3500 ft/sec (approximately 110 kft altitude) and the barometric altimeter data is employed to update the state vector starting at 2500 ft/sec (altitude approximately 80 kft). These data are not processed during the transonic (mach-jump) region from vehicle mach 1.6 to mach 1.1 (between approximately 66 kft and 52 kft altitude). The barometric altimeter data weighting is a linear function of altitude above the runway altitude from a minimum value of 16 ft at zero altitude above the runway with a slope equal to 10 percent of altitude above the runway, 1 sigma. The data edit criteria is 8 sigma.

The MSBLS system is comprised of onboard transceivers and two ground stations located very near the runway. One ground station, located approximately 3 kft from the end (threshold) of the runway, provides a measurement of elevation angle to the vehicle. (The measurement is actually the wedge angle of a scanned beam from the ground station and, thus, not a true elevation angle.) The second ground station, located at the far end of the runway, provides a measurement of range (via a transponder) to the vehicle and a measurement of (lateral) azimuth angle (via a scanned beam) to the vehicle. Acquisition of these data are limited by the geometrical volume coverage of the ground station scanned beams as well as onboard software testing to assure that the data being received does actually come from the ground sites associated with the particular runway chosen for landing.

Upon satisfactory acquisition of MSBLS range and azimuth data, the onboard navigation filter modes from the three-state configuration to the single-state configuration. The execution rate is also doubled to once per 1.92 seconds for high speed processing of the MSBLS data throughout the remainder of the trajectory. Should either (or both) of MSBLS range or azimuth data be labelled as invalid for three consecutive filter cycles, the filter will automatically transition back to processing of TACAN and/or barometric altimeter data, but remain in the single-state configuration.

The weight applied to MSBLS range data is 24 ft, 1 sigma. The weight applied to MSBLS angular data is 3 mrad, 1 sigma. The data edit criteria for MSBLS data is 20 sigma.

Unique Design Features

Several unique design features are incorporated into the data filtering process. Most of these features are related to the three-state configuration. Although redundancy management (RM) of multiple data sources has been used for some time, it is believed that the orbiter navigation represents certainly one of the first times that RM techniques have been applied to state vectors and several technical issues have been resolved. Even though the navigation filter maintains three state vectors, the onboard guidance and control functions are supplied with a single state at all times. This single position and velocity vector is obtained via a selection process applied to the three states on a component by component basis. Since each of the individual states are associated with a particular IMU, the IMU status (i.e., good, failed, etc.) is obtained from the IMU RM function and used to indicate the status of each state vector. If all IMUs are indicated as valid the middle value of each of the components of the state vectors is chosen as the selected state. If only two IMUs are indicated as good then the average value of the two state vectors form the selected state vector. If only a single IMU is indicated as good, then the corresponding position and velocity vector comprises the selected state.

Without regard to the current number of good state vectors, there is only a single covariance matrix maintained by the filter. This covariance matrix is propagated by employing the selected state vector, and thus not truly associated with any one of the individual three states although it is employed for measurement updating for all three states. The measurement partials vector is also computed based on the selected state. The measurement gain vector is then computed from the measurement partials vector and the single covariance matrix, and applied to update each of the individual state vectors. The measurement residual, however, is computed for each individual state vector so that a single measurement may cause different updates to the individual states.

The fact that measurement updating is accomplished with a peculiar blend of selected quantities and individual quantities does not seem to have caused any particular performance difficulties, although several theoretical questions about this kind of process still remain unresolved.

Another variation from theoretical consideration lies in the fact that the measurement partials vector computation and the filter covariance matrix are generally not in time synchronization with each other. The covariance matrix is always propagated to fixed time points. The partials vector is computed at the measurement data time (usually a short interval in the past from the time of covariance matrix propagation) but is applied to the covariance matrix to obtain the gain vector without regard to time synchronization.

Due to the extremely accurate nature of the MSBLS data and the expected state errors resulting from prior processing of TACAN and barometric altimeter data, the position-velocity correlation components in the covariance matrix are zeroed before initial processing of MSBLS data. This process permits correction of the state position components without inducing an (initial) artificial transient into the velocity components. Any velocity errors are subsequently corrected when the position-velocity correlation elements of the covariance matrix are reconstructed by the filter.

Known Design Limitations

The only identified design limitation is that large state (position) errors combined with initial TACAN data processing when the vehicle is almost directly above the TACAN ground sites can cause an increase in state error. This is a geometry problem whereby the direction of correction is computed erroneously. This problem is solved with appropriate crew procedures related to initiation of TACAN data processing.

Simulations have shown that the state divergence is a temporary phenomena, and the filter does recover with continued data processing and measurement data geometry change. The magnitude of the state error transient and the convergence time, however, are sufficiently large to compromise crew and vehicle safety criteria.

State Update With Ground-Computed Data

The onboard navigation state vectors may be updated with ground-computed data in three distinct ways. During orbital coasting flight a complete M50 coordinate system position vector, velocity vector and associated time tag may be uplinked during periods of communication with the ground mission control system. The uplinked state vector is propagated forward or backward until time synchronization with current (ongoing) onboard navigation state time is achieved. The propagated vector is then used to reinitialize each of the three state vectors to a common value prior to normal cyclic processing.

The whole vector update capability is commonly used during the mission to correct error growth over long time periods of onboard state propagation. In particular it is the primary technique employed a short time before deorbit burn ignition to provide a highly accurate position and velocity vector for use during and after the maneuver.

Provided that a sufficient ground radar tracking arc exists, a second type of update capability exists during periods of coasting flight. This update capability is called a "delta-T" update, and is employed to correct downtrack position errors. The underlying assumption is that the vehicle trajectory is known and that the onboard position and velocity vector is a solution to the vehicle motion at some time, t, which isn't the value of time tag that the onboard navigation has currently associated with that position and velocity vector. Thus a downtrack position error can be interpreted as a time tag error. The value of time tag adjustment is computed by ground navigation and relayed via voice communication to the crew. The crew then manually enters the adjustment value into the onboard computer.

The delta-T update has been employed in a variety of pre-flight simulations. However the navigation errors in actual flight have been sufficiently small that this update capability has not yet been employed.

The third type of state update capability is provided for use during periods of non-coasting flight, specifically for the post-blackout, low altitude flight regime. This update is called a delta-state update. Position and velocity differences between the ground navigation and onboard navigation are computed in the runway coordinate system (1). If the computed differences are large enough to violate vehicle safety standards, these differences are uplinked to the onboard navigation and added directly to the existing onboard position and velocity vectors. No time tag is associated with this update; the assumption being that the rate of error growth is small enough that any errors remaining after the update is incorporated into the state will be small enough that the solution satisfies safety criteria. The post-update state errors are then reduced via normal onboard measurement data processing.

Like the delta-T update, the delta-state update has not yet been used during actual flight.

During actual flight the processing of TACAN data has proven sufficient to reduce onboard navigation errors to well within acceptability criteria. The only state updates performed during flight have been whole vector updates (pre-deorbit burn) planned as part of the normal mission timeline.

Reliability and Stability of the Filter

The reliability and stability of the onboard navigation filter has proven to be quite remarkable. It has been through several years of repeated testing in an extremely large number of man-in-the-loop and non-real-time nominal, stress and worst case failure mode cases. This filter is neither perfect nor magic; however, the combination of inherent design capabilities, crew procedures, and ground monitor functions result in a high level of reliability. The previously identified design limitation is not a problem unless large onboard failure conditions occur in conjunction with repeated violations of well defined ground controller and onboard crew procedures.

This author has been repeatedly requested to supply test case conditions for which a navigation system failure will occur so that orbiter crews can experience (via simulation) the effects of navigation failure cases. This author has met with little success in supplying such test case inputs when considerations of no more than two failures can be specified (fail-operational, fail-safe coverage) and normal ground and onboard procedures are followed.

PREDICTED AND OBSERVED FLIGHT PERFORMANCE

Figures 1 through 12 present a comparison of predicted and observed performance for shuttle flights 1 through 4 (4). The coordinate system employed is referred to as "earth-relative UVW." The origin is at the (true) vehicle location; the U-axis is along a radius from the center of the earth, positive upward; the W-axis is in the direction of the vehicle's angular momentum vector, relative to earth center; and the V-axis completes the right hand system, in the direction of vehicle motion relative to the surface of the earth.

These data show the generally excellent performance of the on-board navigation system. No unplanned ground updates have been performed. The filter converged to the measurement data as expected.

The importance of TACAN data to the performance of onboard navigation is illustrated in Figure 2 near 130 kft altitude where the predicted performance curves drop sharply to about 2 kft error.

Stabilization of the altitude channel prior to barometric altimeter data processing is illustrated in Figure 1 above approximately 85 kft altitude. The influence of barometric altimeter data processing is shown by the decrease in predicted error from 85 kft to 70 kft altitude.

The importance of MSBLS data processing is exhibited in predicted performance for all state components in Figure 7 through 12. On some flights MSBLS data was not acquired until below the altitude expected from preflight predictions. Thus, flight performance lies outside predicted performance until MSBLS data processing was actually initiated.

Plot Legend

On each plot the heavy line is the preflight predicted performance envelope. It was derived from Monte Carlo simulation studies. The flight number is the label for each of the curves obtained from actual flight data. Where it appears within the plot scale, the straight line segments labelled "G" are the constraints defining acceptable performance boundaries.

FIGURE 1

POSITION ALTITUDE

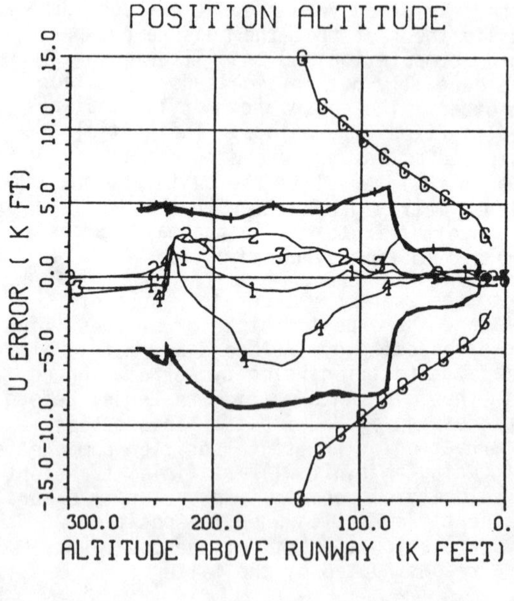

FIGURE 2

POSITION DOWNRANGE

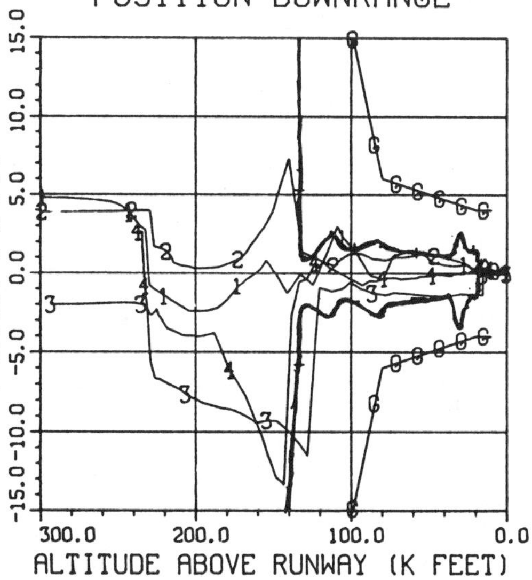

FIGURE 3

POSITION CROSSRANGE

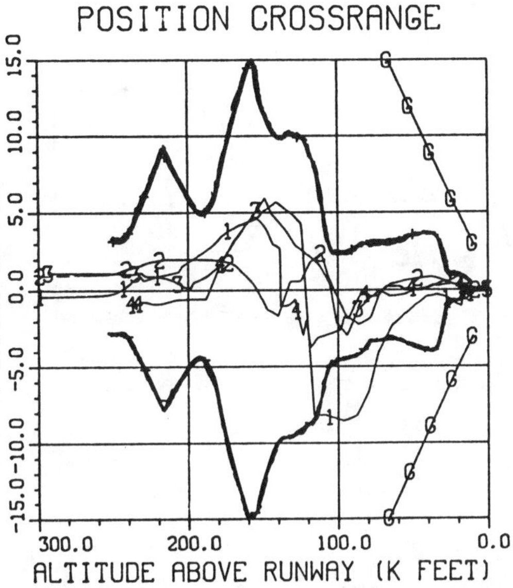

FIGURE 4

VELOCITY ALTITUDE

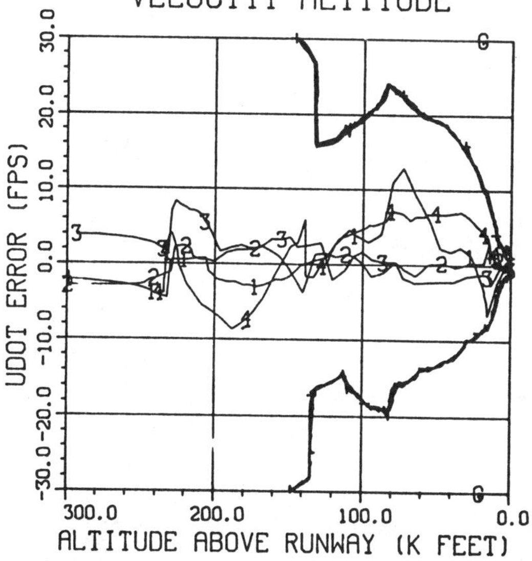

FIGURE 5

VELOCITY DOWNRANGE

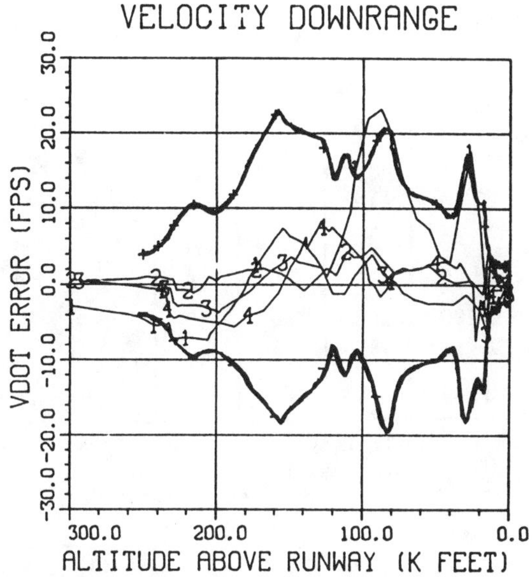

State Vector Accuracy Requirements

Guidance-determined navigation accuracy requirements are presented in Table 1. Although the requirements are expressed only at descrete altitudes, the agreement with guidance experts is that those points are connected with straight lines. These requirements resulted from performance simulations and are considered to be "soft" requirements. If navigation errors remain within these limitations then navigation is considered to be satisfactory; however violation of the requirements does not necessarily imply violations of vehicle safety criteria. An additional constraint imposed upon navigation is that the change in downrange velocity error not exceed 1 ft/sec^2. This requirement, too, is a "soft" requirement. It applies only during the entry guidance phase, from 400 kft altitude to approximately 80 kft altitude, and is construed to exclude transient values.

Anomalies Observed and Corrective Action

On flight 1 a bias of approximately 1.3 degrees was observed in the TACAN bearing data (5). The source of this bias was determined to be a difference in the way in which ground station alignment reference is determined and established and the definition of true north used in both onboard and ground navigation for the shuttle. The existence of this computational difference was documented in February 1981, shortly before flight 1

FIGURE 6
VELOCITY CROSSRANGE

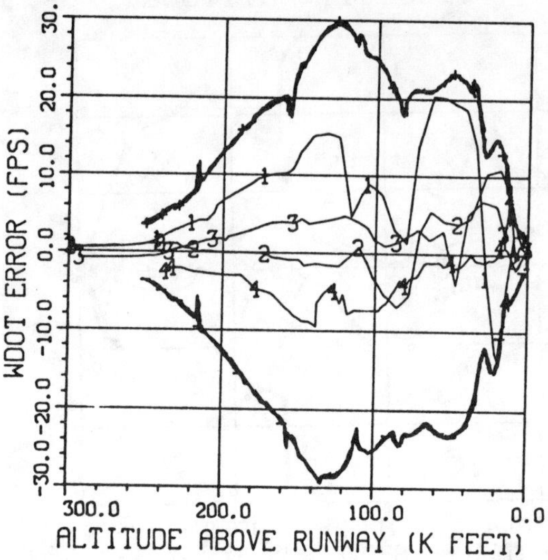

FIGURE 7
POSITION ALTITUDE

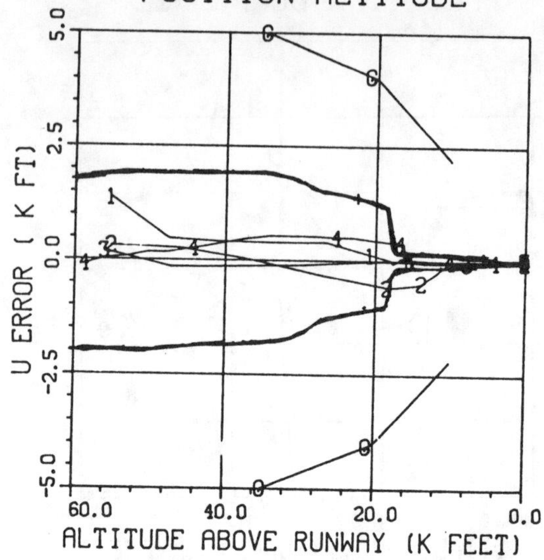

Table 1 - Navigation Performance Constraints

ALT / STATE COMP.	150 Kft	130 Kft	80 Kft	20 Kft	10 Kft	Touch-Down
U*	18	12	8	4	2.2	5
V	90	30	6	4	4	200
W	90	30	18	5	3	30
UDot**	100	100	50	30	30	1.5
VDot	175	175	60	40	40	***
WDot	175	175	70	40	40	1.5

*Position constraints in feet
**Velocity constraints in ft/sec
***No explicit value expressed; must be "reasonable"

FIGURE 8
POSITION DOWNRANGE

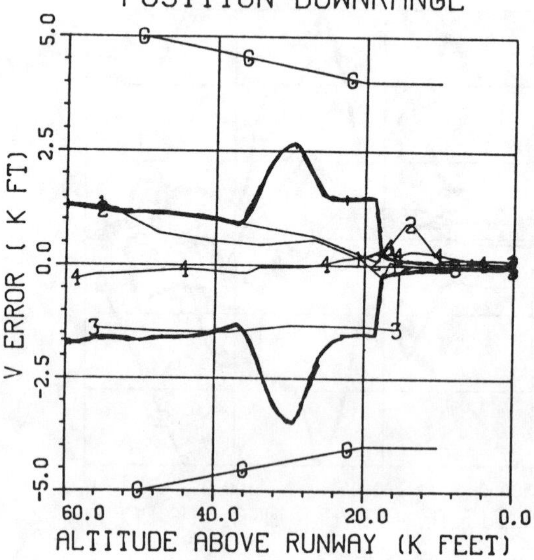

(6). The effect of this bias on the position vector is shown on Figure 3 near 120 kft altitude in the trace for flight 1. The sudden change in the error curve at that point is actually due to the fact that (until flight 3) two values of measurement weight were applied to the TACAN bearing data. At ranges larger than 150 n.m. a weight of .1 radian was used; at ranges less than 150 n.m. a weight of .02 radian was used. The change the error curve shows the filter convergence to the biased data when the .02 radian weight was employed. For the second flight, the TACAN ground station alignment was changed. Beginning with the third flight, the TACAN bearing data is always weighted at .1 radian, 1 sigma.

Although not explicitly evident in this flight data, TACAN bearing data has also exhibited other anomalous behavior (5). Either because of low signal strength or secondary reflection (multipath) reception, at least one of the onboard TACAN transceivers has demonstrated a "40 degree glitch" anomaly on flights 1, 3, and 4. This anomaly is characterized by measurement data which shows errors that are multiples of 40 degrees. In several cases the anomaly lasted long enough that the onboard TACAN redundancy management (RM) function declared a failure in a particular unit although that unit operated properly before and after the anomaly. Flight software changes are developed to specifically check for 40 degree data anomalies.

FIGURE 9

POSITION CROSSRANGE

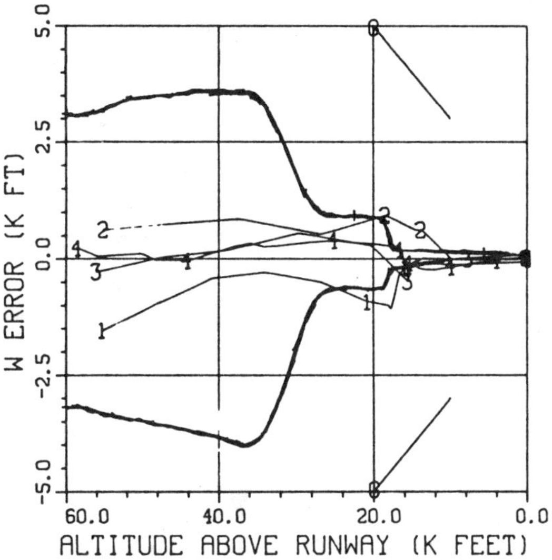

FIGURE 11

VELOCITY DOWNRANGE

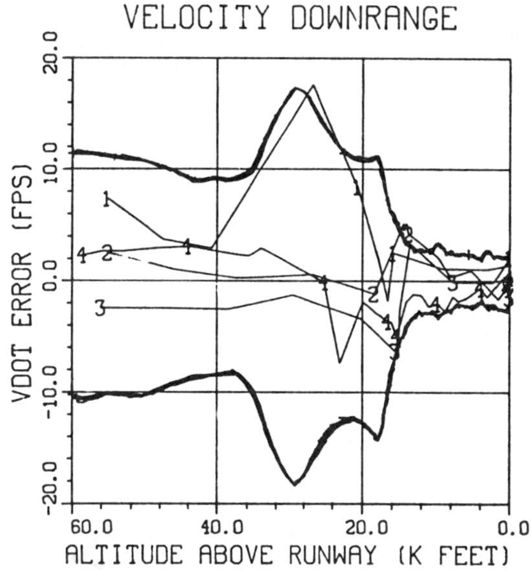

FIGURE 10

VELOCITY ALTITUDE

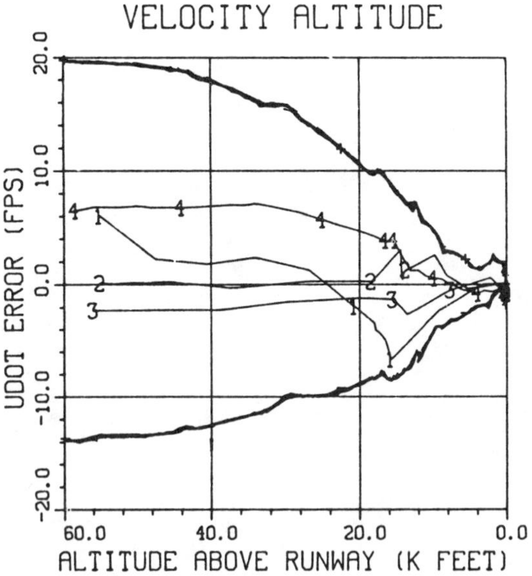

FIGURE 12

VELOCITY CROSSRANGE

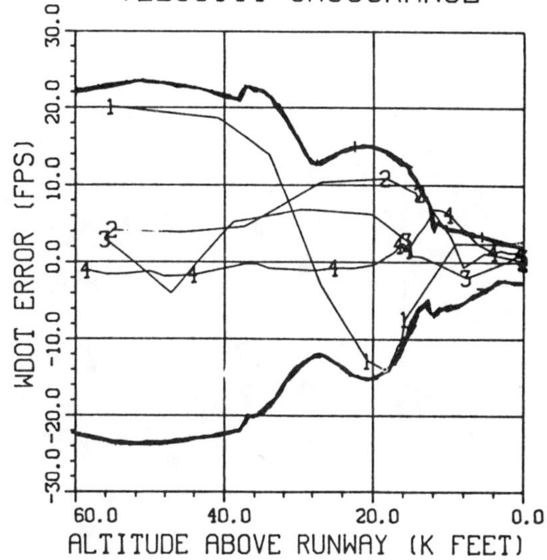

The barometric altitude measurement has also been observed to exhibit anomalous behavior near 85 kft altitude. This behavior has been attributed to the location of the ADTA probes, airflow around the vehicle, and vehicle attitude changes at that time. Current crew procedures are to not permit altimeter data processing until after this point in the trajectory although the filter weighting of the data minimizes any impact to the state vector.

MSBLS azimuth data have shown a variable, bias-type error anomaly very near vehicle touchdown. The filter response to these data is a lateral position error with a peak amplitude of near 50 ft; large enough to exceed the 30 ft navigation state error budget at touchdown. The source of this error has been determined to be a cross-polarization error, and hardware modifications to the MSBLS ground station azimuth antenna are in progress.

MSBLS elevation data also exhibits erratic behavior at low elevation angles due to multipath propagation. An onboard software change has been implemented to declare the data invalid for measurements less than 1.2 degrees.

CONCLUSIONS

The onboard navigation function has been observed to perform in agreement with design objectives and preflight predictions. Excellent performance has been obtained for all flights. Sufficient capability and design flexibility has permitted accommodation of data anomalies with parameter value changes rather than equation alteration.

ACKNOWLEDGEMENT

The work reported here has been supported by the NASA, Johnson Space Center, under contract NAS9-16105.

REFERENCES

1. "Entry Through Landing Navigation FSSR Input," MDTSCO Transmittal Memo 1.4-MPB-1530, J. J. Ewell Jr., R. C. Pocklington, J. K. Bhatt, N. L. Pierce, R. G. Peterson, R. E. Pritchard, 15 September 1978.

2. "Space Shuttle Operational Level C Functional subsystem Software Requirements (FSSR) Document, Guidance, Navigation, and Control, Part B, Entry Through Landing Navigation," STS 81-0004, Rockwell International Shuttle Orbiter Division, 1 May 1981.

3. "Uniform Representation of the Gravitational Potential and Its Derivatives," S. Pines, AIAA Journal, Vol. 11, No. 11, November 1973.

4. "STS-1, 2, 3, 4 Entry State Vector and Preflight Prediction Performance Comparison," MDTSCO 1.4-TM-D1730-728, S. Sanchat, 24 September 1982.

5. "Quick-Look Results for STS-1 Entry Onboard Navigation," MDTSCO 1.4-TM-D1530-28, J. J. Ewell Jr., 22 April 1981.

6. "Space Shuttle TACAN Evaluation for Nominal Entry to Edwards AFB, California," H. J. Leister, HW Air Force Communications Command/FFNM, Scott AFB, Report 80/66S-200, 15 February 1981.

SATELLITE AUTONOMOUS NAVIGATION - STATUS AND HISTORY

M. A. Chory - Senior Staff Engineer, TRW Control Systems Engineering

D. P. Hoffman - Project Manager, TRW Defense Projects Division

J. L. LeMay - President, St. Joseph Sciences, Inc.

ABSTRACT

Autonomy of operations for satellites operating in earth orbit is a technology area which has received increased attention in the last 5-10 years. The advances in onboard computing systems have facilitated implementation of some of these concepts. One aspect of autonomy is highlighted in this paper-that of satellite navigation. Mechanization of this function in a spacecraft design implies the ability to determine position and velocity of the vehicle onboard in real time. Autonomous navigation in its purest sense would have this implementation independent of external sources, however, some variations in this strict definition are possible and indeed may be more practical.

This paper introduces the concept of satellite navigation and highlights some of the key work done in the evolution of the technology. Contemporary designs which have been or are being developed through flight implementations are discussed and their performance capabilities summarized. Fully autonomous systems are assessed as having capabilities in the range of a few kilometers while the GPS (Global Positioning System) which is partially autonomous can attain accuracies of tens of meters or better. Programs such as the Strategic Defense Initiative (SDI) which are considered the current drivers to future navigation needs are also discussed to project activities which should motivate advancement of the state of this art in the near term time frame. Throughout the discussions and descriptions, references from the open literature are cited as aids to those interested in detailed research and to acknowledge previous contributions to this important field.

INTRODUCTION

Autonomous operation of satellites has obvious advantages over the traditional dependence on ground support. Increased space traffic alone suggests that, in time, autonomy at some level will be necessary to reduce the loading of ground resources. There are also the considerations of the investment required to support operations of large numbers of satellites and the geopolitical concerns of maintaining remote tracking stations. All of these contribute impetus to a trend toward satellite autonomy. Finally, certain satellite missions may dictate autonomy due to the environments in which they must operate.

All satellites require some form of navigation. This function lends itself to self-contained operation. In spite of this, however, the development of autonomous navigation for satellites has been sporadic and no operational systems are known to exist today with this function.

What is autonomy for satellite navigation? In the strict sense, a system is autonomous if it is self-contained in the satellite, operates in real time, is nonradiating, and does not depend on ground support. The third of these characteristics and, to some extent, the fourth are motivated by a desire for operation in a hostile environment. Some of the desirable features of autonomous navigation systems identified above are achievable without all of these characteristics. Indeed, such systems are, generally speaking, easier to implement. We will distinguish, in the sequel, between full and partial autonomy with regard to satellite navigation systems.

Our purpose in this paper is to present an overview history and status of satellite autonomous navigation. We have attempted to make the presentation useful in a short space through the use of key references. No claim of completeness can be made, but we are hopeful that the major aspects of this interesting and important subject have been addressed. Through the references, readers should be able to access the additional levels of detail required. The focus is on military activities since these have been major applications drivers. However, it is recognized that NASA also has made major contributions to the overall technology.

A historical summary of satellite autonomous navigation developments is given first. This includes a discussion of most of the viable concepts which have been proposed/developed over the years. Most developments have been in the technology section of government funding.

Following the historical summary, a discussion of representative contemporary systems is given. These are satellite autonomous navigation systems currently in concept/development/operational stages. Performance comparisons are highlighted in this discussion where possible.

The final discussion projects applications and requirements for autonomous navigation. Here, it is noted that SDI holds the potential for being the strong

Reprinted from *Rec. IEEE Position Location and Navigation Symp.*, pp. 110–121, 1986.

application driver previously lacking for this function. It promises to bring focus to the sporadic technology-driven developments of the past two and a half decades and lead to operational flight implementations.

DEVELOPMENT HISTORY

A history of satellite autonomous navigation system development has been determined to a large extent by the technological advancements in the constituent supporting areas. Satellite navigation is the process of determining the position and velocity vectors of a satellite as functions of time. Our interest is in earth-orbiting satellites and the major supporting areas are motion models, navigation measurement types/sensors, mathematical algorithms, and computers.

Motion models for earth-orbiting satellites are characterized by long periods between applications of non-conservative forces. Gravity (including geopotential harmonic terms for precision) is the dominant force and accurate models were quickly developed with early satellite data. Atmospheric drag can be significant at low altitudes but very rapidly becomes negligible and can be ignored at altitudes above about 1000 km. At higher altitudes, especially geosynchronous and beyond, the gravity effects of the sun and moon become significant. Finally, solar radiation pressure can be important depending upon the satellite configuration. On the whole, satellite dynamics are well understood and do not pose implementation limitations at this time.

Navigation measurements for satellite navigation are of two generic types-vector and scalar. It is well known that scalar range tracking data from earth-fixed locations are very effective for satellite orbit determination/navigation. Indeed, these are the mainstay for ground-based systems. Our emphasis here is on autonomous navigation for which ranging data are difficult to realize.

The most easily implementable satellite-based measurements have been found to be of the vector type. This explains the parallel development of inertial attitude reference systems along with satellite navigation. It also explains the great interest in the development of sensors capable of making satellite-based scalar measurements effective for navigation. The determination of measurement types effective for autonomous navigation and the development of sensors capable of making such measurements are quite separate matters. The following discussion will show this distinction.

Navigation algorithms have undergone an interesting historical evolution and one could say that we have arrived at a point we could describe as generic standardization. These developments have closely paralleled the explosive growth in the capability of satellite-borne computers. This is partly coincidental and partly not. The reality is that non-implementable algorithms are almost always set aside until adequate computing capability is available in a practical implementation. To an extent this was the case with Kalman filtering which is now the accepted method of formulation in navigation and attitude reference designs.

The history which follows for satellite autonomous navigation treats the interrelationships among the subsystems involved on an implicit rather than explicit basis. A summary of the historical evolution is shown in Table 1.

Satellite navigation done on an autonomous basis has been with us at some level of development since virtually the beginning of the space age. The appeal of self-contained satellite navigation and a definition of autonomy have been given in the introduction. The first major satellite autonomous navigation program is considered to be the Air Force Program 283 which took place during 1963 and 1964. The contractor was United Aircraft Corporation in Farmington, Connecticut. Technical monitoring was done by The Aerospace Corporation. It was at this time that it became widely appreciated that inertially-referenced horizon scanner data are sufficient for satellite navigation. The Program 283 concept consisted of an attitude reference system (3 strapdown gyros and a gimballed star tracker), a horizon scanner (which effectively measures local-vertical direction), a flight computer, and a clock.

Extensive analysis and simulations were carried out in the 283 program to predict system performance. Accuracy was limited largely by horizon scanner errors (instrument and phenomenology), especially bias errors in the pitch component (in the orbit plane) of the local-vertical direction. Such errors translate directly into navigation errors, one minute of arc giving two kilometers of error at low satellite altitudes. Although never accepted as the formal Air Force position, the consensus at the time was that the 283 system's predicted overall position accuracy was about two kilometers. It was not appreciated until much later that the 283 navigation loop algorithms [1, 2], developed in an ad hoc manner, were in fact a suboptimal version of what one would have obtained using Kalman filtering techniques. This was a preview of things to come.

Program 283 included some component laboratory tests funded by the contractor and the Air Force. Development beyond the laboratory phase did not occur. It was too early for consideration of strapdown star sensors; hence, gimbal readout errors were found significant for the attitude reference design. The strapdown inertial measurement unit (accelerometers were included) was viewed as high-risk because the technology was in its infancy. This is ironic since Program 283 proved to be the pilot for strapdown inertial navigation systems and the continued development by STL (Space Technology Laboratories-now TRW) resulted in it being flown on Apollo missions. The 283 accelerometers were never intended for satellite navigation usage but were included to support a dual-mode use, the other mode was to be boost navigation.

Table 1. Satellite Autonomous Navigation History

DATE (S)	AIR FORCE PROGRAM	MEASUREMENT TYPE (S)	MAJOR DEVELOPMENTS	REFERENCES
1963-64	283	HORIZON SCANNING	NAVIGATION SYSTEM STRAPDOWN IMU TECHNOLOGY	1, 2
1964	SSGS	HORIZON SCANNING; KNOWN/UNKNOWN LANDMARK TRACKING	VOLUMINOUS PERFORMANCE PREDICTIONS	3 - 8
1965	ULTRA	UNKNOWN LANDMARK TRACKING	ULTRA INSTRUMENT (PARTIAL COMPLETION)	NONE
1967-74	SPARS*	STRAPDOWN SLIT STAR SENSORS & STRAPDOWN GYROS	PROTOTYPE FOR DMSP	9
1970	STRETCH	HORIZON SCANNING	MORE PERFORMANCE PREDICTION	AEROSPACE CORP LIBRARY
1970-73	ANT	KNOWN/UNKNOWN LANDMARK TRACKING INTERFEROMETER CONCEPTS	SYSTEM LEVEL REFINEMENTS; PARTIAL SENSOR DEVELOPMENT	10
1973-75	HANS	54 CONFIGURATION COMBINATIONS	PERFORMANCE PREDICTIONS; COMPARISON; SEXTANT, PASSIVE RANGING RECOMMENDATIONS	11
1976-79	HAARS*	STRAPDOWN MOSAIC (INCLUDING CCD) STAR SENSORS & STRAPDOWN GYROS	STRAPDOWN STAR SENSOR DESIGN SYSTEMS CONCEPT WITH 3 STAR SENSORS WITH AND WITHOUT STRAPDOWN GYROS	16
1977-81	ANARS	SEXTANT	FLIGHT TEST UNIT; ON-ORBIT PERFORMANCE	12 - 15
1979-85	MADAN	HAARS ATTITUDE SYSTEM, HORIZON SCANNING/OTHER NAVIGATION SENSORS	MADAN NAVIGATION SYSTEM DEVELOPMENT	17, 18

*ATTITUDE REFERENCE SYSTEMS

Program 283 ended, but interest in low-altitude satellite autonomous navigation continued. Measurement types which respect the strict definition of autonomy proved to be scant indeed for near-earth satellites. In addition to the earth limb, earth fixed landmarks were convenient reference points and interest quickly focused in this direction. The contract vehicle for these studies was the Air Force Standardized Space Guidance System (SSGS) which took place in 1964.

The SSGS program was broader in scope but had autonomous navigation for satellites in near earth orbits as its centerpiece. Four contractors were involved: IBM, STL, North American, and Sperry. Basic horizon-scanner systems were much in evidence in these studies but accuracy-limited. The culprit was, and is, the basic limitation in earth-limb definition. The best wavelength for good definition is 14-16 microns [3] but the definition is insufficient for some applications. The concepts at the time used the horizon scanner system for autonomous navigation as a reliable but not-so-accurate sure thing. The rationale was that horizon scanners were onboard for other purposes as often were inertial attitude reference packages. For better navigation accuracy, other measurements were needed and the emphasis was on earth landmark tracking.

Two categories quickly emerged-known landmark tracking and unknown landmark tracking. As the names imply, the former refers to the use for navigation of earth-fixed reference points of known location while the latter does not require that such reference points be of a priori known location but only earth-fixed. In either case a landmark sensor is required to provide measurements of the pointing vector between the satellite and the landmark. If such a measurement vector is inertially referenced (with an attitude reference package for example) then satellite navigation can be accomplished. As an aside, concepts which utilize earth features which are not literally points may also be included in this general concept.

For a time, unknown landmark tracking was a concept with some controversy. The basic observability of such a system is not apparent and a geometrical demonstration of the process proved elusive. To this day a simple demonstration is not known to exist. The closest is a theorem out of MIT at the time of SSGS which supports observability of some but not all of the satellite states. The Air Force made requests for simple observability demonstrations but no one produced a simple, yet convincing argument. However, the developments at North American proved to be sound. In fact, as time passed, it became apparent that known and unknown landmark tracking are quite similar. For example, one can use coarse initial satellite position to initialize landmark coordinates. Then, both satellite navigation states and landmark coordinates are estimated in the navigation algorithm. A real issue presents itself in this context. The continual addition and deletion of landmark coordinates in a navigation state vector is an implementation complexity at best and an unacceptable suboptimality at worse.

What emerged from SSGS with regard to satellite autonomous navigation were three basic types of sensing: horizon scanning, known landmark tracking, and unknown landmark tracking. The use of atmospheric refraction effects on stars is viewed as a variation of horizon scanning. At this time algorithms using Kalman filtering became standard. Emphasis was on low-altitude applications and no serious consideration was given to lunar sensing concepts or passive ranging techniques. Hence, the systems considered were of the vector measurement type and this generated interest in inertial attitude reference systems. Onboard computers were experiencing exponential growth but were not yet able to easily incorporate the navigation algorithms. As for accuracy, that seemed to be in the eyes of the beholders, but the general range of performance error was 0.2-2.0 kilometers. Technical papers on SSGS-type navigation systems soon appeared [4-8].

SSGS was followed in 1965 by a small Air Force program called ULTRA (Unknown Landmark Tracker). The contractor was North American and the goals were to develop the unknown landmark concept more completely and to produce a prototype sensor. The program was short-lived and the algorithms were not fully developed. A planned aircraft test with ULTRA never took place and ULTRA went on the shelf.

At about this time (1965) there was a significant amount of effort industry-wide on satellite autonomous navigation studies. These emphasized horizon scanning and known/unknown landmark tracking systems. In the meantime, the Air Force had begun program 681D-Advanced Space Guidance. This program actually was to carry satellite autonomous navigation activities from approximately 1965 to 1974. Interest by the Air Force in autonomous navigation significantly waned in the late 60's and emphasis shifted to the SPARS (Space Precision Attitude Reference System) portion of Program 681D. SPARS had competitive multiple-contractor programmatic phases studying various attitude determination concepts. The Lockheed/Honeywell team emerged as the developer of a system using strapdown gyroscopes and strapdown slit-type star sensors. [9]. The program was an unqualified success and the SPARS was subsequently flown on the DMSP (Defense Meteorological Satellite Program).

In 1970 the Air Force funded an internal program at the Aerospace Corp. called STRETCH (Space Technology Requirements and Engineering Tests of Component Hardware). STRETCH included many activities on attitude determination and included more definitive work on horizon scanning satellite autonomous navigation systems. STRETCH produced a 10 volume final report and several symposia on the results [accessible from the Aerospace Corp. library under STRETCH title].

After STRETCH, the Air Force had a revitalized interest in satellite autonomous navigation and the Autonomous Navigation Technology (ANT) program was begun in 1970. Competitive contractor efforts continued for three years. Again the systems were horizon and landmark sensing but interferometric techniques for landmark tracking were also introduced [10]. This technology program improved understanding of these systems and included some laboratory demonstrations of subsystems. The results were compared with Air Force-provided standard trajectories for blind Monte Carlo comparisons. What emerged was a better understanding at the system level; however the landmark tracking sensor developments were not completely successful. On the other hand, horizon scanning systems were, in fact, a reality. Horizon scanner data combined with SPARS data, for example, could produce a navigation system. All that was needed was to combine these data in an on-board computer.

For mission and other reasons, an interest emerged in the Air Force for satellite autonomous navigation at high satellite altitudes (synchronous and beyond). An extensive study called HANS (High Altitude Navigation Study) was conducted by The Aerospace Corporation in early 1973 under Air Force direction to examine all conceivable configurations (a total of 54) of navigation systems for high altitude satellites [11]. This study was briefed to industry and set the stage for significant future contractual developments. Several recommendations from the study were made including the development of passive ranging techniques as well as the development of a space sextant. Both provided scalar measurement data not dependent on attitude reference systems. Competitive contractor efforts followed during 1973-75. Passive ranging proved illusive but the sextant development lead to a major new program. The program was designated ANARS (Autonomous Navigation and Attitude Reference System) led by the space sextant development by the Martin Company during 1977-81 [12-15]. Because of the implementation concept (gyros were included), the system could also be used in an attitude reference mode. This was not, however, the thrust of the sextant program. The sextant program was a satellite autonomous navigation concept with a wide range of altitude applications since it used the lunar limb as a reference. The space sextant has been developed and flight tested and has established the soundness of sextant data for navigation [15].

In the meantime, studies were made of various attitude reference systems at high altitudes. This program was called HAARS (High Altitude Attitude Reference System), and took place during 1976-79. HAARS did not have new startling concepts but did introduce the use of mosaic sensors for stellar data [16]. This was significant in that the idea of strapdown star sensors with wide field of views gave the advantage of strapdown combined with vector data. This was different, for example, from the slit crossing device used in SPARS.

CCD (Charged Coupled Device) technology was progressing to the point where its use could be exploited for attitude determination in wide-field-of-view strapdown star trackers. Several sensors could provide satellite attitude accurately without gimbals. In fact, the field of views involved were sufficiently large that a single device could provide three axes of inertial attitude knowledge.

The HAARS attitude reference system using CCD sensors developed into an Air Force-sponsored navigation system called MADAN (Multimission Attitude Determination/Autonomous Navigation) by adding usually-available earth horizon scanning data. MADAN concepts emphasized applications at high altitudes but could be used at any altitude and moreover left the door open for the use of other navigation sensors instead of, or in conjunction with, horizon scanners. MADAN development was conducted by TRW [17,18] during 1979-85.

CURRENT ACTIVITIES

The contributions and activities highlighted in the preceding section have formed the basis for work currently being done in the area of satellite navigation. This section summarizes the key efforts which comprise the current system level work as known to the authors. In some measure, this takes the form of an update to the survey paper prepared in 1984 [19]. Specific descriptive detail of contemporary implementations is included primarily through use of extensive references in these discussions.

Since the preparation of [19] there have been three major changes affecting satellite navigation. The first is the advent of the Strategic Defense Initiative (SDI) research as a driving force in the applications area. Second is the de-emphasis and phase out of the contractual efforts on the Space Sextant technology demonstration and MADAN programs. Lastly is a great body of work which has appeared in the open literature on the Global Positioning System (GPS) now also being referred to under the program name Navstar. The new GPS material has focused on performance at altitudes above that of the Navstar constellation and also on results obtained from flight experience on LANDSATs 4 and 5. Although navigation using GPS is not autonomous, this system concept has gone the farthest in the implementation and testing arena. It provides a comparison base for other concepts in terms of performance, complexity, and vulnerability.

Figure 1 is an updated version of the performance summary contained in [19]. The survey work of Ferguson and Tapley [20] tends to corroborate these data. Technology projections by NASA [21], the Air Force [22] and Hummel [23] were also used to project these performance predictions. The implementations concepts are grouped into four general categories: 1) direct earth horizon sensing, 2) inferred earth horizon sensing through stellar refraction, 3) satellite crosslinks and 4) Navstar (GPS). These approaches are fully autonomous with the exception of Navstar. Since the advent of SDI, emphasis has centered on satellite

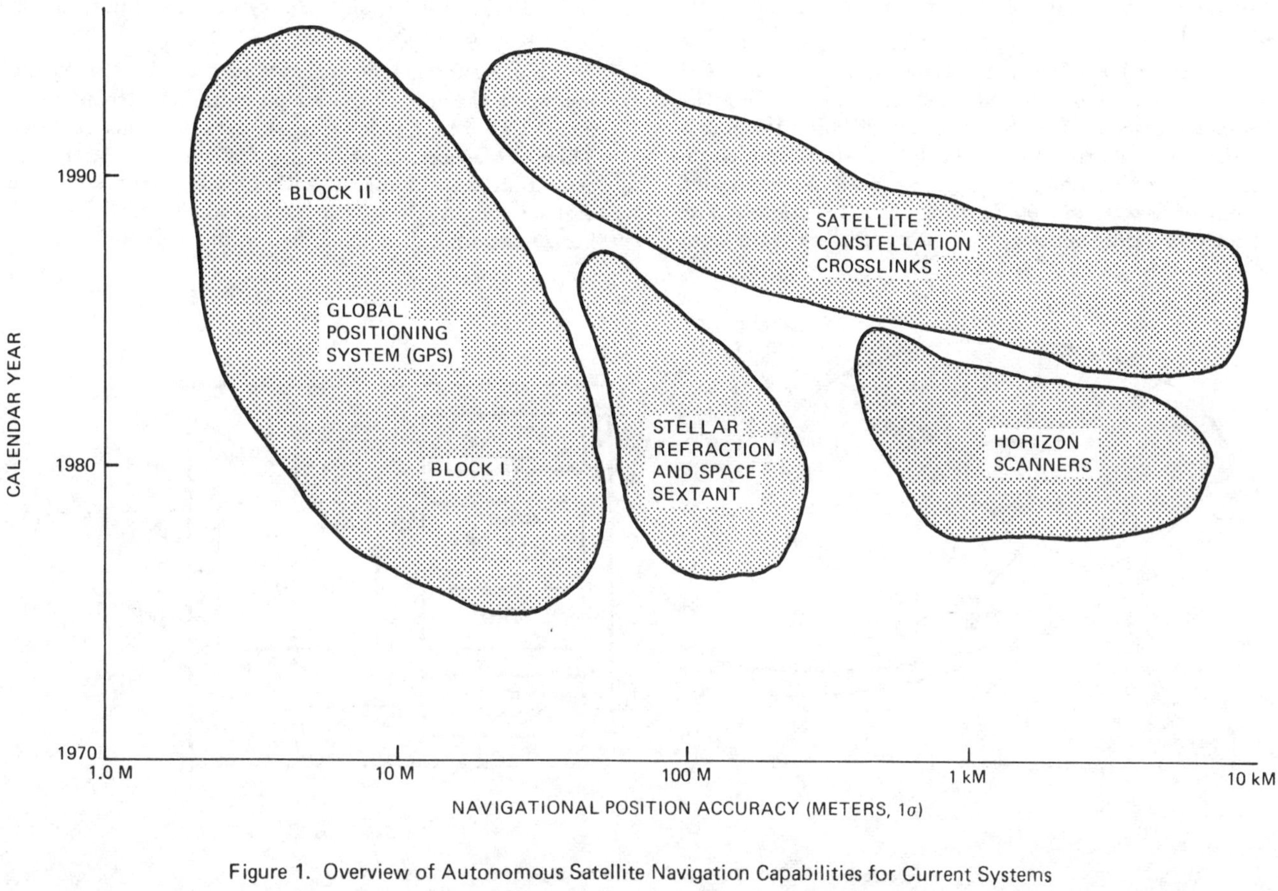

Figure 1. Overview of Autonomous Satellite Navigation Capabilities for Current Systems

crosslinks and stellar refraction with GPS as a backup system. Renewed interest in making GPS autonomous has also surfaced. A brief description of each of the above noted navigation techniques follows.

Electro-optical Autonomous Navigation Techniques

A typical system concept for electro-optical (EO) autonomous navigation is shown in Figure 2. These systems typically provide 1) spacecraft attitude using a stellar-aided inertial sensing system and 2) spacecraft position and velocity. As discussed in the historical section, many schemes have been employed using earth and/or lunar horizon sensing (for example MADAN and Space Sextant).

The MADAN system concept provided an example of how to solve the autonomous navigation problem by two techniques: 1) earth horizon sensing, or 2) stellar refraction measurements. The significant limiting factor in the earth sensor schemes is the accuracy of the measurement of the direction of the satellite-to-earth vector. Figure 3 taken from [21] gives expected earth sensor accuracies up to the year 2000 as projected by NASA. The MADAN goal, given 1980's horizon scanning technology, was 1.8 km over an extended range of altitudes. Aerospace Corporation studies have shown the feasibility of this autonomous navigation scheme using actual DMSP Star sensor data, earth sensor data and gyros. Positional accuracies of the order of 7.6 km were obtained [24].

The stellar refraction scheme also uses the CCD star sensor as a navigation sensor. Figure 4 shows the geometry associated with the technique when the star is in the satellite orbit plane. Analytical studies by Hummel [23] and Gounley [25] project position estimation errors of less than 100 m using this technique. The effect of satellite altitude and star sensor random error on navigation accuracy were calculated by Hummel as shown in Figure 5 [23]. Figure 6 of [21] gives expected star tracker accuracies up to the year 2000. IBM has shown the feasibility of the star refraction scheme by processing real satellite data [26]. Positional accuracies within 750 meters were achieved. In a more recent study, Draper has processed nighttime observations of starlight refraction from the satellite HEAO-2 (High Energy Astronomy Observatory) to profile the stratosphere (density, temperature and pressure) and to estimate variability around the world [27]. These models were used in a navigation simulation to predict performance. Performance results on the order of 40m-185m (one sigma) were obtained using the HEAO-2 data for low altitude satellite orbits (<2000 km) [28].

The space sextant uses star measurements and measurements of the moon's bright limb. Martin-Marietta has built a proto-flight model and some limited orbital testing has been completed. The principle of operation is described in [12] - [14]. Navigation accuracies of 244 m are projected.

Although development has proceeded in fits and starts, the EO techniques continue to be pursued because of their potential for a high degree of autonomy. MADAN and Space Sextant are no longer being funded. NASA has produced a solid-state CCD star tracker (STELLAR) which, though tested successfully in the laboratory, has not been qualified for space [21]. Ball, Perkin-Elmer and Honeywell are pursuing the CCD technology [23].

Some surveillance concepts for the SDI are expanding upon the use of the stellar refraction technique by also using the mission sensor data for this function. These IR sensors provide data of the type needed for refraction measurements without significant additional cost or complexity. This is akin to the use of existing horizon scanner data embodied in previous concepts.

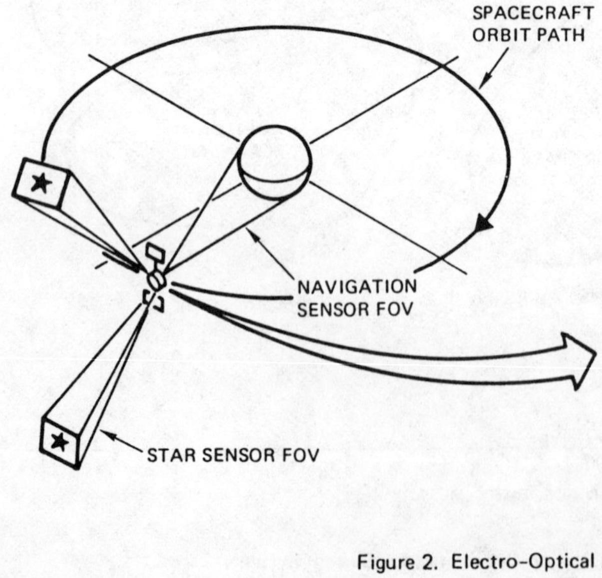

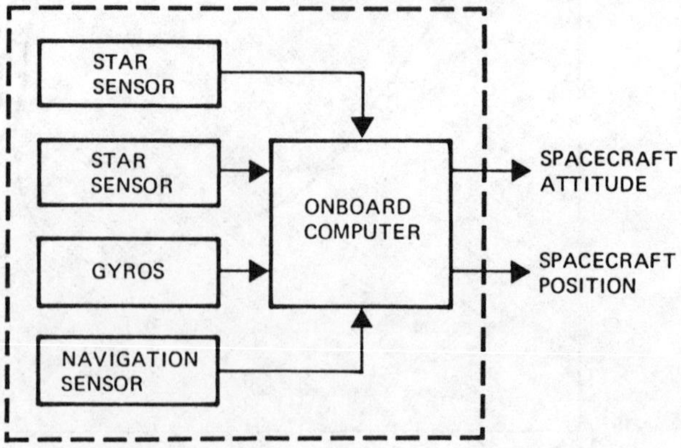

Figure 2. Electro-Optical Autonomous Navigation Overview

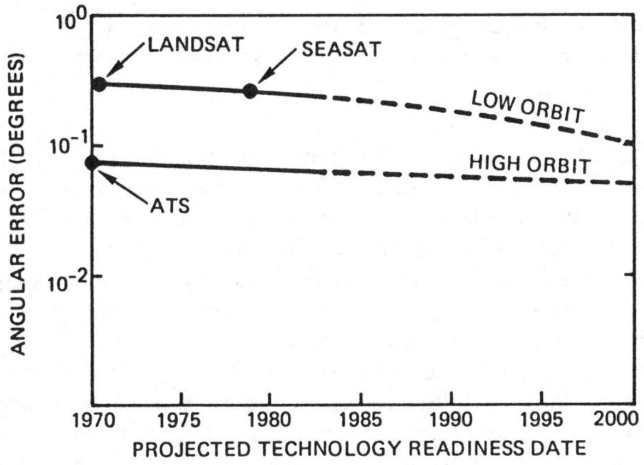

Figure 3. Horizon Sensor Angular Error

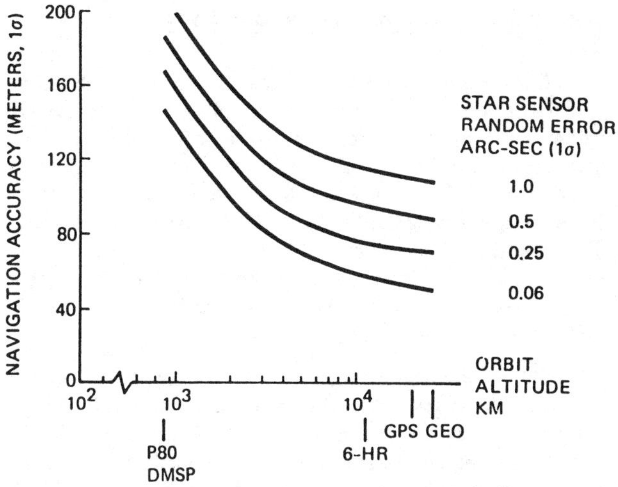

Figure 5. Effect of Orbit Altitude on Navigation Accuracy Using Stellar Refraction Measurements

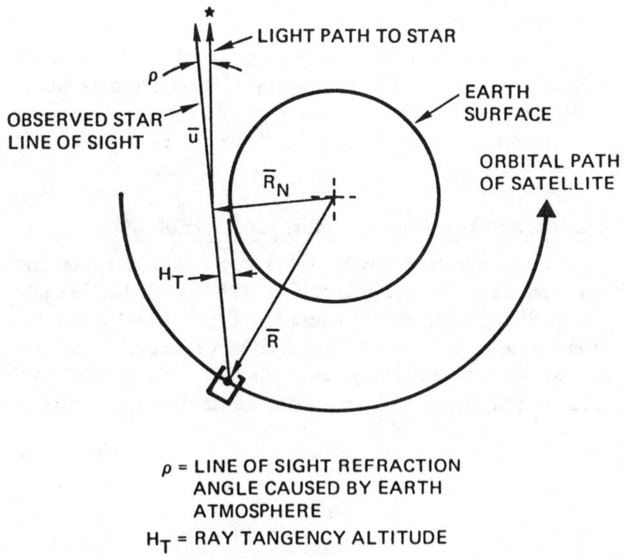

Figure 4. Atmospheric Refraction of Star Sighted Through the Earth's Atmosphere

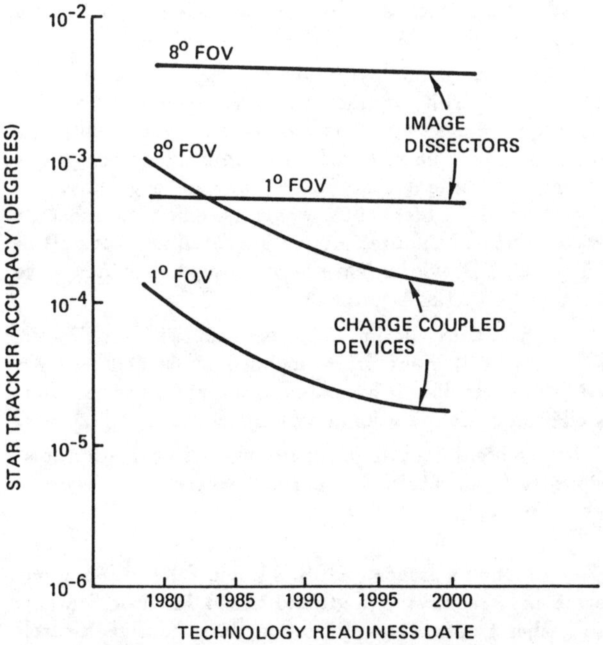

Figure 6. Projected Star Tracker Accuracy

Satellite Crosslinks

Measurements made through the observation of, or interaction with, other satellites can improve the ephemeris estimate of an autonomous navigation system. The concept, performance and status were explained in [19]. The details of designs using crosslink data for navigation are very application dependent. The accuracies achievable depend on the available satellite sensors for the specific constellation. Theoretically, system performance on the order of GPS performance could be achieved using satellite crosslinks with ranging measurements.

Until GPS, the Joint Tactical Information Distribution System (JTIDS) had been the only program actively pursuing algorithm development for navigation using crosslink ranging. As described in the next section, the SDI concept could make great use of a crosslink sharing system for navigation. Its application is such that large groups of satellites are in constant communication with one another for operational reasons. Autonomous satellite navigation using crosslinks is therefore being pursued by SDI programs and also by GPS [29].

NAVSTAR: Global Positioning System

Background - The Global Positioning System (GPS) is currently under development by Rockwell under DoD Sponsorship to provide world-wide navigation capability for ground, air, naval and space vehicles. Originally, [30], the system was to consist of 24 satellites, with eight satellites in each of three orbit planes. The original system would have provided global coverage with at least four and more often, six to nine satellites in view. Accuracies were predicted to be in the 10 meter range.

Budget constraints in the 1978 and 1979 period resulted in reduction of the satellite constellation from the planned 24 to the minimum 18 required for global three-dimensional (latitude, longitude, altitude) capability. Three satellites are deployed in each of six orbital planes, spaced by 120 deg in each plane. The orbital planes are equally spaced 60 deg in longitude and inclined to the equator at 55 deg. The orbits are circular at 20,183 km, resulting in orbit periods of 11h 57 min 58.35 sec (one-half sidereal day).

Current plans [31] and [32] call for 18 primary satellites in orbit, augmented by three on-orbit spares. On-orbit spares will improve system availability by reducing the time required to resume full coverage following a satellite outage. Before the Challenger disaster, the 21-satellite block buy was proceeding on schedule with launch of the first production satellite (Block II) in 1986 [32]. Previous Navstar prototype satellites were launched on Atlas boosters.

A test constellation is in operation with six Block 1 satellites in 12 hour orbits inclined at 63 degrees. An additional satellite is available to maintain the test constellation until operational buildup commences [32].

The identification of future satellite navigation missions and the status of current receiver hardware is given in [33].

Autonomy Issues - GPS is not a fully autonomous system. It requires: 1) ground based Monitor Stations to gather transmitted navigation data from each satellite; 2) processing at the Master Control Station to determine updated satellite ephemerides and clock offsets; and 3) uplink of the updates from upload stations. However, the availability of stable atomic clocks and the ability to store precise long-term ephemerides on board each GPS satellite reduces the need for frequent ground intervention. Autonomous operation without degraded performance for up to a week is projected when the system is fully operational. Additionally, for other satellites to use GPS, they must rely on the radiated signals from the GPS constellation.

Ananda et al [29] have established the conceptual feasibility of GPS autonomous navigation using satellite-to-satellite range measurements. User accuracies can be maintained for 180 days without any contact with the ground. Implementation feasibility has been determined by Rockwell [34]. These changes have not been incorporated in the Block II design concept however.

Performance - Navigation accuracy using GPS depends on the product of the basic system range error and the Geometrical Dilution of Precision (GDOP) parameters. Low values of the GDOP parameter indicate good geometrical distribution of the satellites being used for the navigation solution. The geometry is best in low earth orbit where nine or more GPS satellites are in view. Numerous studies have been conducted that characterize satellite navigation performance from low earth orbit to geosynchronous [35-39]. Positional accuracies in the 5 to 30 meter range can be achieved for orbital altitudes less than GPS altitude. In a geosynchronous orbit, the GPS satellites are visible only when they are on the opposite side of the earth (see Figure 7). For this reason, these satellites can track only one or two GPS satellites (Figure 8) most of the time. The accuracy is degraded to approximately 100 meters (Figure 9, [39]). Figure 9 summarizes the positional accuracy for various orbits.

GPS can also be used for on-board attitude determination [40- 42]. The feasibility of interferometric attitude determination and the concept of operations for this function are explained in the references previously noted.

Hardware Implementation and Verification

The experience with GPS receivers in space has been on Landsat 4 (1982) and 5 (1984) with the Magnavox GPSPAC [33,43-45] using the Block 1 test constellation. The objectives of the LANDSAT experiment were to validate the accuracy and reliability of the GPSPAC and to determine on-board satellite navigation accuracy.

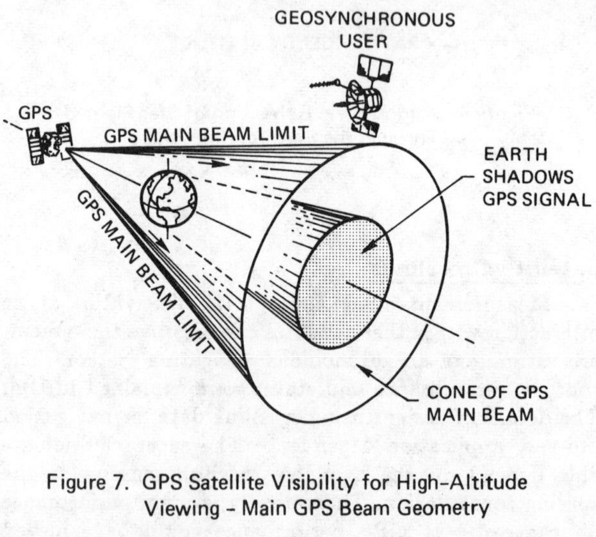

Figure 7. GPS Satellite Visibility for High-Altitude Viewing - Main GPS Beam Geometry

A GPSPAC description, operational history of the experiment, and analytical and on-orbit performance are described in the references. The position uncertainty for any revolution was found to be between 5 and 50 meters when four GPS satellites were in view. Texas Instruments also has a space GPS receiver that has been proposed for a shuttle experiment [46].

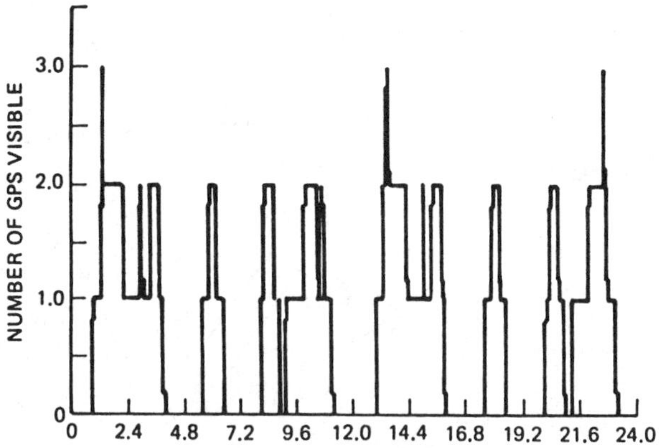

Figure 8. Geosynchronous User Satellite - Number of GPS Satellite in View

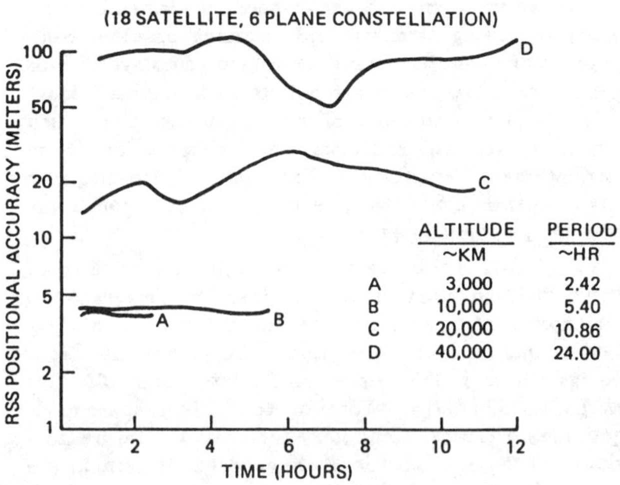

Figure 9. Positional Accuracy Using GPS for Various Satellite Altitudes

PROJECTED NEEDS/ APPLICATIONS

The historical perspective of satellite autonomous navigation reviewed previously has highlighted the evolution of the concepts of implementation and technology developments. An assessment of the current status of the current driving forces is developed in this section. The assessments are subjective and therefore provide but one possible view of the projected needs.

Advent of the Strategic Defense Initiative (SDI) research was in 1983. Its attendant various space based elements represents the major current push to spacecraft autonomy overall and navigation in particular. Certain of the scientific space programs related to earth sciences also serve as drivers for this technology but generally to a lesser degree. Astronomical, space science and commercial programs, by contrast, currently represent only a minor impetus to the incorporation of satellite navigation autonomy.

Scientific satellite programs typically involve no more than two or three vehicles which must be operated in an integrated fashion. Many times a program will consist of only a single satellite. In this context, the influences of transfer or sharing of real time data involving navigation based information are minimal. Post facto knowledge is usually adequate. When navigation is required for these applications, it is more directed toward enhancing the ability of the satellite to perform mission tasking more simply and/or with reducing the needs for extensive ground facilities and staffing. An example would be the targeting of a mission sensor to a series of terrestrial targets for earth resources survey. The presence of onboard navigational data would allow use of summary level time-tagged geographic data to be fed to the satellite. Onboard processing would then be used to establish pointing commands for the sensor and/or spacecraft to achieve the desired target coverage. Unless sensor fields of view are very small, the navigational accuracy required to perform this type of targeting is on the order of kilometers. These types of applications in general are not considered at this time to be the overriding motivation for satellite autonomous navigation.

The SDI environment differs markedly from that described above for scientific satellite operations. In this application area, there is great necessity for information sharing and integration in real time both within constellations of similar functions as well as among differing constellations. For this integration to be possible, it is necessary that the data be in a common base or at least in one that is well defined and controlled by each data source. The SDI mission is one in which a variety of space assets are utilized in a restricted time frame and during a period when the operational environment may well be very hostile. This implies fast reaction, limited availability of external data sources, and possible breakup of a nominal infrastructure. These considerations force implementations which can operate with a minimum of interaction with the ground and still retain a high degree of readiness and an ability to carry out tasks in a highly dynamic environment.

Figure 10 shows a representative context for SDI operations. Space based assets are used to provide surveillance, battle management, weapon deployment, and kill assessment. These functions are distributed according to some established architecture. This architecture is likely to involve groupings of spacecraft (constellations) with similar functions, for example surveillance.

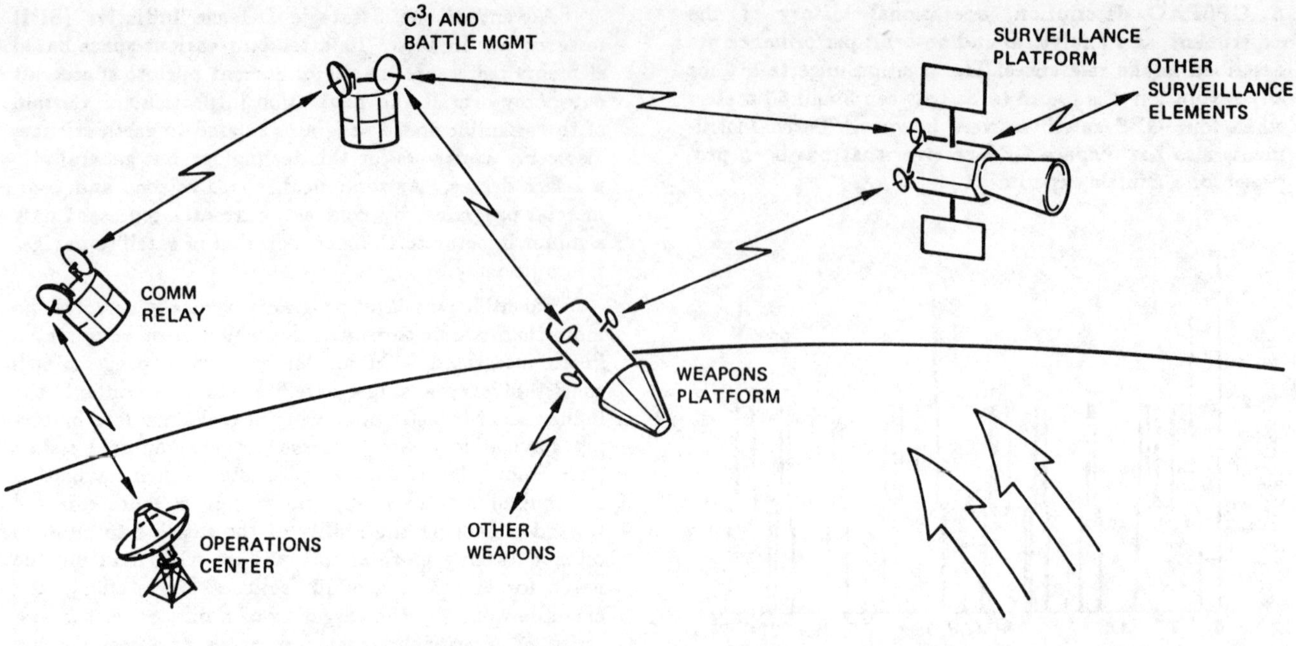

Figure 10. Possible Model for Strategic Defense Architecture

Each surveillance satellite collects data from its own complement of sensors. These data must be processed into a form that can be shared with other members of the constellation which may be covering the same or adjacent spatial areas. What is desired from the surveillance function is the definition of a trajectory for each of a potentially very large number of detected objects. Line-of-sight vectors, and for some implementations ranging data, from multiple surveillance sources must be integrated to derive the highest confidence and most accurate trajectory data. This necessitates placing all of the observation data in a common coordinate reference. Navigation and attitude reference information from each surveillance platform is needed to accomplish this. To achieve the trajectory accuracies required (small fractions of kilometers), observation data from several sources must be integrated and smoothed. This implies a significant level of data transfer within the surveillance constellation. It also implies real time navigation data from each source to levels of 10-100 meters or better for some applications.

Surveillance data is then transferred to the battle management function which in turn assigns weapons based upon data from these orbital platforms as to their positions in the attack space. This too dictates autonomous navigation within the weapons platform to accuracies of the same order as the surveillance requirements. Deployment of weapons can also involve transfer of navigational data from platform to weapon for initialization, again enforcing the need for an autonomous navigator onboard these satellites.

Operations among differing satellites as noted could take place in an environment where the elements themselves are being attacked and jamming or other communications countermeasures are being employed. This environment establishes a requirement to navigate accurately under conditions of evasive maneuvers which implies a transient short term as well as a long term navigational capability. External electro-magnetic environments can also preclude reliance upon some externally supplied navaid sources.

The SDI requirements, plus the fidelity required of the navigation function, are considered to represent the major motivating forces to satellite autonomous navigation at this time. Future Military Space Systems Technology Plans (MSSTP) such as the latest one (1983/84) as discussed in [22] will continue to highlight space navigation as a critical technology for projected future missions. Defense applications are expected to remain the drivers for this capability.

ACKNOWLEDGEMENTS

The information of this paper was generated through reviews of the applicable literature, personal contacts, and general knowledge of the subject matter gathered from experience in the field. As a result, it reflects the work of a large number of individuals and organizations. These contributions are all gratefully acknowledged. In addition, the authors would like to thank M. Hager-Kullmann for her help in preparing the manuscript.

REFERENCES

[1] United Aircraft Corporation Report, "Mathematics of the UA Space Navigation System," April 17, 1963.

[2] J. L. LeMay, "Synthesis Procedure for Gain Selection in the UAC SNS Mathematical Loops," The Aerospace Corporation Report No. ATN-64(4533-40)-1, 2 August 1963.

[3] W. G. McArthur, "Horizon Sensor Navigation Errors Resulting From Statistical Variations in the CO_2 14-16 Micron Radiation Band," paper presented Ninth Symposium on Ballistic Missile and Space Technology, San Diego, California, August 1964.

[4] J. S. Meditch, J. L. LeMay, and J. P. Janus, "Analysis of an Horizon Scanner Autonomous Orbital Navigation System," Pan American Congress of Mechanical and Electrical Engineering and Related Branches, Mexico City, Mexico, October 1965.

[5] G. M. Levine, "A Method of Orbital Navigation Using Optical Sightings to Unknown Landmarks," AIAA/ION Guidance and Control Conference, August 1965.

[6] N. F. Toda and F. H. Schlee, "Autonomous Orbital Navigation by Optical Tracking of Unknown Landmarks," AIAA/JACC Guidance and Control Conference, August 1966.

[7] J. F. Bellantoni, "Unidentified Landmark Navigation," AIAA Journal, No. 18, 1967.

[8] W. L. Brogan and J. L. LeMay, "Orbit Navigation with Known Landmark Tracking," Proc. 1968 Joint Automatic Control Conference, June 1968.

[9] W. R. Davis and J. A. Miller, "SPARS - A Completely Strapdown Concept for Precise Determination of Satellite Vehicle Attitude," presented to the Spacecraft Attitude Determination Symposium, The Aerospace Corporation, September 30 to October 2, 1969.

[10] D. H. Aldrich, "Interferometer Landmark Tracker Navigation System," IBM Electronics Systems Center, NAECON 1974 Record, p. 592.

[11] J. L. LeMay, et al., "High Altitude Navigation Study," The Aerospace Corporation, El Segundo, California, Report TR-0073 (3491)-1, June 29, 1973.

[12] H. A. Garcia and W. J. Owen, "Design and Analysis of a Space Sextant for High Altitude Navigation," AIAA Journal of Spacecraft and Rockets, Vol 13, No. 12, December 1976.

[13] "Design and Laboratory Testing of Self-Contained High Altitude Navigation System, Phase 1 - The Space Sextant Autonomous Navigation Attitude Reference System (88-ANARS)," Phase 1 Final Report, MCR-77-196, Martin Marietta Aerospace, Denver, Colorado, 1977.

[14] W. J. Owen and R. A. Booker, "A Precision Sensor for Spacecraft Navigation and Attitude Determination," proceedings of the ION National Aerospace Meeting, April 1977.

[15] R. A. Booker, "Space Sextant Autonomous Navigation and Attitude Reference System - Flight Hardware Development and Accuracy Demonstration," Rocky Mountain Guidance and Control Conference, paper AAS 78-124, March 1978.

[16] N. Westheimer, "High Altitude Attitude Reference System Development Study," Final Report AFAL-TR-79, Rockwell International Corporation, May 1979.

[17] N. P. Laverty, K. J. McAloon, J. L. Roberts, and I. J. Williams, "Multimission Attitude Determination and Autonomous Navigation (MADAN)," American Astronautical Society, Paper No. AAS-80, Annual Rocky Mountain Guidance and Control Conference, February 17-21, 1980.

[18] "Multimission Attitude Determination and Autonomous Navigation (MADAN)," Final Report, TRW, AFSTC-TN-85-01, August 1985.

[19] M. A. Chory, D. P. Hoffman, C. S. Major, and V. A. Spector, "Autonomous Navigation -- Where We Are in 1984", AIAA Guidance and Control Conference, Seattle WA, August 20-22, 1984.

[20] J. R. Ferguson and B. D. Tapley, "Autonomous Spacecraft Navigation using Current and Planned Sensor Data", Technical Report AD-B070-035L, Applied Research Laboratories, University of Texas at Austin, 7 December 1982.

[21] "NASA Space Systems Technology Model," Volume IIA, Space Technology Trends and Forecasts, NASA, January 1984.

[22] E. J. Pelka, et al, "The Military Space System Technology Plan - A Guidance Navigation and Control Perspective from Technology Workshop 3", IEEE PLANS 1984.

[23] S. G. Hummel, "Spacecraft Autonomous Navigation Using Stellar Refraction Measurements", AIAA Guidance and Control Conference, Seattle WA, August 20-22, 1984.

[24] "DMSP Autonomous Navigation," Aerospace Corporation Quarterly Technical Report, April 1982.

[25] R. Gounley, R. White and E. Gai, "Autonomous Satellite Navigation by Stellar Refraction", AIAA Guidance and Control Conference, Seattle WA, August 20-22, 1984.

[26] International Business Machines, "Autonomous Ephemeris Technical Operating Report", CDRL 215A2, USAF Space Division, 10 June 1983.

[27] R. L. White, F.A. Barnes, and S. J. Brodeur, "HEAO-2 Observations of Stellar Atmosphere Refraction," Advances in Remote Sensing Retrieval Methods, Williamsburg, VA, 30 October - 2 November 1984.

[28] R. L. White, S. W. Thurman, and F. A. Barnes, "Autonomous Satellite Navigation Using Observations of Starlight Atmospheric Refraction," Institute of Navigation Annual Meeting, 25-27 June 1985.

[29] M. P. Ananda, et.al., "Autonomous Navigation of the Global Positioning System Satellites", AIAA No. 84-1863-CP, AIAA Guidance and Control Conference, Seattle, WA, August 20-22, 1984.

[30] B. S. Parkinson and S. W. Gilbert, "NAVSTAR: Global Positioning System - Ten Years Later", Proceedings of the IEEE, Vol.71, No.10, October 1983.

[31] T. A. Stansell, "Civil GPS from a Future Perspective," Proceedings of the IEEE, Vol.71, No. 10, October 1983.

[32] J. P. Porter and W. A. Hite, "Overview/Current Status of the NAVSTAR Global Positioning System", IEEE PLANS 1984.

[33] B. L. Masson, M. P. Ananda, and J. Young, "Functional Requirements of Next Generation Spaceborne Global Positioning System (GPS) Receivers," Aerospace Corporation, 1983.

[34] A. Codik, "Autonomous Navigation of the GPS Satellites: A Challenge for the Future", Proceedings of the National Technical Meeting, Institute of Navigation, 1985.

[35] J. E. Farr, "Space Navigation Using the NAVSTAR Global Positioning System (GPS)," Guidance and Control Conference, Vol 39, AAS, 1979.

[36] P. Axelrad, "Near Earth Orbit Determination and Rendezvous Navigation Using GPS", M.S. Thesis, Dept. of Aeronautics and Astronautics, MIT, February 1986.

[37] G. Matchett, "Navigation Accuracy Issues for Near-Earth Orbital Users of GPS," AAS & AIAA, Annual Meeting on Space Enhancing Technological Leadership, AAS, 1980.

[38] P. S. Jorgensen, "Autonomous Navigation of Geosynchronous Satellites Using the NAVSTAR Global Positioning System", IEEE National Telesystems Conference, 1982.

[39] M. P. Ananda and P. S. Jorgenson, "Orbit Determination of Geostationary Satellites Using the Global Positioning System", Aerospace Corporation, 12 March 1986.

[40] J. F. Ellis and G. A. Creswell, "Interferometric Attitude Determination with the Global Positioning System", AIAA Journal of Guidance and Control, Vol 2, No.6, 1978.

[41] K. M. Joseph and P. S. Deem, "Precision Orientation: A New GPS Application," International Telemetering Conference, San Diego, CA, October 1983.

[42] W. S. Burgett and S. D. Roemerman, "The Development and Applications of GPS Determined Attitude", IEEE National Telesystems Conference, San Francisco, 1983.

[43] P. S. Jorgensen, "Navigating Low Altitude Satellites Using the Current Four NAVSTAR/GPS Satellites", Global Positioning System, Vol 2, Institute of Navigation, pp 112-121.

[44] W. P. Birmingham, B. L. Miller, and W. L. Stein, "Experimental Results of Using the GPS for Landsat 4 Onboard Navigation", Global Positioning System, Vol 2, Institute of Navigation, pp 231-238.

[45] H. Heuberger, "Performance of the GPS Package on LANDSAT 5", IEEE PLANS 1984.

[46] J. G. Peters, "Feasibility of a GEOSTAR Experiment Onboard the Space Shuttle," Report for Naval Surface Weapons Center, Applied Research Lab - University of Texas, ARL-TR-83-18, 1983.

Part 7
Introduction to Animal Navigation

7.1 The Problem

MANY animals can find their way home after being displaced hundreds of meters (e.g., dogs) to hundreds of kilometers (e.g., pigeons). Most homing pigeons will find their loft after being displaced 100 km, even when artificially deprived of some sensory information; many will return from 400 km. Other animals (e.g., birds, salmon, turtles, and whales) migrate seasonally over long distances, some for thousands of kilometers. A few individuals will return to the precise site of their departure. Whether they are homing or migrating, these animals navigate by methods that are understood in general while slowly being revealed in detail. These animals are the ultimate neural-net navigators, using data from several sensors. They are born with wired algorithms and, perhaps, an innate data base. They add to the data base and may modify the algorithms by learning the characteristics of their surroundings with their sensors. From a human viewpoint, they appear to navigate. Their navigation mechanism evolved millions of years ago and have adapted to ice ages, the precession of the Earth's spin axis, the migration of the poles, and many reversals of the magnetic field. The need to adapt to these changes and the survival value of variability in migration routes may explain why so little data are innate.

My interest, for 30 years, has been to find out whether their technology is applicable to long-distance navigation by man. In the pre-microchip days of the early 1960s, I was impressed by the tiny volume of the birds' sensors and computers in comparison to the airborne computers of the day (50 lb for 4000 sixteen-bit words) but concluded that their sensor-processing algorithms were inaccessible to man. That conclusion is unchanged today.

7.2 Experiments

During the past hundred years, controlled experiments have been conducted with caged and free-flying birds. In the 1950s and 1960s, animal navigation experiments were often done by nontechnical biologists who attributed incredible accuracies to birds and fish on long-range migrations. In the 1970s, there was an explosion of work on bird navigation, much of it done by quantitatively minded behavioral biologists. An enormous amount of drudgery has been invested in breeding, feeding, testing, and observing tens of thousands of birds, thereby creating a primitive understanding of the sensors used by migrating or homing birds. The reprint by Baker, pp. 421–434, describes how the data are collected. Besides the tests with caged birds (e.g., in planetariums and Helmholz coils) and the homing experiments that he describes, free-flying birds have been observed by radar and by eye, and banded birds have been observed at discrete points in their travels. Baker uses the term "goal orientation" synonymously with "navigation" in the sense that a displaced goal-orienting bird would adjust its heading toward its original destination. He uses the term "orientation" in the sense of not changing heading (whether inborn or learned) even when the bird is displaced. Birds unfortunately do not cooperate with the experimenter and perform badly in captivity. Hence, it is difficult to design experiments that determine which cues they use.

Fish experiments are far more difficult than bird experiments, though the economic benefits are greater because of the food value of migrating fish (e.g., salmon, tuna, and herring). It is known that at least 10% of salmon released in certain rivers return to them. A smaller number return to nearby streams. This is evidence of purposeful swimming, not random search. A few acoustic transponder tests have been done on free-swimming fish, telemetering depth, position, and heading for a few tens of miles offshore. The consensus of the literature is that fish detect the odor of their stream but there is no data about how they get close enough to detect the odor and the direction of increasing concentration. In many species, celestial cues are ruled out because of the turbidity of the water and the depth at which the fish swim.

7.3 Mission Phases

(a) *Take-off.* As hormone levels rise in a migrating bird, it stores fat, becomes restless, and eventually takes off. Birds do not fly in "instrument" weather when the horizon and ground are invisible. When looking over water or over the top of an overcast (from a high-ground roost), a bird will wait restlessly until the migrating urge becomes so strong, it goes anyway.

(b) *Initial departure heading.* Homing birds estimate the initial direction toward their loft based on observations made during the outbound leg. Except for "racing" pigeons, which are trained to follow a specific route (and no other), birds 50–100 km from home show initial dispersions of $\pm 60°$ from the direction toward home and do not follow direct routes. As a homing bird explores the release area, it adjusts its heading closer toward home. If it is deprived of outbound information and hence is unable to determine the correct flight direction, it will commence a search pattern. If it searches in vain, it will eventually lose its homing urge and roost anew.

Migrating birds could not possibly calculate a precise heading toward last year's wintering or breeding area, thousands of kilometers away. Given the large wind-drift angles at which they fly, a precise heading would do them no good. Even if they could estimate their latitude from the elevation of the noon Sun (discussed below), they have no way to estimate longitude. Hence, first-year migrators follow adults or, if they are migrating alone, depart at an inborn azimuth (south, southeast, or southwest for northern-hemisphere birds in fall; north, northwest, or northeast for the same birds in the spring). That the heading is programmed is proven (in storks, crows, starlings, etc.) by the fact that artificially displaced migrators continue on the same heading rather than "recalculate" their heading and change course to the "original" destination.

After takeoff in clear whether, a bird with a camera strapped to its head circles slowly while staring at the horizon approximately 10 times per minute before assuming its initial heading. The initial heading (called the "vanishing bearing" by modern bird biologists) is established relative to terrain cues and their observation of celestial bodies, as described below. It depends on the direction of the wind arriving at the bird's loft and on prominent landmarks near the point of departure.

(c) *Speed and energy*. Much is known about the optimum flying speed for minimum power consumption and for maximum range. Overland migrators do not store large reservoirs of fuel; they stop to feed, often for several days at a time. By contrast, birds migrating more than 2000 km over water put on 30 to 40% of their body weight as fat, as the migrating season approaches. They thus have sufficient fuel without feeding en route. Airspeeds range from 6 m/s for small birds to a reported 15 m/s (seems high) for large ones. The key problem the bird must solve is to fly in a wind whose direction is variable and whose magnitude is comparable to its own speed of flight. Hence the bird must either accept large along-track and cross-track dispersions or correct for drift. Because many birds spend most of their time at home searching for food, migration does not substantially increase their daily fuel consumption.

(d) *Altitude*. Radar data show that homing birds fly at an altitude of 50–300 m, whereas long-distance migrators fly at 100–2000 m. Some studies show a correlation between flying altitude and wind; others do not. Some birds seem to observe their own ground speed and drift angle. I suspect they test various altitudes to find the fastest ground speed, at least over land. Overland migrators stop and wait during periods of low visibility and high winds. Large birds, migrating over land, soar in thermal updrafts rather than fly a direct heading, thus complicating their navigation. Even among those that migrate in flocks, some soar in groups, some individually. Overwater losses undoubtedly increase if the birds are caught in foul weather.

(e) *En-route navigation*. Homing birds navigate with respect to the terrain and its odors, using the Sun and stars as heading cues. As noted above, young migrating birds fly at an inborn heading (called the "preferred compass direction" by modern biologists). If the bird is artificially displaced, it will continue to fly at the inborn heading (though it is inappropriate for the displaced location) and will nest in a new area.

Migrating birds can probably sense their approximate latitude ($\pm 5°$) from the peak elevation of the Sun. Experiments to determine if they can also measure longitude have all been negative; indeed, there is no conceivable mechanism by which they could do so. Hence, they apparently navigate in one dimension. When the migrating urge strikes, they notice their latitude and adopt an inborn heading, as described above. Long-range migrators establish their heading with reference to the Sun, stars, and perhaps magnetic information. Radar data at sea show birds drifting with the wind. These birds are uniformly distributed over hundreds of miles of ocean and show no evidence that they are position-fixing. Their dispersions increase under an overcast. However, there is some evidence that birds flying over land will compensate for a visually observed drift angle; when blown off course, some will fly back to it. Most likely, they are seeking the mainstream of a migrating flock. Statements that birds off course will adjust their heading to intercept the intended night's stop are unverifiable. Some radar data show that birds follow great circles on east–west migrations in Eurasia, avoiding tundra; other observations show birds favoring routes over land.

An unusual example is the migration of plovers from Canada to Argentina. In their first year of life, they migrate over land via Central America and return on the same route. In later years, they migrate south via the Caribbean Sea and the Antilles Islands, returning via Central America. I think it is more likely that they have two inborn headings than that they infer the Atlantic headings from their experiences on the Central American route. The Central American route for first-year migrants may have evolved to protect young birds who are not yet capable of sustained overwater flight.

(f) *Destination*. Long-distance migrators probably do not aim for a specific site. At least on their first flight, they hold heading and fly until their hormone levels fall or the climate is appropriate, whereupon they search for a nest (southbound) or their home area (northbound). Having chosen a nest, many are able to remember the route next year, to the breeding area and back to the wintering area. However, if they are blown off course, they will search for landmarks (including associated odors and peak Sun elevation) on the original route. If they do not find a site that matches their memorized data, they will nest or breed in another area. Many become lost and never find mates. If a long-distance migrator happens to arrive within 100 km of last year's nesting area, it might recognize the terrain and return to its former nest. If the return dispersion is larger, the bird will probably nest anew.

Homing birds appear to recognize their loft and the landmarks surrounding it, by visual cues and perhaps by odor. Perhaps one-third of pigeons, fitted with translucent contact lenses that are supposed to prevent them from focusing beyond 6 m, can approach from 15 km to within 1 or 2 km of home.

7.4 Animal Sensors

Animal sensors are just beginning to be understood:

(a) Birds certainly observe the Sun and perhaps correct for the Sun's azimuth with a short-term internal clock. Some measurements of the accuracy with which birds can be trained to feed show their internal clock to be accurate to ± 15 min, which would permit them to hold a true heading $\pm 5°$ when the Sun is in view. A 15-min clock error would be consistent with a 200-nm east–west error, if the bird were navigating by the Sun or stars. Steering a true heading by the Sun would perhaps require storing the absolute azimuth of sunrise or sunset and a clock correction thereafter, a remarkable feat if it were possible. Since bird's eyes operate independently (for objects beyond a few meters), their field of view is larger than a hemisphere; they can observe the Sun while flying at most azimuths. Their judgment of latitude is probably on a par with that of Indian Ocean and Pacific Island human navigators (Part 4.1, pg. 147) who associated one peak Sun elevation with their home and another with their destination. Those primitive people did not think of latitude as a coordinate on a spherical Earth and neither do birds. Birds may also use the Moon as an azimuth reference and as an indicator of latitude change during the course of a year, though data are scanty. It is incredible to expect a bird to calculate absolute latitude from the Moon's complex motion.

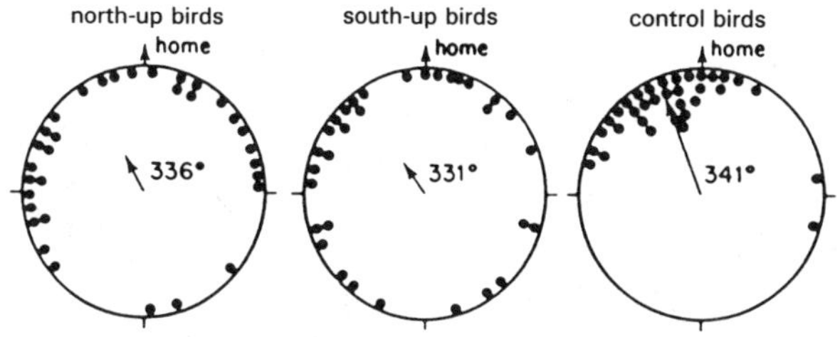

Fig. 7-1 Evidence for a bird's ability to detect the Earth's magnetic field [from (Baker, 1984), based on experiments by Visalberghi and Alleva, 1979]. "North-up" and "south-up" birds wear magnets on their heads. The controls wear nonmagnetic weights that do not disturb the Earth's magnetic field. When released under the Sun, all birds show homeward orientation but those wearing magnets show more scatter in their vanishing bearings. The vector shows the mean vanishing bearing.

(b) Birds can apparently find the axis of rotation of the night sky and can therefore hold heading on a clear night without knowing time. Once they have learned the sky, they can find the axis of rotation with substantial portions of the sky obscured. Experiments show that they can remember the sky for at least 6 months, plenty of time to allow for overcast weather. Birds are known to observe the Sun's azimuth at sunset. That observation might be the initial condition for a rapid solution for the axis of the night sky. It would be valuable to do an experiment to measure how long it takes a bird to solve for the axis of rotation. Birds observing a nonrotating sky fly random headings. Experiments by Emlen in a planetarium showed that birds can find the actual axis of rotation of the night sky, even if it does not pass near Polaris. This ability allows them to adapt to the 26,000-year precession of the Earth's spin axis.

(c) There is some disagreement in the experimental literature about whether birds use landmarks. I am certain that homing birds do use them; migrating birds use occasional landmarks for en-route cues and detailed landmarks when they reach their approximate destination. They evidently memorize what they see, feel, and smell in long-term, nonvolatile storage ("photographic memory"). Migrating birds probably follow flyways because they have learned key landmarks on earlier migrations, while following older birds.

(d) There is major disagreement about whether birds sense the Earth's magnetic field. Caged birds in an artificial magnetic field show random behavior which, by elaborate statistical analysis, sometimes has a slight correlation with the magnetic field. In some experiments, free-flying birds wearing head-magnets become confused; in other experiments they are unaffected by the magnets. Some experiments on caged birds seem to show that they sense the component of the Earth's field along their horizontal body axis; some show that they sense the vertical component. Still others purport to show that the bird senses the magnetic dip angle, hence the possibility of measuring magnetic latitude. Figure 7-1, from Baker (1984), shows the results of a series of experiments with homing pigeons. Birds were fitted with head-magnets; some with the north poles "up," some with the north poles "down." They were transported from their home loft to a release site (the distance was not stated but was probably 50 km). When released, all groups departed at the same mean vanishing bearing but magnet-carrying birds had much larger bearing dispersions than did control birds. The results imply that the birds were initially confused by the magnetic fields. In other experiments, birds were transported to a release site in closed containers in which the vertical component of the Earth's magnetic field was reversed by a Helmholz coil. When released, the experimental birds departed with a 30° average offset from the heading of the control birds, which were transported in closed containers exposed to the normal Earth's field. Walcott (1989) described experiments with matched homing pigeons in which some appear to sense a magnetic anomaly, though most do not. He speculated that birds may sense the horizontal spatial gradient of the magnetic field. Data such as these are the sole, if indirect, evidence that birds are sensitive to magnetic fields.

If a bird could measure the magnetic field, I doubt that its sensor would be more accurate than 5% (3° latitude in midlatitudes). The data-processing algorithms would have to adapt to migration of the Earth's magnetic poles and to occasional magnetic reversals. No magnetic sense organ has ever been identified in an animal, despite an intensive search.

(e) Homing birds sense the odor of the surrounding area and associate it with a location, as a rattlesnake associates the infrared image (from its pits) with the optic image (from its eyes) as described by Newman and Hartline (1982). Figure 7-2, by Keeton as reported in (Aidley, 1981), shows the effect of odor. Homing pigeons that lived in cages with normal airflow were moved to "deflector cages" in which the apparent wind direction was rotated almost 90° from the true direction. When released (at an unstated distance but probably 50 km from home), their vanishing bearing was also rotated 90°, though the birds eventually found their way home. In other experiments, birds were taken to a release point in sealed containers. The controls were supplied with outside air on the outbound trip; the experimentals were supplied with "pure" bottled air. Upon release, the two groups started home on radically different vanishing bearings, though the homing speeds and percentage returns were the same for both groups. Those experiments "prove" that birds use odor to navigate. Yet, other experiments on birds with plugged or anesthetized nostrils have shown successful homing, though sometimes taking longer than with an unimpeded sense of smell.

Although it appears that birds map odors into geographic

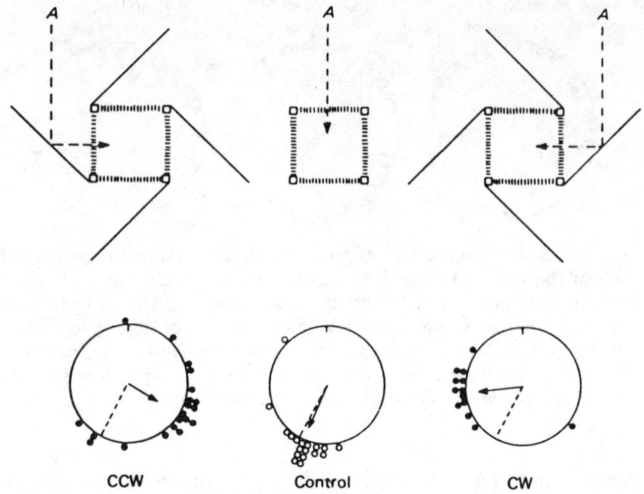

Fig. 7-2 Evidence for a bird's ability to use odors for navigation [redrawn by Keeton from Baldaccini *et al.*, 1975 as reported in (Aidley, 1981)]. Winds from the north, carrying odor "A," enter the control loft from the north but enter the other lofts from the east and west because of the deflectors. The three circles show the angular distribution of the vanishing bearings of birds raised in these lofts and released north of home. The control birds oriented properly southward toward home. By contrast, the clockwise birds, that had experienced "A" winds from the east, oriented more westerly, while the counter-clockwise birds, that had experienced "A" winds from the west, oriented easterly.

locations, several biology texts [see (Yap, 1970), for example] say that birds' brains lack highly developed olfactory regions. This is the typical evidence for olfactory cues to navigation.

(f) Birds can sense the apparent vertical (resultant of gravity and their own acceleration) by measuring the forces on their viscera, perhaps more accurately than mammals can. The visceral muscles act as an analog accelerometer, whose accuracy is unlikely to exceed 5%. However, more precise analog biological force sensors do exist. For example, the human ear detects a variation of 10^{-9} of atmospheric pressure, but it does so for an oscillatory input to a tuned cavity, not for a steady pressure. Behavioral scientists have proposed that, if steady-state visceral forces could be sensed to 0.1%, birds could resolve the measured forces through visually observed attitude and measure their approximate change in position. However, no evidence has ever been found of inertial navigation abilities in birds, even as coarse as 10% of distance traveled. Experiments are difficult to devise.

(g) Other tests have shown that birds sense, and still others have shown that they do not sense, low-frequency (1- to 10-Hz) sounds and the direction of polarization of visible and ultraviolet light.

Long-distance migration is almost always in flocks. I believe that birds migrating in a group make collective decisions about heading, thus reducing sensor errors from the one-bird standard deviation (σ) to $\sigma/\sqrt{N}$ where N is the number of birds in the migrating flock. Thus, a flock of 25 birds may hold heading with one-fifth of the error of any single bird in the flock flying alone. I know of no experiments that illuminate the short-term control dynamics of several birds correcting course, as a migrating group, to arrive at a collective course.

7.5 Conclusion

I conclude that the mechanism of animal navigation is a complex integration of many senses whose details are unlikely to be discovered soon. I see no hope soon of being able to write an algorithm to process the sensor data as the bird does. I conclude that long-distance birds have dispersions of hundreds of miles, flying by heading, time, and visual observations until their local senses (terrain, odor, Sun elevation) take over. If they find themselves within a familiar area, they might return to last year's nest; if not, they will nest anew. Birds on long overwater flights probably have huge losses at sea (I know of no data on percentage arrivals). By contrast, short-distance homing pigeons appear to fly by sight and odor, recognizing the area surrounding their loft from past flying experience.

The engineer is far from being able to profit from the technology of bird or fish navigation. The animal evidently uses multiple sensors, fuses the sensor outputs in its brain, and modifies the algorithm according to circumstances. The animal learns as it gains experience, even changing the mix of sensors used, thus making testing difficult. It associates topography, odors, and the position of heavenly bodies with its route, and remembers the patterns for years. I suspect that its brain contains random-number generators that cause variability of routes. The bird is the ultimate neural-network, expert-system navigator but its algorithms have not yet been studied. Indeed, algorithmic studies are just beginning of simple worms, whose brains contain perhaps 300 neurons (versus 3×10^7 in a bird). I view the bird as an automaton. Griffin (1984) has speculated that a bird's brain is not as determinate as I have implied. He suspects that animals can make conscious "freewill" decisions. Since we can neither ask them about their choices nor infer consciousness from the wiring diagrams of their brains, such speculations are at present undefinable and unverifiable. I wonder how complex a brain must get before it is not deterministic and before it is conscious of itself.

The engineer will profit once direct access is obtained to the animal's brain and its wiring diagram is understood. The engineer wants to know the nature of the pre-wired instructions and data, the method of loading sensor data into the brain, and the reconnection of pathways that results from the bird's experience. Experiments at Caltech in the 1970s [see Pettigrew's chapter in (Schmidt-Koenig and Keeton, 1978)] showed how difficult it is merely to hold the spherical shape of an excised bird's eye, on which the image of the Sun was being mapped, much less keeping a bird's excised brain alive.

Knowledge of the structure of the bird's brain and the algorithms for data processing will be applicable far beyond the technology of navigation. Study of this extraordinary phenomenon has barely begun.

References and Bibliography

Aidley, D. J. (Ed.), *Animal Migration.* Cambridge: Cambridge University Press, 1981. Routes and energy expenditure for bird and fish migration.

Baker, R. R., *Bird Navigation: The Solution of a Mystery?* New York, NY: Holmes and Meier Publishers, Inc. 1984. Observations and test procedures for measuring the navigation senses of homing and migrating birds. His Chapter 3 is reprinted in Part 7, pp. 421–434.

Eastwood E., *Radar Ornithology.* London: Methuen, 1967. Elementary rador theory, pictures of bird echoes, and results of observations in Europe and North America: bird counts and altitudes.

Gauthreaux, S. A. (Ed.), *Animal Migration, Orientation, and Navigation.* San Diego, CA: Academic Press, 1983. Distinguishes among landmark pilotage, compass orientation, and navigation. Wind direction and drift angle. Sensing of Sun, stars, polarized light, and odor. Energy consumption on flights. Imprinting of migrating route. Preprogrammed azimuth. Kenneth Able's chapter on two-dimensional navigation. No proof of magnetic sense.

Griffin, D. R., *Animal Thinking.* Cambridge, MA: Harvard University Press, 1984. Speculates on a continuum of consciousness from bacteria to man. Innumerable examples of "intelligent behavior" in insects, birds, bats, porpoises, and primates. Does not define experiments that can determine if behavior is "conscious."

Mead, C., *Bird Migration.* New York, NY: Facts on File, 1983. Good summary of the state of knowledge of bird navigation. History of the discovery of migration. Routes taken by individual species.

Newman, E. A. and P. H. Hartline, "The infrared vision of snakes," *Scientific American*, vol. 254, pp. 116–127, March 1982. Describes pit and eye sensors, their mechanical alignment, and the mapping of both images onto the same brain cells.

Schmidt-Koenig, F. W. and W. T. Keeton (Eds.), *Animal Migration, Navigation and Homing.* Berlin: Springer-Verlag, 1978. Forty-one papers emphasizing navigation. Mostly birds; some invertebrates and fish.

Walcott, C., "Show me the way you go home," *Natural History*, pp. 40–46, Nov. 1989. Brief survey of homing pigeon experiments, concentrating on magnetic field sensitivity. Compares identical breeding stocks raised in identical cages a few miles apart. When flown from a known magnetic anomaly, one group is disoriented while the other is not. Nevertheless, the author believes birds use magnetic cues

Yapp, W. B., *The Life and Organization of Birds.* London: E. Arnold, 1970. Traditional bird biology. Small olfactory lobes; "usually said they have no sense of smell, probably wrong."

Major types of experiment

R. Robin Baker
Reader in Zoology, University of Manchester

3.1. Homing experiments

If navigation is an integral part of exploration, it follows that experiments on navigation should be an attempt by the experimenter to mimic the exploration process. At the same time, by controlling the amount and types of information available to the animal at different stages during 'exploration', the experimenter can aim to find out when the different types of information are most important for successful navigation.

Effectively, this is what displacement–release or 'homing' experiments try to do (Fig. 3.1). Animals are taken from their home range, displaced beyond the limits of their familiar area, and released. Usually, at some stage during the experiment, they are deprived of particular information (e.g. by preventing them from seeing the route over which they are being displaced). At the same time, 'control' individuals are treated in exactly the same way except that they are not deprived of the information being evaluated. Some measure of navigational ability is then made and the performances of control and experimental animals compared to see if the deprivation has had any influence.

Reprinted with permission from R. R. Baker, *Bird Navigation: The Solution of a Mystery?*, New York: Holmes and Meier Publishers, Inc., 1984, ch. 3, pp. 19–32.

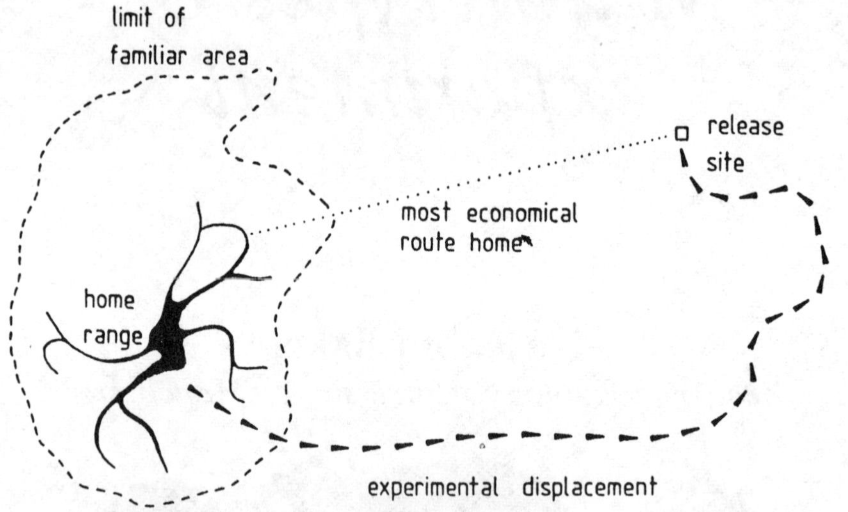

Fig. 3.1 Displacement–release experiments mimic natural exploration [From Baker (1981a)]*

3.2. Homing success

In the early days of experiments on navigation, the measure of performance most widely used was homing success. The proportion of released birds that returned home and the speed with which they did so was used as the measure of navigational efficiency. Homing success was abandoned as the primary measure, however, when it was realized that most often it failed to provide critical evidence for even the existence of navigation.

When a group of animals is released away from home, a lucky few might return successfully, not because they could work out the direction of home from the release site but because by chance they happened to set off in the correct direction. One of the characteristics of homing experiments is that the proportion of birds that returns home decreases as the distance of release increases (Fig. 3.2). This effect could be due to any number of factors, few of which relate to navigational ability. For example, the further an individual has to travel, the less motivated it may be to attempt to home. Similarly, the further it has to travel over unfamiliar terrain, with no knowledge of where to seek shelter or hide, the more prone it may be to predation. More importantly, however, the further from home the release site, the less likely is an animal to arrive back in its familiar area purely by chance.

Wilkinson (1952), for birds, and Saila and Shappy (1963), for salmon, developed mathematical models in an attempt to work out what proportion of individuals would return home and at what range of speeds if they had no navigational ability at all but instead employed various

* For the reference list see *Bird Navigation: The Solution of a Mystery?*

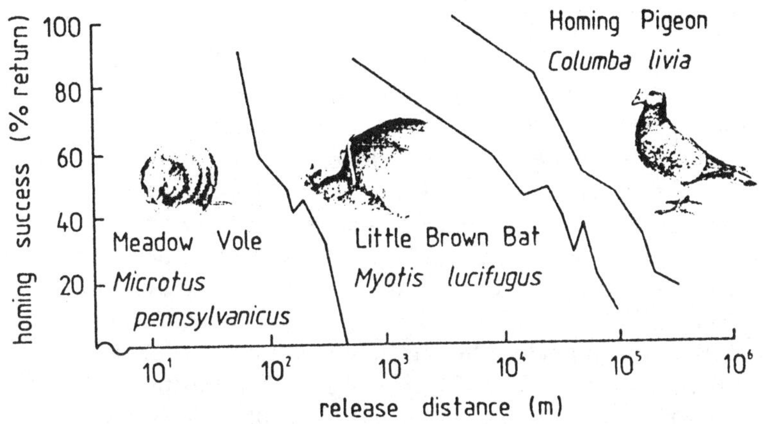

Fig. 3.2 For birds, as for other animals, homing success declines with increase in release distance

[From Baker (1981a), after Bovet]

patterns of random search. The close fit between observed homing performances and the predictions of their models led to a revolution in experimental design. No longer was it enough simply to release a group of animals and to wait for them to return. Some measure had to be found of the ability of birds to recognize home direction at the release site. After all, if it were found that more individuals set off in the home direction than in any other, random search could not be involved no matter what the characteristics of homing performances. For example, even if no birds returned home, an ability to recognize home direction from the release site could be inferred if animals set off in the correct direction. If we were to displace snails and release them 100 km from home, they may be perfectly able to work out the direction of home and set off in that direction, but we may doubt that any would actually return.

3.3. Vanishing points

In homing experiments on pigeons, the measure that is now universally adopted in measuring their ability to solve navigational problems is the compass bearing of their vanishing point. Pigeons are released singly, from the hand, and then watched through binoculars until they vanish from sight. The compass bearing of the last sighting before the bird vanishes is recorded. The next bird is not released until the previous bird has vanished and successive birds are faced in different directions to be released from the hand. The use of vanishing points requires that a release site has equally good visibility and open aspect in all directions. Pigeons are a convenient size for recording vanishing points but the same measure has been taken on much smaller birds, such as swallows.

3.4. 'Pointing' home

There are a number of disadvantages in having to release the experimental animal and watch it disappear in order to measure its 'answer' to the navigational problem set by the experimental conditions. One of these disadvantages is that unless the individual homes successfully, it cannot be used again. Another is that during the homeward flight it gains unknown experience that may influence its performance on any subsequent flights. The major disadvantage, however, is that it is usually difficult and often impossible to manipulate experimentally the information available to the bird at the release site. There would be a tremendous advantage in being able to restrain the animal at the release site and have it 'point' towards home rather than having to watch it set off in that direction.

Of all experimental animals, humans are the most convenient in this respect. They can be asked literally to point towards home. Even more conveniently, they can also be asked to state or write down their estimate of the compass direction of home to see if this differs from the direction in which they pointed. It is very difficult to separate these two different estimates of home direction using other animals, particularly birds. Yet, given both of these measures, a great deal could be inferred about the navigational process.

Animals other than humans have been restrained at the release site and, in effect, asked to point toward home. Most often, this technique involves the use of a so-called orientation cage to restrain the animal. Such orientation cages have been used with particular success for mice (Fig. 3.3) but have also been used for birds (e.g. bank swallows (*Riparia riparia*) – Sargent 1962). Unfortunately, homing pigeons have so far proved to be totally unco-operative in orientation cages. The latest attempt to design a cage in which a pigeon will indicate home direction is being made by the Frankfurt group.

Measures taken in an orientation cage are either the frequency of movements or the lengths of time spent in different directions. These measures are then processed in some way to calculate the mean direction of the animal's movements. This mean direction is then taken to be the animal's estimate of home direction.

3.5. Circular statistics

It is not enough, of course, simply to collect a number of estimates of home direction, whether from vanishing point bearings or from a set of mean directions in an orientation cage, and to judge them by eye for evidence of an ability to recognize home direction. Some statistical technique is required to calculate the probability of a given set of estimates arising by chance. Unfortunately, as estimates of home direction are angles and thus restricted to values on a circle of 360 degrees, the more familiar statistical techniques are not applicable. Instead, we have to use a group of statistical

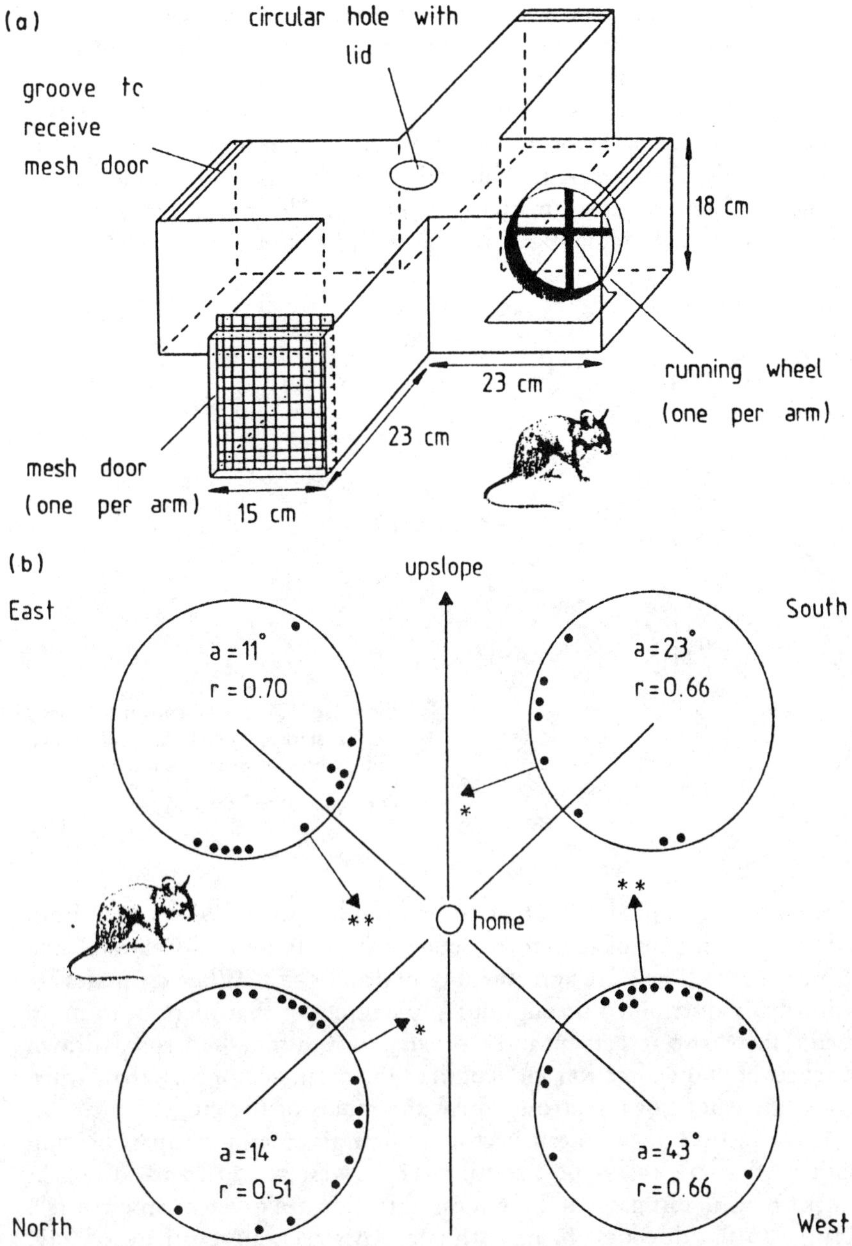

Fig. 3.3 The use of an orientation cage to study navigation by a small mammal

At the test site, an animal was introduced into the perspex cage (a) through the circular hole. The time spent in each arm was recorded and used to calculate a directional preference. When these preferences are plotted (b), they show that the mice are pointing more or less toward their place of capture ('home').

[From Baker (1981a), after Mather and Baker]

tests known collectively as circular statistics. All of the tests used most commonly in orientation research are described by Batschelet (1981). Here I shall give only the basic information required to be able to interpret the diagrams scattered through the rest of this book.*

The first step in analysis of a set of directional estimates (e.g. Fig. 3.4) is to calculate their mean vector. It is usual when illustrating such a data set to show the individual estimates of direction as points around a circle relative to the direction being estimated (e.g. home). The mean vector is then shown as an arrow radiating from the centre of the circle.

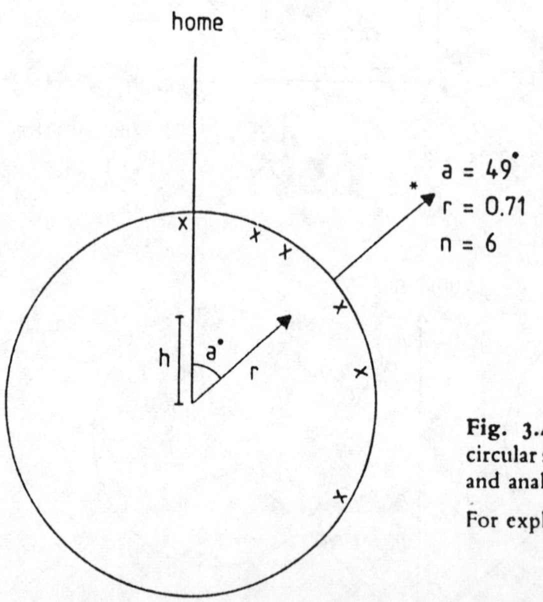

Fig. 3.4 Graphical conventions and circular statistics used in the presentation and analysis of orientation data

For explanation, see text.

A mean vector has both direction ($a°$) and length (r), where the length indicates the amount of agreement between the individual estimates. When all estimates are identical, agreement is total and $r = 1$. When estimates are uniformally distributed throughout 360 degrees so that there is no mean vector, there is no agreement and $r = 0$. By convention, the arrow is drawn to reflect both the direction and length of the mean vector such that, when $r = 1$, the length of the arrow equals the radius of the circle.

The direction of the mean vector is often given as a compass bearing with north = 0°, east = 90°, south = 180°, west = 270°, and so on. At other times, all estimates and the mean direction are given as an error ($e°$) relative to the direction being estimated. An accurate estimate of, say, home direction is thus given as $e = 0°$. An estimate ninety degrees clockwise of the correct direction would be $e = +90°$, or simply $e = 90°$, whereas an estimate ninety degrees anticlockwise of the correct direction would be $e = -90°$. An estimate precisely opposite to the correct direction would be $e = 180°$.

* Citations such as "this book" and specific chapters refer to *Bird Navigation: The Solution of a Mystery?*

If a uniform set of estimates of direction gives a mean vector with $r = 0$, we can check for non-uniformity in the data by evaluating, in effect, whether the r that we calculate for our data is significantly greater than zero. The statistical test for this is known as Rayleigh's z-test which, to be precise, tells us whether our data show a unimodal departure from uniformity. There are other tests which allow for other, such as multimodal, departures, but these are less commonly used.

The z-test tells us whether there is significant agreement among the estimates of direction in a given experiment, but it does not tell us whether agreement is over the correct direction. After all, it would be possible for a group of pigeons, for example, all to depart, one after the other in precisely the same direction and for us to calculate a mean vector of $r = 1$. These data would be highly significant by the z-test, yet their mean error could be $e = 180°$. Obviously the z-test by itself is not enough to tell us whether a given set of estimates indicates an acceptable ability to recognize home direction among our experimental group. To do this, we need a test that takes into account the mean error.

The first step in testing for homeward orientation is to apply a modification of the Rayleigh test, known as the V-test. Effectively, this test examines the length of that component of the mean vector that lies in the home direction. This component, usually known as the homeward component, is represented in Fig. 3.4 by h and can vary from $+1$ (when all estimates are exactly in the home direction) to -1 (when all estimates are exactly in the opposite direction).

The V-test tells us whether there is a significant vector in the expected direction. It does not, however, tell us whether the mean error is significantly different from 0° (Aneshansley and Larkin 1981). The final step in evaluating homeward orientation, therefore, after establishing significant non-uniformity by either z- or V-tests, is to determine the 95 per cent confidence limits of the mean error. If these include 0° (e.g. when $e = 30° \pm 35°$), an ability to recognize home direction by the group being tested can be accepted. If not (e.g. when $e = -30° \pm 15°$), then homeward orientation has not been established.

3.6. Separating goal orientation from orientation

So far we have considered the evaluation of homeward orientation in purely statistical terms, once the data have been collected. However, even if such data are consistent with homeward orientation from a particular release site, this is not enough to demonstrate that goal orientation is in operation.

The earliest experiments on homing pigeons often used release sites on a narrow range of compass bearings from the home loft. From the loft used by Kramer at Wilhelmshaven on the north German coast, for example, it was much more convenient to release birds from the south than from the north. As pointed out by Matthews (1968), however, such experiments do not separate goal orientation from simple orientation. Suppose that, for some unknown reason, homing pigeons from a particular loft always fly to the northwest upon release. If such pigeons were only ever released from sites at different distances to the southeast of their loft, the impression gained would be of accurate homeward orientation at all distances. Not until the birds were released from sites to the northwest would it become clear that they were employing orientation, not goal orientation. It has become standard practice, therefore, in experiments on goal orientation, to release groups from at least two (opposite) directions from home, and preferably from four.

It might seem unlikely that any animal, when displaced from home and released, would ever simply orientate on a particular compass bearing, unrelated to the direction of home, unless trained to do so as in the case of racing pigeons. Yet this is precisely how some birds do behave. When mallard (*Anas platyrhynchos*), for example, are displaced from Slimbridge, England, and released, they fly almost always to the northwest (Matthews 1968), no matter what the direction of 'home' (Fig. 3.5). Mallard from other areas also show a preferred compass direction, though not always to the northwest (Matthews and Cook 1982). Wallraff (1978b) examined data from a variety of pigeon lofts in the United States and Europe and found that birds from each loft had their own, apparently loft-specific, preferred compass direction, or PCD as he termed it. When released the birds showed mean vanishing bearings that were a compromise between their PCD and the direction of home.

In the case of mallard, the PCD was maintained beyond the vanishing point for not more than about 20 minutes and most ducks had landed or broken away from their original direction within 16 km of the release site. Even then, there was no evidence that orientation gave way to goal orientation, for subsequent recoveries were scattered at random (Matthews 1968). In the case of homing pigeons, orientation in the PCD does seem eventually to give way to goal orientation, for many birds still return home, albeit with reduced homing speeds (Matthews 1968). In the study by Michener and Walcott (1967), in which pigeons were trained from one direction and then released from another (e.g. Fig. 2.3), orientation in the training direction was sometimes maintained for up to 80 km before goal orientation became evident.

3.7. Testing for orientation

There is still no good explanation for why birds such as mallard and homing pigeons should sometimes orient in a PCD following displacement rather than towards home (but see Chapter 11). At first sight, such behaviour is inconvenient, requiring caution in the design and interpretation of experiments. In another sense, however, it is convenient because it

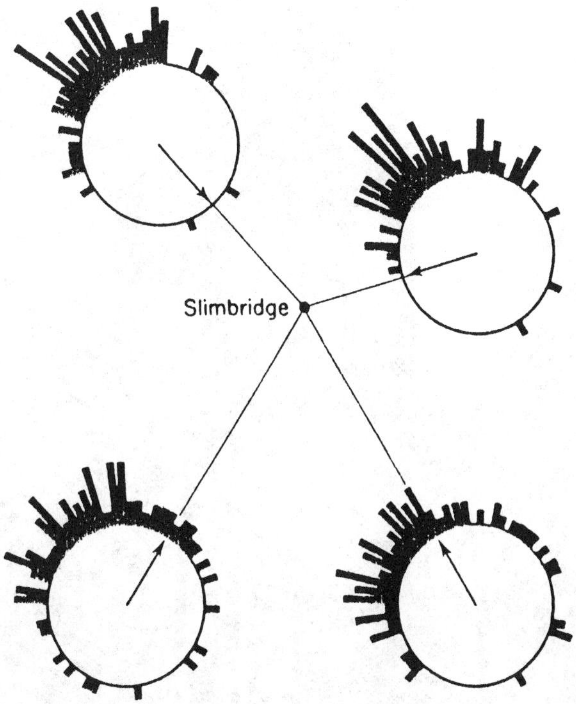

Fig. 3.5 When mallard (**Anas platyrhynchos**) are displaced from Slimbridge, southwest England, they fly to the northwest irrespective of the direction of 'home'

The shortest spoke represents the vanishing bearing of one bird.

(Re-drawn from Matthews 1968.)

(a)

(b)

(c)

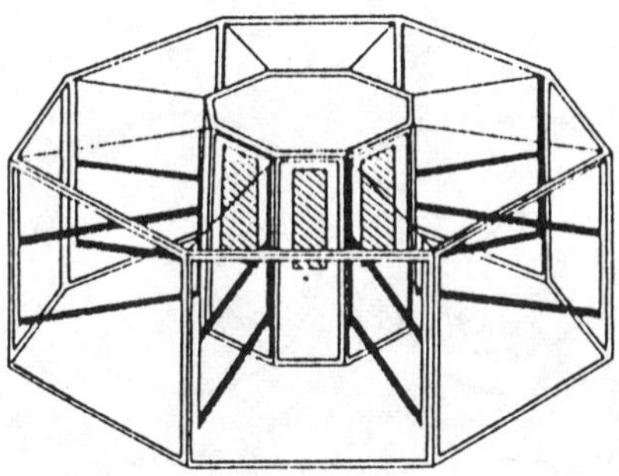

Fig. 3.6 The use of caged migrant birds to study migratory orientation.

Caged birds of migrant species, such as continental European starlings, **Sturnus vulgaris**, when housed in special orientation cages (as in (a)), may be used to study migratory orientation and timing. The caged starling in (b) becomes restless at about the same time as its free-living counterparts on the telephone wires are migrating. It will also orient in the same compass direction as they migrate. Orientation mechanisms may then be studied by, for example, altering the magnetic field through the cage by means of Helmholtz Coils as in (b). (c) the octagonal cage developed by Merkel and Fromme (1958) and later used extensively by Wiltschko.

Each time the bird lands on a perch the event is recorded automatically on paper tape. Data are later analysed by computer. (Photos by Gerlinde Orth)

offers an experimental situation in which compass orientation may be studied. As we shall see in the next chapter, there are many ways in which a bird could make use of an ability to recognize compass direction in solving problems of goal orientation. Indeed, paradoxical though it may seem, there are some forms of navigation that can be achieved entirely through detection of compass direction. A vital step, therefore, in trying to understand bird navigation is to study the ways in which birds detect compass direction. Experiments on how they manage to orient in their PCD is one method available, but there are others.

The most widely used of all methods of experimenting on compass orientation is to restrain a migrant bird in an orientation cage (Figs. 3.6 and 3.7). Throughout most of the year, such birds behave much like any other caged bird. At those seasons that the free-living members of their species are performing their annual migrations, however, caged birds become extremely restless. The pattern of this migratory restlessness, or *Zugunruhe* as it is often called, mimics in many respects the pattern of migratory activity in wild birds. For example, nocturnal migrants when caged are restless at night (Fig. 3.8). Longer-distance migrants show more restlessness than shorter-distance migrants (Chapter 12). Most important of all as far as orientation is concerned, and certainly the most unexpected, is that the restlessness is oriented in the migration direction traditional for the species concerned (Chapter 7).

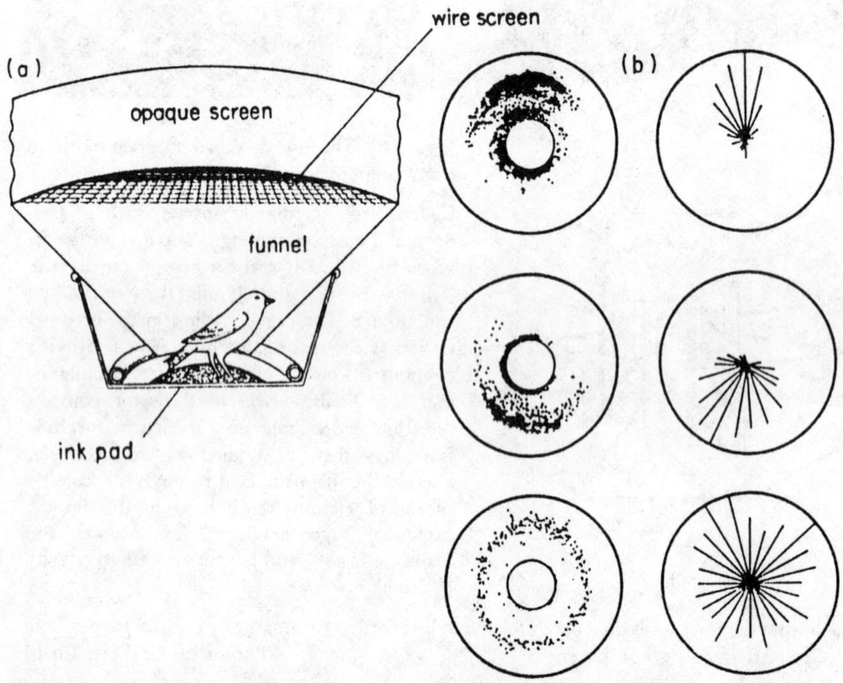

Fig. 3.7 The Emlen funnel introduced by Emlen and Emlen (1966)

When the bird jumps up the side of the cage (a), footprints are produced on the blotting paper lining the funnel. Footprints are then transcribed into vector diagrams as in (b). More recently, ink has been abandoned and the blotting paper replaced by 'Tippex' paper, which scratches easily when the bird jumps. Vector diagrams are produced by counting the number of scratches in each direction. [Compiled and re-drawn from Schmidt-Koening (1979)]

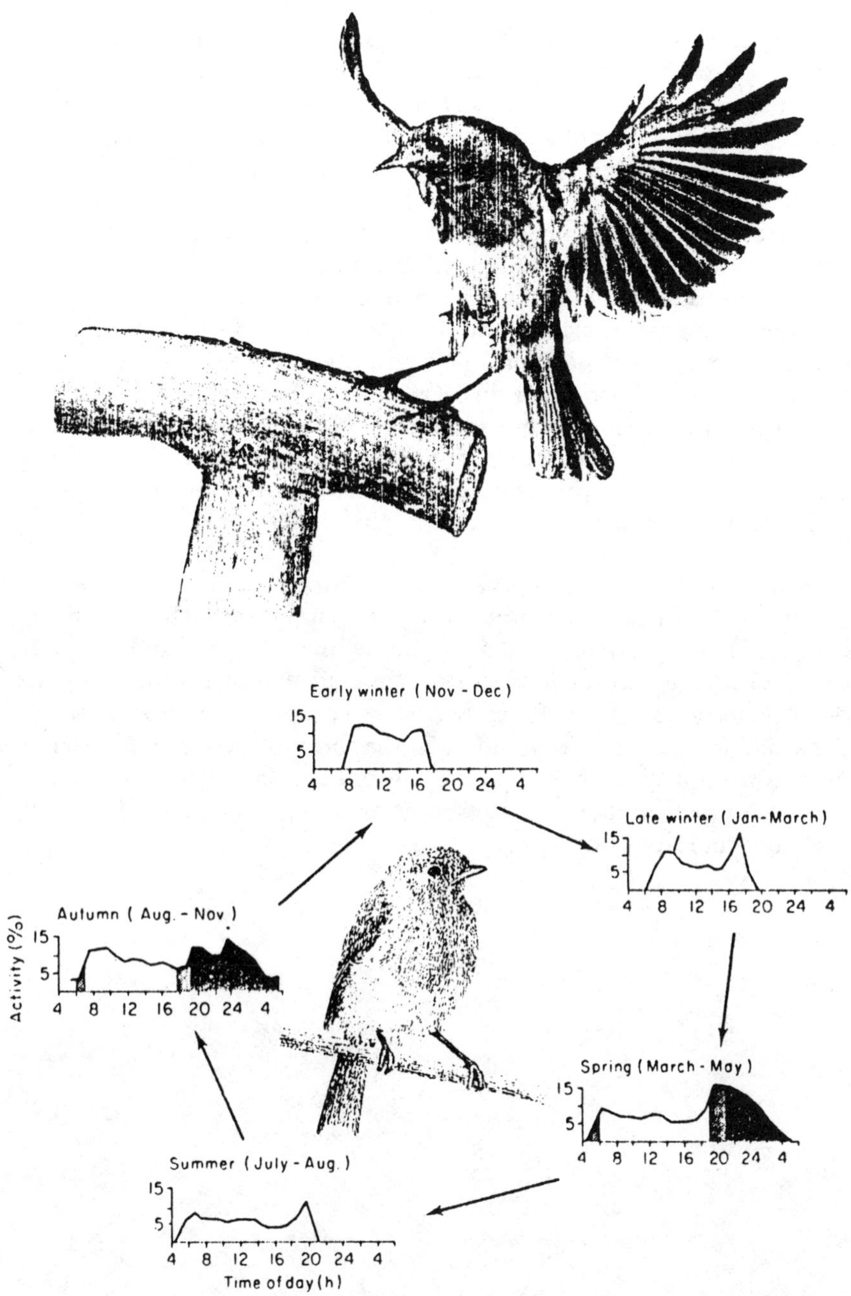

Fig. 3.8 Annual cycle of daily activity by caged European robins (**Erithacus rubecula**)

Diurnal activity is unshaded; nocturnal activity, black; dusk and dawn activity, stippled. During the migration seasons, caged robins are active at night when free-flying birds are migrating.

[From Baker (1978), after Merkel. Photo by Ronald Thompson, courtesy of Frank W. Lane]

This compass orientation by caged migrants during migratory restlessness was first noticed as long ago as 1949 by Kramer (1949) during experiments on caged passerines. His observation opened the way for subsequent generations of researchers to manipulate available compass cues and begin to unravel the compasses that are the basis of migratory orientation.

There are, however, disadvantages in being restricted to a study of compass orientation only during periods of migratory restlessness in caged birds. Not least, it means that there are long periods during the year when no experiments can be carried out. Kramer himself reduced this constraint (e.g. Kramer 1953) by devising an experiment to test for a sun compass in starlings (*Sturnus vulgaris*). For this, Kramer first trained a bird to take food from a pot in a particular compass direction. Then he began to manipulate the apparent position and movement of the sun (see Chapter 6). Training procedures similar in principle have been used to study the magnetic orientation of salamanders also (Phillips and Adler 1978).

In this chapter the major types of experiment used in the study of orientation and goal orientation have been surveyed. Such experiments have yielded a myriad of details. Sometimes results from different experiments are consistent with one another; at other times they are downright contradictory. If ever there were a danger of being unable to see the wood for the trees, it would be in the forest of orientation research. Now is the time, before we become engrossed in the detailed findings of particular experiments, to step back, to seek a few general principles, and to ask: how could birds navigate?

Part 8
Navigation in the 21st Century

8.1 Trends

PROJECTIONS of navigation technology are feasible, 30 years ahead, because the large investments preclude rapid change. The investment in a radio navaid includes:

—transmitter/receiver stations, ground-based and space-based,

—monitor stations (some of which may be "differential") and satellite-control stations. Differential ground stations are described in Part 3.6, pg. 68, and Part 4.8, pg. 155,

—propagation measurements and prediction algorithms,

—clever data-reduction algorithms for on-board computers and ground computers,

—the dissemination of charts, station locations, and ephemerides,

—extensive documentation and public information, and

—receivers, purchased by users from many competitive companies in many countries.

Radio systems tend to remain operational for 50 years, though the vehicle-borne receivers are constantly improved. Receiver manufacturers introduce the latest computers, analog chips, and digital microcircuits to reduce cost, add features, and increase reliability. Ground stations are usually upgraded every 10 to 20 years, but the signals in space remain unchanged.

I forecast the following trends, common to all vehicles:

(a) The distinctions among navigation, position location, and surveying will continue to diminish as satellite navigation aids are used for surveying and as data processing speeds increase. Navigation results must be available in real time as explained in Part 1 ("time delay," pg. 1). By contrast, the best surveying accuracy requires that the same radio signals be observed for several hours, sometimes several days.

(b) Navigation technology for the next 30 years will be dominated by the GPS-Glonass. The success of these navaids will depend on:

—continued military willingness to deploy the operational satellites, maintain the ground stations, and spend money for supporting staff and documentation:

—the willingness of the operator to assure continuous service, preferably without deliberately degraded accuracy. This may require that GPS be transferred to the FAA or Coast Guard when it is operational, as were earlier military navigation aids (e.g., Loran and Omega). The huge cost of developing the satellites will, by then, have been paid out of the military budgets. I do not know how Glonass is managed by the Soviet bureaucracy.

—availability of receivers in a variety of costs and accuracies. Widespread adoption of GPS seems assured with the marketing of single-channel C/A-code receivers for 3000 1989-dollars. A U.S.-Soviet agreement in 1989 and small-power desires for independence will probably assure the manufacture of dual-mode receivers.

—the willingness of the GPS operator to broadcast GPS status. Status is now broadcast every 30 s to 12.5 min, too infrequently to monitor time-critical users such as an aircraft executing an instrument landing. At least one industry committee [see (Kalafus, 1989) in the Bibliography for Part 3] is studying a network of independent monitors, perhaps at differential GPS stations, that broadcast status information frequently enough to announce a failure within 10 s.

—America's continued dominance of navigation technology and its ability to establish worldwide standards. It is likely that European and Japanese users will design multipurpose navigation sets that receive Soviet navaids as well as GPS, to reduce their dependence on foreign military services.

(c) If GPS and Glonass are successful, there will be intense pressure by governments to decommission most of their other navigation aids and by large users to buy receivers of fewer different types. Offsetting this pressure, many users of older navaids will exert pressure to delay decommissioning until installed receivers are scrapped. Some of the radio bands allocated to navigation would then become available for other purposes. Whoever wins, total reliance on GPS or any single radio aid is probably not wise.

(d) The next advance beyond GPS will be the nav–com satellite, which will be used by the land, marine, and aviation communities except for the most security-conscious military vehicles. In the past, navigation and communication systems were designed and procured by different bureaucracies. Equipment was often produced by different suppliers. Now that satellites can provide both services worldwide, many organizations are proposing combined satellites (see sections 3.4 and 4.10). Government-owned nav–comm satellites could create instant air-traffic control systems in undeveloped areas. Commercial owners of nav–comm satellites are already renting services to oil platforms, truck fleets, and railroads. In a debt-ridden United States, they might rent services to the FAA.

A major issue is whether the new satellites will be specialized by medium (e.g., land, sea, air) as are mobile communication services. Another issue is how vehicle transmitters and satellite operators will be licensed, a necessary measure to allocate the limited spectrum. Vehicles moving across national boundaries (e.g., aircraft, ships, and intra-European trucks) will demand areawide compatibility. There is no danger that *all* satellite communications will be centralized; it appears that satellites serving fixed users will continue to be operated by organizations related to the national telephone companies. City navigators will probably not use satellites, as described earlier, because of the large number of subscribers and the availability of inexpensive alternatives.

(e) Integration of multiple navigation aids will continue as a way to calculate the best estimate of position and velocity. On-board computers are inexpensive, as is on-board software whose development cost can be amortized among many customers. Two forms of navigation integration are apparent:

—multisensor navigation is in widespread use. It combines fixes from 2 or more navigation aids, usually in a Kalman filter;

—hybrid navigation is coming into use. It combines individual measurements from 1 or more navigation aids (e.g., solving for the intersection of a Loran and Omega hyperbola or a DME circle with a Loran circle). An example of a hybrid system is described by Enge and McCullough (1988–89) in the Bibliography of Part 4.

(f) Integration of navigation, communication, flight control, and other functions in common on-board avionics will continue. Multipurpose displays and controls will grow, as described in Part 5.5, pg. 240. Where sales volume justifies the investment in custom microchips, then cost, accuracy, and availability will benefit. In the 21st century, the sharing of circuits by radio sensors in the same band may reverse the current trend toward distributed computing (1 computer per traditional subsystem).

(g) The reliability of equipment and software will continue to improve as custom microchips are used, as proven software routines are reused, as hardware packaging becomes standardized (especially for the military), and as realistic schedules and budgets are allocated for software testing.

(h) Expert system software will continue to be developed, for example:

—software that detects failures, isolates them to a reconfigurable box, reselects modes, and switches among boxes (''pilot's associate'');

—software that replans missions in real time, for land vehicles, ships, aircraft, and spacecraft, allowing for terrain, navaids, intervisibility, Sun angle, and hostile activity;

—software that monitors and eventually controls air traffic; and

—software that can create a map from sensor observations and calculate an optimum route for an unmanned vehicle (especially a land robot).

(i) Absolute time will be regularly disseminated with an error of 0.1 μs via the GPs and Loran navaids. This will be of benefit to scientists and navigators. It wil permit one-way ranging to an accuracy of 30 m.

(j) The Sarsat–Cospas rescue satellite system (Part 3.6) will acquire new uses and its positioning accuracy will improve. A new class of Sarsat-compatible locator beacons will be sold that permits more precise positioning and adds 406 MHz to its service frequencies. Governments have begun to mandate their use by high-risk travelers to save money for searches. For example, American fishing boats have been required to carry beacons since August 1989.

8.2 Cartography

The Introduction to Part 2 describes the diminishing distinctions between navigation and geodetic sensors.

Digital maps will displace paper charts on large ships and aircraft, perhaps in the form of optic disks. The Defense Mapping Agency (DMA) issued a standard for color UTM maps on compact disk in 1989. They issued more than a thousand sheets of 1:50,000 topographic maps, mostly of Western Europe. With this inventory of maps, mobile CD-readers and CRTs will soon be introduced into ground vehicles, beginning with military, emergency, and commercial fleets. Those nations that standardize on formats will benefit by integrating maps into their land, air, and sea transportation networks. If competitive American companies refuse to use a common standard for civil digital maps, they will be forced to adopt foreign standards, as they did for videotapes. Some trends in digital maps will be:

—creation of local digital data bases by local specialists;

—invention of map-matching algorithms that recognize radar and optic fixes, terrain profiles, and road networks;

—further automation and computer-generated cues for vehicle crews, especially for single-place automobiles and aircraft;

—on-board data-base managers that store and manipulate data about street networks, route structures, navigation stations, fixes, waypoints, satellite orbits, etc. In the past, little on-board data were stored in a navigation system since sensor data were discarded after passing through the recursive filters.

8.3 Land Vehicles

The Introduction to Part 3 describes the current status of land navigation.

Low-cost electronics are already permitting the introduction of navigation systems into emergency vehicles (fire engines, police cars, ambulances) and into truck fleets. Land navigators will soon be installed in rental cars and personal cars, starting as luxury items. Civil land navigators will probably depend on navigation satellites *outside cities*, using known road locations to improve accuracy. Where communication to a dispatch center is needed, highway-vehicle operators in the United States will probably pay to route calls via commercial communication satellites that specialize in mobile or land-mobile service. Europe and Japan will establish government-owned (or licensed monopoly) communication (or nav–comm) satellites that offer standardized services to their citizens. Foresighted designers are collecting navigation and communication data that will allow them to build inexpensive, clever land navigators and communication systems.

In cities, two trends are emerging. One is a Loran-based system that reports via existing cellular and emergency radio links. The other, more widespread, is a self-contained dead-reckoning system. As noted in section 3.4, city navigators use a compass or differential-wheel rotation and odometers with electropost updates, manual updates, or expert system map updates using the known street configuration to improve accuracy between updates (see Part 3.4). I suspect that electroposts are likely to be placed in traffic-light stanchions. In Europe and Japan, they are likely to be government-owned; in the U.S., they are likely to be privately owned and to pay rent to the cities. Everywhere, receivers are likely to be purchased competitively by the users. Many vehicles may soon carry compact-disk readers for entertainment and map-reading. City users will have communication links, for personal messages and traffic updates, via cellular radio. Once navigation is feasible in a city, using software unique to that city, automated routing to a desired destination will follow. If routing is to account for temporary traffic conditions, the vehicles will have to receive external information, provided by a public or private traffic-management organization.

High-value *military off-road* vehicles will carry GPS and inertial navigators because many of them need attitude and heading as well as position and because GPS service might be lost in wartime. *Military on-road* vehicles will use low-cost dead-reckoning and GPS. On battlefields, where commanders must know the location of their forces, PLRS will be the preferred navaid for at least the next 10 years. Beyond that, it is not clear whether JTIDS and EPLRS will co-exist (with data gateways between them) or evolve into a new, unified system that serves troops, land vehicles, and aircraft.

Inertial surveying equipment, with and without GPS, will continue to replace theodolite-distance measurements because of its cost and speed advantages.

The American Association of Railroads and the Railway Association of Canada are developing a navigation system as part of a comprehensive train-control system (designed by ARINC) whose objectives are to assure dispatchers that the main lines are clear and to guide trains through switches. Each locomotive will dead-reckon between electroposts located at junctions. An electropost will radio its identification code to a train when it passes. Each train will radio its position, at intervals of several minutes, on 900-MHz radio links to existing receivers located every 30 mi, trackside. These receivers will be gateways to the railroad's microwave, fiber-optic, or telephone communication network. Tests began in 1988. Another system, developed by the Burlington-Northern Railroad, is also under test.

Several commercial navigation services are planned. One offers to track mobile transponders in the same way as Sarsat tracks emergency beacons. Another relays vehicle-measured Loran positions to a dispatch center. Land-mobile satellite bands have been assigned at L-band: 1530–1560 MHz downlink and 1630–1660 MHz uplink. If these new mobile bands are privatized, as appears likely in the United States, then competing noninteroperable navigation and mobile radio services will spring up for land, sea, and air, perhaps also using air-mobile L-band frequencies.

The future navigation of unmanned land vehicles is described at the end of the Introduction in Part 3.3, pg. 66.

8.4 Ships

The Introduction to Part 4 describes the state of the art of ship navigation. The era of the celestial fix has ended. Sextants are being stored in moldy suitcases next to slide rules. Small ships already rely on Loran, Transit, the magnetic compass, and depth sonar. They will continue to do so until some of these radio aids are displaced by GPS. Lifeboats now carry sextants and Sarsat beacons. I suspect that portable satellite receivers will replace the sextant, even there, since crews will lose the ability to do celestial navigation.

The use of Loran-C may be cut back as GPS satellites come into service. The reprint paper by Woodman *et al.* in Part 4, pg. 221, describes European plans to alter their northeastern chains when U.S. funding ceases.

Large ships will use these same radio aids, supplemented by the following devices. Gyrocompasses, derived from inertial navigation technology, will still be their heading references. Doppler sonar may displace the electromagnetic log. Multi-sensor ship navigators have already been installed in some naval vessels but it is not clear whether dead-reckoning will be of value on *civil ships* of the 21st century, given the precision and continuity of satellite-based radio aids. However, use of inertial navigators will continue to increase in the *naval fleet* since they also measure attitude and heading and hence are useful for fire control and to initialize aircraft and missiles. All large ships will continue to rely on radar for collision avoidance and for navigation near land.

Submarines are used for military purposes, underwater rescue, oceanographic and archaeological exploration, repair of cables (as fiber-optic cables proliferate), and repair of oil platforms, for example. *World-ranging submarines* will depend on their inertial navigators (SINS) at depth. Their position updates will be by GPS near the surface, and perhaps by bottom-mapping (terrain correlation or pictorial) under water. It is possible that the matching of gravity gradients will be used to navigate a ship in real time, in the next generation.

Doppler sonar will give accurate velocity fixes, as it does now on a few ships. Omega will fade from service as its poor accuracy is no longer acceptable, even underwater. *Short-range submersibles,* operating from a mother ship, will depend on dead-reckoning by gyrocompass and electromagnetic log, updated by precisely located acoustic transponders on the ocean floor.

8.5 Aircraft

The Introduction to Part 5 describes the current state of the art of aircraft navigation. The upgrade of the North American air-traffic control system in the next century will affect it as follows:

—eventual decommissioning of Vortac navaids, to save money. The FAA and military may agree on assured GPS service and status broadcasts. The FAA may also adopt Loran.

—possible decommissioning of the en-route surveillance radars, thereby placing full reliance on GPS, Loran, and communication links;

—establishment of two or three communication satellites to replace the FAA's en-route radio network;

—establishment of combined nav–comm satellites to perform all traffic-control services. Such a system would handle 50,000 aircraft that are in-flight over the United States at any time in the next century. The economics are being studied by organizations in many countries, as described in section 8.1.d.

—resolution of the inherent conflict between collision-avoidance systems (that request on-board-derived maneuvers) and the air traffic control system (that is predicated on ground controllers knowing all maneuvers in advance); and

—resolution of the inherent conflict between air-derived optimum climb-descent profiles and the ground assignment of altitudes.

European countries, East and West, will upgrade their traffic-control systems in the 1990s as the common market strengthens and as the Soviet bloc opens to the West. Radar coverage of the Mediterranean will increase. Western Europe's traffic coordinator, Eurocontrol, will be strengthened. European airspace (for civil and military aircraft) has denser traffic than does North America. Hence, its traffic-control agencies are motivated to use clever software and the latest hardware, perhaps surpassing the United States in automation. They may direct aircraft to climb and descend in "tubes," rather than clearing all airspace below or above an aircraft. This will place severe requirements on horizontal and vertical navigation. The Soviet Union is already upgrading its own system to conform to ICAO standards. Soviet-bloc countries are opening airlanes to Western traffic for the first time.

Civil committments are being made to VLF-Omega, despite its severe limitations, suggesting that an entrenched user base will develop before equivalent-cost GPS receivers become available. The North American Loran chains will be used by general aviation for en-route navigation and nonprecision approaches. Civil commitments are also being made to GPS. ARINC is preparing interface standards for GPS and combined GPS-Glonass sets. At least one airline has ordered GPS-Glonass receivers for its Boeing 747 aircraft. These aircraft will constitute another entrenched user base in the 21st century. GPS and Loran may become the only publicly funded U.S. radio aids in the 2010s.

As air-traffic density increases in East Asia, South America, and (much later) Africa, new regional traffic-control systems may arise, emulating Eurocontrol. Those regions have the opportunity of investing directly in nav–comm satellites since they have little investment in conventional traffic control equipment.

Positive control, implying communication and navigation, will certainly be implemented over the North Atlantic and North Pacific before the turn of the century, to serve at least 200 aircraft that are in the air over each ocean at any time. Proposals have already been made by the FAA, Mitre, ARINC, ICAO, NASA, and INMARSAT for communication and nav-comm satellites. L-band satellite-to-aircraft frequencies have been assigned. A single satellite system may serve all mobile users; land, sea, and air. Massoglia (1989) has reviewed the status of oceanic air traffic control and has forecasted proposed L-band communication, via satellites, of on-board-derived position and of clearances.

The ICAO issued a report by the prestigious Committee on Future Air Navigation Systems that recommended the retention of local VHF and radar navigation and communication and the addition of a worldwide system. The report recommended GPS-Glonass as the worldwide navigation system and satellite L-band (1.545–1.555 GHz and 1.646.5–1.656.5 GHz) as the worldwide communication system. The report expects that the worldwide system will be the only one in many parts of the world (e.g., oceans). The report expects that it will supplement local systems (for en-route navigation, nonprecision approaches, and essential communications) in Third-World countries where local nav–comm aids are often not working.

Whether or not domestic navigation is based on GPS, military restrictions will be imposed on aviation during national emergencies. Traffic would be restricted (e.g., visual-flight-rules prohibited) and selected navaid stations would be shut down. Plans for doing so are already in place in the United States [see the FAA reference (FAA, 1987) to SCATANA in the Part 5 Introduction, pg. 229] and, presumably, in other industrial countries.

U.S. military aircraft will adopt GPS, alone or as an update for self-contained inertial and radar navigation. GPS is less vulnerable to enemy attack than are Omega or Loran; destroying 24 satellites is vastly more expensive than destroying 60 radio-emitting ground stations. Inertial and radar navigation are required for security and for weapon delivery. Fighter aircraft will still use multimode radar to find targets. Battlefield-support aircraft will probably use JTIDS communications if terminal costs fall (Part 3.5); navigation may be from JTIDS, GPS, or both. The future of specialized military navigation is described in the Introduction to Part 5.

For landing, MLS will gradually displace ILS as old aircraft are retired and MLS receiver costs fall. It will probably dominate the Category II and Category III markets. Meanwhile, ILS is doubling the number of channels by reducing the spacing from 100 kHz to 50 kHz. GPS might become a landing aid in remote areas, using a portable ground station that broadcasts differential measurements and instantaneous status on an existing radio frequency. Thus, civil and military organizations could have instant instrument landing facilities.

Carrier landings will continue to be made with the landing aids discussed in Part 5.4, SPN-46 gradually replacing SPN-42. There are not enough carrier-based aircraft to justify the development of a wholly new system for many decades.

Figure 5-6, pg. 238, shows the antenna complement of an airliner of the late 1990s. It illustrates the antennas needed to accommodate many navaids and communication systems for en-route flight, collision avoidance, and landing. Probably no single aircraft would carry all these antennas. The figure explains the motivation to design integrated navigation aids that require fewer antennas. It also shows that, even with integrated navaids, coverage and shadowing will always require multiple antennas.

8.6 Spacecraft

The Introduction to Part 6 describes the current state of the art of spacecraft navigation. Because of the discontinuous coverage of STDN and SCN, navigation has depended on mathematical models to interpolate and extrapolate measured state vectors. However, a revolutionary change is occurring; GPS and TDRS now permit continuous state vector measurements in low and medium Earth orbits.

Publicly owned spacecraft will continue to use NASA's STDN-TDRS or the Air Force's Satellite Control Network. Both networks are being upgraded to reduce operating costs and raise the communication bandwidth. The military will use both GPS and SCN. *Privately owned* spacecraft will continue to be navigated by the users' own ground stations. NASA will rent tracking and telemetry services to private operators. The trend toward lightly manned ground-based control rooms will continue, with staffs being augmented during critical mission phases.

Spacecraft at altitudes below approximately 6000 nm can measure their on-board state vector continuously with GPS or TDRS. They have already demonstrated 5- to 80-m real-time position accuracy on orbit and are expected to reach centimeter accuracy in the 1990s, after postflight analysis. Multichannel GPS receivers will be needed in space. GPS covers the entire low-orbit region and many GPS satellites are visible simultaneously from an orbiting spacecraft. For example, a 6-channel GPS receiver is being developed for the Topex-Poseidon geodetic satellite [see the reprint by Carson in the Introduction to Part 6, pg. 317, and the reference by Yunck (1988) in the Introduction to Part 2].

TDRS measures to better than 100-m accuracy and covers nearly the entire low-orbit region. However, TDRS does not give instantaneous position fixes as does GPS. Several hours are required for the ranging data to be processed on the ground, administratively approved, and, if necessary, uplinked to the spacecraft. This delay is tolerable for most satellites (e.g., geosynchronous communication satellites) but not for others (e.g., manned spacecraft). It will undoubtedly be reduced as ground-computer speeds increase and procedures are refined.

To predict the trajectory in the past, between discontinuous fixes, or in the future, mathematical models must be used. The accuracy of prediction is limited by the errors in measuring the initial state vector, by the extent of modeling of the gravity field of the Earch and other celestial bodies, and by the measurement of forces due to venting and the firing of rocket motors. In low orbit, unpredictable atmospheric drag is the largest limit on prediction. These nongravitational forces could be measured by a low-range accelerometer whose bias is so low that it could be left "on" indefinitely without accumulating large bias errors. Regrettably, no such instrument exists. In higher orbits, the initial measurement error dominates. Fourth-order gravity models predict within a few tens of meters per revolution and can execute in real time; hence they are adequate for most purposes. A future distributed avionic system could have one computer that is dedicated to orbit prediction.

GPS and TDRS will reduce a spacecraft's dependence on ground tracking. The long-sought goal of navigational autonomy or near autonomy is within reach.

Sensors, computers, and data buses for on-board navigation and control will continue to be integrated with other spacecraft avionics. Functions that are now performed on the ground will, in part, be performed on-board by "expert-system" software. Examples are attitude determination of unmanned spacecraft, failure detection, failure correction, and mission replanning. Autonomous navigation will probably first be introduced on military spacecraft, for example, for automated rendezvous and docking.

Evolutionary upgrades will be made to the Shuttle in new inertial navigation sets and computers, improved displays, and added software for direct ascent, automatic rendezvous, and docking. In a decade, the Shuttle will be re-engined and its avionic system completely refurbished, probably with distributed microcomputers and new sensors.

The National Aerospace Plane is probably being designed with avionics that are an outgrowth of the Space Shuttle's, including planned improvements that NASA could not afford to fund.

Deep-space navigation will probably not undergo a radical change in the next decade. However, planetary surface explorers will contain dead-reckoning software and may use nav–comm satellites that orbit the planet. Explorers will be heavily autonomous in navigation, mission replanning, failure detection, and reconfiguration. The Moon, and later Mars, will be equipped with surface relay stations and land lines for navigation and communication. Guidance systems will be needed for a cable-laying robot and for a lunar-surface launch sled, which may return payloads to Earth.

The Space Station will probably make occasional stationkeeping maneuvers, which will be calculated on the ground. It will eventually contain its own targetting software for launching lunar and planetary spacecraft. It may track debris, predict their orbits several revolutions ahead, and either make an evasive maneuver (for a large object) or deploy a shield (for a small object). Kessler (see Part 6 Bibliography), pg. 331 predicted a 0.1 probability of collision with an object larger than 1-m diameter in 10 years and 0.3 probability of collision with debris of all sizes. Hence, a new form of space collision-avoidance will evolve, in which ground-tracking of large objects and on-board tracking of small objects predict collisions. Good orbit-prediction accuracy will reduce the false-alarm rate. Space rescue vehicles and ambulances will be equipped with rendezvous and de-orbit navigation systems, as described in Part 6.4.

8.7 Animal Navigation

The current revolution in biochemistry may affect navigation. Animals are the ultimate "natural intelligence" navigators but we have no idea how they process their sensor data. By the end of the 21st century, some of their techniques of sensing and pattern-recognition may be applied to man-made navigation devices.

Author Index

Subject Index

1. The authors of reprints are indexed on page 441.
2. Acronyms are indexed in the glossary, pages xi–xii.
3. The numbers in italics refer to the Introduction at the beginning of each Part of the book. Other numbers refer to the reprints.
4. Authors of references are listed in the Bibliographies after each Introduction and each Reprint. They are not indexed.

B

D

E

F

M

Q

R